Push your Career Publish your Thesis

Science should be accessible to everybody. Share the knowledge, the ideas, and the passion about your research. Give your part of the infinite amount of scientific research possibilities a finite frame.

Publish your examination paper, diploma thesis, bachelor thesis, master thesis, dissertation, or habilitation treatises in form of a book.

A finite frame by infinite science.

Infinite Science
Publishing

An Imprint of
Infinite Science GmbH
MFC 1 | Technikzentrum Lübeck
BioMedTec Wissenschaftscampus
Maria-Goeppert-Straße 1
23562 Lübeck
book@infinite-science.de
www.infinite-science.de

© 2018   Infinite Science Publishing
University Press and
Academic Printing

Imprint of Infinite Science GmbH,
Technikzentrum | MFC 1
Maria-Goeppert-Straße 1
23562 Lübeck, Germany

Cover Design and Illustration: Uli Schmidts, metonym
Editorial and Copy Editing: University of Lübeck

Publisher: Infinite Science GmbH, Lübeck, www.infinite-science.de
Printed in Germany, BoD, Norderstedt

ISBN: 978-3-945954-47-8

Bibliografische Information der Deutschen Nationalbibliothek:
Die Deutsche Nationalbibliothek verzeichnet diese Publikation in der Deutschen Nationalbibliografie; detaillierte bibliografische Daten sind im Internet über http://dnb.d-nb.de abrufbar.

Student Conference Proceedings 2018

7th Conference on Medical Engineering Science
3rd Conference on Medical Informatics
1st Conference on Biomedical Engineering

Lübeck, March 7-9, 2018

Editors in Chief

T. M. Buzug, H. Handels, S. Klein

Associate Editors

C. Debbeler, K. Gräfe, J.-H. Wrage, S. Venker

Editors

H. Botterweck, T. M. Buzug, G. Buntrock, C. Damiani, F. Ernst, S. Fischer, H. Gehring, T. Gutsmann, J. Haase, H. Handels, M. Heinrich, C. Hoffmann, R. Huber, C. Hübner, G. Hüttmann, J. Ingenerf, N. Jochems, M. Kleemann, S. Klein, M. A. Koch, M. Leucker, N. Linz, K. Lüdtke-Buzug, A. Madany Mamlouk, T. Martinetz, A. Mertins, Y. Miura, J. Modersitzki, R. Moll, S. Müller, A. Neumann, J. Obleser, M. Ratecas, R. Rahmanzadeh, P. Rostalski, M. Ryschka, A. Schweikard, F. Spitzenberger, R. Wendlandt, C. Wendt

Exhibitors and Sponsors

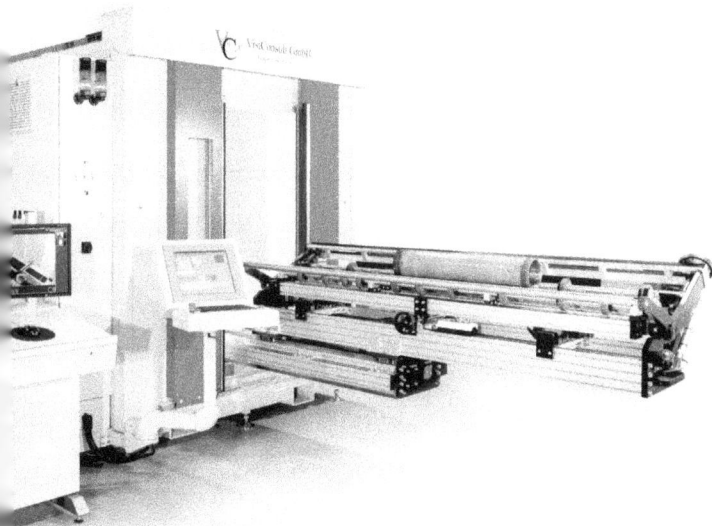

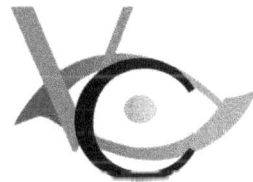

Students surprise themselves when they work with us.

Will you?

Working at Philips is
more than just a job. It's about
people, passion, innovation
and... the fun!

senTec

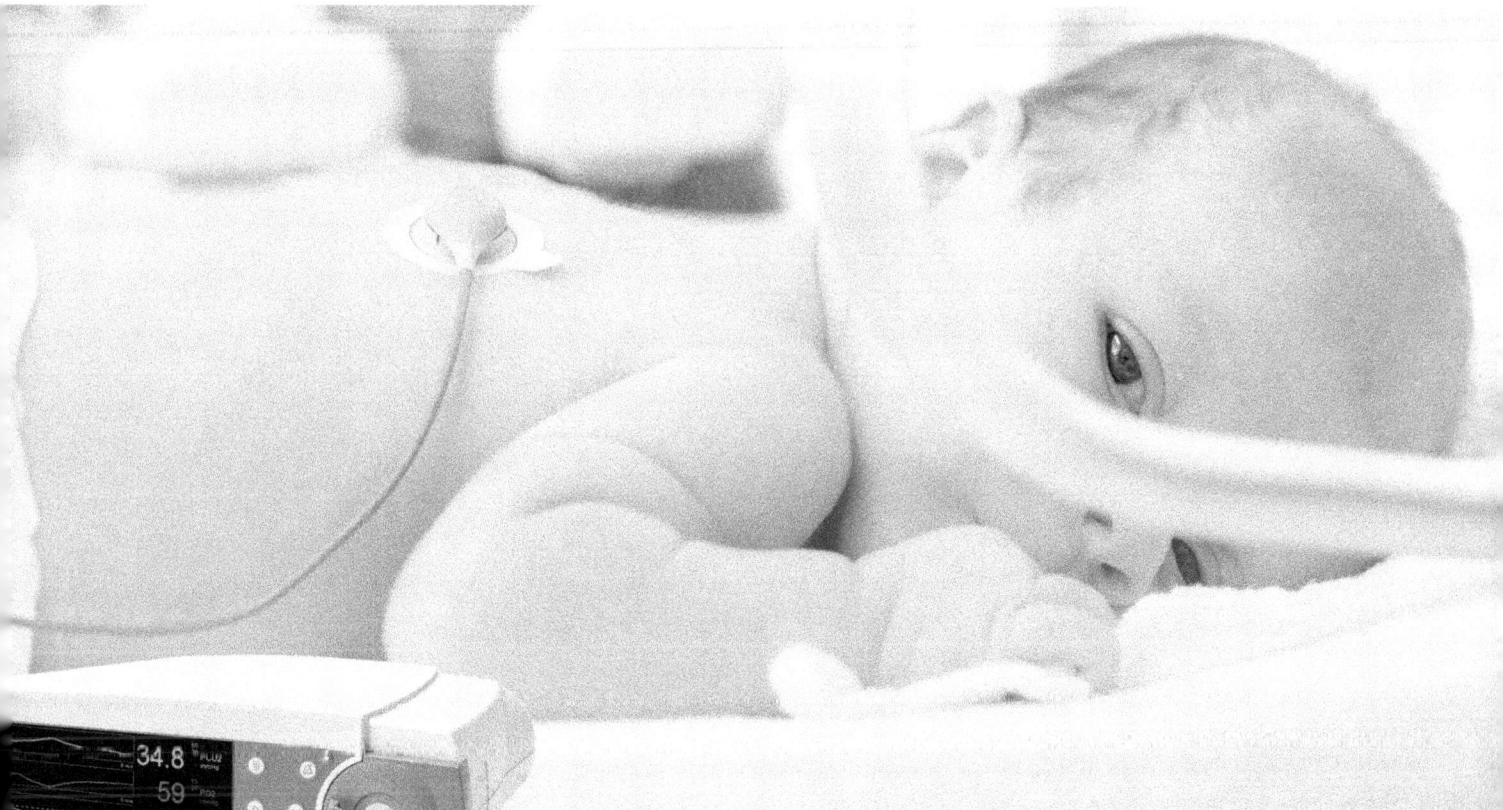

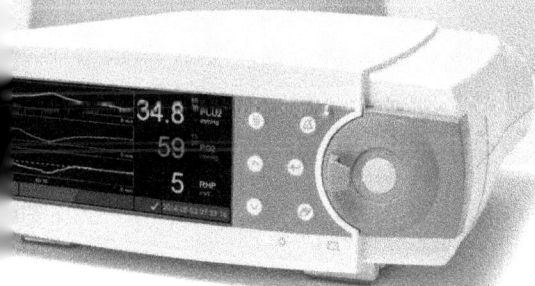

Advancing Noninvasive Patient Monitoring™

SenTec ist ein junges, innovatives und international tätiges Unternehmen im Bereich Medizintechnik.

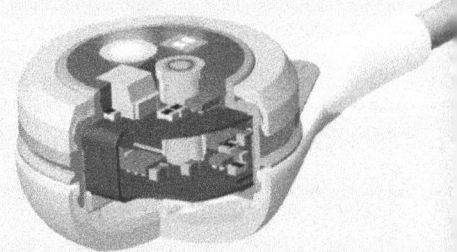

Mit unseren engagierten Mitarbeitern in Rostock und in der Schweiz, entwickeln und produzieren wir unsere high-tech Produkte weltweit. Unsere SenTec DigitalMonitoringSystems werden zur nicht-invasiven und kontinuierlichen Überwachung von Patienten eingesetzt – sei es zur Überwachung der Atmung von Patienten mit schweren Lungenerkrankungen oder der Beatmung von Frühgeborenen.

Getrieben von unserem hohen Anspruch an die Qualität und Zuverlässigkeit unserer Produkte, streben wir mit unseren Forschungs- und Entwicklungstätigkeiten stetig nach neuen innovativen Messtechnologien.

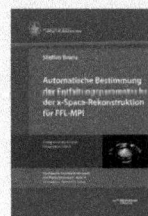

Sponsors

Angebote der IHK zu Lübeck für Studierende

Erst das Studium und dann... Über den besten Studienplatz, interessante Praktikumsplätze, Finanzierung und Förderung sowie die Möglichkeiten, sich selbstständig zu machen und von gestandenen Unternehmerinnen und Unternehmern zu lernen, unter www.Mein-Unternehmen-Zukunft.de

Best of ...

Karrieretag

Der Karrieretag ist eine gemeinsame Veranstaltung der Universität zu Lübeck, der Fachhochschule Lübeck, der BioMedTec Management GmbH sowie der IHK zu Lübeck und richtet sich an Studentinnen und Studenten, Absolventinnen und Absolventen. Beim Karrieretag stellen Sie die Weichen für Ihre berufliche Zukunft und knüpfen Kontakte zu zukünftigen Arbeitgebern. Mehr unter: www.ihk-sh.de/karrieretag

Praktikumsbörse

Die IHK-Praktikumsbörse www.praktikum-sh.de bietet Schülern und Studierenden die kostenlose Möglichkeit, Praktikumsplätze bei Unternehmen in Schleswig-Holstein zu finden.

Beratung StudiLe

Das Studium mit integrierter Lehre (StudiLe) ist ein duales Studienmodell, welches eine betriebliche Ausbildung mit einem Bachelorstudium an der Fachhochschule Lübeck verbindet. In circa vier Jahren können somit zwei berufsqualifizierende Abschlüsse erworben werden. Nähere Informationen finden Sie unter www.studile.de.

Informationen für Studienabbrecher

Die IHK zu Lübeck engagiert sich im Netzwerk „Zweifel am Studium?" und steht Studienabbrecherinnen und Studienabbrechern für individuelle berufliche Beratungsgespräche zur Verfügung. Eine erste Orientierungsberatung zur Studiensituation und dem persönlichen Profil bietet die Handwerkskammer Lübeck im Studentenwerk auf dem Campus der Lübecker Hochschulen an: www.kursaenderung-ins-handwerk.de

Existenzgründung und Unternehmensförderung

Ob Sie eine Firma gründen, die Wettbewerbsfähigkeit Ihres Unternehmens für die Zukunft sichern oder die Nachfolge regeln möchten – wir stehen von Anfang an an Ihrer Seite und begleiten Ihr Unternehmen von der Gründung bis zur Übergabe. Mehr zu finden unter: www.ihk-sh.de/basisinfos

Speziell für Studierende bieten wir Beratungstage zur Existenzgründung auf dem Campus und eine individuell zugeschnittene Finanzierungs- und Förderberatung.

Bonus?
Punkte!

Das Bonusprogramm der Techniker

Ganz gleich, wie Sie ins Schwitzen kommen: Wer sich für seine Gesundheit einsetzt, wird belohnt. Dabei motiviert das TK-Bonusprogramm nicht nur mit Geld, sondern auch mit gesunden Extras. Ich berate Sie gern:

Annika Naber
Hochschulberaterin
Tel. 01 51 - 46 13 01 51
annika.naber@tk.de

INTER-
NATIONALER
MARKT

SOLIDES
WACHSTUM

✳ COHERENT.

OFFENE
KOMMUNIKATION

PARTNER-
SCHAFTLICHER
UMGANG

ARBEITEN BEI
COHERENT ...

TECHNISCHE
INNOVATION

VIELES SPRICHT
DAFÜR!

STABILES
TEAM

REGIONALE
VERBUNDENHEIT

WERTE

STANDORTE

Arbeiten bei Coherent –
Vieles spricht dafür!

Coherent und ROFIN sind jetzt zusam-
men das weltweit größte Laserunter-
nehmen. Mehr als 5.000 Mitarbeiter in
über 40 Ländern arbeiten an führenden
Photonik-Lösungen für industrielle, wis-
senschaftliche und medizinische Anwendungen.

>> *Wir sind erfolgreich, weil wir nicht nur
unsere Produkte kontinuierlich weiterent-
wickeln, auch unsere Mitarbeiter werden
immer besser. Dass wir dabei auf den Ein-
zelnen eingehen, ist für uns ebenso selbst-
verständlich, wie der wertschätzende Umgang miteinander, der die Atmosphäre bei
Coherent prägt.* <<

• Produktion
• Sales/Service

Santa Clara

Deutschland: Lübeck, Hamburg, Göttingen, Mainz, Dieburg, Kaiserslautern,
Gilching, Günding, Starnberg, Freiburg, Overrath und Zorneding

Coherent LaserSystems GmbH & Co. KG

Seelandstraße 9 ▪ 23569 Lübeck
www.coherent.com

✳ COHERENT.   ✳ COHERENT. | DILAS   ✳ COHERENT. | rofin   ✳ COHERENT. | NUFERN

When care demands more

SPEED

is everything.

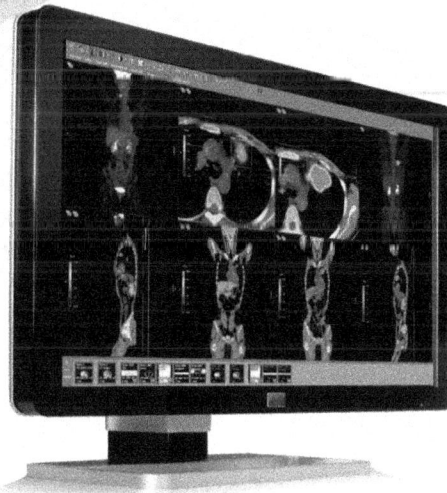

Sie programmieren gerne, kennen sich mit IT aus und interessieren sich für medizinische Anwendungen?
Sie suchen Dynamik und denken und arbeiten ergebnisorientiert?
Dann sind Sie bei uns genau richtig!

An unserem Standort in Berlin beschäftigen wir mehr als 30 Mitarbeiter. Die Entwicklung der Software wie auch Global Customer Support befinden sich hier. Wir suchen zum nächstmöglichen Zeitpunkt mehrere:

- **Software Engineers C++**

- **Application Engineers (Clinical Systems)**

Unser Angebot:
- Unbefristete Festanstellung
- Individuelle und flexible Arbeitszeitmodelle (Teil-/Vollzeit)
- Einstellung von sowohl Berufseinsteigern als auch Berufserfahrenen
- Arbeit in einem hochmotivierten und kreativen Team mit flachen Hierarchien
- Innovatives, zukunftsorientiertes Branchenumfeld
- Kostenlose Getränke, frisches Obst und ein attraktives Gehalts- und Urlaubspaket

Visage Imaging verfügt über 25 Jahre Erfahrung im Bereich der medizinischen Bildverarbeitung und ist ein etablierter Arbeitgeber mit hohem Qualitätsanspruch. Haben wir Ihr Interesse geweckt? Dann werden Sie Teil eines Technologieführers und schauen Sie bei uns vorbei.

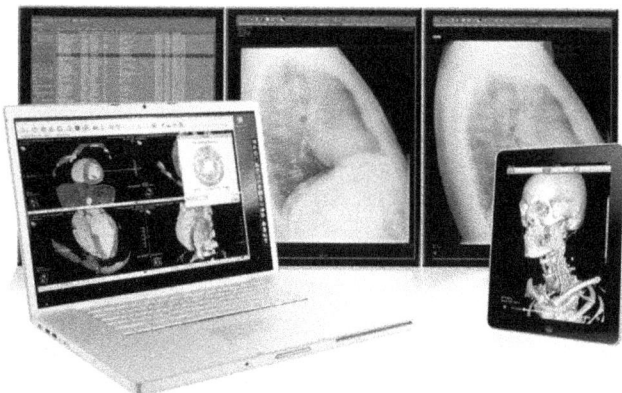

VISAGE IMAGING®

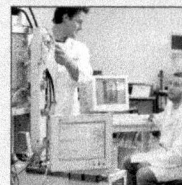

Student Conference
Proceedings 2018

Scientific Program Committee

Botterweck, Prof. Dr. Henrik	Department of Applied Natural Sciences, Lübeck University of Applied Sciences
Buzug, Prof. Dr. Thorsten M.	Institute of Medical Engineering, Universität zu Lübeck
Buntrock, PD Dr. Gerhard	Institute for Software Engineering and Programming Languages, Universität zu Lübeck
Damiani, Dr. Christian	Medical Sensors and Devices Laboratory, Lübeck University of Applied Sciences
Ernst, Prof. Dr. Floris	Institute for Robotics and Cognitive Systems, Universität zu Lübeck
Fischer, Prof. Dr. Stefan	Institute of Telematics, Universität zu Lübeck
Gehring, Prof. Dr. Hartmut	Department of Anesthesiology, University Medical Center Schleswig-Holstein, Lübeck
Gutsmann, Prof. Dr. Thomas	Research Center Borstel, Leibniz-Center for Medicine and Biosciences
Haase, Dr. Jan	Institute of Computer Engineering, Universität zu Lübeck
Handels, Prof. Dr. Heinz	Institute of Medical Informatics, Universität zu Lübeck
Heinrich, Prof. Dr. Mattias	Institute of Medical Informatics, Universität zu Lübeck
Hoffmann, Dr. Christian	Institute for Electrical Engineering in Medicine, Universität zu Lübeck
Huber, Prof. Dr. Robert	Institute of Biomedical Optics, Universität zu Lübeck
Hübner, Prof. Dr. Christian	Institute of Physics, Universität zu Lübeck
Hüttmann, PD Dr. Gereon	Institute of Biomedical Optics, Universität zu Lübeck
Ingenerf, Prof. Dr. Josef	Institute of Medical Informatics, Universität zu Lübeck
Jochems, Prof. Dr. Nicole	Institute for Multimedia and Interactive Systems, Universität zu Lübeck
Kleemann, Prof. Dr. Markus	Clinic for Surgery, University Medical Center Schleswig-Holstein, Lübeck
Klein, Prof. Dr. Stephan	Medical Sensors and Devices Laboratory, Lübeck University of Applied Sciences
Koch, Prof. Dr. Martin A.	Institute of Medical Engineering, Universität zu Lübeck
Leucker, Prof. Dr. Martin	Institute for Software Engineering and Programming Languages, Universität zu Lübeck
Linz, Dr. Norbert	Institute of Biomedical Optics, Universität zu Lübeck
Lüdtke-Buzug, Dr. Kerstin	Institute of Medical Engineering, Universität zu Lübeck
Madany Mamlouk, PD Dr. Amir	Institute for Neuro- and Bioinformatics, Universität zu Lübeck
Mertins, Prof. Dr. Alfred	Institute for Signal Processing, Universität zu Lübeck
Miura, PD Dr. Yoko	Institute of Biomedical Optics, Universität zu Lübeck
Modersitzki, Prof. Dr. Jan	Institute of Mathematics and Image Computing
Moll, Dr. Ralf	Department of Applied Natural Sciences, Lübeck University of Applied Sciences
Müller, Prof. Dr. Stefan	Medical Sensors and Devices Laboratory, Lübeck University of Applied Sciences
Neumann, Dr. Alexander	Institute of Medical Engineering, Universität zu Lübeck
Obleser, Prof. Jonas	Department of Psychology, Universität zu Lübeck
Rafecas, Prof. Dr. Magdalena	Institute of Medical Engineering, Universität zu Lübeck
Rahmanzadeh, Dr. Ramtin	Institute of Biomedical Optics, Universität zu Lübeck
Rostalski, Prof. Dr. Philipp	Institute for Electrical Engineering in Medicine, Universität zu Lübeck
Ryschka, Prof. Dr. Martin	Depart. of Electrical Engineering and Computer Science, Lübeck University of Applied Sciences
Schweikard, Prof. Dr. Achim	Institute for Robotics and Cognitive Systems, Universität zu Lübeck
Spitzenberger, Prof. Dr. Folker	Depart. of Applied Natural Sciences, Lübeck University of Applied Sciences
Wendlandt, Dr. Robert	Clinic for Orthopedic and Trauma Surgery, University Medical Center Schleswig-Holstein
Wendt, Dipl.-Ing. Christian	Department of Applied Natural Sciences, Lübeck University of Applied Sciences

Conference Chairs

Thorsten M. Buzug (Chair), Institute of Medical Engineering, Universität zu Lübeck; **Heinz Handels (Chair)**, Institute of Medical Informatics, Universität zu Lübeck; **Stephan Klein (Chair)**, Medical Sensors and Devices Laboratory, Lübeck University of Applied Sciences; **Hartmut Gehring (Co-Chair)**, Department of Anesthesiology University Medical Center Schleswig-Holstein, Lübeck

Local Coordination

Christina Debbeler, Institute of Medical Engineering, Universität zu Lübeck; **Ksenija Gräfe**, Institute of Medical Engineering, Universität zu Lübeck; **Jan-Hinrich Wrage**, Institute of Medical Informatics, Universität zu Lübeck; **Silke Venker**, Lübeck University of Applied Sciences; **Gisela Thaler**, Institute of Medical Engineering, Universität zu Lübeck; **Susanne Petersen**, Institute of Medical Informatics, Universität zu Lübeck

Preface and Acknowledgements

After the great success of the previous meetings from 2012 to 2017, the Student Conference 2018 shows continuing growth both in quality and quantity of scientific contributions. In this year, the 7th Student Conference on Medical Engineering Science is hold together with the 3rd Student Conference on Medical Informatics and the 1st Student Conference on Biomedical Engineering.

The organization team of the Institute of Medical Engineering, the Institute of Medical Informatics, and the Program Coordination for Biomedical Engineering at the University of Applied Science Lübeck in cooperation with the Chamber of Industry and Commerce (IHK) Lübeck and the Life Science North Management GmbH, the North German Life Science Cluster Agency, has spared no effort to provide an excellent conference, where master students of the campus present their recent research results to a broad public of academics and industry.

The contributions show, how new approaches and methods in medical engineering and medical informatics can advance medicine, health, and health care. Moreover, this conference offers a good opportunity for both students and companies to get in touch at the Recruiticon, i.e. a satellite recruiting fair with industrial exhibition. Students from the Life Sciences programs present their results from projects carried out at the laboratories, clinics and institutes of Lübeck's Universities, in international research facilities, or research-oriented industrial companies. The conference focus has been placed on topics from medical engineering and medical informatics. The interdisciplinary field of medical engineering has been established at the Lübeck University of Applied Sciences for decades and Medical Engineering Science (Medizinische Ingenieurwissenschaft – MIW) is an important bachelor and master program at the Universität zu Lübeck as well. Both universities jointly offer the international master degree course Biomedical Engineering (BME). Furthermore, in the young master program Medical Informatics (Medizinische Informatik – MI) the 3rd Student Conference on Medical Informatics is integrated as an important element where project results in the emerging field of digital medicine are presented by the students.

As Conference Chairs, we thank all the people who worked with enthusiasm and dedication to make the conference a successful event. We want to thank the companies who support the meeting. Moreover, our thanks go to Infinite Science for producing these proceedings and organizing the Recruiticon meeting supporting the student conference. Personally and on behalf of all colleagues of the Student Conference Committee, we especially want to thank Christina Debbeler and Ksenija Gräfe from the Institute of Medical Engineering, they have been the central contact points for all questions of students and the program committee members as well as Jan-Hinrich Wrage from the Institute of Medical Informatics editing the proceedings. Their in-depth overview of all details of this event is the key to the success of the Student Conference 2018.

Lübeck, March 7-9, 2018

Prof. Dr. Thorsten M. Buzug
Chair of the 7th Student Conference
on Medical Engineering Science

Prof. Dr. Heinz Handels
Chair of the 3rd Student Conference
on Medical Informatics

Prof. Dr. Stephan Klein
Chair of the 1st Student Conference
on Biomedical Engineering

Exhibitors and Sponsors

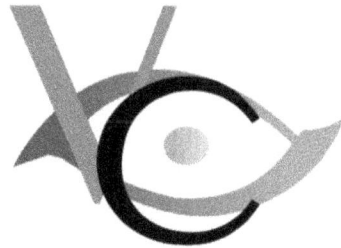

VisiConsult
X-ray Systems & Solutions

PHILIPS

senTec

Dräger

EUROIMMUN
a PerkinElmer company

Sponsors

Contents

Biomedical Engineering

Biomedical Optics

Biochemical Physics

Image Processing

Medical Imaging

Signal Processing

Safety and Quality

E-Health

1

Biomedical Engineering

Development of an Appropriate Temperature Profile for Brazing a 5-pole Feedthrough of Pacemakers

M. Ziauddin [1], H. Kalb [2], and S. Müller [3]

[1] Biomedical Engineering, University of Applied Sciences Lübeck, .ziauddin@stud.fh-luebeck.de
[2] Biotronik, Nürnberg, hermann.kalb@biotronik.de
[3] Department of Applied Natural Sciences, University of Applied Sciences Lübeck, stefan.mueller@fh-luebeck.de

Abstract

The quality of a feedthrough depends on the materials used for its brazing and the selection of a best-suited brazing temperature profile. Metal and ceramic joint quality is influenced significantly by the brazing duration above the melting point of the brazing material. In case of implantable devices (pacemakers in this case), the controlled use of bio-active metals (e.g. Copper) is also important. In this experiment a comparative analysis of different brazing temperature profiles is made and the effect of these profiles on the formation of the Titanium-Ceramic ($Ti-Al_2O_3$) and Niobium- Ceramic ($Nb-Al_2O_3$) joints is studied. Furthermore, the effect of brazing on the flow of Gold (Au) and Copper(Cu) along the feedthrough pins is studied by means of a Digital Light Microscope.

1 Introduction

A (hermetic) feedthrough (Fig. 1) in a pacemaker provides, the electrical and mechanical isolations, of the internal circuitry from the external environment. To achieve these characteristics, feedthroughs have to be both mechanically and electrically stable. Moreover, as component of an implantable device, it has to be bio-compatible and corrosion resistant as it cannot be excluded with absolute certainty that the feedthrough gets in contact with body fluids. A feedthrough consists of several components (Fig. 2), usually a metal flange provides the supporting base of the feedthrough. This flange is built in a metallic housing of a pacemaker. Inside the flange, the electrical signal carrying pins are installed with the help of isolation ceramic [1]. Pins are isolated from each other by ceramic and are brazed with a brazing material to form a joint with the ceramic. The ceramic itself, carrying the pins, is also brazed to the metal flange.

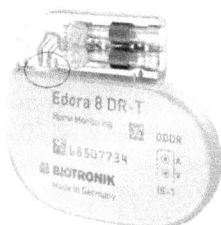

Figure 1: A 5-pole feedthrough (marked in a circle) in a pacemaker.

Ceramic has become a material of choice in electronic

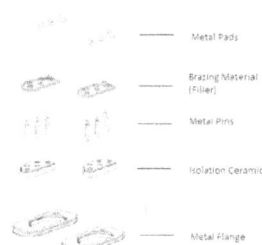

Figure 2: Components of a 5-pole feedthrough of pacemakers.

industries where combined properties of products, like electrical conductivity and electrical isolation are desired [2]. In most applications, ceramic-metal joints are limited to a temperature below 1000^0C [3],[4]. In the case of feedthrough, ceramic-metal joint can be exposed to a higher temperature while the feedthrough is welded to the metal housing of pacemaker. It is desired that feedthrough joint must withstand any failure because of high temperature during welding process, to fulfill such requirement high temperature ($>1000^0C$) brazing is preferred for the feedthrough's production. For high temperature applications, the selection of ceramic and metal type as well as brazing material are important. The brazing cycle of a ceramic-metal joint usually consists of three independent parts [5]; Preheating, Brazing and Cooling (Fig. 3). The preheating temperature is usually less than the melting temperature of the brazing material and the duration of preheating can be varied based on size of the materials to be brazed. In the Brazing portion, the temperature reaches the melting temperature of the brazing material and its

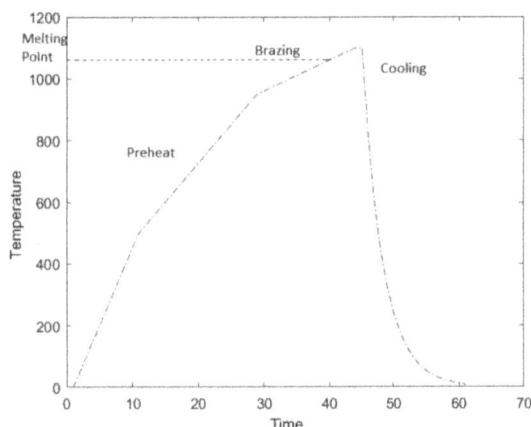

Figure 3: A typical brazing temperature profile of a feedthrough.

duration is kept as minimum as possible. A prolonged duration in this segment may affect the ceramic-metal joint quality [6].

In this study, gold (Au) was selected as the brazing material. Several temperature profiles, comprising of different segments of the brazing profile was compared with regards to visual inspection criteria (standard procedure of manufacturer to check brazed feedthrough in production) in order to select the best brazing profile for brazing the 5-pole feedthrough in a double chamber brazing fixture.

2 Material and Methods

Niobium (Nb)pins and Titanium (Ti) flange was used in the experiment. Gold is selected as brazing material because of its biocompatibility and its non-corrosive nature in body environment [7]. The pads are made of nickel with a copper alloy foil as its base to join the pads with the pins in brazing process and itself a non bio-compatible material for the implant, it is intended to control the overflow of copper along the pins. Ceramic and metals used for feedthrough are listed in Table 1.

Table 1: Feedthrough materials list.

Description	Materials
Pins	Nb (99.50 wt%)
Flange	Ti (100 wt%)
Ceramic	Al2O3 (99 wt%)
Brazing Material	Au (99.95 wt%)
Pads	Nb (90 wt%)

2.1 Ceramic Joint Assembly

The surface of ceramic is coated with a metal layer before assembling for brazing as it is a requirement for ceramic brazing to receive a good joint between brazing material

and ceramic [8]. For the first step of the assembly process, the metal flange is inserted in the brazing block and then the isolation ceramic is inserted inside the Ti flange. Nb pins are inserted into the ceramic holes, after that, a Au ring is placed in between the flange and ceramic, Au disks are placed between pin and ceramic as the brazing material. Finally, Ni pads are placed on the top the pins. A fully assembled brazing block is shown in Fig. 4.

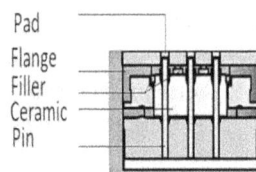

Figure 4: All the components of feedthrough are assembled inside the brazing fixture.

2.2 Heating and Temperature Profile

Heating is done in several stages. In the beginning, brazing block is heated up to 600^0C and then in the preheat segment, block is preheated at a rate of 10 units/time. In the next heating segment, temperature is ramped up at a rate of 5 units/time. Brazing starts when temperature reached 1063^0C [9]. In the next heating segment, temperature is gradually increased at a rate of 3 units/time and finally during the very high temperature segment, temperature is slowly increased at a rate of 1 unit/time [Table 2]. To investigate the temperature profiles, different heating time was applied during each segment therefore the brazing temperature changed for each profile as seen in Fig. 5.

The brazed samples are then inspected under a Digital Light Microscope (Keyence VHX 5000, Keyence Corporation, Japan) to check brazing quality, particularly, to determine the extent to which gold and Cu flow downward along the pin surface.

Table 2: Different temperature profiles.

Ramp (temp. unit/time)	10	5	3	1
Profile	Seg. 1 [time]	Seg. 2 [time]	Seg. 3 [time]	Seg. 4 [time]
1	33.2	19.6	11.7	4
2	34.7	19.6	11.7	4
3	36	27.6	0	0
4	37	27.6	0	0

3 Results and Discussion

3.1 Brazing Quality

In profile 1, the preheat segment (segment 1 and 2 in this case) is short, the brazing segment, segment 4, tempera-

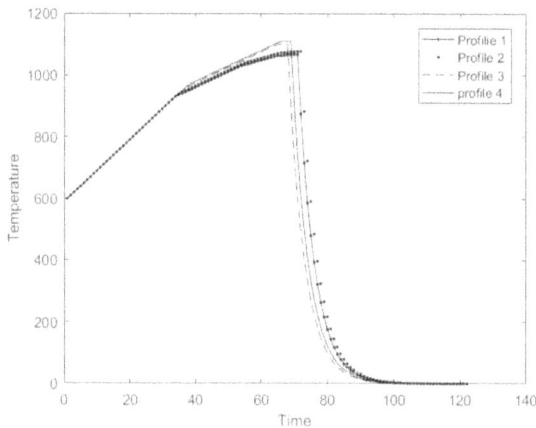

Figure 5: Different temperature profiles.

ture just reached the brazing point and then cooling started. Therefore, in all brazed samples the gold rings were not melted and pins were not attached to the ceramic, Fig. 6a. In profile 2, preheat temperature is increased (segment 1) and other segment settings are unchanged. The brazing samples of this profile shows that all gold rings are melted and attached the ceramic to the metal flange, but still not all pins are attached to the ceramic, Fig. 6b. In profile 3, the preheat segment (segment 1) is prolonged even more and the second segment is increased as well. To limit the over heating, the high temperature regions are removed in this profile. All the samples, brazed in this profile, are without any melting failure, Fig. 6c. In the profile 4, the preheat segment (segment 1 in this case) was increased further than the profile 3, and other segments remain unchanged. Every samples of this brazing profile were brazed well and an overflow of gold on the flange is visible in the earth pin of feedthrough as a result of very high temperature, Fig. 6d.

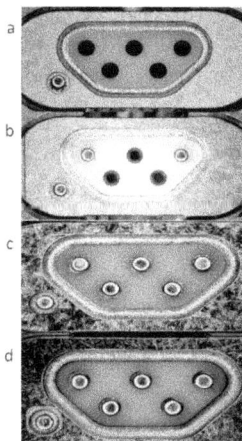

Figure 6: Feedthrough samples brazed at different temperature profiles, a) sample of profile 1: all gold rings unmelted b) sample of profile 2: two gold rings melted c) sample of profile 3: all gold rings melted d) sample of profile 4: all gold rings melted.

3.2   Gold downflow

Only the samples of profile 3 and profile 4 are fully brazed, therefore samples of these profiles are examined for gold downflow along the pin surface. Brazing samples of profile 3 does not show any downflow of gold along the pin surface, Fig. 7a, whereas in the samples of profile 4 gold flows down along the pins, Fig. 7b.

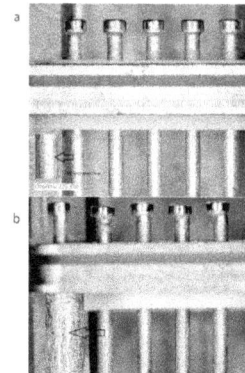

Figure 7: Gold down flow along the pins of the brazed samples; a) sample of profile 3: No gold flow along pins b) sample of profile 4: gold flows down along the pins.

3.3   Copper downflow

In samples of profile 3, Cu flow is observed only in the ground pin Fig. 8a, in profile 4 samples, Cu downflow is visible in all pins of the feedthrough, Fig. 8b.

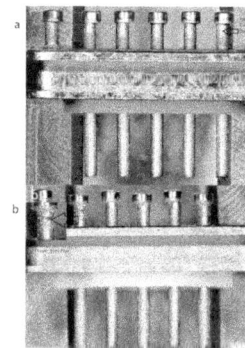

Figure 8: Copper downflow along the pins of the brazed samples;a) sample of profile 3: No copper flow along pins b) sample of profile 4: copper flows down along the pins.

4   Conclusion

Temperature profile with a short preheat segment and then slow heat rate up to the brazing point melts the brazing material of feedthrough and make joints between the ceramic and metal flange, ceramic and pins, pads and pins without any copper downflow and gold downflow along the pins.

Temperature profile with a long preheat segment with further increment of its temperature during segment 4 brazes the feedthrough components but causes the gold and the copper downflow along the pins, which violate the brazing criteria of the feedthrough.

Brazing was done with only one particular supply lot of the pins and ceramics, result of the experiment may change with the change of the supplies (pins and ceramics). Future experiments are in the planning phase to validate the findings of this experiment with different supply lots and to check the mechanical and electrical isolation property of the feedtrhough.

Acknowledgement

The work has been carried out at BIOTRONIK SE & Co. KG, Nürnberg and supervised by Prof. Dr. Stefan Müller Department of Applied Natural Sciences, University of Applied Sciences Lübeck.

5   References

[1] Robert E. Kraska, Frank J. Wilary and Joseph F. Lessar, *Hermetic electrical feedthrough assembly*. Patent US4678868 A.

[2] H. P. Martin, A. Triebert and B. Matthey, *Ta-Ni braze for high temperature stable ceramic-ceramic junctions*. Löt 2013.

[3] J. A. Fernie, R. A. L. Drew, K. M. Knowles, *Joining of Engineering Ceramics*. International Materials Review vol. 54, 2009.

[4] M. R. Rijnders and S. D. Peteves, *Joining of alumina using a V-active filler metal*. Scripta Materialia vol. 41, pp. 1137-1147, 1999.

[5] G. Waning and Bielefeld, *Protective atmosphere in heat treatment furnace*. Löt 2013.

[6] Yongtong Cao, Jiazhen Yan, Ning Li, Yi Zheng and Chenglai Xin, *Effects of brazing temperature on microstructure and mechanical performance of Al2O3/AgCuTi/Fe-Ni-Co brazed joints*. Journal of Alloys and Compounds vol. 650, pp. 30-36, 2015.

[7] ET Demann, PS Stein and JE Haubenreich, *Gold as an implant in medicine and dentistry*. Journal of Long Term Effects Medical Implants, Chapter 15, pp. 687–698, 2005.

[8] Osamu Kano and Atsuo Senda, *Method of metallizing ceramic material*. Patent US 4795658 A.

[9] M. H. Sloboda, *Industrial gold brazing alloys*. Gold Bulletin vol. 4, no. I, March 1971.

Requirements for Tapes for Sensor Attachment to the Human Skin

F. Eckardt [1], I. Menn [2] and S. Klein [3, 4]

[1] Medizinische Ingenieurwissenschaft, Universität zu Lübeck, franziska.eckardt@student.uni-luebeck.de
[2] SenTec GmbH, Rostock, ingolf.menn@sentec.com
[3] Medical Sensors and Devices Lab, Fachhochschule Lübeck, klein@fh-luebeck.de
[4] Institute for Electrical Engineering in Medicine, Universität zu Lübeck

Abstract

Sensor attachment tapes must maintain a tight fit of the sensor in many different circumstances (such as bodily movements, in incubators) so that reasonable measurements can be taken without interruptions. In order to be able to develop comprehensible uniform measurements of the required bond strength, a literature search was carried out on the basis of the corresponding DIN standards. For the examination and further development of medical tapes, a biological assessment of medical devices must always be carried out in accordance with DIN EN ISO 10993. However, there are standards only in industry that determine the adhesive strength of tapes, which after extensive inspections largely can be adapt to medical technology.

1 Introduction

For the further development of the current patient applicators of the company SenTec GmbH, reproducible methods are to be found under consideration of national and international standards, with which the adhesion of tapes on the skin can be determined. In the application, the tape should adhere very well on the one hand to allow a long-lasting accurate measurement with the respective sensor and on the other hand, the skin (especially those of neonates) neither injure during detachment nor irritate to much during the measurement. In order to be able to define the boundary conditions of an adequate test first, the current requirements for a tape for clinical use have to be determined from standards in advance.

To measure the transcutaneous oxygen and carbon dioxide partial pressure ($tcpO_2$ and $tcpCO_2$), sensors are used for non-invasive monitoring. In order to attach them airtight to the skin, retaining rings are glued to the skin and the sensors are clicked into it. As a tape for the retaining rings adhesive tapes are used with adhesion by pressure. These must retain their adhesive power in many different climatic conditions and with different skin types and conditions, for a tight fit of the sensors. The $tcpCO_2$ measurement is used in anesthetics, intensive care units, emergency departments, general practitioners, neonatal care, sleep diagnostics, pulmonology, home care and, nowadays, sports medicine and physiology [1], [2].

2 Material and Methods

For this work, only the currently valid DIN standards are considered more closely. The tapes must be biocompatible, non-irritating, painless removable and sensitive according to DIN EN ISO 10993. Furthermore the tapes must be breathable, germ-free and the adhesive power must be even when using disinfectants.

2.1 Biocompatibility according to DIN EN ISO 10993 - Biological evaluation of medical devices

All medical devices must have been subjected to a biological assessment in accordance with DIN EN ISO 10993. Because there must be no disadvantages to the patient in the application, which prevent the health and recovery considerably. For this reason, possible test methods are described in several parts. Especially importan for tapes is part 10 - test for irritation and skin sensitization important. In addition, all medical devices are subject to part 1, which describes the evaluation and testing as part of a risk management system.

2.1.1 Part 1 (2017): Evaluation and testing within a risk management system (Draft)

During a risk management process, medical devices are assessed for potential biological risks to humans. For this purpose, this section outlines general procedures that are applicable to any medical device (including active, non-active, implantable and non-implantable) in direct or indirect contact with the patient and which directly or indirectly come into contact with the attending physician. For this purpose, general principles for assessment within a risk management procedure are described, a general classification of medical devices by type and duration of body contact is shown and reference is made to the evaluation of available data from all sources, paying particular attention to gaps in the available datasets which must be closed during the evalution. In addition, this part provides a basic framework by which the biological safety of the medical device can be assessed. Fig. 1 shows a systematic approach for a biological evaluation as it can be performed according to the standard [3].

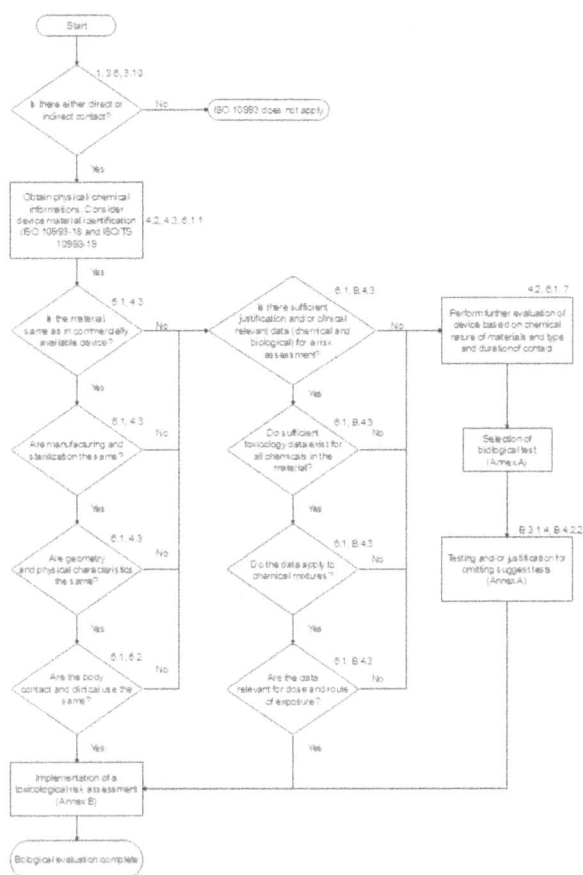

Figure 1: Systematic approach to a biological evaluation from DIN ISO EN 10993-1:2017.

2.1.2 Part 10 (2014): Tests for irritation and skin sensitization

Potential hazards during contact between humans and chemical substances exiting the medical device must be asses. In this part these assessment was made by the irritation of the skin, the mucous membrane and the eyes or a sensitization of the skin. Pre-test considerations for irritation, including in silico and in vitro dermal exposure, shall be performed after this part. It describes the conduct of in vivo tests and describes key factors for interpreting the results (Table 1). In addition, special tests are described

in which the medical devices come into contact with areas other than the skin [4].

Table 1: Classification for skin reaction

Reaction	Numerical grading
Erythema and eschar formation	
No erythema	0
Very slight erythema (barely perceptible)	1
Well-defined erythema	2
Moderate erythema	3
Severe erythema (beet-redness) to eschar formation preventing grading of erythema	4
Oedema formation	
No oedema	0
Very slight oedema (bareley perceptible)	1
Well-defined oedema (edges of area well-defined by definite raising)	2
Moderate oedema (raised approximately 1 mm)	3
Severe oedema (raised more than 1 mm and extending beyond exposure area)	4
Total possible score for irritation	8

2.2 Adhesive power

The tapes must not injure the skin of the patient when peeling, which is often the case with the sensitive skin of neonates. Therefore it is the best not to work with tapes, because in premature babies the skin is not fully developed and it can come to bruising and open wounds. There is dermal instability, the cornea is underdeveloped, and the epidermis and dermis are not yet firmly attached. At an age of two weeks, however, the skin of premature babies corresponds to mature babies [5].

The same tape may be used for a maximum of 24 hours. Especially in premature babies the measuring points are changed every one to two hours, with healthy skin every two to six hours. Because the sensors have to heat the measuring range between 41 and 44 °C to get the best possible results [1]. The change avoids skin burns, but also means that more tapes are used for fixation. Often two tapes with retaining rings are applied onto the patient and then changed between these places. Thus, the bond strength must always be given at elevated temperature, when changing the sensors, with common body movements, when changing clothes and body wash. In addition, the adhesiveness is influenced by the following factors: hair, use of creams, lotions, makeup and contaminated surfaces. On the other hand, there are more uncontrollable factors: skin breathability, sensitive, elastic and rough surfaces, cell renewal, low surface energy, climate, age and health of the patient, dieting and secretion production. To measure the bond strength, some standards from the industry can be adapted, which are briefly named in the following chapters.

2.2.1 DIN EN 1939: 2003 - Self adhesive tapes - Determination of peel adhesion properties

DIN EN 1939 specifies four methods for measuring the bond strength [6]:

1. measurement of the bond strength on stainless steel at an angle of 180°,

2. measurement of the bond strength on the own backside at an angle of 180°,

3. measurement of the bond strength on double-sided coated and transferable adhesive tapes at an angle of 180°,

4. measurement of the adhesive force of the release material on adhesive tape at an angle of 180°.

At the second and third method, other materials can be used for the test plates, which come closest to the later application.

2.2.2 DIN EN 1943: 2003 - Self adhesive tapes - Measurement of static shear adhesion

To measure the ability of pressure-sensitive adhesive tapes not to detach under constant load acting parallel to the surface of the adhesive tape and the carrier, seven methods are defined in DIN EN 1943 [7]:

1. measurement of the shear resistance on a standardized, vertically arranged steel test plate,

2. measurement of shear resistance on a vertically arranged test plate covered with a standard NIST SRM 1810A hardboard,

3. measurement of the shear resistance on a vertical test plate covered with a hardboard for which agreements have been made between the buyer and the seller,

4. measurement of the shear resistance of fiber-reinforced adhesive tape applied to a standardized, horizontally arranged steel test plate,

5. measurement of the shear resistance of fiber reinforced adhesive tape applied to a horizontally arranged test panel covered with a standard NIST SRM 1810A hardboard,

6. measurement of the shear resistance of fiber-reinforced adhesive tape applied to a horizontally arranged test plate covered with a hardboard for which agreements have been made between the buyer and the seller,

7. measurement of the shear resistance on a vertically arranged, standardized test plate made of steel at elevated temperature after 10 min waiting time.

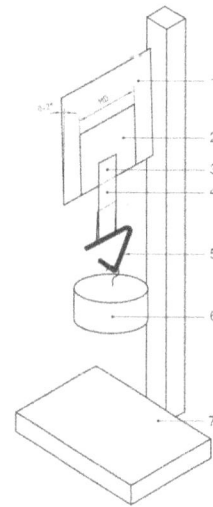

Figure 2: Schematic diagram from DIN 1943 for determining the shear resistance according to methods 1, 2, 3 and 7. 1: test plate, 2: steel plate or hard cardboard 3: test surface, 4: test strip, 5: clamp or hook, 6: test mass, 7: base plate with timepiece.

2.2.3 DIN EN 1945: 1996 - Self adhesive tapes - Measurement of quick stick

In the DIN EN 1945 a method is described to measure the initial tack of an adhesive tape, which is continuously withdrawn at an angle of 90° from a plate [8].

2.2.4 DIN EN 12023: 1996 - Self adhesive tapes - Measurement of water vapor transmission in a warm humid atmosphere

The DIN EN 12023 specifies a method for measuring the permeability of water vapor under test conditions. For this purpose, the adhesive tape separates the moist atmosphere present in an external container (93% R.F.) from a test cell with dry air. Then the mass of the test cell, which has diffused through the tape, is measured. The water vapor permeability is given as the increase in mass per unit time and per unit area of the test cell opening [9].

2.2.5 DIN EN 12024: 1996 - Self adhesive tapes - Measurement of resistance to elevated temperatures and humidity

The standard DIN EN 12024 specifies a method for determining the behavior of adhesive tapes at a defined temperature and humidity. Therefore the adhesive tape is exposed in a test environment of an elevated temperature and humidity for a certain time and then examined the existing effects on the physical or chemical properties [10].

3 Results and Discussion

A biological assessment of medical devices according to DIN EN ISO 10993 must always be implemented before a product is launched on the market. Due to the fact that the adhesive tapes and retaining rings are obtained from other companies, this assessment is also carried out there.

For its own further development, above all, the desired bond strength must be tested, which can be adapted for these purposes on the basis of the cited standards from industry.

To determine the bond strength, the first three methods (in particular the first method) of DIN EN 1939 can be used. The currently used adhesive tapes can be measured as a reference in order to subsequently determine measured values that can still be classified as sustainable when peeled off but still have enough adhesive power. For many patients the removal of tapes from the skin is unpleasant to painful and can lead to bruising and injury, especially in sensitive skin, such as premature and mature babies.

In order to be able to measure the effective load on the tape and thus possible displacements, the first, second, fourth, fifth and last method can be used from DIN EN 1943. The body movements and the sensor itself forces act on the retaining ring with tape, which can be affected to the bond strength and can lead to a possible premature detachment. Above all, the last method will be of greater interest, because in this method the influence of increased ambient temperature is taken into account, because with the $tcpCO_2$ measurement, the measuring point has to be heated to 41 to 44°C. The fourth and fifth methods use fiber-reinforced adhesive tapes and a heavier mass. These methods could be carried out on a trial basis with smaller masses, since the adhesive tapes used certainly do not endure such large forces.

The DIN EN 1945 can also be used to obtain an assessment of the quick stick, which is still rated as bearable when removing the tape for a patient.

With the DIN EN 12023 the permeability of water vapor can be measured. This is of greater interest because secretions and sweat escape through the skin, which can lead to a moist, warm environment under a tape. With the help of DIN EN 12024 the resistance to elevated temperature (the measuring point must be heated between 41 and 44°C) and air humidity can be measured, so that can be determined as a dimension in later measurements where it does not come to the release of the tape.

4   Conclusion

For the measurement of bond strength, the next step is to develop tests based on the corresponding standards. In addition, user surveys should be carried out in order to further define criteria for the development of the retaining rings and to connect these with the adhesion measurements. In this way values can be determined which can later be used as evaluation points for new tapes. With the use of new materials, a biological assessment must always be made in order to rule out later reactions to the patient. The preliminary considerations and tests can also be adapted to other uses of tapes, e.g. for insulin pumps, fixation plasters for venules and other sensors.

Acknowledgement

The work has been carried out at SenTec GmbH and supervised by the Institute for Electrical Engineering in Medicine, Universität zu Lübeck and the Medical Sensors and Devices Lab, Fachhochschule Lübeck.

5   References

[1] M. Stücker, U. Memmel, P. Altmeyer, *Transkutane Sauerstoffpartialdruck- und Kohlendioxidpartialdruckmessung - Verfahrenstechnik und Anwendungsgebiete*. In: Phlebologie 4/2000, F.K. Schattauer Verlagsgesellschaft mbH, Stuttgart, pp. 81–91, 2000. Available: http://www.schattauer.de/t3page/1214.html?manuscript=1065PB [last accessed on 2018-01-08].

[2] D. Teising, H. Jipp, *Neonatologische und pädiatrische Intensive- und Anästhesiepflege - Praxisleitfaden*. Springer Verlag, Berlin, Heidelberg, 2016.

[3] DIN Deutsches Institut für Normung e.V., *DIN EN ISO 10993-1:2017-04 - Draft - Biological evaluation of medical devices - Part 1: Evaluation and teting within a riskmanagement system*. Beuth Verlag GmbH, Berlin, 2017.

[4] DIN Deutsches Institut für Normung e.V., *DIN EN ISO 10993-10:2014-10 - Biological evaluation of medical devices - Part 10: Tests for irritation and skin sensitization*. Beuth Verlag GmbH, Berlin, 2014.

[5] H. Traupe, H. Hamm, *Pädiatrische Dermatologie*. Springer Medizin Verlag, Heidelberg, 2006.

[6] DIN Deutsches Institut für Normung e.V., *DIN EN 1939:2003-12 - Self adhesive tapes - Determination of peel adhesion properties*. Beuth Verlag GmbH, Berlin, 2003.

[7] DIN Deutsches Institut für Normung e.V., *DIN EN 1943:2003-01 - Self adhesive tapes - Measurement of static shear adhesion*. Beuth Verlag GmbH, Berlin, 2003.

[8] DIN Deutsches Institut für Normung e.V., *DIN EN 1945:1996-04 - Self adhesive tapes - Measurement of quick stick*. Beuth Verlag GmbH, Berlin, 1996.

[9] DIN Deutsches Institut für Normung e.V., *DIN EN 12023:1996-11 - Self adhesive tapes - Measurement of water vapor transmission in a warm humid atmosphere*. Beuth Verlag GmbH, Berlin, 1996.

[10] DIN Deutsches Institut für Normung e.V., *DIN EN 12024:1996-07 - Self adhesive tapes - Measurement of resistance to elevated temperature and humidity*. Beuth Verlag GmbH, Berlin, 1996.

Optimization of the Inspiratory and Expiratory Controller of an Innovative Anesthesia Device

S. Schmees [1], M. Meyer [2], P. Rostalski [3]

[1] Medical Engineering, Universität zu Lübeck, steffen.schmees@student.uni-luebeck.de
[2] Drägerwerk AG, Lübeck, marius.meyer@draeger.com
[3] Institute for Electrical Engineering in Medicine, Universität zu Lübeck, philipp.rostalski@uni-luebeck.de

Abstract

Anesthesia devices are subject to ever increasing demands. To accomplish this, new prototypes have to be designed and tested. Additionally, for mandatory ventilation a control design is necessary to supply the patient with a precise pressure and volume. For the design a proper system model has to be found. These models differ from patient to patient. Therefore this work focuses on a controller, which can be switched for the two patient categories "pediatric" and "adult". In order to do this, two system models for these categories were experimentally determined and integrated into the closed loop control design. Afterwards the newly designed controllers have to be validated within their assigned patient category. The results of the experiment show an improvement for both patient categories, especially for the adult class.

1 Introduction

Mandatory ventilation in modern anesthesia devices is usually done with a positive pressure, contrary to the physiological respiratory drive. Because of these non-physiological conditions the anesthetist should be able to choose exact inspiratory and expiratory pressure settings to spare the lung from ventilator induced lung injuries, such as a barotrauma or volutrauma [5]. To comply with the requirement, a robust control system is needed. Commonly in anesthesia devices a proportional, integral (PI) closed loop controller is used. For control design the lung has to be modeled. In analogy to electrical circuits the lung model can be simplified to the physical sizes pressure P, volume V and flow $\dot{V}$ over the constant parameters resistance R and compliance C [6]

$$R = \frac{\Delta p}{\Delta \dot{V}}, \ C = \frac{\Delta V}{\Delta p}. \qquad (1)$$

The equivalent circuit diagram in Fig. 1 shows the simplification of the lung as a series connection of a resistance and a capacitor. Considering that, mandatory ventilation can either be done with a pressure or volume controlled ventilation (PCV, VCV). In Fig. 2 the time dependent PCV curve can be seen. It should be noted, that the airway pressure (PAW) has a rectangular shape. Furthermore the PAW has an offset, the positive end expiratory pressure (PEEP). This is necessary to protect the alveoli from collapsing [6]. This ventilation mode results into a decelerating flow profile. The area under the flow curve corresponds to the applied tidal volume. The purpose of this paper is to improve the inspiratory and expiratory controller of an innovative anesthesia device, with a controller, which can be switched for the two patient categories, pediatric and adult. This should improve the PCV.

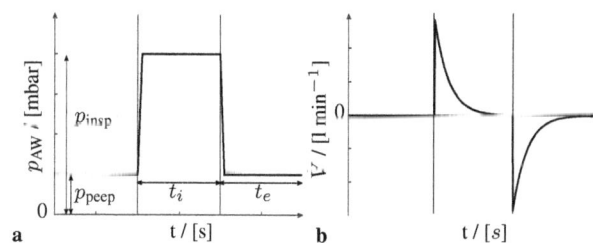

Figure 2: (a) Pressure and (b) flow curve dependent on time for the pressure controlled ventilation [2]

2 Material and Methods

2.1 Anesthesia Device

The prototype used is a half-closed anesthesia device. It operates with a circuit system to reuse a part of the expiratory air. In consequence of this recirculation of

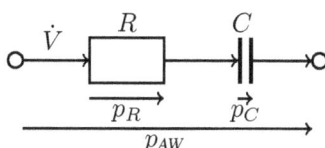

Figure 1: Simplified electrical circuit diagram [5]

the expiratory branch into the inspiratory, the anesthetic agents can be used in a resource-conserving manner. To prevent a hypercapnia of the patient, the carbon dioxide is removed with soda lime and oxygen is added externally. The prototype uses a centrifugal compressor (blower) to generate a pressure which supplies the patient with the necessary volume. For ventilation control, two pressure sensors (inspiratory and expiratory) and one flow sensor are used. Additionally a breathing bag as a gas reservoir is given, to prevent a dependency of the respiratory minute volume on the fresh gas flow. Furthermore an adjustable pressure limiting valve (APL) ejects the excessive gases [2]. The evaluation of the implemented controller is done with a rapid control prototyping system. The control algorithms are implemented as a real-time application in a Simulink® (MathWorks GmbH, Massachusetts, United States) model. Subsequently the Simulink Coder generates the code to send it via ethernet to the real-time target machine Speedgoat (Speedgoat, Liebefeld, Schweiz), which communicates with the I/O, like the sensor and actuator unit of the connected prototype. Additionally the Simulink-Real-Time Explorer is used to design a graphical user interface (GUI), to display or modify parameters.

2.2 System Identification

In a previous work, the system identification of the overall anesthesia device, was done without any test lungs. Instead the inspiratory and expiratory connections were short-circuited by combining them with tubes in a closed y-piece. To improve the system modeling of the overall system, a more realistic model should be selected. Therefore in this work two glass test lungs in combination with a resistance are used for the system identification. The lungs correspond to a pediatric ($C = 20$ ml mbar^{-1}, $R = 19.4$ mbar s l^{-1}) and an adult lung ($C = 49.6$ ml mbar^{-1}, $R = 4.91$ mbar s l^{-1}). The two system models are used to design a controller, which can be switched for the two patient categories. With the simplification in Fig. 1 and Eq. (1), the system can be modeled by a first order differential equation. The control system is implemented in discrete-time with a sampling interval of 10 ms and hence, the z-transformation is used to obtain the discrete-time system model, leading to the system transfer function [4]

$$G(z) = \frac{b_1 z^{-1}}{1 - a_1 z^{-1}}. \tag{2}$$

To identify the system behavior the expiratory pressure sensor is being evaluated after the blower is changed stepwise, to receive a step response. For the inspiratory system model a positive and for the expiratory a negative step is used. The identification is done with the System Identification Toolbox (MATLAB®), with an ARX-method. It uses a least-squares method, which minimizes the quadratic error between the theoretical model and the experimental measured data in an iterative approach [1]. To obtain a more robust estimation, different step sizes were used and the coefficients

b_1 and a_1 of the corresponding transfer functions were averaged respectively. The results of the averaged transfer functions for the model with the pediatric lung G_{ped} and the adult lung G_{ad} are shown in Table 1, whereby the unit delay in (2) is negligible. Additionally the transfer function G_y for the model with the closed y-piece is appended.

Table 1: Y-piece, pediatric and adult transfer functions for the inspiratory and expiratory model

Inspiratory model	Expiratory model
$G_{\text{y,insp}}(z) = \frac{0.123z}{z-0.878}$	$G_{\text{y,exp}}(z) = \frac{0.062z}{z-0.938}$
$G_{\text{ped,insp}}(z) = \frac{0.076z}{z-0.924}$	$G_{\text{ped,exp}}(z) = \frac{0.053z}{z-0.949}$
$G_{\text{ad,insp}}(z) = \frac{0.036z}{z-0.966}$	$G_{\text{ad,exp}}(z) = \frac{0.035z}{z-0.966}$

Fig. 3 shows an example of the step responses for the models of the two patient categories and a simulation of the system models respectively. A PEEP = 5 mbar and $p_{\text{insp}} = 20$ mbar were used. It should be noted that the measured data have a deviation of the set point in all measurements. The discrepancy of the modeled data to the set point are minor, but the rise time is shorter in the adult models than in the measured data. Additionally the measurements exhibit a strong oscillation, especially with the inspiratory adult model. To avoid the set point deviation and oscillations a control design is necessary.

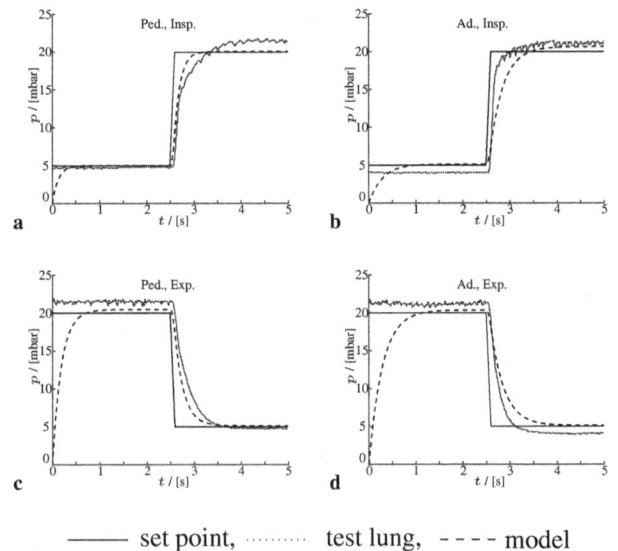

Figure 3: Measured data for an open-loop control with PEEP = 5 mbar and $p_{\text{insp}} = 20$ mbar. Additionally the system modelling and the set point curve is attached. (a) Inspiration with pediatric test lung, (b) Inspiration with adult test lung, (c) Expiration with pediatric test lung, (d) Expiration with adult test lung.

2.3 Control Design

Based on the system models respectively a PI-controller for the inspiration and expiration, with different gains, was designed (Fig. 4). The user shall now be able to choose with a switch between the both patient categories and in further consequence the according controller. The controllers contain a proportional gain K_P to get a quick reaction of the control variable to the input signal and an integrator with the gain K_I to reduce the remaining set point deviation [3]. The output limit of the blower can lead to a wind-up effect. Therefore a back-calculating anti-wind up with the gain K_W is integrated in the control design [7]. To improve the control behavior of the inspiratory controller a feed forward control is added to this controller. This concept is done with an inversion based design of the estimated system models [3]. Additionally a switch signal decides between the expiratory and inspiratory breathing phase and switches the controller, respectively. To avoid larger steps between both PI-controllers the output of the controllers follows a tracking signal. The difference between the output and the tracked signal is led back to the integrator input (bumpless transfer). Furthermore the inspiratory integrator is reset to zero in the expiratory breathing phase and analogue the expiratory integrator is reset to the PEEP. With these adjustments both integral components would not discharge at the change of the breathing phase and consequently the controllers should be faster. The PI-controllers were first tuned in a simulation with Simulink® with the according estimated transfer functions and afterwards determined experimentally.

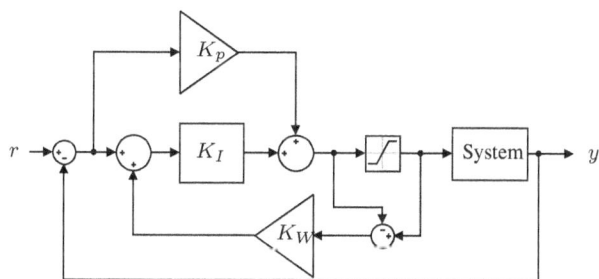

Figure 4: Closed loop PI-controller with anti-wind up compensation

3   Results and Discussion

To evaluate the control design, the results are compared to the previously implemented controller. This controller uses a similar control design, but does not include the integrator reset and was only tuned for the closed y-piece. For evaluation, the mean pressure values and standard deviations of the inspiratory and expiratory breathing phase are compared with measurements of different step sizes (table 2). One measurement lasts 38 s and the inspiratory and expiratory breathing phases last 6 s each. To calculate the mean, the signals after the settling time were summarized and averaged out for the corresponding controller. Additionally, the following conditions were given and should be adhered: For the inspiration and expiration, a tolerance of the settings

Table 2: Settings for the pressure controlled ventilation

PEEP / [mbar]	P_insp / [mbar]
5	7; 10; 15; 20; 25
10	12; 15; 20; 25; 30
15	17; 20; 25; 30; 35
20	21

about 10 % of the set point and a standard deviation of $\pm$ 2 mbar is allowed respectively.

3.1   Pressure Controlled Ventilation

The measurements of a PCV with a PEEP of 5 mbar and an inspiratory pressure of 20 mbar are shown in Fig. 5. It can be seen that in the inspiration phase the controller developed based on the pediatric model and the previous controller have a similar rise time. In contrast to that the controller developed based on the adult model is faster. In the expiration phase both switchable controllers have a smaller rise time. This means that the controller for the patient cat-

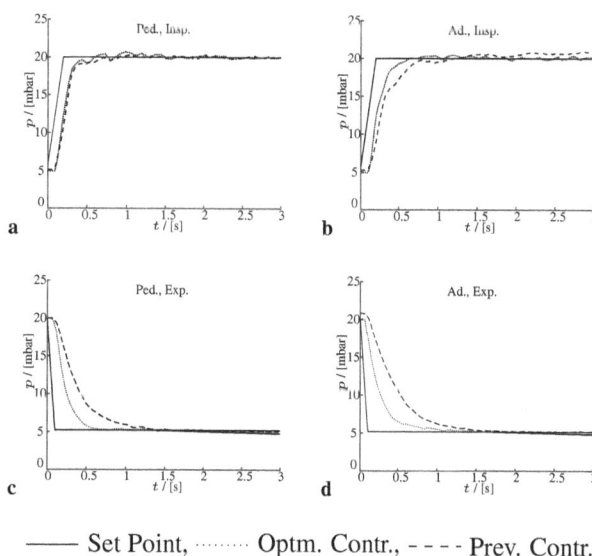

—— Set Point, ········ Optm. Contr., – – – – Prev. Contr.

Figure 5: Closed loop control step response for the optimized and previous controller, for comparison the set point is attached. (a) Inspiration with pediatric test lung, (b) Inspiration with adult test lung, (c) Expiration with pediatric test lung, (d) Expiration with adult test lung.

egory pediatric could barely be improved in the inspiration breathing phase, in contrast to the adult patient category. One reason could be that the system models of the pediatric test lung and the closed y-piece are more alike than the adult test lung. In all measurements the expiratory breathing phase could be improved, this is owed to the reset of the expiratory integrator. These results can also be confirmed in table 3. Thereby the mean time constants T for a step response of first order for both controllers are summarized. This constant corresponds to the time after 63 % of the set point is reached [3].

Table 3: Time constants of a control path of first order for the optimized and previous controller.

	Optm.	Prev.
$T_{\text{Ped,Insp}}$ / [s]	0.254	0.293
$T_{\text{Ped,Exp}}$ / [s]	0.217	0.351
$T_{\text{Ad,Insp}}$ / [s]	0.265	0.373
$T_{\text{Ad,Exp}}$ / [s]	0.222	0.441

In Fig. 6 the deviation of the mean values from the set point $\Delta p = p_{\text{set}} - p_{\text{mean}}$ and the standard deviations of the measurements are summarized. The dotted line at zero shows no deviation and the enveloping dotted line includes the given 0.2 mbar standard deviation limit. The first four values before the vertical dotted line are the deviations of the mean value from the PEEP and the remaining from the p_{insp}. It can be seen that for the pediatric test lung, the mean values do not differ much from the set point. However the standard deviation for measurements of high pressure leaves the limiting area of 0.2 mbar standard deviation for both controllers. This means that the controller oscillates too much at high pressure. For the adult test lung the mean values of the adult controller do not differ much from the set point, except at a PEEP of 15 mbar. In contrast to that, the mean values of the previous controller have a larger scattering of up to 1 mbar especially for high pressure values. Likewise the requirements on the standard deviations can not be met in a lot of cases.

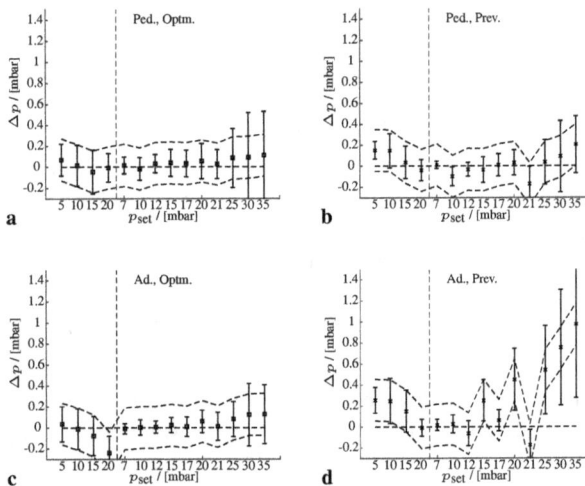

Figure 6: Summarized mean values and standard deviations of the optimized (optm.) and previous (prev.) controller for the two patient categories. (a) Optimized controller for pediatric lung, (b) Previous controller for pediatric lung, (c) Optimized controller for adult lung, (d) Previous controller for adult lung.

4 Conclusion

This work shows that the previous controller could be improved, especially the expiratory controller is faster for both patient categories. This is achieved by the reset of the integrator to the initial PEEP value. Furthermore the optimized controllers are reasonable in the case of an adult test lung, because the adult inspiratory controller has a shorter rise time and smaller oscillations. This is due to the fact that the adult system model fits the adult test lung better, so that the feed forward has a higher impact in the closed loop control. On the other hand the system model of the pediatric test lung and the y-piece are similar, therefore there were no improvements due to the new controller.

In a further experiment the tightness of the prototype should be verified. Thereby a leakage of $5.27\,\mathrm{l\,min^{-1}}$ at a pressure of 40 mbar could be observed. This high leakage indicates further non-linearities and makes the closed loop control more difficult, because the linearity assumption of the system model does not hold. In order to get further improvement a tighter system could be necessary. Furthermore a more sophisticated controller could be used. For example model predictive control whereby the system model contains the inspiratory as well the expiratory breathing phase.

Acknowledgement

The work has been carried out at Drägerwerk AG, Lübeck, and supervised by Institute for Electrical Engineering in Medicine, Universität zu Lübeck.

5 References

[1] C. Bohn, H. Unbehauen, *Identifikation dynamischer Systeme - Methoden zur experimentellen Modellbildung aus Messdaten.* Springer Vieweg, vol. 1, pp. 55–75, 2016.

[2] R. Kramme, *Medizintechnik.* Springer-Verlag, vol. 4, pp. 477–493, 2011.

[3] J. Lunze, *Regelungstechnik 1 - Systemtheoretische Grundlagen, Analyse und Entwurf einschleifiger Regelungen.* Springer Vieweg, vol. 10, p. 213, 2014.

[4] J. Lunze, *Regelungstechnik 2 - Mehrgrößensysteme, Digitale Regelung.* Springer Vieweg, vol. 8, p. 213, 2014.

[5] J. Werner, *Biomedizinische Technik - Automatisierte Therapiesysteme.* Walter de Gruyter GmbH, vol. 9, pp. 155–182, 2014.

[6] L. Ziegenfuß, *Beatmung.* Springer Medizin, vol. 5, pp. 30-35, 2013.

[7] S. Tarbouriech, G. Garcia, J.M. Gomes da Silva, I. Queinnec, *An Overview of Anti-windup Techniques. In: Stability and Stabilization of Linear Systems with Saturating Actuators.* Springer, vol. 1, p.267–281, 2011.

Detection of Patient Ventilator Asynchrony based on Surface Electromyography

F. Spieß [1], J. Graßhoff [2], E. Petersen [2], and P. Rostalski [2]

[1] Medizinische Ingenieurwissenschaft, Universität zu Lübeck, felix.spiess@student.uni-luebeck.de
[2] Institute of Electrical Engineering in Medicine, Universität zu Lübeck,
{j.grasshoff, eike.petersen, philipp.rostalski}@uni-luebeck.de

Abstract

In assisted ventilation, the air support provided by the ventilator device is controlled according to patient demand by recording and analyzing physiological signals. As a non-invasive, time performant acquisition procedure, detection of respiratory muscle potentials via *surface Electromyography* (sEMG) represents a promising approach. Misinterpretation as well as inadequate factorization of selected parameters lead to discrepancies in the operational state between patient and ventilator, called *Patient Ventilator Asynchrony* (PVA). These distortions are attended by physiological miserable outcomes [7]. Several processing steps must be applied to extract triggering events for change in breathing phases out of raw sEMG signals. Four patterns including different approaches for each processing segment were investigated. As major influence the results provided, irregular detection behavior due to noisy shape of processed signals came into view. So, careful preprocessing should be of central interest concerning further research based on these findings.

1 Introduction

During assisted ventilation procedures, the air supply from the ventilator device is regulated by the patient demand. This is achieved by recording and analyzing of physiological outputs representing breathing effort. Several mechanical as well as bioelectrical signals have been investigated concerning this application [1]. Well established treatments tracking patients breathing demand are predicated on direct measurement of the airflow initiated by the patient. Due to the prolonged delay between the start of patient demand and development of considerable signals amplitudes, such practices are incapable of tracking patients breathing effort in real-time. Furthermore, subjects suffering from respiratory muscular weakness cannot initiate recognizable flow amplitudes reliably. Another pneumomechanical signal is originated in invasive measurement of esophageal pressure (Pes) by balloon catheters [3]. In respect to the invasive nature as well as sensitivity towards systematical or spontaneous distortions, its applicability under clinical circumstances cannot be ensured generally.

To get direct access to patients breathing effort, excluding a maximum amount of periphery physiological system components, the detection of myoelectric potentials of respiratory muscles has gained increasing attention the last decade, expressed as *neurally adjusted ventilator assist* (NAVA), [7]. The most common procedure consists of invasive recording of electrical potentials in the diaphragm muscular system, referred as *Electrical Activity on Diaphragm* (also Edi or EAdi) [4]. This signal is recorded by inser-

tion of an electrode in the crus of the diaphragm by oral or nasal catheter. In contrast to invasive measurement procedures, detection of myoelectric potentials by electrodes placed at the body surface, known as *surface Electromyography* (sEMG) represents a promising approach for detection of breathing demand. The signal acquisition principle is depicted in Fig. 1.

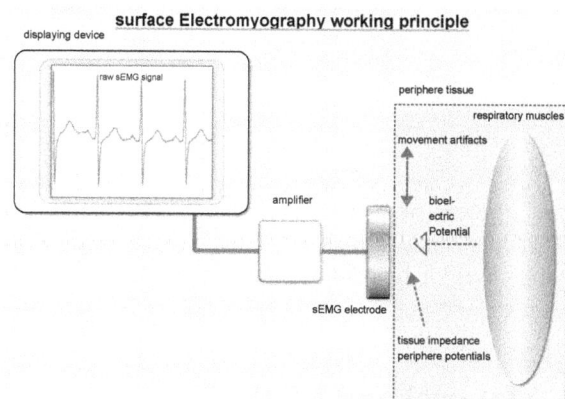

Figure 1: Working fundamentals of sEMG recordings. Bioelectric potentials of respiratory muscles are recorded by a bipolar electrode undergoing distortion processes. Afterwards, the signal gets amplified before further processing is applied.

Placement of electrodes at the body surface implies specific distortions originating from bioelectric potentials of periphery muscle groups, predominantly cardiac interferences, tissue-intrinsic impedance as well as relative movement processes. Main task with respect to these limitations consists of specific filtering, preventing the relatively low amplitude breathing activity patterns while ensuring sufficient exclusion of highly dominant cardiac signal oscillations [8]. Nonlinearity of respiratory and cardiac motions accuse overlapping spectral harmonics. Conservative spectral filtering techniques are not capable of separating those components without losses. More advanced strategies based on Wavelet transformation or Empirical Mode Decomposition are currently under investigation showing promising performance [9].

The whole breathing activity cannot be traced back to a single action potential. Respecting that, a transformation representing the longterm variation of the signal, suppressing the highly frequent single potential oscillations the raw sEMG recording consists of, is generated. The processing pattern is visualized in Fig. 2. Multiple different approaches like simple linear envelope functions, entropy or wavelet based strategies exist [2], [6]. Concerning changepoint detection, several procedures ranging from constant amplitude thresholds [4] to statistical feature based approaches [5] have been implemented. During the research activity concerning this publication, it is the proclaimed pursuit to investigate the ability of sEMG based activity detection strategies with respect to synchrony to patient demand. As reference signal representing patients breathing activity, EAdi measurements were taken into account.

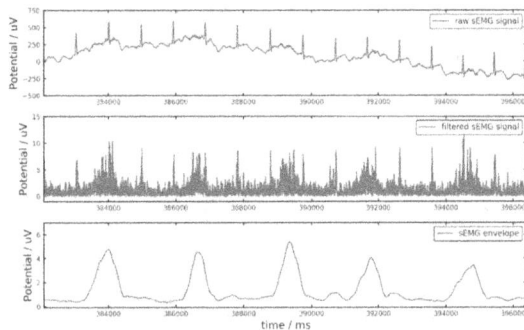

Figure 2: Steps of signal preprocessing. Top figure: raw sEMG signal, middle: high pass filtered and rectified signal, bottom: envelope function via moving median

2 Material and Methods

2.1 Processing Patterns

To get a valid estimation about the influences of the several processing steps, each one of them has been varied. First, the preprocessing of raw sEMG signals was investigated. On the one hand, a zero phase, fourth order butterworth high pass filtering with 50 Hz cutoff frequency, followed by a linear envelope generation via moving median of 402 ms was put up. To assess the preprocessing strategy, mainly

the importance of the balance between ECG removal and respiratory signal preservation, a linear envelope based on a moving variance calculation was set up on a pre-filtered signal developed by Dräger. As reference for the envelope formation, a proprietary procedure invented by Dräger was simultaneously examined. Additionally, two different approaches to detect start and end of breathing activity were compared. One strategy sets a constant threshold as onset detection criterion, whereas offset value is represented by decreasing of envelopes value below 70 percent of the current maximum of inspiration phase [4]. This strategy will be named *constant onset detection*, if not stated otherwise. Its manifestation on a sEMG envelope function is visualized in Fig. 3.

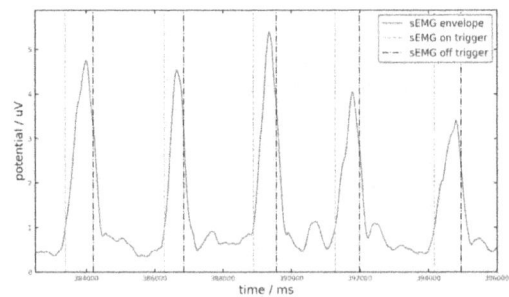

Figure 3: detection of trigger events representing changepoints in breathing phase via constant threshold application

In contrast, a dynamic procedure extracts the mode of the distribution of sEMG values between two consecutive maxima by gaussian kernel density estimation. The onset value is recorded by marking the chronologically last point in current interval, this mode value is passed in ascending manner [5]. This technique will be referred as *dynamic onset detection*.

To summarize the different combinations of processing elements, four patterns were investigated. (1): high pass filtering followed by median envelope including constant onset detection. (2): moving variance envelope function settled on the artifact reduced signal invented by Dräger including constant threshold detection. (3): high pass filtering followed by median based envelope generation attended by dynamic onset detection. (4): Dräger envelope function in combination with constant onset detection.

2.2 Patient Ventilator Asynchrony

To get an estimation of the different appearances of patient ventilator asynchrony, each on-cycling and off-cycling detection event was compared to its time dependent closest EAdi equivalent, defining a 200 ms time interval as synchrony tolerance. Afterwards, the recorded events were categorized as synchronous or asynchronous. The distinction between the several manifestations of asynchrony strongly varies in literature. Basically, three most likely types can be pointed out. First, the so called missed efforts, taking place when the analyzing device misses a patients effort of breathing. These are named *missed triggers* and are recorded when no EMG trigger can be found within the tolerance interval of an EAdi correspondent. On the other hand, sometimes multiple triggering events are detected within one synchrony-interval, called *extra triggers*. Another regularly occurring incidence is the so called *auto triggering*. Those events take place when breathing changepoints are sampled automatically by the device without patient demand, so anywhere outside of the synchrony interval around any EAdi changepoint in our case.

The data source the different approaches were tested on was put up at the university hospital of Monza, Italy, and contains 415 datasets recorded during clinical daily routine. Most of the files contain the desired signals. Nevertheless a pre-selection is inevitable to exclude statistical distortions due to examination of non interpretable intervals. As first step, the amount of patient files were reduced to 115 sets. For the final examination, four representative data frames were chosen. Those were selected with respect to signal quality still seen as interpretable while containing distortions regularly occurring in the entire recordings set. All algorithms were computed using Python programming language.

3 Results and Discussion

To parameterize the appearance of the already named forms of asynchrony, the following detection parameters were extracted for onset as well as offset detection events:
I. The entire amount of sEMG as well as EAdi based triggers including on-cycling as well as off-cycling incidents. II. The synchrony index, described as the amount of synchronous EMG triggers in relation to all EAdi events. III. The missed trigger quotient was calculated counting EAdi triggers not attended by an sEMG correspondent within the tolerance interval in relation to the whole amount of EAdi based detections. IV. The auto trigger ratio, computed by counting sEMG triggers not settled in synchrony interval divided by the whole number of sEMG triggers. V. The extra or double trigger ratio expresses the number of EAdi triggers having more than one sEMG trigger event in relation to the whole amount of EAdi events.
As a second interpretation criterion, the distributions of the time delays between EAdi and sEMG based events where plotted for the whole set of the final four patients recordings as well as for every single dataset. The distributions were generated in two different fashions. To get an estimation about the synchrony behavior, trigger delays got sampled towards the closest EAdi correspondent. Furthermore, every sEMG event settled in a constant time interval of ten seconds towards any EAdi trigger was plotted. This basically provides information about the general correlation between the two acquisition strategies. Exemplary, such a distribution for one patient record is visualized in Fig. 4. These parameters provided following results, as Table 1 implies: Generally, independent of the procedure applied, main issue is located at the missed as well as auto trigger ratio, listed

Table 1: synchrony performance of different patterns. Procedures numbered equivalently to chapter 2.

procedure	sync. on/off	missed on/off	auto on/off
(1)	0.259/0.228	0.74/0.792	0.659/0.638
(2)	0.354/0.301	0.645/0.608	0.501/0.608
(3)	0.202/0.339	0.756/ 0.79	0.695/0.65
(4)	0.221/0.452	0.676/0.631	0.617/0.348

as *missed on/off*, *auto on/off* in Table 1. According to visual inspection of the signals, longterm amplitude variation as well as less articulated activity regions, so locally bad conditioned SNR of the envelope functions, get instantaneously visible. These distortions lead to inappropriate detection of activity by constant threshold procedures. In case of the dynamic threshold generation, debugging showed high sensitivity towards intervals showing multi-modally distributed values, so segments including several signal oscillations. The mode computed by kernel density estimation did not appear in the hole section in these cases, consequently, triggering did not take place. The extraction of the envelopes maxima appears to be critical concerning this behavior. Choosing less restrictive constraints increases probability of including noisy side peaks, leading to rise of auto trigger rate. More exclusive parameter selection creates already named multi-modally distributed intervals, accusing missed trigger efforts. Several different filtering strategies of the maxima detection as well as moving average calculation of the mode values did not provide appropriate improvement concerning performance.

In contrast to the regularly occurring auto and missed triggers, double triggers did not appear during examination of the four datasets and only took place a few times while iterating over the entire dataset of 115 files. Only found during application of constant thresholding, this rare incidence takes place if signals amplitude oscillates between values above as well as below onset threshold multiple times within one synchrony tolerance interval. Because of the general smoothness of the applied envelope functions, abrupt changes in signal leading to multiple triggering events are that unlikely in appearance.

Highest synchrony rates, expressed as *sync. on/off* in Table 1, are settled at the processing chain including the prefiltered signal provided by Dräger, originating from the highly pronounced activity regions with general smooth appearance, as visual inspection suggests. Also the constant threshold detection settled on the Dräger own envelope provided high synchrony during off-cycling detection. That, again supported by visual inspection, implies the general reliability of the envelope function and marks the constant threshold on-cycling detection as less suitable for real-time detection based on that envelope signal.

Main goal concerning the interpretation of the obtained distributions was to visualize systematical relations between sEMG guided activity detection in relation to EAdi based events. The shape of the distribution provides information about the systematical dependencies between the two methods. A certain appearance of the distribution expresses the general form of correlation between the different approaches of activity detection, as shown in Fig. 4.

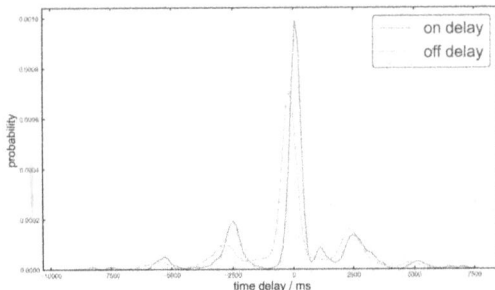

Figure 4: Trigger delay distribution for one patient file with constant time window of 10 seconds relatively to each EAdi trigger. Major fraction of sEMG based events are settled within 500 milliseconds around the EAdi correspondents, like in approximately 75 % of the cases.

4 Conclusion

Having regard to the results provided by the examination parameters, several procedure dependent as well as systematic conditions appear to have major impact on detection performance. These limitations can be overcome either by improvement of the detection algorithms or by generating envelope functions providing well suited SNR allowing more rudimentary detection strategies to work out efficiently. As become instantaneously visible by inspecting the signals and trigger events, it is intuitive to mention the envelope function as indispensable element of the processing pattern. Main task to improve those conditions is settled in an effective preprocessing of EMG signals, excluding cardiac potentials without of removing excessive amount of respiratory activation patterns. Working on less well behaved signals, the detection algorithms need to be modified to take these constraints into account, making them more computationally expensive so less real-time performant and interpretable. As shown working on reliably pre-filtered signals invented by Dräger, it is possible to implement a real-time performant preprocessing strategy providing an ECG removed sEMG signal providing well suited SNR by preventing respiratory potentials. Resulting in an easy to interpret envelope function, this allows rudimentary detection algorithms to work out efficiently.

So as the main suggestion concerning further research activity based on these findings, a thoroughly implemented preprocessing and envelope formation should be of central interest. Current approaches dealing with wavelet based or novel time-frequency resolved strategies like *Empirical Mode Decomposition* [9] show promising results concerning artifact removal. The more well behaved and interpretable the envelope function is, the simpler and comprehensible the detection strategies can be, leading to well ana-

lyzable and reproducible results. Alternative approaches of envelope generation using entropy dependent signals, particularly *Sample Entropy* [6] has been tried because of decreased necessity of ECG removal but was not chosen as final processing element due to high computational costs. So improvement of algorithmic efficacy is also recommended because of promising performance.

Acknowledgement

The work has been carried out at Drägerwerk AG & KGaA, Lübeck, Germany and supervised by the Institute of Electrical Engineering in Medicine, Universität zu Lübeck, Germany.

5 References

[1] D. R. Hess, *Respiratory Mechanics in Mechanically Ventilated Patients*. Respir. Care 2014, vol. 59, pp. 1773-1794, 2014.

[2] C. Kowalski, *Detection of Patient Ventilator Asynchrony using surface Electromyography*. Masters Thesis, Drägerwerk AG & KGaA, Institute of Electrical Engineering in Medicine, 2017.

[3] J. Graßhoff, *Model-Based Estimation of Respiratory Effort using Esophageal Pressure*. Masters Thesis, Drägerwerk AG & KGaA, Institute of Electrical Engineering in Medicine, 2016.

[4] C. Sinderby et al, *An automated and standardized neural index to quantify patient-ventilator interaction*. Critical Care 2013, vol. 17, 2013.

[5] L. Estrada, A. Torres, L. Sarlabous, and R. Janee, *Onset and Offset Estimation of the Neural Inspiratory Time in Surface Diaphragm Electromyography: A Pilot Study in Healthy Subjects*. IEEE journal of biomedical and health informatics, 2017.

[6] P. Zhou and X. Zhang, *A Novel Technique for Muscle Onset Detection Using Surface EMG Signals without Removal of ECG Artifacts*. Physiol. Meas., 2014, pp. 45-54, 2014.

[7] H. Yonis et al, *Patient-ventilator synchrony in Neurally Adjusted Ventilatory Assist (NAVA) and Pressure Support Ventilation (PSV): a prospective observational study*. BMC Anesthesiology, 2015, vol. 15, 2015.

[8] R. Chowdhury et al, *Surface Electromyography Signal Processing and Classification Techniques*. Sensors 2013, vol. 13, pp. 12431-12466, 2013

[9] X. Zhang and P. Zhou, *Filtering of surface EMG using ensemble empirical mode decomposition*. Med. Eng. Phys. 2013, vol. 35, pp. 537–542, 2013.

Critical evaluation of gaps within the cobas m 511 hematology analyzer parameter portfolio

H. Kettner [1] and S. Klein [2]

[1] Biomedical Engineering, University of Applied Sciences Lübeck, hanna.kettner@stud.fh-luebeck.de

[2] Medical Sensors and Devices Lab, University of Applied Sciences Lübeck, klein@fh-luebeck.de

Abstract

In this paper the parameter portfolio of the cobas m 511 integrated hematology analyzer was investigated by performing a gap analysis and evaluating the clinical benefit of the missing parameters, compared to the Sysmex XN-Series. The literature research revealed that the clinical benefit for measuring the immature platelet fraction (IPF) is not completely proven. The first documented benefit of IPF was rebutted by recent studies. For the measurement of the immature reticulocyte fraction (IRF) and the immature granulocytes (IG) the clinical benefit seems evident according to the current literature. The study concludes, that the development of the parameters IG and IRF would be beneficial according to the current state of science and would complete the parameter portfolio of the cobas m 511 while the benefit of an invention of IPF is in doubt. For all three parameters follows that the development of the state of science should be considered.

1 Introduction

The blood can give hints in the diagnosis of diverse diseases and blood counts are a standard diagnostic instrument. In the field of hematology, currently the company Sysmex is the market leader with 63 % market-share in Germany [1]. At the moment Roche Diagnostics is about to place the integrated hematology analyzer cobas m 511 in the german market which uses the Bloodhound® technology. In this study, the parameter portfolio of the cobas m 511 will be investigated and the relevance of the missing parameters compared to the Sysmex XN analyzer series shall be checked for several clinical indications. The results can be used for decicion about an invention of the parameters for the cobas m 511.

1.1 The cobas m 511 hematology analyzer

The cobas m 511 integrated hematology analyzer from Roche Diagnostics is a digital morphology analyzer, cell counter and classifier which prepares, stains and analyses blood samples using the Bloodhound® technology (Fig. 1). The Bloodhound® technology represents the core intelligence of the cobas m 511 and is a unique approach in the field of digital hematology. In this approach the results of the morphology analysis of the cells are presented in detailed images which if necessary can be analysed and interpreted on a computer screen (Viewing station). With this technique, in many cases, manual microscopic review by the medical technologists can be eliminated.The workflow of the cobas m 511 consists out of five main steps. First, 30 microliters of blood are aspirated from the sample tubes. Then a defined blood volume of $1\mu l$ is printed onto a de-

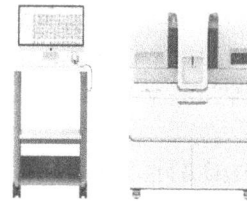

Figure 1: The cobas m 511 integrated hematology analyzer of Roche Diagnostics including the Viewing Station monitor [2].

fined area on a slide forming a monolayer of cells. The printed monolayer is then stained using the Romanowsky staining method, which is required for differentiating the cells. For location and counting of the cells multispectral imaging is used. Finally, the results including the morphology images are displayed in the viewing station. For the multispectral imaging a 5x objective lens and a 50x objective lens are used. After staining, the slide is moved to the 5x objective lens. There, the cells are located and pictures of the cells are taken. With the help of different colors of a LED light source (colors red, yellow, green, blue), red blood cells, nucleated red blood cells, white blood cells and platelets can be differentiated. As the printing volume as well as the area is known, absolute cell counts can be determined. With the 50x objective lens and four colors of light the system performs an analysis of the cells. Using multispectral imaging all reported parameters can be determined or calculated. For example by measuring the light absorption at multiple points within one cell, the received information about cell height and cell hemoglobin can be converted into the mean corpuscular volume (MCV) and the mean corpuscular hemoglobin (MCH) [3].

2 Material and Methods

2.1 Gap analysis of the cobas m 511

As a first step of this work, a gap analysis was performed. Therefore, the parameter portfolio of the cobas m 511 hematology analyzer was compared with the parameter portfolio of the Sysmex XN-9100 analyzer, as Sysmex is the market leader in the field of hematologically in-vitro diagnostics and has the broadest hematology portfolio on the market [1].

2.2 Evaluation of the missing parameters

After the gap analysis, the missing parameters were categorized into groups: Parameters, that are further discussed, parameters that can not be implemented technically on the cobas m 511 and parameters that have another nomenclature or are offered with alternative determination options on the cobas m 511. For the parameters that are further discussed, the clinical evidence and medical benefit was investigated by the assessment of clinical studies. It was searched for indications where the presence of these parameters is beneficial for diagnosis. Therefor clinical studies were used to verify or disprove the benefit of measuring a parameter in a certain indication.

3 Results and Discussion

3.1 Gap analysis: Parameters of the cobas m 511 and the XN-9100

For both analyzers, the following parameters are available [4]: Leukocytes: White blood cell count (WBC), neutrophil percent and count (%Neut/ #Neut), lymphocyte percent and count (%Lymp/ #Lymph), monocyte percent and count (%Mono/ #Mono), eosinophil percent and count (%EOS/ #EOS), basophil percent and count (%BASO/ #BASO); Erythrocytes: Red blood cell count (RBC), hematocrit (HCT), MCV, red blood cell distribution width (RDW), red blood cell distribution width - standard deviation (RDW-SD); Hemoglobin: Hemoglobin concentration (HGB), MCH, mean corpuscular hemoglobin concentration (MCHC); Reticulocytes: Reticulocyte percent and count (%RET/ #RET); The results show that both analyzers provide a complete blood count (CBC) and 5 part differential and thereby cover all the standard parameters needed.

Nevertheless the parameter portfolio of the cobas m 511 lacks several parameters (see Table 1).

3.2 Evaluation of the missing parameters

Among the missing parameters are several, that are not relevant for discussion, as they can not be implemented technically on the cobas m 511, including for example low flu-

Table 1: Identified gaps within the parameter portfolio of the Roche cobas m 511 and the Sysmex XN-9100

Parameter	Category	Roche	Sysmex
WBC-BF	Leucocytes	-	✓
MN % and #		-	✓
PMN % and #		-	✓
IG % and #		-	✓
RBC-BF	Erythrocytes	-	✓
NRBC % and #		✓	-
NRBC		-	✓
PLT	Thrombocytes	✓	-
MPV		✓	-
IPF % and #		-	✓
PDW		-	✓
P-LCR		-	✓
PLT-I/ PLT-F/ PLT-O		-	✓
HGBr	Reticulocytes	✓	-
IRF		-	✓
LFR/ MRF/ HFR		-	✓
RET-HE	Stem cells	-	✓
HPC% and #	and others	-	✓
TC-BF #		-	✓

orescent reticulocytes (LFR), medium fluorescent reticulocytes (MFR), high fluorescent reticulocytes (HFR), which are measured by using a flow based measurement method based on impedance. Furthermore, White blood cell count - body fluid (WBC-BF), Red blood cell count - body fluid (RBC-BF) and total nucleated cell count - body fluid (#TC-BF) can be neglected, as body fluid samples are not applicable on the cobas m 511. Another group of parameters to be neglected are those for which cobas m 511 uses other nomenclature or offers alternative determination options. These parameters are optical, impedance based and fluorescence based platelet count (Plt-O, Plt-I, Plt-F) for which the cobas m 511 offers the parameter Plt. The lack of the parameter platelet distribution width (PDW) can be overcome with the parameter mean platelet volume (MPV), which is a platelet index that also increases during platelet activation. Platelet larger cell ratio (P-LCR) is a ratio parameter that can be calculated out of the results of the cobas m 511 if required. Out of the remaining parameters (mononuclear percent (%MN), mononuclear count (#MN), polymorphonuclear percent (%PMN), polymorphonuclear count (#PMN), haematopoietic progenitor cell percent and count (%HPC, #HPC), immature platelet fraction (IPF), immature reticulocyte fraction (IRF) and immature granulocytes (IG) the research parameters IPF, IRF and IG are worth a closer examination, as these parameters measure precursor cells that can give important conclusions about the presence of specific diseases. The several indications for the three parameters can be seen in Table 2.

3.2.1 Evaluation of the immature platelet fraction

Immature platelets are one- to two-day-old thrombocytes [5]. As shown in Table 2, several indications for the mea-

Table 2: Relevance of the parameters IRF, IPF and IG shown with several studies verifying or refuting their benefit for indications

Indication of parameter		confirmed by	limited/refuted by
IPF	Differential diagnosis of thrombocytopenia	[6]	[11]
	Monitoring of thrombocyte generation after chemotherapy or BMT	[7]	
	Determination of necessity and timing of platelet transf.	[8]	
	Prediction of bleeding risk in ITP	[9]	
	Indication of ACS	[10]	[12]
IRF	Engraftment prediction after hematopoietic SCT	[8]	
	Prediction of BMR in patients with leukemia after chemotherapy	[13]	[14]
IG	Marking of inflammation	[15]	
	Prediction of infection or positive blood culture	[16]	
	Prediction of sepsis	[16]	
	Prediction of SAP	[17]	
	Monitoring of inflammation after liver transplantation	[18]	

surement of the immature platelet fraction can be found. The most important indication area is the differential diagnosis of thrombocytopenia. The IPF was identified to offer the possibility to distinguish between the different possible reasons for the thrombocytopenia. It was found to increase in case of a consumptive cause of thrombocytopenia like autoimmune thrombocytopenic purpura (AITP) and acute thrombotic thrombocytopenic purpura (TTP). IPF was at normal level at aplastic causes of thrombocytopenia like in bone marrow failure (BMF) [6]. Additional to this indication, the IPF measurement can be used to monitor the thrombocyte generation after chemotherapy or bone marrow transplantation (BMT) or determine the necessity or the timing of platelet transfusion [7] [8]. Besides, IPF was found as an important predictor of significant bleeding risk in patients with idiopathic thrombocytopenic purpura (ITP) [9]. Moreover, it was proven, that the IPF is increased in acute coronary syndromes (ACS) [10].

Contrary to the above mentioned papers latest studies limit the value of IPF as parameter to distinguish ITP from BMF by detecting increased IPF values for patients with BMF as well as for patients with ITP. The reasons found for that phenomenon are first, that IPF is increased in patients with low platelet count independent of underlying diseases and second, that thrombocytopenia is a disease of corrupt megakaryopoiesis and not only of platelet destruction [11]. In addition it was found out, that the IPF is not associated with the prevalence or the extend of coronary artery dis-

ease and should therefore not be used as marker of coronary atherosclerosis [12]. Summing up, according to our results, the clinical benefit of the parameter IPF is not proven finally, and therefore the parameter is not yet disadvantageous to be missed in the parameter portfolio. What stands out is, that especially the recent studies doubt the additional benefit of the parameter. In 2017 IPF was found increased in patients with BMF as well as in patients with ITP [11], which questions the clinical benefit recorded in the studies of 2004 [6].

3.2.2 Evaluation of the immature reticulocyte fraction

The literature research shows, that IRF, as well as IPF, are engraftment predictors and that IRF can be used as predictor of neutrophil recovery after hematopoietic stem cell therapy (SCT) [8]. Furthermore, IRF was identified as predictor of bone marrow recovery (BMR) in patients with acute leukemia after chemotherapy [13]. The literature research yields limits for the indication of the IRF, as erroneously elevated IRF values were detected in leukemia patients. Reason for that is the measuring method, in which leukocytes of leukemic patients are insufficiently stained and misidentified as immature reticulocytes [14]. Finally, the clinical benefit of the parameter IRF seems to be evident as all the studies found prove its value. Nevertheless, the competitive advantage of the Sysmex analyzer over the cobas m 511 concerning this parameter is limited by the danger of wrong identified immature reticulocytes. As the cobas m 511 uses a different measuring technology, the problem of wrong classified IRF is not likely. Therefore, the potential advantage of the cobas m 511 over the Sysmex XN-series should be checked.

3.2.3 Evaluation of the immature granulocytes

The literature research shows several indications for the measurement of IG, which is a parameter showing active immune reaction. It was identified as new marker of acute inflammation [15]. It was determined as better predictor of infection or positive blood culture than the WBC count and comparable to the absolute neutrophil count. Furthermore, it was identified as specific predictor of sepsis [16]. In addition, IG promises to be a good predictor of severe acute pancreatitis (SAP) [17]. After Liver Transplantations, IG seems useful as monitoring parameter of inflammation [18]. Concluding, the measurement of IG promises to have a clinical benefit according to the present state of the studies.

4 Conclusion

The study concludes, that the development of the parameters IG and IRF for the cobas m 511 would be beneficial according to the current state of research. The benefit of IPF as part of the portfolio is in doubt. For all of the three parameters follows, that the development of the state of science should be considered.

Acknowledgement

The work has been carried out at Roche Diagnostics Deutschland GmbH, Mannheim and supervised by the Center of Biomedical Engineering, Fachhochschule Lübeck. I would like to express my special thanks to Tobias Stumpf, product manager in the field of haematology at Roche Diagnostics Deutschland GmbH.

5 References

[1] VDGH Verband der Diagnostica-Industrie, *IVD Monitor 2014*, 2015.

[2] Roche Diagnostics International Ltd, *cobas.com, cobas m 511*, August 2017. [Online]. Available: http://www.cobas.com/home/product/hematology-testing/cobas-m-511.html. [Access 15th January 2018].

[3] Roche Diagnostics International Ltd, *cobas.com, cobas m 511*, August 2017. [Online]. Available: http://www.cobas.com/home/product/hematology-testing/cobas-m-511.html. [Access 15th January 2018].

[4] C. TODAY, *CAP TODAY, An interactive guide to labratory software and instrumentation "Sysmex America, XN-9100, Hematology analyzers, November 2017"*, November 2017. [Online]. Available: http://www.captodayonline.com/productguides/instruments/hematology-analyzers-november-2017/sysmex-america-xn-9100-hematology-analzyers-november-2017.html. [Access 12th January 2018].

[5] Sysmex Deutschland GmbH, *IPF (Immature Platelet Fraction) - Die klinische Bedeutung der Bestimmung unreifer Thrombozyten fr die XN-Serie*, Xtra Vol. 18.2, February 2014.

[6] C. Briggs, S. Kunka, D. Hart, S. Oguni and S.J. Machin, *Assessment of an immature platelet fraction (IPF) in peripheral thrombocytopenia*, British Journal of Haematology, Vol. 126, pp.93-99, July 2004.

[7] Y. Abe, H. Wada, H. Tomatsu, A. Sakaguchi, J. Nishioka, Y. Yabu, K. Onishi, K. Nakatani, Y. Morishita, S. Oguni and T. Nobori, *A simple technique to determine thrombopoiesis level using immature platelet fraction (IPF)*, Thrombosis Research, Vol. 118, pp. 463-69, 2006.

[8] I. Morkis, M. Farias, L. Rigoni, L. Scotti, L. Gregianin, L. Daudt, L. Silla and A. Paz, *Assessment of immature platelet fraction and immature reticulocyte fraction as predictors of engraftment after hematopoietic stem cell transplantation.*, International Journal of Labratory Hematology, April 2015.

[9] A. McDonnell, K.L. Bride, D. Lim, M. Paessler, C.M. Witmer and M.P. Lambert, *Utility of the immature platelet fraction in pediatric immune thrombocytopenia: Differentiating from bone marrow failure and predicting bleeding risk*, Pediatric Blood & Cancer, February 2017.

[10] E.L. Grove, A.M. Hvas und S.D. Kristensen, *Immature platelets in patients with acute coronary syndromes*, Thrombosis and haemostasis, pp. 151-6, January 2009.

[11] A. Cybulska, L. Meintker, J. Ringwald und S.W. Krause, *Measurements of immature platelets with haematology analysers are of limited value to separate immune thrombocytopenia from bone marrow failure*, British hournal of haematology Vol. 177, pp. 612-619, April 2017.

[12] M. Verdoia, M. Nardin, R. Rolla, P. Marino, G. Bellomo, H. Suryapranata, G. De Luca und Novara Atherosclerosis Study Group, *Immature platelet fraction and the extent of coronary artery disease: A single centre study*, Atherosclerosis Vol. 260, pp. 110-115, May 2017.

[13] R. Raja-Sabudin, A. Othman, K. Ahmed-Mohamed, A. Ithnin, H. Alauddin, H. Alias, Z. Abdul-Latif, S. Das, F. Abdul-Wahid und N. Hussin, *Immature reticulocyte fraction is an early predictor of bone marrow recovery post chemotherapy in patients with acute leukemia.*, Saudi medical journal; Vol. 35, pp. 346-9, April 2014.

[14] J. Huh, H. Moon und W. Chung, *Erroneously elevated immature reticulocyte counts in leukemic patients determined using a Sysmex XE-2100 hematology analyzer*, Annals of Hematology; Vol. 86, pp. 759-62, October 2007.

[15] C. Briggs, S. Kunka, H. Fujimoto, Y. Hamaguchi, B. H. Davis und S. J. Machin, *Evaluation of immature granulocyte counts by the XE-IG master: upgraded software for the XE-2100 automated hematology analyzer*, Labratory Hematology, pp. 117-24, 2003.

[16] M. A. Ansari-Lari, T. S. Kickler und M. J. Borowith, *Immature Granulocyte Measurement Using the Sysmex XE-2100: Relationship to Infection and Sepsis*, American Journal of Clinical Pathology, Vol. 120, pp. 795-799, November 2003.

[17] M. Lipinski und G. Rydzewska, *Immature granulocytes predict severe acute pancreatitis independently of systemic inflammatory response syndrome*, Przeglad Gastroenterologiczny Vol.12, pp. 140-44, June 2017.

[18] S. Shiga, H. Fujimoto, Y. Mori, T. Sakata, Y. Hamaguchi, F. S. Wang, Y. Inomata, K. Tohyama und S. Ichiyama, *Immature Granulocyte Count after Liver Transplantation*, Clinical chemistry and labratory medicine, pp. 775-80, September 2002.

Directional Hearing in clinical everyday life with the ERKI system of the company Auritec

M. Keßler [1], and J. Mody [2]

[1] Medizinische Ingenieurwissenschaft, Universität zu Lübeck, marlene.kessler@student.uni-luebeck.de
[2] Auritec Medizindiagnostische Systeme GmbH, Hamburg, vertrieb@auritec.de

Abstract

ERKI (recording of directional hearing at children; Erfassung des Richtungshörens von Kindern) is the only approved clinical system that automatically verifies directional hearing. It is now possible to obtain valid information about the ability to localize short auditory events. This paper combines various studies on the localization of sounds, in particular results with the ERKI system, to get an overview about this topic. Important findings are standard variation and typical anormal forms of directional hearing and how Cochlear Implants and Auditory Processing Disorder affect the ability of directional hearing.
All in all it is clear that directional hearing is one of the basic auditory abilities of people in everyday life. But more studies (e. g. with the ERKI system) will be needed in the future to understand the processes between ears and brain. And get more information how hearing loss affects directional hearing.

1 Introduction

Currently, the focus of hearing loss and deafness treatment is on speech understanding at rest and noise. Together with the technical possibilities, speech understanding has dramatically improved due to the high quality of hearing aid/implant fitting in the last few years. However, the world of hearing is more than just understanding speech. So another very important skill is directional hearing. This means the ability to localize sound source with respect to its direction and distance.

Studies found two main mechanisms in the human auditory pathway: The level and time differences between the two ears, so-called Interaural Level Difference (ILD) and Interaural Time Difference (ITD). Therefore, good sound localization requires both ears. With ITD, a sound reaches one ear earlier than the other, and through this short delay the direction of the sound source can be perceived. With ILD, the sound level on one ear is higher than on the other ear (due to the head shadow effect). So humans are able to exactly locate a sound source from the front up to 1 degree with these two mechanisms [1].

This paper discusses how human, especially hearing impaired people (CI and APD), can locate sound sources.

2 Material and Methods

With the ERKI system it is possible to measure the sound localization between -90 degree (left) and +90 degree (right) within a range of 5 degree with real and virtual sound sources, the setup you can see in Figure 1. The signal can be selected between white noise, pink noise and one speech signal. The signal is 300 ms long to prevent head movements during the measurement. The patient uses a control dial to indicate the direction where he/she assumes the sound source and then confirms it. The next test signal is automatically generated after patient's response.

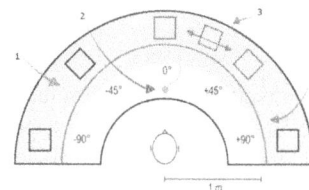

Figure 1: Setup of ERKI: 1: coverage 2: control dial 3: virtual sound source 4: LED-strip

Most measurements with the ERKI-system done by Plotz and Schmidt at the IHA Oldenburg [7], [9], [10], [4].
Other authors for example Zheng [12] and Litovsky [6] have dealt with directional listening with bilateral (on both sides) cochlear implants. The results are taken into account in the following.
In the future, it may be possible to detect another disease more effectively by precisely determining the directional hearing like auditory processing disturbances (APD). Important publications written by Kunze [5] and Böhme [2].
From 1960 onwards, research [1], [8] has been carried out on the topic of directional hearing and first insights have been gained. Today's knowledge has based on these findings and further detailed studies have been designed.

3   Results and Discussion

The following section will discuss standard values and typical forms of directional hearing. Also two possible fields of application will be demonstrated, cochlear implants (CIs) and Auditory Processing Disorder (APD).

3.1   Standard values of directional hearing

To check directional hearing, a standard value must be determined for studies or diagnostics in hospitals. Comprehensive research has led to similar results. These measurement results are discussed with respect to the following ERKI system [11].

The evaluation of direction of sound sources is best at a frontal angle (0 degree), due to the general orientation of sensory perception towards the front. Accuracy decreases once the sound source moves to the sides.

Figure 2 presents the standard values of directional hearing in the grey shaded area [11]. Normal directional hearing occurs at the center when the maximum deviation of the direction indication is maximal 10 degree. The deviation increases slightly to the sides (-/+90 degree) and is laterally maximal 20 degree in normal range.

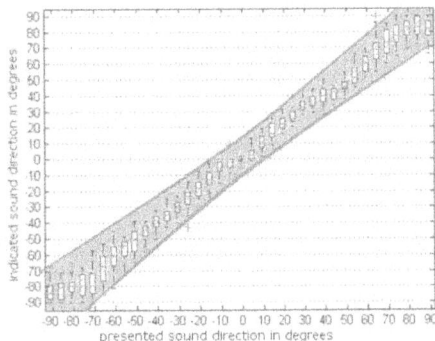

Figure 2: Study results of standard values of directional hearing with the ERKI system with test persons between 6 - 37 years. The grey area displays normal hearing [11].

3.2   Different forms of directional hearing

1. **Optimal directional hearing:** The direction is correctly indicated over the whole spatial width (left, right, center) Central: max. deviation 10 degree; lateral: max. deviation 20 degree (see Figure 3, 1. case; [10]).

2. **Shifting to the center:** Central direction is correctly indicated, the extreme sides are shifted to the center. This means that signals from the sides are heard incorrectly and more shifted to the center. A "squeezing" takes place towards the middle. Spatial perception does not take place in its entirety (see Figure 3, 2. case; [10]).

3. **Lateral shift:** A distinction is made between the left and right sides. Signals from the center are perceived

as coming from the side. This lead to a lateral division. There is binaural hearing but a monaural perception. Furthermore, signals directly from the front are assigned to one side/ear (typical: inexperienced bilateral CI carriers) (see Figure 3, 3. case; [10]).

4. **Combination of 2. and 3. case:** Signals mostly from the central area (approx. -/+20 degree) are perceived as having shifted to the center (0 degree). Lateral signals (> -/+20 degree) are perceived as having shifted to the sides.

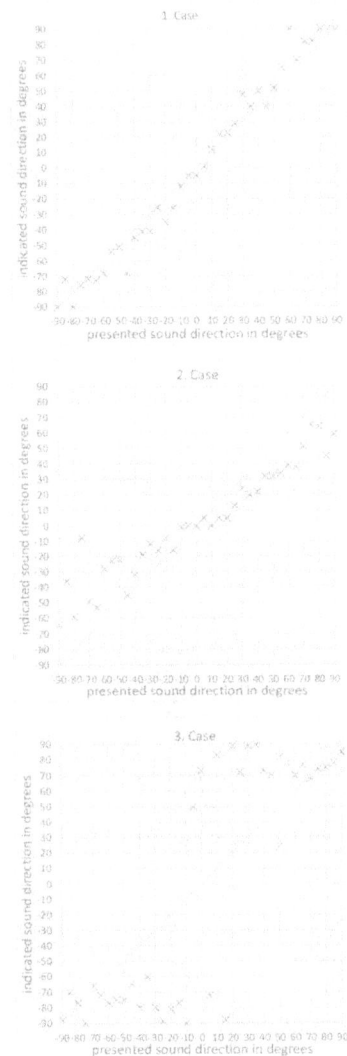

Figure 3: Three different forms of directional hearing measured with ERKI (test persons: children 8/9 years) 1. case: optimal/normal directional hearing, 2. case: Shifting to the center, 3. case: Lateral shift, monaural perception [10].

3.3   Directional hearing with Cochlear Implants (CI)

In a CI treatment, hearing should be as close as possible to normal hearing. This applies not only to linguistic performance but also to the general acoustic perception of space.

The localization of sound sources is an important parameter here. However, until now there has hardly been any possibility to examine directional hearing in the daily clinical routine. With ERKI, this is possible now.

3.3.1 Monaural CI treatment

At a monaural CI-fitting a clear deviation from the normal directional hearing can be detected, an exact spatial allocation is hardly possible. As can be seen in the following study (Figure 4), there is a lateral displacement to the CI side. [3]. Generally speaking monaural hearing is rather diffuse and shallow [8].

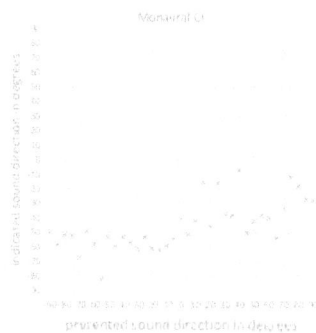

Figure 4: ERKI results of test persons (N= 9, f= 4, m= 5, median: 49 years) with monaural CI treatment (CI: -90 - 0 degree; deaf: 0 - +90 degree). No directional hearing is possible. Localization to the CI-side [3].

3.3.2 Bilateral CI treatment

In bilateral CI care, perception typically changes significantly. Depending on the type of hearing loss and hearing aid supply, directional hearing has already been severely restricted before a CI treatment.

At the beginning of the bilateral CI treatment no directional hearing is visible (Figure 5 top), usually the first step is a lateralization (Figure 5 bottom). In Figure 5 on the bottom the sound direction is divided mostly between left and right. Signals from the center (0 degree) are still mostly shifted to the sides. Complete side errors can also occur in parts. A both-sided but one-ear perception of sound sources exists, like in Figure 3, 2. Case. After a period of familiarization with bilateral CI, binaural processing usually occurs. In Figure 5 bottom it already started a little bit. The answers of the patient are beginning to follow the diagonal line. Study results displace a better perception of sound sources compared to the beginning of CI treatment [9]. The localization accuracy is lower, above an angle of -/+45 degree, probably due to characteristics of the CI-processors [7].

Thus, ERKI provides an insight into the binaural processing of CI patients. Whether there is a connection between the right and left side or whether only separate listening is possible. The extent to which additional funding and clinical care could help here, has to be subject to further studies and no clear study results are available yet [6], [12].

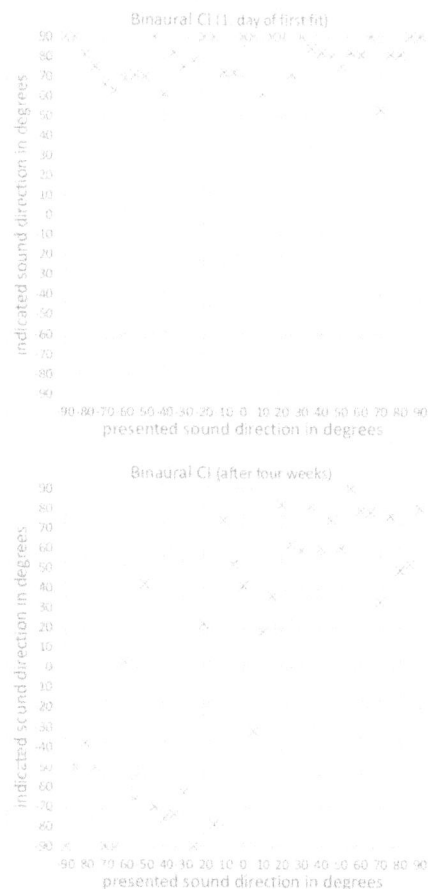

Figure 5: ERKI results of a child with binaural CI treatment (male, 11 years, right 2. CI with 11, left 1. CI with 6). Top: results at first day of fitting, bottom: results after four weeks wearing bilateral CIs. The results illustrate the change of directional hearing from a complete lateralization to the right side to a more "balanced" localization [9].

ERKI is a good opportunity to test the hearing ability of patients during CI treatment. Thus deficits and, above all, progress are made visible. For example, in the case of a bilateral CI treatment after the first adjustments, after three and six months the directional hearing can be checked with ERKI. In this way, the positive results can be tracked and deficits can soon be detected. Should the results fall short of expectations, steps in therapy can be initiated immediately. (e. g. balancing in CI-Map)

3.4 Auditory Processing Disorder (APD)

Generally speaking, APD refers to a variety of disorders that affect the way the brain processes auditory information. But patients have still a normal sound audiogram. The diagnosis of APD is usually difficult to determine in everyday clinical life, as the degree of manifestation varies strongly from one patient to the other. On the whole, studies did not show aspects that were common to all affected patients [2]. The lack of a uniform diagnostic picture of APD, is also reflected in the results of directional hearing. So an APD may be present, but is not reflected in the directional hearing and results of ERKI are within a normal range as well. The guidelines of the DGPP (Deutsche Gesellschaft für Phoniatrie und Pädaudiologie; German Society for Phoniatrics and Pediatrics) (as of 2015) recommend that APD patients make a "Binaural Interaction Test" to check the directional hearing ability among other tests. So localization capability is only one test area in APD diagnostics. Nevertheless, a disturbance of the binaural processing is usually present, which is indicated by an anormal directional hearing. Studies have demonstrated that children with APD have significantly poorer directional hearing than a normal hearing control group. More than 70 percent of APD children displayed deficits in directional hearing [5].

With the automatic ERKI system testing directional hearing can be accomplished under 10 minutes, which plays an important role in clinical life and patient concentration. In addition, no speech signal is required, so that speech processing and language skills do not have any influence on the results. By checking the localization performance, an important component of everyday hearing is mapped. In this way, substantiated results about the binaural auditory pathways can be determined for each patient and a very important component of the auditory processing can be tested. In order to get a more precise picture of directional hearing of APD patients, comprehensive studies are still lacking but they will certainly be available in the future.

4 Conclusion

In summary, it should be noted that directional hearing could be a great opportunity to improve hearing diagnosis and treatment. It can provide more and deeper information about the patient's hearing status and can help to understand the hearing processes. With regards to the technical advances, this can improve hearing in normal situation for hearing impaired people like CI- or APD-patients. But clear is also that directional hearing and ERKI, as diagnostic tools, are still at the beginning. There must be much more studies to prove the benefits and needs for patients and clinical everyday life.

Acknowledgement

The work has been carried out at AURITEC Medizindiagnostische Systeme GmbH and supervised by Obleser at the Institute of Psychology, Universität zu Lübeck. I would like to thank J. Mody, owner and CEO of Auritec, for the opportunity to do my internship at Auritec.

5 References

[1] Blauert, J. (1974): *Räumliches Hören*. Stuttgart: Hirzel (Monographien der Nachrichtentechnik).

[2] Böhme, G. (2008): *Auditive Verarbeitungs- und Wahrnehmungsstörungen (AVWS) im Kindes- und Erwachsenenalter. Defizite, Diagnostik, Therapiekonzepte, Fallbeschreibungen*. 2., vollst. überarb. und erg. Aufl. Bern: H. Huber.

[3] Brandt, J.; Knief, A.; Schmidt, K.; Am Zehnhoff-Dinnesen, A.; Plotz, K. (2017): *Poster: Lokalisation bei Cochlea-Implantat Patienten am erweiterten Mainzer-Kindertisch (ERKI-Setup) in Abhängigkeit der Mikrofoneinstellung des Sprachprozessors*. DGA Aalen, 2017. IHA; Universitätsklinikum Münster.

[4] Knief, A.; Deuster, D.; Rosslau, K.; Demir, M.; Am Zehnhoff-Dinnesen, Antoinette; DGPP|Akademie für Hörgeräte-Akustik (Hg.) (2014): *Entwicklung des Sprachverstehens und des Richtungshörens nach bilateraler Cochlea-Implantat-Versorgung*. Universitätsklinikum Münster.

[5] Kunze, S. (2010): *Richtungshörvermögen und Stapediusreflexe von Kindern mit und ohne AVWS*. LMU, München. Soziale Pädiatrie und Jugendmedizin.

[6] Litovsky, R. Y.; Parkinson, A.; Arcaroli, J. (2009): *Spatial hearing and speech intelligibility in bilateral cochlear implant users*. In: Ear and hearing 30 (4), S. 419–431.

[7] Plotz, K.; Schmidt, K.; Schoenfeld, R.; Loewenheim, H.; Bitzer, J. (2016): *Entwicklung des Richtungshören nach Cochlear Implant – Vergleich realer und virtueller Schallquellen*. DGPP. JADE Hochschule; IHA; HNO-Universitätsklinik EMS Oldenburg.

[8] Röser, D. (1960): *Die zentralen Vorgänge beim Richtungshören*. In: Archiv f. Ohren- Nasen- u. Kehlkopfheilkunde 177 (1), S. 57–72.

[9] Schmidt, K.; Decker, A.; Bohnert, A.; Rader, T.; Läßig, A.; Plotz, K. (2017): *Poster: Erhebung der Lokalisationsleistung von Kindern Erhebung der Lokalisationsleistung von Kindern mit einem Cochlea-Implantat am erweiterten Mainzer-Kindertisch (ERKI-Setup)*. DGA Aalen. JADE Hochschule; IHA. Oldenburg; Klinik für HNO-Heilkunde, Universitätsmedizin Mainz.

[10] Schmidt, K.; Plotz, K. (2016): *Poster: Lokalisation virtueller Schallquellen mit einem automatisierten Erweiterungsmodul am Mainzer-Kindertisch -ERKI-*. 5th European Pediatric Conference Berlin. Link: http://pedconfberlin2016.com/en/poster/katharina-schmidt (12.08.2016)

[11] Schmidt, K.; Plotz, K.; Bitzer, J.; Kissner, S.; Heimann, J.; Nienaber, M.; Seinecke, B., T.: Jade Hochschule, IHA Oldenburg, Deutschland (2015): *Lokalisation von realen und virtuellen Schallquellen am Mainzer Kindertisch*. DGPP. JADE Hochschule; IHA. Oldenburg.

[12] Zheng, Y.; Godar, S. P.; Litovsky, R. Y. (2015): *Development of Sound Localization Strategies in Children with Bilateral Cochlear Implants*. In: PloS one 10 (8), e0135790.

A model to estimate pressure within a syringe from the corresponding force applied on the piston

N. Manjunath Swamy[1]

[1] Biomedical Engineering, University of Applied Sciences, Lübeck, nithishi.swamy@stud.fh-luebeck.de

Abstract

During anesthesia, pressure sensors for determining excessive injection pressure are often disposed and must be replaced after each application. A feasible solution is to use reusable force sensor to reduce cost of usage. In this paper, a model to estimate pressure within a syringe from the force applied on the piston is developed. Data from force and pressure sensors is sampled by a microcontroller and sent on to a PC for analysis. It is observed that as force increases pressure rises and using this relationship suitable trendlines (models) are developed. Accordingly, polynomial models n = 2 and 3 are chosen. The models overestimates pressure for most readings giving an early warning to stop unsafe injections. However, there are cases of underestimation condition which are dangerous to patients undergoing anesthesia. A safe model is one which always estimates higher pressures. Thus, by using the models force sensors may be used for pressure estimation.

1 Introduction

Insertion of syringe into the human body is considered as a "fundamental skill" and medical professionals and students require years of practice and experience to familiarize themselves on its proper use. Also in order to use the syringes effectively and safely without causing unnecessary damage to the tissues, doctors and medical students must be aware of the force exerted on the syringe and the corresponding pressure of the fluids inside the syringe [1]. Thus, these factors strongly create the need to design a system to compute force and pressure values to aid medical practitioners in making knowledgeable decisions on how to proceed with an injection instead of subjective evaluation ("syringe feel method") [2]. According to the study carried out by A. Hadzic et al., safe injections are possible at pressures < 4 Pounds per square inch (psi). However, pressures > 20 psi at the onset of injection lead to severe fascicular injury and persistent neurological defects. Therefore such high pressures are considered unsafe and must be avoided [3].

Pressure sensors used to determine high injection pressures during aneasthetic drug administration are often disposed. A probable solution is to use a reusable force sensor. Literature relates to the use of force-sensitive resistors (FSR) [4] and among others an optical fiber sensor to measure injection pressure [5]. This paper describes a model to estimate pressure within a syringe from the measured amount of force exerted on the piston. By using the model , a capacitive force sensor may be used to estimate pressure from the corresponding force applied instead of a pressure sensor. If the model indicates pressure > 20 psi at the onset of injection, such injections must be paused to ensure that force applied meets the required injection pressures.

Capacitive force sensor is selected over FSR because of the following reasons [6]:
1. capacitive sensor technology is more stable with regard to repeatability and durability when compared to resistive technology. Also it can measure low levels of force accurately.
2. capacitive force sensor is less susceptible to failure even when multiple loads are applied on them .
Use of optical fiber sensor describes a new method for real-time continuous monitoring of injection pressure at needle tip [5].

2 Material and Methods

2.1 Experimental Setup

Fig. 1 shows the setup to measure force and pressure. Capacitive force sensor is placed on the piston of B Braun 20 ml syringe using a double sided tape and MPX 2300 DT1 pressure sensor is connected to its nozzle. It must be noted that pressure within the syringe is developed by having a closed system and the experiment is carried out by compressing air regardless of the injection speed.

2.1.1 Force Measurement

Fig. 2 represents the scheme for force measurement. Capacitive sensor measures force applied on the syringe up to 45 N. It is combined with an interface board which offers analog output for immediate Data Acquisition (DAQ) integration and an Inter-Integrated Circuit (I2C) based interface for integration into embedded systems [7]. The Microcontroller unit (MCU) receives sensor data using I2C commu-

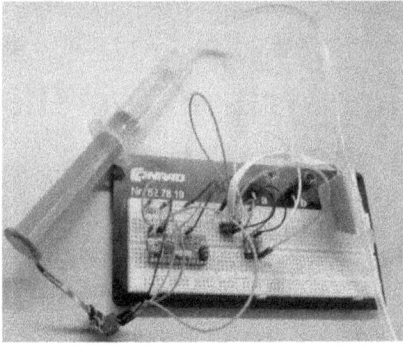

Figure 1: Setup to measure force and pressure values from force and pressure sensor, respectively.

nication which is displayed on personal computer (PC) for visual interpretation. This is achieved by using an Universal Asynchronous serial Receiver and Transmitter (UART) communication between MCU and PC [8]. MCU transmits the sensor outputs via Universal Serial Bus (USB) which are finally received by the HTerm terminal program on PC in digits.

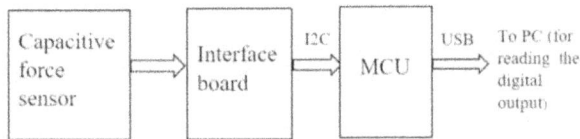

Figure 2: Block diagram to measure force applied on the force sensor.

2.1.2 Pressure Measurement

Fig. 3 represents the scheme for pressure measurement. MPX2300DT1 pressure sensor is a low cost, miniature differential pressure sensor consisting of a silicon piezo resistive sensing element which detects changes in pressure [9]. It offers an analog output voltage which corresponds to these changes in pressure values. However, before converting these analog values into digital, they must be amplified accordingly so as to fit the input voltage range (0V to the selected reference voltage) of the Analog to Digital Converter (ADC) in the MCU.

For the purpose of amplification, AD8226 Instrumentation amplifier with a single gain resistor, R_G, is used and suitable gain is calculated based on the following factors:
1. sensitivity of the pressure sensor;
2. maximum pressure intended to be measured (approximately 40-45 psi);
3. reference voltage of ADC.

Based on these factors, it is estimated that a gain of 24 provides the required signal amplification. Accordingly, R_G value is calculated, as in (1) [10].

$$R_G = \frac{49.4 \, k\Omega}{G - 1} \tag{1}$$

Where:
G is the estimated gain.
The amplified signal is then fed to the appropriate single ended input channel of the ADC. Thus ADC converts the analog signal into its corresponding digital values through successive approximation, as in (2) [8].

$$ADC = \frac{V_{in} 1024}{V_{ref}} \tag{2}$$

Where:
V_{in} is the input voltage on the selected input pin;
V_{ref} is the reference voltage to which ADC assigns maximum value (since the resolution of the ADC is 10 bit [8], the maximum value in this case is 0x3FF). The internal reference voltage of 1.1 V is used here.
After the ADC conversion has been successfully implemented, the digital values are transmitted via USB using UART communication which are received by HTerm terminal program on the PC (in digits) for further analysis and evaluation.

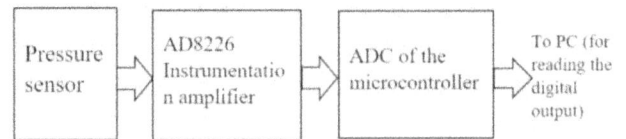

Figure 3: Block diagram to measure pressure within a syringe.

For meaningful interpretation of force and pressure, the digital values are converted into their respective analogue values i.e. pressure in psi and force in N. Since the capacitive sensor is uncalibrated, force values are represented in digits.

2.2 Implementation

Five data sets of force in digits and pressure values in psi are obtained and the number of measurements for all 5 data sets is varied from 1200 to 2000. For each measurement the value of force (x) is inserted into models under consideration. The corresponding estimated pressure (y) is calculated which is compared with data obtained from experimental setup and the deviation is given in the form of RMSE which is calculated, as in (3)

$$RMSE = \sqrt{\frac{\sum_i^n (\hat{y}_i - y_i)^2}{n}} \tag{3}$$

Where:
i is the pointer which points at each measurement;
n is the number of measurements for which the experiment is carried out;
ŷ-y is the difference between estimated pressure and actual pressure value.

3 Results and Discussion

A typical valid analog output range from the force sensor is 0.5 V (no force) to 1.5 V (maximum force), corresponding to the 0x0100 to 0x2FF (9 bit) digital output range. Also, the ADC output in case of pressure measurement typically ranges from 0 to 0x3FF which is then converted into psi units [8].

3.1 Model Selection

Fig. 4 shows a plot of force over pressure for five data sets along with the models to estimate pressure (y) from known amount of applied force (x). It can be seen that the pressure rises as the force applied on the piston increases. Using this relationship suitable trendlines/models are developed to fit all data sets combinedly. However there is a portion of force range for which the pressure does not increase. This is the static frictional force range. Once the applied force exceeds this range, pressure rises considerably. Accordingly, force range from 256 to 560 for the given set of readings do not cause an increase in pressure. But, the models estimate high pressures for this range and hence the resulting pressure is different from the actual value. So a threshold must be set to disregard frictional force. From experimental readings, it is noted that a force of less than or equal to 420 does not cause an increase in pressure and therefore the threshold is set to 420.

Table 1 shows RMSE obtained by applying polynomial of order n = 2 to 6, exponential and power models on all five data sets. Table 2 represents RMSE values obtained by ignoring force and pressure values below 420. The rmse decreases in comparision to that depicted in Table 1.

By ignoring data below 420, any models under consideration could be used for pressure estimation. However, even though polynomial models n = 4 to 6 estimate higher pressures for most data sets, they have a decreasing pressure behavior for force values from 890 to 940. As a result the models represent an underestimatation condition for such high forces and hence are not considered safe for pressure estimation. Out of the remaining models, polynomial model n = 2 and 3 are chosen because they overestimate pressure for most of the readings in comparison to exponential or power models. Also, they do not represent a decreasing pressure behavior in contrast to other polynomial models.

Table 1: RMSE values for five data sets from different models, in psi

Data	Poly					Exp	Pow
	n=2	3	4	5	6		
1	3.23	2.82	2.72	2.74	2.63	4.11	2.15
2	1.90	1.69	1.39	1.37	1.49	2.39	3.48
3	2.92	2.46	2.08	2.09	2.03	3.70	4.15
4	2.40	2.53	2.60	2.60	2.60	2.20	2.08
5	1.22	1.22	1.15	1.13	1.16	1.51	1.88

An overestimation condition gives an early warning to prevent any unsafe injections. Contrary to it, an underestimation condition poses significant risk to patients as the injection is preceded with higher pressures than estimated thereby unnecessarily damaging the tissues.

3.2 Evaluation

Polynomial model n = 2 is applied on data obtained from similar capacitive force sensor and the RMSE values are calculated accordingly. From Table 3, it can be seen that the model shows minimum deviation from the actual data thereby indicating that it fits to the readings obtained from these sensors.

Table 2: RMSE values obtained by applying the set threshold, in psi

Data	Poly					Exp	Pow
	n=2	3	4	5	6		
1	3.06	1.61	2.45	2.54	2.45	3.85	1.90
2	1.32	0.99	1.01	0.97	0.93	1.85	2.95
3	2.37	1.49	1.69	1.69	1.53	3.18	3.64
4	1.93	2.02	2.09	2.21	2.16	1.74	1.63
5	0.96	0.73	0.66	0.83	0.82	1.18	1.56

Table 3: RMSE values calculated by applying polynomial model n = 2 on readings obtained from similar force sensor

data	RMSE, in psi
1	1.60
2	1.930
3	2.006
4	3.064
5	2.759

4 Conclusion

A relationship between force and pressure is established and suitable trendlines are developed to fit all five data sets. Accordingly, polynomial models of n = 2 and 3 are chosen for pressure determination. The RMSE values decreases by ignoring the static frictional force range. Validity of the model on independent readings obtained from other capacitive force sensor demonstrated repeatability of results. With the help of the models, reusable force sensors may be used to estimate pressure instead of a pressure sensor which is disposed in most cases. The models over estimates pressure for most of the readings and gives an early warning to stop any unsafe injections. However there are also cases of under estimation which poses a risk to the patient as more pressure is being delivered than estimated and the models fail to identify such high pressures . A safe model is one which always has an over estimation condition. However even such models are not perfect as they will inherently have a high RMSE value.

As future work, aim is to carry out experiments with fluids, syringe of different sizes and by considering injection

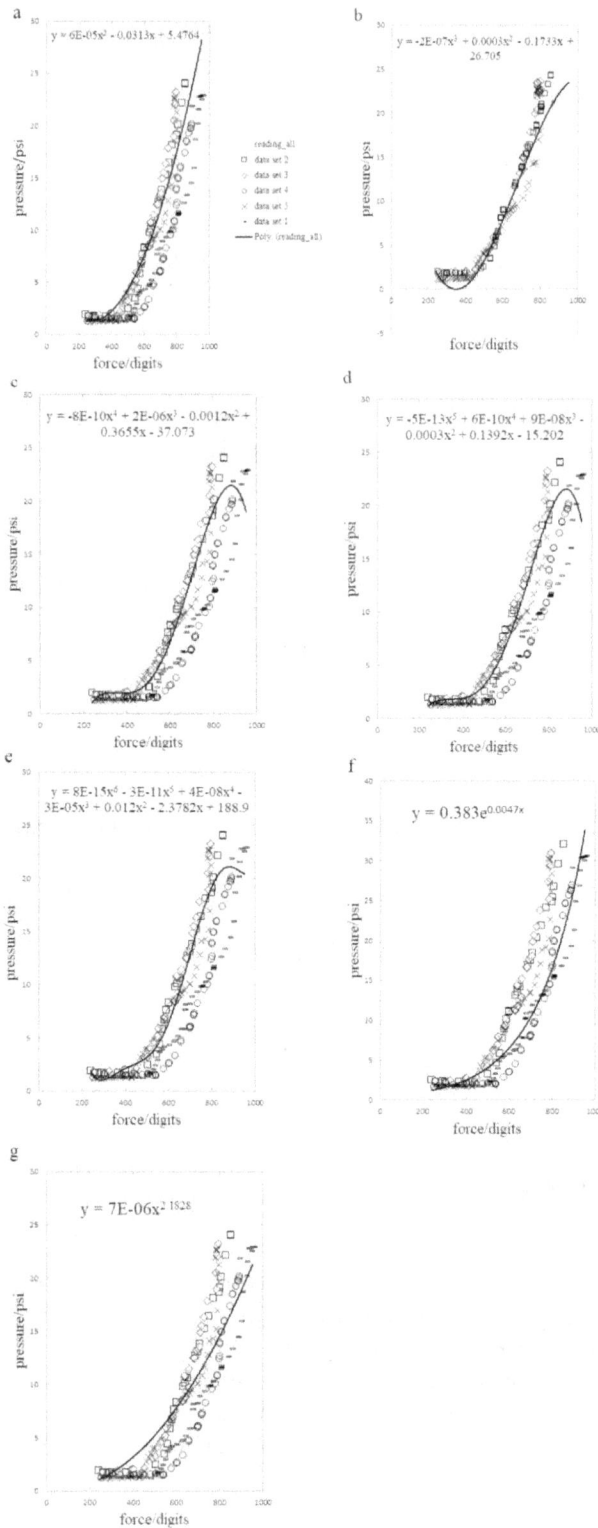

Figure 4: A plot of force over pressure for five data sets with models: (a)-(e) Polynomial n=2 to 6; (f) Exponential; (g) Power.

Acknowledgement

The work has been carried out at Hotswap Deutschland GmbH and supervised by S.Müller, Department of Applied Natural Sciences, University of Applied Sciences, Lübeck.

5 References

[1] LaCourse, J. R, P. McWilliam and inventors, *System to measure forces on an insertion Device*. United States patent application No. 13/135,152, 2012.

[2] R. Claudio et al., *Injection pressures by anesthesiologists during simulated peripheral nerve block*. Regional anesthesia and pain medicine, vol. 29, no. 5, pp. 201-205, 2004.

[3] A. Hadzic et al., *Combination of intraneural injection and high injection pressure leads to fascicular injury and neurologic deficits in Dogs*. Regional anesthesia and pain medicine, vol. 29, no. 5, pp. 417-423, 2004.

[4] C. Johnstone, A. Pawa, D. Onwochei, R. Vargulescu, C. Razavi, *Stop before you block and injection pressure - Juggad innovation meets regional anesthesia*. The European society of regional anesthesia, 2017.

[5] C. Quadri, A. Saporito, *Real-time continuous monitoring of injection pressure at the needle tip in regional anaesthesia: description of a new method*. The European society of regional anesthesia, 2017.

[6] R. Puers, *Capacitive Sensors: when and how to use them*. Sensors and Actuators A: Physical, Elsevier Publishing, 1989.

[7] Force sensor datasheet, *Single tact miniature force Sensors*. Available:https://www.robotshop.com/media/files/pdf2/singletact_datasheet.pdf [last accessed on 2017-11-27].

[8] Microcontroller data sheet, *ATmega48A/PA/88A/PA/168A/PA/328/P*. Available: www.trassare.com/wp-content/uploads/2017/07/ATmega328_datasheet.pdf [last accessed on 2017-11-22].

[9] Pressure sensor data sheet, *MPX2300DT1, 0 to 40 kPa, differential compensated Pressure sensor*. Available: https://www.nxp.com/docs/en/datasheet/MPX2300DT1.pdf [last accessed on 2017-12-21].

[10] Instrumentation amplifier data sheet, *AD8226 wide supply range, rail-to-rail output Instrumentation amplifier*. Available: https://www.nxp.com/docs/en/datasheet/MPX2300DT1.pdf [last accessed on 2017-12-21].

speed in order to investigate any considerable changes in model behavior.

Investigation of OCT phase fluctuations correlated with neuronal activity and measurements of mechanical and optical properties in rodent brain tissue

Tabea Kohlfaerber [1], Norbert Linz [2], Pin-Chieh Huang, [3] and Stephen A. Boppart [4],

[1] Medizinische Ingenieurwissenschaft, Universität zu Lübeck, tabeakohlfaerber@student.uni-luebeck.de

[2] Institute of biomedical optics, Universität zu Lübeck, linz@bmo.uni-luebeck.de

[3] Department of Bioengineering, Beckman Institute for Advanced Science and Technology, University of Illinois at Urbana-Champaign, phuang16@illinois.edu

[4] Department of Electrical and Computer Engineering, Bioengineering Program, College of Medicine, Beckman Institute for Advanced Science and Technology, University of Illinois at Urbana-Champaign, boppart@illinois.edu

Abstract

The neural activity in rodent brain slices was investigated using phase-resolved Optical Coherence Microscopy (OCM). The brain tissue was excited by adding KCl and a significant difference between phase fluctuation during periods of neuronal activity and non neuronal activity was shown. Additionally, mechanical properties and optical properties of the brain tissue were measured with a different OCT system. For the characterization of different mechanical properties, a magnetic coil was integrated into a magnetomotive OCT system. The Young's Modulus, which provides information about the stiffness of the tissue, was measured for different parts of the brain. It was shown that the thalamus is the stiffest region in the brain. The refractive index of the tissue was measured in dependence on the compression. The refractive index increased non-linear with increasing pressure from 1.32 to 1.50 after 60% compression.

1 Introduction

Traditional methods for measuring electrical activity in neurons *in vitro* are based on contrast-enhancing dyes, like voltage- or calcium-sensitive fluorescence dyes [1]. However, the use of dyes has several disadvantages, such as photo-bleaching or toxic effects. Optical coherence tomography (OCT) is a noninvasive and label-free approach, that is based on detecting optical changes in neuronal tissue [2]. It was shown that these changes are intrinsic and accompanied by functional activity. These changes are correlated to three separate factors: changes in the membrane potential [1], changes in the local refractive indices of the membrane or the cytoplasm, and with the geometry of a cell [2].

The detection of the changes in back scatted light requires an imaging system with high resolution, fast acquisition rates and high sensitivity to small optical changes, which is provided by OCT. It has been demonstrated that changes in OCT intensity are related to neuronal activity in the brain [1]. Recently, it has been shown that the activity in a complex neuronal structure, like in the isolated Drosophila central nervous system can be directly detected using phase-resolved measurements [3]. In this work, a phase-resolved OCM has been used to detect patterns of neuronal activity. Besides the neuronal activities, the mechanical and optical properties of the mouse brain were studied in this work. It is well known that mechanical factors play an important role in regulating brain activity. Nevertheless, most research efforts have focused on the biochemical or electrophysiological activity of the brain [4]. In general, mechanical properties (like the stiffness of the tissue) and optical properties (like the refractive index n) can support the clinical diagnosis between normal and malignant tissue. In this paper dependence of n due to tissue compression was examined. Such studies could be helpful in correcting image distortion caused by injury, surgery or disease [5]. A 1310 nm OCT system was used to measure both n and the stiffness of the brain tissue. This required an extended setup as a Magnetomotive Optical Coherent Elastography (MM-OCE).

The OCM, the OCT and the MM-OCE used for the experiments were custom-made.

2 Material and Methods

2.1 Imaging systems

Two different OCT imaging systems were used for the experiments in this work. A Spectral Domain (SD-) OCT system was used to image the structure of brain slices as well as for the measurements of optical and mechanical properties. For the latter, the OCT-system was extended to a Magnetomotive (MM-) OCE system. The light from the laser source is split by a fiber coupler into sample and reference arm. The back scattered light from the sample interferes with the reflected light from the reference path and the interference signal is measured by an optical spectrometer. The system has a superluminescent diode as light source with a centered wavelength at 1310 nm and a bandwidth of 170 nm . This leads to an axial and lateral resolution of ~6 µm and ~16 µm (full width at half maximum (FWHM)). A 1024-pixel linescan camera was used in the spectrometer providing an optical imaging depth of 2.2 mm.

MM-OCE is a functional extension of OCT that employes "magnetic" nanoparticles (MNPs) as dynamic imaging contrast agents [6]. For the MM-OCE, an electromagnetic coil was placed in the sample arm of the 1310 nm spectral domain OCT imaging system. The coil had a inner diameter of 2 mm to enable the optical beam to pass through, for imaging the sample located below the coil. The magnetomotive force acting on the MNPs is proportional to the "square of the magnetic field". To obtain a sinusoidal displacement of the MNPs, a unipolar square root sinusoidal voltage,

$$V(t) = V_0 \sqrt{\frac{1 + sin(2\pi f_b t)}{2}} \quad (1)$$

was applied to the coil through a computer controlled power supply. Here, V_0 is the peak voltage and f_b is the applied modulation frequency [7]. This way tissue containing MNPs can be triggered externally for sinusoidal movements.

A second OCT system, the spectral domain (SD) -OCM system was used for the investigation of phase fluctuations correlated with neuronal activity. It is based on a Michelson interferometer with a 50/50 coupler as shown in Fig. 1. A superluminescent diode with a centered wavelength at 860 nm and a bandwidth of approximately 80 nm was used as light source, giving a calculated axial resolution of 4 µm (FWHM) and a lateral resolution of 0.6 µm. The used objective has a numerical aperture of NA = 0.6 (LUCPLFLN40X, Olympus, NA = 0.6). The inverted construction of the microscope enables to place the sample in an imaging dish with its surface in upwards direction suitable for living cells and tissue. A galvanometer-based 2-axis optical scanner was used for scanning across the sample. The interference signal from reference and sample arm is detected by a spectrometer, operated at an A-scan rate of 60 *kHz*.

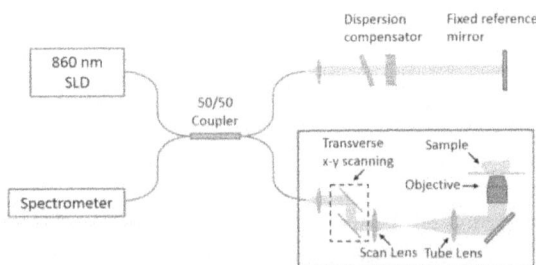

Figure 1: Schematic of the inverted spectral-domain optical coherence microscope (SD-OCM)

2.2 Measurements

Both systems were used to investigate *ex-vivo* mouse brain tissue slices. Fig. 2 a) shows an *en face* of the whole brain slice. a) was imaged with the 1310 nm OCT system. b) were imaged with the 860 nm SD-OCM System. The box in a) indicates the section of a part of the cortex, shown in b), where the cell bodies (dark regions), as well as the myelinated axons (the bright lines) can be seen clearly.

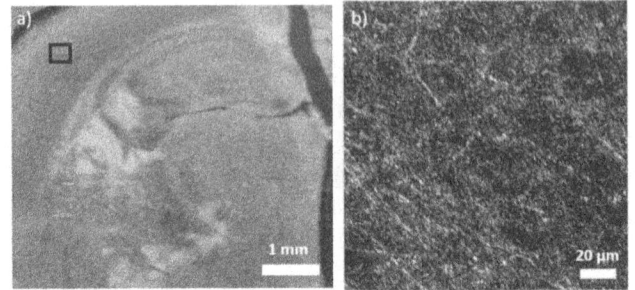

Figure 2: *en face* of a brain tissue slice. a) *en face* the whole left side of the brain slice measured with the 1310 nm OCT-System. The box indicates the section shown in b), detected with the SD-OCT system at 860 nm wavelength. b) shows a 160x160 µm *en face* image of a part of the cortex

2.2.1 Measurement of neuronal activity

For the measurement of neuronal activity, 500 µm and 300 µm thick brain slices from up to six week old mice were used. The slice was fixed in an imaging dish. To keep the tissue alive, it was placed in a pre-oxygenated artificial cerebrospinal fluid (aCSF). The tissue was stimulated by applying 100 µl KCl in a concentration of 66 mM to the imaging dish. This way, KCl depolarizes neurons and triggers an measurable action potential.

The neuronal activity was measured, by the 860 nm SD-OCM system, that enables phase detection. The OCT beam was positioned over a cell body prior to KCl application. During the measurement, the beam was fixed and remained within the cell of interest over the entire measurement time. A M-mode dataset with 512 A-scans, 512 frames and a linescan rate about 20 *kHz* were acquired, leading to a recording time about 13 s. Multiple datasets were recorded to measure an extended period of time.

Tetrodotoxin (TTx) served as a control measurement to KCl. TTx is an aminoperhydroquinazoline poison, which inhibits the firing of action potential in neurons. 1 µM of TTx was dissolved in artificial cerebrospinal fluid (aCSF) and applied. It binds to the voltage-gated sodium channels and blocks the passage of sodium ions into the neurons.

2.2.2 Measurement of mechanical properties

The 1310 nm OCT system was used to measure the elasticity of the biological sample. Assuming the sample has negligible viscosity, the Young's Modulus E [6]

$$E = \frac{4f_0^2 \pi^2 mL}{S} \qquad (2)$$

was calculated. However for the calculation, it is necessary to know the natural frequency f_0, the thickness L, the cross-section S and the weight m of the sample. For measuring f_0 of the brain, each brain slice was soaked in a magnetic nanoparticles (MNPs) -solution (incubation solution (aCSF mixed with iron oxide nanopowder in a concentration of $2\,mg/ml$). MNPs can respond to a magnetic field and serve as force transducers [6]. If the tissue undergoes a forced vibration oscillating at the natural frequency of the tissue, the sample would exhibit the largest displacement and hence, a resonant peak at f_0 would be observed.

2.2.3 Measurement of optical properties

The 1310 nm OCT-system was also used to measure the refractive index of the tissue. A flat glass coverslip was attached to the top of the biological sample and fixed. The refractive index n was then calculated by comparing the of the optical thickness d_{tissue} of the brain slices and the physical thickness d_{airgap}.

$$n = \frac{d_{tissue}}{d_{airgap}} \qquad (3)$$

First, d_{tissue} was measured directly as thickness of the slice in the OCT image. The physical thickness d_{airgap} of the brain tissue was measured using the thickness of the air gap between the bottom and the coverslip. To measure the dependence of n under compression, the coverslip was pressed down to apply uniformly compressive strain. The brain tissue was compressed by 24, 45, 57, 71, 76 and 85 percent with respect to the initial optical thickness.

3 Results and Discussion

In the following section, the results of the different experiments are presented and dicussed. The data was analyzed using Matlab (Mathworks 2017). Matlab and Image J was used to visualize the images.

3.1 Phase fluctuation correlated with neuronal activity

The raw phase data was evaluated by an unwrapping algorithm, which eliminated phase wrapping and allowed the phase values to go beyond the normal 2π limits. Furthermore a background was always recorded and subtracted during data processing with Matlab. The standard deviation of "phase" was calculated. Fig. 3 shows strong fluctuations in "phase" after applying KCl to the sample. Phase data are plotted over the number of frames, which correlate to the time. Each frame represents 25.6 ms. Hence, Fig. 3 shows the phase over 1.92 minutes.

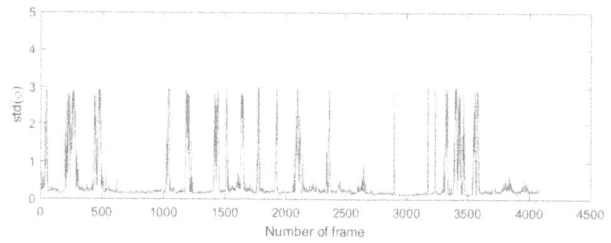

Figure 3: Phase fluctuation after KCl application.

Fig. 4 shows the comparison of KCl and TTx application. The phase after the application of TTx, shown in Fig. 4 b) stays noisy, but shows a significant distinction to a) the standard deviation of the phase during neuronal activity. Phase fluctuations are introduced by the relative displacements between scatterers, and the transient nanometer-level changes in neuronal thickness occur during action potential.

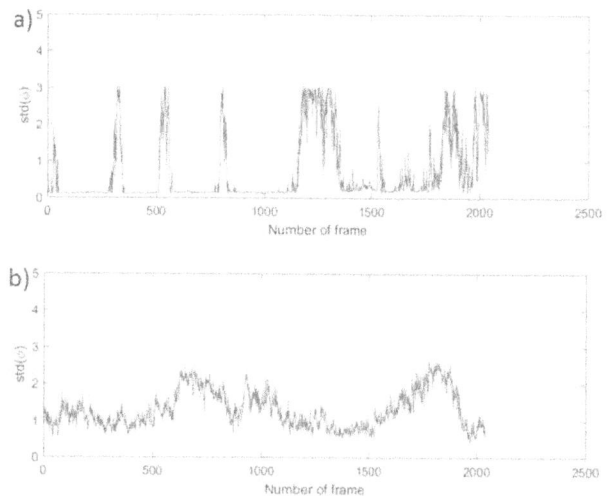

Figure 4: Phase fluctuation after KCl (a) and TTx (b) application. Both graphs show the standard deviation of the phase.

3.2 Mechanical properties

The Young's Modulus was measured in three different brain regions using the MM-OCE system: the thalamus, the cortex, and the hippocampus. 2 mm thick brain slices were used. The measured resonant frequencies of the hippocampus, cortex and thalamus are $f_{hippocampus} = 245\,Hz$, $f_{cortex} = 190\,Hz$ and $f_{thalamus} = 185\,Hz$. Using the weight and volume of each cross section, Young's Moduli were calculated for each brain region: $E_{hippocampus} = 7.60\,kPa$, $E_{cortex} = 6.74\,kPa$, $E_{thalamus} = 11.19\,kPa$. This implies that the thalamus is the stiffest region. The difference between hippocampus and cortex is not significant. In total, five different measurements were made. In every measurement the results varied by $\pm 1.5\,kPa$, which amounts to ± 20 percent. Here, we quantify the temporal-resolved characteristics of the magnetomotive vibrations instead of the absolute displacement amplitudes. This is because the amount of the MNPs diffused into the brain varies across different regions. The differences in the MNP concentration can be related to many factors such as tissue dependent diffusion of MNPs, the biological functions, or properties of the cells and tissues [7]. In each measurement, the thalamus always shows a higher value for the Young's Modulus. There is no significant distinction between hippocampus and cortex.

3.3 Optical properties

Fig. 5 shows the change of the refractive index n of the hippocampus under compression with respect to the initial optical thickness. n increased non-linear with increasing compression strain. It was shown that n increased significant over 25% compression to 60% compression and then reached a constant value.

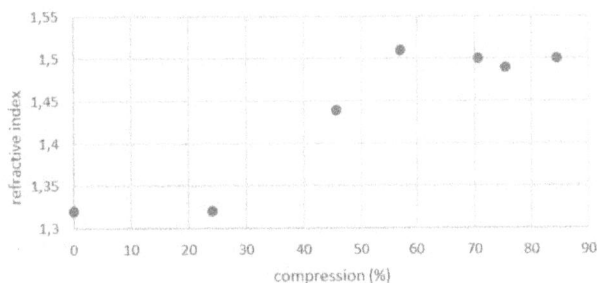

Figure 5: Refractive index of the hippocampus, measured under compression in seven steps

The increase of n from 1.32 to 1.50 caused by compression is significant (13.6 %). Although the n measured here is different from that reported in the literature ($n = 1.38$) [5], the general trend which shows a positive correlation between n and compression still agrees with the previous study. This knowledge may lead to an improvement of the correction of imaging distortion of biological sample that could be compressed due to injury, surgery or disease. Additionally, the information on how n changes with regard to physical deformation may be used to characterize or potentially diagnose diseases which are related to tissue deformation, such as hydrocephalus and brain tumors [5].

4 Conclusion

Two OCT systems were used to investigate mechanical and optical properties in rodent brain tissue. A 860 nm OCM system, with its high phase sensitivity capability, was used to measure phase fluctuations correlated to neuronal activity *in vitro*. It was shown that neuronal activity is correlated to phase fluctuations in the OCT data. Standard deviation of the phase showed a significantly different signal during neuronal activity compared to no neuronal activity after application of TTx.

For measuring mechanical properties, in particular the Young's Modulus, a 1310 nm OCT system was used for MM-OCE. The parameter indicates, that there is a difference in the stiffness of different brain regions. A significant distinction of Young's Modulus was measured between the region of the thalamus on one hand and the hippocampus and cortex on the other hand. inconsistent distribution of the MNPs in brain tissue, as well as the boundary condition issues that might be a result of the small thickness of the slices, it is debatable whether or not MM-OCE at the low excitation frequency ($<1kHz$) should be applied to brain tissues.

OCT is furthermore well suited to study optical properties of the brain like the refractive index. The experiments in this work showed a significant increase in the refractive index with compression of the tissue.

Acknowledgement

The work has been carried out at the Biophotonics Imaging Laboratory, Beckman Institute for Advanced Science and Technology, University of Illinois at Urbana-Champaign and supervised by the Institute of Biomedical Optics, Universität zu Lübeck.

5 References

[1] B. W. Graf, T. S. Ralston, H.-J. Ko, and S. A. Boppart *Detecting intrinsic scattering changes correlated to neuron action potentials using optical coherence imaging*. Opt. Express Vol. 17, Issue 16, 13447-13457, 2009.

[2] M. Lazebnik, D. L. Marks, K. Potgieter, R. Gillette, and S. A. Boppart, *Functional optical coherence tomography for detecting neural activity through scattering changes*. Opt. Letters, Vol. 28, Issue 14, 1218-1220, 2003.

[3] M. Q. Tong et al, *OCT intensity and phase fluctuations correlated with activity-dependent neuronal calcium dynamics in the Drosophila CNS*. Biomed. Opt. Express 8(2), 726-735, 2017.

[4] A. Goriely et al, *Mechanics of the brain: perspectives, challenges, and opportunities*. Springer Biomechanics and Modeling in Mechanobiology, 2015.

[5] J. Sun et al, *Refractive index measurement of acute rat brain tissue slices using optical coherence tomography*. Opt. Express 20(2): 1084–1095, 2012.

[6] P. Huang et al, *Magnetomotive Optical Coherence Elastography for Magnetic Hyperthermia Dosimetry Based on Dynamic Tissue Biomechanics*. IEEE J. Sel. Top. Quantum Electron., Vol. 22, Issue 4, 2015.

[7] A. Ahmad et al, *Mechanical contrast in spectroscopic magnetomotive optical coherence elastography*. Phys. Med. Biol. 60 6655, 2015.

Analysis of blood pressure data of children receiving general anesthesia to investigate the feasibility of creating age-specific reference values

R. Hillgruber [1]

[1] Biomedical Engineering, University of Applied Sciences Lubeck, rosa.hillgruber@stud.fh-luebeck.de

Abstract

Reference values for mean arterial blood pressure (BP) exist only for children undergoing inhalational anesthesia (IH). This pilot study creates nomograms, which show age-dependent reference BP values for children receiving total intravenous anesthesia (TIVA), IH and procedures with mixed techniques (MIVA) using a cohort of children, undergoing general anesthesia at a single center. With ethical approval, BP data were extracted from a vital signs database and age dependent BP nomograms were created. Data from 3913 cases were available: 2334 cases with TIVA, 847 cases with IH, and 732 cases with MIVA. Median BP varied from 46 mmHg for neonates to 72 mmHg at 18 years of age during TIVA, 44 mmHg to 71 mmHg for patients receiving IH, and 43 mmHg to 74 mmHg for MIVA, respectively. The median BP was higher with TIVA than with IH, demonstrating an anesthesia-specific BP effect. Future work includes repeating the study with a larger sample size, improved artifact rejection, and procedure-specific subgroup analyses.

1 Introduction

General anesthesia is required for most pediatric surgeries, regardless of the age of the patient. In the past, inhalational anesthesia (IH) has dominated the practice of pediatric anesthesia [1]. However total intravenous anesthesia (TIVA) has been shown to reduce some of the undesired side-effects of general anesthesia, including postoperative nausea and vomiting [2], and perioperative respiratory adverse events (PRAE); therefore, its use is becoming more common in everyday clinical practice [1].

Given that one of the fundamental tenets of anesthetic care is to maintain and monitor intraoperative physiological homeostasis, vital signs monitoring is a significant aspect of perioperative anesthetic care: this includes frequent measurement of blood pressure (BP), heart rate (HR), blood oxygen saturation (SpO_2), and exhaled carbon dioxide concentration ($etCO_2$). Reference values for non-invasive BP (NIBP) for healthy, non-anesthetized children are available [3], and a recent multi-center retrospective cohort study added nomograms of NIBP observed in children during maintenance of general anesthesia with IH using anesthetic vapor [4]. However, no reference values are currently available for children undergoing TIVA.

At BC Children's Hospital (BCCH), TIVA is the most commonly used type of general anesthetic (> 60%). Due to differences in pharmacology between these modalities, it is possible that TIVA results in different intraoperative BP profiles than those observed during IH. This provides us with an opportunity to evaluate the findings of [4], with a different type of anesthetic. Additionally, identification of cutoff values from these data might help anesthesiologists guide their anesthetic management, as BP is a key surrogate indicator of organ perfusion [5]. Finally they could be used for the setting of age-appropriate blood pressure alarm limits.

The purpose of this study was to develop age-specific mean arterial BP nomograms for children undergoing general anesthesia: TIVA, solely IH, or in combination, in which an inhalational induction of anesthesia was followed by intravenous maintenance (MIVA).

2 Material and Methods

With the approval of the UBC/Children's & Women's Health Center of BC Research Ethics Board [H17-03317] a retrospective observational cohort study was performed. It made secondary use of data, extracted from a local vital signs database, which collected data from the operating room (OR) multi-parameter patient monitoring systems (Datex AS/3 GE Canada, Mississauga, ON). This database contains approximately 40,000 procedures conducted between January 2013 and December 2016. In this pilot study, we limited ourselves to data collected between January and September of 2016.

The vital signs database is de-identified, meaning it does not include any personal health information, including the age of the patient. Therefore, cases needed to be matched with the Operating Room Scheduling Office System (ORSOS), in which the date and time of the surgery (time when the patient entered and exited the OR), the procedure location, and demographics, including the age of the patient, were available.

2.1 Inclusion and exclusion criteria

Data were included from all children (age <19 years), who underwent general anesthesia in the BCCH ORs during January - September 2016, for whom vital signs data, including NIBP, were available; procedures requiring cardio-pulmonary bypass (having a threshold of a body temperature under 33 °C in the 10th percentile of available temperature data) as well as cases which could not be matched to a patient, were excluded from the analysis.

2.2 Case matching

A probabilistic matching algorithm was used to match vital signs data to a patient record in ORSOS, required to add the age to the BP data. Due to the fact that vital signs are only recorded when sensors are connected to the patient, as well as that the times provided in ORSOS are documented manually, the start and end time of available vital signs might not be exactly the same as times provided in ORSOS. However, since the pulse oximeter is typically the device which is connected to the patient first upon entry to the OR, the time of the first SpO_2 measurement was used as a surrogate for the case start time. Similarly, the pulse oximeter is the last sensor removed, as its module accompanies the patient during transport to the anesthetic care unit. Therefore, the time of the last SpO_2 was used as the case end time.

While this method was successfully used in many previous studies [6], factors such as a shifted wall or patient monitor clock, a transcription error from the OR time sheet, or a delay between room entry/exit and application of pulse oximeter can result in differences in reported times. A 25 min time frame at the start as well as a 30 min time frame at the end of the surgery was allowed for the matching process.

2.3 Determination of case type

The type of anesthetic technique was determined by analyzing the minimal alveolar concentration (MAC) of inhaled anesthetic vapor. Cases were defined as either:

a) TIVA, where the sum of all observed MAC values was zero, since no anesthetic vapor was administered (1 % tolerance);

b) IH, in which the MAC value is greater than or equal to 0.3 for at least 70 % of the case time; or

c) MIVA, in which procedures were started with an inhalational induction of anesthesia leading to a MAC >0.2

for part (<20 %) of the procedure duration, but with MAC at zero for most (>50 %) of the procedure duration.

Cases not falling in any group were excluded from further analysis.

2.4 Data analysis

BP artifacts were removed using algorithms implemented in Matlab (version R2016b, The Mathworks, Natick, MA); first, physiologically impossible values were excluded, e.g. negative values or where the expected relation of systolic > mean > diastolic pressure was not observed; next a 5-point moving median filter was applied to the signal which was sampled at 0.1 Hz, resulting in a 50 second averaging window. From all cases, in which NIBPs were available, 20 randomly sampled values, with replacement, were extracted.

Data were presented as median with interquartile range (IQR) ($q_1 - q_3$) and range [min-max]. BP data were tabulated and aggregated over cases, from which nomograms for mean BPs were created. Data were grouped by age using the following cutoffs (each excluding the previous categories): <1 month, <3 months, <6 months, <1 year, and then continuing in one year intervals until <19 years. For presentation, box plots with population percentiles were created. BP distributions between the three anesthetic techniques were compared using Wilcoxon rank-sum test; Bonferroni corrections were applied whenever multiple comparisons were performed (adjustment of p-value: p / number of simultaneously performed statistical tests).

3 Results

Between January and September 2016, 8058 surgeries were captured in ORSOS. However, for this preliminary study we limited ourselves to cases which were assigned to a main OR (ORs 1 to 7, and the Dental OR). This reduced the number of potential matches to 5531 cases; excluded cases included medical imaging, and other offsite locations. With the currently available matching algorithm (see section 2.2), 4850/5531 (88%) cases were matched successfully to vital signs data, and of these 4040/4850 (83%) cases were successfully assigned to one of the three anesthetic case types. Finally, excluding cases of patients older than 19 years of age, or without non-invasive BP, 3913 cases remained in the presented analysis.

The median (IQR [range]) age of the children in this cohort was 6 (3-12 [0-18]) years. The distribution of anesthesia type was: TIVA, 2334 cases (60%); IH, 847 cases (21%); and MIVA, 732 cases (19%). Median BP by technique were TIVA: 64 mmHg, IH: 63 mmHg and MIVA: 64 mmHg; when TIVA vs. IH, TIVA vs. MIVA, and MIVA vs. IH were compared, all comparisons were statistically significant (p<0.001).

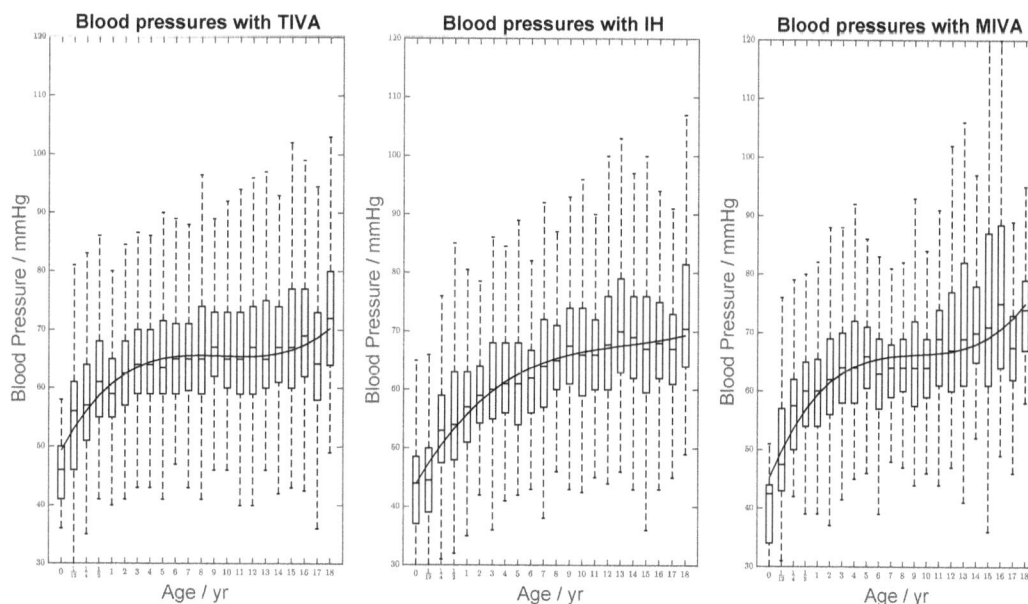

Figure 1: Blood pressure nomograms for the anesthetic types TIVA, IH and MIVA for ages between 0 and 19 years, shown using boxplots, with extending to the last datum within 1.5 IQR. Finally, a cubic regression of median BP values is overlaid.

BP increased by age (see Fig. 1); median (IQR [range]) mean BP of children receiving TIVA was 46 (41-50 [36-69]) mmHg at birth, and increased to 72 (64-80 [49-108]) mmHg at age 18. BP of children receiving IH were 44 (37-48.5 [27-76]) mmHg at birth, increasing to 70.5 (64-81.5 [49-109]) mmHg; while BP of children with MIVA were 42.5 (34-44 [29-51]) mmHg and 74 (67-79 [58-110]) mmHg in those ages respectively.

When fitted using cubic regression, the median BP, with correlation coefficients of $R^2 = 0.893$ for TIVA, $R^2 = 0.966$ for IH, and $R^2 = 0.894$ for MIVA, and a = age in $years$, can be described as

$$BP_{TIVA}(a) = -0.009 \cdot a^3 - 0.35 \cdot a^2 + 4.7 \cdot a + 45, \quad (1)$$

$$BP_{IH}(a) = -0.004 \cdot a^3 - 0.22 \cdot a^2 + 4.2 \cdot a + 40, \quad (2)$$

$$BP_{MIVA}(a) = -0.011 \cdot a^3 - 0.44 \cdot a^2 + 5.8 \cdot a + 40, \quad (3)$$

Differences in median BP, for children with a clinically significant difference in BP ($\geq$ 5 mmHg) between groups, split by anesthetic type and age are summarized in Table 1. Median BPs of patients from all anesthetic groups younger than one month as well as IH patients with an age between 1 to 3 months were more than 5 mmHg lower compared to the next age group respectively.

Patients with IH generally had a lower median blood pressure than patients with TIVA or MIVA within the same age category; significant results ($p < 0.05/3$) were found for all anesthetic types between ages of 0 and 5 years, for TIVA vs. IH at ages 6, 13, 14, 16, and 17, and for MIVA vs. IH at ages 9-11 and 14-16 years.

4 Discussion

With 3913 analyzed cases of BP data this pilot study demonstrates, as already known [3], an increase in median BP by age. It adds reference BP for children during TIVA, which were reported to be ranging from 46 (41-50) mmHg at birth to 72 (64-80) mmHg at age 18. The cubic median fittings in (1)-(3) show that the BP increase by age of patients while the slope of TIVA is less sharp than for IH or MIVA patients. This may indicate a pharmacodynamic difference between the three anesthetic modalities. The comparison of median BP in Table 1 demonstrates that particularly for children between 0 and 5 years, there exist differences in BP by the type of anesthetic provided. Clinically significant results (a difference $\geq$ 5 mmHg), where only found in TIVA vs. IH for the ages < 3 month, < 1, and 13 years, MIVA vs. IH for the ages < 1, 5, and 16 years, and TIVA vs. MIVA for the ages < 3 month, and 16 years

4.1 Comparison with the literature

Compared to the recently published results [4], where BP values between 33 mmHg at birth to 67 mmHg at the age of 18 years were reported, our results show a slightly higher BP in children receiving IH: 44 mmHg at birth to 70.5 mmHg at the age of 18 years. Differences, particularly at the lower end is clinically significant, and may reflect differences in institutional practices.

4.2 Limitations

The generalizability of our data may be limited, both as data are only from a single center (BCCH) as well as

Table 1: Median BP by age group, *p*-values are from the comparison using Wilcoxon rank sum test of age group differences. BP values of clinically significance between the three anesthetic groups as well as compared to the next and previous age group of the same anesthetic type respectively are highlighted in bold

Age	Patients			Median			*p*-value		
---	TIVA	IH	MIVA	TIVA	IH	MIVA	p(TIVA,IH)	p(MIVA,IH)	p(TIVA,MIVA)
<1 m	6	28	2	**46**	44	**42.5**	0.001	0.006	<0.001
<3 ms	17	29	11	**56**	44.5	**47.5**	<0.001	<0.001	<0.001
<6 ms	21	40	16	57	**53**	**57.5**	<0.001	<0.001	0.628
<1 yr	39	35	27	**61**	54	**60**	<0.001	<0.001	0.003
5	197	38	42	63.5	**61**	**66**	<0.001	<0.001	0.001
13	88	37	31	**65**	**70**	69	<0.001	0.330	<0.001
16	110	59	27	**69**	**68**	**75**	0.008	<0.001	<0.001
17	46	32	19	**64.3**	67	**67.5**	<0.001	0.374	<0.001
18	15	10	10	**72**	70.5	**74**	0.489	0.051	0.238

due to the unique anesthetic practice predominantly used at BCCH. Additionally, the sample of children, while large (>5,000) did not capture large number of infants, neonates from patients older than 16 years due to their low prevalence in the surgical population espiacially for the MIVA and TIVA groups.Applying our methodology to the larger cohort available will overcome this problem.

Next, the matching algorithm, in its current stage of implementation, discarded a high number of potential cases. This includes instances, in which two surgical teams switched their physical location, but where this change was not updated in ORSOS. Improvements in the probabilistic matching are likely increasing the cohort size, as well as the reliability of the match.

4.3   Future work

To increase the representativeness of our results all available data from the vital signs database in the study period should be used in the analysis. Additionally, in order to compare the results of this study with the available reference values [4], a distinction between the preparation phase (time before the surgical procedure starts), and the procedure phase needs to be made. Such data was not available in the currently available ORSOS export but can be requested from an analysis in BCCH's performance metrics and reporting department. Next, subgroup analyses by sex, as well as by procedure group should be performed. Finally, nomograms and analyses of the systolic blood pressures should be completed as these remain in common clinical use, despite their known methodological limitations.

5   Conclusion

This study demonstrated a difference in observed BP with anesthetic type, whereby median BP with TIVA was slightly higher than BP with IH. Whether this difference is clinically significant, thereby motivating a potential switch in anesthetic technique, remains to be evaluated in a more robust fashion.

Acknowledgement

The work was performed at the Research Institute, BC Children's Hospital, Vancouver, Canada, and was supervised by Dr. M. Görges, Department of Anesthesiology, Pharmacology & Therapeutics, University of British Columbia (UBC) and by Prof. Dr. S. Müller, Department of Applied Natural Sciences, University of Applied Sciences Lübeck. The author would like to thank Dr. S. Whyte for his clinical guidance in protocol development, as well as Mr. N. West for editorial assistance.

6   References

[1] G. R. Lauder, "Total intravenous anesthesia will supercede inhalational anesthesia in pediatric anesthetic practice," *Paediatric Anaesthesia*, vol. 25, pp. 52–64, Jan 2015.

[2] J. Lerman and M. JÖhr, "Inhalational anesthesia vs total intravenous anesthesia for pediatric anesthesia," *Paediatric Anaesthesia*, vol. 19, pp. 521–534, May 2009.

[3] H. K. Neuhauser, M. Thamm, U. Ellert, H. W. Hense, and A. S. Rosario, "Blood Pressure Percentiles by Age and Height From Nonoverweight Children and Adolescents in Germany," *Pediatrics*, vol. 127, pp. 978–988, Apr 2011.

[4] J. C. De Graaff et al., "Reference Values for Noninvasive Blood Pressure in Children during Anesthesia: A Multicentered Retrospective Observational Cohort Study," *Anesthesiology*, vol. 125, pp. 904–913, Nov 2016.

[5] G. Thomas and V. Duffin-Jones, "Monitoring arterial blood pressure," *Anaesthesia & Intensive Care Medicine*, vol. 16, pp. 124–127, Mar 2015.

[6] M. Görges et al., "Developing an objective method for analyzing vital signs changes in neonates during general anesthesia," *Paediatric anaesthesia*, vol. 26, pp. 1071–1081, Nov 2016.

Development of an electromagnetic compatible SpO2-simulator for the use with pulse oximeters

S. Sharif [1], S. Puttfarken [2], and S. Müller [3]

[1] Biomedical Engineering, University of Applied Sciences Lübeck, soeren.sharif@fh-luebeck.de
[2] Drägerwerk AG & Co. KGaA, Lübeck, stefan.puttfarken@draeger.com
[3] Fachbereich Angewandte Naturwissenschaften, University of Applied Sciences Lübeck , stefan.mueller@fh-luebeck.de

Abstract

Pulse oximeters, like any other medical device, run through various tests to succeed in different registration procedures. One is about ensuring the electromagnetic compatibility. Objective of this paper was the development of a device which is able to confirm that a pulse oximeter is electromagnetic compatible. The developed SpO2-simulator allows the simulation of one certain SpO2-value and various different pulse rates while reducing the used electronics to a minimum of a simple timer based pulse generator. Achieving the main function by means of using an artificial finger and changing its volume using a pneumatic system, the simulators electromagnetic compatibility is ensured. Other comparable, available devices are often based on directly modifying the pulse oximeters signal, which is significantly incompatible to electromagnetic radiation. First tests produced results similar to other research on calibration devices for pulse oximeters with the advantage of making electromagnetic compatibility tests possible.

1 Introduction

In today's medical patient monitoring pulse oximetry is one of the standard methods, due to its fast, non-invasive and easy measurement. Pulse oximetry is an optical process based on the Beer-Lambert Law. The value of interest is the ratio of oxygenated hemoglobin to oxygenated and de-oxygenated hemoglobin, i.e. the functional oxygen saturation, SpO2. The measured values are achieved by using two LED's of different wavelengths (red and infrared) and analyzing the different transmission in blood for both. Therefore a finger sensor is emitting light on the one side of the finger and detecting the transmitted light on the other side. To get rid of constant tissue-absorption (DC-component) only pulsatile parts (AC component) of the transmitted light is used for calculation of the R-value as shown in equation (1).

$$R = \frac{(AC/DC)_{red}}{(AC/DC)_{infrared}} \qquad (1)$$

The calibration of a pulse oximeter is done by using a calibrating curve created by comparing the R-values to the results of a CO2-meter during a measurement on human subjects [1].

During registration procedures a manufacturer has to ensure that his medical device, in this case a pulse oximeter, is electromagnetic compatible, i.e. the device is measuring the correct values even when it is exposed to electromagnetic radiation. The standard calibration procedure is not applicable for this, due to the need of human subjects and a CO2-meter which might also not be electromagnetic compatible. To solve this problem a device is needed which on the one hand replaces a human subject, i.e. simulates SpO2-values and on the other hand is electromagnetic compatible. There is already a lot of research on alternative calibration procedures for pulse oximeters to get rid of the need of human subjects and some devices are already available. Even if those calibration devices are not licensed for official registration procedures they facilitate a manufacturer to perform calibration testing before performing expensive, official registration procedures.

The reference [2] has as objective to develop a novel calibration technique using an "artificial finger with a variable spectral-resolved light attenuator in conjunction with an extensive clinical database of time-resolved optical transmission spectra of patient's fingers...". The same approach was continued in [3] to a first prototype with good results compared to the standard calibration procedure. A different approach is described in reference [4], there "a programmed, motorized, dual-axis system was designed to provide certain calibration values as specified by the user through a computer interface". The OxSim OX-1 by Pronk Technologies [5] and some devices by Fluke Corporation based on the US Patent 5.348.005 [6] are already available. Both are converting the light emitted by the pulse oximeter to modulated signal and those back to modulated light flashes which are detected and analyzed by the pulse oximeter.

All those devices and research have in common that they do not focus on the electromagnetic compatibility, this is where the approach of this work begins. While the main focus is proving electromagnetic compatibility not calibration there is no need for adjustable SpO2-values. Objective is to develop a simulator which is able to simulate exactly one value but ensuring that this value is not affected if the device is exposed to electromagnetic radiation. To achieve this the simulation is based on an artificial finger which achieves its main function without the use of electronics, but with a pneumatic volume changing finger.

2 Material and Methods

2.1 Development of the SpO2-simulator

The SpO2-simulator being developed here will provide several sub functions to achieve its main function, the simulation of saturation and pulse rate, when being attached to a pulse oximeter. Fig. 1 shows how the sub functions are connected with each other to build the complete SpO2-simulator.

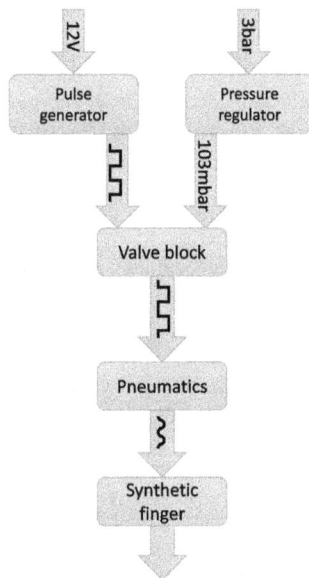

Figure 1: Block diagram of the SpO2-simulator

2.1.1 Pressure regulator

The pressure regulator is used to decrease the input air pressure, for this purpose a precision pressure regulator by WatsonSmith was used [7]. Variation of the decreased pressure will not change the pulse rate, but the measured saturation. To achieve reproducible results one should ensure that the input air pressure for the valve block, i.e. the decreased output of the regulator, will not change. For the upcoming measurements the pressure was regulated to 103mbar.

2.1.2 Valve block

The valve block is used to modify the constant air pressure to a periodic changing air pressure, changing stepwise between 103mbar and 0mbar. The speed of the periodic changes correlates to the measured pulse rate. Therefore

it is possible to change the measured pulse rate by changing the speed of the periodic changes. For the performed measurements a intern developed solenoid valve block was used.

2.1.3 Pulse generator

The pulse generator is based on a NE555 by Unisonic Technologies Co., Ltd [8], which is an integrated circuit timer.

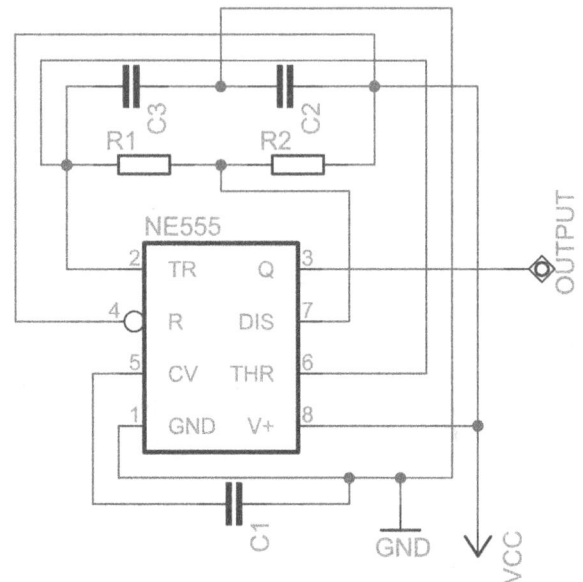

Figure 2: Electric circuit of the pulse generator. Circuit is based on the timer *NE555*. R1 & R2 are used to vary the pulse.

Fig. 2 shows the electric circuit to control the Period, T of the timer. This circuit is based on typical applications as they are mentioned in the product specification [8]. The most relevant parts of the circuit are the capacitor C3 and the two resistors R1 and R2. C3 is charging through both resistors R1 and R2, but discharging only through the resistor R1. As soon as C3 is charging above 2/3VCC an internal transistor is turned on and C3 begins discharging. The internal transistor controls the output of an internal flip-flop. This way the circuit has a periodic changing high/low output. The capacitor C2 connected between GND and VCC provides functional stability of the circuit, while capacitor C1 prevents swinging. Both C2 and C3 are $0,1\mu F$ and C1 is $1\mu F$.

The values of the resistors are kept variable as changing these defines the Period, T of the pulse generator and therefore the simulated pulse rate. In the following Measurements different Settings are uses, which can be seen in Table 1.

As shown in (2) the pulse rate in beats per minute is the reciprocal value of the Period, T multiplied with 60 seconds [8].

$$f\,[bpm] = 60 \cdot \frac{1}{0.693\,(R_2 + 2R_1) \cdot C_3} \qquad (2)$$

Table 1: Implemented settings and pulse rates

Setting	R1/MOhm	R2/MOhm	Pulse/bpm
I	3,0	2,1	107
II	1,5	3,9	125
III	3,0	6,2	71
IV	1,5	2,1	170

2.1.4 Pneumatics

The combination of pulse generator and valve block, connected to a const. air pressure, is generating a pulsatile, rectangular changing air pressure. The flow generated by this changing air pressure flows through a pneumatic system (Fig.3) before it reaches the artificial finger. Whenever the valve block is open the air pressure of 103mbar generates a flow, which is filling the artificial finger with air. When the valve block is closed the air pressure is 0mbar and the air flows back out of the artificial finger. This generates a pulsation in the artificial finger. To achieve a pulsation similar to a real pulse a check valve, CV is added to the pneumatics. This way the air flow from the valve block to the artificial finger is passing two equal throttle valves, R1 and R2 and in the opposite direction the air flow passes only the throttle valve R1, leading to a reduced flow in this direction.

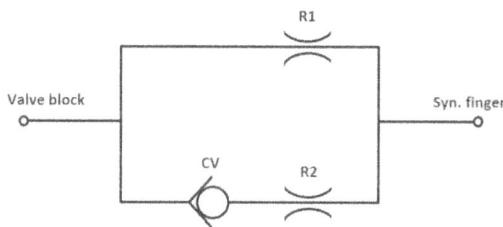

Figure 3: Pneumatic layout: The check valve, CV allows only air flow in valve block -> syn. finger direction.

2.1.5 Artificial finger

The artificial finger (Fig. 4) consists of a tough, non-transparent plastic giving the finger its shape and prevents of artifacts due to outer light sources, combined with a flexible and transparent plastic allowing the finger to pulsate and the light of the oximeter to pass through the finger. A tube allows the air to flow in and out of the finger.

2.2 Evaluation of the SpO2-simulator

The evaluation of the SpO2-simulator is done by comparing it with the OxSim OX-1 by Pronk Technologies [5]. Both, the SpO2-simulator and the OxSim OX-1 were tested in long-term tests over a period of 12 hours. For each test a standard pulse oximeter with a finger sensor is attached to the device under test and the SpO2- and pulse rate-values were measured. In this case the measurements were performed using a *Masimo M-LNCS Y-I Multisite Sensor* attached to a *Dräger Infinity Delta Monitor* [9] under room

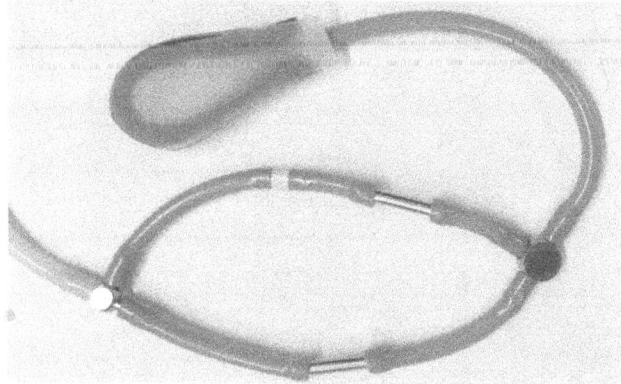

Figure 4: Artificial finger and pneumatics, allowing pulsation due to a change of air pressure.

temperatur conditions. During measurement it is ensured that the outer light sources are not changed and the device under test is not moved to prevent measuring artifacts. For the SpO2-simulator the test is repeated for each setting, while for the OxSim OX-1 only one setting with a pulse rate of 80bpm and a saturation of 85% is used.

As basis for the evaluation of the SpO2-simulator the sample standard deviation (ssd) is calculated for each measurement and compared to the ssd calculated (3) for the measurement with the OxSim OX-1. To calculate the ssd equation (3) is used.

$$ ssd = \sqrt{\frac{1}{N} \sum_{i=1}^{N} (x_i - \bar{x})^2} \quad (3) $$

3 Results and Discussion

The results presented in the following sections are based on the methods described before. The developed SpO2-simulator is compared to the OxSim OX-1 with focus on the deviation of measured SpO2- and pulse rate-values.

3.1 Evaluation of the SpO2-simulator

Fig. 5 shows the results for all settings and the OxSim OX-1 achieved in the long-term measurement. The upper graph is showing the change of pulse over time and the lower graph is showing the change of saturation over time.

The mean values as well as the sample standard deviations (ssd) for both, the pulse rate and the saturation are presented for all settings and the OxSim OX-1 in table 2. A slight difference between the measured mean pulse rates (see table 2) and the calculated pulse rates (see table 1) can be seen for all settings except of setting III and the reference OxSim OX-1.

3.2 Discussion

The SpO2-simulator was developed to measure simulated SpO2- and pulse rate-values with an oximeter, while ensuring electromagnetic compatibility. Fig. 1 shows the

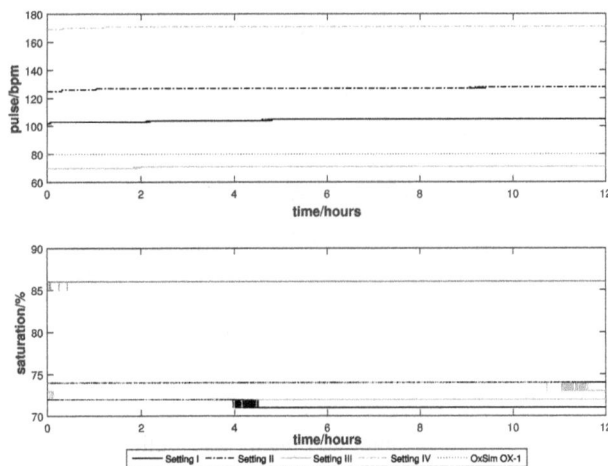

Figure 5: Top: Change of pulse over time - Bottom: Change of saturation over time

Table 2: Measured pulse rates and saturations

Setting	Pulse/bpm		Saturation/%	
	Mean	ssd	Mean	ssd
I	105	0,9	74	0,0
II	127	0,9	71	0,4
III	71	0,3	74	0,5
IV	171	0,5	72	0,1
OxSim OX-1	80	0,0	86	0,0

separate parts of the developed device in an overview. By achieving the simulators main function without using electronics, but pneumatics and placing the electronic pulse generator outside the artificial finger, negative electromagnetic effects are avoided.

Comparing the sample standard deviation of each setting with the OxSim OX-1 the latter shows smaller deviation for the pulse rate and the saturation. Only single settings show completely constant results, i.e. $ssd = 0,0$ for either the pulse rate or the saturation. With the highest deviation of $ssd = 0,9$ for the pulse rate and $ssd = 0,5$ for the saturation all settings show results that support the workability of the developed sensor, as the ssd is calculated for the complete 12 hours and to obtain useable results regarding electromagnetic compatibility way shorter measurements are performed. This statement is also supported by the investigation of the graphs in Fig. 5. Changes only occur in short, defined regions, most time during the period of 12 hours the values do not change at all and thus facilitate measurements for ensuring electromagnetic compatibility.

4 Conclusion

Due to the developed SpO2-simulator it becomes possible to evaluate the measuring quality of pulse oximeters and, what is the most important achievement, evaluating the electromagnetic compatibility of pulse oximeters or devices with integrated pulse oximeter interfaces, e.g. some anesthesia and artificial respiration devices. This feature is pro-

vided by none of the available SpO2-simulators for pulse oximeters.

Providing electromagnetic compatibility due to giving up using electronics as the main principle to simulate SpO2-values is straight forward and allows a simulation that is independent of the used sensor type.

While the SpO2-simulator is not offering an option to change the simulated saturation it features changeable pulse rates.

In future experiments a redesign of the artificial finger could be investigated in terms of decreasing the deviation of the simulated pulse rate- and SpO2-values. Beside of this a future version with adjustable saturation would further improve the workability of the simulator.

Acknowledgement

The work has been carried out at Drägerwerk AG & Co. KGaA, Lübeck and supervised by the University of Applied Sciences Lübeck. In addition I would like to thank S. Puttfarken and Prof. S. Müller for their supervision.

5 References

[1] A. C. Guyton, J. E. Hall, *Textbook of medical physiology*. 11. ed., Elsevier Saunders, Philadelphia, Pa., 2006

[2] Ch. Hornberger, Ph. Knoop, W. Nahm, H. Matz, E. Konecny, et al., *A prototype device for standardized calibration of pulse oximeters*. In: Journal of clinical monitoring and computing 16, Grin Publishing, pp. 161–169, 2000.

[3] B. Weber, *Direkte Spektralmodulation mittels Mikrospiegelarray zur Wiedergabe zeitaufgelöster Fingertransmissionsspektren für die Pulsoxymeterkalibration*. Inauguraldissertation, Universität zu Lübeck, 2017.

[4] D. Reid, H. Robertson, M. Sonntag, *Pulse oximeter calibratior*. Bachelorthesis, Worcester Polytechnic Institute, 2011.

[5] *Technical specifications for OxSim OX-1*. Pronk Technologies Inc.

[6] E. Merrick and P. Haas, *Simulation for pulse oximeter*. Bio-Tek Instruments Inc., Winooski, Vt. United States Patent, Patent Number: 5.348.005, Date of Patent: Sep.20 1994

[7] *Manostat Precision Pressure Regulator Series Manually Operated, datasheet*. IMI Watson Smith Ltd.

[8] *NE555 linear integrated circuit, single timer, product specification*. Unisonic Technologies Co., Ltd, Document number: QW-R106-001.G

[9] *Infinity delta and delta XL patient monitors, datasheet*. Draeger Medical Systems, Inc., Document number: 90 51 660

Development of a Simulation Module for a Synchronized Multi-Robot System Using the Simulation Software KUKA.Sim Pro for Supporting Process Validation

M. Stender [1]

[1] Medical Engineering Science, University of Luebeck, mareike.stender@student.uni-luebeck.de

Abstract

Process automation is getting more and more important in numerous sectors of industry and research. Especially repetitive and monotonous tasks as well as processes in extreme or dangerous ambient conditions are predestined for automation. Driven by the high complexity of nowadays automated systems just as by their expensive realization, simulations are included in the planning process of such systems. In this work the use of a single simulation module for simulating several scenarios of multi-robot systems is validated by realizing the complex simulation of the multi-robot system "medusa robot". A simulation software which enables implementations of realistic robot movements is used to create this simulation module and a simple application example is analyzed with respect to robot interactions and risks of collision. Advantages and boundaries of the developed simulation module and a short outlook complete this paper by comparing the simulated features of the medusa robot to two already existing medical robot systems.

1 Introduction

Nowadays the hard tasks in the automotive industry as well as in other industrial and research sectors are fulfilled by automated manufacturing lines or stand-alone systems including robot support. By delegating repetitive and exhausting work to machines and robots the efficiency of the whole production process is increased and especially in case of work in exposure to chemical substances the automation is beneficial with respect to occupational health and safety.

The design of complex automated solutions is simplified by the integration of simulations of single process steps or the entire process. Simulations enable the visualization of the whole sequence and a rapid modification without wasting much time and hardware.

The benefits of realizing a simulation module, as it is made in this work, are the expandability and reproducibility of it. It is possible to easily adapt the module to several scenarios of industry or research. The simulated features of process variants can then be readily compared. Driven by this fact the simulation module for the multi-robot system medusa robot is created, which can not only be used for the medusa robot but also, for example, for the simulation of an elaborated pharmaceutical handling robot sequence.

The special feature of the medusa robot is the way of interaction of the included robots with each other while fulfilling their tasks synchronized without collisions.

The simulation module is established in the KUKA.Sim Pro [1] environment and validated relating to reachabilities, risks of collision, and possible synchronization of the included robots by simulating simple handling processes.

With the outcome of a first basic simulation module, the features of the simulated medusa robot are compared to existing multi-robot systems in the medical sector like the CellPro [2] and MiroSurge [3] system.

2 Methods

The multi-robot system medusa robot was patented by Daimler AG in 2014 [4] and is part of the research project 3DProCar [5]. It is built of seven KUKA robots, one large KR1000 titan [6], four KR3 AGILUS R540 [7] and two KR6 AGILUS R900 [8] based on the described concept in [4]. The mounting for the KR3 and KR6 robots is realized by a baseplate of aluminium [9].

The settings for the baseplate and the positions of the small-sized robots on it are based on researches relating to the stability of the whole system, the reachabilities of the small-sized robots and the limited load of the KR1000 robot. The structure of this multi-robot system is shown in Fig. 1.

The simulation software KUKA.Sim Pro is developed by KUKA. It enables the simulation of any kind of KUKA robot and the emulation of diverse robot systems. This software also provides the opportunity to generate the final programming for the robots by using the tool KUKA.OfficeLite [10] directly during building the simulation.

The development of a simulation starts with choosing the robots for the simulation from the library called "ecatalog" in KUKA.Sim Pro. The ecatalog contains several other devices e.g., conveyors, linear units and different devices as well which can be flanged to the robots. The inserted

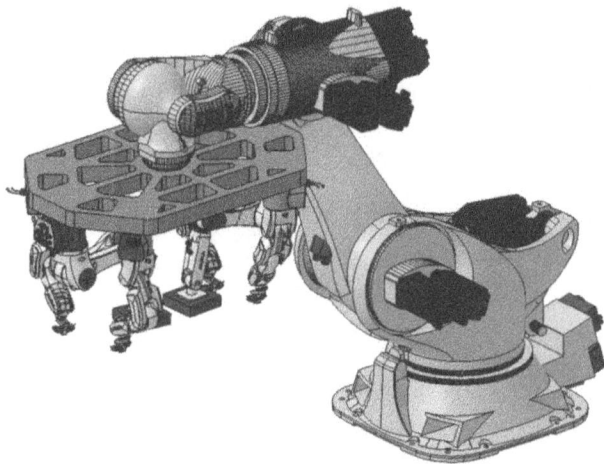

Figure 1: The shown multi-robot system called medusa robot is constructed of seven KUKA robots, one KR1000 titan [6], four KR3 AGILUS R540 [7] and two KR6 AG-ILUS R900 [8], [9].

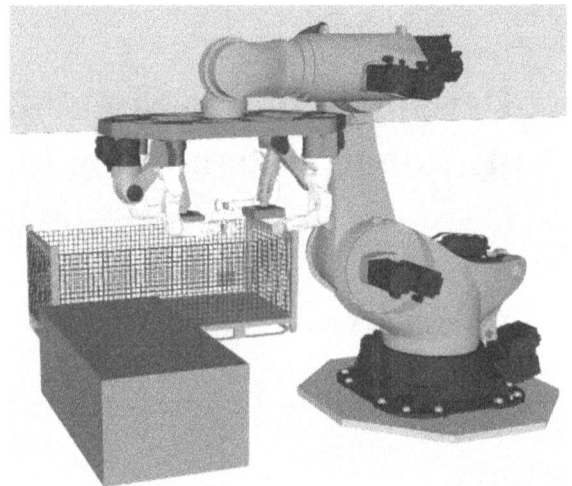

Figure 2: In this figure, the Medusa robot, aimed at inserting a block into a tray, is shown as an example of simulation modules.

robots have the same individual axes boundaries of movement as the real robots. Grippers also have to be inserted of the ecatalog or as .stp files which can be created in a CAD (computer-aided design) software. The inserted grippers have to be flanged to the robots.

Thereafter, all robots of the system have to be taught. Teaching is the process of delivering the operation sequence of desired points to the robot. Every desired axes point is saved in the shape of an angle property. In this way all the robot axes included in the system have to be taught. To support a synchronized movement of the robots the same velocity has to be chosen for all included robots. Also, delays of milliseconds or seconds in the sequence of desired points can be inserted.

To enable the inter-robot communication, inputs and outputs in the routines of the robots have to be set. That means that e.g., robot 1 waits for an input (true or false) of robot 2. Robot 2 sets the output after fulfilling its task to true or false. Thus, robot 1 gets the input for which it has waited and is able to start the next task.

By starting the simulation the joint points are converted to the path for every single robot and the robots move synchronized without collisions and interact with each other with a realistic cycle time.

In the case of the medusa robot the seven included robots are inserted out of the ecatalog. The baseplate, infrared emitters and needle grippers are inserted as given templates which are created in the CAD software CATIA [11] (Version: V5 R19).

Relating to the research project 3DProCar the task of the medusa robot is to handle and treat a soft textile which shall be preformed before it will be consolidated in a further step. The handling of the textile is performed by the needle grippers which are flanged to the KR3 robots. The infrared emitters which are flanged to the KR6 robots apply heat to the textile in order to prepare the textile for the following treatment step.

To flange the small-sized robots KR3 and KR6 to the baseplate they have to be connected and parented to the baseplate. Thereafter, the baseplate with the small-sized robots on it is connected to the KR1000 robot. The needle grippers and the infrared emitters are flanged in the same way to the KR3 and the KR6 robots.

A detailed operation sequence is developed in the context of this work. This operation sequence includes considerations about the possibility of simultaneous movement of the small-sized robots in order to reduce the cycle time which is an important requirement in the system engineering.

Moreover, calculations of the reachablities of the small-sized robots are included in the developed operation sequence.

Due to the fact that every robot owns six independent axes, 36 particular axes have to be taught in the case of the medusa robot. This means that every axis of a robot has to be moved separately in order to reach the desired points. The special feature of the developed simulation module is the interaction of the included robots.

A screenshot of a simple simulation including the medusa robot is shown in Fig. 2. As a placeholder for various possible objects a simple block is shown which the medusa robot has to grab using the gripper of the flanged KR3 robots. After gripping the block, the medusa robot carries it to the tray and inserts it there.

3 Results

The developed simulation module for the multi-robot system medusa robot ends up in several advantages for improving the system engineering of complex projects and process validation. As a result of the simulations realized with the simulation module, the needed parameters of the project of [4] can be defined and the limits of the project can be determined without the risk of collision of real robots. Including the real robot characteristics like size, velocity, and axes angles in the simulation module an investigation of the

Figure 3: This photo shows the CellPro system [2]. It is a system of two synchronized robots which are controlled by a panel outside the housing. The main benefit of Cell-Pro is to fulfill tasks without contaminations risks caused by human intervention.

reachabilities, the risks of collision, and the real cycle time of the robots in the range of milliseconds is possible. The possibility to include various devices, and objects as CAD files increases the validity of the simulation. Nevertheless, there are limits in the simulation module, for example, the examination of dynamics and forces which affect the baseplate of the medusa robot. But by using a software for analyzing the behaviour of the material of the baseplate this point is verifiable. Depending on the complexity of the process which shall be simulated the planning of the operation sequence becomes more complicated, especially because every single axis of the included robots have to be taught while avoiding collisions.

Due to the high modularity of the developed simulation module, it is applicable in several sectors of industry and research. To show the possible application of the simulation module for simulating medical robotic systems a comparison with two already used systems, the CellPro and the MiroSurge, is summarized in the following paragraph.

CellPro

The robot isolater called CellPro shown in Fig. 3 is develeloped by Shibuya Hoppmann. The main benefit of this system is the replacement of human operators in sterile-manufacturing processes because humans carry the risk of contaminating sensitive products. The robots act in an isolated room. To adapt the system to the task area the number of robots in the isolated room is configurable. The system is controlled by a remote control from outside the housing. A function to train precise tasks which the robot duplicates based on the operators movement is integrated in the system. The several robots work interactively together like the human hands to perform the desired process sequence. Cell and tissue processing, handling of radioactive products, and researches of new virus are tasks which can be fulfilled by the CellPro [2].

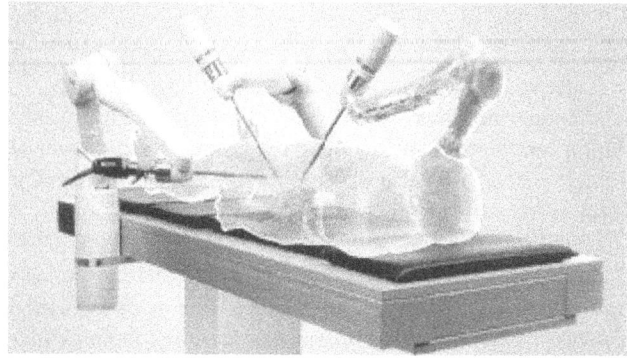

Figure 4: Shown is the minimally invasive robotic surgery called MiroSurge. It includes one robot guiding the endoscope (left) and two robots with laparoscopes (right) which can be attached to the surgical table [3].

MiroSurge - Telemanipulation in Minimally Invasive Surgery

The telemanipulation system MiroSurge is developed by the German Aerospace Center to increase the quality of minimally invasive surgery. The MiroSurge system arrangement adresses challenges for the surgeon like lost hand-eye coordination and missing direct manual contact to the operation area. The system includes a master console with a 3D-display and two haptic devices. The surgeon sits in this master console to operate the patient by moving the haptic devices. Furthermore, this system includes three robots. One of them holds an endoscopically instrument and can guide a stereo video laparoscope. The other two robots carry surgical instruments equipped with miniaturized force/torque sensors. With the development of the MiroSurge system the minimally invasive surgery gets more precise. This higher precision enables various more complicated applications, for example, in the cardiosurgery.

Compared to these both systems the simulation module provides performance characteristics of the medusa robot which show several similarities and differences. Both Cell-Pro and MiroSurge use manipulator systems, meaning that the included robots do not work autonomously. Contrary to this in the simulation the robots are following the program and the process is running without intervention of an operator. Besides interrupting the whole simulation there is no opportunity to change or stop the routine while running. In the industrial use there is the advantage that no human intervention is necessary for running the routine. In the medical sector every task needs several possibilties of controlling and interrupting in a case of an emergency situation during the surgery.

Relating to the number of robots and the expandability for other applications e.g., more complicated tasks the simulation module for the medusa robot delivers the highest flexibility. The medusa robot currently includes six small robots. By using lighter robots the system would be more expandable. The CellPro system as well includes an expandability with more robots but the space in the isolated room limits this fact. The MiroSurge is designed for use of three robots. Generally it is possible to use the simulation module for the medusa robot for simulating the two presented systems.

To realize the processes of the CellPro system the simulation module has to be reworked relating to the room in which it acts. By inserting a simple block in the dimension of the isolater this is possible. Thereafter, four of the six small-sized robots of the medusa robot in the simulation module can be demounted. On the other hand the process of the CellPro system can be extended to more complicated process of preparation of pharmaceuticals or cell manipulation by using the four other robots of the medusa robot.

The MiroSurge system as well is simulateable by using the simulation module for the medusa robot. Three of the six robots of the medusa robot can be used as it is developed in the MiroSurge system. One of them can handle the endoscopically instrument and two of them can carry surgical instruments. Furthermore, the three other robots of the medusa robot can handle instruments of the lung ventilator or the anaesthetic machine.

4   Conclusion

In the last years the simulation of robot based systems became more important. Especially in relation to process validation simulations are inevitable. By using simulations time and material can be saved. Moreover, limitations and challenges are already detectable in an early state of development and engineering.

Driven by this motivation a simulation module for the medusa robot is developed which enables the simulation of a multi-robot systems. By using this it is possible to realize hypothetical settings and to analize the behaviour of the multi-robot system in extreme situations e.g., the accuracy of repetition during fast movement in a confined space.

Thus, the usage of a medusa-like robot system and the module could be validated for an application in the pharmaceutical and the medical industry. Potential limitations and advantages could be identified.

To conclude, the usage of the simulation module for the multi-robot system medusa robot in the medical sector is possible. There are, however, some aspects which have to be explicitly researched. As an example, the required space of the industrial robots limit the possible use of the simulation module in the medical sector. Furthermore, the accuracy of repetition and the precision of the robot movement have to be analyzed and the hygiene specifications in the medical sector have to be considered.

Nevertheless, the comparison of the simulation module for the medusa robot with the already applied medical robotic systems CellPro and MiroSurge shows an promising possibility to use the newly developed simulation module for promoting the research and development of medical multi-robot systems.

Acknowledgement

The work has been carried out at IBG Technology Hansestadt Lübeck GmbH and was supervised by Dr.-Ing. S. Keipert-Colberg. Also it was supervised by Prof. Dr.-Ing. A. Schweikard, Institute for Robotics and Cognitive Systems, University of Luebeck. This project has been partly founded by the research project 3DProCar (support code: 02P14Z027) [5]. This research and development project is funded by the Federal Ministry of Education and Research (BMBF) in the framework concept "Innovations for the Production, Services and Work of Tomorrow" and funded by the Energy and Climate Fund and supervised by the Project Management Agency Karlsruhe (PTKA).

5   References

[1] KUKA Industries GmbH Headquarters. *Technology KUKA.Sim.* Location: Augsburg.

[2] Xact consortium, 2013. *Expert cooperative robots for highly skilled operations for the factory of the future.* P. 15. University of Patras.

[3] U. Hagn, et al. 2009. *DLR MiroSurge: a versatile system for research in endoscopic telesurgery.* In: International Journal of Computer Assisted Radiology and Surgery. [online]. March 2010. Volume 5. Publisher: Springer-Verlag. pp 183–193. ISSN 1861-6429

[4] Daimler AG, 2013. *Verfahren zur Herstellung eines Bauteils aus faserverstärktem Kunststoff und Vorrichtung zur Durchführung des Verfahrens.* Inventor: Dipl.-Ing. (FH) Eckhard Reese. 11. April 2013. DE 10 2012 019 958 A1.

[5] M. Gude. *FOREL-Newsletter.* TU Dresden, no. 2, 2015.

[6] KUKA Roboter GmbH, 2016. *KR 1000 titan, KR 1000 L750 titan.* Version: Spez KR 1000 titan KR C4 V5. Location: Augsburg.

[7] KUKA Roboter GmbH, 2016. *KR 3 AGILUS.* Version: Spez KR 3 AGILUS V3. Location: Augsburg.

[8] KUKA Roboter GmbH, 2017. *KR 6 R900 sixx.* Version: V24.1. Location: Augsburg.

[9] D. Bünning, 2018. *Entwicklung einer Mehrfach-Roboter-Kombination für eine Positionieraufgabe.* [bachelor thesis.] Fachbereich Maschinenbau und Wirtschaft, Fachhochschule Lübeck.

[10] KUKA Roboter GmbH, 2011. *KUKA.OfficeLite 8.2.* Version: KUKA.OfficeLite 8.2 V4 en. Location: Augsburg.

[11] CENIT. *Newsletter CATIA-V5R19. NEUERUNG IN CATIA V5 R19.* Location: Stuttgart

Robust evaluation of timing parameters in the gait cycle

I. Ryan [1], R. Wendlandt [2],

[1] Biomedical Engineering, University of Applied Sciences Lübeck, imogen.ryan@stud.fh-luebeck.de
[2] Clinic for Orthopedic and Trauma Surgery, University Medical Centre Schleswig-Holstein, Campus Lübeck, wendlandt@biomechatronics.de

Abstract

Observation of gait can be used to diagnose certain conditions or problems in regards to the human locomotor system. It can provide information about a person's health and help identify the state of his or her disease or condition. Therefore, acquiring accurate and reliable timing parameters of gait data is important to achieve this. An evaluation script was developed to analyse gait data and extract accurate timing parameters. This script was written to analyse data not only from healthy patients, but also from patients with locomotor problems or neurological illnesses, where the data is much more erratic. The final script proved to be reliable for a large number of data sets, meaning it can be used to extract the timing parameters from patients' gait data and use these parameters to diagnose or to do further studies.

1 Introduction

Human walking is a locomotive method using two legs, where at least one foot is in contact with the ground at all times [1]. Whereas gait can be defined as the manner of walking, or how a person walks. The gait cycle begins with the 'initial contact' of one foot with the ground (the reference foot) and ends when this same foot contacts the ground again. It has two phases, the swing phase and the stance phase, also known as the support phase.

Fig.1 shows the different events of the gait cycle. They follow this order:

Initial contact, reference foot

Toe-off, opposite foot

Initial contact, opposite foot

Toe-off, reference foot

Initial contact, reference foot

Fig.1 also shows the double support and single support timing. Initial contact of the opposite foot occurs while the reference foot is still on the ground, from this moment, until the toe-off of the reference foot, there is a period of double support. The time between the initial contact of the reference foot and the toe-off of the opposite foot, is also a period of double support. The times in between the periods of double support are the periods of single support, the timing of the reference foot single support corresponds to the opposite foot swing phase.

The stance phase of the gait cycle usually amounts to 60% of one cycle, and the swing phase is usually 40%. This is dependent on the speed of walking, the faster the pace the

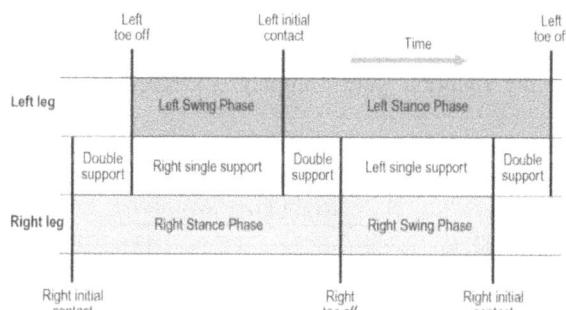

Figure 1: The gait cycle, beginning with right initial contact, of the left leg and right leg [1].

longer the swing phase and the shorter the stance phase, until the transition to running where there is no double stance phase at all [1].

A person's gait can provide information about his or her health and can be used to diagnose a neurological disease or muscular and skeletal problems. A completely healthy person will have low variability in his or her gait. The walking speed, the step length, the step rate and the gait timing will all stay relatively steady, and within normal ranges, during walking [5]. In healthy adults the ratio between the step length and step rate (normalised for height) is constant from person to person of around 6.5 mm/(step/min) [3] . In Fig. 2, which shows gait data from a healthy person, this steadiness can be clearly seen. The gait timing can be easily derived from the data, with clear initial contact and toe-off events as the maxima and minima of the heels' position data.

Illnesses, such as basal ganglia disorders like Parkinson's and Huntington's disease, can result in visible abnormal

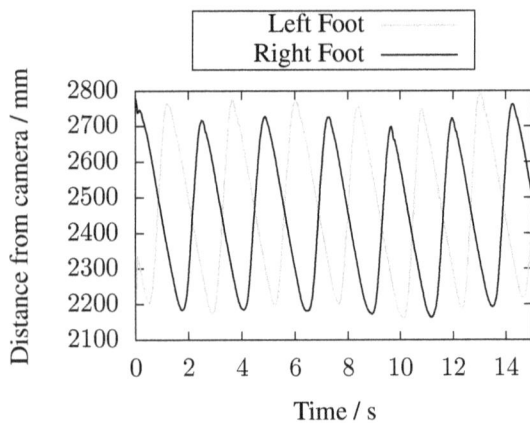

Figure 2: Gait data from a healthy subject.

gait patterns such as lurching, difficulty turning, imbalance, slow and shuffling gait [4]. It is thought that the basal ganglia are important in initiating and regulating motor programs for balance, gait and the sequencing and fluidity of movement [6]. Chorea, abnormal involuntary movement, is also present in these diseases, which cause more difficulties and abnormalities in the gait cycle. Skeletal or muscular problems can result in avoidance of pain and compensating movements, where the patient is correcting for the problem by having abnormal gait. Limping is a broad example of this.

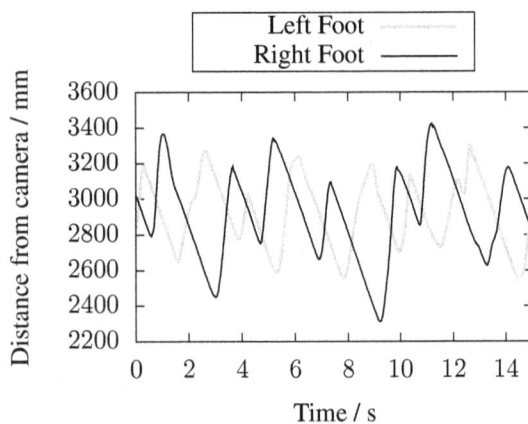

Figure 3: Gait data from a patient with Huntington's disease.

Fig. 3 shows gait data from a patient with Huntington's disease. The abnormalities of the gait make it very difficult to identify exactly where the initial contact and toe-off events of the reference foot and opposite foot are located. In comparison to Fig. 2 it is clear the patient does not have normal gait timing or even normal order of the gait events, occasionally with one foot taking two steps while the opposite foot remains in the stance phase (double steps). Therefore, gaining accurate gait timing from data such as this, is a demanding task.

As technology and science advances, the diagnoses of these locomotor problems and neurological illnesses are improv-

ing. Gait is being used as a measure to identify, diagnose and classify them, by measuring patients' gait cycle and then analysing it. Measures such as the stride length (m), cadence (steps/min), velocity (m/s), stride time (s), single and double limb support times (s) and the walk ratio are all important in the analysis of patients' gait. They are used to assess his or her levels of neuromotor control and to acquire other informative gait parameters, which can lead to the distinction between different diseases and the stage of the disease [4]. Gaining accurate data on the gait timing of all patients is therefore important, no matter how abnormal their gait is. To achieve this, an evaluation script was developed to analyse raw gait data, identify where the abnormalities are, exclude them and give accurate time points for the initial contact and toe-off points in the patient's gait cycle. For the evaluation script to be fully functional, it must be reliable for all gait data and must identify all abnormalities, not only some.

2   Material and Methods

2.1   Raw Data

Gait data used in this project was acquired from the motion and gait analysis laboratory of the University Medical Centre Schleswig-Holstein, Campus Lübeck. A study had been carried out in this lab on 6 patients with Huntington's disease, who were due to receive deep brain stimulation surgery. Their gait was tested in three different scenarios; before surgery, after surgery with the stimulator turned on and after surgery with it turned off. This gave a total of 24 different data sets to use; to create and test the evaluation script.

Gait data from healthy subjects was also available to use, for comparison.

The testing set-up in the lab included a treadmill, active LED markers, a 3D motion analysis camera and a computer. The camera used was the Motion Analysis System AS 200 from LUKOtronic. The markers were attached to the patient's body at 16 different positions; at both heels, both knees, both hips, both upper ankle joints, the spine on the left and the right side, both elbows, both shoulders, at the sacrum and at the lower neck (C7 vertebrae).

The patients walked on the treadmill and the 3D camera was set up directly behind them, about 2.5 meters away. There was a handrail for the patients to hold on their left side, this was required as some of the patients had difficulties standing alone. The treadmill was set to a speed of 2.5 km/h and, based on their capabilities, the patients walked for between 35 seconds and 2 minutes. Due to the varying severities of the patient's illnesses, they could not all walk for a full 2 minutes.

Data for the X, Y and Z-directions were acquired from each of the markers on each patient. The data files had varying levels of abnormalities which gave a good scope to work with. The number of abnormalities in the 24 data sets was first evaluated manually to obtain baseline data.

2.2 Data Analysis

The raw data was analysed in a program called DIAdem by National Instruments. It was clear that the data file from the heel marker in the X-direction was the best to use for this analysis. It was the closest available data set to the toe, which gives more accurate data on the toe-off and initial contact time points [2] . This data file gave a graph with time on the x-axis, and distance from the heel to the camera on the y-axis.

An evaluation script was developed to take the raw data and output the usable timing points of the gait.

2.2.1 Evaluation Scripts

Three versions of the evaluation script were created before it was finalised. Each version had a different approach to identifying the abnormalities and each had varying levels of success.

The first version followed the method found in [2]. Firstly, frequency based filtering was carried out on the raw data, then the second derivative of data was calculated. This resulted in the minimum and maximum peaks corresponding to the initial contact and toe-off points respectively.

The second version used a pattern finding command and the pattern of the abnormalities, to identify the abnormalities within the data and exclude them from the usable steps.

The third and final version uses digital filtering and double step detection to identify the real steps within the raw data.

2.2.2 Workflow

The following outlines the workflow of the final evaluation script and describes how the raw data is being analysed:

1. The raw data was filtered digitally at first. Fast Fourier Transforms (FFT) were used to calculate the step frequency of both the feet. 2.8 times the mean step frequency between the two feet was then used as the limit frequency to filter the data with an infinite impulse response, Bessel, low pass filter.

2. The maximum and minimum peaks of this filtered data were detected.

3. The filtered data was checked for any double steps. This was implemented by checking within each step of one foot, for the presence of two or more initial contact events of the opposite foot.

4. The steps in the opposite foot which would be affected by the double steps in the reference foot were identified.

5. The second step of the double step was detected by identifying the step before it, meaning both steps involved in the double step will be removed.

6. Lighter filtering was carried out on the raw data. One sixth of the sampling frequency used during data acquisition, was taken as the limit frequency. The sampling frequency in this case was 50 Hz, therefore 50/6

Hz was taken. The same digital filter was used as before. The lighter filtering was used to accurately identify the time points, as the heavy filtering altered the initial contact and toe-off points so the stance and swing phase were almost equal.

7. The maximum and minimum peaks were detected on the lighter filtered data.

8. The two sets of filtered data were matched up. The usable steps within the initially filtered data were identified within the lightly filtered data. The time points from these steps in the lightly filtered data were used and outputted as the results, showing on the raw data which steps are usable and their timing parameters.

3 Results and Discussion

Table 1 shows the total number of abnormalities in each data set, and the number of abnormalities that the final version of the evaluation script detected and therefore excluded from the results.

The final evaluation script detected all the abnormalities that were present in each data set, and only outputted actual usable steps. Therefore this version of the script is functioning completely and can be relied on.

Table 1: Effectiveness of the Final Evaluation Script

Data Set	No. of Abnormalities	No. detected
1	34	34
2	0	0
3	0	0
4	1	1
5	4	4
6	0	0
7	3	3
8	0	0
9	2	2
10	1	1
11	1	1
12	1	1
13	19	19
14	24	24
15	19	19
16	4	4
17	2	2
18	5	5
19	19	19
20	6	6
21	2	2
22	2	2
23	0	0
24	0	0

Table 2 shows the results from all 3 versions of the evaluation script with a selection of some of the worst data sets (i.e. the data sets with the highest number of abnormalities). This comparison shows that the code evolved as it progressed, and how it's effectiveness changed.

Table 2: Comparison of Evaluation Script

| | Abnormalities Detected | | |
Data Set	Version 1	Version 2	Version 3
13	11	5	19
15	17	6	19
18	5	2	5
19	16	6	19
20	6	6	6

Version 2 worked well with the data sets which weren't so severely affected. Most small abnormalities, such as small stumbles and trips, were detected and could then be excluded. However, with the data sets with more abnormalities, version 2 struggled to detect the multiple double steps that occurred. This can be seen in Table 2, where it only functioned fully with data set 20. The other problem which made version 2 unreliable, was the number of completely usable steps being detected as abnormal steps, leaving some data sets with barely any usable data.

Version 1, as can be seen from Table 2, worked much better than version 2. It was very successful in detecting the double steps in the gait data, however it still missed the occasional abnormality. It is vital that all the double steps are detected, otherwise the detection of the affected steps on the opposite foot would not function. This is because the data sets from the right and left heel are not directly connected. Therefore, even with one missing double step, the results are rendered unreliable.

Version 3 was fully functional and therefore reliable. The pattern finding element from version 2 was removed, and no usable steps were being detected as abnormal steps. Only the actual abnormal steps were being detected. The section of the code, which detects the double steps, functions completely with the wide range of data sets available. The filtering using FFT's , which was also used in version 1, is vital to ensuring the functionality of the script. This is clear from Table 2, as both version 1 and version 3 used this filtering method and are the most successful versions.

The data that was available had an impact on how the analysis developed. As the test set-up that was used had the camera directly behind the patient, an LED marker on the toe was not detectable. The best alternative to the toe marker was the heel marker, as it followed the same pattern of movement. However, the time-point of it's initial contact is slightly later, and it's toe-off time-point is slightly earlier. This might be the reason the method used for version 1 of the evaluation script, which was based on [2], was not as effective, since data from a toe marker was used for this analysis.

4   Conclusion

Both acquiring and analysing gait data from patients with neurological diseases is a difficult task. The data resulting from their walk can be confusing and chaotic, making the analysis and extraction of useful information challenging.

The three methods used to attempt to analyse the data had varying levels of success. It is clear a 100 % success rate is required for the method to be reliable, and to allow the user the confidence that the results are correct. The final method used was able to identify the usable steps within the gait data, and exclude the trips, drags and double steps, on even the most severe data set.

Useful information can then be extracted from this data, such as the gait timing, initial contact, toe-off, swing and stance phases, which can be used to learn more about locomotor problems and diseases.

This automatic classification of the abnormalities within gait data, and the automatic acquisition of the timing parameters, is now a trustworthy method to replace manual evaluation of the data. And it provides a quicker and more accurate approach to obtaining information from patient's gait.

Acknowledgement

The work has been carried out in the Biomechanics Laboratory, Clinic for Orthopedic and Trauma Surgery, University Medical Center Schleswig-Holstein, Campus Lübeck.

5   References

[1] M. Whittle, *An introduction to gait analysis.* Butterworth-Heinemann, 2007.

[2] Clinical Gait Analysis, *Toe-Off.* Available: http://www.clinicalgaitanalysis.com/faq/toe-off.html [last accessed on 2017-12-20].

[3] V. Rota, L. Perucca, A. Simone and L. Tesio, *Walk ratio (step length/cadence) as a summary index of neuromotor control of gait: application to multiple sclerosis.* In: International Journal of Rehabilitation Research, vol. 34, pp.265–269, 2011.

[4] A. Delval, P. Krystkowiak, J. L. Blatt, E. Labyt, K. Dujardin, et al., *Role of hypokinesia and bradykinesia in gait disturbances in Huntington's Disease. A biomechanical study.* In: Journal of Neurology, vol. 253, pp. 73–80, 2006.

[5] N. Sekiya, H. Nagasaki, *Reproducibility of the walking patterns of normal young adults: test-retest reliability of the walk ratio(step-length/step-rate).* In: Gait and Posture, vol. 7, pp. 225–227, 1998.

[6] J. M. Hausdorff, M. E. Cudkowicz, R. Firtion, J. Y. Wei and A. L. Goldberger, *Gait Variability and Basal Ganglia Disorders: Stride-to-Stride Variations of Gait Cycle Timing in Parkinson's Disease and Huntington's Disease.* In: Movement Disorders, vol. 13, no. 3, pp. 428–437, 2011.

Investigation of corona treatment to enhance adhesion of silicone on PUR-films

M. Saathoff [1], C. Koester [2], C. Wendt [3]

[1] Biomedical Engineering, University of Applied Sciences Lübeck, malte.tamme.saathoff@fh-luebeck.de
[2] BSN medical an Essity Company, Hamburg, carmen.koester@bsnmedical.com
[3] Institute of Applied Sciences, Fachhochschule Lübeck, christian.wendt@fh-luebeck.de

Abstract

In today's medicine silicone dressings represent a significant part in the treatment of difficult, slow-healing wounds. Their skin-friendly appearance allows atraumatic dressing change. During production of thermoplastics with silicone coating, adhesion promoters are used. Some of these chemical primer cause allergies and are harmful to health, which is why their application in wound care is gladly dispensed with. In this project it was investigated whether a non-chemical corona plasma pre-treatment of a polyurethane (PUR) film improves its adhesion properties in a way that it can replace a classic silicone primer. For this, three different samples were prepared and a test method was developed to examine the anchoring. It was found that corona treatment improves the adhesion properties of PUR-films for medical silicone, but does not achieve the results of silicone primer. During the implementation, some improvement potentials were discovered, which did not fully expose the process as a potential alternative.

1   Introduction

In modern medicine silicone wound contact layers (SWCL) take a major role in the treatment of severely exuding and slow- healing wounds [1]. They are characterized by their skin-friendly adhesion, gentle fixation and an atraumatic dressing change in the acute and further developed wound care treatment [2]. The silicone layer of advanced wound dressings must be deep-seated in the plaster composite. This is necessary to avoid detachment of material which may possibly remain in the wound, for example during dressing changes. Some wound dressings should be permeable to liquid and air, while the gentle surface and skin-friendly adhesion must be maintained even with strong cross-linking and anchoring of the silicone layer [3],[4].

Depending on the medical application of the dressings, different support materials are selected. In this examination a polyurethane film was coated with silicone. For a clean non-detachable bond between the silicone and the support material, the surface energy of the substrate should exceed the surface energy of the silicone by a about $2 - 10 \frac{mN}{m}$ [5]. The literature value of the surface energy of PUR is $32 \frac{mN}{m}$ [5], silicones about $24 \frac{mN}{m}$ [13]. The difference in surface energy is already sufficient for an adhesive coating. However, a wider gap between the surface energies and the generation of reactive groups on the substrate surface can increase the anchoring. To raise the surface energy of the substrate and therefore improve the connection of PUR and silicone, so-called adhesion promoters are used [8]. Some of these adhesion agents work chemically, others physically or mechanically [6]. All procedures change the surface texture and raise the surface energy of the material to be coated [6]. Some of these chemical mediators trigger allergies, which should be avoided in medical applications [7]. Corona treatment is a physical pre-treatment procedure for substrates to be coated. It increases the surface energy and adhesion properties of polymers like PUR [5]. In addition to using ambient air as process gas in atmospheric pressure, oxygen containing functional groups (e.g. -C-O, -C=O) as well as nitrogen containing groups (NH_2, N-C=O etc.) can be added on polymeric surfaces [5],[8],[9]. These active groups allow the silicone to anchor better to the surface [9]. Is the corona pre-treatment of a polyurethane film a method to raise the connection properties and ensure the silicone adhesion to the plaster composite? Are the results comparable or even better than current methods in the production process involving the usage of primer?

2   Material and Methods

To produce atraumatic soft silicone wound care dressings, a SWCL must be integrated into the product. In some cases, the SWCL consists of a high elastic film coated with silicone, which is later on laminated with further components of the plaster. Figure 1 shows the schematic composition of the plaster.

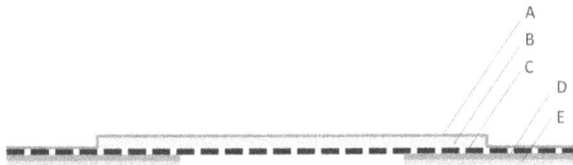

Figure 1: Schematic of plaster composition (A)- PUR Cover, (B)- Foam pad, (C)- Anti microbial fabric, (D)- PUR-film, (E)- silicone layer

2.1 Material

To investigate the influence of an adhesion promoter on anchoring the silicone to the PUR-film, three samples are examined. The first film is wetted with a common silicone primer. This is used in classical production and serves as a reference value for an optimal result. The second film is corona pre-treated. It is the film of interest to evaluate the differences between the primer pre-treated film. A third film is not pre-treated but also investigated as a reference for a worst-case product, in which the silicone is anchored without any promoter.

Polyurethane Film
The silicone coated layer in the plaster composite is a high elastic polyurethane film. PUR can be varied within a wide range. Depending on the degree of cross-linking and added components, thermoset, thermoplastics or elastomer can be fabricated [10]. In this process, a thermoplastic PUR was used. This is characterized by its freedom from plasticizer, tightness against liquid media at the same time high water vapour permeability and high mechanical strength. It is also skin-friendly, heat-resistant and cold-flexible [11]. All coated films had a thickness of $60 \pm 6\,\mu m$.

Silicone
The used silicone is a two component RTV (Room-Temperature-Vulcanizing) biocompatible silicone. Components A/B are mixed homogeneously in a ratio $A : B = 51 : 49$ to make the silicone layer a little soft and sticky. RTV means after mixing the components, the platinum catalyst in component A starts cross-linking at room temperature. The pot life at $23\,°C$ is 73 minutes. At temperatures around $85\,°C$ the silicone vulcanizes rapidly [14].

Primer
The used primer is a specially formulated silicone primer for use with platinum-cured systems where conventional silicone primers are insufficient. It is dripped onto the film and glazed by deflection rollers before coating.
The amount of primer is about $4\,\frac{ml}{m^2}$.

2.2 Methods

Corona preparation
The film is processed externally for the corona pre-treatment. The film is unwound and passed over deflection rollers through a discharge zone (gap between roll and electrode is 1.5 mm) with a speed of $10\,\frac{m}{min}$ at room temperature. After the discharge zone the film is rewound. Five samples of the film are treated with different energies ($100, 150, 200, 250, 500\,W \cdot \frac{min}{m^2}$) and the surface energy at each sample is examined with test ink (DIN ISO 8296). This indicates the surface energy depending on running or staining on the substrate. It is known that after some time, the surface energy decreases [13], so three different storage periods (one, three and five days) were set. After storage time, the surface energy is retested and the results compared with the directly measured samples and an untreated sample. The untreated PUR-film has a tested surface energy of $36\,\frac{mN}{m}$, which is almost the literature value of $32\,\frac{mN}{m}$ [5]. A corona treatment processed with $250\,W \cdot \frac{min}{m^2}$ lead to the best surface energy directly after processing ($72\,\frac{mN}{m}$) and after all storage times ($46\frac{mN}{m}$). As a result a roll of 100 m PUR-film is processed with $250\,W \cdot \frac{min}{m^2}$ to be then coated in-house with silicone with the method described below. The time between pre-processing and coating is about four hours.

Coating of material
In the production process, the two silicone components are mixed homogeneously with a static mixer in a ratio $A : B = 51 : 49$ and applied to the PUR-film by means of a coating bar in the desired thickness. To check if the silicone layer has the desired mass the weight of a sample is measured before and after coating. The gap of the coating bar is adjusted until the area density of $200\,\frac{g}{m^2}$ is reached. In a subsequent heating section of about $85\,°C$ the silicone cross-links to some extent before it is covered by a release liner for better storage and post-processing. All films were coated with the same area density of silicone.

Test method
To test the anchoring of silicone to the PUR-film and to have an in-process quality control, a procedure for the assessment was developed. For this purpose, the release liner is removed and the silicone layer of the substrate to be examined is exposed. The sample is then attached to a lab bench with the silicone site up, fixed with two strips of adhesive silicone tape (roll width 50 mm) spaced 50 mm apart, generously overlapping the width of the sample on both sides. A schematic view can be found in figure 2.1. As seen in figure 2.2, a third strip of tape is now weighed and then placed in the longitudinal direction on the silicone surface so that it glues on the first and second strip of tape. This creates an area of $0.0025\,m^2$.

With a pinch roller (3 ± 0.05 kg) is now driven back and forth five times over the pattern without exerting additional pressure by hand. After rolling, was waited one minute to give tape and silicone time to connect. As shown in figure 2.3, the third strip of adhesive tape was carefully pulled of at 180 degrees. To assess the silicone release and thus the adhesion to the substrate, the third strip of the silicone tape was weighed again after removal. The difference to the initial value was noted.

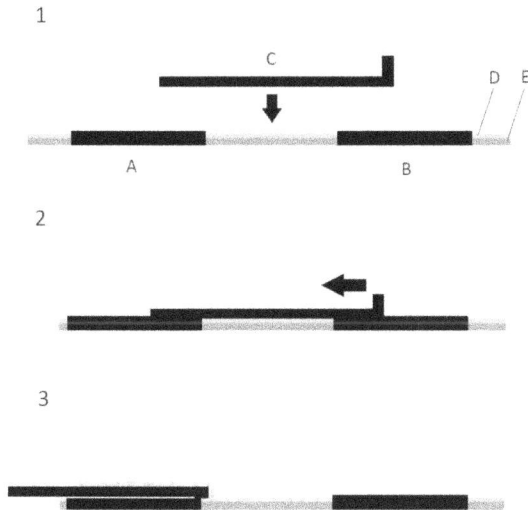

Figure 2: Schematic of test method, 1- first stripe of tape (A) and second stripe of tape (B) are applied to silicone layer (D) coated on PUR-film (E), 2- third stripe of tape (C) is attached, wait one minute, 3- pull off at 180 degrees, removed silicone is weighed

3 Results and Discussion

3.1 Results

The results of the investigation are shown in the tables 1,2 and 3. Each sample was tested for five times to compensate for measurements errors and material variations. Subsequently, the mean value of the measured weight is generated. The difference in weight leads to a degrading silicone mass. This is extrapolated to one square meter by means of the defined test area ($0.0025\,\mathrm{m}^2$) and then percentage mass removal is calculated. Figure 3 shows a photo of the test strips. The release of silicone is already recognizable on two of the strips.

3.2 Discussion

The results show that the corona-pretreated PUR-film has clearly better adhesion properties to silicone than an untreated PUR-film. The mass of removed silicone dropped from $37.92\,\frac{\mathrm{g}}{\mathrm{m}^2}$ to $12.64\,\frac{\mathrm{g}}{\mathrm{m}^2}$. Even if the surface energy

Table 1: Untreated PUR film

Weight before execution [g]	0.6432
Weight after execution [g]	0.7380
Difference [g]	0.0948
$\frac{\mathrm{g}}{\mathrm{m}^2}$	37.92
Removed silicone	19 %

Table 2: Corona pre-treated PUR-film

Weight before execution [g]	0.6238
Weight after execution [g]	0.6554
Difference [g]	0.0316
$\frac{\mathrm{g}}{\mathrm{m}^2}$	12.64
Removed silicone	6 %

Table 3: Primer pre-treated PUR-film

Weight before execution [g]	0.6392
Weight after execution [g]	0.6394
Difference [g]	0.0002
$\frac{\mathrm{g}}{\mathrm{m}^2}$	0.08
Removed silicone	0 %

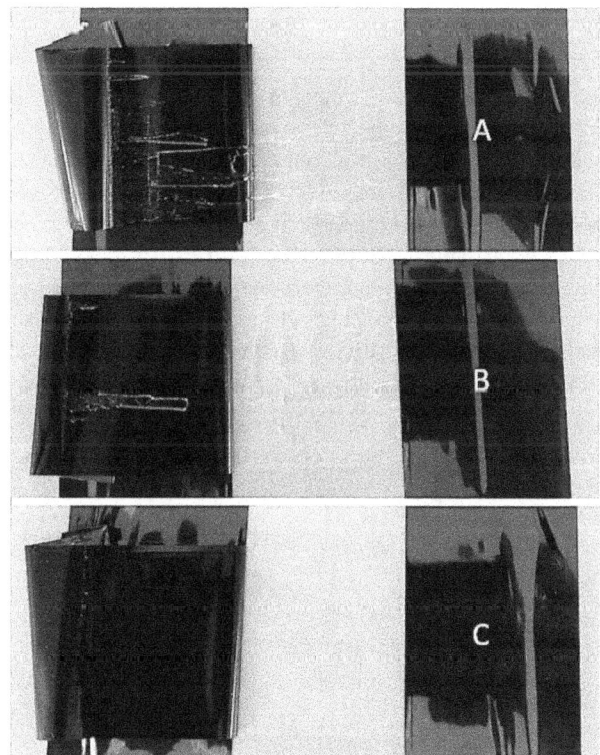

Figure 3: Pictures of (A)- Untreated film (B)- Corona pre-treated film (C)- Primer pre-treated film after execution

of PUR allows a silicone coating, the anchoring on an untreated film is not sufficient: the test method provokes a detachment of silicone. Even though this is not comparable to a normal application on a patients skin, high demands are placed on the product, which makes an untreated PUR film insufficient for medical use.

In addition, the results show that a primer-pretreated film provides adhesion properties that cannot be achieved with the performed corona treatment. In this series of experiments, only the surface energy is measured after the corona treatment; what influence volatile active groups and radicals have is not determined. Furthermore, it is not measured or determined which volatile active groups are formed and to what extent and duration they are present. Because of the external pre-processing, a time loss of four hours before coating occurs, which already has an influence on the presence of active groups and radicals. An inline corona treatment would shorten the time between stimulation and coating, which should further improve the adhesion properties of the silicone. Discarding the procedure as an alternative coating method can not be judged at this point, as there is still considerable room for improvement. Initial results from a clinical study show that the plasters of primer and corona treated PUR-films do not vary greatly, whether the manufacturing process is therefore changed depends significantly on the costs and the order situation and cannot be further assessed here.

4   Conclusion

The applied corona treatment of the PUR-films for medical use does not provide the desired adhesion-promoting properties as the classic silicone primer does. However, the influence of volatile active groups resulting from corona treatment has not been considered and investigated. They are only available for a limited period of time, which requires immediate post-treatment coating for optimal results. Whether the process must be rejected as an alternative pretreatment depends on further tests, which also take into account the presence of active groups on the substrate surface and the time between treatment and coating.

Acknowledgement

The work has been carried out at BSN medical an Essity Company, Hamburg and supervised by the Institute of Applied Sciences, Fachhochschule Lübeck.

5   References

[1] R. White. *Evidence for atraumatic soft silicone wound dressing use*. Wounds UK, Aberdeen, 2005.

[2] BSN medical. *Cutimed Siltec Sorbact*. Available: http://www.bsnmedical.de/produkte/wund-und-gefaessversorgung/produktew/p/cutimed-siltec-sorbact.html [last accessed 2018-01-08].

[3] A. Hansen. *Fast cure-high moisture vapour transmission rate adhesives improve wound care*. Adhesion Age 46 (22–25), 2003.

[4] A. Derbyshire. *Using a silicone-based dressing as a primary wound contact layer*. British Journal of Nursing vol. 23, No 20, 2014.

[5] Tantec A/S, Knowledgebase, Denmark. *What is surface treatment?*. Available: https://www.tantec.com/what-is-surface-treatment.html [last accessed 2018-01-08].

[6] L. A. Bloomfield. *Primer system for bonding conventional adhesives and coatings so silicone rubber*. Int. J. Adhesion & Adhesives 68 (239-247), 2016.

[7] BSN medical. *Safety data sheet primer*. Manufacturers and suppliers may not be named for reasons of data protection.

[8] V. Seitz, K. Arzt, S. Mahnel, C. Rapp, S. Schwaminger, et. Al. *Improvement of adhesion strength of self-adhesive silicone rubber on thermoplasic substrates*. Int. J. Adhesion & Adhesives 66 (65-72), 2016.

[9] K. G. Kostov, A.L.R. dos Santos, R.Y. Honda, P.A.P Nascente, M.E. Kayama et. Al. *Treatment of PET and PU polymers by atmospheric plasma generated in dielectric barrier discharge in air*. Journal of Surface and Coatings Technology vol. 204 (3064-3068), 2010.

[10] Z. Zhang, O. Ortiz, R. Goyal, J.Kohn. *Biodegradable Polymers*. Handbook of Polymer applications in medicine and medical devices (303-325), 2013.

[11] BSN medical, *Product specification PUR-film*. Manufacturers and suppliers may not be named for reasons of data protection.

[12] D.M. Brewis, R.H. Dahm. *A rewiew of electrochemical pretreatments of polymers*. Int. J. Adhesion & Adhesives 21 (397-409), 2001.

[13] G. Habenicht. *Kleben: Grundlagen, Technologien, Anwendungen*. 4. erweiterte Auflage, Springer-Verlag Berlin Heidelberg, 2002.

[14] BSN medical, *Product specification silicone*. Manufacturers and suppliers may not be named for reasons of data protection.

Study of Antimicrobial Effects of the Fungal Toxin Candidalysin by Atomic Force Microscopy

C. Borchert [1], C. Nehls [2], T. Gutsmann [2]

[1] Medizinische Ingenieurwissenschaft, Universität zu Lübeck, cborchert@student.uni-luebeck.de
[2] Division of Biophysics, Research Center Borstel, cnehls@fz-borstel.de, tgutsmann@fz-borstel.de

Abstract

Candidosis is the mycosis of *candida albicans*, a human pathogenic fungi that produces various peptides, one of them being Candidalysin, which seems to play a major role during the infection. With atomic force microscopy, Candidalysin induced effects on membranes were analyzed to understand its mechanics. Whole cells and bacteria as well as reconstituted membranes were measured and provided information regarding binding and membrane damaging activities. The experiments with bacteria showed that *Escherichia Coli* tends to resist the permeation of Candidalysin. In contrast the experiments with the reconstituted membranes showed their disintegration. Lipopolysaccharide may be the cause of this observation. To uncover possible resistance mechanisms, further experiments have to be executed on models which resemble the structure of natural membranes more and more.

1 Introduction

candida albicans is usually a benign member of the human microbiota, but can be the reason for a mucosal infection. During invasion of cells the peptide Candidalysin is secreted, which is responsible for the human pathogenic effects [1]-[3]. An existing interaction model for human membranes describes that Candidalysin binds to a membrane and if it overcomes a certain concentration, intercalates into the membrane. After intercalation a structural change and building of lesions is described, which leads to the disintegration of the membrane [4].

The described interaction model may be adapted for bacterial membranes. Measurements of bacteria, human cells and reconstituted membranes exposed to Candidalysin are supposed to provide new knowledge of the peptide interaction with membranes. This knowledge may reveal a way to prevent or treat candidosis.

Measurements of bacteria and red blood cells (RBCs) give an overview of Candidalysin induced effects. Possible antimicrobial effects on bacteria and cell damaging effects on RBCs are analyzed by surface measurement with atomic force microscopy (AFM). In addition reconstituted membranes, consisting of different lipid compositions, deliver information on binding and permeating processes.

2 Material and Methods

2.1 AFM

The part in contact or interaction with the sample is the cantilever, a tiny bar, that functions as a spring. A tip is attached to the end of the cantilever. Forces, that appear between tip and sample, like van der Waals forces, lead to a deflection of the cantilever [5], [6].

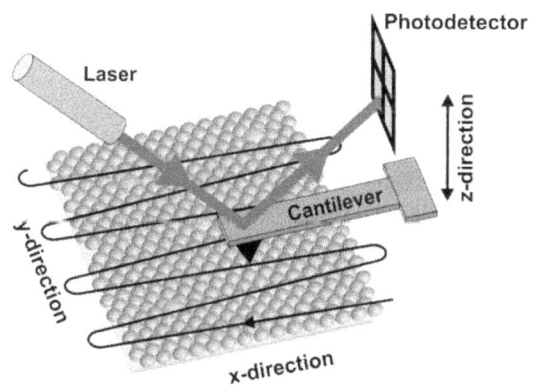

Figure 1: Functional design of an Atomic Force Microscope. Due to forces occurring between tip and sample, the cantilever is deflected during the scan. This deflection is detected by the reflection of a laser beam on a photodiode. According to a feed back loop the space between tip and sample is kept constant.

A laser beam is focused on top of the cantilever and the deflection of the cantilever is detected by the dislocation of the beam from the middle of a four-segment photodiode (Fig.1). In contact mode the deflection is kept zero by feed back control of the z-direction piezoelement. This gives a topographic image of the sample surface. In AC-mode the cantilever is stimulated to oscillate near its resonance frequency, which is determined by thermal fluctuation [6]. In this work the MFP-3D by Asylum Research was used with various cantilevers.

2.2 Bacteria & Red Blood Cells

The idea was to prove Candidalysin induced effects by measuring increased surface roughness on bacteria and RBCs. Therefore four samples of *Escherichia Coli* were prepared. One with a Candidalysin concentration of 128 μg/ml, one with 256 μg/ml and two controls without Candidalysin. The samples were fixated with a 4 % Paraformaldehyde solution. With the AFM single bacteria were localized, vertically aligned and then measured in contact mode in air with a CSG10 Cantilever by NT-MDT. 1 μm x 1 μm images of the bacterial surface were taken with high resolution to determine the surface roughness. The natural curvature of the bacteria increases the measured roughness. To receive useful and comparable data of surface roughness the natural curvature has to be subtracted (Fig.2A/2C). The result of this subtraction is a flattened picture, which shows only the surface roughness (Fig.2B/2D).

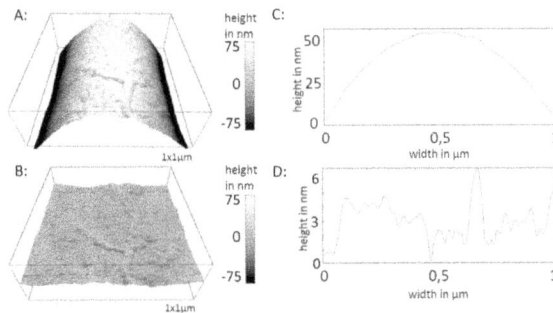

Figure 2: Determination of the surface roughness by the subtraction of the natural curvature. A: Original hight image containing an overlay of the surface roughness and the natural curvature. B: Surface roughness. C: Cross section through the Original image with a maximal value of 50 nm. D: Cross section through the surface with a maximal value of 6 nm.

The preparation of the RBCs included concentrations of 1 μg/ml, 5 μg/ml and 10 μg/ml Candidalysin plus two controls without Candidalysin. The measurement plan was equal to the bacteria. Cantilevers with a spring constant higher than 0,1 N/m did not work on the prepared RBCs. The OMCL-RC800PSA-1 and the OMCL-TR400PSA-1 by Olympus were suitable in terms of the spring constant.

2.3 Solid Supported Membranes

On a reduced model like reconstituted membranes Candidalysin induced effects are easier to determine. To measure lipid membrane models with AFM, vesicles from these lipids were prepared first. For a reconstituted bacterial membrane the lipids Phosphatidylethanolamine (PE) and Phosphatidylglycerol (PG) were composed in a molar ratio of 4:1 (PE:PG). For the human model it was pure 1,2-dioleoyl-sn-glycero-3-phosphocholine (DOPC). These vesicles are spread onto the substrate. The used substrate was mica, a sheet silicate, that builds atomically flat sheets. The formed sensitive bilayer is surrounded by liquid buffer

and therefore measured in AC-mode.

Many membranes, consisting of several molecular components, separate into coexisting phases. Usually regions in rigid phase are higher than regions in liquid or gel phase. These regions existing in different phases are called domains [7] and can be measured with AFM.

With usual measurement, domains are the only indication of a membrane. PE:PG membranes show domain formation. But in many cases the domains are not locatable and stable. Additionally pure DOPC does not form domains. So to verify the integrity of the membrane, force plots were recorded. Under continuous measurement the peptide was added to identify changes of the membrane structure over time. Concentrations of 30 μg/ml, 100 μg/ml and 300 μg/ml were used.

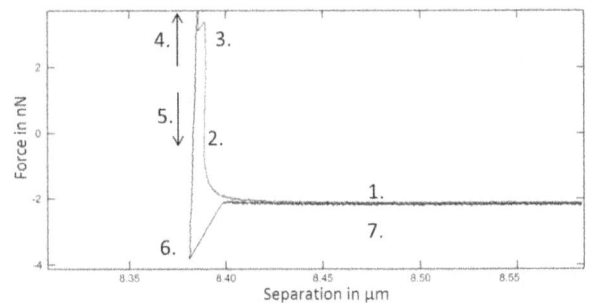

Figure 3: Force plot with a membrane breakthrough (3.) and the typical adhesive force peak (6.). 1. engaging the membrane, 2. deflection on the membrane surface, 3. membrane breakthrough, 4. deflection on the substrate, 5. deflection cutback, 6. adhesion-force-peak, 7. withdrawing.

For the force plots the cantilever engages the sample surface. When the tip is in direct contact with the surface it drives on until a adjustable force is reached, then it withdraws. The plot shows the increase of force on the first contact with the membrane surface. When the membrane is not able to withstand the pressure any further a breakthrough occurs, which identifies the membrane as such (Fig.3). If there is no breakthrough it is unlikely that there is a membrane.

3 Results and Discussion

3.1 Bacteria

Due to a systematic error, control 1 stands out of the collected data and therefore has to be excluded from the surface roughness evaluation. On this sample, further named aggregated sample, the bacteria were compact like a carpet and physically supported each other. Thereby the natural curvature in the measured area was decreased and the surface roughness clearly smaller in comparison with the other samples.

The measurement of the bacterial surface roughness shows no significant difference between the Candidalysin exposed samples and the remaining control. Though the control seems to be a bit smoother (Fig.4). Overcorrection of the natural curvature in the marginal areas of the 1 µm x 1 µm images would make their evaluation inaccurate. 0,5µm x 0, 5 µm cutouts from the center of the 1 µm x 1 µm images were evaluated. Additionally recorded 0, 5 µm x 0, 5 µm images from the same spot delivered values, that were a bit smaller (Fig.4). The bacterial membranes are intact and no clear effect of peptide can be seen in the roughness of the bacteria. This is not the cause of technical limits but other structures, like residual fimbriae or flagella, lying on the bacterial surface have bigger effect on its roughness than binding peptides (Fig.2A/B). The Candidalysin exposed bacteria showed aggregation in contrast with the control. The aggregated sample, former control, showed aggregation too, but in another way. The low surface roughness proves that (Fig.4, aggregated). The aggregates were one bacterium high, so they lie side by side. The aggregates of the Candidalysin exposed samples were up to 600 nm high and disordered. One bacteria is about 150 nm high. The bacteria lay one above the other. This is an indication, that the Candidalysin binds to the bacteria and makes them sticky. Another argument for this is, that during the measurement the cantilever more often stuck on the Candidalysin exposed samples than on the control and the aggregated sample.

A Minimum Inhibitory Concentration (MIC) test with the used bacteria and the maximal used peptide concentration showed no effect. Thus the MIC is above 256 µg/ml.

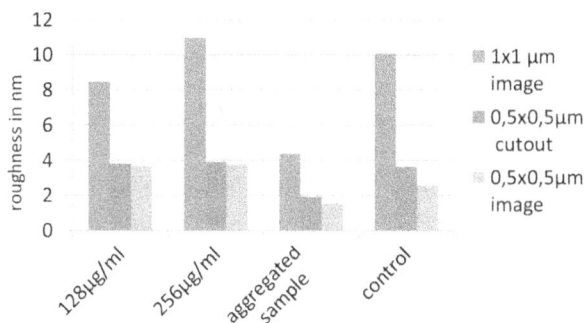

Figure 4: The diagram shows the roughness of 4 samples. The images, delivering data for the 1 µm x 1 µm bars, were affected by artifacts in the margin areas. The 0,5µm x 0, 5 µm cutouts take their data from the center of the 1 µm x 1 µm images. The aggregated sample stands out with its smoothness. But between the control and the Candidalysin exposed samples is no significant difference.

3.2 Red Blood Cells

RBCs are difficult to measure, due to preparation the samples were covered with salt crystals of the buffer. These crystals had to be washed off carefully to not damage or wash away the RBCs. If the samples do not dry long enough they get shifted by the cantilever. The choice of the right cantilever is essential. Because the edges of the RBCs are high and sharp, a soft cantilever with a relative low spring constant is needed. Also the tip has to be sharp and high

enough. The so far used cantilevers that were soft enough had pyramidal tips, that caused artifacts (Fig.5).

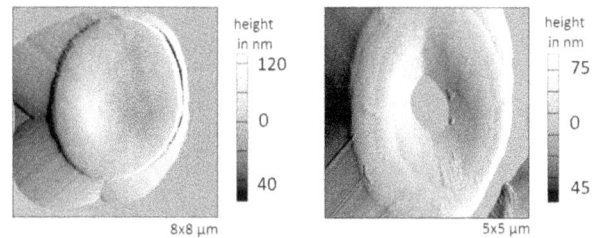

Figure 5: The 8 µm x 8 µm image shows the upside of a RBC (left). The edges show diagonal ramps according to the aligned cantilever tip. The 5 µm x 5 µm image shows a RBC with a deeper dent (right). Because the cantilever does not reach the deep parts the center appears plane.

The pyramidal tip has a sharpened peak, which is not high enough to be the first point in contact at sharp and high edges, like these of the RBCs (1 µm and higher). This causes problems in terms of surface roughness measurement, because the dents on the RBCs are not equally deep. The dent of the RBC (Figure 5 left) is flat enough to be completely imaged. But most other RBCs have deeper dents (Fig. 5 right), what leads to a contact with the diagonal edges of the pyramidal tip prior to the tips peak. Therefore deeper parts can not be measured and will be imaged even.

3.3 Solid Supported Membranes

Reconstituted membranes made of PE:PG had a higher rigidity (breakthrough at about 5nN) than the reconstituted membranes made of DOPC (breakthrough at about 2nN). Still both were disintegrated by a Candidalysin concentration of 100 µg/ml (Fig. 6). Although the concentrations of 300 µg/ml and 30 µg/ml showed no visible effects, it is to presume, that concentrations of 100 µg/ml and above will disintegrate such membranes. The reason that these concentrations showed no effects can be various. First the peptide needs to diffuse to the measured area. If the time is too short the peptide does not arrive the right spot during measurement. Second, residual vesicles may bind most of the peptides. And third, the 30 µg/ml concentration might be too low, to have a visible effect.

4 Conclusion

With the used concentrations, Candidalysin was not able to damage *Escherichia Coli* effectively. Nevertheless the aggregation of the bacteria implies that the peptide binds to the bacterial surface and makes them stick together, what may be enough to neutralize them. Force plots may deliver a proof by measuring increased adhesive power.

For the membrane models Candidalysin disintegrated both, the human and the bacterial model. A decreased concentration should be measured to determine differences. Lipopolysaccharide (LPS) is the first structure the peptide

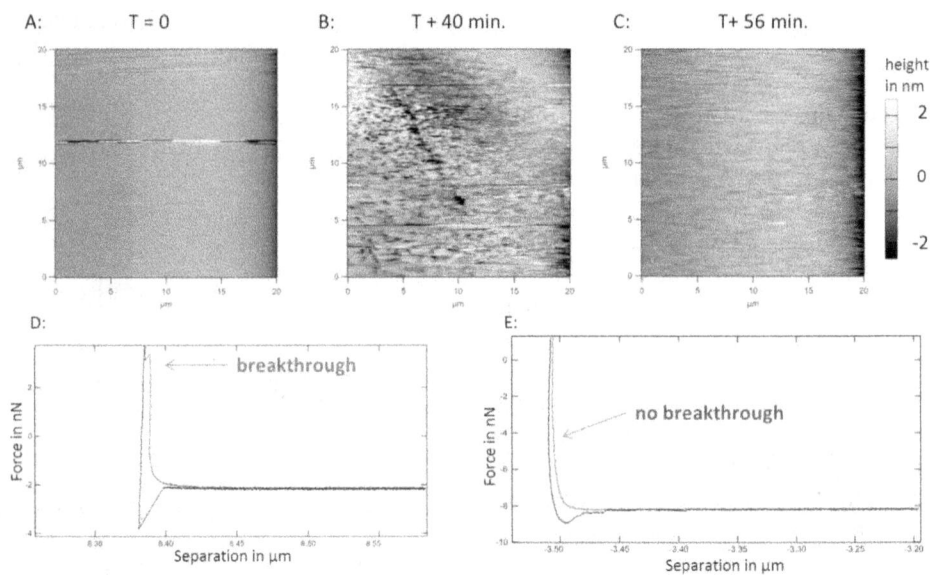

Figure 6: Exemplary measurement of PE:PG membrane with addition of 100 µg/ml Candidalysin. A: Zero point with addition of the peptide. B: After 40 minutes the peptide binds and intercalates. C: After 56 minutes no further reaction, the membrane has disintegrated. D: Force plot of the intact membrane, before zero point. E: Force plot of the disintegrated membrane, 60 minutes after zero point.

is encountered with, so it is a good idea to have a look at membranes containing LPS. Candidalysin might damage the inner bacterial membrane but is not able to overcome the LPS layer.

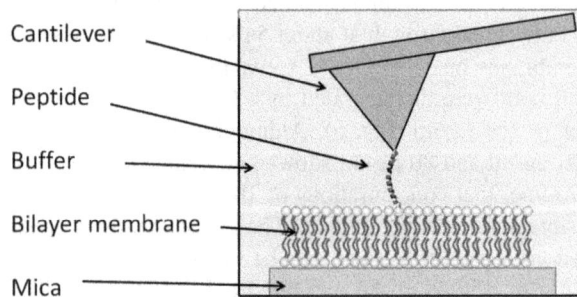

Figure 7: Scheme of a binding measurement with a helical peptide bound to the cantilever tip.

Additional binding measurements are planned to exactly determine binding affinities for different lipids and their binding forces. In these measurements the peptide is directly bound to the cantilever tip and then binds to a membrane (Fig.7). By pulling the peptide, binding forces between peptide and membrane will be measured. Effects like binding and permeabilization were already proved with techniques like FRET (Förster resonance energy transfer) and planar membranes [4]. By high resolution imaging the AFM delivers complementary data to support such measurements. The measurement of forces is a big strength of the AFM and no other technique is able to measure this so precisely. Therefore the measurements of forces, like bind-ing forces or adhesive forces should move into focus.

Acknowledgement

The work has been carried out at the Division of Biophysics, Research Center Borstel and supervised by Prof. Dr. Thomas Gutsmann and Dr. Christian Nehls. We are grateful for excellent technical support by Kerstin Stephan.

5   References

[1] D. L. Moyes, et al. *Candidalysin is a fungal peptide toxin critical for mucosal infection.* Nature, 2016.

[2] G. D. Brown, et al. *Hidden killers: human fungal infections.* Sci. Transl Med. 2012.

[3] D. Wilson, J. R. Naglik, and B. Hube. *The missing link between candida albicans hyphal morphogenesis and host cell damage.* PLoS pathogens, 2016.

[4] J. Wernecke, *Biophysical charakterisation of the fungal peptide toxin ECE1-III and its interaction with lipid membranes.* PhD thesis, Universität zu Lübeck, 2016.

[5] G. Binning, C. Quate und C. Gerber, *Atomic Force Microscope.* 1986.

[6] E. Meyer, *Atomic Force Microscopy.* University of Basel: Progress in Surface Science, vol. 41, pp. 3-49, 1992.

[7] R. Lipowsky and R. Dimova, *Domains in membranes and vesicles.* Journal of Physics: Condensed Matter, vol. 15, no. 1, 2002.

The Hemodynamic Response Function in Functional Magnetic Resonance Imaging: Variability Between Brain Regions and Differences Between Rats and Humans

B. Akinola [1], J. Baudewig [2], S. Boretius [2], and M. A. Koch [3]

[1] Biomedical Engineering, University of Applied Sciences Lübeck, benson.olumide.akinola@stud.fh-luebeck.de
[2] Funktionelle Bildgebung, Deutsches Primatenzentrum GmbH , {JBaudewig,SBoretius}@dpz.eu
[3] Institute of Medical Engineering, Universität zu Lübeck, koch@imt.uni-luebeck.de

Abstract

Measurement of the time to peak in functional magnetic resonance imaging (fMRI) hemodynamic response yields information about the neural dynamics of brain activity. In fMRI experiment neural activity occurs in the millisecond range but measured indirectly with the hemodynamic response. The general linear model is used to statistically estimate activated brain regions to this response. In this study, the hemodynamic response function and its temporal parameters for a short visual stimulus in humans and a long stimulus in rats and humans was estimated by fitting the convolution of a gamma function and the stimulus time course to the hemodynamic response function (HRF) of activated regions. Results show a method to estimate a temporal parameter (time to peak) of the HRF for different sampling procedures in a short stimulus event-related fMRI. Variation of the HRF across species and brain's functional areas (visual and motor cortex) using long stimulus was observed.

1 Introduction

Functional magnetic resonance imaging (fMRI) is a non-invasive method of brain imaging to map and localize brain regions (functional areas) e.g. vision, motor and cognitive functions. It is an indirect measure of the neural activities using a blood oxygenation level dependent (BOLD) contrast [1]. fMRI has a spatial resolution of ≈ 1 mm with a compromising temporal resolution [2], because it measures hemodynamic response (blood flow) to neuronal activity. How neurons inform the vasculature in order to trigger an increased blood flow is not fully understood yet.

In fMRI experiments, the neural activity is calculated as statistical values. Statistical methods compare time course of signal intensities with the modeled response using the hemodynamic response function (HRF) of each voxel describing the regions of neural activity. The statistical maps of these activated regions overlaid on the brain anatomical images relate brain areas to functions. Therefore, the shape and timing of the BOLD response (HRF) have an huge impact on the results of the statistics.

The shape and timing of the HRF varies across species and within the subject's brain regions [3]. Time to peak of 2.49 s was measured in the somatosensory cortex [4] and 2.73 s in the motor cortex [5] for a 330 μs electrical stimulation of rats' forepaw. In a human fMRI experiment, TTP of 5.1 s for 1 s visual stimulus [6] and a delay of 5.43 s for 100 ms visual stimulus [7] was measured. Variability was seen across age groups with 5.38 s in adults for a 1 s motor task

[8], TTP of 4.94 s to the effect of sampling was recorded also for a 1.5 s motor task [9]. An increase in TTP for increasing stimulus length was observed in a motor, visual and auditory fMRI experiment [10]. TTP of HRF was also influenced by order of stimulus presentation [11].

The HRF's temporal parameters have a significant delay in short stimulus event-related designs, therefore stimulus duration is important to measure the time to peak. The paper describes a modeling procedure to estimate the TTP of the BOLD response of activated voxels obtained by the general linear model (GLM) and estimates variation of the HRF and its peak latency across brain regions for short and long stimulus presentation in rat and human subjects.

2 Material and Methods

The expected BOLD response is the convolution of the stimulus time course with an impulse response function

$$x(t) = s(t) * h(t) = \int_{-\infty}^{\infty} s(t-u)h(u)du. \qquad (1)$$

Where, $x(t)$ is the BOLD response, $h(t)$ an impulse response function, $s(t)$ is the stimulus time course. The BOLD signal is sampled with a period of TR (repetition time) to extract volumes of images for activation maps.

2.1 Statistical Analysis of the BOLD HRF

The General Linear Model (GLM) predicts the contributions of stimulus to image voxels and estimates the temporal properties of the BOLD signal. The GLM is defined by (2) for a voxel's time series as

$$Y = Xb + e. \qquad (2)$$

Where, X is the design matrix of predictor time courses, b is a vector of beta weights estimating the contribution of a predictor, e is error values, Y is the measured signal, each column represents a single voxel.

The impulse response function $h(t)$ describes the hemodynamic response. The impulse response function used combines two gamma functions [12] expressed by (3) as

$$h(t) = A\left[\frac{t^{\alpha_1-1}\beta_1^{\alpha_1}e^{-\beta_1 t}}{\Gamma(\alpha_1)} - c\frac{t^{\alpha_2-1}\beta_2^{\alpha_2}e^{-\beta_2 t}}{\Gamma(\alpha_2)}\right], \qquad (3)$$

where A is the amplitude, α determines the shape, β controls the scale, c controls the ratio of the undershoot, t is the reference time and Γ is a gamma function. The initial values of the gamma function parameters [12] are $\alpha_1 = 6$, $\alpha_2 = 16$, $\beta_1 = \beta_2 = 1$ and $c = 1/6$. Hence, the statistical estimation of the parameters of the BOLD signal is highly dependent on the correctness of the design matrix. The time to onset and peak determines the HRF's temporal behavior.

2.2 Experimental Paradigm

Data from a previous fMRI study from the Deutsches Primatenzentrum functional imaging department, Göttingen of measurements in medetomidine-anesthetized rats and human subjects were utilized.

2.2.1 Rat fMRI data

The functional MRI data measured for 6 rats was provided. The rats were anesthetized with medetomidine and placed in a 9.4 T MR system (Bruker Ettlingen) with GE-EPI sequences, (TR 1500 ms, TE 15 ms, flip angle 90°, 30 contiguous coronal slices 0.5 mm thick, in-plane resolution 0.2 × 0.2 mm², 220 repetitions). Electrical stimulation (9 Hz, 3 mA, 0.3 ms pulse width) of the forepaws of the rats during 3 blocks of a 30 s stimulation period with 3 blocks of rest (330 s total for each run) was done and T2*-weighted images were acquired for 15 runs, 220 volumes each.

2.2.2 Human fMRI data for long stimulus

The T2*-weighted MR images with 129 volumes were acquired on a 3 T Magnetom Spectra (Siemens, Erlangen), EPI sequences with a TR = 2000 ms, 29 transverse slices of 4 mm thickness, (2 ×2 × 4 mm³ voxel size) for 6 subjects. Three of the subjects performed a motor task. The subjects focused on a screen in the MR scanner during the resting period (8 blocks) and in the 8 blocks of stimulation, tapped their fingers. Three subjects performed a visual experiment with a 12 s visual stimulus projected across the left and right fields of a screen during MR acquisition.

2.2.3 Short stimulus and TR optimization

For a valid estimation of the HRF, a visual experiment using a full field stimulation (checkerboard, 2 Hz) was presented for 5 s to a human subject and 15 s rest period for TR range of 100 ms, 250 ms, 500 ms, 1000 ms, 2000 ms with 20°, 30°, 45°, 60° and 70° flip angle respectively (flip angle derived by an Ernst angle calculator, *mritoolbox.com/ErnstAngle.html*). A single slice covering the calcarine suclus for each run was acquired.

2.3 Data Analysis

Data were analyzed for activation, the time course of the activated voxels in the somatosensory area (S1FL) for rats, the visual cortex (V1) for both short and long stimulus, the supplementary motor area (SMA) and the primary motor cortex (MC) in human subjects were defined as the region of interest (ROI) using the BrainVoyager QX2.8 (Brain Innovation, Maastricht, The Netherlands) fMRI analysis tool. The designed model used the time course in the fMRI experiments to estimate the BOLD response for each experimental paradigm. The BOLD impulse response function was fitted using MathWorks MATLAB R2015a (Natick, MA, USA) to the measured rat and human HRF to investigate the GLM analysis' assumption, the time to peak and the variation of the HRF within species and brain systems. The *"findpeaks"* and the *"resample"* routine functions of The MathWorks MATLAB and Statistics Toolbox 2015a (Natick, MA, USA) were also used in the estimation of the time to peak of the extracted and the fitted HRF data.

3 Results and Discussion

The time to peak is estimated for the measured 5 s visual fMRI experiment. TTP for the event-related averaging of the extracted ROI's data and the gamma function fit model's HRF are shown in Table 1.

Table 1: TTP of BOLD response in ROI for 5s visual stimulus and TTP of the gamma (γ) fit model for the TR ranges

TR (ms)	TTP (s)	γ fit TTP(s)
2000	11.20	10.72
1000	9.34	7.92
500	7.53	7.57
250	6.48	7.20
100	9.12	8.10

3.1 Influence of Temporal resolution

The Fig.1–5 show the average BOLD responses of activated voxels within the defined ROI (visual cortex) for the measured data and the double gamma HRF fit for the 5 s visual stimulation using different sampling procedures. The acquisition points are shown and the BOLD response's shape is influenced by the parameters of the impulse response function in the fitted estimate. The TTP estimation in Table 1

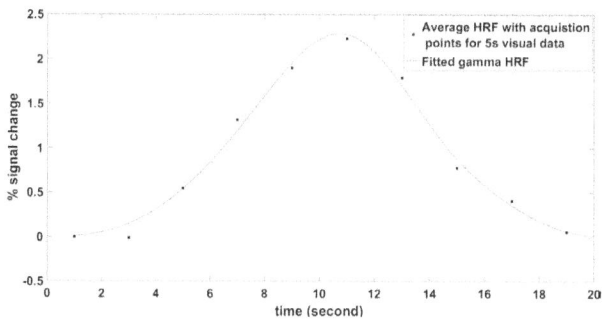

Figure 1: Average HRF for activated voxels in visual cortex ROI for 2000 ms TR with estimated HRF fit.

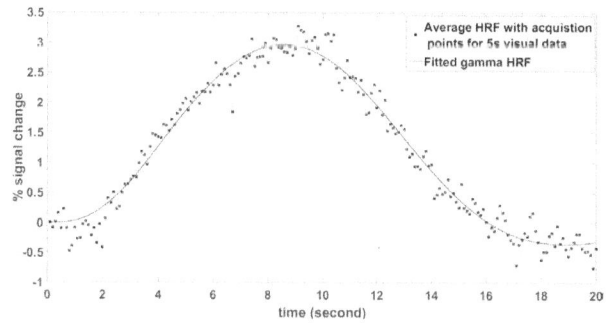

Figure 2: Average HRF for activated voxels in visual cortex ROI with estimated HRF fit, TR 1000 ms.

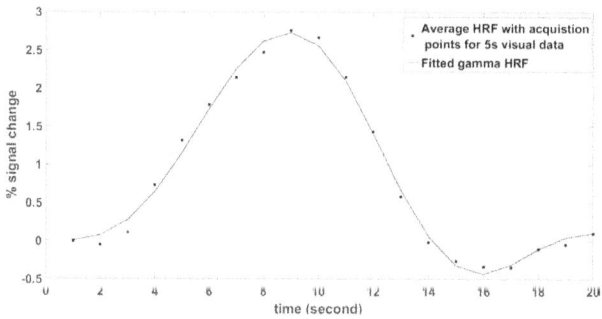

Figure 3: Average HRF for activated voxels in visual cortex's ROI (solid squares) and estimated fit, TR 500 ms.

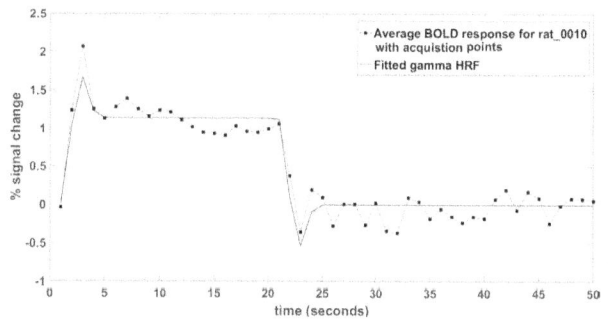

Figure 4: Average HRF for activated voxels in visual cortex's ROI (solid squares) and estimated fit, TR 250 ms.

shows a difference in peak time latency across the TRs. The time to peak for the averaged BOLD response and for the fitted model for the rat data is indicated in Table 2. The TTP

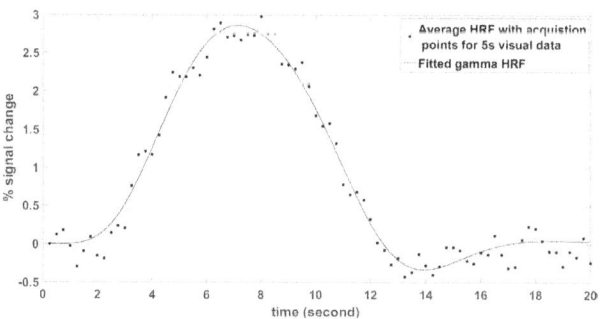

Figure 5: Average HRF for activated voxels in visual cortex's ROI (solid squares) and HRF fit, TR 1000 ms.

for human subjects' SMA and motor cortex average HRFs is shown in Table 3. Table 4 shows the TTP for the primary visual cortex (V1) for the 3 human subjects for a 12 s visual stimulation.

Table 2: Latency of the HRF from activated voxels and the fitted (γ) model in the rat's ROI.

Subject	TTP (s)	γ fit TTP (s)
rat_008	3.35	3.29
rat_009	2.77	2.87
rat_010	2.87	2.92
rat_014	3.19	3.28
rat_015	3.03	3.05
rat_016	3.13	3.15

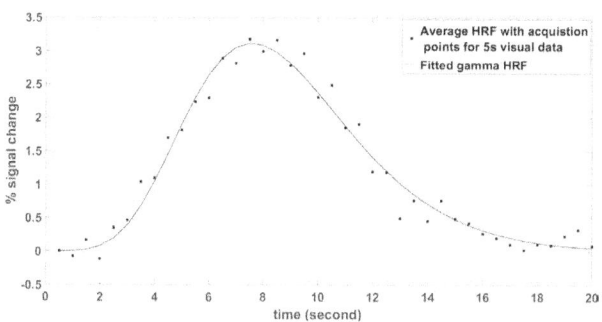

Figure 6: Average HRF of rat data with estimated HRF fit.

Table 3: TTP of average hemodynamic response and gamma (γ) fit in each ROI for 12 s motor task

Subject	ROI	TTP (s)	γ fit TTP (s)
H0105	SMA	4.19	4.33
	MC	5.34	6.05
H0107	SMA	4.10	5.03
	MC	6.08	5.82
H0109	SMA	3.77	7.06
	MC	4.16	6.88

Table 4: TTP of average HRF in ROI for 12 s visual stimulation of visual cortex and TTP of the gamma (γ) fit

Subject	ROI	TTP (s)	γ fit TTP (s)
H0106	LV1	6.80	6.79
	RV1	6.74	7.04
H0108	LV1	4.63	5.80
	RV1	6.08	7.06
H0110	LV1	7.91	7.41
	RV1	7.97	7.22

3.2 Fitting and variance across species

For the different experimental paradigms, there is a need for
different GLM models. This is indicated by the varying delay in both humans and rats' HRF. The block design model
for rat's activation would not accurately predict activations
in the human data due to this difference. TTP was faster
in the rats' somatosensory cortex Fig. 6 than in the human's motor cortex with a mean time to peak 3.09±0.18 s
and 5.19±0.96 s, respectively. The gamma function fits the
short stimulus optimally (Fig.1–5), hence a short stimulus is
suitable to measure the HRF's TTP for this model. The long
stimulus HRF showed a difference in TTP between modalities (visual and motor cortex) with a mean TTP 6.68±0.34 s
in the human visual cortex. Also, within-subject variability
was measured in Table 3 for the SMA and the motor cortex with a mean TTP of 4.02±0.22 s and 5.19±0.96 s respectively. Comparison of the HRF's variation and its TTP
estimation in rats and humans for short stimulus would be
carried out in future experiment.

4 Conclusion

The time to peak was estimated in rats and humans brain regions to show variability between their HRF. Future experiment would estimate HRF variation across different modalities using shorter stimulus (< 5 s) for measuring the time
to peak parameter and its difference between rat and human
subjects using the same experimental paradigm.

Acknowledgement

The work has been carried out at the Funktionelle Bildgebung, Deutsches Primatenzentrum GmbH, Goettingen and
supervised by Prof. Martin Koch at the Institute of Medical
Engineering, Universität zu Lübeck.

5 References

[1] S. Ogawa et al., *Intrinsic Signal Changes Accompa-
nying Sensory Stimulation: Functional Brain Mapping
with Magnetic Resonance Imaging*. Proceedings of the
National Academy of Sciences of the United States of
America vol. 89, no.13, pp. 5951–5955, 1992.

[2] R. S. Menon and S. G. Kim, *Spatial and Temporal Limits in Cognitive Neuroimaging with FMRI*. Trends in
Cognitive Sciences, vol. 3, pp. 207–216, 1999.

[3] D. A. Handwerker, J. M. Ollinger and M. D'Esposito,
*Variation of BOLD Hemodynamic Responses Across
Subjects and Brain regions and Their Effects on Statistical Analyses*. NeuroImage, vol. 21, pp. 1639–1651,
2004.

[4] A. C. Silva, A. P. Koretsky and J. H. Duyn, *Functional
MRI Impulse Response for BOLD and CBV Contrast
In Rat Somatosensory Cortex*. Magnetic Resonance in
Medicine, vol. 57, pp. 1110–1118, 2007.

[5] J. A. De Zwart et al., *Temporal Dynamics of the BOLD
FMRI Impulse Response*. NeuroImage vol. 24, pp.
667–677, 2005.

[6] H. L. Liu and J. H. Gao, *An investigation of the Impulse
Functions for the Nonlinear BOLD Response in Functional MRI*. Magnetic Resonance in Medicine, vol. 18,
pp. 931– 938, 2000.

[7] J. M. Ford, M. B. Johnson, S. L. Whitfield, W. O.
Faustman and D. H. Mathalon, *Delayed Hemodynamic
Responses in Schizophrenia*. NeuroImage, vol. 26, pp.
922–931, 2005.

[8] T. Arichi et al., *Development of BOLD Signal Hemodynamic Responses in the Human Brain*. NeuroImage,
vol. 63, pp. 663-673, 2012.

[9] F. M. Miezin, L. Maccotta, J. M. Ollinger, S. E. Petersen and R. L. Buckner, *Characterizing the Hemodynamic Response: Effects of Presentation Rate, Sampling Procedure, and the Possibility of Ordering Brain
Activity Based on Relative Timing*. NeuroImage vol. 11,
pp. 735–759, 2000.

[10] D. A. Soltysik, K. K. Peck, K. D. White, B. Crosson
and R. W. Briggs, *Comparison of Hemodynamic Response Nonlinearity Across Primary Cortical Areas*.
NeuroImage, vol. 22, pp. 1117–1127, 2004.

[11] F. H. Lin et al., *fMRI Hemodynamics Accurately Reflects Neuronal Timing in the Human Brain Measured
by MEG*. NeuroImage, vol. 78, pp. 372–384, 2013.

[12] M. A. Lindquist, J. Meng Loh , L. Atlas and T. D.
Wager, *Modeling the Hemodynamic Response Function
in FMRI: Efficiency, Bias and Mis-modeling*. NeuroImage, vol. 45(1 Suppl), pp. S187-98, 2009.

Development and evaluation of a test and training environment for endovascular aortic procedures using rapid prototyping as part of NAV-EVAR project

J. Bouchagiar [1], A. Höfer [2], M. Horn [2], and M. Kleemann [2]

[1] Medizinische Ingenieurwissenschaft, Universität zu Lübeck, juljan.bouchagiar@student.uni-luebeck.de

[2] Universitäres Gefäßzentrum (UGZ), Bereich Gefäß- und endovaskuläre Chirurgie, Klinik für Chirurgie, Universitätsklinikum Schleswig-Holstein-Campus Lübeck, Markus.Kleemann@uksh.de

Abstract

This review provides an overview about constructing and building of a training plattform to gain rapid prototyping experience of arterial vessel models. The target is to build a prototype which will allow development and testing of new surgery techniques. In a second step a dummy for training purposes will be developed. First a list of requirements was defined where the anatomical correctness of the aorta were included. In addition upcoming questions were answered regarding other necessary anatomical structures and orientation landmarks. The dummy has a modular setup with interchangeable 3D printed anatomical models. Due to the modularity it is possible to simulate different pathological aortic aneurysm.

1 Introduction

The most common cause of death in industrialized countries are cardiovascular diseases e.g. globally more than 200 million people are currently suffering from peripheral occlusive disease (PAD) [1]. Over the last decade the treatment of endovascular diseases became more important as more accurate diagnostic methods and treatments were developed [2].

For complexity reasons this paper is focused on the aortic aneurysm. The aortic aneurysm is an enlargement of the aorta. It is defined as a dilation of 1.5 times the normal size. The aneurysm itself causes no symptoms only when it ruptures. This kind of rupture results in massive internal bleeding with little chance of survival [3].

At the beginning of last century, first X-ray based diagnostic methods for humans were introduced. In principle the techniques didn't evolve much. The usage of a contrast agent with all its known disadvantages is still very common. The same applies to endovascular treatments which make use of X-ray. It became an important imaging tool because of the major challenge to visualize the exact position of an inserted catheter in the vessel [4].

Due to the relevance of cardiovascular diseases the number of surgeries is rising and so the exposition rates of radiation to the medical staff is rising too. First cases of brain tumors of the performing personnel are reported. Especially left sided tumors are common which can be related to the fact that often the X-ray device is located on the left-hand side [5].

The second big disadvantage of the common imaging principles is the usage of contrast agents. Due to the good contrast, density iodinised contrast agent are used which can cause allergic reactions. Besides that another big disadvantage is, that it can decrease the renal function [2]. Especially a lot of patients suffering from vascular diseases show also impaired renal function. Already 40 % of all patients with abdomic aortic aneurysm are suffering from impaired renal function as some studies reveal [6] [7].

Currently there are only existing some vessel models of the aorta for training purposes [8] [9]. As part of the Navigated Contrast-Agent and Radiation Sparing Endovascular Aortic Repair (NAV-EVAR) project, this paper focuses on building a complete test dummy for more realistic impression as testing platform for the development of new endovascular imaging tools to get away from X-ray based processes for detection of aortic aneurysms and providing orientation support for the catheter treatments.

2 Material and Methods

The goal is to manufacture a prototype dummy with the possibility of simulating the treatment of aortic aneurysms and practising its treatment. It should generate experience for further development of dummies which shall be used as training environment for inexperienced surgeons. The purpose with this training dummy to expand the training of new surgeons. The current medical training is in Germany defined by the Medical Associations and is a guided training during the surgeries [10]. Furthermore a second type of dummy should be constructed for developing new navigation techniques and treatments to be flexible for changing requirements in the future.

2.1   Definition of the prototype

At first the prototype had to be defined. Due to the requirement for a test environment for practising surgeries the anatomical correctness of the aorta and the aneurysm is the most important quality requirement. With regard to material selection it was sufficient that the model of the aorta is waterproof, to be able to simulate blood flow, and as low-priced as possible to reduce the costs. It is positioned and installed inside a human modeled body. For further development and testing the whole system should be modular to allow the exchange of different parts or the addition of new ones like other landmarks if new navigation methods it require.

In the following sketch all outgoing vessels are drawn.

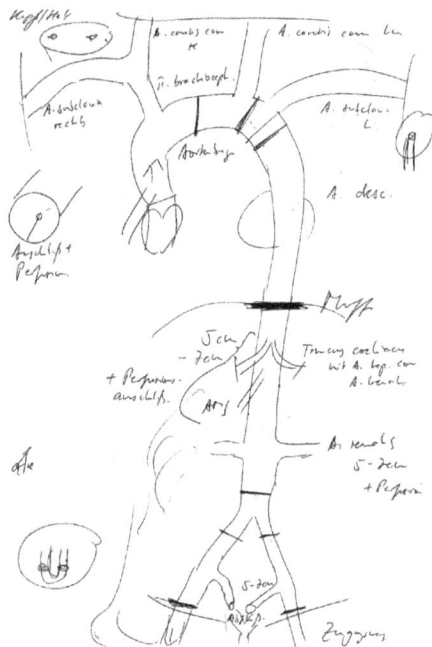

Figure 1: All needed outgoing vessels from the Aorta model are drawn as it is been seen distal, mostly all vessels starting from the heart till inguinal arteries, without coronary arteries. Possible cutting positions to achieve the modularity are indicated by the strokes.

As seen in Fig. 1 the following outgoing vessels are included into the model due to the anatomical correctness: brachiocephalic artery with right subclavian artery (till end of arm stump), right carotid artery (till end of neck stump), left carotid artery (till end of neck stump) left subclavian artery (till end of arm stump), descending thoracic aorta, abdominal aorta, celiac trunk with common hepatic artery and splenic artery, superior mesenteric artery, renal artery, common iliac artery, internal iliac artery, external iliac arteries and common femoral artery. The length of the anatomical correct part of all visceral arteries seen from aorta is planned with 5-7 cm. The figures behind some vessels are giving the required length of anatomical correctness of this vessel, seen from the aorta.

Due to the requirement of perfusion, all vessels are connected to a hosepipe system to enable the circulation of

fluid. The inlet should be at the origin of the aorta. The pipe system can be connected to an external pump. The access to the vessels shall be possible through the groin, the arm stumps or the neck stump. For the groin vessel should be an included access structure with an angle of 30-40 degr. The diameter of the access pipe allows devices up to 8 mm to be inserted. It is also possible to preinstall these devices so that the system is permanently equipped as shown in Fig. 2.

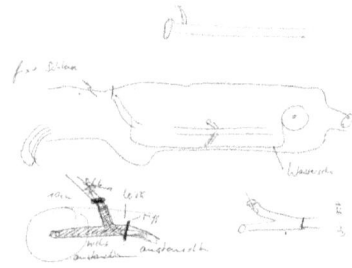

Figure 2: Approximate position of the aorta in the dummy and in the lower left corner the mounting of the catheter at the inguinal artery.

In order to simulate different aneurysm positions and types it is necessary to be able to exchange the aorta or parts of it. For orientation purposes the model has also built in the spinal column from the thoracic vertebra III until the beck bone, as landmarks for further X-Ray imaging. To make it also CT usable it is necessary to avoid using any metal objects which will deteriorate the results.

2.2   Construction of the prototype

To maintain the data anatomically correctly we had to select one patient with an aortic aneurysm. Wchose a patient from an anonymised database with an infrarenal aortic aneurysm. From the CT data we developed the anatomic structure of the aorta and the spine by the 3d software Slicer 4.8.0 as shown in Fig. 3.

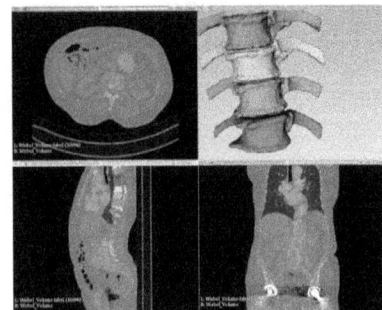

Figure 3: Different perspectives of vertebrae Th III - Th XII derived from the CT data using slicer 4.8.0. Clockwise from upper left corner: longitudinal, 3D, sagittal, transversal.

To enable mounting purposes the model was equipped with fixations. For this the CAD software Catia V15 was used. Different fixations were created in the computer model. They were printed together with the model in order to achieve better structural stability and easier mounting to achieve the required modularity. The whole model is entirely constructed in Catia and shown in Fig. 4.

Figure 4: Entire construction of the dummy using Catia V15. All anatomical parts and all parts needed for perfusion and fixing the models at the anatomical correct position are shown.

The model was printed in the facilities of the company Medizinische Modellbau Manufaktur GmbH in Wildau. Processing and assembly of the parts was also done by the company. For the bone structures polylactic acid (PLA) was chosen, for the aorta silicone and for the pipe connections VeroBlue. PLA was chosen as material because of the low price and easy handling. VeroBlue had to be waterproof and able to be printed with the available printers. This choice represents the most economical compromise. The material for the pipe connections had also to be waterproof and durable this is why VeroBlue was used.

3 Results and Discussion

The current status is shown below.

3.1 Results

Fig. 5 illustrates the upper part of the spinal column. It is mainly for orientation and fixation purposes during the testing. The function of the spinal canal is to guide the returning fluid from those thoracic vessels which are connected to the aorta. These are brachiocephalic artery, left carotid artery and left subclavian artery.
The whole spinal column is mounted into the body of the dummy as shown in Fig. 6. Due to the early status of the project it was not possible to show more parts of the testing environment as only a few parts were completed.
The supporting structure is a prefabricated body in which accessing holes were already cut. The skin material is silicone and is also prefabricated and pulled over the supporting body. This is shown in Fig. 7.

3.2 Discussion

The needed modularity could implemented and had to be tested if it can be used as intended. The durability of the individual components such as flanges, models and supporting structures has still to be verified.

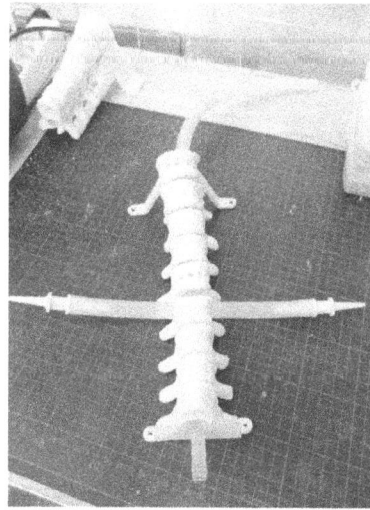

Figure 5: Assembled cranial view of spinal column with the hosepipe in the spinal canal for the returning liquid from the thoracic vessels which are connected with the aorta model.

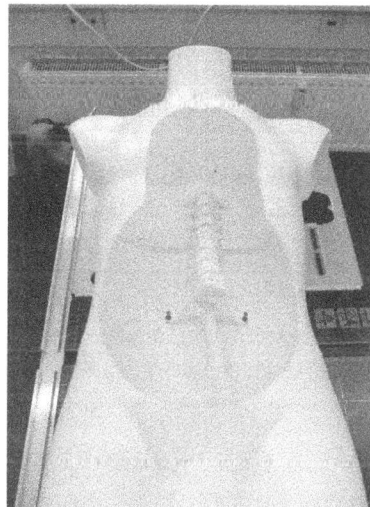

Figure 6: Assembly of the spinal column into the dummy body. Only the thoracic vertebrae of the printed spinal column is mounted.

Also further test surgeries have to be conducted with the dummy in order to find out how realistic the selected materials behave in comparison to the natural human counterparts and whether it will be necessary to print new models using more real life materials. No kind of statement was possible at this stage.
Further usage should show if the dummy is capable for the development of new surgery techniques and if necessary to be modified.

4 Conclusion

Development and manufacturing of the dummy prototype is just completed. Although final acceptance test has not yet been conducted we gained a lot of experience. We succeeded in standardizing several connecting parts, which are

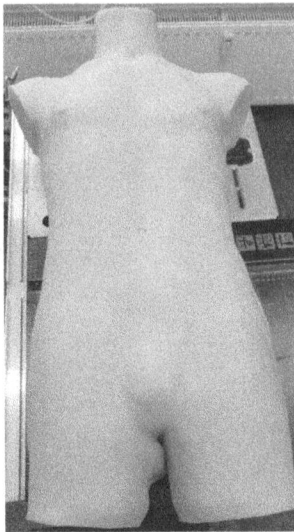

Figure 7: Appearance of the final dummy with pulled over silicone skin.

very important for the required modularity. It should be possible to easily exchange the different aneurysm models and practise their on surgery techniques.

However a CT measurement of the entire dummy is highly recommended in order to verify and define the differences between the manufactured parts and the patient dimensions. During the 3D printing process it was experienced that the printing materials get heated up and become softer. This can lead to deviations from the original anatomical structure. Knowing this it is now very important to further investigate how endovascular surgical procedures with humans can be simulated in the testing environment and how natural the materials behave.

Also further research should show if such an artificial environment is a proper alternative to cadaver studies or animal testing.

As a further step and in order to test alternative navigation methods this dummy could be modified to make it ultrasound capable. Also for other methods further modifications are possible. For example, metal dots can be installed as landmarks. The necessary modifications of the dummy shall be defined in a separate research and development project. At this moment the creation of a laboratory testing environment is planned in order to test the established navigation methods as described in study [2].

Acknowledgement

The work has been carried out at UKSH (Universitätsklinikum Schleswig Holstein) in Lübeck in cooperation with the University of Lübeck. Technical support was provided by c/o Medizinische Modellbau Manufaktur GmbH in Wildau.

This project was sponsored by the Federal Ministry of education and research. FKZ: 13GW0228A

Figure 8: Federal Ministry of Education and Research logo

5   References

[1] F. Fowkes. R. Gerald, *Comparison of global estimates of prevalence and risk factors for peripheral artery disease in 2000 and 2010: a systematic review and analysis*, The Lancet, 382.9901, 2013.

[2] M. Horn, J. Peter Goltz, E. Stahlberg, N. Papenberg, F. Ernst, M. Kleemann, *Endovascular interventions proceeded under contrast agent and radiation sparing using navigation and imaging techniques for holographic visualisation*, European symposium on vascular biomaterials 2017, 2017.

[3] D. Reiche, *Lexikon Medizin*, Elsevier Health Sciences, 2003.

[4] M. Horn, J. Nolde, J. Goltz, J. Barkhausen, W. Schade, C. Waltermann, *An Experimental Set-Up for Navigated-Contrast-Agent and Radiation Sparing Endovascular Aortic Repair (Nav-CARS EVAR)*. Zentralblatt für Chirurgie, 140.5, pp. 493-499, 2015.

[5] A. Roguin, J. Goldstein, O. Bar, J. A. Goldstein, *Brain and neck tumors among physicians performing interventional procedures.*, The American journal of cardiology 111.9, pp. 1368-1372, 2013.

[6] S. R. Walsh,Y. T. Tang, J. R. Boyle. *Renal consequences of endovascular abdominal aortic aneurysm repair*, Journal of Endovascular Therapy, 15.1, pp. 73-82, 2008.

[7] R. K. Greenberg, T. Chuter, M. Lawrence-Brown, S. Haulon, L. Nolte, *Analysis of renal function after aneurysm repair with a device using suprarenal fixation (Zenith AAA Endovascular Graft) in contrast to open surgical repair*, Journal of vascular surgery, 39.6, pp. 1219-1228, 2004.

[8] S. Mafeld, et al, *Three-dimensional (3D) printed endovascular simulation models: a feasibility study*, Annals of translational medicine, 5.3, 2017.

[9] I. O. Torres, N. De Luccia, *A simulator for training in endovascular aneurysm repair: The use of three dimensional printers*, European Journal of Vascular and Endovascular Surgery, 54.2, pp. 247-253, 2017.

[10] Bundesärztekammer, *Facharzt Weiterbildung*. Available: http://www.bundesaerztekammer.de/fileadmin/user_upload/downloads/pdf-Ordner/Weiterbildung/MWBO.pdf [last accessed on 2018-02-07].

Comparison between Piezoelectric Actuator and Brushless Motor for Insulin Delivery

N. Tadrisi Parsa [1], C. Wuertele [2], S. Wuerzburger [3], and M. Ryschka [4]

[1] Biomedical Engineering, University of Applied Sciences Lübeck, nasrin.tadrisiparsamoghadam@student.fh-luebeck.de
[2] Department of Hardware Development, Roche Company, Mannheim, christian.wuertele@roche.com
[3] Department of Hardware Development, Roche Company, Mannheim, steffan.wuerzburger@roche.com
[4] Department of Electrical Engineering, University of Applied Sciences Lübeck, martin.ryschka@fh-luebeck.de

Abstract

Diabetes occurs when the pancreas does not produce enough insulin (a hormone that regulates blood glucose). In some cases the daily injection of insulin is vital. For avoiding several injections per day and having more precision controlled insulin delivery, medical technology has introduced insulin pumps. These devices are connected to the patient and can deliver the insulin continuously. This paper describes the application of piezoelectric micropumps and briefly compares it with brushless micropumps which are currently on the market.

keywords: Insulin pumps, Piezoelectric micropumps, DC/DC converter, Brushless DC micropumps

1 Introduction

Diabetes is one of the most important chronic diseases. It is a metabolic illness created by an insufficient generation of or shortage counteraction to insulin. For a diabetic person, it may be necessary to measure the blood glucose regularly and some patients have to inject insulin every day. In recent years, many studies have been conducted to make this procedure easier for the patients and permit them to have more participation in the daily activities with a risk reduction of long-term side effects. As a result, different types of insulin pumps have been introduced which infuse insulin continuously and accurately by using cannulas which are placed in subcutaneous tissue.Technology has introduced piezoelectric micropumps which are controlled by an electronic board without using the conventional electromagnetic energy which is used in brushless DC motors. In the following article, we will discuss the implementation of piezoelectric micropumps and compare it with brushless micropumps for insulin pumps.

2 Materials and Methods

2.1 Piezoelectric micro pump

When the piezoelectric elementary cells are applied with the external electric field, a mechanical strain will be produced by these cells (crystal) which is called piezoelectric effect. This effect is usually used for piezoelectric sensors and motors in different shapes of actuators. The pump which is used in this research is Micropump mp6 from Bar-

tels microtechnik with unimorph actuator which has active and passive valves. This is a combination of piezo disk and metal disks. By applying voltage the piezo disk will move; therefore, the active valves are controlled with electrical voltage by contraction and expansion of the valves and the passive valves control the direction of the injection. The principle behind these pumps is delivering the drugs

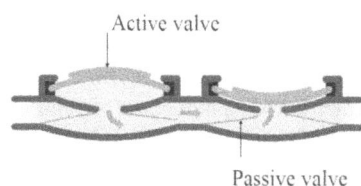

Figure 1: Piezoelectric Valves

without using the common motor pumps. The closer look at the functional principle of the pump reveals two pumps behind each other. This results in a higher output pressure and a more reliable function. For controlling the active valves (actuators) we need a high supply voltage which is the most important challenge in designing the driver. The drug delivery in this system is controlled by the frequency and V_{p_p} voltage which will be explained in the following.

2.1.1 Mathematical description

Polarized piezoelectric materials are characterized by several coefficients and relationships. In simplified form, the basic relationships between the electrical and elastic properties can be represented as follows:[9]

$$D = dT + \epsilon^T E. \tag{1}$$

$$S = s^E T + d_t E. \tag{2}$$

where d represents the piezoelectric constants, T the mechanical stress, E electrical field, S the mechanical strain, ϵ^T the Permittivity, s^E Compliance or elasticity coefficient, and D is the electric displacement.[6]

2.1.2 Driver board

The board is responsible for driving the micropump. It consists of power management, a microcontroller which produces control signals, DC/DC converter and serial to parallel converter part.

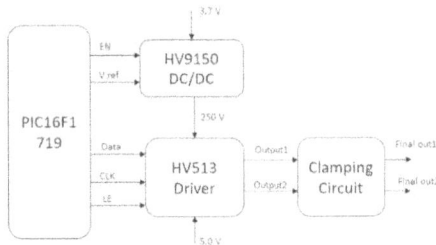

Figure 2: Driver board diagram

2.1.3 Controller

The microcontroller PIC16F1719 is used as an important part of the system to produce the control signals and to communicate with other parts of the board like sensors and push buttons for controlling the speed and time of drug delivery. controlling signals and data that comes from PIC microcontroller, control the high voltage IC and change the output voltages. In the following, these signals are described more precisely. Generally, the microcontroller sends five control signals to the high voltage part of the circuit: Enable signal, Reference voltage, Data, Clock and Latch signal. These signals control the VP_P and the frequency of the final high voltage driving signals to the mp6 micro pump.

Enable Signal: Is used to control HV9150 (DC/DC) converter.

Reference Voltage: Is generated by microcontroller to control the boosted output of HV9150.

Data Signal: This signal caries 8 bits of data which is produced by micro controller and goes to the high voltage driver IC (HV513) to change into the parallel signals with high voltage output for the piezoelectric micropump (up to 250V).

Clock: This signal controls the data transformation from the micro controller to the inner shift register of IC HV513. This procedure happens during the rising edge of the clock signal.

Latch Signal: When the latch goes high, the data will transfer from shift register to the output. Consequently, by controlling the latch signal, the frequency of the output signal will be changed. This means that by changing the latch signal the speed of drug delivery will change.

2.1.4 High voltage signal production and sensors

The HV9150 is a high output voltage DC/DC converter that has a built-in charge pump converter and a linear regulator for a wide range of input voltage.[7] The output voltage is controlled by two main signals. Enable signal which is the controller enable pin (EN), serves two main purposes. The most obvious function is to turn on and off the controller, and the other function is to act as a trigger to activate the device to accept external voltage reference.[7]. A reference voltage which is the second important signal, controls the output voltage (Vp_p). We can use internal or external V_ref and in this board, the externals V_ref is used for changing and controlling the Vp_p manually. By changing the Vp_p the speed of micro pump will change and we can control it linearly. The input signal for this IC is 3.7 V of lithium battery and the output can be regulated to 250 V.

The HV513 is a low-voltage to high-voltage serial-to-parallel converter with eight high-voltage push-pull outputs. This device is designed to drive capacitive loads such as piezoelectric transducers. It can also be used in any application requiring multiple high-voltage outputs with medium-current source-and-sink capabilities.[3] In this circuit, we only need two output signals because the micropump has two actuators. Data from microcontroller goes to the shift register of HV513 by CLK controlling signal and by latch control transfers to the output. As a result, we can control the frequency of the output by these controlling signals and consequently control the speed of the micro pump.

additionally, an EEPROM is used for data storage and a temperature sensor is used for monitoring and controlling the drug delivery more precisely which are connected to the microcontroller with the I2C interface.

2.1.5 Clamping circuit

A clamper is an electronic circuit that fixes either the positive or the negative peak excursions of a signal to a defined value by shifting its DC value. A diode clamp (a simple, common type) consists of a diode, which conducts electric current in only one direction and prevents the signal exceeding the reference value; and a capacitor, which provides a DC offset from the stored charge.[5] A Zener diode is also used to regulate the voltage. The output of high voltage fed the clamping circuit to fix the signal to -50 V. This voltage is required for the mp6 micropump.

2.2 Brushless micropumps

Brushless DC motors are simple, magnets attached to a shaft are pushed and pulled by electromagnetic fields that are managed by an electronic speed control. Electrical energy is converted to mechanical energy by the magnetic attractive forces between the permanent magnet rotor and a rotating magnetic field induced in the wound stator poles.[8] in a brushless DC motor, the rotor is the magnet and the stator is made up of the coils. The speed of the motor can be controlled by the microcontroller with switching the GPIO pins. The feedback of the rotor position is done by three Hall sensors installed in the motor. Microcontroller sends signals to the coils to drive the motor due to the signals from Hall sensors. As the brushless DC motor needs high current, we can not connect the microcontroller to the motor directly because it will harm the microcontroller. Consequently, MOSFETs are used to solve this problem. The diagram is shown in figure 3. In principle,

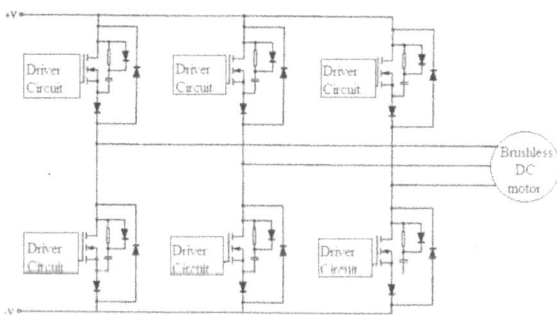

Figure 3: Brushless DC motor Diagram

for effective usage of the motor, the output of this motor should be extended. Therefore, 2 gears have been used to fulfill this issue. At the end, around 1000 rotation of the

Figure 4: Extending the rotation by using gears

motor, we will have one rotation in the biggest gear which causes 1.2 mm of movement of the syringe. As a result, the minimum amount of delivery is 0.025 μl per minutes.

3 Results

3.1 Piezoelectric

At the starting point, the microcontroller sends enable and reference voltages to the DC/DC converter to produce the high voltage signal. By setting the high voltage output, HV513 can be ready to get signals from PIC which are data with clock control and latch signal. Figure 5 RC filter is designed for reducing the audible noise. The input of this filter is given by the high voltage driver and make the sharp edges of the output smoother and rounded. Figure 6 When one of the parallel outputs is on the other one is off and vice versa.

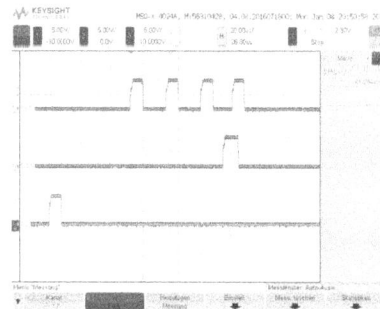

Figure 5: Piezoelectric Control signals. From the top: (CLK, Data, LE signal)

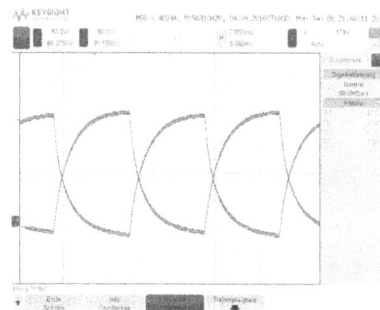

Figure 6: Output voltages

3.2 Brushless

As it can be seen from the figure 7, hall sensors, detect the position of the rotor and send some signals to the microcontroller. Based on the signals from hall sensors, the microcontroller decides which coils should be driven and sends voltages to them. By controlling the speed of sending signals, the speed of rotation will be controlled and consequently, the speed of delivering the drug.

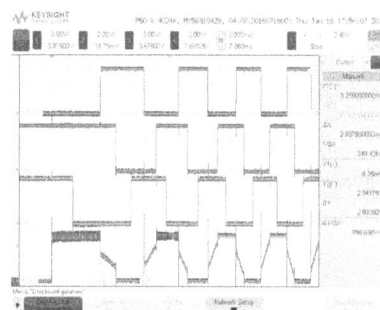

Figure 7: BLDC motor_hall signals. From the top: Hall1, Hall2, Hall3 and Motor 1 signals.

It should be noticed that the hall sensors are located in a position that they always send two signals on, or two signals off to the microcontroller. By defining this signals in the microcontroller, it will decide which motors should have signals and in which direction. Figure 8 shows the output voltages of the coils. It is obvious that the outputs are not completely rectangular, because of the stored energy and inherent property of the coils.

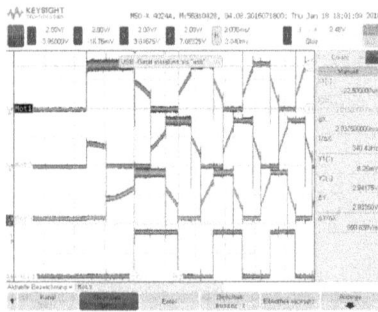

Figure 8: BLDC motor_ Motor signals. From the top: Mot1, Mot2, Mot3 and Hall1 signals.

4 Discussion

In this part, we will discuss the benefits and also the drawbacks of piezoelectric micropump and its driver board and compare it with brushless micropump which is currently used in typical insulin pumps.

Generally, the main advantages of piezoelectric micropumps are the fast response and simple structure. The voltage and frequency can be changed linearly when the motor is running, so the speed of delivering can be changed precisely when the system is working. The weight and size of the pump are considerable for designing the wearable drug delivery pumps. In mp6 pump, due to the passive valves the micro pump can only pump in to one direction. However, they need high actuation voltage which makes the designing of the driver more complicated. Therefore, the power consumption would be high (around 400 mW). Figure 9 shows the power consumption of the pumps.

It should be also considered that the high voltage has a certain risk for the patient; therefore, a special safety concept must be foreseen.

In brushless DC motors, the main advantage is the simple implementation and controlling. As it is shown in figure 9, the power consumption of brushless motor is less than 100 mW.(For measuring the power consumption, the consumption of the whole board in the running mode and standby mode has been measured for 30 seconds each, and the minimum and maximum values are subtracted for both pumps). This low power consumption can save the battery power

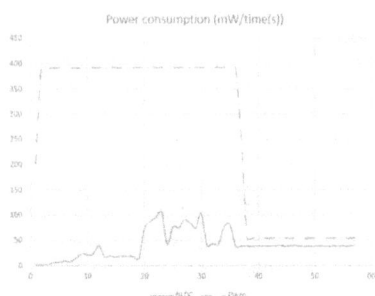

Figure 9: BLDC(typical insulin pump) and Piezoelectric power consumption

for a long time. This pump is not directly in contact with

the drug and the voltages in the electronic board is so low and it cannot harm the patient which is a great advantage for this pump. The minimum drug delivery is 0.025 μl per minute which is much less than the piezoelectric micropumps; however, the maximum amount of drug delivery is only around 3 ml per minute which is much less than the maximum amount of drug delivery in the piezoelectric pump. The major problem of this implementation is using gears which extend the rotation of the motor but it also wastes the energy. At the end the typical values of the piezoelectric micropump and brushless micropump is shown in table 1;

Table 1: Properties of Piezoelectric and BLDC pumps

Pump type	Piezoelectric	BLDC
Dimensions	$30x15x3.8mm^3$	$55x40x19.8mm^3$
Weight	$2g$	$16g$
Power consumption	$< 400mW$	$< 80mW$
Max volume flow	$7ml/min$	$3.18ml/min$
Min volume flow	$1\mu l/min$	$0.025\mu l/min$

Acknowledgement

The work has been carried out at Hardware development department, Diabetes care, Roche Company, Mannheim, as a part of study program of Biomedical Engineering, Fachhochschuele Luebeck. I would like to express my special thanks to Christian Wuertele and Steffan Wuerzburger who were my supervisors at Roche Company.

5 References

[1] Diabetes Fact Sheet, WHO, November 2008(http://www.who.int/mediacentre/factsheets/fs312/en/).

[2] Zhang Feng, Fu-Ho Lee, "Microchip Technology Incorporated 'Piezoelectric Micropump Driver Reference Design".

[3] Datasheet hv513, http://www.microchip.com/mymicrochip/ filehandler.aspx?ddocname=en570612.

[4] Operating manual for Micropump Mp6 and controller.

[5] $wikipedia, clampingcircuits$, www.en.wikipedia.org/wiki/Clamper_%28electronics%29

[6] IEEE Standard on Piezoelectricity. 1987. ANSI/IEEE, Std. 176. Morgan Matric Inc. Piezoceramic Databook. 1993. Morgan Matroc Inc., Electroceramics Division.

[7] Datasheet HV9150, http://ww1.microchip.com/downloads/en DeviceDoc/20005689A.pdf

[8] http://ww1.microchip.com/downloads/en/AppNotes/00857B

[9] https://www.piceramic.com/en/piezo-technology/fundamentals/

Investigating *Nanoparticles' Permeability* using a Tumor-Microenvironment-on-a-Chip

A. Cameron[1], H. Wang [2] and C.-X Zhao [2]

[1] Biomedical Engineering. Lübeck University of Applied Sciences, anna.patricia.cameron@stud.fh-luebeck.de

[2] Australian Institute of Bioengineering, University of Queensland, {haofei.wang, z.chunxia@uq}@uq.net.au

Abstract

Effective accumulation of nanomedicine at the tumor site remains a significant challenge for nanoparticle-based drug delivery, but little is known about nanoparticle rigidity's effect on extravasation and accumulation- mainly due to the difficulties of studying these factors in vivo. This article reports a Tumor-Microenvironment-on-a-Chip, consisting of a tumor microenvironment channel, one blood vessel channel and a lymph channel. The tumor microenvironment was constructed using agarose– its porosity imitating ECM selectivity– to facilitate examination of nanoparticle' shell rigidity, flowrate and interstitial pressure on ECM permeability. A difference between the "soft" Liposome and "hard" PLGA nanoparticles was established; PLGA exhibiting a diffusion permeability coefficient (P_d) $\sim$100x, and permeability $\sim$10x slower than Liposome. Flowrate and permeability were shown to have a parabolic relationship, whereas, interstitial pressure and permeability have a linear relationship. Interestingly, it was shown that flowrate has the greatest effect on the permeability of the ECM.

1 Introduction

Microfluidic devices provide a platform to visualize a biological process. A device providing insight into tumor biology is referred to as a Tumor-on-a-Chip. Current, Tumor-on-a-Chip technology has been developed to study nanoparticles and tumor representation; such as the size and design of the particles, and the formation of 3D spheroids and multicellular tumor spheroids.

Although, these advances are necessary- little effort has been put into developing the microenvironments which execute the transport of the nanoparticles to tumor spheroids. Researchers working with microchips have chosen arbitrary values, close to what they expect in vivo, for flowrate- for instance Albanese's ground breaking work [1]. Yet, exploration has been limited in detailing the relationship between particle rigidity, flowrate, interstitial pressure and permeability. The importance of selecting the correct particle or value cannot be understood unless the possible error is known.

This paper proposes a simple Tumor-Microenvironment-on-a-Chip, composed of three microchannels in Fig. 1. With the outer, interchangeable, channels representing the capillary and lymph vessels, and the inner channel the extracellular matrix (ECM). Low Melting Point Agarose provides an ECM substitute- with its porosity mimicking the ECM selectivity. Nanoparticles can flow from the capillary through the ECM channel and out into the lymph— much like the true biological process. Using the device,

the effect of nanoparticle rigidity on permeability could be measured— by calculating permeation speed of "hard" Poly(lactic-co-glycolic acid) (PLGA) and "soft" Liposome with the confocal microscope. Then, PLGA was placed under hydrostatic pressure to measure permeation speed with differing interstitial pressure. Lastly, the dependence of permeability on flowrate was found, by varying the flow of nanoparticles into the capillary channel using a syringe pump.

2 Material and Methods

The design of both the microfluidic chip, and the designated nanoparticles' drew heavily on previous work. A group protocol had been developed for the production of microfluidic chips and was adhered to- although, this protocol is not documented scientifically, Stroock's review encompasses each process [2]. The nanoparticles' were created to be of similar size.

2.1 Nanoparticles

Liposome nanoparticles were suspended in Phosphate-buffered Saline following previous protocol [3] and PLGA particles were suspended in water [4]. Both particles were approximated at 50 nm in size measured via, Dynamic Light Scattering (DLS) and Transmission Electron Microscopy (TEM). Nanoparticles were labeled with DiI-fluorescent carbocyanide dye- for fluorescent microscopy.

2.2 Microfluidic Chip

To fabricate the microfluidic chip shown in Fig. 1, photolithography was used to fabricate an SU8 mask in conjunction with soft lithography to mold the chip. Glass slides were spin coated and bonded to the PDMS chip using plasma coating. Each production technique is outlined in Stroock's overview of microfluidic systems.

2.3 Agarose ECM

Low Melting Point Agarose was diluted with both 0.1 M Phosphate Buffered Saline and deionized water to create a 2 % gel. It could be pipetted into the chip's middle channel after being liquefied and allowed to cool to bodily temperature at 37 °C. Water-based agarose was used in PLGA testing, and PBS-based agar was used in Liposome testing- to avoid osmotic pressure.

2.4 Interstitial Pressure Control

Tubing was added to the inlets and outlets of the capillary and lymph channels, to create hydrostatic pressure– shown in Fig. 4C. The height of the tubing directly correlates to the applied channel pressure. The difference between the capillary and lymph channel pressure, represents the applied interstitial pressure.

2.5 Flowrate Control

A Harvard Apparatus Model 11 Syringe Pump was used to create flow in the capillary channel. The pump was connected to the inlet and set to the specified flowrate ($6.75\,\mu l/\mathrm{min}$-$2.25\,\mu l/\mathrm{min}$)– excess liquid from the outlet was removed periodically.

2.6 Zeiss 710 Confocal Microscopy

A Zeus Confocal Microscope was used for fluorescent intensity imaging within regions of interests (ROI). Using 10 second time intervals to gather data for processing. The layout of the ROI's within the microfluidic channels, was reproduced following Ho Yan Teck's methodology [5]. The areas were approximated at similar sizes.

3 Results and Discussion

Confocal microscopy was used to gather the fluorescent intensities of the ROI over time. Ho Yan Teck's method was followed. The fluorescent intensity over time was gathered from the tumor microenvironment channel. These intensities were then normalized to the area and averaged. The data could then be approximated by a line of best-fit. The slope of this line, average fluorescence over time, represented $\frac{dI_{avg,gel}}{dt}$ in (1) and (2).

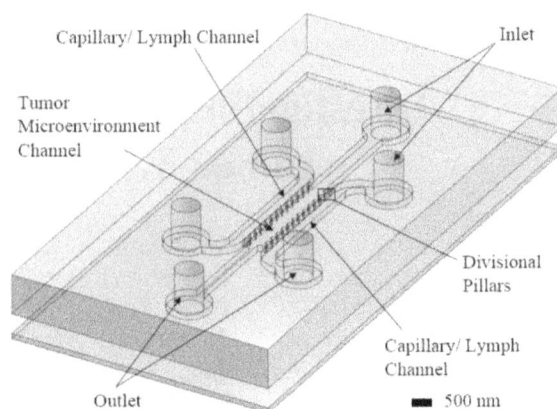

Figure 1: Schematic of Tumor-on-a-chip Microenvironment.

Equation (1) was used to calculate the diffusion permeability coefficient (P_d) [5].

$$P_d = \frac{A_{gel}(\frac{dI_{avg,gel}}{dt})}{w_{capillary}(I_{capillary} - I_{lymph})_{t=0}}. \quad (1)$$

Parameter A_{gel} was the average area of the Agarose ROI. $w_{capillary}$ represented the width of the capillary channel, and $I_{capillary}$ was the initial fluorescent intensity when the channel was filled with nanoparticles. I_{lymph} defined the fluorescent intensity within the lymph channel with no Nanoparticles present.

Equation (2) was used to calculate the permeability [6].

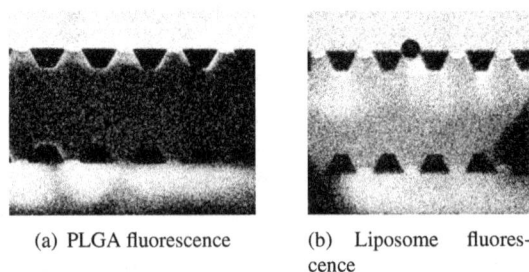

(a) PLGA fluorescence

(b) Liposome fluorescence

Figure 2: Fluorescent permeation of the tumor microenvironment with (A) PLGA and (B) Liposome.

$$P_d = \frac{1}{2(I_{capillary})}(\frac{dI_{avg,gel}}{dt})(\frac{A_{capillary}}{P_{pillar}}). \quad (2)$$

A correction factor of the permeation area is described by $\frac{A_{capillary}}{P_{pillar}}$. $A_{capillary}$ was the total area of the nanoparticles in the capillary— from the first pillar to the last, multiplied by the width of the channel. P_{pillar} represented the perimeter of this area that is open for permeation- with $100\,\mu m$ between pillars shown in Fig.1.

The microfluidic chip was designed to isolate the ECM interaction- to gather permeation data. The design incorporated three channels- the capillary, ECM and lymph. The

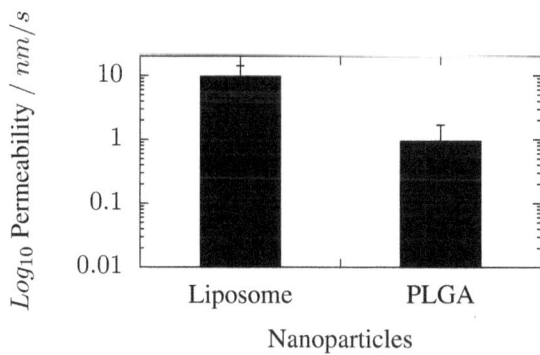

(a) Nanoparticle' Permeability through agarose

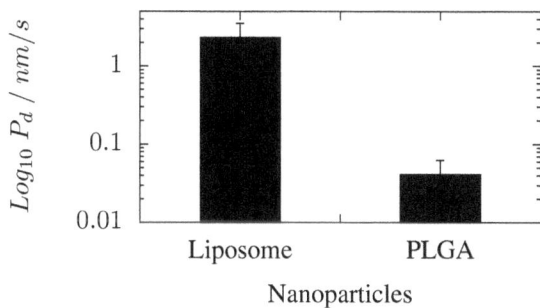

(b) Nanoparticle P_d through agarose

Figure 3: (A) Decreased permeability of the rigid PLGA nanoparticle. (B) Lower P_d of the PLGA in comparison to Liposome nanoparticles.

ECM was filled with Low Melting Point Agarose to create a permeable structure. The pillars were designed to confine the agarose to the middle channel via surface tension. The inlet and outlet of each channel allows the addition of pressure and flow; necessary components for modeling the true conditions of the capillary and lymph.

3.1 Nanoparticle Rigidity

With the variety of nanoparticles being produced, most can still be separated into "hard" and "soft" particles. Although, there are uptake statistics, little is known about particle rigidity and ECM penetration. To test this "hard" PLGA and "soft" Liposome were used, to permeate agarose; allowing calculation of the P_d and permeability. Liposome, from Fig. 3A and 3B, was found to have a P_d of $2.33\,\text{nm}\cdot\text{s}^{-1}$ and permeability of $9.74\,\text{nm}\cdot\text{s}^{-1}$. The same figure's show that PLGA has slower results with a P_d of $0.042\,\text{nm}\cdot\text{s}^{-1}$ and permeability of $93.8\,\text{nm}\cdot\text{s}^{-1}$.

Not only were the particles distinguishable in their permeation but also in their fluorescence. Shown in Fig. 2, Liposome (Fig. 2B) moved in a flood-like fashion, whilst PLGA (Fig. 2A) only speckled the agarose. Literature stated that 2 % agar had a pore-size greater than that of our 50 nm particles- we theorize that the hard-shelled PLGA lacked the elasticity to flood the porous structure, instead it was forced pore-to-pore. This slowed P_d and permeation time.

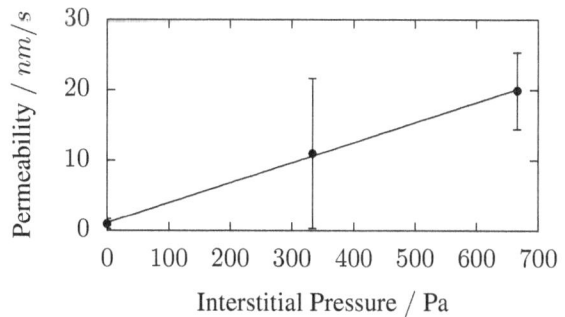

(a) Nanoparticle' permeability with increasing interstitial pressure

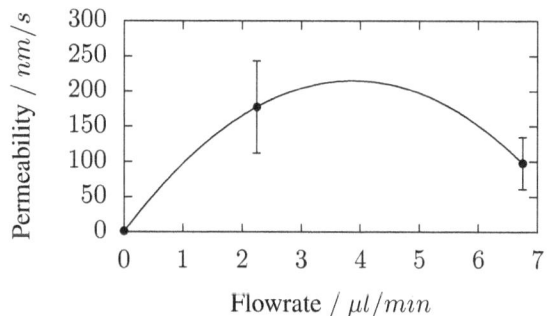

(b) Nanoparticle' permeability with increasing flowrate

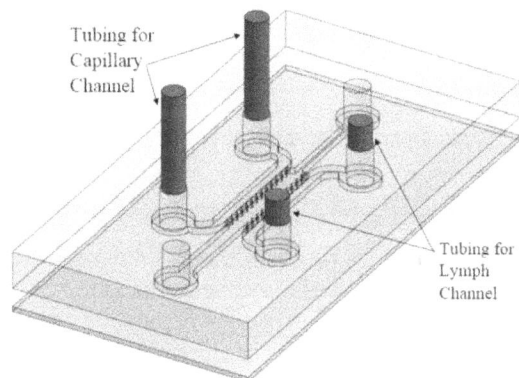

(c) Tubing to add interstitial pressure

Figure 4: (A) Linear relationship between permeability and interstitial pressure. (B) Parabolic relationship between permeability and flowrate.(C) Schematic of the interstitial pressure setup.

3.2 Interstitial Pressure and Flowrate

Tumor targeting technology capitalizes on the enhanced permeability and retention effect (EPR) to deliver specific sized nanoparticles to tumor cells. A key component of the delivery time of the nanoparticles is the interstitial pressure and flowrate of the capillary. To mimic in vivo data, values were based off Murine tumors; with an average capillary velocity of $0.75\,\text{mm}\cdot\text{s}^{-1}$–$6.75\,\text{mm}\cdot\text{s}^{-1}$ [7], pressurized to 2666 Pa [8]– the lymph was approximated at a quarter of the flowrate and pressure. This velocity can be translated to a flowrate of $5.63\,\mu l\cdot\text{min}^{-1}$–$50.6\,\mu l\cdot\text{min}^{-1}$ in a 0.50 mm wide by 0.25 mm deep channel with an interstitial pressure of 2000 Pa.

Fig. 4A shows the linear relationship of interstitial pressure and permeability, with a (root mean squared error) RMS of 0.22313. This relationship is expected, as the greater the force applied through the ECM, the faster the particle permeability. Fig. 4B suggests that flowrate has a clear parabolic relationship, with a RMS of 1. Given the correct flowrate, nanoparticles could reach a maximum permeability– in our device the peak flowrate would be $4\,\mu l/min$. This is probably reminiscent of the shear force which removes particles from the ECM diffusion interface if the flow is too fast. Whereas, if the flow is too slow the particles reach the interface later. Therefore, a balance between both the shear and the delivery time would create a more efficient system.

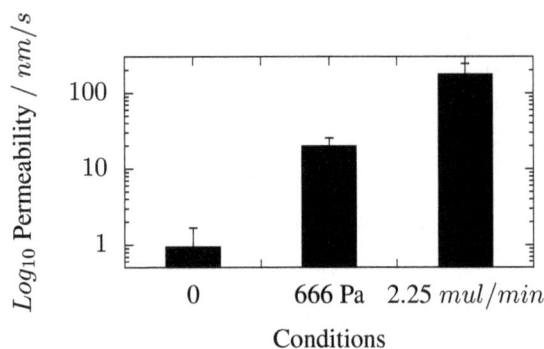

Figure 5: Nanoparticle permeability under differing conditions.

Most interestingly, in Fig. 5, when conditions are close to true biological conditions, flowrate has the greatest effect on permeability. Often flowrate in microfluidic chips is set at a simple, obtainable value. This research shows that more emphasis must be put on the convection processes- specifically flow rate- occurring within the microfluidic chip.

4   Conclusion

This study showed the viability of a three-channel microenvironment for exploring the permeability of ECM to Nanoparticles. By measuring permeability and diffusion permeability coefficients (P_d) it was deduced that rigid particles' travel slowest in the ECM. This is likely due to the particles' lack of elasticity, causing them to be bumped pore-to-pore. Interstitial pressure showed a direct relationship to permeability; the greater the interstitial pressure the greater the force pushing the Nanoparticles across the ECM. Flowrate showed a parabolic relationship, indicating a maximum permeability could be reached. A flowrate should be selected which balances shear force and particle speed; To avoid removal of particles from the ECM interface before they can permeate, whilst delivering them as quickly as possible to the same interface. When using values close to those of the true biological conditions it was found that

flowrate had the greatest effect on permeability. Indicating, that future research should emphasize the use of correct flowrates' to collect feasible data.

Acknowledgement

The work has been carried out at the Australian Institute of Bioengineering, and supervised by Dr. Ralf Moll, Universität zu Lübeck. This work was supported by the Australian Research Council (ARC) Future Fellowship Project (FT140100726) and 2016 UQ Foundation Research Excellence Award fund. Acknowledging financial support from the award of the ARC Future Fellowship (FT140100726). The work was performed with the Queensland node of the Australian National Fabrication Facility, a company established under the National Collaborative Research Infrastructure Strategy– providing nano and micro-fabrication facilities for Australia's researchers.

5   References

[1] Albanese, Alexandre, et al. *Tumour-on-a-chip provides an optical window into nanoparticle tissue transport.* Nature communications, vol. 4, pp. 2718, 2013.

[2] Stroock, Abraham D., and George M. Whitesides. *Components for integrated poly (dimethylsiloxane) microfluidic systems.* Electrophoresis, vol. 23.20, pp. 3461-3473, 2002.

[3] Ran, Rui, Anton PJ Middelberg, and Chun-Xia Zhao. *Microfluidic synthesis of multifunctional liposomes for tumour targeting.* Colloids and Surfaces B: Biointerfaces, vol. 148, pp. 402-410, 2016.

[4] Baby, Thejus, et al. *Fundamental studies on throughput capacities of hydrodynamic flow-focusing microfluidics for producing monodisperse polymer nanoparticles.* Chemical Engineering Science, pp. 169, 2017.

[5] Ho Yan Teck, et al. *A facile method to probe the vascular permeability of nanoparticles in nanomedicine applications.* Scientific Reports, vol. 7, pp. 707, 2017.

[6] Adamson, Roger H., and Fitz-Roy E. Curry. *Determination of Microvessel Permeability and Tissue Diffusion Coefficient of Solutes by Laser Scanning Confocal Microscopy.* Journal of biomechanical engineering, vol. 127.2, pp. 270-278, 2005.

[7] Kelly-Goss, Molly R., et al. *Dynamic, heterogeneous endothelial Tie2 expression and capillary blood flow during microvascular remodeling".* Scientific reports, vol. 7.1, pp. 9049, 2017.

[8] Williams, S. A., et al. *Dynamic measurement of human capillary blood pressure.* Clinical science, vol. 74.5, pp. 507-512, 1988.

Gain-Scheduled PI Controller Design for PEEP-Valve Control in an Anaesthesia Device

D. Kleinewalter [1], J. Börner [2], T. Rahlf [2] and P. Rostalski [3]

[1] Medical Engineering, Universität zu Lübeck, dennis.kleinewalter@student.uni-luebeck.de
[2] Drägerwerk AG & Co. KGaA, Lübeck, Jonas.Boerner@draeger.com, Till.Rahlf@draeger.com
[3] Institute for Electrical Engineering in Medicine, Universität zu Lübeck, philipp.rostalski@uni-luebeck.de

Abstract

Anaesthesia devices are commonly used medical instruments in hospitals all over the globe. Therefore, the optimization and development of new methods and implementations is necessary to guarantee the survival of patients in all imaginable scenarios. This paper examines a state-of-the-art controller design and implementation for PEEP-valve control within ventilation in an anaesthesia device, manufactured by Drägerwerk AG. Since assurance of the patients safety is the most important and inevitable requirement, the main objective for the controller design is to facilitate a suitable and secure expiration for the patient. Regarding this issue, an improvement of the current implemented open-loop control is aspired. To ensure the endeavoured objective, a the system identification, a controller design and its implementation is conducted. This concludes in an evaluation with an emphasis on stability and robustness. Overall, the evaluation exhibits an improvement of the former open-loop control and complete this papers targets.

1 Introduction

Modern anaesthesia devices allow patients in diverse health conditions to be ventilated effortlessly and painlessly during a condition of unconsciousness. There are three categories of respirators regarding the reuse of the exhaled gas, which are determined as opened, semi-closed and closed systems. In closed systems a complete reuse of the ventilated gas is possible, whereas in semi-closed systems only a fraction of the gas is recycled. In open systems the breathing gas is completely removed from the system. The anaesthesia device utilized in this elaboration, is a semi-closed system. With this device several ventilation modes can be used. The most common modes are pressure controlled (PC) and volume controlled (VC) as well as spontaneous respiration (SPN) of the patient [2]. The pressure controlled ventilation mode is shown in Fig. 1, including the expiration time (t_e) and the inspiration time (t_i) as well as the pressure and flow. Within various ventilation modes, the "positive end expiratory pressure", PEEP, is one of the main set values. As the name indicates, the PEEP is the pressure after an expiration of a patient. It is the residual pressure, even after a full expiration to avoid atelectasis. In general the lung is characterised by different capacities and volumes, which include the total lung capacity, the vital capacity, the inspiratory capacity, the functional residual capacity and the tidal capacity together with the residual volume, the expiratory reserve volume, the tidal volume and the inspiratory reverse volume [2]. In addition to the lung characteristics, the lung is described as a flexible and elastic organ. In this

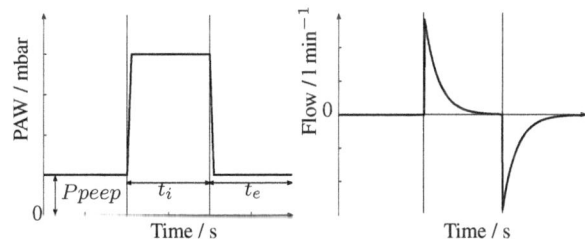

Figure 1: pressure controlled ventilation scheme and associated flow similar to [1]

context the elasticity is not only defined by its elastic filaments but rather through the surface forces of the alveoli as well as their connection to the surrounding pulmonary tissue [2]. Hence, the relation between the volume variation ΔV and the pressure variation Δp conforming to [2] is defined as the compliance:

$$C = \frac{\Delta V}{\Delta p}, \qquad [C] = \frac{\text{ml}}{\text{mbar}}, \qquad (1)$$

which implies the smaller the compliance C, the bigger the variation in pressure Δp needed to provide a specific volume to the patient. Moreover, the pressure Δp is related to the flow rate $\Delta \dot{V}$ according to [2] by a resistive element:

$$R = \frac{\Delta p}{\Delta \dot{V}}, \qquad [R] = \frac{\text{mbar}}{\frac{1}{s}}. \qquad (2)$$

With both equations (1) and (2) the lung is characterised and can be represented in an electric equivalent circuit as shown

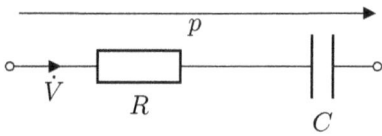

Figure 2: electric equivalent circuit for lung characteristics according to [5]

in Fig. 2, where according to the similarities in the given equations the following analogy has been made: voltage corresponds to pressure, current corresponds to flow rate and capacity corresponds to compliance.

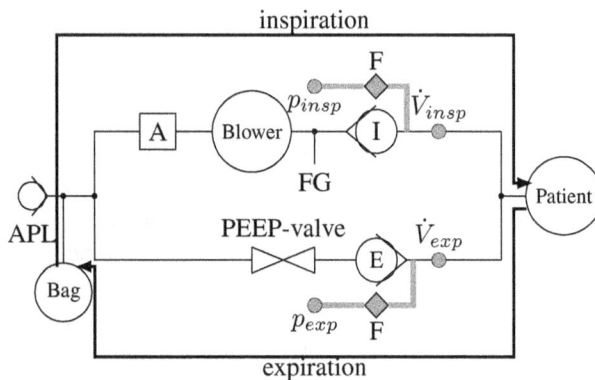

Figure 3: pneumatic scheme of an anaesthesia system similar to [1]

Furthermore, the pneumatic circuit of an anaesthesia device is shown in Fig. 3. As Fig. 3 indicates, there are two branches for each breathing cycle, inspiration and expiration. Both branches include flow sensors ($\dot{V}_{insp}$ and $\dot{V}_{exp}$) as well as pressure sensors (p_{insp} and p_{exp}), a manual ventilation bag and the APL, which limits the pressure of the whole system in SPN and hot-wired otherwise. Within the inspiratory branch, the air flows through a carbon dioxide absorber (A), then through the blower, which delivers the air to the patient and afterwards through the fresh gas pipe (FG). Continuative, there is a inspiratory check valve, as well as a filter. The expiratory branch includes an expiratory check valve and the PEEP-valve, which in the aspired controller design is the regulated actuator.

2 Material and Methods

The following paragraph illustrates the elaborated materials and methods including a system identification, the controller design and its implementation.

2.1 System Identification

The system identification process is performed with different measurements for various patient compliances and resistances. Typical combinations of these parameters for different patients can be seen in Table 1. The system identification is executed with the System Identification Toolbox of

Matlab®. As an input, the toolbox uses a randomly selected expiration phase of a resistance/compliance combination as well as a given step function from 15 mbar to 5 mbar. The duration of this single expiration is 2 s. The toolbox estimates a second order discrete time transfer function for every resistance/compliance sequence as in [3]:

$$G(z) = \frac{b_0 + b_1 z^{-1} + b_2 z^{-2}}{1 - a_1 z^{-1} + a_2 z^{-2}} \tag{3}$$

, which define the different system models. A linear second order model is chosen, because of an easy and fast calculation and further processing. The coefficients for the numerator b_0, b_1, b_2 and denominator a_0, a_1, a_2 are estimated through the process. For standardization the coefficient a_0 is according to [3] equal to 1, the coefficient b_0 equal to 0 since the system possesses no feedthrough. The calculations facilitate therefore models with four free coefficients. Overall, 20 different models were estimated with the toolbox. The plants are used for a PI controller tuning process in a simulation model.

Table 1: Typical nominal value combinations for resistances ($\frac{mbar}{\frac{l}{s}}$) and compliances ($\frac{ml}{mbar}$) taken from [4]

patient	Compliance	Resistance
adult	70-100	2-4
small children	20-40	20
infants	10-20	20-30
newborn	3-5	30-50

2.2 Controller Design

The main objectives for the controller design include a minimal undershoot, so that the ventilation not detracts the patient. A faster response time as the former adaptive open-loop control as well as minimal permanent deviation from the set-point are aspired. The proportional part of the designed controller enables a fast response time, the integral part eliminates steady state errors. With the system models estimated with the System Identification Toolbox of Matlab®, a basic PI-Controller is tuned. In Fig. 4 (a), the general closed-loop control circuit is shown. The circuit includes the model $\frac{Y(z)}{U(z)}$, a PI-Controller and the feedback. As an input, the controller gets the set-point signal during expiration. The output of the controller is defined by the determined pressure for the PEEP-valve. In Fig. 4 (b), the inner elements of the controller, shown in Fig. 4 (a), can be seen. It contains an anti-wind-up element with gain K_b, a saturation as well as proportional gain P and integral gain I. The gain K_b is set to $K_b = 1$ and the limits for the saturation are set to appropriate values for the given pressures. The controller is then tuned heuristically with different lung characteristics to the given requirements. As a result, the identified parameters are stored into lookup tables for the integral and proportional gain.

The above described PI-Controller is now extended by a gain scheduling aspect with regard to the patients lung characteristics. The new design is shown in Fig. 5. Note, that in

(a) general discrete PI-Controller scheme

(b) inside of the PI-Controller

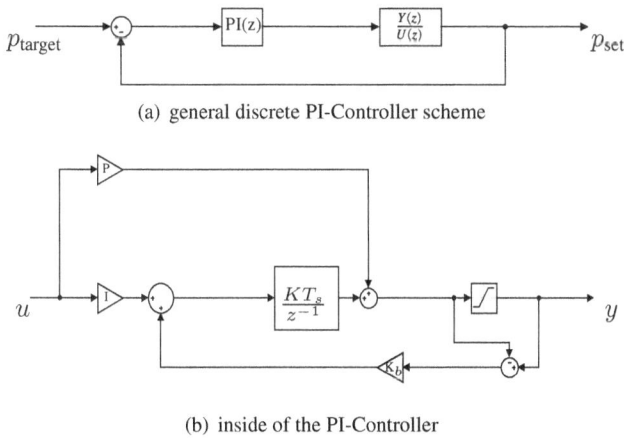

Figure 4: PI-controller scheme

the simulation process a time-discrete varying transfer function is used to model different patients instead of the fixed model plant. The inner elements of the PI-Controller block remain equivalent to the block visualized in Fig. 4. Within a further extension to the previous general PI-Controller design, a suitable initial condition as well as a reset for the integrator are derived. The reset of the integral gain is represented by a boolean signal, which determines whether the patient inhales or exhales. As long as the patient inhales, the integrator is set to the initial condition specified by:

$$I_0 = \max \left(\frac{\Delta p}{2} + \text{PEEP}, \text{PEEP} + 10 \right) \quad (4)$$

where, Δp is the deviation between the maximal inhaling pressure and the PEEP. Equation 4 ensures an appropriate initial condition of the integral gain for various steps, lung characteristics and ventilation modes. Furthermore, the current estimation of compliance and resistance is inserted into the lookup tables for the proportional and integral gain. The lookup tables are heuristically tuned with the above mentioned method. The result is a 5×4 matrix for compliance and resistance of a patient, which is deposited in the lookup tables. These tables interpolate, if not an exact specified value is given as an input. With this design, the controller is flexible and adaptive for different situations. It facilitates one single PI-Controller for various compliance/resistance combinations respectively various patients in an uncomplicated way.

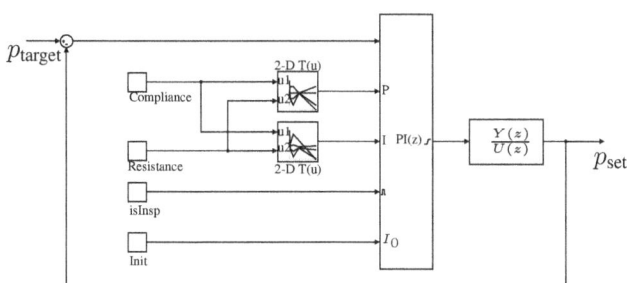

Figure 5: gain scheduled PI-Controller design

2.3 Implementation

The presented gain-scheduled PI-Controller (s. Fig. 5) is implemented with Matlab/Simulink® within the model for the ventilation control of the anaesthesia device. The model is used to generate executable code for a hardware-in-the-loop simulation. This code is transferred through a rapid control prototyping system onto the device to analyse the implementation. This development method empowers an analysis and manipulation during run-time. It is possible to visualize and measure every single signal of the model blocks as well as to modify every constant. With the help of the real-time system, the implemented controller is verified and evaluated in comparison to the prior adaptive open-loop control in the next section.

3 Results and Discussion

The implemented gain scheduled PI-Controller is evaluated against the prior adaptive open-loop control for various patient characteristics. The evaluation is managed with single expiration phases viewing the set-point, the controller and the adaptive open-loop control process. Fig. 6 shows the process for an adult lung with a compliance $C = 50 \frac{\text{ml}}{\text{mbar}}$ and a resistance $R = 5 \frac{\text{mbar}}{\frac{l}{s}}$. The visualized expira-

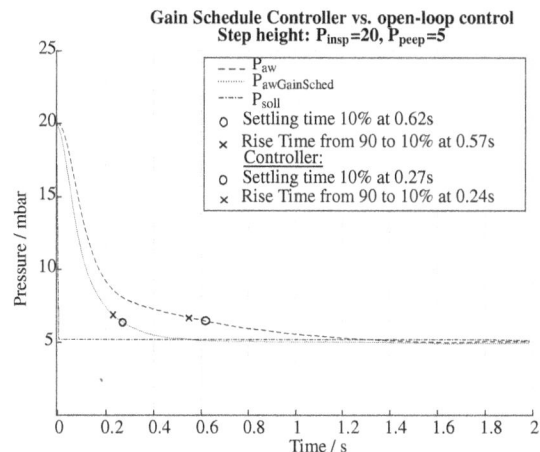

Figure 6: single expiration phase for an adult patient $\left(C = 50 \frac{\text{ml}}{\text{mbar}}, R = 5 \frac{\text{mbar}}{\frac{l}{s}} \right)$ and a pressure controlled ventilation

tion is measured within a pressure controlled ventilation mode. Both implementations reach the set-point specification during expiration. The step height is determined by the inspiration pressure ($P_{\text{insp}} = 20 \, \text{mbar}$) and the PEEP ($P_{\text{peep}} = 5 \, \text{mbar}$). However, the gain-scheduled controller is remarkably faster than the open-loop control. The 10 % settling time is reached after 0.27 s, whereas in contrast the 10 % settling time for the open-loop control is reached after 0.62 s. This is 350 ms faster than the open-loop control. Furthermore, the rise time of the closed-loop controller is in the same range ahead of the open-loop control (with ca. 300 ms). In contrast to that, the open-loop control includes

Table 2: Evaluations of the open-loop (OL) against the closed-loop (CL) controller for different patient characteristics and a step height from 15 mbar to 5 mbar and a pressure controlled ventilation. Parameters with $[C] = \frac{ml}{mbar}$ and $[R] = \frac{mbar}{\frac{l}{s}}$

parameters	adult (C= 50, R= 5)	infant (C= 10, R= 12)	infant (C= 10, R= 19)	child (C= 20, R= 19)	adult (C= 50, R= 46)	selftestlung
settling time OL (in s)	0.64	0.41	0.42	0.51	0.47	0.46
settling time CL (in s)	0.36	0.26	0.39	0.31	0.36	0.36
rise time OL (in s)	0.59	0.36	0.38	0.47	0.43	0.41
rise time CL (in s)	0.32	0.33	0.35	0.28	0.33	0.32
Overshoot OL (in %)	0.0	0.0	0.0	0.21	4.68	0.0
Overshoot CL (in %)	2.01	0.0	0.0	0.0	3.69	0.0
Undershoot OL (in %)	2.89	9.82	4.52	0.22	0.0	5.66
Undershoot CL (in %)	2.93	3.81	4.61	0.98	0.0	4.13
Resistance OL (in mbar/l/s)	16.08	31.04	41.93	37.16	61.88	24.26
Resistance CL (in mbar/l/s)	14.08	29.62	41.61	34.07	61.99	22.54

a zero undershoot and an 1.76% overshoot though the gain-scheduled controller a minimal undershoot of 1.98% and a zero overshoot. This undershoot, respectively overshoot is hardly noticeable in Fig. 6. Continuing, the overall system resistance is about $2 \frac{mbar}{\frac{l}{s}}$ lower. This means a much easier expiration for the patient resulting e.g. in a better mucus extrusion. Both implementations reach a nearly zero steady state error. As a conclusion the gain-scheduled controller achieves the desired requirements for a healthy adult. Moreover, the implemented gain-scheduled PI-controller is much more responsive to new inputs, e.g. another PEEP value. This enables the anaesthesia device to respond fast to a change in the patients state. A further evaluation of the closed-loop controller examines different lung characteristics (different patients) and another step height. Table 2 exhibits the different evaluations for 10% settling time, rise time from 90% to 10%, over- and undershoot as well as the systems resistance for the adaptive open-loop control and the gain-scheduled controller. The measurements are received with a pressure controlled ventilation with a step height from 15 mbar to 5 mbar. The controller design managed to have a faster response time in almost every scenario than the prior adaptive open-loop control. Especially the 10% settling time is in every case faster, with the addition that the set point is followed more precisely. Furthermore, the system resistance for the controller is in most cases lower than the resistance with open-loop control. This allows an easier expiration for the patient. The overshoots vary with different patient characteristics, so that in general the overshoot is less than 4% for all patient characteristics. In addition, the undershoots are in general less than 5% for every patient. This globally small margin shows the performance of the closed-loop control. In addition, the controller is able to achieve a better performance also to an artificial lung for a selftest (s.Table 2), which include non-linearities and a resulting difficult task for the controller. Overall, the set-point is reached faster and with less edges, so that the patient is able to respire easily during the ventilation.

4   Conclusion

Summarizing, the designed gain-scheduled PI-Controller for PEEP-valve control produces an overall better performance than the former adaptive open-loop control. The controller performance is robust and stable in different pa-

tient scenarios, due to the adjustable controller parameters in combination with the measured patient characteristics. Furthermore, the closed-loop controller creates no periodic oscillations and especially produces minimal undershoots in the range of under 5%. This enables the straight forward implementation and good solutions for various patient characteristics as well as for different step heights. Notably, for an adult patient, the considerably faster response time improve the anaesthesia devices ventilation. With this short evaluation in the section above a general advantage of a controlled PEEP-valve for expiration could be revealed. Further tests and more patient scenarios need to be evaluated to confirm the presented performance enhancement. This includes also evaluations for leakage and other difficult environments for the anaesthesia device. Furthermore, different evaluations for other ventilation modes such as volume controlled ventilation or supported ventilation are necessary. As a main consequence, the designed and implemented controller facilitates a superior expiration for the patient in comparison to the open-loop control.

Acknowledgement

The work has been carried out at Drägerwerk AG & Co. KGaA, Moislinger Allee 53-55, 23558 Lübeck and supervised by the Institute for Electrical Engineering in Medicine, Universität zu Lübeck.

5   References

[1] R. Kramme, *Medizintechnik*, 4. edition, Springer Medizin Verlag, 2011.

[2] R. Larsen, T. Ziegenfuß, *Beatmung*, 5. edition, Springer Medizin Verlag, 2012.

[3] J. Lunze, *Regelungstechnik 2*, 8. edition, Springer Vieweg, 2014.

[4] W. Oczenski et al., *Breathing and mechanical support*, Blackwell Science, 1997.

[5] J. Werner et al. *Automatisierte Therapiesysteme*, Reihe Biomedizinische Technik no. 9, Walter de Gruyter GmbH, 2014.

Cell Adhesion on Titanium Surfaces after Modification by Plasma Electrolytic Oxidation and Sol-gel Coating

M. Bodrova [1], A. Kopp [2]

[1] Biomedical Engineering, University of Applied Sciences Lübeck, mariia.bodrova@stud.fh-luebeck.de
[2] Meotec GmbH & Co. KG, alexander.kopp@meotec.eu

Abstract

Titanium is one of the most commonly used materials for implants in trauma applications due to its low density, high corrosion resistance and biocompatibility. Nevertheless, engineers try to modify titanium surface in order to improve wear and corrosion protection or to reduce the possible infections which might happen in the human body after the implant is introduced. In this study titanium samples treated by plasma electrolytic oxidation (PEO), including silver additives, and by PEO along with sol-gel dip-coating were examined using scanning electron microscopy (SEM), energy dispersive spectrometer (EDS), contact angle measuring system and cell culture testing. It was demonstrated that material, containing silver on its surface due to PEO coating process, might be considered as a good candidate for orthopedic applications and further investigations.

1 Introduction

Titanium (Ti) alloys are widely used in medical field, for example for hip and knee replacement or for dental purposes. The reasons for that are such titanium features as low density, high corrosion resistance, biocompatibility and good mechanical properties [1], [2]. However, specific surface modifications, such as plasma electrolytic oxidation (PEO), could be used to further improve the material properties. PEO is known as industrial surface treatment for metals to form ceramic-like porous coatings offering improved biocompatibility, enhanced wear and corrosion protection [2], [3]. H. Habazaki et al. reported in their experiments that the wear resistance was remarkably improved after applying of PEO for titanium alloy [4]. Corrosion examinations made by D. Krupa et al. showed that PEO modification improved the corrosion resistance of titanium [5].

Adhesion of microorganisms to orthopedic trauma hardware is a critical factor in implant-associated infections [6]. Previous studies showed that the presence of hydrophobic coatings for implants eliminated the clinical signs of infection in vivo in a large animal infection model [6]. Also silver is a recognized antimicrobial agent, which encourages the reduction of microbial contamination when the implant is integrated into living tissue [7]. Therefore silver additives or hydrophobic coatings might be investigated as possible candidates to make the material less dangerous with respect to surrounded tissue and to prevent an infection.

In this work titanium samples were modified by PEO or by PEO followed by sol-gel dip-coating and then analyzed regarding coating thickness, surface morphology, wettability, cytotoxicity and cells proliferation.

2 Material and Methods

2.1 Sample preparation

30 Titanium Grade 4 discs 2 mm thick and 18 mm in diameter were divided into three groups, named P01, P02 and P03. All samples were coated applying plasma electrolytic oxidation technique by using an alternating current power supply (Aixcon, Stolberg, Germany) and function generator (ISO-TECH 25MHz True Dual Channel Arbitrary Waveform Generator AFG-21225, ISO-Tech Kunststoff GmbH, Ahaus, Germany) at a frequency of 200 Hz. During the PEO process the Ti samples were used as anodes and a stainless steel plate as cathode. Hypered electrolyte with silver elements was used during PEO for the samples of group P03 and without silver for groups P01 and P02. The chemical composition of the electrolyte was: sodium calcium edetate, ammonium dihydrogen phosphate, phosphoric acid, ammonia solution and a silver containing component. Starting with 10.5 V_{pp} until the sparks appear, the samples were kept afterwards in the electrolyte at 1.5 V_{pp} for five minutes. The temperature was kept at 20°C using a cooling thermostat (LAUDA Alpha RA 8, LAUDA-Brinkmann).

After PEO all samples were rinsed with distilled water, followed by ethanol and finally dried at room temperature.

P02 samples were additionally treated by dip-coating into a fluoride-based sol-gel (purchased from WiBOTec-Surfaces GmbH & Co. KG, Vettweiss, Germany) stirred for five minutes with the magnetic stirrer before usage. The specimen was hanging in the beaker inside the glass chamber connected to a pump (Mini Laboratory Pump – VP 86, VWR International) which created a vacuum. The chamber was connected via a peristaltic tubing pump (REGLO Analog

ISM 795C, ISMATEC, Wertheim, Germany) with the sol-gel solution container. Once the pressure inside the chamber reached 150 mbar, sol-gel was slowly poured into the beaker by tubing pump with the flow rate of 38 ml/min. By the time the sample was completely covered by the solution, tubing and vacuum pumps were switched off and the sample was lifted up with withdrawal speed of 3 cm/min by means of a motorized unit programmed by programmable controller (MCC-2: Programmable Two-Axes Stepper Motor Controller, Phytron, Gröbenzell, Germany). Finally, the P02 samples were treated in the oven for 30 minutes with $110°C$.

2.2 Characterization methods

2.2.1 Coating thickness

The thickness of the samples was first examined with a handheld coating thickness measurement device (Isoscope FMP10, HELMUT FISCHER GmbH, Sindelfingen, Germany).

2.2.2 Surface morphology

The surface and cross-sectional chemical characterization of PEO- and sol-gel-treated samples were investigated using Scanning Electron Microscope (SEM, Philips XL30, Philips GmbH, Hamburg, Germany) equipped with an energy dispersive spectrometer (EDS, Octane Plus, EDAX, Weiterstadt, Germany) with the acceleration voltage of 20 kV. One sample from each group was examined by scanning its surface from the top and corresponding EDS spectra were obtained.

2.2.3 Wettability

The surface wettability was quantified by drop shape analysis. The contact angle measurements were performed for three samples per each group, using a contact angle measuring system (Vision GmbH). Distilled water droplets of approximately 3 μL volume were placed onto the samples surface using a manual microsyringe. Images of the horizontal view of the drops were captured by a camera (VRMC-16/M, VRmagic, Mannheim, Germany) and analyzed using a software - Image Access Standard (Imagic, Glattbrugg, Switzerland). The contact angle was measured 5 times for each sample and the average values were calculated.

2.2.4 Cytotoxicity and cells proliferation

Direct and indirect cell culture testing for cytotoxicity and proliferation of cells were carried out in the Research Laboratory of the Department of Plastic Surgery and Heavy Burns of BG University Clinics Bergmannsheil, Ruhr-Universität Bochum. Mouse fibroblast cell line L-929 was cultured in Minimum Essential Medium (MEM) supplemented with 10% fetal bovine serum, penicillin/streptomycin (100 U/μL each) and L-glutamine in a final concentration of 4 μM. Positive and negative controls were performed by growing the cells on plastic and

by cell cultivation on RM-A, a polyurethane film containing 0.1 % zinc diethyldithiocarbamate (ZDEC) (Hatano Research Institute, Japan), respectively. Cell proliferation test (XTT) was investigated via Cell Proliferation Kit II (XTT) (Roche Diagnostics) and Cytotoxicity test (LDH) was examined via LDH-Cytotoxicity Assay Kit II (BioVision). Both tests were done according to the manufacturer's instructions. Quantification of the results was performed by measuring the absorbance with a Elx808 Ultra Microplate Reader (Bio-Tek Instruments GmbH).

3 Results and Discussion

3.1 Thickness measurement

The thickness of the coating after treatment was between 10 and 15 μm for P01 and P02 specimens. Additional applying of the dip coating into the sol-gel for P02 samples didn't contribute to any measurable changes of the coating's thickness. In contrast, the coating thickness for P03 specimens was in the range of 7-9 μm.

3.2 Surface morphology

SEM images for all three sample types (Fig. 1) reveal that the surface of the samples contain a lot of pores, which is characteristic for PEO coating [2]. Darker regions were observed on the surface of P02 samples under SEM (Fig. 1b). EDS spectrum says that in that area the content of fluorine in one chosen spot (24,83% w) is higher than in the spot outside this area (0,56% w). This could therefore be a small inhomogeneity of the sol-gel coating.

EDS spectrum for P02 (Fig. 2) of the area shown in the Fig. 1b presents the components derived from the electrolyte (e.g. carbon, phosphorous, oxygen), sol-gel (fluorine) and the substrate (titanium). Fig. 3b shows the cross sections of P02 sample. EDS map showed that the fluorine is mostly contained inside PEO pores. This image was done on the sample's edge, while on the flat plane no similar situation was found, which might be because of surface destruction during grinding made before SEM analysis. In contrast, the pores of P01 (Fig. 3a) and P03 were not filled with any substance containing fluorine neither on the edges nor on the flat planes. As a result, it can be assumed that the sol-gel might fill the pores on the coating's surface. It is not visible on SEM images of the cross section if there is a sol-gel also outside the pores on P02. Hence, it is hard to estimate the thickness of the sol-gel coating.

3.3 Wettability

Water contact angle on the surface of the P01, P02 and P03 samples were obtained as $57° \pm 2.73°$, $119° \pm 1.55°$ and $44° \pm 2.49°$ respectively, indicating that PEO surface itself and PEO surface with silver additives exhibit hydrophilic behavior, while PEO surface treated with the sol-gel solution is hydrophobic.

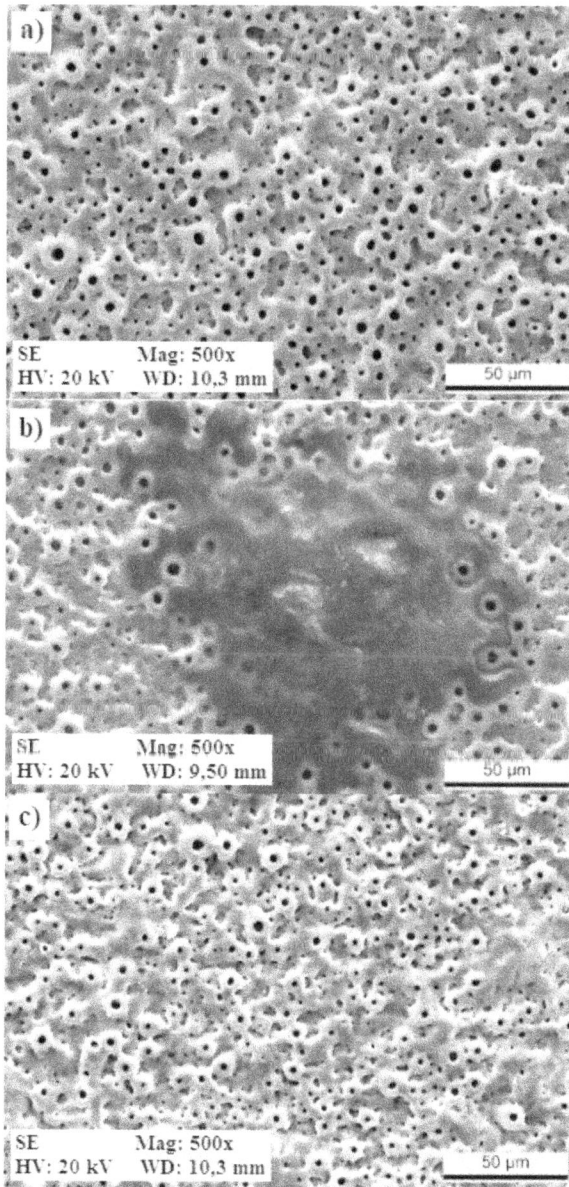

Figure 1: SEM image of top surface of: a) P01, b) P02 and c) P03.

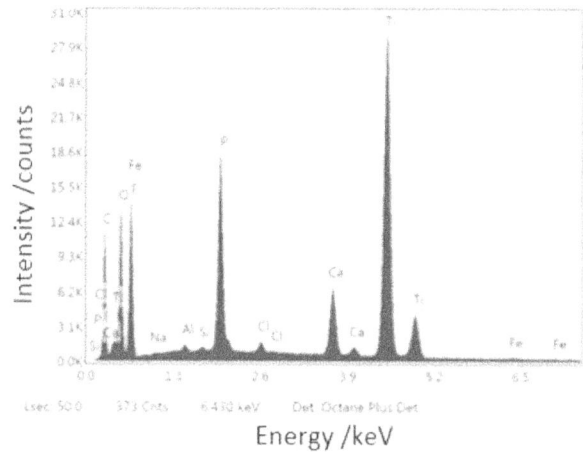

Figure 2: EDS spectrum for P02.

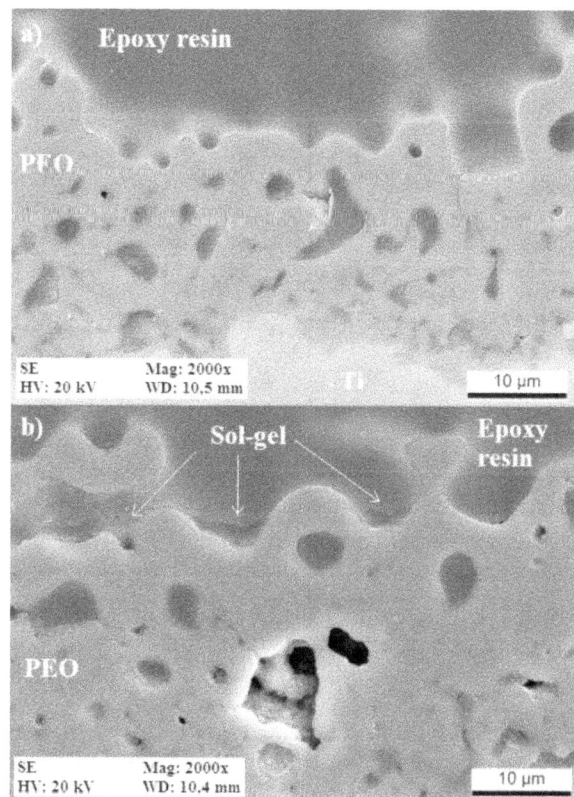

Figure 3: Cross section SEM Image of: a) P01 and b) P02.

3.4 Cytotoxicity and cells proliferation

Fig. 4 presents the results of the direct tests for cytotoxicity and cell proliferation. Both direct and indirect tests in terms of cytotoxicity showed that P03 samples were less toxic to the cells in comparison with positive control, whereas titanium itself and P01 samples were almost on the same level as the positive control. In contrast, P02 samples were 25% more cytotoxic than positive control in direct test and nearly as cytotoxic as positive control in indirect test.

Proliferation test estimated that the cells prefer to grow rather on P01 specimens than on P02 or P03. Indirect test showed high proliferation level for all specimens in question.

4 Conclusion

In this work PEO and dip-coating into sol-gel solution were applied as a surface treatment for titanium samples, which were afterwards investigated regarding the coating's composition, water repellency, cells proliferation and cytotoxicity.

Specimens with silver showed low rate of cell proliferation in direct cell culture testing. This might be an advantage for titanium bone plates, for example, during extraction because the surrounding tissue or tendons probably will not adhere to the plate's surface and will make the surgery easier and safer. On top of that, these samples re-

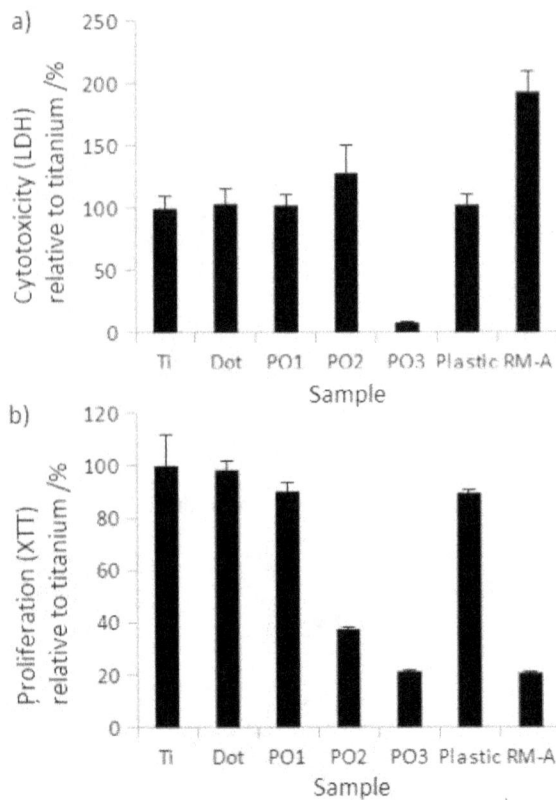

Figure 4: Direct tests of: a) Cytotoxicity and b) Proliferation.

vealed the smallest level of cytotoxicity in comparison with other specimens. Cytotoxic compounds are not favorable for implants, because they can influence on the living cells causing necrosis, apoptosis or inflammatory reactions of the surrounding tissue [8].

Sol-gel coatings appeared to be even more cytotoxic than titanium itself. But it showed hydrophobic behavior and moderate cell proliferation in comparison with titanium in direct testing.

Summarizing, PEO modified titanium with silver particles showed promising results as a material for orthopedic implants and further research in this area should be encouraged.

Acknowledgement

The work has been carried out at Meotec GmbH & Co. KG (Aachen, Germany) and supervised by Dr. C. Damiani, MSGT Lab, University of Applied Sciences Lübeck.

5   References

[1] M. Manjaiah, R. F. Laubscher. *Effect of anodizing on surface integrity of Grade 4 titanium for biomedical applications*. Surface and Coatings Technology, vol. 310, pp. 263–272, 2017.

[2] H. F. Nabavi, M. Aliofkhazraei, A. S. Rouhaghdam. *Electrical characteristics and discharge properties of hybrid plasma electrolytic oxidation on titanium*. Journal of Alloys and Compounds, vol. 728, pp. 464–475, 2017.

[3] X. Lu, M. Mohedano, C. Blawert, E. Matykina, R. Arrabal, K.U. Kainer, M.L. Zheludkevich. *Plasma electrolytic oxidation coatings with particle additions – A review*. Surface and Coatings Technology, vol. 307, part C, pp. 1165–1182, 2016.

[4] H. Habazaki, T. Onodera, K. Fushimi, H. Konno, K. Toyotake. *Spark anodizing of β-Ti alloy for wear-resistant coating*. Surface & Coatings Technology, vol. 201, issue 21, pp. 8730–8737, 2007.

[5] D. Krupa, J. Baszkiewicz, J. Zdunek, J. Smolik, Z. Słomka, J. W. Sobczak. *Characterization of the surface layers formed on titanium by plasma electrolytic oxidation*. Surface & Coatings Technology, vol. 205, issue 6, pp. 1743–1749, 2010.

[6] T. P. Schaer, S. Stewart, B. B. Hsu, A. M. Klibanov. *Hydrophobic polycationic coatings that inhibit biofilms and support bone healing during infection*. Biomaterials, vol. 33, issue 5, pp. 1245–1254, 2012.

[7] E.I. Zamulaeva, A.N. Sheveyko, A.Y. Potanin, I.Y. Zhitnyak, N.A. Gloushankova, I.V. Sukhorukova, et al. *Comparative investigation of antibacterial yet biocompatible Ag-doped multicomponent coatings obtained by pulsed electrospark deposition and its combination with ion implantation*. Ceramics International, vol. 44, issue 4, pp. 3765–3774, 2018.

[8] D. Granchi, E. Cenni, G. Ciapetti, L. Savarino, S. Stea, S. Gamberini, et al. *Cell death induced by metal ions: necrosis or apoptosis?* Journal of materials science: materials in medicine, vol. 9, pp. 31–37, 1998.

Development of Subject Specific Musculoskeletal Model of the Lower Extremity after Total Knee Replacement

U. Malik[1,2], M. Kebbach[2,*], R. Wendlandt[3], D. Klüß [2], R. Bader [2]

[1] Biomedical Engineering, University of Applied Sciences Lübeck, usman.malik@stud.fh-luebeck.de

[2] Biomechanics and Implant-Technology Research Laboratory, Department of Orthopaedics, Rostock University Medical Center, Germany, * maeruan.kebbach@med.uni-rostock.de

[3] Biomechanics Laboratory, University Medical Center Schleswig-Holstein, Germany robert.wendlandt@uk-sh.de

Abstract

Multibody simulation can be used for musculoskeletal modeling to compute joint angles and biophysical quantities, e.g. muscle forces, during dynamic activities of daily living. In this present study, a generic musculoskeletal model of the lower limb was scaled to the subject-specific morphology of a male patient treated with an instrumented total knee replacement (TKR) in publicly available dataset. To obtain joint angles of the lower right extremity an inverse kinematic analysis of a squat was implemented. Maximum values for the flexion angle of the hip, knee and ankle joint amounted to $60.57°$, $92.1°$ and $61.3°$ respectively. Subsequently, an inverse dynamic model was developed in order to compute muscle forces during the squat movement. This study provides ground work for further full lower limb musculoskeletal modeling to incorporate a more detailed knee model.

1 Introduction

Knee contact forces essentially occur during carrying out activities of daily living. Therefore, knowledge of knee biomechanics plays a vital role for pre-clinical investigation of new materials and designs, understanding implant failure mechanisms and improvement of functional outcomes after total knee arthroplasty (TKA). Electronic force sensors have been utilized to measure joint contact forces of the knee in vivo [1],[2]. Measurement procedures resulted in the release of extensive databases of joint forces which were recorded for telemetric endoprosthetic patients during in vivo loading [2]. Based on this methodology, the OrthoLoad Project by Bergmann et al. [3] and the "grand challenge competition predicting in vivo knee loads" by Fregly et al. [4], were derived. These datasets are recognized widely as gold standards for validation of measurements procedures of in vivo knee contact. These initiatives have stimulated the musculoskeletal modeling community to improve the clinical utility of their computational models through extensive validation.

Apart from joint loading patterns, these procedures revealed individual variations in terms of difference in anatomy and joint kinematics [4]. Furthermore, information provided through such datasets is limited, available for selected number of patients and for particular type of endoprostheses under controlled conditions. Force generated through muscles or transmitted through ligaments are not included. For conclusive information on the conditions of individual patients as well as biophysical quantities (muscle and ligament forces) computational musculoskeletal models were developed in the past [5]. For example, Arnold et al. [6] have proposed a dynamic full lower-body musculoskeletal model (Fig. 1).

Validation of musculoskeletal modeling is required before use in clinical practices [4]. In this context, Fregly et al. [4] initiated the grand challenge competition for in vivo prediction of knee loads for computational model's evaluation and validation. The data set includes motion capture data, ground reaction forces (GRFs), electromyography (EMG) signals, knee contact force data and geometric models for knee implant, bone geometrical models as well as pre- and post-operative computed tomography (CT) scan data of patients who received instrumented total knee implants.

The objectives of this study are: a) to develop a generic inverse kinematic musculoskeletal model of the lower limb based on motion capture data, which are capable of estimating joint angles; and b) to perform an inverse dynamic analysis to estimate muscles forces in vivo by using joint angles as input from inverse kinematic analysis and GRFs.

2 Material and Methods

The motion of a patient with specific implant were taken into account for the development of patient specific lower extremity musculoskeletal model.

2.1 Experimental Data Collection

The data used in this study were part of the fourth grand challenge competition to predict *in vivo* knee loads [4]. The data were obtained from a male subject (age: 88yr, height: 168 cm, and weight 66.7 kg) who received an instrumented total knee replacement (generation 1 tray design (eKnee)) on the right knee [2].

2.2 General Musculoskeletal Modeling

AnyBody Modelling system (AMS) version 6.0.4 (Any-Body Technology, Aalborg, Denmark) was used for development of the subject-specific musculoskeletal model. The musculoskeletal model was extracted from Anybody Managed Model repository (AMMR V 1.6.3) based on the database Twente Lower Extremity Model (TLEM 1.1). The lower limb model was actuated by 38 muscle units [7] and each muscle is divided in different muscle line of action based on muscle morphology, with approximately 159 muscle elements for each lower extremity [7]. Muscles were modelled as three element Hill-type model [8] with default properties in AnyBody software.

2.3 Subject Specific Modeling

Patient weight, length and marker trajectories for a squat trial are taken into account for developing a subject specific musculoskeletal model. Marker trajectories from the data set were overwrite using MOKKA, 3D Motion and Kinetic Analyzer, version 0.6.2 (Laboratory of Movement Analysis and Measurement, Lausanne, Switzerland), to edit and visualize motion capture data [9]. Marker positions were

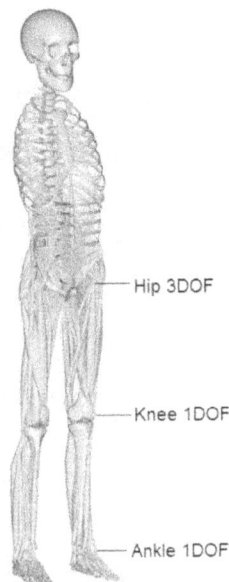

Figure 1: The human musculoskeletal model from Anybody Managed Repository was the basic platform used to develop subject specific models in present study . The body model has 15 segments and 5 degrees of freedom for each lower extremity.

applied and implemented to a custom script for motion capture model provided by Anybody. Length-Mass scaling law was used to optimize the local marker coordinates, model parameters and segment lengths according to the patient data. All bone segments were scaled linearly to fit the generic model data to the available patient specific dataset. Linear segment scaling law [10] is used to scale the length of the lower body segments, the pelvic bone width, and the foot lengths. This optimization algorithm minimized the difference between model markers and experimentally recorded marker positions during one frame of a squat movement by using the method of Andersen et al. [11].

3 Results and Discussion

3D musculoskeletal modeling is opening new possibilities, in terms of movement simulation and understanding of the interaction between joints, ligaments, and muscles. Moreover, the use of instrumented prostheses now offers a unique chance to validate the models in terms of forces estimation during a movement.
The goal of this study was to present musculoskeletal modeling framework based on subject-specific motion capture as an input to inverse kinematics-based method that concurrently predicts knee joint angles. The inverse dynamic based method permitted the estimation of quadriceps muscles forces, while still employing motion capture and force plate data as GRFs.
The estimated quadriceps muscles forces during squatting show good agreement with literature data [12],[13] and supports the validity of our model.

Joint Angles: An inverse kinematic model was developed to compute joint angles of the lower right extremity by using marker trajectories of a squat motion as input. Fig.2 presents the results of the inverse kinematic analysis of the right lower extremity during a squat movement. Each squat trial took approximately 2.7 to 3.1 seconds.
For one squat trial, there were no crucial changes in hip abduction angle and external rotation. The hip flexion showed a maximum value of 60.6° and a starting angle of -12.02°. At the knee joint, the maximum flexion angle amounted to 92.09°. Ankle flexion showed a value between 20.1° and 61.3°.

Muscle Forces: Subsequently, an inverse dynamic analysis was perfomed to compute muscle forces by using the obtained joint angles and available GRFs as input. Quadriceps muscles forces during squat trial were obtained using an inverse dynamics based analysis in the AnyBody modelling system (Fig. 3).
Inverse dynamic analysis showed a large range of the M. vastus lateralis force with maximum value of 1130 N. Maximum muscle forces for M. vastus intermedius , M. rectus femoris and M. vastus medialis were 324 N, 283 N and 532 N respectively.

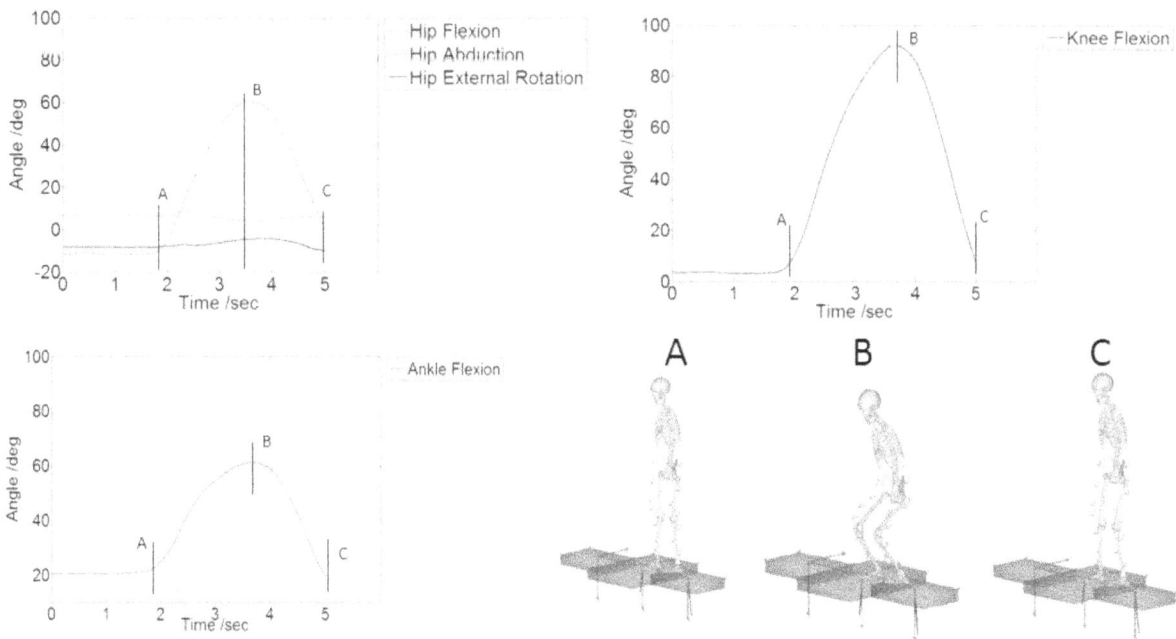

Figure 2: Results of inverse kinematic analysis during a subject specific squat motion. Marker trajectories are input to an inverse kinematic analysis to obtain joint angles for hip, knee and ankle. Joint angles plotted relative to one squat trial at three different positions (A) initial position, (B) middle position, and (C) end position.

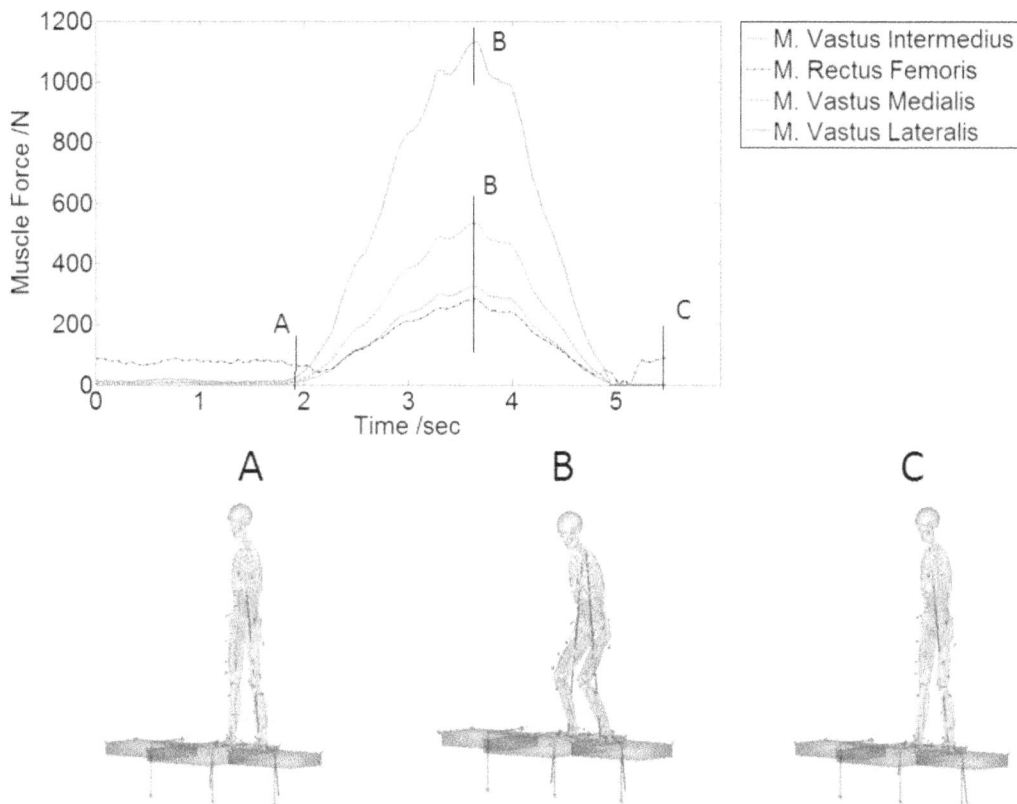

Figure 3: Inverse dynamic analysis performed with subject specific squat motion. Ground reaction forces and computed joint angles as input. Calculated forces for M. rectus femoris, M. vastus intermedius, M. vastus medialis and M. vastus lateralis plotted relative to one squat trial at three different positions (A) initial position, (B) middle position, and (C) end position.

This study has several limitations that are worth discussing. First, we represent the knee joint as hinge joint and the knee contact forces were underestimate in our model. However, the use of 12DOF knee model with deformable contact for the accurate estimation of knee joint contact forces during squat motion will be developed in further studies.

Second, for obtaining reliable subject-specific model prediction, better scaling of musculoskeletal model to the subject-specific anatomy will be carried out in further studies using new advanced scaling morphing techniques.

Third, ligaments were not incorporated in our present lower extremity model. Further research will address more detail incorporation of ligamentous restraints in knee model. Fourth, inverse dynamic-based analysis did not include muscle synergy models in the prediction process for the forces in antagonistic muscles.

4   Conclusion

In conclusion, a musculoskeletal model is utilized for patient specific architecture of the lower limb to predict joint angles using an inverse kinematic model. Furthermore, M. quadriceps forces were predicted by performing inverse dynamic analysis. This study provides ground work for further musculoskeletal modeling of the lower limb incorporating more detailed knee models. Integration of subject specific bone models and total knee implants as well as a knee contact model and ligament structures will be done in further studies.

Acknowledgment

This study was carried out at and supported by the Biomechanics and Implant Technology Research Laboratory, FORBIOMIT, Rostock University Medical Center. We wish to thank B. J. Fregly, Ph.D., D. D. D'Lima, MD, Ph.D. and colleagues as the organizers of the fourth "Grand Challenge Competition to Predict in vivo Knee Loads" for making such invaluable experimental data publicly available.

5   References

[1] Kutzner, I., Heinlein, B., Graichen, F., Bender, A., Rohlmann, A., Halder, A., and Bergmann, G., "*Loading of the Knee Joint During Activities of Daily Living Measured In Vivo in Five Subjects*," J. Biomech., 43(11), pp. 2164–2173, 2010.

[2] D'Lima, D. D., Townsend, C. P., Arms, S. W., Morris, B. A., and Colwell, C. W., "*An Implantable Telemetry Device to Measure Intra-Articular Tibial Forces*," J. Biomech., 38(2), pp. 299–304, 2005.

[3] Bergmann, G., "Orthoload.com," Charite Univ. Berlin, http:// www.orthoload.com, 2008

[4] Fregly, B. J., Besier, T. F., Lloyd, D. G., Delp, S. L., Banks, S. A., Pandy, M. G., and D'Lima, D. D., "*Grand Challenge Competition to Predict In Vivo Knee Loads*," J. Orthop. Res., 30(4), pp. 503–513, 2012

[5] Erdemir, A.D , McLean, S., Herzog, W., and van den Bogert, A. J., "*Model Based Estimation of Muscle Forces Exerted During Movements*," Clin.Biomech. (Bristol, Avon), 22(2), pp. 131–154, 2007.

[6] Arnold, E. M., Ward, S. R., Lieber, R. L., and Delp, S. L., "*A Model of the Lower Limb for Analysis of Human Movement*," Ann. Biomed. Eng., 38(2), pp. 269–279, 2010

[7] Horsman, K., Koopman, H., Helm FC, Prosé, LP., ,"*Morphological muscle and joint parameters for musculoskeletal modelling of the lower extremity*," Clin Biomech (Bristol, Avon). 22(2):239-47, 2007.

[8] Zajac, F. E., 1989, "*Muscle and Tendon: Properties, Models, Scaling, and Application to Biomechanics and Motor Control*," Crit. Rev. Biomed. Eng., 17(4), pp.359–411.

[9] Armand S., "Biomechanical ToolKit: Open-source framework to visualize and process biomechanical data," Comput Methods Programs Biomed. Apr; 114(1):80-7, 2014.

[10] Rasmussen, J., Zee, M. de, Damsgaard, M., Christensen, S. T., Marek, C., and Siebertz, K., "*A General Method for Scaling Musculo-Skeletal Models*," 2005 International Symposium on Computer Simulation in Biomechanics, Cleveland, OH, 2005.

[11] Andersen, M. S., Damsgaard, M., MacWilliams, B., and Rasmussen, J., "*A Computationally Efficient Optimisation-Based Method for Parameter Identification of Kinematically Determinate and Over-Determinate Biomechanical Systems*,"Comput. Methods Biomech. Biomed. Eng., 13(2), pp. 171–183, 2010.

[12] Nagura T., Matsumoto H., Kiriyama Y., Chaudhari A., and Andriacchi P.T., "*Tibiofemoral Joint Contact Force in Deep Knee Flexion and Its Consideration in Knee Osteoarthritis and Joint Replacement*," Journal of Applied Biomechanics, 22:305-313, 2006

[13] Innocenti B., Pianigiani S., Labey L., Victor J., and Bellemans J.," *Contact forces in several TKA designs during squatting: A numerical sensitivity analysis*,"Journal of Biomechanics 44, pp. 1573–1581, 2011.

Magnetic fields generated by planar coils with potential usage towards medical applications

A. Medrea [1], P. Klemm [2], T. Senkbeil [2] and P. Cörlin [2]

[1] Biomedical Engineering, University of Applied Sciences Lübeck, elena.alexandra.medrea@stud.fh-luebeck.de

[2] Trafag GmbH sensors & controls, {philippe.klemm, tobias.senkbeil, philipp.coerlin}@trafag.de

Abstract

Magnetic fields have played a crucial role in medicine for a long time. They have been utilized for various applications. In this research study planar coils were investigated which can be used to produce small magnetic fields in the range of a few mT. The magnetic fields of planar coils with different numbers of windings and different arrangement of the generator coil were directly measured with a Gaussmeter. In addition to that the sensitivity of two types of planar coils with different generator windings was investigated by using a torsion torque bench. It has been concluded that these planar coils can be effectively used for the purpose of creating magnetic fields for medical applications such as fracture and bone-healing.

1 Introduction

In 1820, Danish scientist Hans Christian Oersted observed that when electric current was passed through a coil of wire, the needle of a compass lying nearby deflected [1]. This happens because when electric current is passing through a conducting coil, a magnetic field is produced around the coil, and the phenomenon is called electromagnetism [1]. Through his observation Oersted discovered that magnetism and electricity are related phenomena [1]. The region of space around a magnetic object within which the effects of magnetic forces can be detected is called magnetic field. The Standard International (SI) unit to measure the magnetic field is Tesla (T). To develop a quantitative understanding of the strength of magnetic field, the strongest pulsed non-destructive magnetic field produced is in a laboratory at National High Magnetic Field Laboratory's Los Alamos, New Mexico, USA and its magnitude is 100 T. This is of similar magnitude as the magnetic field strength of the average white dwarf star. The earth itself behaves like a big magnet and has its own natural magnetic field also known as the geomagnetic field that ranges from 25 μT to 65 μT. On the other hand, the strength of a typical refrigerator magnet is around 5 mT.

Nowadays, magnetic fields have found tremendous usages in the medical field. A widely known application is the Magnetic Resonance Imaging, operating in the range of 1.5 T up to 7 T. Magnetic Resonance Imaging is a non-invasive imaging technique revolutionizing the imaging technique in the past three decades [2]. Another medical application making use of magnetic fields is Transcranial Magnetic Stimulation. Transcranial Magnetic Stimulation is also a non-invasive technique used to stimulate the cortex of the human brain in conscious individuals using magnetic fields in the range of a few T. Several studies have shown that Transcranial Magnetic Stimulation might have therapeutic effects such as having beneficial effects on major depression or serving as a therapeutic application in Parkinson's disease [3, 4].

There have been some studies that have shown that certain magnetic fields might have a positive influence on the skeletal system [5]. Static magnetic fields operating in the range of thousandths of T were used for various experiments. In one study, the healing process of a fracture of a rabbit's radius was monitored for a duration of 4 weeks under an exposure in the range of 22 mT to 26 mT magnetic fields. Afterwards it was shown that thicker and stronger trabecular bone was found around the fractured area [6]. Moreover, it has been shown that long term exposure to 1.5 mT magnetic fields may be useful in treating osteoporosis [7].

In this paper, a set of planar coils is introduced, whose objective is to generate and sense magnetic fields of up to a few mT. Because of their small size in the range of few millimeters, they could be implanted and be used as potential tool of physical therapy promoting fracture and bone-healing. The study described in the present paper has been divided into two parts. For the first part, the magnetic fields generated by planar coils with different number of windings and different winding arrangements were directly measured with a Gaussmeter. During the second part of the study, the sensitivity of planar coils with two different numbers of generator windings was examined by using a torsion testing bench.

2 Materials and Methods

The experimental set-up for sensitivity measurements consisted of a torsion testing bench *Inspekt T-500H* manufactured by *Hegewald & Peschke Meß- und Prüftechnik GmbH*. A stainless steel shaft made of *X45CrMo4* with a diameter of 25 mm and a length of 220 mm was mounted on the testing bench. A bidirectional torque could be applied. Applying torsion stress to this shaft induces a direction dependent change of the magnetic properties within the shaft, which may be sensed with a thoroughly designed coil setup. The coil design is illustrated in Fig. 1(a). The printed circuit board (PCB) is comprised of four sensing coils positioned on the outer parts of and a generator coil positioned in the middle of the coil package. When a voltage U is applied to the generator coil, an electric current $I = \frac{U}{R}$ flows through the coil, where R is the resistance of the coil. This flow of current induces a magnetic field in the vicinity of the generator coil. The strength of the magnetic field is measured in terms of the magnetic flux density. Magnetic flux density is defined as the amount of magnetic flux in an area taken perpendicular to the direction of magnetic flux, where magnetic flux represents the number of magnetic field lines passing through a surface such as a loop of wire. The magnitude of the magnetic field generated by a steady current I is given by the Biot-Savart law as given below:

$$B = \mu_o n I$$

where, B = magnetic flux, μ_o = permeability (inductance per unit length) of the medium where we are measuring the fields , n= number of windings of the coil. In this case, the permeability of the medium air is 1. The unit of the magnetic flux density is T.

The magnetic field generated by the generator coil is guided by a ferrite core into the shaft material as illustrated in Fig. 1(b). The setup comprising of the ferrite core with 5 small legs and the associated coils is mounted on the shaft as shown in Fig. 1(c). While applying torque to the shaft, compressive and tensile stresses arise which are orthogonal directions of stress as illustrated in the Fig. 1(c). These stressed states cause the magnetic properties of the shaft to change, and this change is manifested in the change of the distribution of the magnetic flux. The changes that occur in the magnetic properties of the shaft are measured as changes of the induced voltages of the sensing coils.

In addition, a Gaussmeter *FH 54 Gauss-/ Teslameter* manufactured by *Magnet-Physik Dr.Steingroever GmbH* was used to directly measure the magnetic fields generated by the planar coils. The effects of the positioning of the generator area was also studied in this experiment. The generator area was positioned either on top or at the bottom as illustrated in the Fig. 1(d).

2.1 Planar coils design

The coils used for the experiment are made of multiple sheet-like layers stacked on top of each other where copper wiring was embedded in an epoxy matrix. The sensing coils and the generator coils have different number of windings. In this article, the number of windings of the different coils were normalized to one specific planar coil

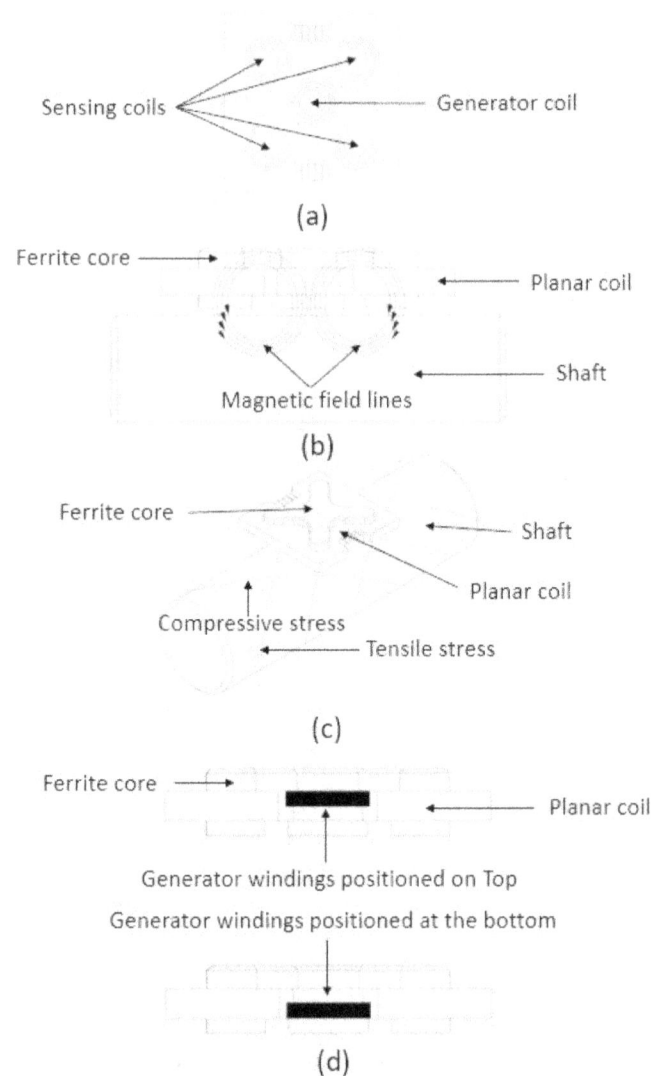

Figure 1:   (a) Coil package comprising four sensing coils and one generator coil. (b) Ferrite core placed on coil. Coil mounted on shaft. Magnetic fields generated by the generator coil guided into the shaft material. (c) Ferrite core placed on coil. Coil mounted on a shaft. Torque applied to the shaft. Two orthogonal stresses arise - compressive and tensile stresses. (d) Generator windings positioned on top of the printed circuit board (PCB). Generator windings positioned at the bottom of the PCB.

setup. The normalized values are shown in Table 1. The windings of the generator coil can be placed either in every layer (EL), at every second layer (ESL), on top (T) or at the bottom (B) as illustrated in the Table1.

3   Results, Discussion and Conclusion

3.1   Results

The generator coils were driven by an alternating current voltage where the maximum voltage amplitude and

Table 1: Number of windings and distribution of windings within the coil

Coil Id	Normalized sensor windings	Normalized generator windings	Generator windings distribution
a1	standard	standard	EL
a2	standard	+40%	EL
a3	standard	-20%	EL
a4	standard	-40%	EL
b1	standard	-50%	ESL
b2	standard	-30%	ESL
c1	standard	-50%	T/B
c2	standard	-30%	T/B
d1	+14%	standard	EL

frequency were kept constant. In the first part of the experiment, the effect of varying the number of windings of the planar coil on the magnetic field was studied. The magnetic flux density was directly measured with the Gaussmeter. The magnetic field strengths were recorded with a tolerance limit as illustrated by the vertical lines of the Fig. 2(a). It can be observed from the Fig. 2(a) that as the number of windings for planar coil systems a1 to a4 increased, the magnetic fields decreased and vice versa. The reason behind that was because the voltage was kept constant and as the number of windings increased the length increased. Because of an increase in the length of the wire the resistance increased. According to the Ohm's Law and at constant voltage, an increase in resistance leads to a reduction of current. According to the Biot-Savart law, if there is a current drop the magnetic field will decrease. Generally, a smoother transition between magnetic field strengths of coils with different windings would be expected, a variation in production accuracy may have led to the observed sharp change between planar coil system a1 and a3.

Furthermore, the effects of the positioning of the generator area was also studied in this experiment. The results of this part of experiment has been shown in the graph of Fig. 2(b). The dark grey vertical bars illustrate the magnetic fields when planar coils with standard-50% windings were used and the light grey vertical bars illustrate the magnetic fields when planar coils with standard-30% windings of series b and c as named in Table 1 were used. It can be observed from Fig. 2(b) that the intensity of the magnetic fields when the generator area was positioned on the top the magnetic field was the smallest and when the generator area was positioned at the bottom the magnetic field intensity was the highest. On the other hand, positioning the generator area at every second layer did yield a magnetic field in the middle of the aforementioned two geometries. Furthermore, it was observed from Fig. 2(b) that the coils designed with only standard-50% windings (dark grey vertical bars) generated higher intensity of magnetic fields as compared to the coils designed with standard-30% windings (light grey

vertical bars). The reason behind that was as the number of windings increased the length increased and subsequently resistance increased at constant voltage.

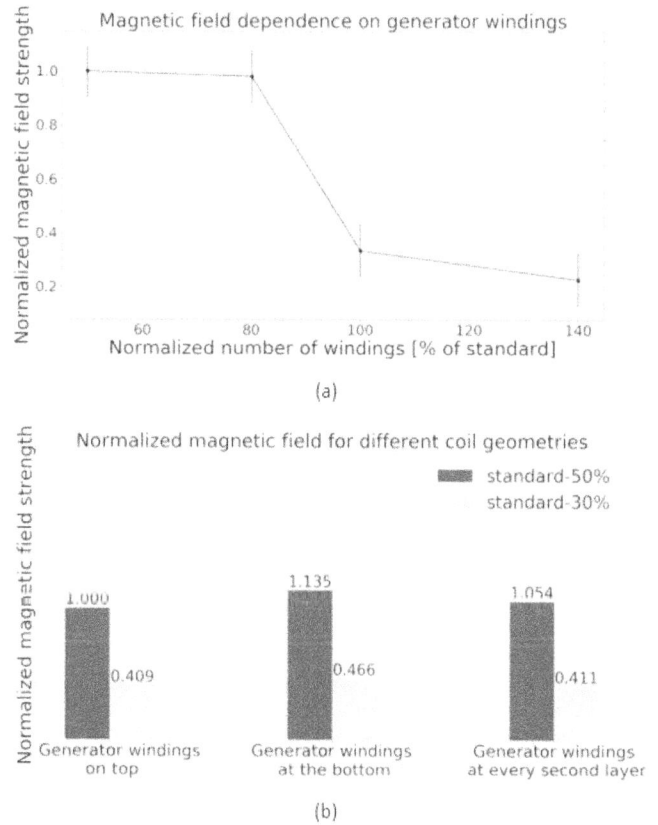

(a)

(b)

Figure 2: (a) Magnetic field strength as function of generator windings for planar coil systems a1 to a4. (b) Magnetic field intensity for different planar coil geometries.

For sensitivity measurements planar coil with a standard number of generator windings were used. Two variants of planar coils with different sensing windings were investigated, named a1 and d1 in Table 1. In Fig. 3 the sensor output normalized to the output of the sensor with planar coil system a1 is shown as a function of torque applied to a steel shaft, as described in Sec. 2. When no torque is applied, both planar coil systems show an output of zero % Full Scale (%FS). Increasing the applied torque to maximum load leads naturally to a sensor output of 100 %FS for planar coil system a1, which has the lower amount of sensing windings. For the same torque the planar coil system d1 with higher amount of windings shows an increased sensor output of 124 %FS. The sensor output is a measure of the voltage induced in the sensing coils. As the induced voltage of a coil is given by:

$$U = -N \frac{d}{dt} \int_A \vec{B} d\vec{A} \approx N \dot{B} A$$

with N the number of windings, $A = \pi r^2$ the inner area of the coil and $\dot{B}$ the time derivative of the magnetic field, the general observed behavior is expected [8]. However, the quotient of induced voltages $U_1/U_2 = 1.36$ is higher than the quotient derived from the number of windings $N_1/N_2 = 1.15$. We argue that this ratio is due to slight distance variations of shaft to planar coil system due to an imprecise mounting system.

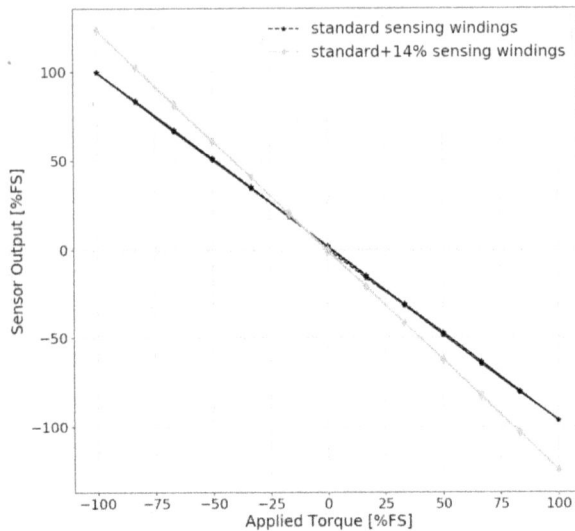

Figure 3: Coil sensitivity as function of sensing winding. The induced voltage in the sensing coils is shown as function of torque applied to a steel shaft mounted in a distance of 1mm to the planar coil system. The voltage induced in the sensing coils with a lower amount of windings (black) due to the applied torque is lower than the voltage induced in the coils with a higher amount of windings (grey).

4 Discussion and Conclusion

Numerous studies and applications have demonstrated successful implementation of magnetic field in medicine. Low intensity magnetic fields have found tremendous use in bone healing and regeneration process. In this study, we demonstrated and described an experimental process of using planar coils to generate low intensity magnetic fields. The coils used in this experiment are small, occupy a small area and are able to generate low magnetic fields in the range of a few mT. These properties could be exploited in the medical field for improving the process of fracture healing and bone regeneration. Moreover, they could be adopted into a low-cost, minimally invasive medical tools for preventing osteoporosis and improving bone healing process. They can either be applied externally in a non-invasive way or implanted internally with minimal invasion since they occupy a small area. It would be advantageous to administer them internally for subjects suffering from chronic pain due to osteoporosis that require long term treatment.

Acknowledgement

This work has been carried out at Trafag GmbH sensors and controls, Unterensingen, Germany and supervised by Prof. Dr. M. Ryschka, University of Applied Sciences Lübeck.

5 References

[1] N. Kipnis. Chance in science: The discovery of electromagnetism by H.C. Oersted. Science and Education, 14(1), 1-28. https://doi.org/10.1007/s11191-004-3286-0, 2005.

[2] A. Kangarlu. Physics of High-Field Magnetic Resonance Imaging and Applications to Brain Tumor Imaging. Handbook of Neuro-Oncology Neuroimaging (Second Edition). Elsevier. https://doi.org/10.1016/B978-0-12-800945-1.00021-5, 2016.

[3] A. Pascual-Leone, B. Rubio, F. Pallardó, & M. D. Catalá. Rapid-rate transcranial magnetic stimulation of left dorsolateral prefrontal cortex in drug-resistant depression. Lancet (London, England), 348(9022), 233-7. Retrieved from http://www.ncbi.nlm.nih.gov/pubmed/, 1996.

[4] H. R. Siebner, C. Mentschel, C. Auer, & B. Conrad. Repetitive transcranial magnetic stimulation has a beneficial effect on bradykinesia in Parkinson's disease. NeuroReport, 10(3), 1999.

[5] Y. Wang, & Q. Qin. Computer Methods in Biomechanics and Biomedical Engineering A theoretical study of bone remodelling under PEMF at cellular level, 5842(January). https://doi.org/10.1080/10255842.2011.565752, 2011.

[6] G. K. Bruce, C. R. Howlen, R. L. Huckstep. Effect of a static magnetic field on fracture healing in a rabbit radius: preliminary results. Clin Orthop Relat Res.;222:300-306, 1987.

[7] V. Akpolat, M. S. Celik, Y. Celik, N. Akdeniz, & M. S. Ozerdem. Treatment of osteoporosis by long-term magnetic field with extremely low frequency in rats, 25(August), 524-529. https://doi.org/10.1080/09513590902972075, 2009.

[8] J. D. Jackson, C. Witte, M. Diestelhorst and K. Müller. Klassische Elektrodynamik. Berlin, Boston: De Gruyter. Retrieved 23 Jan. 2018, from https://www.degruyter.com/view/product/211200, 2013.

Transferring a Deep Cityscape Synthesis Approach to the Medical Domain

D. Weller [1], L. Hansen [2], M. Blendowski [2], and M. Heinrich [2]

[1] Medizinische Ingenieurwissenschaft, Universität zu Lübeck, david.weller@student.uni-luebeck.de

[2] Institute of Medical Informatics, Universität zu Lübeck, {Hansen, Blendowski, Heinrich}@imi.uni-luebeck.de

Abstract

Supporting medical professionals in the evaluation of medical image data is a main emphasis of today's medical image processing [1]. Recently, neural networks have shown to be very successful in computer assisted classification. Yet, the foundation for a neural networks capability to analyse images, is an extensive training with a suitable amount of data. In terms of medical imaging, legal requirements restrain the amount of accessible training data. A solution for this issue could be the usage of synthetic generated images, to increase the volume of possible training data. In 2017 Q. Chen and V. Koltun presented a new approach for the artificial synthesis of photorealistic images. We use this concept on a collection of CT images and show that generating results with a promising quality is possible, but a more extensive training and further preprocessing of the data has to be done.

1 Introduction

An adequate amount of training data is crucial for a neural networks' capability of learning a complex task like content based image classification and recognition [2]. This dependency on suitable training data can be a restriction for successful training and therefore limit the performance of a network. Using not only real world, but also synthetic training data is a promising approach for tackling this issue [3]. In the field of captcha recognition the training of a network, exclusively using synthetic data, has already succeeded in state-of-the-art performance [4]. Realistic features are the fundamental requirement for such synthetic data. Generating artificial but realistic data, however is a challenging task. The following example illustrates this problem: Generating photorealistic images is feasible for skilled painters, who have good knowledge of the object they want to draw. Beside rendering an exact copy of an object they have studied, artists can also draw diverse instances of such an object that are all equally realistic. Consider an artist who specialises on painting buildings. It would be a simple task for him, to study one specific building and then draw a number of variations of it, by changing size, position and style of some key features like windows and doors. As long as these changes stay within a reasonable limit, all of these paintings could be considered as realistic. In contrast, a computer can store and reproduce the image of one building exactly, but the rendering of varying, photorealistic instances of this building would be a much more complex task that leads to central problems of computer graphics and artificial intelligence. In 2017 Q. Chen and V. Koltun presented a new approach for the photorealistic image synthesis, using a single feed forward neural network [5]. Based on much more complex, feature based loss function, their model is capable of synthesising a diverse collection of images, for one given input. The network of the original publication was trained on a cityscape dataset of urban scenes and their semantic label images. Motivated by these results, we transferred and trained this network architecture on a CT dataset. Our motivation is to synthesise a diverse collection of images, from one semantic label map. This feature might be useful, for artificially increasing the amount of training data that is required for neural network based image classification.

2 Material and Methods

The synthesis of an image is built up from a semantic label map, that is drawn from a segmentation image S and serves as the input for the network. Such a label map has the form of a third order tensor $\mathcal{L} \in \{0,1\}^{i,j,k}$, where i,j are the pixel resolution and k matches the number of segmented classes. A one-hot-vector of length k assigns a class to each pixel, so that the sum over all k elements with the coordinates i,j always equals one: $\sum_k \mathcal{L}_{i,j}(k) = 1$. For this work, the semantic label maps are based on the segmentation of a CT data-set, that includes 13 different classes. The original segmentation of our training set contains 11 internal organs and a void class for all pixels that are not part of these organs. We add another class for pixels, that show tissue or bones, but are not part of the 11 organ classes. Otsu's threshold based segmentation was used to add the additional background tissue and bone class [6]. The network is trained to synthesise grayscale images, corresponding to a given label map $\mathcal{L}$. Chen and Koltun defined three characteristics such a network architecture must fulfil:

- *Global Coordination:* Global consistency must be ensured for a realistic synthesised object. For our work this means a consistent level of vascularisation in a lung, without abrupt changes. In our model, the synthesis begins on a very low resolution of 4×4 pixel and is then doubled with each section of the network, the label map is passed through. Therefore widespread structures can be initialised at lower resolution levels and progressively refined with each module.

- *Memory:* For a convincing appearance, the generated images need to have high resolution. Especially with respect to a realistic synthesis of medical image data, the resolution must be fine enough, to feature the specific texture of various organs. With our model, the resolution of the images is dependent on the depth of the network. The network is designed as a cascade of modules, each one using the output of the last module as input and increasing the resolution by one octave.

- *High resolution:* High resolution images occupy millions of bits of memory, even if they are compressed with state of the art techniques. Aiming for a good reproduction of learned examples (memorisation) and an acceptable quality of results for new inputs leads to a large-scaled network architecture. Our network contains $140M$ trainable parameters, due to this precondition.

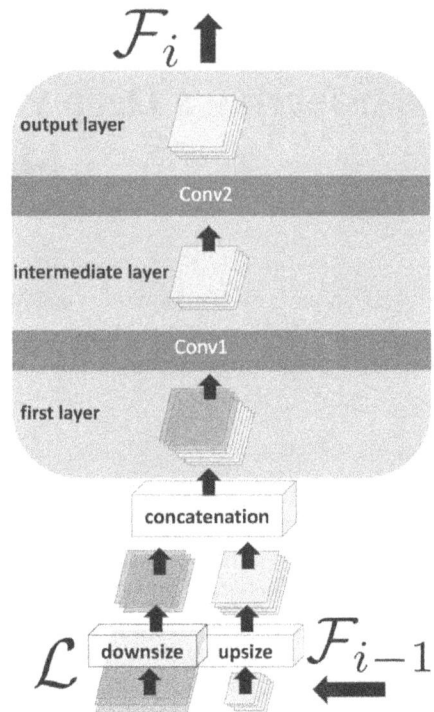

Figure 1: Basic concept of a single module m_i, used in the CSR cascade. A concatenation of a label map $\mathcal{L}$ and a feature map $\mathcal{F}_{i-1}$ serves as input for a module m_{i-1}. Within this module a new feature tensor $\mathcal{F}_i$ is generated as the output of the module. The same preprocessing, involving an upsampling and a concatenation, is applied prior the consecutive module m_i.

2.1 Architecture of the network

The network is built as a cascade of refining modules m_i, each one receiving the output of a module m_{i-1} as input and refining the resolution by one octave. Therefore we denote this architecture as CSR (Cascaded Refinement Modules). A first module m_0 receives the semantic label map $\mathcal{L}$, downsampled to a resolution of 4×4 pixels as input and generates a feature tensor of the same resolution, that has d channels. A resize, that lifts up the resolution by one octave, follows. In the next module, a concatenation of this feature map and the downsampled label map, takes place. From this concatenation, a feature map is drawn again. A visualisation of the structure is shown in figure 1. This procedure applies for all consecutive modules: A concatenation of the upsampled feature map and the downsampled label map serve as the input for module m_i. Within this module a new feature map is generated. Three layers build up such a module: An input layer, an intermediate layer and the output layer. The input layers channel number is the sum of the k label map channels and the d feature layers, according to previous concatenation. The resolution i, j stays the same for all layers within a module. Between each layers a 3×3 convolution is applied. The number d of feature maps is 1024 for the first 5 modules and reduced to 512 for the last 3 modules. The last module contains 27 feature maps. A final 1×1 convolution generates 9 output images from this feature map.

2.2 Training and Image Synthesis

The training is executed in a supervised fashion on a data set that contains actual CT-images I and their corresponding label maps $\mathcal{L}$. Two networks are used for the training: The CSR modules, whose parameters θ are actually learned during the training and a pretrained VGG-19 network from wich we obtain our loss. The CSR cascade uses the semantic label map $\mathcal{L}$ as input and produces an image $g(\mathcal{L}, \theta)$ as the output. A simple metric for the calculation of the loss would be the Manhattan Distance between an output image $g(\mathcal{L}, \theta)$ and the original image I: $\sum_{i,j} |I_{i,j} - g(\mathcal{L}; \theta)_{i,j}|$. This metric could penalise completely realistic synthesised images, therefore we use a more complex loss function, because a semantic label image does not assign to just one corresponding image. The following example illustrates this problem: We want to train a CSR cascade for synthesising photorealistic images and use a data set, where cars are a class of $\mathcal{L}$. A label map of an image that depicts a blue car is taken from this data set. If the CSR cascade synthesis a perfectly realistic image from this input but colourises the car red, a Manhattan Distance based loss function would indicate a high loss, despite the synthesised image being highly realistic, but an not an exact copy of I. Therefore, a content based loss function is used for this work. The idea of such a loss is, that two images with the same content should trigger a similar activation in a visual perception network

Φ. In such a visual perception network, the first layers Φ_l primarily detect edges and colours, while the activation of deeper layers is dependent on the actual content of an input image. We collect outputs from higher and lower layers of a VGG-19 network, to base our loss on the actual content as well a finer-grained local details. An empirical hyperparameter λ_l, taken from the original publication, balances the contribution of each layer l to the loss. Equation 1 shows this concept.

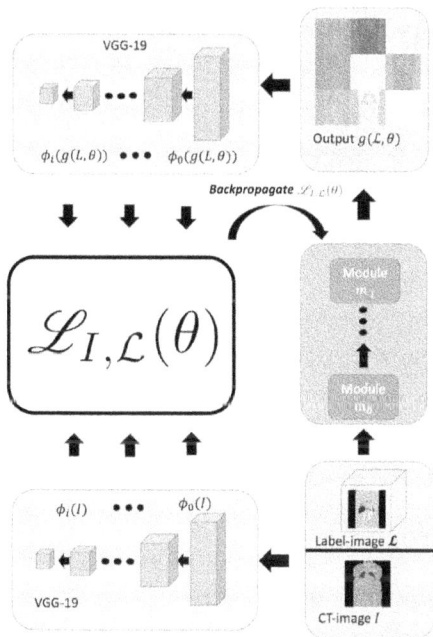

Figure 2: This is the concept of training the CSR cascade. A collection of images is synthesised from an input label map. This collection and the original image are propagated through a VGG-19 network and the resulting feature layer activations serve as input for a more complex loss function. Equation 2 shows this function in detail.

$$\mathscr{L}_{I,\mathcal{L}}(\theta) = \sum_l ||\lambda_l(\Phi_l(I) - \Phi_l(g(\mathcal{L},\theta)))|| \qquad (1)$$

To spread the bets of our network, we go further and apply a collection based loss. From one label map $\mathcal{L}$, a collection of u images is synthesised. We backpropagate the loss $\mathscr{L}$ of the image $g_u(\mathcal{L},\theta)$ that is the current best match for I. For the training, we use CT-images with their corresponding label maps from the VISCERAL (Visual Concept Extraction Challenge in Radiology) data set [7]. The training data consists of 64 preselected coronal thoracical images, from 5 different subjects. Dependent on their position in the sagittal direction, some of the coronal slices depict no or very few of the classified organs. We wanted to focus on the networks' capability to synthesise images showing the total diversity of textures in thoracical organs. Therefore training and test images where chosen that depict multiple organs and not only the background tissue. For testing the performance of our network, we used 5 thoracical images of another subject. Figure 3 shows such tuples. With each

input, we synthesise 9 greyscale images and backpropagate the loss of the best matching one. This loss function is displayed in equation 2. Figure 2 visualises this concept.

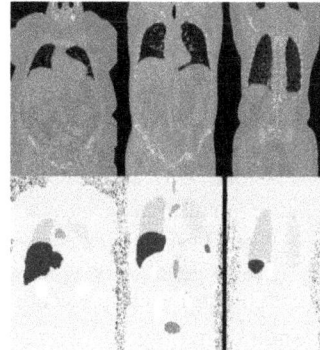

Figure 3: Sample data from our training. The upper row depicts the basic CT-images, the lower row shows a segmentation of this data.

$$\mathscr{L}'_{I,L}(\theta) - \min \sum_l ||\lambda_l(\Phi_l(I) - \Phi_l(g(\mathcal{L},\theta)))|| \qquad (2)$$

3   Results

A result for the synthesis of a whole collection from one input label map can be seen in figure 5. The second image of the last row shows the winner output with the smallest loss. The images show the same basic structures, but with differences in the medium intensity of grey values and contrast. A more detailed view of the image with the smallest loss can be seen in figure 4. This figure compares the winner image, the reference image and its segmentation. It is possible to recognise and distinguish between various organs. Lunges, kidneys and the heart can be recognised, but with less characteristic texture details, compared to the real reference image. Edges and finer details are a lot more blurred and the background tissue does not completely fit the shape of the reference image. Some features that are not even part of separate classes, like the pelvic bones and spine, can be seen to some extent. A further concept for a quantitative evaluation is outlined in section 3.1.

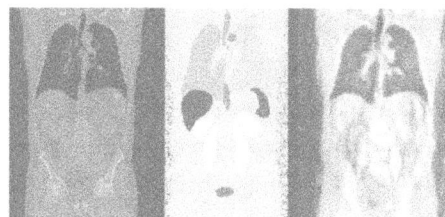

Figure 4: Sequence of original image, the corresponding segmentation and the synthesised image. Multiple organs are recognizable in the synthesised image. To some extent, even pelvic bones and the backbone are depicted.

3.1 Discussion

The appearance of our synthesised images still does not completely match a real CT-image, but the results show some interesting details. First signs of vascularisation and characteristic textures can be recognised, especially in the lunges. Even some content, that is not part of separate classes is present, like the basic structures of pelvic bones and the spine. A main difficulty for the training is the segmentation, that does not distinguish between bones and tissue, but only between organs, background tissue and a total black background. Some subjects show a lot more background tissue, relative to their organ volume, due to their physical condition. This appearance also influences the position and orientation of the organs. The basic segmentation of the background tissue induces a strong blurring that distorts the shape of the synthesised images. A better segmentation of the background tissue should isolate the torso from the black background and reduce the blurring around the boundaries of the body. With a segmentation that includes a bigger number of classes, the CSR cascade should be able to synthesise more separate structures. Furthermore, we expect an increased level of finer texture details with such an improved segmentation. These clearer specifications for the image synthesis should make it easier for the network to learn synthesising smaller texture details. A broader amount of training data might be a solution for the problems, induced by the varying constitution of the subjects. Some images depict tissue brighter than the background. This might be a weakness of the feature based loss function, that does not severely penalise such implausible pixel values, as long as more complex features are depicted adequately. Further experiments with the hyperparameter λ should yield a better balanced loss function. The main motivation behind the synthesis of artificial medical image data is a contribution of synthetic training data to deep learning in medical image analysis. Therefore, testing the performance of a network on the classification of real world medical image data, using only synthetic images as training data, is a reasonable next step for a quantitative evaluation of our results.

4 Conclusion

We have shown that the used network architecture and concept of training are able to produce first results of decent quality. The synthesised images do not completely match real CT-images, but even with limited amount of training data and the rudimentary segmentations, our results show promising features. Especially the networks' capability to depict structures that are not part of separate classes are interesting. Future research will show, to what extent the synthesis of realistic images and an increase of training data can be achieved with this concept.

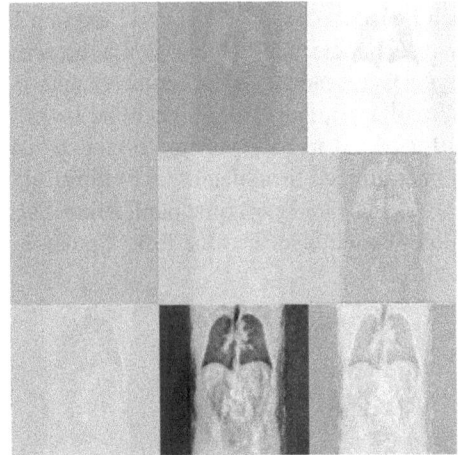

Figure 5: View of a whole collection of images, synthesised from one label map, the image that is the best match for the reference original CT-data the second one of the last row. The images differ primarily in the medium grey values and contrast.

Acknowledgement

This work has been carried out and supervised by the Institute of Medical Informatics, Universität zu Lübeck.

5 References

[1] D. Shen *Deep Learning in Medical Image Analysis.* Available: PMC 2017 June 21.

[2] J. Deng, W. Dong, R. Socher, L. Li, K. Li and L. Fei. Dept. of Computer Science, *ImageNet: A Large-Scale Hierarchical Image Database* Conference on Computer Vision and Pattern Recognition, 2009.

[3] A. Shrivastava,T. Pfister *Learning from Simulated and Unsupervised Images through Adversarial Training.* CVPR 2017, 2017.

[4] T. A. Le , A. G. Baydin, R. Zinkov and F. Wood, *Using Synthetic Data to Train Neural Networks is Model-Based Reasoning* International Joint Conference on Neural Networks (IJCNN), 2017.

[5] Q. Cheng and V. Koltun, *Photographic Image Synthesis with Cascaded Refinement Networks.* International Conference on Computer Vision, 2017.

[6] H. Vala, A. Baxi *A Review on Otsu's Image Segmentation Algorithm* International Journal of Advanced Research in Computer Engineering and Technology Volume 2, 2013.

[7] http://www.visceral.eu/

2

Biomedical Optics

Light-induced Permeabilization of Liposomes

P. Ahrari [1], A. Link [2] and R. Rahmanzadeh [2]

[1] Medizinische Ingenieurwissenschaft, Universität zu Lübeck, paria.ahrari@student.uni-luebeck.de

[2] Biomedizinische Optik, Universität zu Lübeck, {link, rahmanzadeh}@bmo.uni-luebeck.de

Abstract

One technique for the delivery of drugs to the target tissue is the use of liposomes. These serve as a carrier system, in which the drug is encapsulated and introduced into cells and tissue. Subsequently, the charge of the liposome can be released by an external trigger, such as light. In this project, liposomes with two different photosentisizer were prepared and tested. The photosensitizers Chlorin-e6 and benzoporphyrin derivative monoacid were used to observe which photosensitizer achieved better results. The study shows, that both photosentitizers are suitable for opening the liposomes but chlorin-e6 liposomes require a lower irradiation intensity and exposure time to release the liposomes contents.

1 Introduction

For the treatment of many diseases, the transport of therapeutic agents to the target tissue is a key challenge.[1] On the way to their destination, drugs can be untimely expelled from the body through the destructive effects of multiple defense systems and thus can not reach their destination.[2] A potential solution to this challenge are light-sensitive liposomes as carrier-systems. Liposomes which were used since the 1960s, are artificial spherical shaped vesicles composed of phospholipids.[3] The lipids have a hydrophilic head group and a hydrophobic tail group. This amphiphilic character enables lipids to assemble into bilayers with the hydrophobic tails aligning with each other. Liposomes vary in the number of bilayers and their size and may consist of a single (unimellar vesicle) or multiple lipid bilayers (multimellar vesicles). Due to their amphiphilia, liposomes enable the uptake of hydrophobic and hydrophilic molecules and can therefore be used for the encapsulation of substances.[3][4] In order to open the liposomes with light, the principle of Photodynamic Therapy (PDT) is used. PDT is a method of treating tissue alterations such as tumors using light and a photosensitizer (PS). In PDT, the patient is given a photosensitizer. The irradiation with light of a specific wavelength leads to the toxicity of the PS, which reacts with its environment and damages it by the production of singulett oxygen.[5] The damage to cell components leads to necrosis or apoptosis. This principle is used to produce light-sensitive liposomes with a photosensitizer and to open them at the target tissue by irradiation. [5] In order to compare the effects of liposomes with different photosensitizers, liposome with chlorin e6, benzoporphyrin derivative monoacid (BPD) and liposomes without photosentizers are prepared and tested in this study. To test the encapsulation and the release, the fluorescent dye calcein is encapsulated within the liposome. Inside the liposome, the fluorescence of calcein is quenched due to the high concentration. If the membrane of the liposome is permeabilized by irradiation, the calcein is released and the fluorescence increases.[6] This increase of fluorescence intensity is then measured and used to quantify the release rate from the liposome.

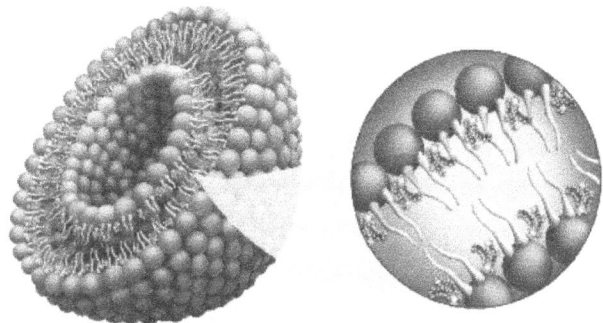

Figure 1: Simplified presentation of a liposome with a hydrophilic head group and a hydrophobic tail group [7]

2 Material and Methods

There are various techniques and different phospholipid compositions for the preparation of liposomes. The selection of the method depends on purpose and criteria such as the physiochemical characterisitics of the liposomes, the costs and the replication facilities. A very common method for liposome preparation is the thin-film hydration procedure or Bangham method [2]. We applied a modified version of this method explained in detail in the following.

2.1 Liposomes

Three different liposomes were produced. Two of these liposomes are light-sensitive, the third is not, since no photosentisizer is added. All prepared liposomes are composed of the lipids 1,2-dioleoyl-3-trimethylammonium-propane (DOTAP), dipalmitoylphosphatidylcholine (DPPC), 1,2-distearoyl-sn-glycero-3-phosphoethanol-amine- N- [amino (polyethylene glycol) -2000] (DSPE2000-PEG) and cholesterol, dissolved in chloroform. Depending on the liposome to be produced, the photosentisizer chlorin-e6 or BPD is added. The entire liposome synthesis occurs in four steps: thin-film preparation, hydration, extrusion and purification.

2.1.1 Thin-film preparation

For the thin-lipidfilm preparation, the lipids and the photosentizer are mixed in a glass tube according to the given quantities in table 1.

Table 1: Components thin-lipid film

Lipide	volume	concentration	molare mass
DOTAP	60 μl	25 mg ml^{-1}	698.5 M
DPPC	300 μl	25 mg ml^{-1}	734.0 M
Cholesterol	38 μl	50 mg ml^{-1}	386.7 M
DSPE2000-PEG	150 μl	10 mg ml^{-1}	2805.5 M

Afterwards either 200 μl BPD or 8,9 μl chlorin-e6 is added. Subsequently, the chloroform is evaporated with uniform rotation under a stream of nitrogen, to produce a homogeneous lipid film. To remove the remaining chloroform, the glass tube is placed in a vacuum excsiccator for 12 hours. Subsequently, a calcein solution is mixed with the quantities given in table 2.

Table 2: Components Calcein solution

	mass
Calcein	0.0748 g
Natriumchlorid	0.0034 g
Aqua dest.	2 ml
NaOH (5N)	approx. 70 μl

2.1.2 Hydration

The lipid film is hydrated after 12 h in the evaporator. For this purpose, 500 ml of calcein solution are added to the film, followed by a 40-minute incubation in a 50 °C waterbath. This is followed by an contrast bath for 120 minutes. The dispersion is transferred to an ice bath with an incubation time of 10 minutes and afterwards to the water bath again for 10 minutes. After the water bath, the dispersion is thoroughly mixed. This process is repeated 6 times, so that the dispersion was incubated for a total of 60 minutes in an ice bath and 60 minutes in a water bath. Hydration forms multilamellar vesicles of the lipids. In order to obtain unilamellar vesicles, the hydration is followed by extrusion.

2.1.3 Extrusion

The extrusion system consists of two Hamilton syringes and a polycarbonate membran with pores of 100 nm diameter.

The dispersion is first drawn up in a Hamilton syringe and then pressed 13 times through the extrusion system which is heated to 50 °C. This is followed by the purification of the liposomes.

2.1.4 Purification

The unilamellar liposomes obtained by the extrusion are purified by gel filtration, the calcein which has not been encapsulated is removed. For this purpose, the dispersion is applied to a Sepharose column CL-4B (Sigma Aldrich). After the dispersion sinks into the gel, the column is filled regularly with phosphate buffered saline (PBS). The dispersion passes through the column and both the purified liposomes and the remaining eluate are collected in a 96 well plate with two drops per well. Subsequently, the liposomes are collected. For this purpose, the calcein fluorescence is measured with a fluorescence spectrometer. The excitation wavelength λ_{ex} used is 485 nm and the emission wavelength λ_{em} is 525 nm. Liposomes from wells which have a similar fluorescence intensity are collected. The liposomes are stored in the refrigerator with exclusion of light.

2.2 Permeabilization

To observe light-induced release from liposomes, they were permeabilized by light irradiation. In order to determine the parameters with which the best release can be generated, various tests were carried out. In these experiments, different irradiances and irradiation times were used. Liposomes containing the photosensitizer chlorin-e6 are irradiated at 660 nm and liposomes containing BPD at 690 nm. The fluorescence intensity was measured before and after the irradiation; the increase in the fluorescence intensity can be used to quantify the release efficiency. As control, liposomes permeabilized with 4 % Triton are used for all experiments.

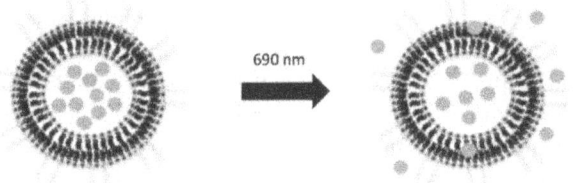

Figure 2: Simplified principle of permeabilization of BPD Liposomes by irradiation

3 Results and Discussion

The permeabilization was investigated after light irradiation. In order to obtain a comparison and to observe the effect of photosensitizers, L1-Calcein Liposomes were first irradiated (prepared without Photosensitizer). This was followed by irradiation of the liposomes with BPD (L1-Calcein-BPD) and Chlorin-e6 (L1-Calcein-Ce6). Each experiment was conducted in multiple times. The error bars represent the standard deviation in the following figures.

Figure 3: Fluorescence intensity of L1-Calcein irradiated at 10 mW/cm^2, 20 mW/cm^2, 40 mW/cm^2 and 80 mW/cm^2 for 1 min, 3 min, 6 min and 12 min

In Fig. 3 the results of the irradiation of the L1-calcein-liposomes are illustrated. Since these were produced without photosensitizer, no calcein is released by the irradiation. Even with increasing irradiation intensity and irradiation time no fluorescence release was observed.

3.1   L1-Calcein-BPD

The L1-calcein-BPD liposomes were irradiated with increasing irradiation intensities and irradiation times. The results of the irradiations are shown in the following figures. The standard deviation may be reduced after further measurements.

3.1.1   Irradiation time

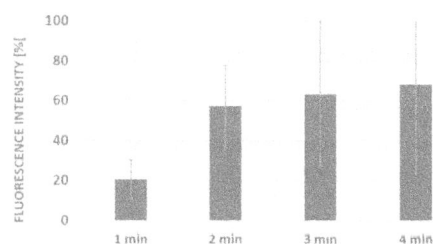

Figure 4: Fluorescence intensity of L1-Calcein-BPD irradiated at 690 nm and 80 mW/cm ^{2} for 1 min, 2 min, 3 min and 4 min

Fig. 4 shows that with an increasing irradiation time (max. 4 minutes) and a constant irradiation intensity of 80 mW/cm^2, the maximum calcein release of 68 % was achieved. A complete release could not be reached.

3.1.2   Irradiation intensity

In Fig. 5 it is shown that the fluorescence intensity increases slightly with increasing irradiation intensity. At constant time of 1 min and 40 mW/cm^2 the fluorescence release is 10 %. At 80 mW/cm^2 radiation intensity 13 % is achieved.

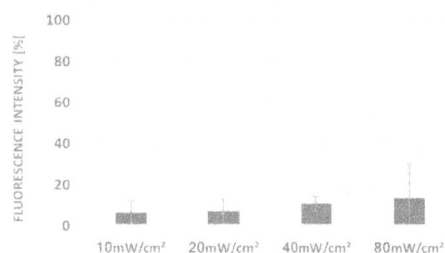

Figure 5: Fluorescence intensity of L1-Calcein-BPD irradiated at 690 nm and 10 mW/cm^2, 20 mW/cm^2, 40 mW/cm^2 and 80 mW/cm^2 for 1 min

3.1.3   Irradiation time and intensity

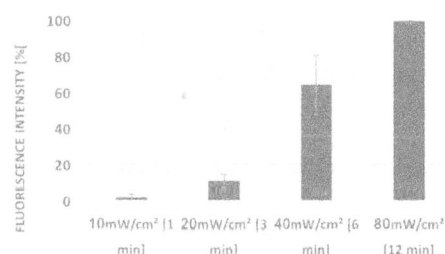

Figure 6: Fluorescence intensity of L1-Calcein-BPD irradiated at 690 nm and 10 mW/cm^2, 20 mW/cm^2, 40 mW/cm^2 and 80 mW/cm^2 for 1 min, 3 min, 6 min and 12 min

The fluorescence intensity in Fig. 6 shows that with increasing irradiation intensity and irradiation time a 64 % release is achieved at 40 mW/cm^2 and 6 minutes. A 100 % release and maximum fluorescence intensity occurs at 80 mW/cm^2 and a 12 minute irradiation time.

3.2   L1-Calcein-Ce6

The L1-calcein-Ce6 liposomes were irradiated with increasing irradiation intensities and times. The results of the irradiations are shown in the following figures. The standard deviation may be reduced after further measurements.

3.2.1   Irradiation time

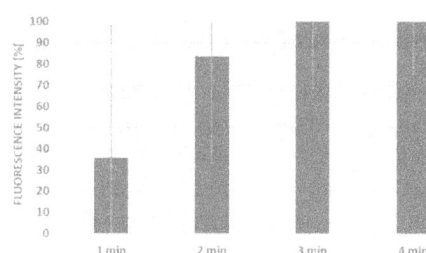

Figure 7: Fluorescence intensity of L1-Calcein-Ce6 irradiated at 660 nm and 80 mW/cm ^{2} for 1 min, 2 min, 3 min and 4 min

Fig. 7 shows that a complete release is achieved within 3 minutes for L1-calcein-Ce6. With an irradiation intensity of 80 mW/cm^2, a fluorescence intensity of 82 % is reached after 2 minutes Irradiation.

3.2.2 Irradiation intensity

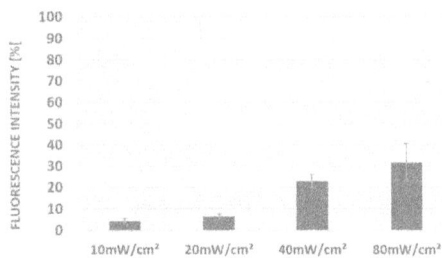

Figure 8: Fluorescence intensity of L1-Calcein-Ce6 irradiated at 660 nm and 10 mW/cm^2, 20 mW/cm^2, 40mW/cm^2 and 80mW/cm^2 for 1 min

Fig. 8 demonstrates that with an increasing intensity to 80 mW/cm^2 and constant irradiation time of 1 min, a 30 % Fluorescence release occurs.

3.2.3 Irradiation time and intensity

Fig. 9 shows that with increasing irradiation intensity and time a complete calcein release occurs. At 40 mW/cm^2 and 6 minutes 90 % of calcein is released. The 100 % release is achieved at 80 mW/cm^2.

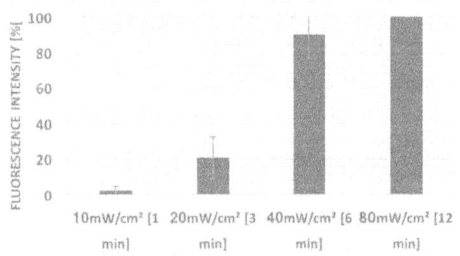

Figure 9: Fluorescence intensity of L1-Calcein-Ce6 irradiated at 660 nm and 10 mW/cm^2, 20 mW/cm^2, 40 mW/cm^2 and 80 mW/cm^2 for 1 min, 3 min, 6 min and 12 min

3.3 Comparison of BPD and Chlorin-e6

The photosensitizer is one of the most important factors in the permeabilization of liposomes. the irradiation of the L1-calcein liposomes proves that no release is possible without photosentizers. In this study we compared the potential of the two Photosentisizers BPD and Chlorin-e6 to open liposomes and to release their cargo. In the case of the L1-calcein-BPD liposomes, an increase in fluorescence occurs during the irradiation time measurements, but complete release does not take place, as in the case of the L1-Calcein-Ce6 liposome. In the irradiation intensity measurements,

an irradiation time of 1 min is not enough for both the L1-Calcein-BPD and the L1-Calcein-Ce6 to achieve a release above 30 %. The Chlorin-e6 liposomes achieved a higher release under the same conditions. With increased irradiation intensity and time, a release rate of 100 % is achieved with both photosensitizers. With Chlorin-e6, however, a lower irradiance and time is needed to achieve the 90 % release.

4 Conclusion

This study demonstrates the importance of photosensitizers in light-induced release from liposomes. Comparing the photosensitizers BPD and chlorin-e6, the liposomes released in both cases. The measurements carried out previously indicate that chlorin-6 requires a shorter irradiation time and irradiation intensity in order to release its contents. Further investigations will characterize this difference in release in more details.

Acknowledgement

The work has been carried out at Institute of Biomedical Optics, Universität zu Lübeck.

5 References

[1] K. A. Carter et al., *Porphyrin–phospholipid liposomes permeabilized by near-infrared light*. Nature Communications,vol.5, 2014.

[2] L. Sercombe, T. Veerati, F. Moheimani, S. Y. Wu, A. K. Sood and S.Hua, *Advances and Challenges of Liposome Assisted Drug Delivery*. Frontiers in Pharmacology,vol.6, pp. 286, 2015.

[3] G Bozzuto and A. Mollinari, *Liposomes as nanomedical devices.* , International Journal of Nanomedicine, vol.10, pp. 975–999, 2015.

[4] M. Alavi, N. Karimi and M. Safaei,*Application of Various Types of Liposomes in Drug Delivery Systems*. Advanced Pharmaceutical Bulletin, vol.7, pp.3–9, 2017.

[5] M. R. Hamblin and T, Hasan, *Photodynamic therapy: a new antimicrobial approach to infectious disease?*. Photochemical and Photobiological Sciences,vol.3, pp.436-450, 2004.

[6] J. R. Lakowicz, *Principles of fluorescence spectroscopy*. Edition 3, Springer Science & Business Media, Baltimore, pp. 59–62, 2006.

[7] Peregrine Ophthalmic. Available: http://www.peregrineophthalmic.com/products/index.html [last accessed on 2018-01-21].

Fluorescence Lifetime Imaging Ophthalmoscopy of the Retinal Pigment Epithelium during Wound Healing after Selective Retina Treatment

A. Hutfilz [1], B. Lewke [2] and Y. Miura [2,3]

[1] Medizinische Ingenieurwissenschaft, Universität zu Lübeck, {alessa.hutfilz, britta.lewke}@student.uni-luebeck.de

[2] Institute of Biomedical Optics, Universität zu Lübeck, miura@bmo.uni-luebeck.de

[3] Department of Ophthalmology, University Hospital Schleswig-Holstein, Campus Lübeck

Abstract

Measurement of the metabolic changes in the wound healing of the human retina after laser treatment is of great interest, though to date there is no detection method. Recently introduced fluorescence lifetime imaging ophthalmoscopy (FLIO) has a potential to diagnose metabolic activation of the retinal pigment epithelium (RPE). In this study RPE-choroid explants from enucleated porcine eyes were irradiated with selective retina treatment (SRT) laser with pulse energies of 80-150 μJ. FLIO was performed 24h and 72h after SRT. The fluorescence lifetime (FLT) of two areas in the RPE wound were analyzed for two spectral channels. Results show a significantly longer FLT in the wound center and a significant decrease of the FLT over time, which showed a dependency on the laser pulse energy especially for the longer wavelength channel. Based on the results, FLIO may function as visualizing method for structural and metabolic change in RPE.

1   Introduction

Fluorescence Lifetime Imaging Ophthalmoscopy (FLIO) holds a new possibility as a method to visualize metabolic changes of the retina [1]. The fluorescence lifetime (FLT) is determined by the average time duration, during which a fluorophore excited to a higher energy level returns to its ground state. Since FLT is the fluorophore-intrinsic value, it can overcome the difficulty in differentiation of the different fluorophores in the retinal tissues, which conventional autofluorescence (AF) measurement struggle with [2]. The retinal pigment epithelium (RPE), a monolayered epithelial at the most outside of the retina, is a crucial integral part of the outer blood retina barrier, and thus the disorder of RPE functionality is related to the pathogenesis of different chorioretinal diseases [3]. Therefore, a treatment to improve RPE functionality is desired as one of the recent therapeutic strategies. One of them is the selective retina treatment (SRT). By using multiple microsecond laser pulses shorter than the thermal confinement and low repetition rates the effects of SRT may be selectively localized to the RPE with the microbubble formation around the melanosomes without thermal diffusion to the surrounding tissues [4][5]. Although metabolic improvement of the RPE is expected by SRT, there is no method to detect metabolic activation of retinal tissues to date, and thus this theory has not been proven yet. Therefore, it has been desired to develop a method to diagnose metabolic changes of human fundus, based on the knowledge that some metabolic cofactors, as flavin adenine dinucleotide (FAD), may change their FLT according to the cell metabolic states [6]. We hypothesized that FLIO can be utilized to detect metabolic alteration of the RPE after SRT. As the first step toward this goal, we investigated the alteration of the FLT of the RPE at the irradiated sites. The RPE cells treated with SRT are killed in a circle form as an area of cell defect of 200 μm diameter, followed by the covering of the defect through the migration and proliferation of surrounding RPE cells. In this study, the RPE in the ex-vivo-RPE-choroid-sclera tissue culture was irradiated with SRT with different laser pulse energies, and the FLT at the laser spots was measured with FLIO over time. With a final Calcein-AM test the viability of the RPE cells and wound closure was verified.

2   Material and Methods

A series of experiments lasts for 5 days. 24h after preparation (day 1) the RPE-choroid-sclera-culture was treated with the SRT (day 2), followed by FLIO measurements after 24h (day 3) and 72h (day 4). In Addition, the RPE viability was tested with calcein-AM at day 4. In the following the detailed methods for each procedure are explained.

2.1   RPE-Choroid-Sclera Culture

As experimental model, enucleated porcine eyes were used, which resembles the human eye in anatomic and physiologic characteristics [7]. For the preparation of RPE-choroid-sclera culture the anterior parts of the eye, lens and the vitreous body were removed, and the RPE-choroid-sclera explant was resected by trepanation as shown in

Fig.1. The RPE-choroid-sclera explant was then cultivated in a Dulbecco's Modified Eagle Medium (DMEM, high glucose, Merck KgaA, Darmstadt, Germany) mixed with 10 % porcine serum, 1 mM sodium pyruvat, and antibiotic antimicrobic agents and maintained in a 5 % CO_2 incubator at 37°C.

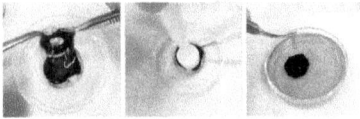

Figure 1: RPE-choroid-sclera-explant preparation.

2.2 SRT Laser System

All RPE-choroid-sclera explants were irradiated with the SRT-Laser system (Medical Laser Center Lübeck GmbH). The setup shown in Fig.2a is a slit-lamp-adapted, frequency doubler Nd:YFL laser with a wavelength of 572 nm and a pulse duration of 1.7 μs. One exposition corresponds to 30 pulses with a repetition rate of 100 Hz and a spot diameter of 200 μm. The pattern by which the RPE was irradiated is a range of laser pulse energies (E_P) from 80 μJ to 150 μJ with increment of 10 μJ, as shown schematically in Fig.2b. For the orientation a marker with 150 μJ was set.

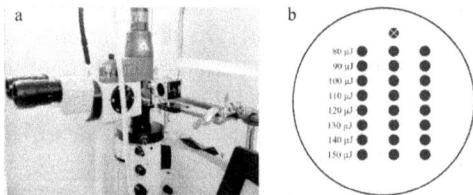

Figure 2: a) SRT setup for ex-vivo irradiation, and b) irradiation pattern on RPE-choroid-sclera explant.

2.3 Fluorescence Lifetime Imaging Ophthalmoscopy (FLIO)

Fluorescence lifetime imaging of the RPE was performed with a FLIO system, a prototype provided by Heidelberg Engineering GmbH (Heidelberg, Germany). The schematic description of the setup is shown in Fig.3. For the FLIO measurement the RPE-choroid-sclera explant was placed in a custom made chamber filled with Dulbeccos's phosphate-buffered saline (PBS). For excitation the object is raster scanned by a 473 nm laser diode with an 80 MHz repetition rate over a 30° field with a focus of 52.00 dpt. By highly sensitive hybrid photon-counting detectors (HPM-100-40; Becker&Hickl, Berlin, Germany) emitted photons were detected in two spectral channels of 498-560 nm (short) and 560-720 nm (long). By means of time correlated single photon counting (TCSPC) modules (SPC-150; Becker&Hickl) the detector signals were registered. Acquired lifetime data were analyzed by Becker&Hickl software (SPCImage 6.4)

by using a biexponential decay model and a binning factor of 0. A decay matrix calculation creates a pseudo col-

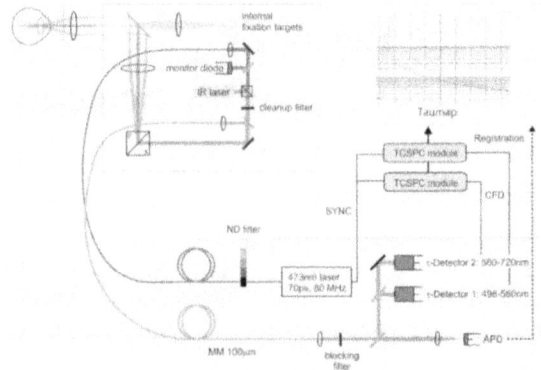

Figure 3: Scheme of FLIO setup [8].

ored image of the FLT. Blue color indicates a long FLT and red/orange color a short FLT. For every spot the data of the section of 20x20 pixels (800 μm x 800 μm) were exported. For each pixel the mean lifetime (t_m) was calculated from the short and long lifetime components (t_1, t_2) and their respective amplitudes coefficients (a_1,a_2) for each spectral channel (Ch1, Ch2), according to the formula:

$$t_m = \frac{a_1 \cdot t_1 + a_2 \cdot t_2}{a_1 + a_2}. \tag{1}$$

2.4 Calcein-AM Viability Test

For measurements of the RPE viability the explants were tested with calcein-AM directly after the 72h FLIO. Calcein-AM diffuse into the cell membrane and convert to calcein by esterase. Calcein fluoresce in the green spectral range by blue light excitation. Dead cells show no fluorescence. Explants were incubated with 5 μM calcein-AM (Thermo Fisher Scientific, Waltham, USA) in DMEM high glucose for 15 min at 37°C and live-dead analysis was performed with fluorescence microscopy. A Nikon Eclipse Ti-E fluorescence microscope with a FITC filter (excitation wavelength of 465-495 nm, dichroic mirror for 505 nm and a barrier filter for 515-555 nm) and a magnification of 4 and 10 were used.

2.5 Statistical Data Analysis

SRT irradiation on one explant was conducted with at least two spots for each energy setting, and experiments were repeated more than three times. To assess the results at both time points statistical analysis by means of Student's t-test was performed and the two-sided P value less than 0.05 was determined as significant. For the evaluation of a feasible E_P-t_m dependency the regression analysis was used and the correlation coefficient R was calculated.

3   Results and Discussion

3.1   FLIO Results

Fig.4a is an exemplary AF image of an RPE-choroid-sclera explant 24h after SRT, representing the fluorescence intensity with gray scale. Fig.4b shows exemplary AF and FLIO images of a RPE-choroid-sclera explant treated with SRT after 24h and 72h. The FLIO image is demonstrated with pseudo color (here shown only with gray scale). Typically the AF at the laser spots becomes unclear over time in both channels, whereas the FLIO image clearly reveal the wound area, even after 72h. Final fluorescence image in Fig.4c of calcein-AM viability test present the wound closure with elongated RPE cells. We found the gradient of the FLT at

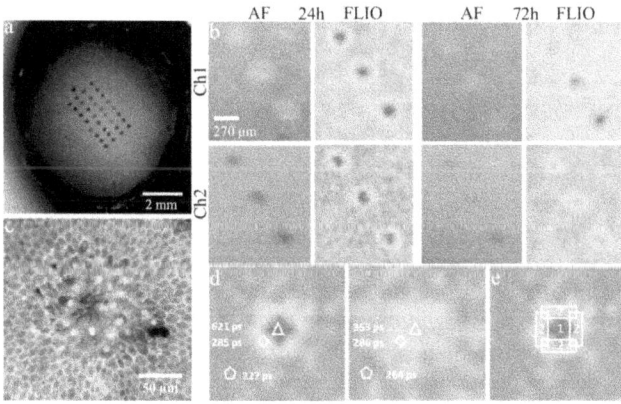

Figure 4: a) Exemplary AF image (Ch2) at 24h. b) AF and FLIO of Ch1 and Ch2 at 24h and 72h after SRT (130-150 μJ). c) Calcein-AM image 72h after SRT. d) FLIO (t_m,Ch2) of 130 μJ wound at 24h (left) and 72h (right). Central area (triangle) with longer t_m than transition area and adjacent RPE (rhombus;pentagon). The gradient became smaller after 72h. e) Selected area for investigation of FLT.

the wound, where the center of the spot shows the longest FLT, which decreases with distance (Fig.4d left) and this gradient become less apparent after 72h (Fig.4d right). After 24h the central area (triangle) has a longer FLT than the transition area (rhombus) and the surrounding RPE (pentagon), whereas at 72h the central area has a decreased FLT than after 24h. The central and transition area show, however, still significantly longer FLT than the one of the surrounding RPE, so that one can differentiate the RPE cells participating in the wound healing process very clearly with the pseudo-color image. These results suggest that FLIO may visualize the wound healing of the RPE more clear than the AF. As the next step, FLT values at the spots were analyzed for different wound areas over time as shown in Fig.4e. The central area is represented within 120 μm x 120 μm and the transition area consisting of 4 sections of 200 μ x 80 μm. The data (t1, t2, a1, a2) from all pixels included in these areas were exported from SPCImage, and t_m was determined for each area for different E_P as well as different time points using the formula shown in (1). The average value and standard deviation of t_m and re-

spective p-values of the Student's t-test are represented in Table 1.   Fig.5 shows the t_m of Ch1 in central area (up-

Table 1: Average values $\pm$ SD of t_m. p: Student's t-test p-values from the comparison between 24h and 72h. R: Correlation coefficients by regression analysis for the relationship between t_m and E_P, shown with respective p-values.

		Area 1 - t_m in ps			Area 2 - t_m in ps		
	E_p in µJ	24h	72h	p	24h	72h	p
Channel 1 t_m	80	1625 ± 384	1771 ± 194	0.10278	996 ± 510	1411 ± 358	0.00234
	90	1512 ± 420	1445 ± 460	0.54716	947 ± 594	1144 ± 471	0.26374
	100	1399 ± 550	1376 ± 449	0.83175	856 ± 544	998 ± 427	0.32671
	110	1458 ± 339	1319 ± 259	0.23077	864 ± 431	838 ± 259	0.87343
	120	1468 ± 354	1119 ± 329	0.05102	787 ± 457	721 ± 197	0.69283
	130	1598 ± 358	1367 ± 372	0.19855	804 ± 405	869 ± 337	0.70698
	140	1718 ± 206	1555 ± 249	0.16160	957 ± 433	954 ± 254	0.98048
	150	1806 ± 223	1512 ± 491	0.10469	1101 ± 510	1097 ± 431	0.98215
	R	0.61 (p=0.11)	0.23 (p=0.21)		0.15 (p=?2)	0.50 (p=0.59)	
Channel 2 t_m	80	848 ± 607	382 ± 73	0.04096	292 ± 84	291 ± 21	0.94790
	90	1038 ± 464	395 ± 47	0.00181	308 ± 59	288 ± 13	0.24459
	100	969 ± 403	390 ± 81	0.00127	286 ± 21	292 ± 25	0.33847
	110	893 ± 410	414 ± 101	0.00574	300 ± 41	294 ± 27	0.66998
	120	910 ± 406	406 ± 72	0.00284	308 ± 53	297 ± 26	0.48899
	130	951 ± 398	403 ± 60	0.00257	297 ± 21	298 ± 19	0.74927
	140	1054 ± 367	425 ± 93	0.00069	332 ± 84	305 ± 31	0.29642
	150	1394 ± 215	478 ± 171	0.00001	324 ± 37	311 ± 41	0.32693
	R	0.66 (p=0.11)	0.83 (p=0.012)		0.71 (p=0.048)	0.94 (p=0.0004)	

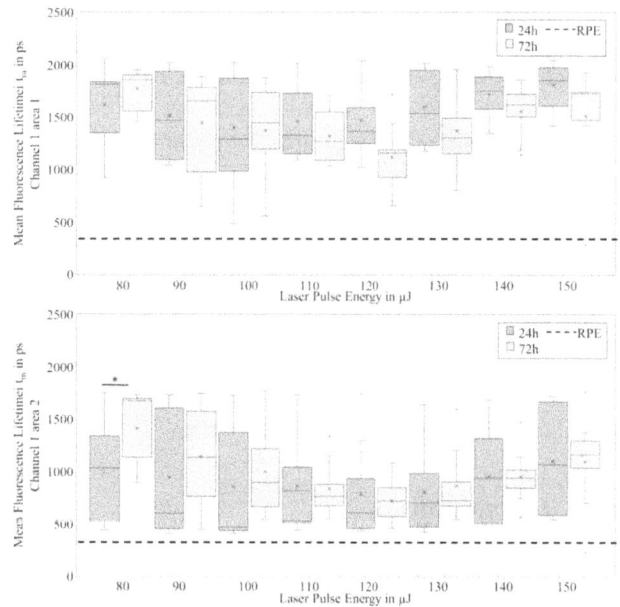

Figure 5: t_m (ps) in Ch1 (shorter wavelength) at the lasered spots with different laser pulse energies (80-150 μJ) at 24h and 72h after SRT for central area 1 (upper) and transition area 2 (lower) with * p<0.05. Dash line (- - -) indicates the level of the average FLT of the surrounding RPE.

per) and transition area (lower) at 24h and 72h for all E_P. t_m among all spots was 1573 ps (24h) and 1433 ps (72h) for area 1, and 914 ps (24h) and 1004 ps (72h) for area 2. Significant difference in t_m was found neither among different E_P nor between time points in both areas. On the contrary t_m of Ch2 showed different behaviors, as shown in Fig.6. Here t_m among all spots was 1007 ps (24h) and

412 ps (72h) for the center, and 306 ps (24h) and 297 ps (72h) for the transition. t_m in area 1 showed a significant decrease after 72h. There was no significant difference in

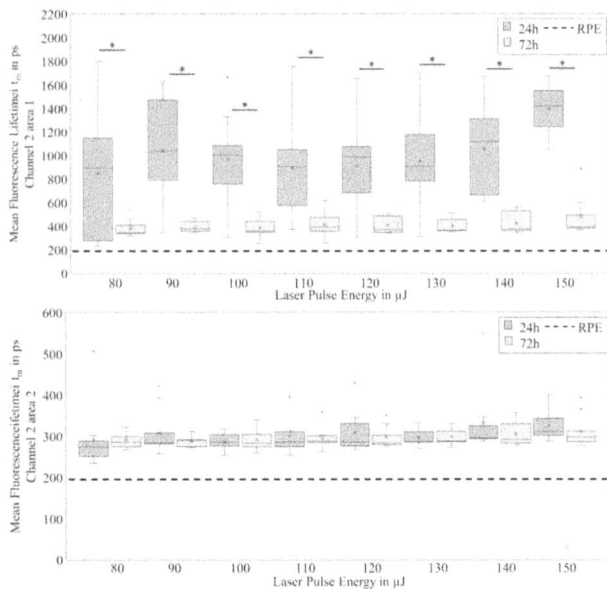

Figure 6: t_m (ps) in Ch2 (longer wavelength) at the lasered spots with different laser pulse energies (80-150 μJ) at 24h and 72h after SRT for central area (1,upper) and transition area 2 (lower) with * p<0.05. Dash line (- - -) indicates the level of the average FLT of the surrounding RPE.

t_m between 24h and 72h in area 2. t_m in Ch2 appeared to have a high energy dependency in both zones for both time points. The dependency between E_P and t_m was evaluated by a regression analysis and resulted correlation coefficients (R), of all t_m-area combination for both channels, is presented with respective p-values in Table 1. Thereby R-values and p-values in both areas at 24h and 72h for Ch1 show no correlation between E_P and t_m. On the other hand, R-values for Ch2 state a clear correlation between E_P and t_m, whereat 72h shows a strong correlation.

3.2 Discussion

At the laser spots, independent of channels, the central area, namely the area of cell defect, shows always long t_m. This is due to the long t_m of the choroid [6]. Interestingly, the transition area, around the laser spots showed also a long t_m, even after 72h, when the wound is almost closed. This change might reflect the metabolic activation of RPE cells, showing metabolic cofactors FAD changed their molecular status with the change of cellular energy metabolism. It is known, that FAD change its FLT dependent of the protein binding states. Free FAD shows a long FLT (2300±700 ps), whereas protein-bound FAD a short FLT (130 ±20 ps for monomeric form and 40±10 ps for dimeric form) [9]. Regarding time-dependent change, there was a significant decrease in t_m in Ch2 in the central area. This result suggests that the t_m of Ch2 may better correlate to the wound closure of the RPE.

4 Conclusion

Results of this study showed that FLIO can clearly visualize the laser spots on the RPE and their wound healing, by indicating not only structural but also metabolic alterations. It may provide us of further insights into the effect of laser treatment of the RPE, including metabolic activation. Further it is planned to investigate the impact of the laser action with FLIO in a larger RPE area around laser spots. In conclusion, FLIO may be a useful noninvasive method to measure the structural and metabolic status of the RPE.

Acknowledgement

The work has been carried out at the Institute of Biomedical Optics, Universität zu Lübeck.

5 References

[1] J. Schmidt et al., *Fundus autofluorescence lifetime are increased in non-proliferative diabetic retinopathy*. Acta Ophthalmologica, vol. 95, no. 1, pp. 33-40, 2017

[2] D. Schweitzer et al., *In vivo measurement of time-resolved autofluorescence at the human fundus*. Journal of Biomedical Optics, vol. 9, no. 6, pp. 1214–1222, 2004

[3] O. Strauss, *The Retinal Pigment Epithelium in Visual Function*. Physiological Reviews, vol. 85, no. 3, pp. 845-881, 2005

[4] R. Brinkmann et al., *In vivo measurement of time-resolved autofluorescence at the human fundus*. Ophthalmologe, vol. 103, no. 10, pp. 839-849, 2006

[5] J. Roider et al., *Selective retina therapy (SRT) for clinically significant diabetic macular edema*. Graefes Archive for Clinical and Experimental Ophthalmology, vol. 248, no. 9, pp. 1263-1272, 2010

[6] D. Schweitzer et al., *Towards Metabolic Mapping of the Human Retina*. Microscopy Research & Techniques, vol. 70, no. 5, pp. 410-419, 2007

[7] Y. Miura, *Retinal pigment epithelium choroid organ culture*. Expert Reviews Ophthalmology, vol. 6, no. 6, pp. 669-680, 2011

[8] C. Dysli et al., *Quantitative analysis of fluorescence lifetime measurements of the macula using the fluorescence lifetime imaging ophthalmoscope in healthy subjects*. Investigative Ophthalmology & Visual Science, vol. 55, no. 4, pp. 2106-2113, 2014

[9] N. Nakashima, K. Yoshihara, F. Tanaka and K.Yagi, *Picosecond fluorescence lifetime of the coenzyme of D-amino acid oxidase*. The Journal of Biological Chemistry, vol. 255, no. 11, pp. 5261-5263, 1980

Ex-vivo Optical Imaging and Measurements of Intrinsic Optical Signals from Porcine Retina with Full-Field Swept-Source Optical Coherence Tomography

S. Burhan [1], C. Pfäffle [2], D. Hillmann [3], B. Kabuth [1], H. Sphar [2], F. Hilge [1] and Y. Miura [2,5], G. Hüttmann [2,4,6]

[1] Medizinische Ingenieurswissenschaft, Universität zu Lübeck, Germany,
{sazan.burhan, bastian.kabuth, felix.hilge}@student.uni-luebeck.de
[2] Institute of Biomedicial Optics, Universität zu Lübeck, Germany, clara-pfaeffle@gmx.de,
{huettmann, miura, sphar}@bmo.uni-luebeck.de
[3] Thorlabs GmbH, Germany, dhillmann@thorlabs.com
[4] Medical Laser Center Lübeck GmbH, Germany
[5] Department of Ophthalmology , Universität zu Lübeck, Germany
[6] Airway Research Center North (ARCN), Member of the German Center for Lung Research (DZL), Germany

Abstract

It has been shown that noninvasive detection of optical path length changes measures neuronal photoreceptor activity of single cones in the living human retina. For a better understanding of the phototransduction of the human eye and the physiological function of the retina, it is necessary to investigate the connection of the observed effects with *ex-vivo* eyes, as these are easier to manipulate and do not produce motion artifacts like *in-vivo* human eyes. Fresh samples of porcine retina were preapared and measured in three different solutions (neurobasal medium with and without porcine serum, phospate buffered saline (PBS)). The culture media should ensure the vitality of the retina for reproducible measurements. It could be proven that even after several hours after the slaughter of the animals the photoreceptor activity could be detected.

1 Introduction

Optical coherence tomography (OCT) offers many advantages for diagnostic in optomalogy and neurology, through the observation of intrinsic optical signals. It can be used to obtain functional information in the retina [1], [2]. There is a great interest in the functionallity and structure of the retina in medical diagnostics, as a variety of diseases can be detected early by targeted observation. In addition, observation of the retinal neuronal activity provides new insights into the visual phototransduction to clarify physiological causes.

Phototransduction is the first step in seeing, whereby photons absorbed by the retinal rods and cones trigger an electrical signal, which is then transmitted to higher-order retinal neurons for processing [3]. Fourier-domain OCT offers the possibility to detect phases of the backscattered light. These phases react to small changes in the optical path length. It is the product of the geometric length of the path light follows through a system, and the index of refraction of the medium through which it propagates.

The optical path length of the outer segment of the photoreceptors is proportional to the time the light need to pass through the cells. Using the phase difference of light backscattered from different depth, small changes in the optical path length (in a nm-range) can be detected in the outer segments (OS), where the phototransduction takes place. Because the changes in the optical path length are small, full-field swept-source OCT is needed to measure reliably changes in the outer segments. FF-SS-OCT enable very high imaging rates of serveral hundred volumens per second. The three-dimensional data acquisition is parallelized axial direction by the fourier domain principle and lateral direction by the simultaneous acquisition of all sampling points.

2 Material and Methods

2.1 Preparation of the Retinal Explants

The porcine eyes were obtained from a small butchery "Schacht" in Bad Oldesloe. The sample preparation was performed within several hours after enucleation. Retinal explants were prepared as shown in Fig.1. The anterior part of the eye, including cornea and lens, and the vitreous body were removed. The tissue sheet including neural retina, the RPE (Retinal pigment epithelium) and the choroid was carefully removed from the sclera, and clamped between the culture rings (minucells and minutissues, Bad Abbach).

The explant in rings was then fixed in the provided holder of the measurement chamber (Fig.2).

The preparation should be done in a dark room to protect the retinal sample from the bright light. The measurment chamber was filled either with a culture medium (neurobasal medium: Thermo Fischer Scientific) or with phosphate buffered saline (PBS) with calcium and magnesium (Sigma Aldrich). The neurobasal medium used in this

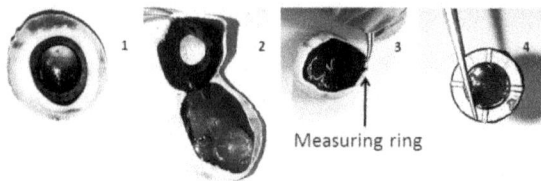

Figure 1: Preparations steps of the porcine retina
The porcine eye before preparations (1). The eye was openend and the vitreous was removed (2). Then the RPE layer was pulled of from the sclera (3) and clamped in a measuring ring (4).

study was without phenol red, because phenol red could affect the optical results. The medium enables the long-term and short-term preservation of homogeneous populations of neuronal cells. To prepare neurobasal medium with porcine serum, 50 ml of 10 % porcine serum, 5 ml of penicillin and streptomycin mixture, 5 ml of sodium pyruvate and 2.5 ml of GlutaMAX^{TM}-I Supplement were added to 450 ml of neurobasal medium [4]. It could be determined by different vitality test that this recipe with neurobasal medium is best suited to keep the porcine sample fresh as long as possible for the measurements [4]. Phosphate buffered saline is a water-based salt solution commonly used in biological research.

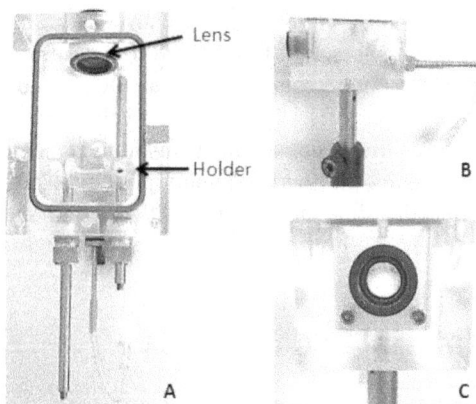

Figure 2: Measurement chamber from different views
The measurement chamber represents an artificial eye in which the retinal tissue is fixed. For an optical adjustment of the focus length, a 14 mm achromat in the front replaces the porcine cornea and lens. The cavity of the chamber was filled with cell culture medium. The measuring rings were pushed into a special holder. The two screws at the rear end of the chamber are used to change the distance between the measuring ring and the lens.

2.2 Full-Field Swept-Source-OCT setup

The used FF-SS-OCT setup is shown in Fig.3. It is based on a Mach-Zehnder interferometer with fiber coupler to generate sample and reference illumination.

The light from a laser source (Superlum Broad sweeper BS 840, 50 nm sweep range, 841 nm central wavelength) is split into two beams for the collimated reference illumination and the collimated illumination of the porcine retina. The backscattered light of the porcine retina is imaged onto a high speed camera (Photron FASTCAM SA-Z), where it superimposed with the reference light. A white LED is used to stimulate the porcine retina. The LED is projected via lenses onto the X-shaped mask. The illuminated pattern is then coupeled into the FF-SS-OCT setup. The swept-source and the camera were synchronized by a trigger signal for each volume acqusition by a minicontroller (Arduino Uno). The setup is described in more details elsewhere [1].

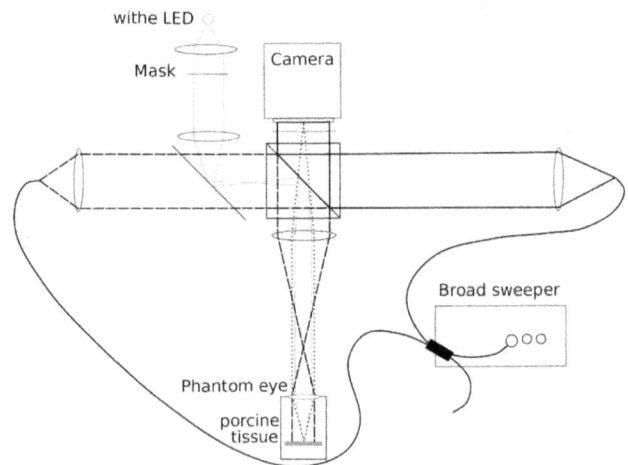

Figure 3: Full-field swept-source OCT setup
Light from the Broad sweeper is split into two beams for the reference illumination (solid line) and illumination of the porcine retina (dashed line). The backsacttered light is imaged onto the camera. The stimulation of the porcine retina is done by a white LED and a mask, which project an "X". The tissue is located in a measurement porcine chamber filled with culture medium.

2.3 Measurements and Processing

The porcine retinal tissue was measured in neurobasal medium in presence of porcine serum and without porcine serum. Comparative measurements were carried out in PBS. First, short series of experiments were performed to evaluate the influence of the medium on measured signals. Subsequently, longer timeseries were taken. To observe changes in the optical path length along the outer segments, phase data were evaluated [1]. A single recorded volume was calculated from 512 containing 640 x 368 pixeals. The field of view was 1.8 mm x 1 mm. The camera acquire OCT-volumes with a rate of 60000 frames per secound. During a measurement 70 volumes were generated with a time interval of 10 ms per volume. The light stimuli started from the 5th volume and lasted over 650 ms. After acquiring all data, the raw images from the camera were transferred to a PC, where processing took place. First the volumetric OCT data for each volume within a time series were reconstructed. Since the porcine retina did not move during the measurements, in contrast to the human eye, motion correction was not necessary. Then the inner segment/outer segment (IS/OS) were segmented and flattenend. A single B-Scan from a volume after flattening is shown in Fig.4. The phase difference between the photoreceptors IS/OS junction and the outer segment tips were evaluating overtime. All processing steps can be found in more detail elsewhere [1].

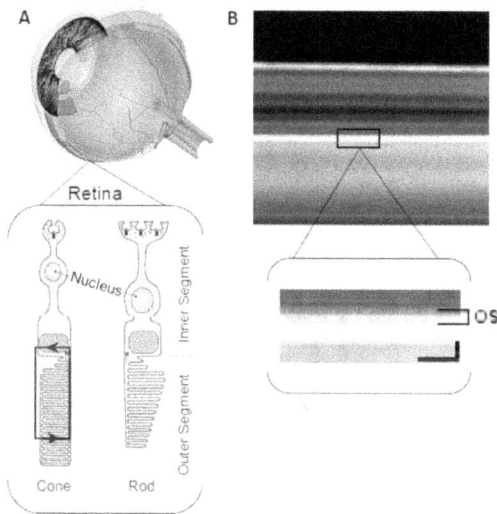

Figure 4: Photoreceptors and B-Scan of a porcine retina (A): Schematic representation of rod and cone photoreceptors. The inner segment (IS) is responsible for protein biosynthesis, while in the outer segment (OS) phototransduction takes places. (B): B-Scan of the porcine retina. Differences in the optical path length in the outer segment (OS) were observed and evaluated.

3   Results and Discussion

The measurements shown that the time after harvesting the eyes plays a major role in photoreceptor activity. A detachment of the retina is an indication for a state of the retina, which is not suitable for measurements (Fig.5). In these eyes mostly a weak or even no signal was detected.

Figure 5: A detached retina
The retina is arched and the individual layers are not clearly separated.

Through a cooled storage of the eyes, the degradation process slowed down. Several hours after the slaughtering, the eyes were fresh enough to observe reaction of the photoreceptors. The optical path length changes were sharply limited to the area of stimulation.

An important role had also be the culture medium. Porcine retina measured in PBS initially almost always showed photoreceptor activity. Samples, which stayed long in PBS, had a detached retina. Therefore PBS is not a suitable culture medium for long term measurements. Fig.6 and Fig.7 show that porcine retina elicits in neurobasal medium a bigger positive response to a light stimulation than in PBS. Neurobasal medium without porcine serum seems to be more suitable for long term measurements. The neurobasal curve in Fig.6 shows a great similarity to the time course of the optical path length changes of human eyes [1]. Both, human eyes and porcine eyes have a more and less linear rise. In the case of neurobasal medium containing porcine

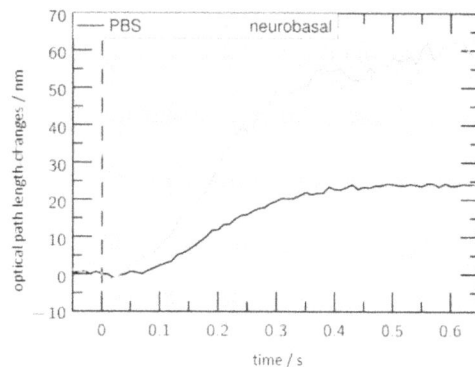

Figure 6: Time course of the optical path length changes
Optical path length changes of the porcine retina after a light stimuli of 650 ms in neurobasal medium without porcine serum (gray line) and in PBS (black line).

serum, effects could not be seen in any measurement series (Fig.7). The porcine serum, which consists mainly of proteins, water and lipids seems to prevent phototransduction. To find a reason for the different behavior, the absorption spectrum of PBS and neurobasal medium with and without porcine serum was measured in a wavelength range of 300 to 1000 nm. The measurements showed no significant difference between the culture media (data not shown). The stimulating light is not absorbed by the media on its way to the retina. The serum probably blocks chemically the phototransduction.

Optimizing the experimental conditions a suitable procedure could be determined to perform reproducible *ex-vivo* measurements of optophysiological effects in porcine retina. At least 70 % of the porcine eyes in PBS and neurobasal without serum showed effects. To further understand the processes of the phototransduction, future measurements could examine the effects of serum or other substances on the retina by adding chemicals to the culture medium. It will give an understanding of the connection the path length changes to fuctional processes in the photoreceptors.

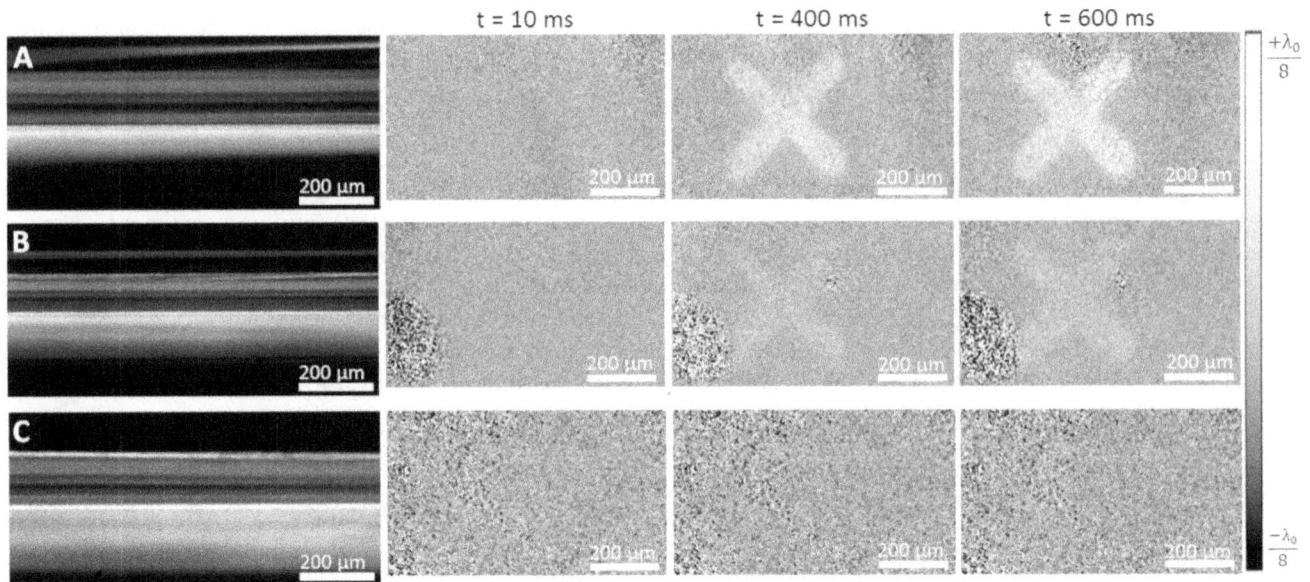

Figure 7: Response from the retina in neurobasal with and without porcine serum and in PBS to a 650 ms stimulus Spatially changes in the optical path lengths of the outer segment were observed from the phase data. Retinal tissue was measured in neurobasal medium without porcine serum (A). During stimulation the signal growed from a small positive response to a clearly visible "X" at about 600 ms. The signal in PBS was considerably weaker (B). No effects could be seen with neurobasal medium containing porcine serum (C). The B-Scans were similar, so the effects are due to the different medium.

4 Conclusion

Even after serveral hours after the slaughter of the animals reactions of the photoreceptors to the light stimuli were observed. In addition, a great influence of the culture medium in the measurement chamber on the results was determined. In neurobasal medium with porcine serum no effects could be detected, while in PBS and in neurobasal medium without porcine serum the retina reacted. We assume that the constituents of the porcine serum interferes with the phototransduction. In PBS the retina detached quite soon from the RPE (Retinal pigment epithelium) layer. Neurobasal medium without porcine serum seems to be suited the best for the measurement of phototransduction with FF-SS-OCT for up to several hours of postmortem time. In conclusion a reproducible *ex-vivo* system to study optophysiological effects with explanted porcine eyes could be developed.

Acknowledgement

The work has been carried out and supervised at the Institute of Biomedical Optics, Universität zu Lübeck. It was funded by the German Science Foundation (DFG HU 629/6-1) "Holographic OCT for functional retinal imaging". We would like to thank all the involved group members.

5 References

[1] D. Hillmann, H. Sphar, C. Pfäffle, H. Sudkamp, G. Franke, and G. Hüttmann, *In vivo optical imaging of physiological responses to photostimulation in human photoreceptors.* Proceedings of the National Academy of Sciences, vol. 113, pp. 13138–13143, 2016.

[2] R. Zawadzki, S. Choi, S. Jones, S. Oliver and J. Werner, *Adaptive optics–optical coherence tomography: optimizing visualization of microscopic retinal structures in three dimensions.* Journal of the Optical Society of America A, vol. 24, no. 5, pp. 1373–1383 , 2007.

[3] P. Liebman, W. Jagger, M. Kaplan, F. Bargoot, *Membrane structure changes in rod outer segments associated with rhodopsin bleaching.* Nature, vol. 251, no. 5470, pp. 31–36, 1974.

[4] B. Kabuth, *Untersuchung von verschiedenen Nährmedien für Schweineretina zur Optimierung der Zellvitalität von Photorezeptoren.* Bachelor thesis, Universität zu Lübeck, 2018.

[5] A. Hendrickson and D. Hicks, *Distribution and Density of Medium- and Dhort-wavelength Selective Cones in the Domestic Pig Retina.* Experimental Eye Research, vol. 74, no. 4, pp. 435–444, 2002.

[6] Y. Xincheng, W. Benquan *Intrinsic optical signal imaging of retinal physiology: A review.* Journal of Biomedical Optics, vol. 20, no. 9, 2015.

Cellular localization of the epithelial cell adhesion molecule (EpCAM) in synchronized cell culture

P. Enzian [1], A. Link [2], and R. Rahmanzadeh [2]

[1] Medizinische Ingenieurwissenschaft, Universität zu Lübeck, paula.enzian@student.uni-luebeck.de

[2] Institute of Biomedical Optics, Universität zu Lübeck, {link, rahmanzadeh}@bmo.uni-luebeck.de

Abstract

The number of cancer patients is increasing and today's treatment options, especially chemotherapy and radiotherapy are associated with several side effects and a low selectivity for tumor cells. Molecular targeted therapies are promising alternative approaches for cancer treatment. In this study, the foundations for a targeting approach for the antigen EpCAM, which is overexpressed in many tumor cells, are established. For this purpose, the localization of EpCAM was investigated with the aid of immunofluorescence staining of MCF-7 breast cancer cells and of synchronized HeLa cervical cancer cells. The synchronization method *mitotic shake-off* for HeLa cells was established. In addition, the intracellular domain of EpCAM was detected in the nucleoli of the two cell lines, which could be a good selective targeting approach for future research.

1   Introduction

The number of cancer patients is increasing due to the rising life expectancy of the population. Breast cancer is a malignant tumor and is the most commonly diagnosed cancer in women worldwide with 23% of total new cancers and 14% of deaths [1]. In cervical carcinoma a total of 4.540 new cases were registered in Germany [2]. In today's clinical practice, malignant tumors are treated by means of surgical operations, radiotherapy and chemotherapy. Especially radiotherapy and chemotherapy are often associated with strong side effects and have a low selectivity for proliferating cells. For this reason, research is carried out on alternative treatment therapies.

The epithelial cell adhesion molecule (EpCAM) is overexpressed in many malignancies especially in epithelial cancers [3]. It has a molecular weight of approximately 40 kDa and is involved in migration, proliferation, adhesion and differentiation [3]. For this reason, EpCAM has been used in other studies for therapeutic research on various epithelial cancers.

In order to gain a better understanding of EpCAM with its extracellular domain (EpEX) and its intracellular domain (EpICD) in proliferating cells, it is important to analyze it during different cell cycle phases. For this purpose, synchronization of cancer cells, can be a helpful method. The proposed technique *mitotic shake-off* allows an enriched cell population in a single stage of the cell cycle [4]. This method has several advantages, e.g. it is based on normal cellular processes and does not require special equipment or reagents. A limitation of the method is the low yield of mitotic cells, which can be collected by shaking off.

Synchronization of the cells could also be carried out by chemical inhibitors. However, these methods have two serious drawbacks: (1) chemical treatment has been implicated in the potential disruption of normal cell cycle regulatory processes, and (2) chemicals may have more than one target in cells, some of which are not fully known [4]. For this reason, only the mechanical method of synchronization is used in this study.

2   Material and Methods

This section presents an overview of the methods focusing on two main parts: immunofluorescence staining and synchronization of cells. First a brief description of the cell cultures is given.

2.1   Cell cultures

The MCF-7 breast cancer cell line was first isolated in 1970 from breast tissue of a 69-year-old Caucasian woman [5]. The MCF-7 cell line developed at the Michigan Cancer Foundation, Detroit, became a standard model in research [5]. Cells were obtained from American type tissue culture (ATCC HTB-22). This adherent cell culture is cultivated in Dulbecco's Modified Eagle's Medium (DMEM) with $4500\,mg/l$ glucose (Sigma-Aldrich, USA) supplemented with 10% fetal bovine calf serum (FBS) (Sigma-Aldrich, USA), 1% Penicillin-Streptomycin (10.000 units/ml Penicillin and $10\,mg/mL$ Streptomycin, Sigma-Aldrich, USA) and 0.1% Insulin (Sigma-Aldrich, USA).

Furthermore, HeLa cervical cancer cell line was used in these investigations. HeLa cells were isolated in 1951 of the

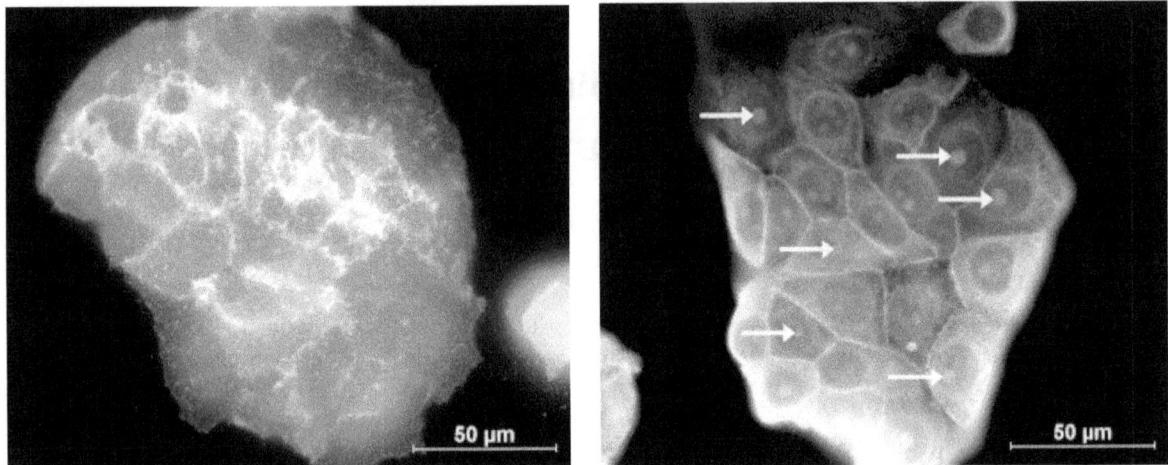

(a) Localization of the extracellular-surface Domain (EpEX).

(b) Localization of the intracellular Domain (EpICD).

Figure 1: Localization of EpEX and EpICD in MCF-7 cells monitored by an inverse fluorescence microscope. (a) After incubation of primary antibody Ber-EP4 (1 : 500) against EpEX and secondary antibody Alexa Fluor 488 (1 : 300) on fixed MCF-7 cells, these were located mainly on the cell surface. (b) Whereas after incubation of the primary antibody clone 4A7 (1 : 100) against EpICD, these were located in the nucleoli (see arrows) of the cell. Furthermore, a staining of the cell membranes was observed.

31-year-old patient Henrietta Lacks [6]. It is the first permanent cell line of human cells used in standard models for cancer research and it was purchased from American type tissue culture (ATCC, No. ACC 57). This cell line is cultivated in DMEM with 4500 mg/l glucose (Sigma-Aldrich, USA) supplemented with 10% FBS and 1% Penicillin-Streptomycin.

Cells were grown in T-75 cell culture flasks in an incubator at 37 °C and a 5% CO_2 concentration. Cells were split at a confluence of about 80%. For this purpose, the medium was removed and the cells were washed three times with Dulbecco's Phosphate-Buffered Saline without calcium chloride and magnesium chloride (PBS). Afterwards 3 ml Trypsin-EDTA were added and incubated for 3 min. For the inactivation of trypsin, medium was added and the suspension was centrifuged for 5 min at $400 \times g$. The supernatant was discarded and the cell pellet was resuspended in 10 ml medium.

2.2   Cell number determination

The cell count is determined microscopically using a hemocytometer. The cell suspension was mixed in a 1 : 1 ration with 0.5% Trypan blue solution. This is used as a vital stain to selectively color dead cells blue. Subsequently 10 µl of this solution were filled in the hemocytometer and counted under a conventional light microscope.

2.3   Immunofluorescence staining

The cellular localization of EpCAM was analyzed by fluorescence microscopy. The images were captured using the Nikon Eclipse Ti inverse fluorescence microscope with NIS-Elements software. All fluorescence images were taken with a 60× oil immersion objective.

The cell density was adjusted to 60.000 cells/ml and 2 ml of the cell suspension were seeded in 35 mm disposable plastic cell culture dishes with a glass bottom and four compartments. Before staining, cells were fixed and permeabilized. For this purpose, the medium was first removed and the cells were washed once with 500 µl PBS and afterwards they were incubated in 4% paraformaldehyde (PFA) for 15 min. To permeabilize the cells, they were treated with 0.25% Triton X-100 for 3 min. Subsequently, the cells were washed three times for 3 min with PBS.

Clone Ber-EP4 (1 mg/ml, Sigma Aldrich, USA) was used as primary antibody against the extracellular-surface domain (EpEX) of EpCAM. Cells were incubated at room temperature for 1 h with clone Ber-EP4 (1 : 500 dilution in PBS and 10% FBS).

For the detection of the intracellular domain EpICD of EpCAM, the primary antibody clone 4A7 (1 mg/ml, Sigma Aldrich, USA) was used. A 1 : 100 dilution in PBS and 10% FBS was prepared and incubated for 1 h at room temperature.

After three times washing with PBS, cells were incubated with the secondary antibody Alexa Fluor 488 of goat anti-mouse IgG (2 mg/ml, Thermo Fisher Scientific). Cells were incubated at room temperature for 1 h with Alexa Fluor 488 (1 : 300 dilution in PBS and 10% FBS).

Before capturing images under the fluorescence microscope, cells were washed three times with PBS and stored in 500 µl PBS. All immunofluorescence stainings were repeated at least three times to obtain representative and reproducible results.

2.4   Synchronization of HeLa cell culture

Synchronized cells undergo the cell cycle in a homogeneous manner, allowing enriched cell populations to be gener-

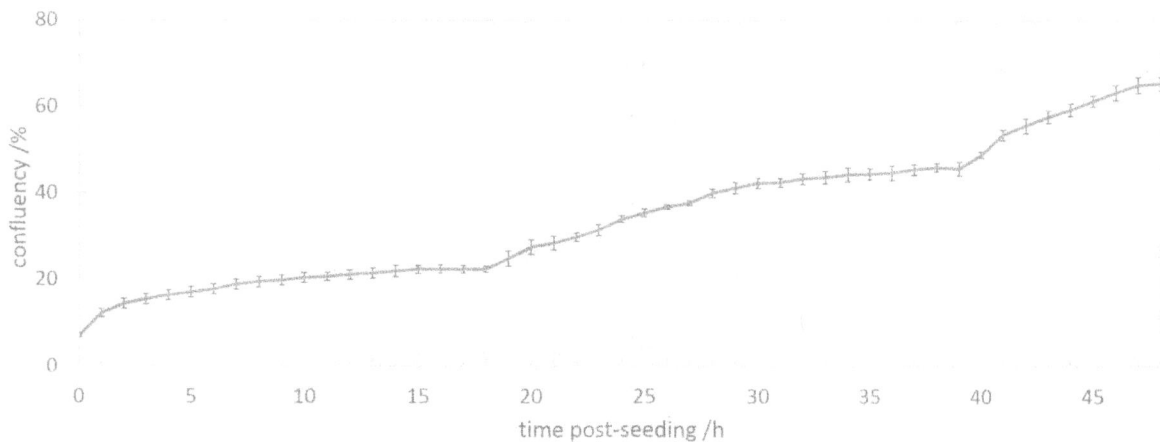

Figure 2: The confluence of synchronized HeLa cells over a period of 48 hours showed increasing values after 18-19h and another one after 38-39h.

ated in a single stage of the cell cycle [4]. The technique *mitotic shake-off* is based on the observation that cells are rounded during mitosis and have fewer points of attachment to the culture flask [4]. Each cell type requires characterization to determine the most appropriate times at which the cells should be harvested after release [4]. The protocol used was modified for our experiments as described below to maximize the number of synchronized cells.

HeLa cells were grown in T175-flasks in an incubator at 37 °C and 5% CO_2 concentration until they reached 70-80% confluency. Subsequently, the growth medium was removed and the cells were washed once with pre-warmed PBS. 6 ml of of 0.25% trypsin solution was added and incubated for 10 min. To shake off loose, rounded cells, the flasks were gently tapped on the bench and then pre-warmed medium was added to inactivate trypsin. The resulting cell suspension was transferred to 50 ml centrifuge tubes and centrifuged for 5 min at $500 \times g$ and room temperature. The supernatant was discarded and the pellet was resuspended in medium and again centrifuged. Afterwards, the cells were counted and cell density was adjusted to 2.5×10^5 cells/ml. Subsequently, 30 ml of cell suspension were transferred into the T175-flasks and incubated for 6 h. The following washing step removes all unattached cells. For this, the cell culture bottles were rinsed twice with 20 ml pre-warmed medium. Subsequently, 30 ml of medium were added and incubated for 10 h. In order to obtain mitotic cells, the cell culture flasks were tapped on the bank and medium was swirled around the flask and then removed by a pipette. The cell suspensions from several flasks were pooled and centrifuged for 5 min at $500 \times g$ to obtain a pellet of mitotic cells. The supernatant was discarded and the cell pellet was resuspended. Finally, the cell density was set for the respective experiments and were seeded in petri dishes.

To determine the confluence and growth characteristics of the synchronized HeLa cells, JuLI™ FL Fluorescence live cell movie analyzer (NanoEnTek Ink.) was used. This is a compact inverted light microscope with two fluorescence channels which was placed in the incubator. This allowed

the observation of HeLa cells under physiological conditions. For this, 2 ml of the cell suspension were seeded with a density of 60.000 cells/ml in cell culture dishes with a glass bottom for microscopic applications. The cells were observed immediately after seeding for a period of 48 h. An image was created every 5 min. In addition, confluence was determined using JuLI™ FL PC for an image based analysis. These investigation were repeated at least three times to obtain representative and reproducible results.

3 Results and Discussion

This section presents the results of the immunofluorescence staining and cell synchronization and first evaluations are delivered.

3.1 Localization of EpCAM in MCF-7 cells

The localization of EpEX and EpICD in MCF-7 cells is shown in Fig. 1. On fixed MCF-7 cells, EpEX was located mainly on the cell surface and bound to the cell membranes, but it could not be observed in the nucleoli, confirming the characteristics of the extracellular domain. Similar EpEX expression was observed in hypopharyngeal carcinoma cells (FaDu) and in colon carcinoma cells [7].

Whereas EpICD was located in the cytoplasm and was also detected in the nucleoli. These results indicated that EpICD, unlike the extracellular domain, can be incorporated into cells.

3.2 Growth properties of synchronized HeLa cells

Fig. 2 shows the confluence of synchronized HeLa cells over 48 h. A strong increase of the area covered by cells to an average of 27% was observed 20 h after seeding. This suggests that mitosis of the cells has taken place in the hour before, as in this phase the cells detached from the petri dishes and re-attached again after mitosis has ended. As a

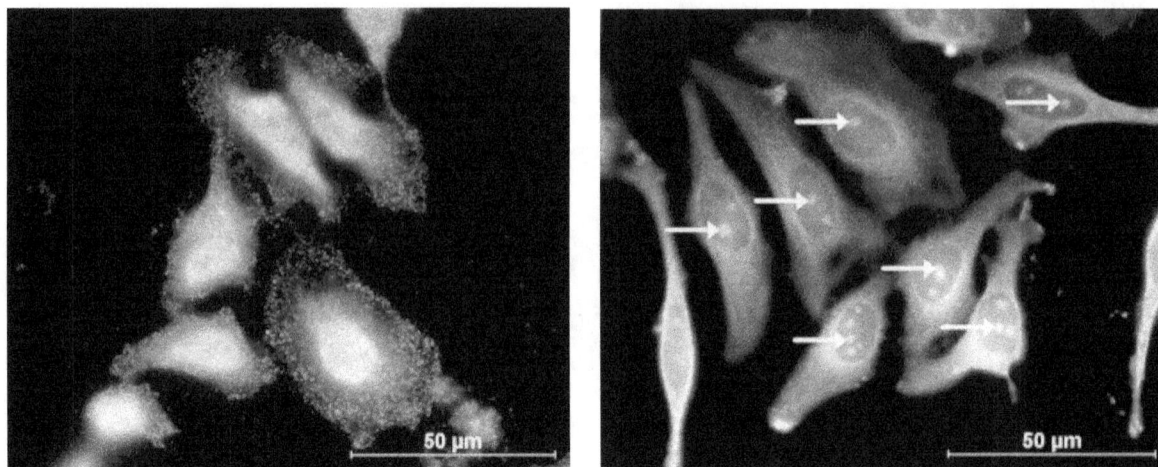

(a) Localization of the extracellular-surface Domain (EpEX).

(b) Localization of the intracellular Domain (EpICD).

Figure 3: Localization of EpEX and EpICD in synchronized HeLa cells was monitored by an inverse fluorescence microscope 12 h after seeding. (a) After incubation of primary antibody Ber-EP4 (1 : 500) against EpEX and secondary antibody Alexa Fluor 488 (1 : 300) on fixed cells, these were located mainly on the cell surface. In addition, amplified signals in the nucleus area were observed. (b) EpICD was detected with the help of primary antibody clone 4A7 (1:100) and secondary antibody Alexa Fluor 488 (1:300). Mainly these were located in the nucleoli (see arrows).

result, this led to an increase in the covered area. The same phenomenon can be observed after about 38-39h.

3.3 Localization of EpCAM in synchronized HeLa cells

The localization of EpEX and EpICD in synchronized HeLa cells is shown in Fig. 3. The images were taken 12 h after seeding in petri dishes. Similar to the non-synchronized MCF-7 cells, EpEX was located mainly on the surface. However, in the synchronized HeLa cells a spotted pattern can be seen on the surface, which was more evenly distributed among the MCF-7 cells. In addition, signals in the nucleus area were detected. As non-synchronized MCF-7 cells, EpICD was observed in the nucleoli and also in the cytoplasm.

4 Conclusion

In this study, synchronized HeLa cell culture was established and first studies on the localization of EpEX and EpICD were implemented. Important properties about the localization of EpCAM have been provided. In particular, the localization of EpICD in the nucleoli may be a promising approach for further research. In order to extend the understanding of the tumor-associated antigen EpCAM, it is important to further study the cell cycle dependency of EpCAM in the nucleoli.

The approach of using synchronized cells offers the opportunity to gain more knowledge into the localization of EpCAM in the different cell cycle phases. Further studies are necessary to determine the effects of EpCAM during mitosis and to characterize them in the different phases to establish it to be a target for selective tumor therapy.

Acknowledgement

The work has been carried out at the Institute of Biomedical Optics, Universität zu Lübeck.

5 References

[1] A. Jemal, F. Bray, M. M. Center, J. Ferlay, E. Ward and D. Forman, *Global cancer statistics*. CA: A Cancer Journal for Clinicians, vol. 61, no. 2, pp. 69–90, 2011.

[2] Robert Koch Institut und die Gesellschaft der epidemiologischen Krebsregister in Deutschland e.V., *Krebs in Deutschland für 2013/2014*. Zentrum für Krebsregisterdaten, vol. 11, pp. 80–83, 2017.

[3] P. A. Baeuerle and O. Gires, *EpCAM (CD326) finding its role in cancer*. British Journal of Cancer, vol. 96, no. 3, pp. 417–423, 2007.

[4] J. Jackman and P. M. O'Connor, *Methods for Synchronizing Cells at Specific Stages of the Cell Cycle*. Current Protocols in Cell Biology, pp. 8.3.1-8.3.20, 2001.

[5] A. S. Levenson and V. C. Jordan, *MCF-7: The first hormone-responsive Breast Cancer Cell Line*. Cancer Research, vol. 57, no. 15, pp. 3071–3078, 1997.

[6] W. F. Scherer, J. T. Syverton and G. O. Gey, *Studies on the Propagation in vitro of poliomyelitis viruses*. Journal of Experimental Medicine, vol. 97, no. 5, pp. 695–710, 1953.

[7] D. Maetzel, S. Denzel, B. Mack, M. Canis, P. Went, M. Benk, O. Gires et al., *Nuclear signalling by tumour-associated antigen EpCAM*. Nature cell biology, vol. 11, no. 2, pp. 162–171, 2009.

In-vivo examination of retinal vessel pulsation during light stimulation by phase-sensitive full-field swept-source optical coherence tomography

F. Hilge [1], H. Spahr [2], C. Pfäffle [2], D. Hillmann [3], S. Burhan [1], and G. Hüttmann [2,4,5]

[1] Medical Engineering, Universität zu Lübeck, felix.hilge@student.uni-luebeck.de
[2] Universität zu Lübeck, Institute of Biomedical Optics, {huettmann,spahr,pfaeffle}@bmo.uni-luebeck.de
[3] Thorlabs GmbH, Germany, dhillmann@thorlabs.com
[4] Medical Laser Center Lübeck GmbH
[5] Airway Research Center North (ARCN), Member of the German Center for Lung Research (DZL), Germany

Abstract

Vascular diseases often affect minor blood vessels at first and cause changes in their stuctural, functional and biomechanical properties. Many of these variations may influence the propagation of the pulse wave through the vessel. Phase-sensitive full-field swept-source optical coherence tomography (FF-SS-OCT) is an imaging technique which can measure the pulsation and the pulse wave velocity of retinal vessels. Here we investigate, whether light stimulus induced vasodilation of retinal micro vessels affects the pulse curves and delay of major arteries and veins. A retinal stimulation unit was developed for the OCT system and pulse curves of a healthy test subject were recorded. Stimulation with intense, flickering light did not affect the amplitude of or delay between arterial and venous pulsation. However, unexpected effects like spatial irregularities in the arterial pulsation and axial retinal motion that was not related to the pulsation of adjacent vessels were observed.

1 Introduction

Optical coherence tomography (OCT) is a three-dimensional, non-invasive and contact-free imaging technique. Using extremely fast phase-sensitive full-field swept-source (FF-SS) OCT very small axial motion within the retina can be detected. This technique was applied measuring the propagation of pulse waves in retinal vessels [1], [2] as well as the response of *single* photoreceptors to incident light [3]. Further, it is known, that retinal vessels are affected by light stimuli [4]. The distance between retinal nerve fiber layer (RNFL) and pigment epithelium (RPE) changes temporally due to pulsation of blood vessels within the neuronal retina. Spatially resolved measurements of the thickness changes allow to follow the pulse waves in arteries and veins. After an arterial pulse wave traverses the capillaries it is believed to enter the vein, which may be the reason for a delay between arterial and venous pulsation. If this is true, the delay may be influenced by the capillary diameter. If the capillaries in a stimulated region of the retina change their diameter as a response to an increased retinal consumption of oxygen, the effect might somehow translate into a change of the observable pulsation of the larger arteries and veins. This effect should be maximized by illuminating a large area. Since vessels especially react to flickering light [4], a pan-retinal stimulation unit with flickering illumination was developed for the OCT setup.

Pulse wave velocity (PWV) depends on the vessel radius r_0. Neglecting viscosity of blood, the PWV within blood vessels is described using by the *Moens-Korteweg equation* [1]

$$PWV = \sqrt{\frac{Eh}{2r_0\rho}} \qquad (1)$$

with the Young's modulus E of the vessel walls, their width h and the density of blood $\rho \approx 1060\,\mathrm{kg\,m^{-3}}$. Using the fact, that a change of r_0 by factor $\alpha = 1 + \Delta r_0/r_0$ modifies h to h/α and assuming a constant Young's modulus, the relation $PWV \propto 1/\alpha$ can be shown. Since Δt is the length of the vessel d divided by PWV, the time delay is linear proportional to α. However, considering micro vessels (1) is not applicable for quantitative computations but most of the relations remain valid qualitatively.

If the predicted effect is indeed observable, the measuring technique used here might be able to diagnose pathologies affecting the ability of capillaries to match their diameter with oxygen consumption *without* necessarily resolving them. Furthermore, the observation of the effect would prove the pulse wave indeed propagates through the capillaries.

2 Material and Methods

The FF-SS-OCT setup is based on the Mach-Zehnder interferometer which can be seen in Fig. 1. The whole setup is described in detail in earlier publications [1] - [3]. The light source (Superlum Broadsweeper 840) has a central wavelength of 841 nm using its maximum sweep range of 50 nm. A beam splitter seperates the emitted light into sample beam and reference wave. The optical power of the sample illumination is 5.2 mW [2]. In order to obtain a collimated illumination of the retina the sample beam is focused into the focal plane of the eye by an achromatic lens. The light backscattered by the retina is imaged onto a camera sensor at tenfold magnification. The collimated reference wave interferes with the backscattered light and is then recorded by a high speed camera (Photron FASTCAM SA-Z) with a frame rate of 60 kHz. Further measuring parameters are listed in Table 1. The recently added components that form the retinal stimulation unit (RSU) are described in the following paragraph.

2.1 Retinal stimulation unit

In order to achieve a stimulation of a large area of the retina, a new stimulation unit was developed and integrated into the OCT setup. The RSU consists of two main components. First, a 1 W-Laser (Lasertack LDM-520-1000-C), which emits at about 520 nm. A stimulation frequency of 10 Hz was chosen due to the fact that earlier research [4] shows this frequency causes distinct vasodilation. The second component is a 45 mm wide, square diffusor plate made of 5 mm thick acrylic glass. It is attached to the aperture of the OCT setup by magnets and then is located right in front of the eye. Hence, it covers the whole field of view of the test subject. A hole with a suitable diameter ensures the diffusor plate does not affect the beam path of the OCT setup. The material the plate is made of is highly scattering. The green laser strikes the plate after passing through a convex lens with a focal length of 30 mm, the diverging light ensures a homogeneous illumination of the diffusor plate.

Since experiments are performed with human subjects, it is required to evaluate possible risks for the test subjects. According to *DIN 60825-1* the maximum radiant exposure of the eye within an exposure time from 10 s to 100 s is $H_{max,ch} = 2.29\,\mathrm{kJ\,m^{-2}}$ for photochemical damage [5]. The shortest and most dangerous wavelength the laser might emit of 518 nm was assumed. The limit $H_{max,therm}$ for photothermal damage is about three times higher at 10 s exposure [5].

The optical power of the scattered light was detected right at the location where the eye of the test subject will be situated. The light sensitive area of the measuring instrument (Newport 1918-C, probe head: 918D-SL-OD3) has a diameter of 10 mm. Thus, the maximum radiant flux in this area is approximately 0.18 J. Regulation DIN 60825-1 requires an aperture of 7 mm for measurements because this diameter is the maximum pupil size [5]. Since the sensitive area of the probe head is larger, the amount of radiant power a

Table 1: List of measuring parameters

Recorded volumes	745
Volume rate	100 Hz
Sweep range	816 nm − 821 nm
Central wavelength λ_0	818.5 nm
Sweep time	0.83 ms
Frames per sweep	50
A-Scan rate	277 MHz
Resolution	$640 \times 360 \times 25$ px
Field of view	2.56 mm × 1.44 mm
Imaging depth	1.75 mm

human retina would have been exposed to is even smaller than the measured value. The measured optical power is about 0.9 mW. An exposure with that amount of power for 100 s would lead to an energy that reaches the retina of 0.09 J which is just half of the energy permitted by DIN 60825-1. The actual values of radiant exposure of the subject's eye are way lower because the measurements never took longer than 15 s. The laser is operated at maximum power to prevent higher optical output power due to electrical malfunctions. An OD2-Filter was used to decrease the radiant exposure by a factor of 100. Nevertheless, the RSU fulfills DIN 60825-1 even *without* the OD2-Filter.

2.2 Measuring procedure

Two large retinal blood vessels (artery and vein) were chosen for the measurements. They are located approx. 1 mm above the optic disc (see Fig. 1, white box in the SLO-image on the right side). They run in parallel. Thus, it can be assumed they are connected by the capillary network feeding and draining the same region. Dilation of capillaries in this region might influence the delay between the arterial and venous pulsation. The pupil of a healthy test subject was dialated using eyedrops. Measurements were carried out either in a completely dark environment or with the RSU.

2.3 Data post-processing and analysis

After the image reconstruction there are further steps to fulfill before the detected phases provide information about change of retinal thickness respectively about vasodilation. Those procedures are briefly described in Fig. 2. Only the two last final steps are depicted here. More detailed information about the previous steps is given in an earlier publication [1]. After the accumulated phase differences between RFNL and RPE are known it is required to select only the phase information of areas that might have been affected by the blood vessels. Hence, just regions close to a vessel are evaluated. Additional selective criteria described below were defined. To improve data quality outliers, whose motion significantly differs from that of the surrounding positions, were excluded from further calculations. Moreover, areas whose axial motion is not affected by the heartbeat are dismissed. Those pixels can be determined by a Fourier transformation of the phase differences with subsequent de-

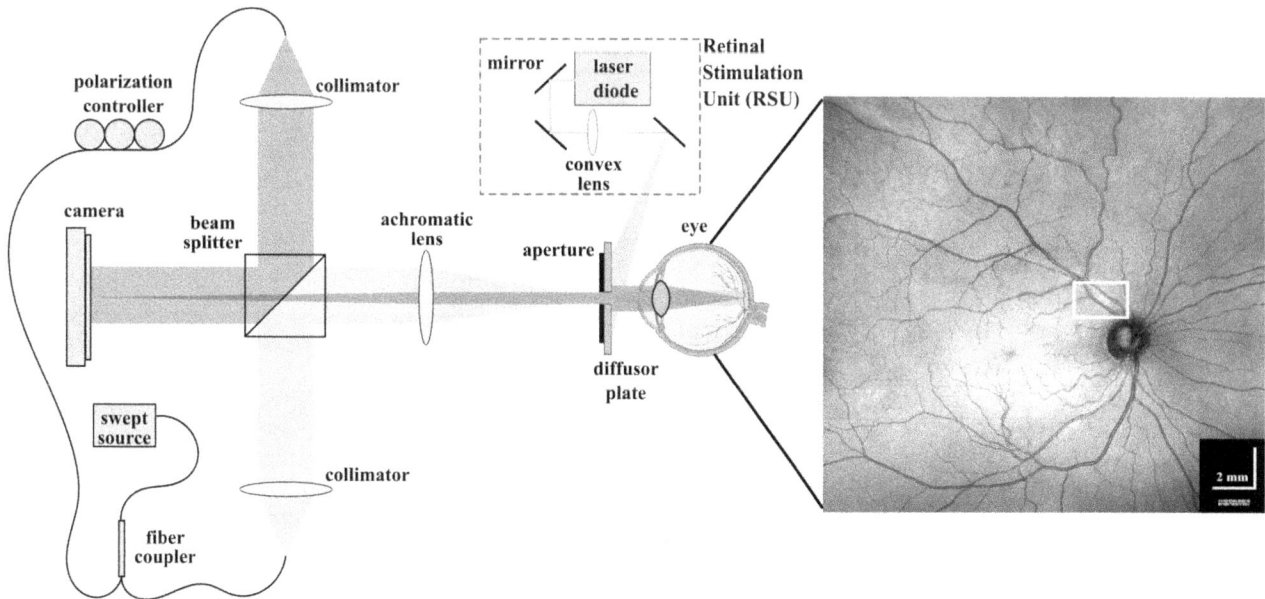

Figure 1: Full-field swept-source OCT setup with retinal stimulation unit (RSU). Examined spot is marked right.

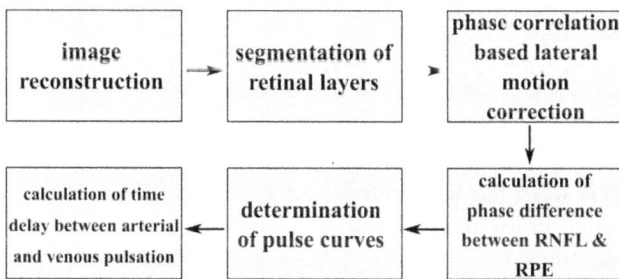

Figure 2: Sequence of post-processing procedures. Further details about the first four steps are given in [1].

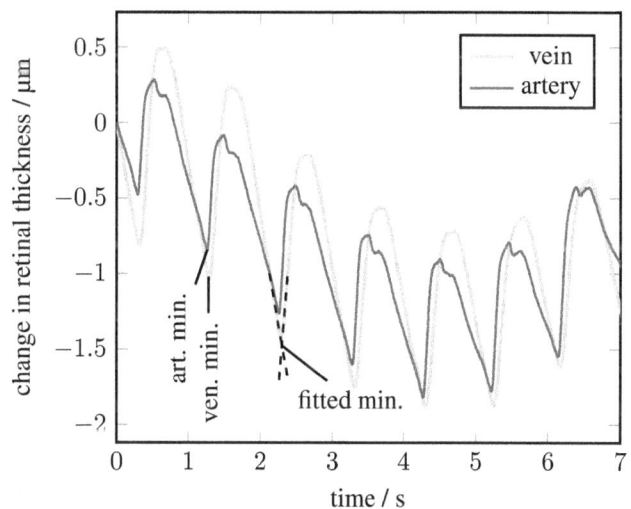

Figure 3: Arterial and venous pulse curves measured without optical stimulation. The intersection of dashed lines is determined to detect the local minimum of the pulse wave with high temporal resolution.

tection of the amplitude of the heartbeat. In a final step, positions suffering from phase wrapping, which is caused by strong motion that exceeds the valid range of phases between $-\pi$ and $+\pi$, were excluded from the data set. Following this, the spatial mean value of the pixels that have not been rejected and are located in close proximity to either the artery or the vein, respectively, were computed to obtain the arterial and venous pulse. Since a phase difference $\Delta\Phi$ of 1 rad corresponds to an optical path distinction of $\Delta z \approx \Delta\Phi\lambda_0/(4\pi n) = 48\,\mathrm{nm}$ [1] the change in retinal thickness is quantifiable. The $n \approx 1.35$ represents the refractive index within the eye. Plotted against time, the characteristic pulse curves of arteries and veins are easy to distinguish (Fig. 3). Finally, two linear functions (Fig. 3, dashed) are fitted along both sides of each local minimum of the pulse curves. Their intersections (Fig. 3, fitted min.) provide local minima with a higher temporal resolution than 10 ms. From these data the time delay between the fitted minimum of the arterial and venous pulsation is computed for every heart beat.

3 Results and Discussion

In Fig. 3 two pulse curves are depicted which were recorded without stimulation. The curves of stimulated measurements look very similar, hence they are not shown. Both curves stand out due to a very high SNR, that was achieved by applying the selective criteria mentioned in section 2.3 without any post-processing. The time delays between arterial and venous pulsation were determined for each of the 7 to 8 cardiac cycles per measurement (Table 2). Computing the last delay was not possible in three measurements, when the last venous pulse could not be recorded completely. Motion artifacts caused some obviously wrong measurements shown in the brackets in Table 2. These values were not in-

Table 2: Time delay of local minima for each heartbeat in ms. Bottom line shows the mean value

beat	stim. 1	stim. 2	stim. 3	dark 1	dark 2
1	33,42	32,74	27,08	30,97	35,34
2	(86,37)	35,74	25,05	29,05	34,55
3	32,62	31,9	26,76	34,31	29,89
4	29,13	34,38	30,32	30,52	35,41
5	29,07	34,59	29,68	30,77	32,23
6	(72,63)	(7,72)	23,61	30,07	31,39
7	(186,36)	22,4	48,46	27,66	31,88
8	25,61	n/a	n/a	n/a	38,35
t_{mean}	29,97	31,95	30,14	30,48	33,63

cluded in the mean value displayed in the bottom line. The results do *not* show any effect of the stimulation on the delay of arterial and venous pulsation.

An insufficient stimulation can be excluded because the human eye is very sensitive to the chosen stimulation wavelength. And the test subject perceived a very bright green light. Since stimulation is expected to cause retinal vasodilation [4] the results shown here imply that capillaries do not affect the pulsation of large retinal vessels. There are three possible reasons for this result. First, the dilation of the capillaries is too small to affect pulsation of large vessels. Another explanation might be, that capillaries indeed change their thickness but without any effect on larger vessels, although the pulse wave propagates through the capillary network. Finally, the pulse wave might not traverse through the capillaries and the reason of venous pulsation is completely different.

Additional effects have been observed. First, there are spatial irregularities in the amplitude of the arterial pulsation (Fig. 4, white arrows), that do not occur along the veins. This effect was already observed in previous publications [1], [2]. Its cause, however, was not determined yet. Spatial differences in the width of the artery wall might cause the irregular change of retinal thickness as well as debris within the blood vessel. Further varification is needed to investigate if the described arterial effect also occurs in a larger group of subjects.

In addition, a large area with weak pulsation that differs highly from the arterial pulse is located above the artery, which can be seen in the top right area of Fig. 4 (dashed oval). This effect increases with longer distance to the artery. Possibly, the pulsation of the choroid beneath the retina influences the measured distance between RFNL and RPE and yields the described results. Whether that effect also occurs within other subjects and its cause should be subject of further research.

4   Conclusion

A change in retinal vessel pulsation due to capillary vasodilation could not be observed. Hence, the proposed idea of a non-invasive examination of the capillary bed by imaging its effect onto the pulsation of larger vessels with OCT might not be as promising as assumed.

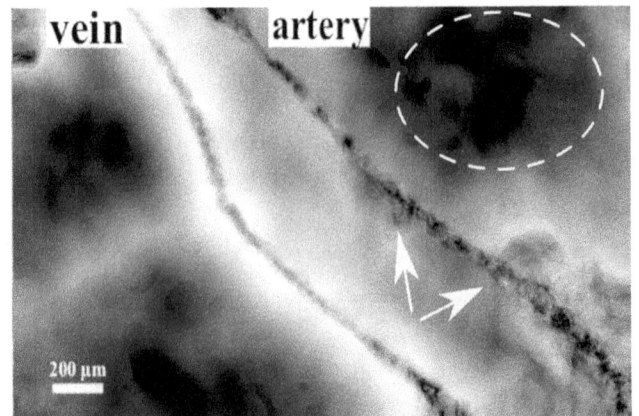

Figure 4: Amplitude of the heartbeat (arbitray units) within the field of view. Inhomogeneous pulsation along the artery is clearly visible (arrows) as well as the dark area with different pulsation behaviour above the artery (dashed oval).

Further research is necessary to explain the origin and potential clinical value of the irregular arterial pulsation. Phase-sensitive FF-SS-OCT is currently the only tool for this task.

Acknowledgement

This work has been carried out and supervised at the Institute of Biomedical Optics, Universität zu Lübeck and was funded by the German Science Foundation (DFG HU 629/6-1 "Holographic OCT for functional retina imaging"). Thanks to all contributing colleagues.

5   References

[1] H. Spahr, *Detektion von retinalen Mikrobewegungen mit phasensensitiver optischer Kohärenztomografie.* PhD thesis, Universität zu Lübeck, 2017.

[2] H. Spahr, D. Hillmann, C. Hain, C. Pfäffle, H. Sudkamp, G. Franke, and G. Hüttmann, *Imaging pulse wave propagation in human retinal vessels using full-field swept-source optical coherence tomography.* Opt. Lett., vol. 40, pp. 4771-4774, 2015.

[3] D. Hillmann, H. Spahr, C. Pfäffle, H. Sudkamp, G. Franke, and G. Hüttmann, *In vivo optical imaging of physiological responses to photostimulation in human photoreceptors.* Proceedings of the National Academy of Sciences, vol. 113, no. 46, pp. 13138-13143, 2016.

[4] K. Polak, L. Schmetterer, and C. E. Riva, *Influence of flicker frequency on flicker-induced changes of retinal vessel diameter.* Investigative Ophtalmology & Visual Scinece, vol. 43, no. 8, p. 2721, 2002.

[5] DIN EN 60825-1:2015-07, *Sicherheit von Lasereinrichtungen - Teil 1: Klassifizierung von Anlagen und Anforderungen (IEC 60825-1:2014)*

Development of an Acoustophoresic Flow Cell for Processing Undiluted Whole Blood at High Flow Rates

G. Bulz [1,2*], T. Mukashev [1,2*], F. Fiedler [2,3*], and S. Müller [2*]

[1] Biomedical Engineering, University of Applied Sciences Lübeck
[2] Medical Sensors and Devices Laboratory, University of Applied Sciences Lübeck
[3] Graduate School for Computing in Medicine and Life Sciences, Universität zu Lübeck
[*] {gordon.bulz, temirlan.mukashev}@stud.fh-luebeck.de, {felix.fiedler, stefan.mueller}@fh-luebeck.de

Abstract

The microfluidic separation of whole blood into its cellular and non-cellular components is of high interest for medical and biomedical applications. One method which allows a continuous processing of blood with a high throughput is the acoustophoresis. It provides the opportunity to quickly analyze blood parameters without the use of disposables. The highest flow rate which enabled the outcome of clinically pure blood plasma from undiluted whole blood so far was $80\,\mu l\,min^{-1}$ [1]. In this paper, we present the design of an acoustophoresic resonator that has the potential to process undiluted blood with a discharge flow of $300\,\mu l\,min^{-1}$, and its development process which followed VDI guideline 2221. This flow cell went through an iterating optimization process of verification by means of fluid-mechanical simulation and adjustment of the parameters.

1 Introduction

During the last decade, the development of lab-on-a-chip devices gained in importance and so the number of microfluidic applications like particle manipulation increased rapidly. One application which is of high importance for medical point-of-care analyses is the separation of whole blood into its cellular parts and blood plasma. To reach this aim, many passive and active continuous separation methods were developed, e.g. methods using the bifurcation law, filtration or magnetic separation [2]. However, the approach which is able to handle the highest flow rates is the acoustophoresis [3]. But still, working with highly concentrated suspensions, like undiluted whole blood, in addition to high flows is always a trade-off in relation to the separation quality. The highest known flow rate which led to the extraction of clinically pure plasma from blood with a hematocrit of $40\,\%$ was $80\,\mu l\,min^{-1}$, with a plasma outcome of $12.5\,\%$ by volume [1]. Our aim is to design an acoustophoresic flow cell which allows the processing of undiluted whole blood at flow rates of $300\,\mu l\,min^{-1}$ to enable spectrometric analyses of lactate and glucose. This value was the maximum discharge flow which could be handled in previous acoustofluidic experiments with diluted blood [4].

1.1 Acoustophoresis

Acoustophoresis is a method where high-frequency ultrasound waves are used to manipulate microparticles within a fluid. Therefore, a piezoelectric ultrasound transducer is attached to a flow cell and so induces mechanical pressure into it. If the width w of the cuboid flow cell channel is equal to a multiple of the half induced wavelength λ according to

$$w = n \cdot \frac{\lambda}{2} = n \cdot \frac{c}{2f}, \qquad (1)$$

where $n \in \mathbf{N} = \{1, 2, 3, ...\}$, c is the speed of sound and f is the frequency, a standing wave with n pressure nodes will be generated inside the channel. The induced acoustic radiation force (ARF) acts on the particles and forces them, due to their acoustic properties, to the nearest node or anti-node. Now a mechanical separation of the particles and their surrounding medium is possible. Fig. 1A illustrates the influence of the ARF on particles which will be focused in a pressure node. Fig. 1B shows the separation process, that is usually made by a $45°$ trifurcation flow splitter, but also every other angle between $0°$ and $90°$ can be used [5]. However, it is worthwhile to use an angle $\leq 11°$ to prevent the creation of turbulences or dead volumes [6].

1.1.1 Acoustic Radiation Force

The ARF is determined by (2). Considering this equation it is clear that F_{AR} is proportional to the volume of the particle V_c, to λ^{-1}, and to the square of the pressure amplitude p_0. Since p_0 results from the transducer, it is directly dependent on the applied peak-to-peak voltage. This leads to the statement that the resulting focusing will improve with an increased frequency f and peak-to-peak voltage U_{pp} [7]. Nevertheless, it is always a compromise increasing them. Commonly used values are in the range of $f = 2\,MHz$ and

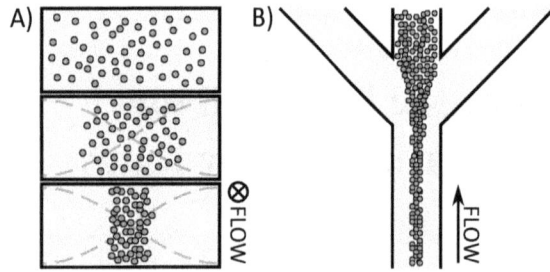

Figure 1: A) schematically shows the cross-section of a microchannel with a width of $\lambda/2$, which is excited by an ultrasonic standing wave. The particles are forced to the center of the channel by the influence of the ARF. B) demonstrates the mechanical separation of the particles going to the middle branch, and the surrounding medium flowing to the lateral branches by means of a $45°$ trifurcation flow splitter.

$U_{pp} = 15\,\text{V}$ [1], [5]. However, the acoustic contrast factor ϕ, see (3), influences the position of the particles along the standing wave. It depends on the density and the compressibility of the particles ρ_c, β_c, and of the surrounding medium ρ_w, β_w, respectively. Equation (4) defines the wave number k. The distance from the node is represented by x.

$$F_{AR} = -\left(\frac{\pi \cdot p_0^2 \cdot V_c \cdot \beta_w}{2\lambda}\right) \cdot \phi(\beta, \rho) \cdot \sin(2kx) \qquad (2)$$

$$\phi(\beta, \rho) = \frac{5\rho_c - 2\rho_w}{2\rho_c + \rho_w} - \frac{\beta_c}{\beta_w} \qquad (3)$$

$$k = \frac{2\pi}{\lambda} \qquad (4)$$

For continuous flow acoustophoresis, the separation quality also depends on the flow $\dot{V}$. This states that the acoustic and the flow force influence each other. So for high flows, the separation quality decreases because the ARF has less time to focus the particles [1]. $\dot{V}$ is calculated according to

$$\dot{V} = A \cdot v, \qquad (5)$$

where A is the cross-section area and v the velocity. Experiments also revealed that an increasing cross-section of the main channel has a negative impact as well [4]. The width-height relationship usually is around 2.5:1 [1], [2]. Also, suspensions with a high particle concentration, like blood, lead to a decreased separation quality [1], [4]. The physiological hematocrit level (Hct) of blood is in a range of 37 % to 54 %. Red blood cells (RBC) have a size of $\sim 7\,\mu\text{m}$ and they tend to migrate to the nodes of ultrasonic standing waves [1], [8]. The speed of sound inside blood is assumed as $c = 1530\,\text{m}\,\text{s}^{-1}$.

2 Material and Methods

The development of an acoustophoresis flow cell with such a high claim is a complex process. Design methodology techniques according to VDI guideline 2221 were applied to find the ideal parameters for the resonator.

2.1 Requirements to the Flow Cell

Initially, a list of requirements for the flow cell was created. This list helped to clarify the task and also was an important guideline throughout the whole development process. Therefore, the internal boundaries were collected using the scenario technique. Additionally, a comprehensive literature research was made to gather relevant information about the topic. The outcome of the most important requirements is listed in Table 1 and specified subsequently.

Table 1: List of Requirements

#	Requirement
1	Continuous flow
2	Processing of undiluted whole blood at flow rates $\geq 300\,\mu\text{l}\,\text{min}^{-1}$
3	Secure and loss-free channeling of whole blood
4	Generation of standing waves inside channel
5	Separation of blood plasma and RBC
6	Observability of separation processes
7	Separate and secure channeling out of blood plasma and RBC without losses
8	Manufacturability of the flow cell using SLE

1: The continuous blood flow will be provided by the use of a neMESYS syringe pump, centoni GmbH, Germany, in combination with a 25 ml glass syringe, ILS, Germany. This ensures pulsation free flows.

2: Since the flow cell shall work for undiluted whole blood with Hct ~ 50 % and a flow rate of $300\,\mu\text{l}\,\text{min}^{-1}$, the channel structures have to be adapted in a way which ensures a good separation quality.

3: For a secure channeling of the blood without losses, a proper connection from the tube system to the flow cell has to be found.

4: The resonator shall be actuated by a 12 mm x 40 mm PZT transducer from PI Electronics with a resonance frequency of 2.125 MHz, which will be supplied by a peak-to-peak voltage of 14 V. To create standing waves inside the channel, (1) has to be fulfilled.

5: The separation of the focused RBC and the plasma shall be made due to branching channel structures.

6: To gain a better understanding of the processes inside the flow cell, they shall be observable with a high-resolution camera.

7: For a separate and secure channeling out of the plasma and the RBC, proper connections from the flow cell to a second tube system have to be made.

8: Another important requirement is that the flow cell shall be manufactured with selective laser-induced etching (SLE). This process allows a production of highly precise ($\sim \pm 1\,\mu\text{m}$) 3D structures with a mean roughness of $\sim 0.2\,\mu\text{m}$ [9].

2.2 Conceptional Design

Based on the requirements list, a function structure was created to visualize the essential functions and their relations.

To find solutions to the functions, diverse problem-solving methods like literature research, the gallery method, brainstorming, and a systematic variation of the parameters were applied. One final concept was chosen by means of a morphological matrix, a table which contains all functions and their associated solutions, from which all solutions were evaluated using a weighted pros and cons list.

2.3 Embodiment Design

According to the basic principles of design theory, the terms clarity, simplicity, and safety were considered throughout the entire design process. This ensures that the technical functions will be fulfilled and that the acoustic resonator will be easy and safe to handle.

The flow cell was constructed by the use of SolidWorks 2014. It went through an iterating process of verification of its fluidic properties with the simulation software ANSYS 18.2, and adjustments of the parameters to optimize the outcome.

3　Results and Discussion

Since the main problem was to process undiluted whole blood at flow rates of $300\,\mu l\,min^{-1}$, parallelization is the best solution to solve the issue of high velocities. According to (5), the velocities inside the channels shrink pursuant to the number of equally sized parallel channels, due to the growth of the cross-section area. The constructed flow cell, as can be seen in Fig. 2, consists of 15 parallel channels. This reduces the initial velocity from $4630\,mm\,min^{-1}$ to $309\,mm\,min^{-1}$, which is around four times lower than the velocity used in [1] in combination with a $2\,MHz$ transducer. In addition, the blood will be processed in five separation steps to ensure an outcome of clinically pure plasma. The flow cell has outer dimensions of $71.7\,mm$ x $10\,mm$ x $7.25\,mm$ (L x W x H). The relatively high thickness shall reduce the heat flow of the thermal energy, which will be a side effect of the oscillating transducer, to ensure that the blood cells will not coagulate within the channel.

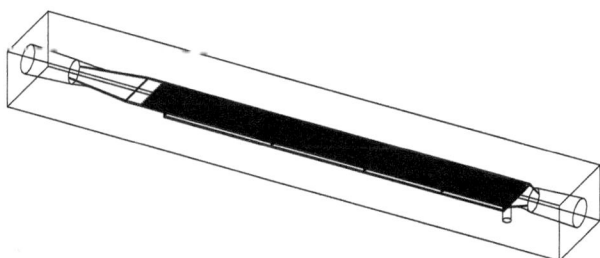

Figure 2: Acoustophoresic flow cell optimized for high flow rates. Its outer dimensions are $71.7\,mm$ x $10\,mm$ x $7.25\,mm$ (L x W x H).

The acoustic resonator consists of 13 parts which can be seen in more detail in Fig. 3. The inlet (a) of the flow cell is conically shaped with a pitch angle of $6°$, so a standard Luer adapter can be used to connect a tube system to the

flow cell. A lofted cut transition zone (b) connects the inlet to the cuboid-shaped main channel (c). It opens with an angle of $10°$. This ensures an even distribution of the blood over the full width of the main channel of $5.68\,mm$, which then splits up into 15 parallel sub-channels (d). These are separated by $20\,\mu m$ thick walls (e) that reach right to the end of the main channel. Every sub-channel has a width of $360\,\mu m$. This equals $\lambda/2$ of the induced frequency, so inside the focusing zone (f) of every sub-channel the RBC will be focused along one central node. The aspect ratio was chosen to be 2:1. After $1\,mm$, the first trifurcation zone (g) occurs to reduce the concentration of the RBC by $\sim 33\,\%$ due to basic fluidic behavior. It has the shape of a straight flow splitter with three equally sized branches. This simple shape is ideal for parallelization purposes. The middle branch leads the RBC directly to a waste outlet (h) perpendicular to the standing wave, the lateral branches lead the rest of the blood along the waste outlet. Now that the concentration is reduced, the remaining cells can be focused properly inside the $12\,mm$ long focusing zone of step 2 before the blood is separated again. Since the concentration of the RBC, the volume itself, and so also the velocity shrink with every step, the distance to the following focusing zones are shortened. After five steps, the cleaned blood plasma passes a second abridged transition zone (i) before it reaches the plasma outlet (j), identical to the inlet. The erythrocytes and parts of the plasma which could not be extracted will flow from the single waste outlets (h) through a collecting area (k) to the vertical collective waste outlet (l).

The fluidic behavior of this flow cell was simulated using the characteristics of blood. Since ANSYS calculates with mass flows instead of flow rates, Table 2 shows the percentual mass flow at the outlets and the recalculation to flow rates. It indicates that $19.1\,\%$ of the total mass flow will reach the plasma outlet (j). This corresponds to $57.2\,\mu l\,min^{-1}$ with an inflow of $300\,\mu l\,min^{-1}$. The loss of $0.1\,\%$ stems from the limited meshing quality.

Table 2: Simulation Results

Opening	Mass Flow in %	Flow in $\mu l\,min^{-1}$
Inlet (a)	100.0	300.0
Step 1	- 38.7	- 116.1
Step 2	- 15.8	- 47.4
Step 3	- 13.6	- 40.8
Step 4	- 11.9	- 35.7
Step 5	- 0.8	- 2.4
Outlet (l)	- 80.8	- 242.4
Outlet (j)	- 19.1	- 57.3
Loss	- 0.1	- 0.3

Assuming a physiological hematocrit level, a reduction of 1/3 on the number of erythrocytes after step 1, and an increment on the clearance with every step, due to a reduction of concentration and velocity, we can expect that the remaining plasma will have a concentration of RBC in a low one-digit percent range.

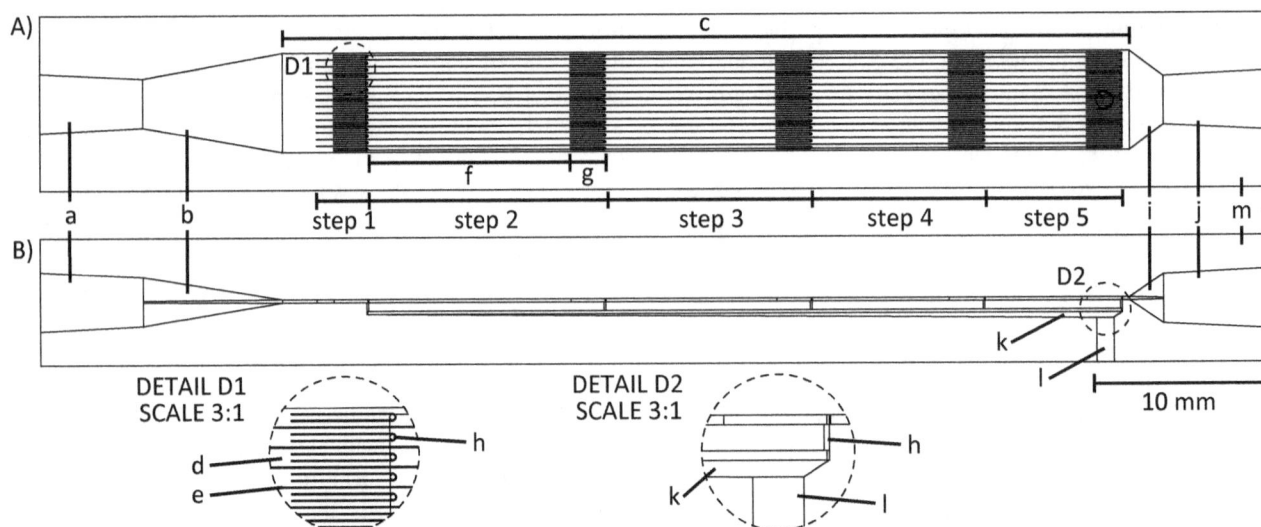

Figure 3: Detailed view on the flow cell: A) top view, B) front view – a: inlet, b: transition zone 1, c: main channel, d: sub-channel, e: separation wall, f: focusing zone, g: trifurcation zone, h: single waste outlet, i: transition zone 2, j: plasma outlet, k: waste collecting area, l: collective waste outlet, m: flow cell body.

4 Conclusion

A new acoustophoresic flow cell was developed, which shall enable the processing of undiluted whole blood at flow rates of $300\,\mu l\,min^{-1}$. It is the first to combine multiple cleaning steps with a high number of parallel channels. The simulated plasma outcome is planned to be clinically pure, and with $\sim 57\,\mu l\,min^{-1}$ more than five times higher than the one mentioned in [1]. This increased outcome could allow new applications based on microfluidic blood separation since the current throughput per minute is in the low microliter range. The simple channel structure facilitates up-scaling to handle even higher flow rates. Nonetheless, in the first place, the resonator has to be manufactured and tested to evaluate the simulation results. Since the design is a compromise between the outcome of blood plasma and its purity, it still can be optimized on the basis of measurement data. The manufacturability of this flow cell was confirmed by the market leader for SLE manufacturing, *LightFab* from Aachen, Germany.

Acknowledgement

This publication is a result of the research project *opLaSens* which was founded by the German Federal Ministry of Education and Research, grant number: 13FH024PX4. The work has been carried out at and was supervised by the Medical Sensors and Devices Laboratory, University of Applied Sciences Lübeck.

5 References

[1] A. Lenshof et al., *Acoustic Whole Blood Plasmapheresis Chip for Prostate Specific Antigen Microarray Diagnostics*. In: Anal. Chem., vol. 81, no. 15, pp. 6030–6037, 2009.

[2] M. Kersaudy-Kerhoas and E. Sollier, *Micro-Scale Blood Plasma Separation: from Acoustophoresis to Egg-Beaters*. In: Lab Chip, vol. 13, pp. 3323–3346, 2013.

[3] D. R. Gossett et al., *Label-free Cell Separation and Sorting in microfluidic Systems*. In: Anal. Bioanal. Chem., vol. 3397, pp. 3249–3267, 2010.

[4] V. Seemann, *Experimentelle Untersuchung der Separation von Erythrozyten und Blutplasma mittels Akustophorese*. Bachelorthesis, University of Applied Sciences Lübeck, 2017.

[5] T. Laurell, F. Petersson, and A. Nilsson, *Chip integrated Strategies for acoustic Separation and Manipulation of Cells and Particles*. In: Chem. Soc. Rev., vol. 36, pp. 492–506, 2007.

[6] J. J. Hawkes and S. Radel, *Acoustofluidics 22: Multi-Wavelength Resonators, Applications and Considerations*. In: Lab Chip, vol. 13, pp. 610–627, 2013.

[7] C. W. Shields IV, D. F. Cruz, K. A. Ohiri, B. B. Yellen, G. P. Lopez, *Fabrication and Operation of acoustofluidic Devices Supporting Bulk acoustic Standing Waves for sheathless Focusing of Particles*. In: J. Vis. Exp., vol. 109, no. 53861, 2016.

[8] D. U. Silverthorn, *Human Physiology – An Integrated Approach*. 7th ed., Pearson, pp. 535–554, 2016.

[9] LightFab GmbH, Aachen, *SLE_3D_printed_glass*. Available: http://www.lightfab.de/index.php/Home.html, [last accessed on 2018-02-06].

A compact handheld OCT-System for homecare applications

M. Münst [1,3], P. Koch [3], H. Sudkamp [3], M. Vom Endt [2,3], J. Franke [2,3], E. Brockmüller [2] and G. Hüttmann [3,4]

[1] Medizinische Ingenieurwissenschaft, Universität zu Lübeck, michael.muenst@student.uni-luebeck.de

[2] Medizinische Ingenieurwissenschaft, Universität zu Lübeck

[3] Medizinisches Laserzentrum Lübeck GmbH.

[4] Institut für Biomedizinische Optik, Universität zu Lübeck, huettmann@bmo.uni-luebeck.de

Abstract

Age related macular degeneration is a leading cause of blindness across developed countries.A homecare device ensuring a short examination intervall will help to improve the treatment of the disease. A homecare device must be small and easy to use. We developed an off-axis full-field time-domain OCT-System that can be used as a handheld device to acquire images of the retina. It is based on an open Michelson interferometer using a free-space emitting superluminescent diode. The system acquires images with 139.000 A-scans/s and measures 235 x 164 x 70 mm.

1 Introduction

Age related macular degeneration (AMD) is a disease of the retina which can lead to vision loss and blindness. It affects the macula which is the central region of our visual field. There are three stages of AMD, early, intermediate and late [1]. Early AMD does not include vision loss and most patients will not notice this stage of the disease, but the occuring drusen under the retina can already be detected with optical coherence tomography (OCT). The intermediate stage includes drusen and pigment changes in the retina but rarely includes vision loss. The late stage includes vision loss and is distinguished in two types, the dry AMD and the more rapid and severe wet AMD, which is responsible for 90% of blindnesses caused by AMD [2]. The wet AMD features neovascular growth to provide nutrients to the retina. These abnormal blood vessels often leak fluid which leads to vision loss. Not all early stages of AMD lead to the late stages. About 5% or 14% of the people with early AMD, depending on whether one or both eyes are affected, will develope a late form of AMD [3]. It is therefore very important to keep track of the developement of the disease, as late AMD is the leading cause of blindness for older people in the United States [4]. In 2010, about 2,07 million Americans suffered from late AMD. Due to the aging population of the US, this number will rise to 3,66 million by 2030 and 5,44 million by 2050 [1].

Wet AMD can be controlled by injecting Anti-VEGF directly into the eye. Vascular endothelial growth factor (VEGF) is a signal protein that stimulates the growth of blood vessels, hence, suppressing it stops the abnormal vessels from spreading. Once wet AMD is treated with Anti-VEGF the symptoms will most likely vanish, but the disease will return in many cases, although it is not clear when. One option is to give these expensive injections regularly to keep suppressing the neovascular growth, accepting the risk of side effects like conjunctival haemorrhage, eye pain and many others. An alternative would be to wait until the patient suffers from vision loss to then treat the patient as soon as possible. Both options are not ideal. A close monitoring of AMD with OCT is a good third option, as it provides information if the disease advances before the patient suffers any further vision loss. Currently, this would imply that the patient visits the doctor at least once a week. Considering the number of patients this workload can not be sustained by medical practitioners across the US. A homecare OCT-System can solve this problem. Much like monitoring blood pressure at home, the patient can measure the current status of the disease at home with a simple and compact device. The measured data will be sent to the doctor and the patient can be called in if further treatment is required. In this paper we will present a compact prototype of an OCT-System which is small enough to be handheld and to be used at home. OCT provides depth information of the analyzed tissue, much like an ultrasonic device but on a smaller scale with better resolution. The most common field of application is the examination of the eye, especially the retina. Diseases like AMD or diabetic retinopathy can be detected with high reliability. Optical coherence tomography is based on low-coherence interferometry and has first been described in 1991 [5]. The development and commercialization has grown rapidly [6] which is why OCT-systems can be found in many clinics today. Due to the sensitivity advantage over time-domain OCT, almost all of these systems are Fourier-domain OCT. These Systems are bulky and range approximately between 35.000 and 120.000 USD in price, while providing high quality images with a big field of view of about 6 x 6 mm. These Systems do not seem appropriate for homecare use. It has been shown however, that these large field of views are not necessary and can be reduced to 2 x 2 mm while still achieving high detection rate of AMD [7]. OCT-Systems currently available do not fulfill the requirements for AMD-Homecare devices and have therefore seeked for a setup that allows to build a small and affordable handheld device fast enough to acquire images of the retina. We have previously described an Off-Axis-Full-Field-Time-Domain OCT-System whose basic principle will be briefly cited as the device introduced in this paper is based on it [8].

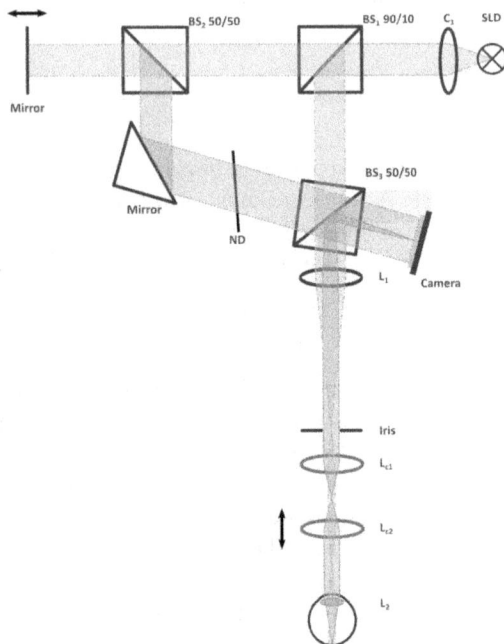

Figure 1: Setup of off-axis full-field time-domain OCT: C_1 is the collimating lens; BS_1, BS_2, and BS_3 are beam splitters; L_1 is an achromatic lens; L_{c1} and L_{c2} are the lenses to compensate for ametropic eyes. L_2 are the refractive elements of the eye; and ND is a neutral density filter. The diaphragm iris is placed in the Fourier plane of both lenses.

As seen in Fig. 1 the beam from the light source is divided into a reference arm (dotted) and sample arm (wider dots) by a 90:10 beam splitter (BS_1). After passing through a second beamsplitter, the light in the sample arm is focused by the lens L_1. It then passes through an iris at the Fourier plane and then passes through another pair of lenses (L_{c1} and L_{c2}) which act as a dioptre compensation for patients with ametropic eyes and can therefore be adjusted while using the device. The light is then collimated by the optics of the eye and illuminates an area on the retina. The retina is then imaged onto the camera. The light in the reference arm also passes through a beamsplitter before reaching the reference mirror. This mirror is mounted on a translational stage to adjust the lenght of the reference arm. To acquire a volume, the length of the reference arm is moved through the measurement range. The beam in the reference arm is then directed onto the camera with an off-axis angle α.

2   Design of the device

In contrast to the previous setup described in the previous section which was based on a Mach-Zehnder interferometer the setup introduced here is based on a Michelson interferometer as this allows a compact design. The following requirements still have to be fulfilled: 1: The light of reference- and sample must fully overlap on the camera, to assure that the image is not vignetted. 2: The reference beam must reach the camera tilted by an angle α. 3: The position of the reference beam on the camera must be stationary and must not depend on the length of the reference arm. 4: The reference beam must show the same orientation as the sample arm. The last condition is necessary due to the fact that we are using a free-space Superluminescent Diode (SLD, Superlum SLD-340-UHP, 100 mW). Although a high radiant flux Φ promises a good sensitivity as can be understood when looking at (1) and (2), the SLD is not fully spatially coherent [10]. This requires for same fractions of the light of reference- and sample arm to be aligned as the contrast would otherwise fade due to the light not interfering. Using a Michelson interferometer it is possible to obtain a higher radiant flux in the eye, as the Mach-Zehnder interferometer has higher losses due to more beamsplitters.

$$Sensitivity = 20 \cdot log_{10} \cdot \sqrt{N_{e,Voxel}} \qquad (1)$$

Where $N_{e,Voxel}$ is the number of electrons collected for each voxel of the sample.

$$N_{e,Voxel} = \frac{\Phi \cdot QE \cdot t_{exp} \cdot T \cdot \lambda_0}{N_{voxel} \cdot c \cdot h} \qquad (2)$$

In this equation, c is the speed of light, and h is Planck's constant. Note that the sensitivity can be increased by increasing the center wavelength λ_0 of the light source, the quantum efficiency QE of the camera, the transmission T between the sample and camera, the incident radiant flux Φ in the imaged field of view, and the exposure time t_{exp}.

Fig. 2 shows the design of the new setup. The light emitted by the SLD is collimated by the lens C_1 and split into reference- and sample arm at the beamsplitter at a ratio of 50:50. The beamsplitter has been rotated by 15° to avoid a prominent reflection on the camera which would already saturate it, making imaging impossible Fig. 3. The path of the reflection is illustrated as the widely dotted line (incident beams in solid, reflected beams in small dotted). The lens in the sample arm has a focal length of 45 mm. Combined with the optics of the eye (f = 17 mm) this results in a magnification of 2,6 on the camera chip. The reference beam mirror is a prism with transparent legs and a reflective hypotenuse. The incident light is refracted at the first leg, as it hits it at an angle of 45°, and is then reflected at the hypotenuse. It exits the prism through the second leg, being refracted parallel to the incident beam. It is then refracted again by the prism P_2 which generates an off-axis-angle of $\alpha = 2,3°$. The reference light is attenuated by a neutral density filter ND. As we demonstrated before [8], off-axis-full-field-time-domain OCT is inherently laterally

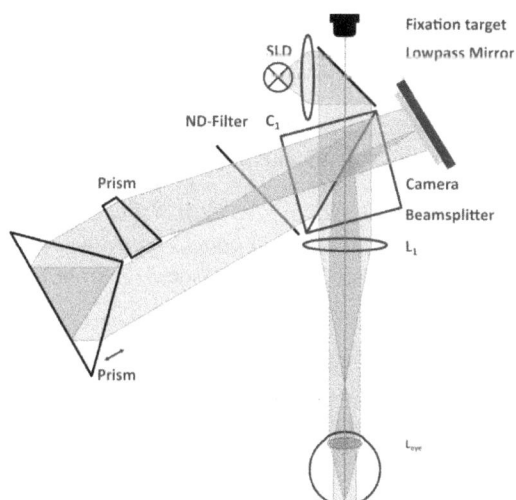

Figure 2: Setup of the off-axis full-field time-domain OCT with Michelson interferometer layout: C_1 is the collimating lens; L_1 is an achromatic lens; L_{eye} are the refractive elements of the eye; ND is a neutral density filter; The first prism is a right-angled prism with reflective hypotenuse; The second prism has a 5° angle to deflect the beam onto the camera.

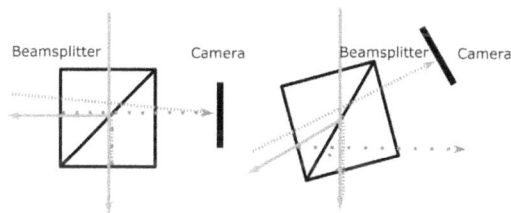

Figure 3: Incident beam (solid), reference beam (small dots) and reflection of the beamsplitter surface (wide dots) illustrated in their path through the orthogonal beamsplitter-setup (left) and the angled beamsplitter (right)

phase stable. This allows us to manipulate the phases numerically after acquiring the image, e.g for correction of defocus. This potential was used to further miniaturize the setup by eliminating the ametropia compensation. As the retina of the eye is imaged onto the camera, OCT images of ametropic patients will be blurred. What has previously been compensated for by a pair of lenses can now be numerically corrected. A fixation target has previously been placed in the same plane as the camera and was therefor imaged onto the retina. Ametropic patiens will therefor also see a blurred image of the fixation target which opposes its cause as the fixation target helps the patient to keep the eyemovent to a minimum. The target should therefor be a clear point to look at. Here we used a laserdiode with a small divergence angle of 12 mrad (DJ532-10, Thorlabs) as a fixation target. The emission of the fixation target is attenuated by a ND-Filter so it can be directed safely into the eye. This target will be seen as a small dot for all patients, no matter where they accumulate to or whatever ametropia they suffer.

The system was designed for a quick assembly without

many adjustments. Only the lightsource needs adjustment, as tolerances of the light exiting the housing, a TO-9 can, are around 150 μm. Everything else is manufactured with high tolerances (around 30 μm) and fixed by dowel pins (H6).

3 Results and Discussion

The dimensions of the whole device excluding the electronics is 235 x 164 x 70 mm. To our knowledge it is the smallest OCT for retinal imaging that includes the lightsource, detector and all the optics in one device. This is small enough to acquire handheld images of a retina (Fig. 4).

Figure 4: The assembled device as seen from above without the lid.

The system acquires 250 images with a rate of 280 Hz resulting in an acquistion time of about 890 ms. The exposure time is 300 μs, which is short enough to avoid fringe washout by unavoidable eye movement. The size of the acquired volume is 4 x 1.5 x 1,9 mm, with a resolution of 620 x 200 A-Scans. This results in 139.000 A-Scans/s which is more than clinical OCT-devices can acquire today. The mean sensitivity was measured by imaging a mirror and results in 65 dB. This does not yet correspond to the theoretical sensitivity which is 71,5 dB. The Scan of a mirror can be seen in Fig. 5.

Figure 5: A B-Scan of a mirror

As no fibers have to be used, the free-space interferometer allows for a quick assembly of the prototype. This is due to the fact that the smallest tolerance, the adjustment of the light source, is about 150 μm. All other parts are alligned through dowel pins and must therefor not be adjusted. The design of the device is compact enough for this to be sufficient to align the reference and sample beam on the camera without losing interference contrast.

4 Conclusion

We have demonstrated a compact off-axis-full-field-timedomain OCT-System based on a Michelson interferometer that can be used as a handheld device. To our knowledge it is the smallest OCT system that includes the lightsource, detector and all the optics in one piece. We believe this setup is simple and small enough to be deployed as a homecare device, primarily for AMD monitoring. AMD monitoring at home will provide a closer examination interval and therefore should guarantee a better outcome of disease treatment, without imposing a higher workload on Ophtamologists and clinics. Upcoming improvements of the device will include electrical integration, as drivers are not yet housed inside the system. In addition, the weight of the system is still too high. This can be reduced by a slimmer design and lighter material of the individual parts.

5 Acknowledgements

This work has been carried out at the Medizinisches Laserzentrum Lübeck GmbH and was supervised by Mr. Hüttmann

6 References

[1] National Eye Institute *AMD Tables*. Available: https://nei.nih.gov/eyedata/amd/tables [last accessed on 2018-10-30]ß.

[2] Augenklinik am Neumarkt. Available: https://www.augenportal.de/augenheilkunde/makula degeneration-amd/makuladegeneration-amd-trockene-feuchte-amd/ [last accessed on 2018-10-30].

[3] National Eye Institute *AMD Facts*. Available: https://nei.nih.gov/health/maculardegen/armd_facts [last accessed on 2018-10-30].

[4] National Eye Institute *Factsheet*. Available: https://nei.nih.gov/sites/default/files/nei-pdfs/NEI_Factsheet_July_2015_v2.pdf [last accessed on 2018-10-30].

[5] D. Huang et al. *Optical coherence tomography*. In: Science, vol 254, pp 1178-1181, 1991.

[6] J. Fujimoto and E. Swanson, *The Development, Commercialization, and Impact of Optical Coherence Tomography*. Investigative Ophthalmology & Visual Science, Vol.57, OCT1-OCT13. doi:10.1167/iovs.16-19963, 2016.

[7] CC. von der Burchard, J. Tode, C. Ehlken, J. Roider, *2mm Central Macular Volume Scan Is Sufficient to Detect Exudative Age-related Macular Degeneration Activity in Optical Coherence Tomography*. Investigative Ophthalmology & Visual Science, Vol.58, 374, 2017.

[8] H. Sudkamp Et Al, *In-vivo retinal imaging with off-axis full-field-time-domain optical coherence tomography*. Optics Letters Vol. 41, Issue 21, pp. 4987-4990, 2016.

[9] E. Couche, P. Marquet and C. Depeursinge, *Spatial filtering for zero-order and twin-image elimination in digital off-axis holography*. Applied Optics Vol. 39, Issue 23, pp. 4070-4075. 2000

[10] C. K. Hitzenberger, M. Danner, W. Drexler, A. F. Fercher, *Measurement of the spatial coherence of superluminescent diodes*. Journal of Modern Optics, vol. 46, no. 12, 1763-1774,1999.

Numerical Simulation of Acoustofluidic Flow Cells with Particles

T. Mukashev [1,2,*], G. Bulz [1,2,*], F. Fiedler [2,3,*], and S. Müller [2,*]

[1] Biomedical Engineering, University of Applied Sciences Lübeck
[2] Medical Sensors and Devices Laboratory, University of Applied Sciences Lübeck
[3] Graduate School for Computing in Medicine and Life Sciences, Universität zu Lübeck
* {temirlan.mukashev, gordon.bulz}@stud.fh-luebeck.de, {felix.fiedler, stefan.mueller}@fh-luebeck.de

Abstract

To get a non-scattering medium for an effective spectroscopic measurement of soluted blood components it is necessary to separate blood plasma and cells. Acoustophoresis is an application of acoustic radiation forces on particles to split plasma and cells up. To achieve a proper separation, several flow cell designs were developed. In this article 3D fluidic simulations of the developed geometries are made with the simulation program ANSYS. As a result for developed designs a comparison between velocity profiles, streamlines and particle tracks was made. We found how pressure, velocity and particle distribution are changing with geometrical characteristics. The next step is to compare experimental and simulation data to validate the results.

1   Introduction

The development of microelectromechanical systems (MEMS) in the recent years shows great performance in the field of medical technology. Some of the MEMS devices are broadly used in the clinic as in vitro diagnostic medical devices to define vital parameters from blood. Acoustophoresis is becoming a widespread method for blood manipulation operated in MEMS and used for the separation of blood components [1]. For instance, the separation of plasma from whole blood was reported to reach high-quality of plasma with less than $6.0 * 10^9$ erythrocytes/l recommended by the Council of Europe [2]. Also the separation of blood components, such as platelets from undiluted whole blood, was reported to achieve a clearance ratio close to 98 % [3]. With small flow rates, $\dot{V}$ used in several papers in the range from $\dot{V}$ = 0.25 µl/min up to $\dot{V}$ = 80 µl/min high separation quality was reached [2], [3]. The objective of this article is to find an optimal value of mass flow rate, optimized design and acoustical parameters which allow decreasing the time consumption without losing separation quality. The previous experiments showed that maximum discharge flow for the acoustofluidic experiments with whole blood was 300 µl/min [5]. As a part of design implementation, we need to find geometrical characteristics which can manage $\dot{V}$ up to 300 µl/min with proper cell separation.

1.1   Governing Equations

The behavior of the flow could be given through the Navier–Stokes equations. They are set of the conservation equations of mass, momentum and energy. Due to the fact that thermal effects are not part of the analysis, energy conservation equation could be neglected. The mass conservation equation can be written as

$$\frac{\partial \rho}{\partial t} + \nabla(\rho U) = 0, \tag{1}$$

where ρ is the density of the medium and U is the velocity vector. The momentum conservation equation has form of

$$\frac{\partial(\rho U)}{\partial t} + \nabla(\rho U \times U) = -\nabla p + \nabla \tau, \tag{2}$$

where p is the pressure and τ is stress tensor which can be defined as

$$\tau = \eta(\nabla U + (\nabla U)^T - \frac{2}{3}\delta \nabla U), \tag{3}$$

where η is the dynamic viscosity of the medium and δ is the strain rate [4].

The forces with the highest effect on the particles are inertial, viscous, drag and acoustic radiation forces. In this article consideration of the acoustic radiation and drag forces will be omitted. To numerically evaluate inertial and viscous forces we can define the dimensionless parameter, Reynolds number Re, which can be written as

$$Re = \frac{\rho v L}{\eta}, \tag{4}$$

where v and L are characteristic velocity and length of the channel, respectively. The physical meaning of Re is a ratio of the inertial to viscous forces so that depending on its value it is possible to neglect with one of the them.

2   Material and Methods

The simulations were done using ANSYS Academic Teaching Advanced CFX, Release 18.2. ANSYS CFX is the solver, used to solve unsteady Navier–Stokes equation in their conservation form. The numerical discretization of the Navier–Stokes equations is done using an element based finite volume method. For spatial discretization were created finite volumes also referred as mesh [4]. Mesh constructed in a way that it satisfies conservation laws.

2.1   Flow Cell Design

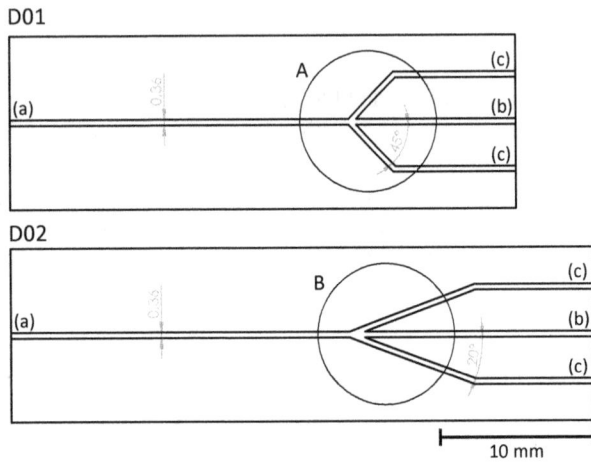

Figure 1: Flow cell designs *D01* and *D02*, top view. Inlet of the flow cell denoted as (a), middle outlet denoted as (b), side outlets denoted as (c). Selected areas "A" and "B" are trifurcation zones for designs *D01* and *D02*, respectively.

The flow cells from Fig. 1 will be named as *D01* and *D02*. The inlet is a part of the flow cell, where the medium and particles enter the domain. An outlet is the part of the flow cell, where the medium and particles leave the domain. The part of the flow cell where the mechanical separation of the flow occurs is the trifurcation zone. The process of separation is caused by pressure diffusion. All the channels presented in Fig. 1 have a width of 360 µm and a height of 180 µm. The main difference between both designs is the opening angle between the middle channel (b) and side channels (c).

2.2   Boundary Conditions

The boundary conditions for the flow and the particles are given in Table 1. For the analysis of mass flow rate were used $\dot{m}_{Inlet} = 0.5$ mg/s, 1 mg/s and 5 mg/s. These values were chosen to observe differences between flows. The boundary conditions for both designs are the same. The injection velocity, v_{inj} is defined as

$$v_{inj} = \frac{\dot{m}}{\rho A}, \qquad (5)$$

where *A* is the area of the rectangular channel.

Table 1: Boundary Conditions for Flow and Particles

Part	Boundary Condition	Value
Flow		
Inlet	Mass flow rate	0.5-5.0 mg/s
Middle outlet	Static pressure	0 Pa
Side outlets	Static pressure	0 Pa
	Static pressure	0 Pa
Wall	No slip wall	$V_w = 0$ m/s
Particles		
Inlet	Injection velocity	7.2-72 mm/s
Inlet	Diameter distribution	6.2-8.2 µm

A particle tracing mode was used to track the behavior of the particles during streaming flow. This is realized using Lagrange particle transport [4]. The morphology of the particles was chosen as *particles solid transport*. The chosen morphology means that the particles are not changing their shape under the action of the flow. The particle's material density was chosen as $\rho = 1120$ kg/m^3, to resemble density of red blood cells. Particle injection regions are subdivided into three equally sized cones. Three cones were enough to fully cover the rectangular shaped channel inlet. They are situated with equal distances to each other. For simplicity reasons, Fig. 2 shows only the middle cone. The angle of the cone is 10 degrees to avoid excessive sticking of the particles to the wall. Each cone inserts 30 particles for steady-state simulation, and one particle every 0.01 s time step over a period of 2 s resulting in 200 particles. For simplicity reasons, the particles are assumed to be spherically shaped.

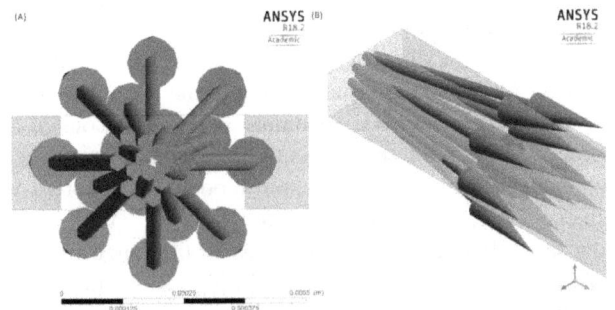

Figure 2: Particle injection region at the inlet. (A) front view in YZ plane. (b) isometric view

To simulate the Non-Newtonian behavior of the blood the Carreau–Yasuda model was used. This model characterizes the dynamic viscosity η as a function of the shear rate $\dot{\gamma}$. It is defined as

$$\eta(\dot{\gamma}) = \eta_H + (\eta_L - \eta_H)[1 + (\lambda\dot{\gamma})^a]^{(n-1)/a}, \qquad (6)$$

where a, n and λ are experimental constants and η_H, η_L high and low shear rate, respectively. The values for experimental constants and shear rates were taken from *Boyd et al.* [6].

3 Results and Discussion

Table 2 gives the mass flow rate distribution between the middle channel and the side channels. It shows that mass flow rate in the side branches for design *D02* is greater than for design *D01*. That means that for the lower angle diffusive forces have greater effect than for the higher angle, hence more flow goes to the side branches.

Table 2: Mass Flow Rate at Different Parts of the Flow Cell

Design	Inlet in mg/s	Middle Channel in mg/s	Side Channels in mg/s
D01	0.5	0.19	0.31
	1.0	0.37	0.63
	5.0	1.83	3.17
D02	0.5	0.17	0.33
	1.0	0.35	0.65
	5.0	1.71	3.29

Fig. 3 compares the velocity profiles of *D01* and *D02* at the beginning of the trifurcation area. The velocity profiles are typically symmetric over the z-axis. Also, they confirm the streamlines distribution from Fig. 4, where maximum velocity in the middle channel of *D02* is higher than in *D01*.

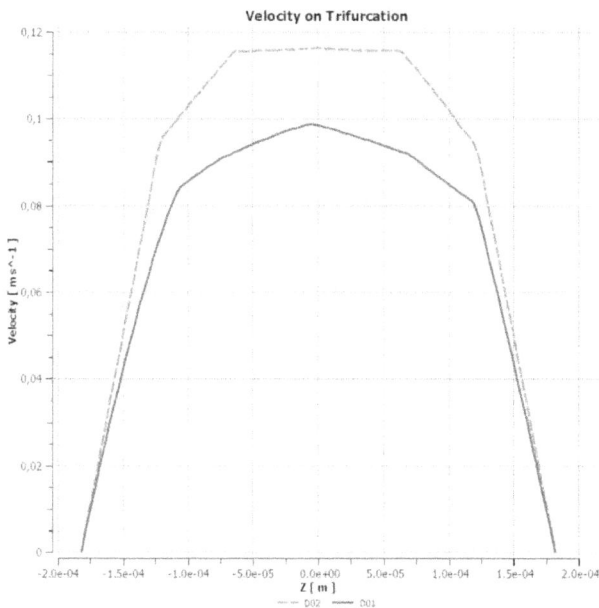

Figure 4: Velocity streamlines at the trifurcation zone.The view made in XZ plane. The legend given in *inverse greyscale*. Mass flow rate value is $\dot{m} = 5$ mg/s.

3.1 Particles

For the mass flow rate $\dot{m} = 5$ mg/s, the particles were unevenly distributed over all three channels. It leaded to $Re = 7$, hence domination of the inertial forces over viscous forces. In the steady type of simulation was decided to show an injection of 90 particles at the same time to observe the effect of the group motion.

Table 3: Particles Distribution Depending on Mass Flow Rate. Steady Type Simulation

Design	Flow Mass flow rate in mg/s	Number of Particles Inlet	Middle Channel	Side Channel
D01	0.5	90	50	40
	1.0	90	51	39
	5.0	90	56	34
D02	0.5	90	44	46
	1.0	90	44	46
	5.0	90	50	40

Table 3 shows relations between mass flow rate and particles allocation for the channels. By analyzing the mass flow rate, it is noted that for $\dot{m} \leq 1$ mg/s, the particles distribution is steady which indicates a balance between inertial and viscous forces which simplifies an analysis. Table 3 and Table 4 shows that for design *D02* more particles are moving to the side branches comparing to design *D01*. Governing

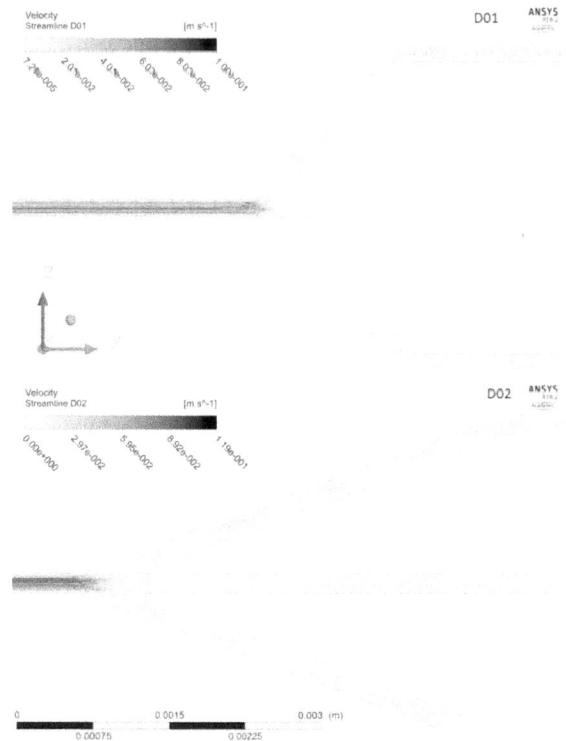

Figure 3: Velocity profile at the beginning of the trifurcation zone over the z-axis. The solid line is the velocity profile of the *D01*. The dashed line is the velocity profile of the *D02*. Mass flow rate value is $\dot{m} = 5$ mg/s

Fig. 4 shows the velocity streamline distribution. In addition, streamlines show the absence of swirls or backflows confirming the laminarity of the flow.

equations have unsteady terms, in order to consider them, we made the transient type of the simulation. For transient

Table 4: Particles Distribution Depending on Mass Flow Rate. Transient Type Simulation

	Flow	Number of Particles		
Design	Mass flow rate in mg/s	Inlet	Middle Channel	Side Channel
D01	5.0	330	237	93
D02	5.0	362	225	137

case, particles entering the flow domain consecutively with a time step of 0.01 s. In general, 600 particles were injected over a time of 2 s. The injections were done consecutively, which means some particles still continued their motion in the domain at the end of the simulation time. Those particles were not taken into account.

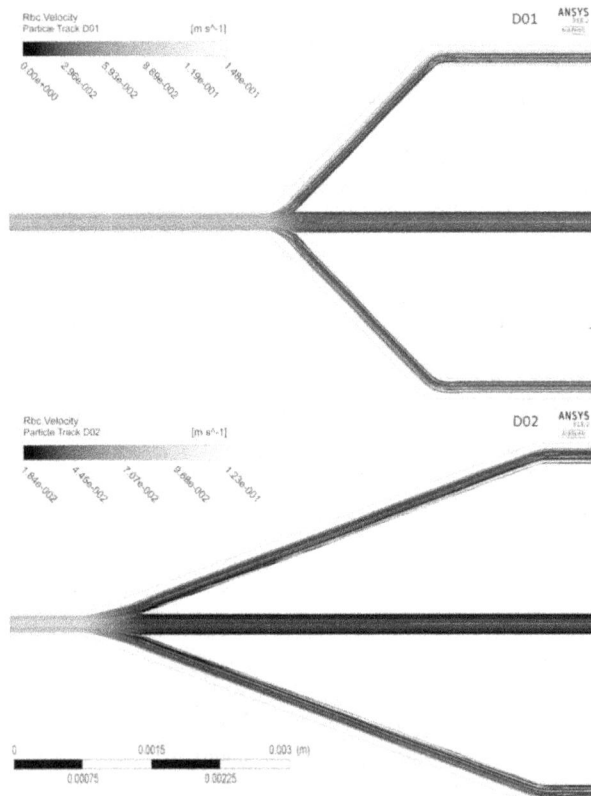

Figure 5: Particles velocity tracks in the trifurcation zone. The view made in XZ plane. The legend given in *greyscale*. Mass flow rate value is $\dot{m} = 5\,\mathrm{mg/s}$

4 Conclusion

In this article, a fluidic simulation of flow cells which are intended to be used in acoustophoresis experiments was shown. Two designs with different geometries were compared using numerical simulation. As a flow behavior analysis, we analyzed velocity streamlines and velocity profile

in the trifurcation zone. As a particle motion analysis, we have done particle tracks simulation. By doing mass flow rate analysis, we found that increasing of the mass flow rate leads to domination of the inertial forces. The particles distribution suggests that design with a smaller angle has better performance.

5 Outlook

For the future, it is planned to modify the simulation by adding acoustic radiation force to analyze the interaction with particles and to compare simulation data with experimental data.

Acknowledgement

This publication is a result of the research project opLaSens which is founded by the German Federal Ministry of Education and Research. Grant number: 13FH024PX4. Images used courtesy of ANSYS, Inc. The work has been carried out at the Medical Sensors and Devices Laboratory, University of Applied Sciences Lübeck.

6 References

[1] A. M. Soliman, A. M. Eldosoky, T. E. Taha, *The Separation of Blood Components Using Standing Surface Acoustic Waves (SSAWs) Microfluidic Devices: Analysis and Simulation.* Bioengineering 4 (2), 2017.

[2] A. Lenshof et al., *Acoustic Whole Blood Plasmapheresis Chip for Prostate Specific Antigen Microarray Diagnostics.* Analytical chemistry 81, pp. 6030–6037, 2009.

[3] J. Nam, H. Lim, D. Kim, S. Shin, *Separation of platelets from whole blood using standing surface acoustic waves in a microchannel.* Lab on a Chip, 11 (19), pp. 3361-3364, 2011.

[4] ANSYS Academic Teaching Advanced CFX, Release 18.2, *Help System, CFX–Solver Theory Guide*, ANSYS Inc., 2017.

[5] V. Seemann, *Experimentelle Untersuchung der Separation von Erythrozyten und Blutplasma mittels Akustophorese.* Bachelorthesis, University of Applied Sciences Lübeck, 2017

[6] J. Boyd, J. Buick, S. Green, *Analysis of the Casson and Carreau-Yasuda non-Newtonian blood models in steady and oscillatory flow using the lattice Boltzmann method.* Physics of Fluids 19, 2007.

[7] F. Petersson, A. Nilsson, C. Holm, H. Jönsson, T. Laurell, *Separation of lipids from blood utilizing ultrasonic standing waves in microfluidic channels.* The Analyst 129(10), pp. 938-943, 2004.

Investigations of Intracellular Transport of Photoimmunconjugates

E. Rybczyk [1], A. Link [2], R. Rahmanzadeh [2]

[1] Medizinische Ingenieurwissenschaft, Universität zu Lübeck, ewelina.rybczyk@student.uni-luebeck.de
[2] Institut für Biomedizinische Optik, Universität zu Lübeck, {link, rahmanzadeh}@bmo.uni-luebeck.de

Abstract

Therapies based on molecular targeting are promising approaches for treatment of cancer and other diseases. Unlike conventional methods, these novel therapies are based on nanoconstructs and light irradiation and allow selective inhibition of cellular molecules such as Ki-67 with low or no systemic side effects. The use of the antibody TuBB-9, which recognizes the antigen Ki-67 highly expressed in proliferating tumor cells, improves the selectivity of this treatment method. The goal of this study is the investigation of various antibody constructs and photosensitizers for intracellular delivery. Fluorescence microscopy investigations and cell vitality tests are performed. This study shows that the TuBB-9 antibody with the attached FITC or Alexa 488 dyes effectively binds to the Ki-67 protein inside cells. It also proves, that the photosensitizer benzoporphyrin monoacid derivative (BPD) has a toxical influence on HeLa cells during irradiation.

1 Introduction

Conventional therapy methods for cancer patients, such as radio- and chemotherapy, show insufficient selectivity against cancer cells and strong side effects [1]. For this reason research is being carried out on alternative, molecular targeted treatment methods for damaging malignant and proliferating cells by using photodynamic therapy (PDT) [2]. Selective inactivation of cellular molecules can be achieved by light irradiation of photosensitizer coupled antibodies. The main challenge with this approach is the choice of the most suitable target protein and the effective transfer of the antibody into affected cells. Ki-67, which is localized in the nucleus of the proliferating cells, seems to be a suitable target protein for this kind of therapy. It is strongly present in malignant cells and is an established prognostic indicator for the assessment of cell proliferation in biopsies from cancer patients. TuBB-9 is a monoclonal antibody that recognizes a physiologically active form of Ki-67. Studies have shown that TuBB-9, conjugated to fluorescein isothiocyanate (FITC) is able to selectively and efficiently eliminate proliferating cancer cells after light irradiation by the production of reactive oxygen species and free radicals. The coupling of conjugates with the nuclear localization sequence (NLS) or encapsulation in liposomes ensure the transport into cells [2].

Figure 1 shows schematically the principle of usage of TuBB-9 antibody, which in combination with FITC and NLS creates the TuBB-9-FITC-NLS construct. For intracellular transport, the construct is taken up into cells via endocytosis. For light-induced release of the encapsulated substances, a photosensitizer binds to the cell membrane

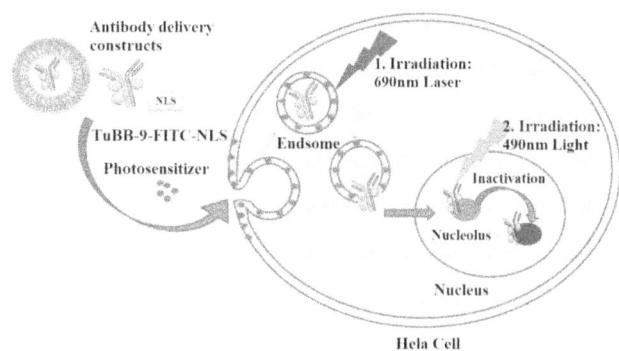

Figure 1: Simplified principle of selective deactivation of the protein Ki-67 after targeting transport of TuBB-9-FITC constructs and light irradiation [2].

and accumulates after endocytosis in the membranes of the endosomal vesicles. During first exposure at a suitable wavelength, the photosensitizer is activated and highly reactive oxygen species are produced. These species destroy the vesicle membrane and the encapsulated construct is released into the cytoplasm of the cell. The second irradiation at a wavelength of 490 nm excites FITC and leads to the inactivation of the protein Ki-67 with subsequent death of the cell [2].

This study focuses mainly on the antibody constructs of TuBB-9 and describes methods for intracellular delivery of the antibody, which include vitality tests and fluorescence microscopy. The toxic influence of the photosensitizer on vital HeLa cells is also investigated.

2 Material and Methods

2.1 Cell Culture

In the context of this study, experiments with human cervical cancer cell line (HeLa) are performed. The HeLa cells are received from American Type Culture Collection (ATCC). They are cultivated in Dulbecco's modified Eagle's medium (DMEM low glucose, Sigma, USA) containing 10% fetal bovine serum (FBS gold, Sigma, USA) and 1% penicillin/streptomycin. This cell line is incubated in a wet incubator with a CO_2 concentration of 5% at $37°C$.

2.2 Antibody Constructs

2.2.1 TuBB-9-FITC

The constructs used in this paper requires preparation of the photoimmunoconjugate of TuBB-9 (an Ki-67 antibody) in advance. To prepare the conjugates, the monoclonal antibody TuBB-9 is labeled with the fluorescent dye fluorescein isothiocyanate (FITC). FITC-conjugated antibodies have an absorption peak at 493 nm and emit fluorescence with a peak around 518 nm [4].

2.2.2 L_1-TuBB-9-FITC and TuBB-9-FITC-NLS

Two different constructs of TuBB-9-FITC are used for the study. The first one involves liposomes, i.e. spherical vesicles synthesized from lipids, which are able to transport large macromolecules intracellularly. The second construct is conjugated with the nuclear localization sequence (NLS) from the SV-40 virus. NLS delivers the light activatable antibody construct, which targets the protein Ki-67 into the cells [2].

2.3 Photosensitizer

HeLa cells are treated with the photosensitizer benzoporphyrin monoacid derivative (BPD), which is activated during the irradiation at 690 nm. This leads to formation of highly reactive oxygen species and permeabilization of endosomal membranes. BPD has a very broad absorption spectrum but only the far-red peak at 690 nm is typically utilized for photodynamic therapy. A good photosensitizer has the ability to preferentially accumulate in the diseased tissue and has a toxic effect on it. In addition, it should have a minimal dark toxicity and thus induces a cell damaging effect only in the presence of light [5].

2.4 Cell Viability Assay

The MTT cell viability assay is a cytotoxicity test that is used to measure the metabolic activity of cells. Based on this activity, the cell vitality as well as the toxic effect of different substances can be determined [6]. When performing this assay, the cells are treated in vitro with the dye MTT(3-(4,5-Dimethylthiazol-2-yl)-2,5-diphenyltetrazoliumbromid). It is a yellow water-soluble dye which is reduced to a blue-violet, water-insoluble formazan. The MTT solution is diluted in a ratio of 1:1.5:1.5 to medium and Dulbecco's Phosphate-Buttered Saline with calcium chloride $CaCl_2$ and magnesium chloride $MgCl_2$ (PBS^+). 100 μl of MTT solution is added per well. After an additional incubation for 2 h, culture medium is removed and the samples are incubated in 100 μl of DMSO (Dimethylsulfoxid) for 30 min on a shaker to dissolve the formazan crystals. Absorbance is measured on a microplate reader (Spectramax M5, Molecular Devices, USA) at 570 nm.

2.5 Fluorescence Microscopy

To analyze the intracellular localization of the photosensitizer and the photoimmunoconjugates, fluorescence microscopic images are taken. Images are captured using Nikon's Eclipse Ti inverse fluorescence microscope and NIS-Elements software. All images are taken with a 60x oil immersion lens (NA: 1.4). The studies are performed on vital as well as fixed and permeabilized HeLa cells.

2.5.1 Microscopy of fixed HeLa Cells

To ensure free access of the antibody to its antigen without irradiation, the cells must be fixed (with 4%-paraformaldehyd) and permeabilized (with Triton X -100 0.25%) before incubation with the primary antibody. This allows a efficient penetration of the antibody into the cell. As a preparation step, HeLa cells are seeded in a concentration of 20.000 cells/ml in 4 compartment glass bottom petri dishes (250 μl per compartment). In the next step, the cells are washed with PBS and incubated in 4%-paraformaldehyde for 15 min. Triton X-100 0.25% is added into each well and the cells are incubated for 3 min. After incubation, the cells are washed three times with PBS. The following constructs and conjugates are examined: TuBB-9-FITC (TuBB-9 concentration: 7.2 mg/ml, in PBS and 10% fetal calf serum), TuBB-9-FITC-NLS (TuBB-9 concentration: 2 mg/ml, in PBS and 10% fetal calf serum) and the TuBB-9 antibody (in PBS and 10% fetal calf serum). HeLa cells are incubated at room temperature without the influence of light for one hour. After the incubation, the sample with TuBB-9 is incubated with a secondary Alexa 488 labeled antibody for 20 min. Alexa Fluor 488 is a sulfonated and chemically modified form of fluorescein. It was designed to address the well known problem of pH sensitive and rapid photo-bleaching fluorescent intensity, common in the FITC dye [4]. In the last step, the cells are washed twice with PBS and afterwards examined in PBS under a microscope.

2.5.2 Microscopy of vital Cells

HeLa cells are seeded at a concentration of 20.000 cells/mL in glass bottom petri dishes (1 ml). In four separate dishes 100 μl L_1-TuBB-9-FITC (78 μl/ml) construct are added, and the cells are incubated for 2 or 24 h. After incubation, culture medium is removed, the samples are washed twice

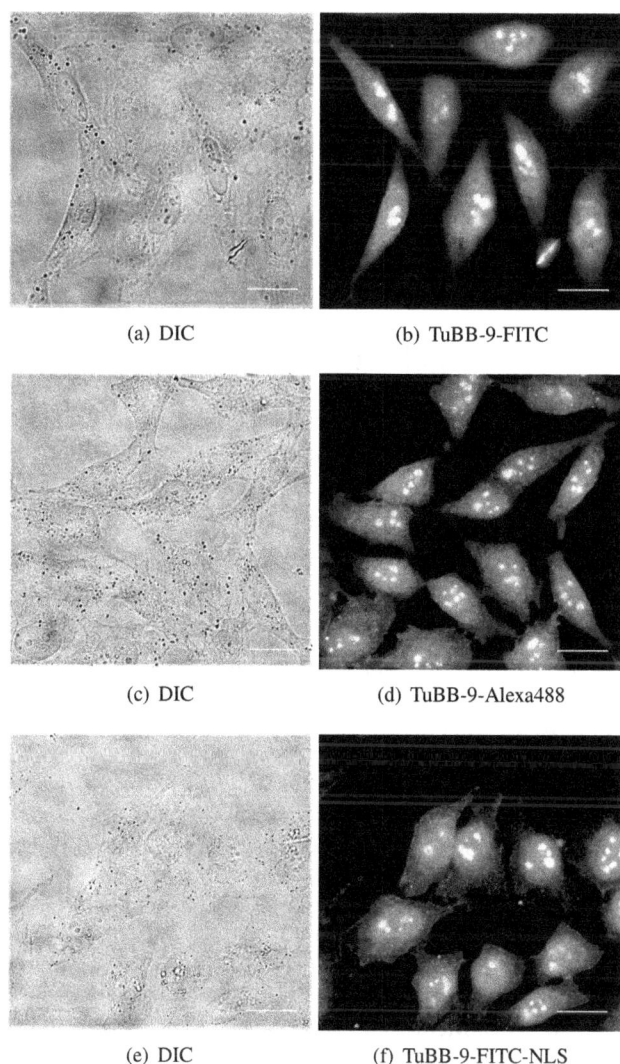

(a) DIC

(b) TuBB-9-FITC

(c) DIC

(d) TuBB-9-Alexa488

(e) DIC

(f) TuBB-9-FITC-NLS

Figure 2: Nucleolar localization of all three TuBB-9 constructs can be observed in fixed and permeabilized HeLa cells. The characteristic spotted pattern of the antibody in the nucleoli can be observed. Size bars: 25 μm.

with PBS$^+$ and then measured in PBS$^+$ under a fluorescence microscope.

2.6 Photodynamic Therapy

PDT (photodynamic therapy) is a minimally invasive therapy, using for the treatment of tissue changes and tumorous disease. This therapy method uses light irradiation of a photosensitizer to create singlet oxygen in the tissue. Cell death (phototoxicity) can be achieved by accumulation of the photosensitizer in the target tissue and by the formation of radicals in a photochemical reaction [7] [8]. To investigate the cytotoxic effect of the photosensitizer, HeLa cells are seeded at a density of 40.000 cells/mL in 96-well plates 200 μl/well. 100 nM BPD is added into each well and the cells are incubated for 20 h. After incubation the samples are washed once in PBS$^+$. The irradiation is performed at 690 nm, according to the absorption spectrum of the photo-

sensitizer. Afterwards, the PBS$^+$ is removed and replaced with fresh medium. 48 hours after irradiation at 690 nm, cell death is evaluated by the cell viability assay.

3 Results and Discussion

3.1 Localization of TuBB-9 Constructs

Cellular uptake and localization of the various constructs are studied on fixed and permeabilized HeLa cells. Fixation and permeabilization treatments remove the molecular barriers, allowing the antibody to reach the nucleus without previous irradiation. Fluorescence images show, that in all three constructs the fluorescence signal within the nucleus is significantly enhanced (see Figure 2). It can be concluded that the TuBB-9 antibody with the attached FITC or Alexa488 successfully localizes the Ki-67 protein and binds to it. This attachment is seen as a spotted pattern in the nucleoli and is characteristic for this antibody.

After demonstrating that the antibody constructs are located on fixed cells in the nucleoli, the next step is the investigation of L$_1$-TuBB-9-FITC construct after an incubation period of 2h and 24 h with vital HeLa cells. The fluorescence microscopy images have shown that the construct in both cases is located outside the nucleus, more specifically it shows predominantly a punctured distribution in the cytoplasm (see Figure 3). This suggests that the antibody constructs are taken up by endocytosis and is trapped in the endosomal vesicles. Due to a low membrane permeabil-

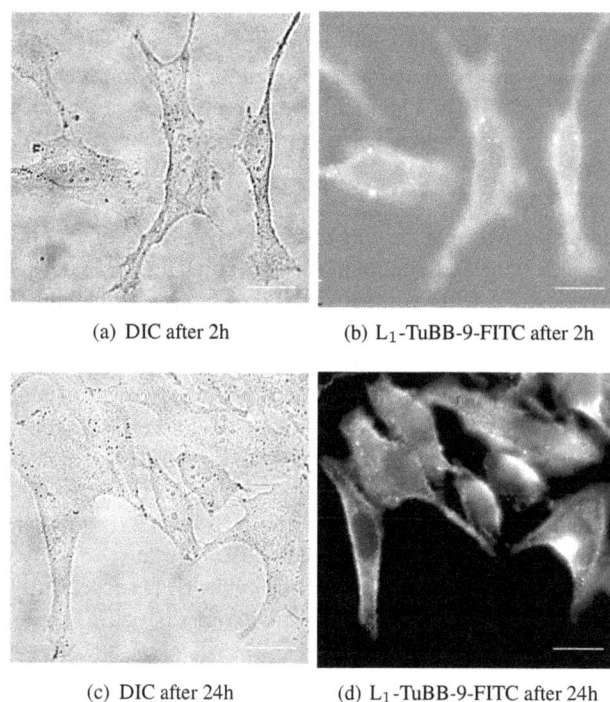

(a) DIC after 2h

(b) L$_1$-TuBB-9-FITC after 2h

(c) DIC after 24h

(d) L$_1$-TuBB-9-FITC after 24h

Figure 3: Localization of L$_1$-TuBB-9-FITC antibody construct in vital cells after incubation for 2 or 24 h. The antibody construct can be observed in the cell but not in the cell nuclei.

ity, the antibody can not escape from the vesicle and enter the nucleus. This effect is much more pronounced after 24 hours of incubation than after 2 hours, where most of the construct is outside of the cell.

3.2 Examination of Cytotoxicity of Photosensitizer

In this section, the cytotoxicity of the photosensitizer on HeLa cells is investigated by means of an MTT assay. The vitality of the samples is measured after different time periods. The cells are divided into two groups: no light irradiation and irradiation with 690 nm laser for activation of the photosensitizer.

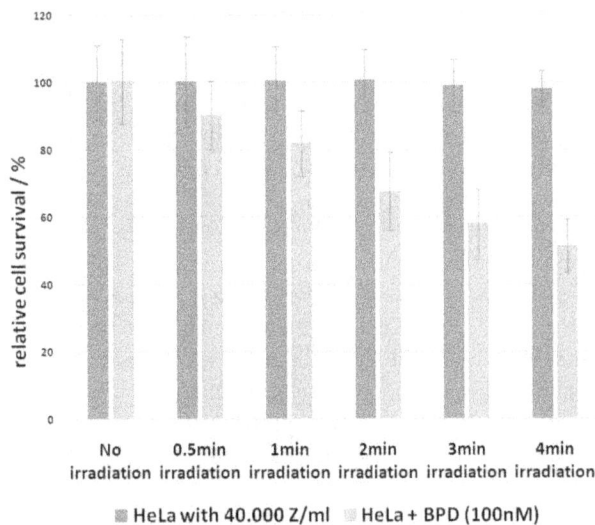

■ HeLa with 40.000 Z/ml ▨ HeLa + BPD (100nM)

Figure 4: Cell vitality after incubation with 100 nM BPD. The cells are irradiated at a wavelength of 690 nm and an irradiance of 8 mW/cm^2 with different irradiation times. Reference form unirradiated cells without incubation with BPD.

Figure 4 shows that the photosensitizer has no significant influence on the cell vitality of the unirradiated samples. This suggests that BPD without radiation has no cell damaging effect. After irradiation, all untreated samples have nearly 100% vitality, so irradiation at 690 nm without BPD is not cytotoxic. However, the cell vitality of HeLa cells incubated with BPD decreases with increasing irradiation time. For example, cell vitality decreases to only 67% for an irradiation period of 2 min and 51% for 4 min. It follows that BPD has a toxic effect on the HeLa cells during irradiation, resulting in cell injury and/or cell death. Because of above properties and a minimal dark toxicity the BPD successfully fulfills the requirements for a good photosensitizer.

4 Conclusion

The TuBB-9 antibody constructs has been proven to be useful for molecular targeted therapies by using photodynamic therapy (PDT), since TuBB-9 is highly effective in binding to an active form of protein Ki-67 [2]. In this study

we showed the localization of the TuBB-9-FITC-NLS conjugates in the cell nucleus in fixed cells and in endosomal compartments in living cells. Furthermore, we tested BPD concentrations which are not cytotoxic for HeLa cells and can be used for the opening of endosomal compartments. However, further research is needed in this field. Studies of cytotoxicity and inactivation of the Ki-67 protein after photochemical internalization will be continued in further studies.

Acknowledgement

The work has been carried out at the Institute of Biomedical Optics, Universität zu Lübeck as a part of a 6 month internship.

5 References

[1] J. Dandler, *Bakteriochlorophyll-Derivate mit Relevanz für die photodynamische Tumortherapie: Lokalisation, Stabilität und Photobiochemie in humanem Blutplasma.* 2008.

[2] S. Wang, G. Hüttmann et al., *Light-Controlled Delivery of Monoclonal Antibodies for Targeted Photoinactivation of Ki-67.* Molecular Pharmaceutics, 1(9):3272-3281, 2015.

[3] R. Rahbari, T. Sheahan et al., *A novel L1 retrotransposon marker for HeLa cell line identification.* National Center for Biotechnology Information, U.S. National Library of Medicine, 46(4): 277–284, April 2009.

[4] F. Wang, W.B Tan, Y. Zhang, X. Fan, M, Wang *Luminescent nanomaterials for biological labelling.* Institute of Physics Publishing, Nanotechnology 17 (2006) R1–R13, November 2005.

[5] F. Fankhauser, S Kwasniewska, *Lasers in Ophthalmology: Basic, Diagnostic, and Surgical Aspects : a Review.* Kugler Publications, ISBN 90 6299 189 0, The Hague, 2003.

[6] T. Mosmann, *Rapid colorimetric assay for cellular growth and survival: application to proliferation and cytoxicity assays.* Journal of immunological methods, 65(1-2):55-63, 1983.

[7] H. Hirschberg, S. Spetalen, et al. *Minimally invasive photodynamic therapy (PDT) for ablation of experimental rat glioma..* US National Library of Medicine National Institutes of Health, 49(3):135-42., June 2006.

[8] O.J. Norum, P.K., Selbo, et al. *Photochemical internalization (PCI) in cancer therapy: from bench towards bedside medicine..* US National Library of Medicine National Institutes of Health, 96(2):83-92, May 2009.

Measurement setup to detect the threshold fluence/abrasion of reflective surfaces

M. Schmidt [1], R. von Elm[2]

[1] Medizinische Ingenieurwissenschaft, Universität zu Lübeck, moritz.schmidt@student.uni-luebeck.de
[2] Coherent LaserSystems GmbH Lübeck, Ruediger.vonElm@coherent.com

Abstract

Reflective materials, which are shot with high laser power, show indications of defects or abrasions after a certain time. Within an experiment with a measurement setup this abrasion of material is supposed to be checked. For this, a laser beam is shot on a reflective sample to receive information about its lifetime and material quality. Therefor, an existing experiment construction is improved with a camera and an objective which enables to view, select or check interesting or conspicuous structures on the sample. This occurred structures will be visually and in written form recorded to show why certain materials get damaged under certain conditions. The final measurement setup also shows which parameters have to be considered to cause a defect or abrasion on a reflective item or surface. The results of this experiment are supposed to enable a potential batch test for all surfaces and materials in the laser manufacturing.

1 Introduction

The laser is nowadays the technical advantage used in the medical and industrial sectors. There are two different laser types. On the one hand the continuous wave laser (CW) and on the other hand the pulsed laser.

By using an instrument or item it abrades. Reflecting surfaces, which are used in the laser manufacturing or in the light deflection, will be worn out e.g. by general usage, heat and power [1][2]. With improvement of material or within the manufacturing process this abrasion can be slowed down. As a result the items receive a better quality and a longer lifetime. The better and more precise the laser and the materials, the better the application in the medical or industrial sectors. Fig. 1 shows the different areas of ablation on a surface [3].

Another important parameter next to the abrasion, regarding to laser and reflecting surfaces, e.g. a mirror, is the reflectivity. It describes the throwback of a light or light waves on a boundary surface, which has a specific index of refraction in consideration of the laws of reflection. This law says that the angle of incidence is equal to the angle of reflection. Furthermore, we can differentiate between a partial-reflective mirror and a high-reflective mirror. At the half-reflective mirror only a part of the light is reflected while the other part passes the boundary surface and spreads in the second medium.

The total-reflective mirror is able to reflect 100% of the light or rather with a little reduction of absorption.

The later damage threshold fluence describes the first abrasion of material. To characterize this abrasion different reflective or half-reflective mirrors, glasses or lenses are shot by a pulsed laser. Within an experiment of this method an approximation to a CW-Laser is supposed to be developed to recognize differences and relations in abrasion of materials.

Figure 1: Example of different threshold fluence and damage radii of the directly ablation of a metal surface on a glass plate [3].

2 Material and Methods

2.1 Reflective Surfaces

A mirror is a reflective surface, which is flat enough to throw back the light in consideration of the laws of reflection. The mirrors in the laser or used for laser beam deflection must tolerate a high power density. That means that they must have a specific low-loss reflection or have to be able to derive the energy resulted by the irradiation. In general households mirrors are coated with aluminium (total-reflective mirrors). This coating of the surface suppresses any significant transmission [4].

2.1.1 Dichroic Mirror

A dichroic mirror which is also called dielectric mirror, is a mirror which is able to reflect just a part of the light spectrum and transmit the other one. It separates the incident light into the wavelength, thus into colors. Dichroic mirrors use the interference of light waves. Either a distributed Bragg reflector who consists of several non-metallic layers or a Fabry-Pérot-Interferometer is used. These mirrors are a special form of an interference filter and basically differentiate from color filters who are used to absorb the light in an area of color ranges.

A special advantage of a dichroic mirror is the low-loss reflection in contrast to metallic mirrors. For this reason they are often used in the laser construction. Even with high laser power, the dichroic mirrors are more effective and resistant against the metallic mirrors. The reflection factor is depending on the wavelength by choosing a suitable number of layers, their thickness and the refractive index. The dichroic mirror is the most important type of mirror used in the laser technology [1][5].

2.2 Experiment Setup

For the experiment a special measuring station is set up to find out why and when an erosion of a reflective material (sample) appears (Fig. 2).

2.3 Laser

The laser which is used in this experiment is a Talisker Ultra Industrial Picosecond Laser average (IR-Laser), developed by Coherent Inc. The maximum power of this Talisker Laser is about 16 W and can be changed by a percentage reduction to a minimum of 0,05 W within the laser. It has a maximum repetition rate of 200 kHz which can be changed by different settings, a wavelength of 1064 nm and generates pulses with a length of 15 ps. Through the whole measurement the maximum repetition rate is used. Table 1 shows the different laser power which is related to the attenuation. The pulse energy is calculated as (1):

$$Pulse\ Energy = \frac{Average\ Power}{Repetition\ Rate} \quad (1)$$

2.4 Scanner And Sample Desk

The pulses are decoupled from the laser and hit directly on a scanner (Arges 2D Scanner Squirrel). The scanner deflects the pulses or laser beam in a 90° angle on a desk, where firstly the position of sample is fixed. At the beginning of the experiment the laser beam is calibrated in x and y direction by the galvanometer mirrors in the scanner. So it was possible to move to every point on the sample and to check the behavior under the laser beam shooting. At the scanner-output a thread is located to mount a lens which can bring the laser beam into focus. For this measurement a f-theta lens with a focal length of 300 mm is mounted.

Table 1: Settings of the laser controller and the resulting power with repetition rate of 200 kHz.

Attenuation [%]	Medial Power [W]	Pulse energy [µJ]
100	0,05	0,25
95	0,97	4,85
90	1,74	8,7
85	2,55	12,75
80	3,41	17,05
75	4,14	19,65
70	4,93	24,65
65	5,71	28,55
60	6,45	32,25
55	7,71	35,85
50	7,94	39,7
45	8,67	43,35
40	9,37	46,85
35	10,11	50,55
30	10,91	54,55
25	11,66	58,3
20	12,42	62,1
15	13,15	65,75
10	14,02	70,1
5	14,76	73,8
0	15,4	77

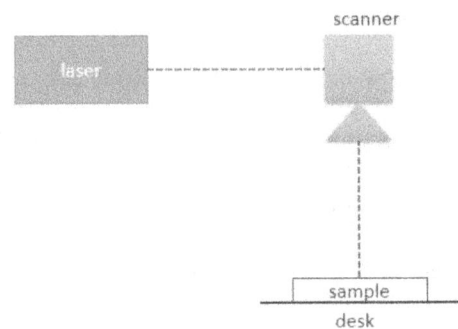

Figure 2: Measurement setup at the beginning of the experiment

Fig. 2 shows the the descriptive development.

The scanner is controlled by a program called InScript and a controller card. This program enables the change of the position of the galvanometer mirrors and it is possible to expose structures or certain points at the surface area of the sample. Furthermore, the controller card is able to connect with the laser, so the laser will only start pulsing if someone starts the program at the computer. The implementations in the program InScript called Jobs and will be executed step by step.

To find out the damage threshold of the material, the first sample which is used for the experiment is a aluminium coated mirror. To avoid further calibrations at each start of the experiments, the mirror is also used to find its right size and position for the deflected laser beam. To guarantee the perfect damage, it is necessary that the reflective sample is in the focal distance from the lens, which focuses the laser beam. The exact location of the focus is determined with a linear stage in z-axis and one waver plate. First, the approximate height is estimated. Afterwards, the exact location is determined by an InScript program and a millimeter dial indicator.

Before shooting the sample with the laser beam, it is checked for defects or contamination under an incident light microscope. After that it is fixed and shot by the laser before it is investigated under the microscope again to compare the results before and after the laser beam shooting. In this state of setup it is still not possible to aim the exact position on the sample again after removing it for the investigation.

For further experiments with the laser the measurement construction is changed to realize more and better results.

2.5 Final Measurement Setup

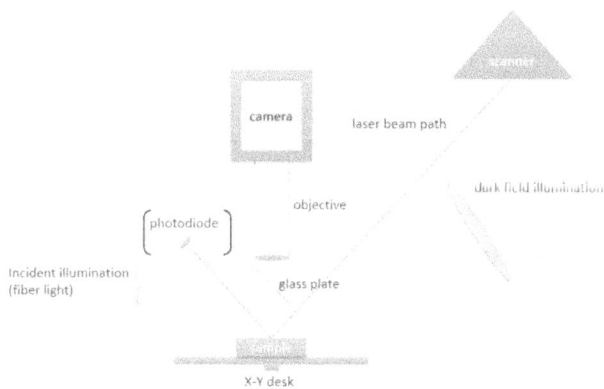

Figure 3: Final measurement setup. The laserbeam hits the sample under an approx. 13° angle. The camera and the objective are above the object and is filming the process.

For this reason a camera (Basler-acA1920-150uc-Python2000) and an objective with a 6 times increased expansion and a working distance of 65 mm (Endmund Optics, Techspec Compact Telecentric Lens, article number: 63743) is attached above the sample and Fig. 3 shwos the designed final setup. Both, camera and objective, enable to present the exact point, where the laser beam hits the sample with a specific angle from the scanner. To change the position of the sample itself the desk, where it is fixed, is able to be moved into x-y direction. First of all, the measurement setup is comparable with a scattered light microscopy, because at the side of the sample an illumination is attached and the scattered light can be detected from the camera. To get an image from an incident light it is necessary to put one glass plate with an angel of 45° between the sample and the objective. Besides, an additional illumination (fiber light) is directed at the glass plate, whose reflection lights on the sample. There it reflects from the sample, passes the glass plate and proceeds into the objective and camera. The problem of this construction is that the glass plate is generating a blurry image. Besides, the laser beam back reflection of the sample is deflecting from the glass into the room.

This reflection is supposed to be detected on a photodiode, which will be used later for the detection of the power dissipation. That is why the same objective is used with an additional inline connection (Endmund Optics, Techspec In-Line Compact Telecentric Lens, article number: 67316). For the first actual measurement in the experiment, a uncoated round glass plate, with a height of 6 mm and a diameter of 19 mm is used. The reflection of the incident light for this glass is approximate 4%. Before using the sample for the experiment it has been cleaned and investigated for any contamination or defects. A large scratch is detected but has not any impact on the further measurements. That is why the sample is fixed on the desk and the camera focus is adapted to the edge of the glass. Because the thickness of this sample is taller than the thickness of the samples used for the settings, it is necessary to lower the distance of the height-adjustable desk and to focus the laser again.

3 Results and Discussion

To avoid major defects, the attenuation of the laser is adjusted to 90%. This attenuation is reduced, rather the laser power is increased during the whole experiment to detect a first defect on the glass provoked by the laser. Even with full laser power no indicators are detected. To control the focus distance of the laser, a paper card is held through the laser beam and the glass. The accuracy of the laser focus is verified, because a few traces of powder from the card arisen on the surface of the glass sample. Even when the card is put under the glass sample, the same phenomenon appears. Anyway the defect of the glass sample occurred just because of the reaction between the laser and the paper card. The threshold fluence of that paper card is located by approx. 85% attenuation.

The next sample which is used in the experiment is a high-reflective (HR) dichroic mirror for a wavelength of 355 nm and high transmission (HT) for 1064 nm at an incident angle of 45°. This mirror has shown the same effects like the experiment with the simple glass plate.

If we compare the back reflection of both samples (glass plate and dichroic mirror) we can find two laser spots on a detector card. This phenomenon is explained by the Fresnel-notation. One part of the laser beam is reflected at the upper boundary surface and the other part is transmitted. The same effect occurs at the backward boundary surface. The reflective laser beam returns to the upper surface, where it escapes out of the medium. A measurement set-up with a non-linear absorption is not possible in this case, because the laser focus diameter of approx. 180 μm is too large, rather the laser power is too insufficient [2].

For the last experiment an aluminum mirror is used. For the beginning the mirror is shot with an attenuation of 95% to relocate the laser focus under the camera. After a certain time a weld penetration occurs at the surface of the mirror and increases and is shown in Fig. 4.

Here we can assume that the aluminum mirror absorbs a part of the laser beam power. The laser power has enough

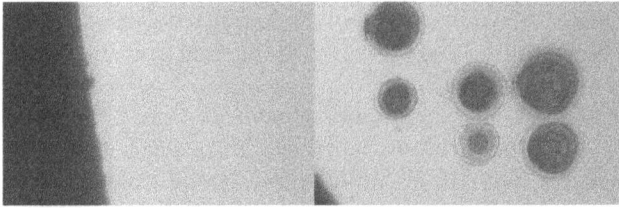

Figure 4: Left side: Sample before testing. Right side: Different locations with the same power over different periods of time (10x magnifaction).

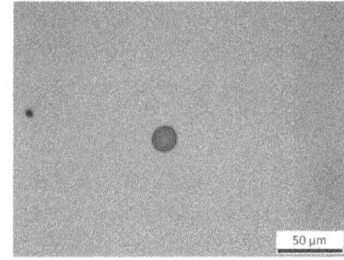

Figure 6: Damagethresholdfluence arised by only one pulse with 38% attenuation under a microscopy with 50x magnifaction (scale 50 μm).

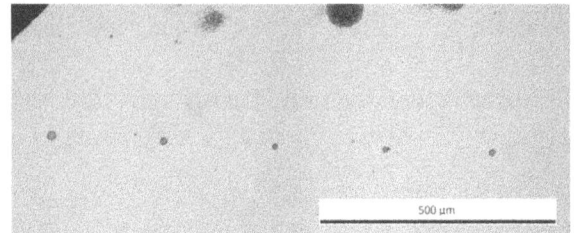

Figure 7: Different positions with a 38% attenuation. Every point act separate and has not the same diameter (scale 500 μm).

energy to get over the threshold fluence and damage the surface of the aluminium mirror. To define the exact threshold, one single pulse with a set pulse energy is shot on the mirrors surface. If the pulse energy is over the threshold, a change on the surface is recognized. The different zones of the laser pulse energy are displayed in Fig. 5.

The area of this threshold is located between 38% and 43% of the attenuation by stepwise increase of the laser energy and is retried at different positions. The extent of the damage is inspected under an incident light microscope Fig. 6. The degree of damage fluctuates at different positions on the surface and shows that the damage threshold at different spots are not same to the laser power. This example is illustrate in Fig. 7.

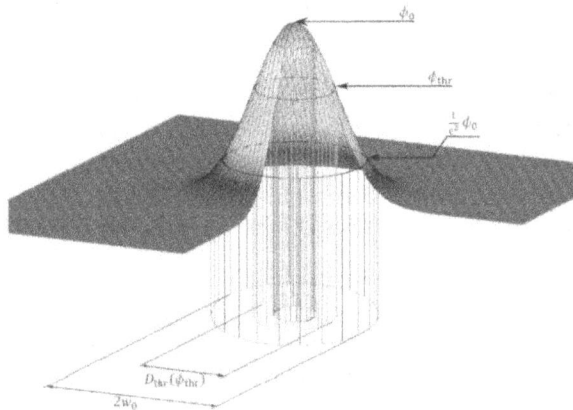

Figure 5: 3D graphic of a Gaussian-shaped laser beam pulse with the maximum fluence ϕ_0, the threshold fluence ϕ_{thr} and the diameter $D = 2w$. Futhermore, the beam waist radius w_0 is defined at the point, where the maximum fluence reduce by $\frac{1}{e^2}$ [3].

4 Conclusion

This described and developed measurement setup shows that it's possible to detect the damage threshold fluence. By changing small components, every material can be checked for its stability and its quality. To get more precise results of the certain materials, it is necessary to use high-resolution images processing systems which can deliver useful and relevant results. Furthermore, this setup can be used for the batch test for all reflective parts which are developed by Coherent Inc. to test their capacity and sort them out when showing any deviations.

Acknowledgement

The work has been carried out at Coherent LaserSystems GmbH & Co. KG, Lübeck and supervised by the Institute of Biomedical Optics, Universität zu Lübeck. Especially I want to thank my supervisor Rüdiger von Elm and his colleagues for their great support.

5 References

[1] Eugene Hecht, *Optics, 6th german editon*. Oldenbourg Wissenschaftsverlag GmbH, München, 2014.

[2] Dr.-Ing. Detlef Breitling, Universität Stuttgart, *Gasphaseneinflüsse beim Abtragen und Bohren mit ultrakurz gepulster Laserstrahlung*. Herbert Utz Verlag, Wissenschaft München

[3] Luigi Slavatore Nobile, *Bachelorarbeit-Untersuchung des Laserablationsverhalen bei der Bearbeitung von dünnen Metallschichten mit Ultrakurzpulslasern*. Hochschule für angewandte Wissenschaften München, 2014.

[4] Available: https://www.rp-photonics.com/mirrors.html [last accessed on 2018.01.12]

[5] Available: http://deacademic.com/dic.nsf/dewiki/327516 [last accessed on 2018.01.12]

Frequency doubling of near infrared sub-nanosecond pulses for two photon microscopy applications

P. Lamminger[1], M. Eibl[2], D. Weng[2], and R. Huber[2]

[1] Medizinische Ingenieurwissenschaft, Universität zu Lübeck, philipp.lamminger@student.uni-luebeck.de

[2] Institute of Biomedical Optic, Universität zu Lübeck, {eibl, daniel.weng, robert.huber}@bmo.uni-luebeck.de

Abstract

Frequency doubling is used to achieve more bandwidth with already existing laser sources and increase the number of different fluorophores that can be excited. Since most lasers for two photon microscopy need a lot of space, the use of fiber lasers, as used in this project, can decrease the spatial requirements and therefore make it more suitable for smaller laboratories. Furthermore, to increase the spectral range of existing fiber lasers, frequency doubling of 1064 nm, 1122 nm and 1186 nm is performed. These wavelengths can be used for two photon excitation in the ultraviolet region for tryptophan. Tryptophan exhibits autofluorescence, which is part of proteins and peptides and can give information about cancer presence. In this study a setup for frequency doubling with a 24% single pass efficiency was designed, evaluated and tested on a two photon excitation fluorescence microscope.

1 Introduction

Fluorescence microscopy and in particular two-photon excitation fluorescence (TPEF) microscopy have shown to be very powerful methods for biomolecular and medical research. TPEF microscopy has higher axial resolution and optical penetration depth compared to one photon fluorescence microscopy. Due to the high photon density needed for TPEF, it is restricted to the focus of the laser beam, which results in good axial resolution with low photodamage and photobleaching, whereas the longer wavelengths result in higher penetration depth of up to 1 mm [1]. The higher intensity needed for TPEF is achieved by short laser pulses, in this case sub-nanosecond pulses.

There are different options to increase the spectral range of already existing lasers and therefore the ability to excite different fluorophores without the need of additional lasers. Examples are the use of an optical parametric oscillator or frequency doubling. Latter was used in this case. The laser used was a fiber based master oszillator power amplifier (MOPA) laser with 1064 nm, 1122 nm and 1186 nm output. Frequency doubling of these wavelengths is further analysed in this study and is particularly interesting for TPEF microscopy, because many substances exhibit autofluorescences in living samples, which are excited by wavelengths in the ultraviolet (UV) region. For example tryptophan, an aromatic aminoacid, is especially interesting, due to its high appearance in most proteins and therefore cancer and skin [2]. In addition, the fluorescence lifetime of tryptophan is strongly dependent on the environment, so fluorescence lifetime imaging (FLIM) can give a lot of informations [4]. Construction of a frequency doubling setup for wavelengths of 1064 nm, 1122 nm and 1186 nm for TPEF mircoscopy is implemented, evaluated and tested in this paper.

2 Material and Methods

In the following frequency doubling, conversion efficiency, the used laser and two photon excitation of tryptophan is explained in detail.

2.1 Frequency doubling

Frequency doubling is only possible with high intensities and in non-centrosymmetric media.

The nonlinear response is derived by the dependence of the polarisation on the eletric field:

$$P = \epsilon_0 \chi E \tag{1}$$

The polarisation is expanded into a taylor series:

$$P = P_1 + P_2 + P_3 + ... \tag{2}$$

$$P = \epsilon_0(\chi_1 E + \chi_2 E^2 + \chi_3 E^3 + ...) \tag{3}$$

$$P = \epsilon_0 \chi E_0 \cos(\omega t) \tag{4}$$

$$P = \epsilon_0(\chi_1 E + \chi_2 E^2) \tag{5}$$

$$P = \epsilon_0(\chi_1 E_0 \cos(\omega t) + \chi_2 E_0^2 \cos^2(\omega t)) \tag{6}$$

with the trigenometric pythagoras and

$$\cos(2\omega t) = cos^2(\omega t) - sin^2(\omega t) \tag{7}$$

it resolves to

$$P = \epsilon_0(\chi_1 E_0 \cos(\omega t) + \chi_2 E_0^2(\frac{1}{2}(1 + \cos(2\omega t)))), \tag{8}$$

with P being the polarisation, ϵ_0 the electric permittivity of free space, χ the electric susceptibility and E the electric field. This means when sending a fundamental wave into a nonlinear crystal two wavelengths can exit after, the fundamental and the frequency doubled. Since χ_2 is very small, the second harmonic (SHG) can only be generated with high intensity (and even higher for higher harmonics). In centrosymmetric medium the generation of P_2 is not possible. Therefore, a medium with birefringence is needed. There are different nonlinear crystals available. To select the suitable crystal for this setup, the SNLO (select nonlinear optics) software is used. The crystals vary in the nonlinear constant, damage threshold and cut angle (different for each wavelength). Also the difference between critical and noncritical phase matching is neccesary. Phase matching is needed, otherwise second harmonic waves generated in each plane of the crystal will interfere destructive. Phase matching is achieved by using birefringence to match the speed of fundamental wavelength and SHG signal. In noncritical phase matching one axis has a changing refractive index with the temperature. In this case a temperature controller is needed. Critical phase matching is achieved by tilting the crystal over a certain angle θ_m (shown in Fig. 1). In fact it is not neccesary to tilt the crystal, but have it cut with the appropriate angle. Critical phase matching was used in this research.

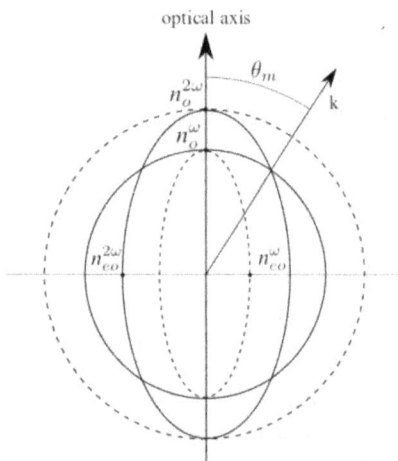

Figure 1: Type 1 critical phase matching. θ_m being the phase matching angle, k the direction of propagation, ω the frequency and n the refractive index for the ordinary (o) and extra ordinary (eo) axis. Phase matching occures, if the refractive index for both, the fundamental and the doubled frequency, match. This happens, if the direction of propagation is in θ_m to the optical axis of the crystal.

2.2   Conversion efficiency

As presented in [5] the conversion efficiency η can be calculated:

$$\eta = \frac{P(2\omega)}{P(\omega)} = 2\left(\frac{\mu_0}{\epsilon_0}\right)^{\frac{3}{2}} \frac{\omega^2 d_{eff}^2 L^2}{n^3}\left(\frac{P(\omega)}{\pi\omega_0^2}\right)\frac{\sin^2\left(\frac{\Delta kL}{2}\right)}{\left(\frac{\Delta kL}{2}\right)^2} \quad (9)$$

with $P(\omega)$ being the power of the fundamental frequency, $P(2\omega)$ the power of the frequency doubled light, d_{eff} the effective nonlinear coefficient, L the length of the crystal, Δk the difference between the wave vectors, ω_0 the beam waist and n the refractive index. Equation 9 shows that focussing on the crystal is increasing the conversion efficiency, since frequency doubling is a nonlinear process. This was further investigated by [5]. Since a low beam waist decreases the Rayleigh length, having a tradeoff between low beam waist and using more crystal length is inherent. With this information different lenses where tested and the most effective one was used for this setup (see section 3).

2.3   Laser

The laser used in this study was a homemade fiber based laser using a MOPA. As seed a premodulated 60 ns pulsed laser diode at 1064 nm with a repetition rate of 100 kHz was used. To generate short pulses an electro-optic modulator is used to supress the light outside of a 250 ps pulse. In the following two single mode ytterbium doped fiber amplifier stages increase the average power of the laser by 20 dB each. The final amplification takes place in the double clad (DC) ytterbium doped fiber stage at the end of the laser, which increases the power by 25 dB. The final peak power is around 4 kW.

To shift the wavelength from 1064 nm to 1122 nm and even further to 1186 nm the pump power has to be increased. This results in raman shift to around 1122 nm and later 1186 nm, which have to be seeded, to be useable.

2.4   TPEF microscopy of tryptophan

Frequency doubling of 1064 nm, 1122 nm and 1186 nm have the advantage of possible two photon excitaion of dyes in the UV region. Here most of the proteins and most important tryptophan are excited. Tryptophan is an aromatic amino acid, which is part of proteins and peptids and therefore more frequent in cancer and skin. In addition the natural answer of the human body against cancer is lowering the resources for cancer. This means the tryptophan level in blood is lower, if it is surrounded by cancerous tissue [7]. Tryptophan has a two photon absorption maximum at 280 nm. This wavelength is achieved with a fundamental wavelength of 1120 nm frequency doubled to 560 nm and used for TPEF microscopy. At 560 nm blood has a high one photon absorption cross section. Therefore, in vivo imaging would have a low penetration depth. The blood absorption curve drops at 590 nm, which makes this wavelength more suitable for in vivo imaging [3]. At the same wavelength the two photon excitation of tryptophan is lower. Reference [6] shows, that two photon excitation has a blue shifted absorption maximum compared to its corresponding one photon spectra. Therefore, a TPEF microscopy test with 1064 nm frequency doubled light was prepared. In [4] it was shown, that bovine serum albumin (BSA) is a simple achievable and well fluorescing sample for two photon excitation at 266 nm.

3   Results and Discussion

In the following the experimental setup, measurements and evaluation of the frequency doubling are presented. The setup and measurements were performed with 1064 nm as fundamental wavelength, unless stated otherwise. Frequency doubling with 1122 nm and 1186 nm have the same setup as with 1064 nm.

3.1   Experimental setup

Fig. 2 shows the setup for frequency doubling of 1064 nm light. Here with the help of manual fiber polarisation controllers the polarisation direction can be changed. The crystal is a 5x3x3 mm (LxHxW) potassium titanyl phosphate (KTP) with type two critical phase matching. In type two critical phase matching two fundamental waves have to be perpendicular to each other. With the help of the polarisation controller the polarisation of the fundamental wave can easily be set to 45° to the crystal axis, which splits the power 50/50 to both orthogonal directions, thus enabling type two phase matching. The KTP crystal has the advantage of a high nonlinear coefficient, which increases the frequency doubling efficiency. An alternative is for example a lithium triborate (LBO) crystal, which has a lower nonlinear coefficient, but higher damage threshold, which is not needed for this setup. For 1122 nm and 1186 nm frequency doubling a 10x4x4 mm KTP is used. The phase matching angle is 23.5° for 1064 nm, 75.4° for 1122 nm and 68.1° for 1186 nm. The crystal was placed in a homemade ring, which was placed in a six-axis kinematic mount. This way the crystal was orientated to achieve maximum frequency doubling.

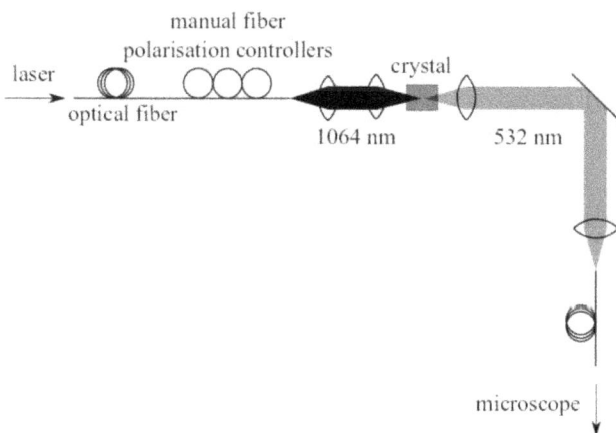

Figure 2: Experimental setup for the frequency doubling process for a wavelength of 1064 nm. The polarisation of the laser light is orientated with the manual fiber polarisation controllers and afterwards focused on the nonlinear crystal and coupled into a single mode fiber.

In [5] focussing on the nonlinear crystal is discussed in detail, resulting in the best outcome with

$$L = 5.68 \cdot z_r, \qquad (10)$$

with z_r being the Rayleigh length and L the crystal length. This means for the best conversion efficiency the right lens must be chosen to adjust the Rayleigh length of the focused beam accordingly, dependent on the collimated beam diameter before the lens, the wavelength and the crystal length. In this setup a lens with a focal length $f = 35$ mm is used for 1064 nm frequency doubing, resulting in a 1 mm Rayleigh length and a $f = 50$ mm lens is used for 1122 nm and 1186 nm frequency doubling resulting in a 2.1 mm and 2.3 mm Rayleigh length.

Behind the crystal the 532 nm light is collimated with a 35 mm lens and with help of a mirror and a 7.5 mm lens the light is coupled into a single mode fiber (SM400). To achieve high coupling efficiency the numerical aperture (NA) of the fiber has to be higher than the NA of the lens. Here a coupling efficiency of 70% was achieved, resulting in an average power of 20 mW with 250 ps pulses and a repitition rate of 100 kHz, which means a peak power of around 800 W.

3.2   Measurements

In Fig. 3 the nonlinear characteristic of the frequency doubling power curve is clearly visible by the fit with the polynomial of the second degree. This measurement was achieved by using a thermal power sensor before and after the frequency conversion. A 40 dB near infrared filter was used behind the crystal to measure only the converted wavelength. This filter also reduced the 532 nm light by 20%, which was corrected in Fig. 3. Also around 20% amplified spontaneous emission (ASE) was measured with a thermal power sensor before the crystal. This was quantified by measuring the peak power via average power times inverse duty cycle and via measuring the exact peak power with a photodiode and an oscilloscope. This discrepancy was also corrected in Fig. 3.

The maximum conversion efficiency achieved was 24%. This percentage has the advantage of not changing the shape of the beam significantly by generating higher modes. These higher modes would cause problems when trying to focus the beam (especially in the microscope). Also 800 W peak power should be sufficient for TPEF microscopy.

Increasing the average laser power with this repitition rate and pulse duration above 120 mW leads to most of the power shifted to a wavelength around 1122 nm, resulting in a lower conversion efficiency. The point of shifting was adjusted via changing the length of the fiber and optimised so 1064 nm and 1122 nm light can be used.

As said before, the setup for 1122 nm and 1186 nm is the same, beside changing the lenses. While measuring the 1122 nm frequency conversion the end of the fiber burned in several times. This happend most likely because the light emitted from the fiber is coming out of the very small area of the fiber core. At the fiber-air transition some of the energy is deposited, resulting in the burned in fiber. By broadening the area at the transition with a coreless (or in this case multimode) fiber end cap, a smaller intensity was achieved. Here a 200 μm multimode fiber was spliced at the end of

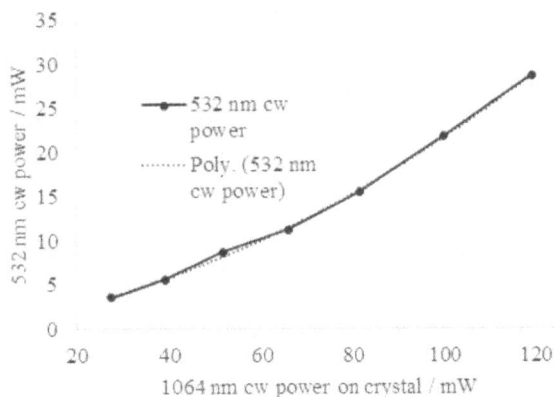

Figure 3: Converted power of 532 nm wavelength as function of the fundamental power of 1064 nm wavelength, with cw being the continuous wave. The fit with the polynomial of the second degree is aligned with the measurement results, which shows the nonlinear behaviour of frequency doubling.

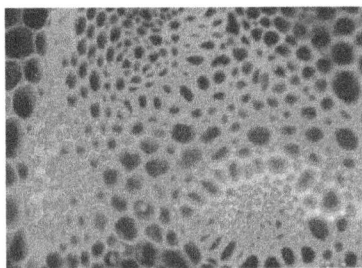

Figure 4: One photon excitation fluorescence microscopy of labelled stem of convallaria majalis. Excitation wavelength was 532 nm. A two photon microscope was used with changed filters, which results in an expected lower resolution than the same image with TPEF and 1064 nm excitation.

the HI1060 fiber, resulting in a three times lower intensity without changing the shape of the beam.

3.3 TPEF Microscope

First of all an image with one photon excitation was taken. For this setup the lenses in the microscope were adjusted for a wavelength of 532 nm. Also the trigger signals for both, the data acquisition and the laser, were synchronised and set to the same pulse generator. The most common sample for 532 nm excitaion is the labelled stem of a *convallaria majalis*, a lily of the valley, which was also used in this test. The result was as expected, lower resolution as a TPEF image, but cellular structures were clearly visible (see Fig. 4).

As stated before, a bright glowing sample, when excited with UV light, with tryptophan is BSA. To investigate this sample an objective for UV-transmission is neccesary. Therefore, a reflecting objtive (Beck Optronic Solutions, Model 5003-000) was used. However, many problems occured, the main ones were focusability, glas fluorescence and problems with the new objective. Most of these where

solved, but resulted in not beeing able to test the tryptophan TPEF. However the converted peak power should be sufficient to excite the tryptophan.

4 Conclusion

The presented setup for frequency doubling of 1064 nm, 1122 nm and 1186 nm is an easy way of implementing frequency conversion and therefore increasing the accessible spectral range of already existing lasers. With the use of fibers an endoscope can be used to bypass the low penetration depth of light, despite the use of two photon excitation. The one photon excitation image gave a proof of concept and the measured intensity of the frequency doubled light is promising for future TPEF images. To learn more about TPEF microscopy with tryptophan, the microscope has to be set up correctly. This would also enable FLIM images with different environments.

Acknowledgement

The work has been carried out at the Institue of Biomedical Optic, Universität zu Lübeck, Luebeck, Germany.

5 References

[1] E. E. Hoover and J. A. Squier, *Advances in multiphotonmicroscopy technology*, Nature photonics, vol. 7, no. 2, pp. 93–101, 2013.

[2] C. Li, C. Pitsillides, J. M. Runnels, D. Côté and C. P. Lin, *Multiphoton microscopy of live tissues with ultraviolet autofluorescence*, IEEE journal of selected topics in quantum electronics, vol. 16, no. 3, 2010

[3] Lin et al., *Imaging leukocyte trafficking in vivo with twophoton-excited endogenous tryptophan fluorescence*, Optics Express, vol. 18, no. 2, 2010

[4] Quentmeier, *Two-Color Two-Photon Excitation of Intrinsic Protein Fluorescence: Label-Free Observation of Proteolytic Digestion of Bovine Serum Albumin*, ChemPhysChem, vol. 10, no. 9-10, 2009

[5] D. L. Robinson and R. L. Shelton, *Frequency doubling conversion efficiencies for deep space optical communications*, TDA Progress Report 42-91, 1987

[6] M. Drobizhev, N. S. Makarov, S. E. Tillo, T. E. Hughes and A. Rebane, *Two-photon absorption properties of fluorescent proteins*, Nature Methods, vol. 8, no. 5, pp. 393-399, 2011

[7] B. Widner, A. Laich, B. Sperner-Unterweger, M. Ledochowski, D. Fuchs, *Neopterin production, tryptophan degradation, and mental depression–what is the link?*, Brain Behav. Immunity, vol. 16, no. 5, pp. 590–595, 2002

Setup and calibration of a commercial optical tweezers system

C. Riesenberg [1], M. L. Torres-Mapa [2], N. Linz [3], D. Heinemann [4], and A. Heisterkamp [2]

[1] Medizinische Ingenieurwissenschaft, Universität zu Lübeck, carolin.riesenberg@student.uni-luebeck.de
[2] Institute of Quantum Optics, Leibniz Universität Hannover, {torres, heisterkamp}@iqo.uni-hannover.de
[3] Institute of Biomedical Optics, Universität zu Lübeck, linz@bmo.uni-luebeck.de
[4] Industrial and Biomedical Optics Department, Laser Zentrum Hannover e.V., d.heinemann@lzh.de

Abstract

Optical tweezers are powerful scientific tools capable of trapping and manipulating microscopic particles and measuring nm-sized movements as well as pN-forces in various applications. To establish an experimental setup for future works, Thorlabs' *Modular Optical Tweezers System* was commercially acquired, set up and characterized. Its trapping capabilities were investigated by performing two different calibration methods: *Equipartition* (EP) and *Power Spectral Density* (PSD). It was shown that the trapping forces are sufficient for further experiments and that the PSD technique is more reliable than the EP technique.

1 Introduction

Focussing of a laser beam through high numerical aperture objectives enables to apply forces to microscopic particles. Provided that these particles are dielectric, transparent for the used wavelength, and have a higher refractive index than the surrounding medium, the gradient forces pull them into an equilibrium position slightly below the laser focus, the so-called *optical trap* [1].

Arthur Ashkin introduced the principle of two-dimensional optical trapping in 1970 [2] and finally developed a three-dimensional trap in 1986 [3]. Nowadays, so-called *optical tweezers* find increasing use in molecular biology, biochemistry and biophysics due to their ability to trap cells and cell organelles, to measure nm-sized movements, and to apply as well as to detect pN-forces [1], [4].

Based on the work of Appleyard et al. [5], Thorlabs developed a low cost optical tweezers system capable of trapping and of additional force measurements [6]. In the present work, the system was firstly set up and adjusted until particles could be trapped reliably in three dimensions (Fig. 1). Furthermore, the system was calibrated for force measurements and different techniques for the calibration were investigated.

2 Material and Methods

In the following sections, the Thorlabs system OTKB/M *Modular Optical Tweezers* [6], including the modules OTKBFM *Back Focal Plane Detection* and OTKBFM-CAL *Force Acquisition*, as well as two calibration methods for the characterization of the optical trap will be described.

2.1 Optical tweezers setup

A schematic drawing of the optical tweezers setup is shown in Fig. 2. The trapping laser is a fiber-coupled laser diode emitting light at 975.5 nm wavelength (Thorlabs BL976-SAG300 and CLD1015). The laser output of the fiber (Thorlabs SM980-5.8-125 and P3-980A-FC-1) is collimated by a triplet collimator (Thorlabs TC06APC-980) and is then expanded to a laser beam diameter of 4.2 mm using two achromatic doublet lenses L1 and L2 (Thorlabs

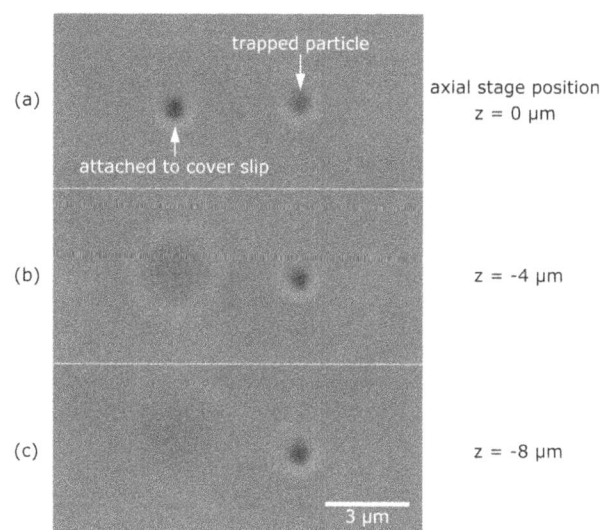

Figure 1: Three-dimensional trapping of a 1 µm silica particle. The particle on the right is held in the optical trap and the particle on the left is attached to the cover slip. Thus, axial movement of the sample stage (a-c) moves it out of focus, whereas the trapped particle remains sharply in focus.

ACN254-050-B, $f = -50\,\mathrm{mm}$ and AC254-150-B, $f = 150\,\mathrm{mm}$). The laser beam is then reflected by a shortpass dichroic mirror D1 (Thorlabs DMSP805R) towards the objective. The objective (Nikon E Plan 100x/1,25 Oil ∞/0,17) focuses the laser beam and creates the optical trap, but it is simultaneously used as an imaging microscope objective enabling observation of the sample via the camera (Thorlabs DCC1240C). The dichroic mirror D1 transmits the white illumination light to the camera and an achromatic doublet lens L4 (Thorlabs AC254-200-A, $f = 200\,\mathrm{mm}$) acts as a tube lens. This way, the optical tweezers' focal plane matches the imaging plane of the microscope.

The condensor (Nikon E Plan 10x/0,25 ∞/-) collimates the laser beam after passing through the sample and a second shortpass dichroic mirror D2 (Thorlabs DMSP805R) reflects it towards a quadrant photodiode (Thorlabs PDQ80A, 150 kHz bandwidth) that is used as a position detector. Therefore, a biconvex lens L3 (Thorlabs LB1027-B) projects the condensor's back focal plane onto the detector. Furthermore, a neutral-density filter (Thorlabs NE06B, $OD = 0.6$) is used to prevent detector saturation.

Beside the hardware, software was supplied by Thorlabs to acquire signals from the photodiode position detector and to control the piezo-driven sample stage (Thorlabs MAX311D/M, 5 nm lateral and 20 nm axial resolution), including a calibration program.

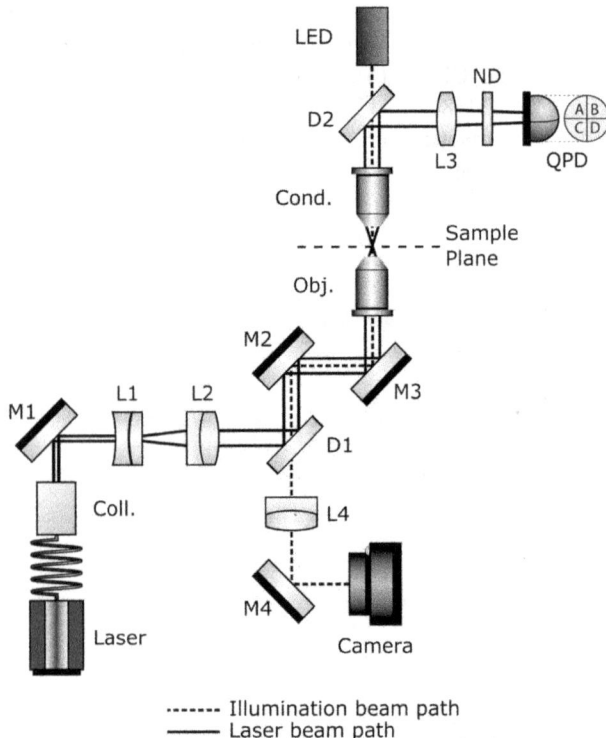

```
------- Illumination beam path
——— Laser beam path
```

Figure 2: Schematic drawing of the optical tweezers and imaging setup. Solid lines indicate the laser beam path starting at the fiber-coupled laser diode (bottom left). A dashed line indicates the path of the LED's white illumination light (top). Coll.: collimator. Obj.: objective. Cond.: condensor. M1–M4: mirrors. L1–L4: lenses. D1–D2: dichroic mirrors. ND: neutral-density filter. QPD: quadrant photodiode.

2.2 Trap stiffness calibration

Besides their trapping capabilities, optical tweezers can be used to measure pN-forces by detecting the displacement of a trapped particle in relation to the trap center. For small displacements x, the trap works like a Hookean spring and the force applied to the particle can be calculated as

$$F = -kx, \tag{1}$$

where k is the so-called *trap stiffness* [1]. This relation shows the importance to detect the particle position with high accuracy, and also to measure the trap stiffness as precisely as possible. For these measurements 1 μm silica spheres in 3 % aqueous NaCl solution were used.

2.2.1 Position detection

One common method to detect the position of the particle is so-called *back focal plane interferometry* [4]. As shown in Fig. 2, a quadrant photodiode is positioned in a plane that is conjugated to the back focal plane of the condensor. Since the laser beam is partly scattered at the trapped particle, the photodiode detects a light pattern caused by interference between scattered and unscattered light. From this interference pattern the position of the trapped particle relative to the trap center can be calculated [1], [6]. Therefore, the first calibration step is to convert the voltage signal of the photodiode to a displacement value.

2.2.2 *Equipartition* (EP) method

For the EP method it is required that the position calibration is performed independently from the stiffness measurement. Therefore, a particle that was attached to the surface of the cover slip was moved through the laser focus along two perpendicular axes while recording the signals S_x and S_y of the photodiode for the x- and y-direction respectively. For small distances from the trap center, the measured voltage has a linear dependence on the particle position [1], [5], [6]. Thus, the conversion factors β_x and β_y in m/V were acquired from the slope of a linear fit in this region.

To measure the trap stiffness, the EP method uses the Brownian motion of a trapped particle. This motion is confined to a small region within the optical trap where the position histogram can be approximated by a Gaussian distribution [7]. After recording the position histogram, the trap stiffness k_i is then calculated using the equipartition theorem

$$\frac{1}{2}k_B T = \frac{1}{2}k_i \langle x_i^2 \rangle, \tag{2}$$

where k_B is the Boltzmann constant, T the absolute temperature and $\langle x_i^2 \rangle$ the statistical variance of the particle position [5], [6].

To assess this technique, both the conversion factor and the trap stiffness were measured for different laser powers of the trapping system.

2.2.3 *Power Spectral Density* (PSD) method

The PSD method also uses the particle's Brownian motion. However, for the evaluation of the trap stiffness the frequency of the particle motion must be observed. Using the quadrant photodiode, a one-sided power spectrum is recorded and fitted by the Lorentzian function

$$S_{VV}(f) = \rho^2 \frac{k_B T}{\pi^2 \mu (f^2 + f_0^2)}, \tag{3}$$

where μ is the drag coefficient

$$\mu = 6\pi\eta a, \tag{4}$$

η the viscosity of the medium, a the particle radius, and ρ the voltage to displacement conversion factor [5], [6]. According to [5] and [6], the trap stiffness k_i is calculated from the known drag coefficient μ and the corner frequency f_0 of the Lorentzian fit:

$$k_i = 2\pi\mu f_0. \tag{5}$$

This calibration method was not only performed for different laser powers, but also for different distances from the cover slip surface. The distance was determined using particles attached to the cover slip as a visual reference point.

3 Results and Discussion

As shown in Fig. 3, the stiffness was determined for different distances from the cover slip at a constant laser power with the PSD method. For both the x- and the y-direction, the stiffness has a maximum at a distance of $10\,\mu\text{m}$ and decreases quickly for smaller as well as for larger distances. For smaller distances, this can be explained by boundary layer effects [5]. Larger distances, on the other hand, cause increasing spherical aberration on the laser beam due to the refractive index mismatch between cover slip ($n \approx 1.5$) and sample medium ($n \approx 1.3$) [1]. Also noticeable is the obvious difference between k_x and k_y, which can only be explained by the laser's polarization [8].

The EP stiffness calibration method was separated into two steps: Firstly, the conversion factor β was determined by moving a particle attached to the cover slip through the laser focus. Because β depends on the laser power (Fig. 4) this measurement was repeated every time the power was changed. The second calibration step, a measurement of the trap stiffness, was then carried out with a different, unattached particle. For the PSD method, only one unattached particle was needed.

The stiffness was determined for different laser powers with the EP method as shown in Fig. 5, and with the PSD method as shown in Fig. 6. The linear dependence can be explained by the dipole model which states that the electrical field of the laser beam $\vec{E}$ induces a dipole moment $\vec{p}$ in the material. According to [9], the potential energy of the induced dipole is

$$U = -\vec{p} \cdot \vec{E} \propto -\vec{E} \cdot \vec{E} \propto -P. \tag{6}$$

Thus, the required energy of a particle to leave the potential well of the optical trap increases with optical power P.

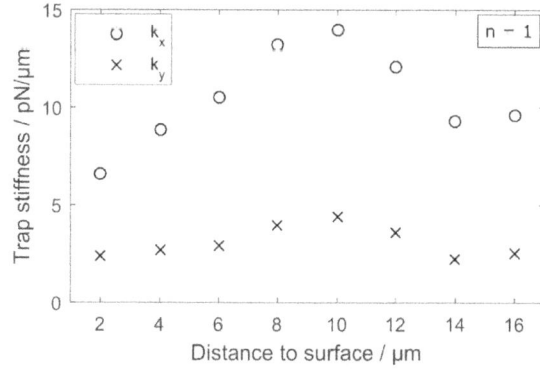

Figure 3: Dependence of the trap stiffness on the distance between the trapped $1\,\mu\text{m}$ silica sphere and the cover slip surface. The stiffness was determined with the PSD method at a constant optical power of $66.6\,\text{mW}$ at the focus. The highest stiffness was achieved at $10\,\mu\text{m}$ distance.

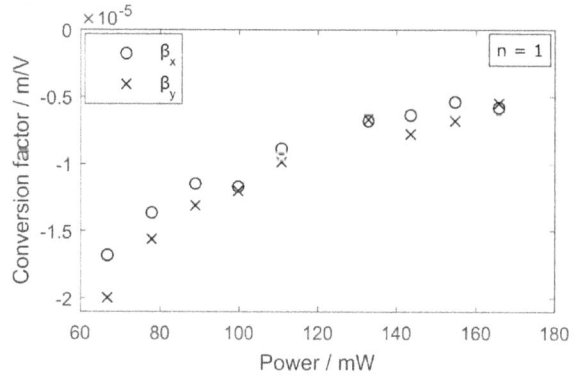

Figure 4: Dependence of the conversion factor β on the optical power of the trap. β was determined by moving a $1\,\mu\text{m}$ silica sphere attached to the cover slip through the laser focus while recording the photodiode's voltage signal.

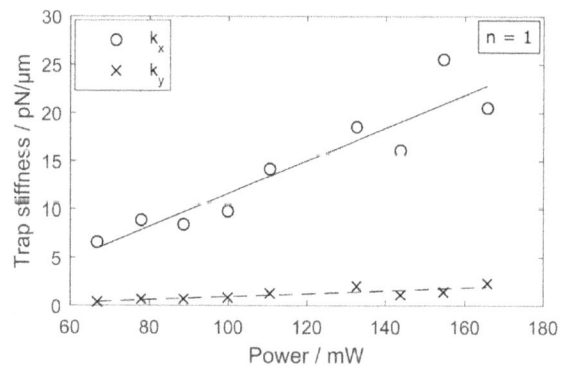

Figure 5: Linear dependence of the trap stiffness on the optical power of the trap, determined with the EP method for $1\,\mu\text{m}$ silica spheres at a distance of $10\,\mu\text{m}$ to the cover slip.

A comparison between the two methods shows that the stiffness values measured with the EP method are smaller than those measured with the PSD method up to a factor of 10. A reason for this could be noise as the EP method makes

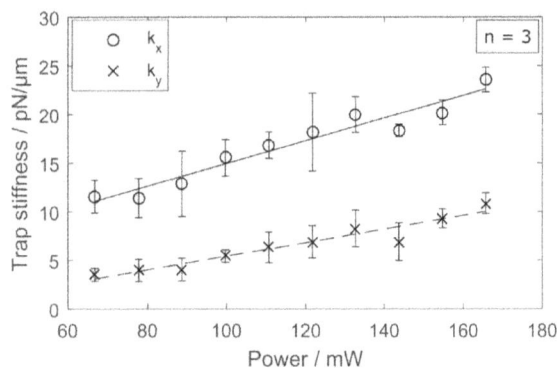

Figure 6: Linear dependence of the trap stiffness on the optical power of the trap, determined with the PSD method for 1 μm silica spheres at a distance of 10 μm to the cover slip.

use of the statistical variance of the particle position, which is increased by any kind of noise. Therefore, the measured stiffness is lower than the actual stiffness [1].

The use of different particles for the two calibration steps could be another point of inaccuracy in the EP method due to slight variations in diameter between individual particles [1]. Furthermore, the attached particle is placed directly on the boundary layer between cover slip and medium. Thus, spherical aberrations that affect the optical trap are not taken into account. It is also difficult to place the particle exactly in the position within the beam path where a trapped particle is located. Another issue could be possible inaccuracies of the sample stage movement.

Although the PSD method is, as shown in the previous section, less prone to error than the EP method, considerable standard deviations between individual measurements were observed. One reason for this is the stiffness's strong dependence on the distance from the cover slip because it is difficult to adjust this distance precisely. Furthermore, variations in particle size could also affect the measured values. The literature considers the PSD stiffness calibration method to be more reliable than the EP method [1], [5], [7], which could be verified by the experiments. Since the voltage to displacement conversion factor is determined simultaneously to the trap stiffness, this method is also more convenient. However, this method requires knowledge about the viscosity of the medium as well as the particle diameter (4), (5). This is especially problematic for biological applications where the exact properties of the medium and the exact dimensions of the trapped object might be unknown. In these cases, the viscosity either needs to be measured as precisely as possible or a different technique like the EP method needs to be used.

4   Conclusion

A commercially acquired optical tweezers system was successfully set up, adjusted and calibrated. The trap stiffness was shown to be linearly dependent on the trap's optical power as well as strongly dependent on the distance of the laser focus from the cover slip. A maximum trap stiffness was measured at a distance of 10 μm. A considerable difference between stiffness values for the x- and y-axis was observed due to the polarization of the laser system. Comparison of the two available calibration methods yields smaller stiffness values for the EP than for the PSD method, likely due to noise influence. It was shown that the PSD method is more reliable and more convenient for future experiments.

Acknowledgement

The work has been carried out at the Lower Saxony Centre for Biomedical Engineering, Implant Research and Development, Hannover, and supervised by the Institute of Biomedical Optics, Universität zu Lübeck.

5   References

[1] K. C. Neuman and S. M. Block, "Optical trapping," *Review of Scientific Instruments*, vol. 75, no. 9, pp. 2787–2809, 2004.

[2] A. Ashkin, "Acceleration and trapping of particles by radiation pressure," *Physical Review Letters*, vol. 24, no. 4, pp. 156–159, 1970.

[3] A. Ashkin, J. M. Dziedzic, J. E. Bjorkholm, and S. Chu, "Observation of a single-beam gradient force optical trap for dielectric particles," *Optics Letters*, vol. 11, no. 5, p. 288, 1986.

[4] J. R. Moffitt, Y. R. Chemla, S. B. Smith, and C. Bustamante, "Recent advances in optical tweezers," *Annual Review of Biochemistry*, vol. 77, no. 1, pp. 205–228, 2008.

[5] D. C. Appleyard, K. Y. Vandermeulen, H. Lee, and M. J. Lang, "Optical trapping for undergraduates," *American Journal of Physics*, vol. 75, no. 1, pp. 5–14, 2007.

[6] Thorlabs, Inc., *OTKB/M - Modular Optical Tweezers System*. Available: https://www.thorlabs.de/thorproduct.cfm?partnumber=OTKB/M [last accessed on 2017-02-21].

[7] K. Berg-Sørensen and H. Flyvbjerg, "Power spectrum analysis for optical tweezers," *Review of Scientific Instruments*, vol. 75, no. 3, pp. 594–612, 2004.

[8] E. Madadi, A. Samadi, M. Cheraghian, and S. N. S. Reihani, "Polarization-induced stiffness asymmetry of optical tweezers," *Optics Letters*, vol. 37, no. 17, pp. 3519–3521, 2012.

[9] S. P. Smith, S. R. Bhalotra, A. L. Brody, B. L. Brown, E. K. Boyda, and M. Prentiss, "Inexpensive optical tweezers for undergraduate laboratories," *American Journal of Physics*, vol. 67, no. 1, pp. 26–35, 1999.

Detection and Removal of Artifacts in Ultra-widefield MHz OCT En Face Images of the Human Retina

K. Rewerts [1], J. Klee [2], J. P. Kolb [3], and R. Huber [3]

[1] Biomedical Engineering, University of Applied Sciences Lübeck, katharina.rewerts@stud.fh-luebeck.de

[2] Medizinische Ingenieurwissenschaft, Universität zu Lübeck, julian.klee@student.uni-luebeck.de

[3] Institute of Biomedical Optics, Universität zu Lübeck, {kolb, robert.huber}@bmo.uni-luebeck.de

Abstract

For a large field of view of ultra-widefield MHz optical coherence tomography of the human retina, relatively long acquisition times are required. Therefore, motion artifacts often occur. The aim is to obtain a diagnostically conclusive image of the retina by registering intact image segments from several data sets of the same patient. Artifact detection is tested on 2D en face images. The detection and classification of strong artifacts works reliably. Microsaccades have not yet been corrected in the 2D images. The necessity of an affine transformation for the purpose of accurate registration is discussed. For the registration a feature-based approach using binary images of the segmented vessel tree is chosen. In the central image region this approach leads to satisfactory results. However, the decrease in image quality towards the outer image area affects the segmentation of the vessels in the periphery.

1 Introduction

Optical coherence tomography (OCT) [1] can be used to non-invasively capture volumetric structures of biological tissue in vivo. This work deals with the registration of images of the posterior segment of the human eye acquired with an ultra-widefield MHz OCT setup based on a Fourier-domain-mode-locked (FDML) laser [2]. The challenge is to qualify this method for clinical use. An essential aspect is to make the examination as pleasant as possible for the patient, that is to omit mechanical immobilization or drug use during the image acquisition process. At the same time, the structures relevant for diagnosis should be displayed clearly and reliably. The observation of the morphology of the posterior ocular structures has a significant importance for the diagnosis of diseases of the eye itself as well as diseases of the whole organism, which cause pathological changes in the posterior eye. For a reliable medical assessment, the image should cover the widest possible area.

The OCT device used here relays the center of rotation of the scanning optics to a pivot point located on the optical axis of the setup. Ideally, the eye is adjusted so that its optical axis coincides with the one of the OCT setup [3]. To achieve the largest possible field of view (FOV), an optimal position for the pivot point has to be determined. For this adjustment a compromise is to be drawn from the following two aspects: Since the scanning angle is limited by clipping of the beam at the iris, the pivot point should be placed as far as possible in the anterior eye. On the other hand, a position deviating from the eye's center leads to an artificial curvature of the retina in the image, since the optical path length in the eye changes with the scanning angle and the retina can move out of the axial image area of the B-scan.

The extremely large FOV facilitates the occurrence of artifacts that can affect the image quality. The transverse scan is accomplished with two galvanometer mirrors that have different pivot points. The adjusted pivot point represents a compromise between the two, so that both pivot points cannot exactly be relay imaged into the eye. The position of the pivot point is thereby spatially unstable, which can cause clipping of the beam leading to shadowing by the iris throughout the image.

As the scanning angle increases, aberrations of the lenses become more and more noticeable. This leads to a decrease of the signal-to-noise ratio. As a result, the peripheral area of the image is darker, the image contrast towards the edges is decreased.

During the image acquisition, the patient must optically fixate a given point and, if possible, not move the eye to avoid image artifacts. This requires a self-control that can hardly be achieved, in particular by diseased or elderly persons, especially since the imaging optics must be brought very close to the eye to be examined. As examples of artifacts caused by an eye movement, the following two are mentioned. The eye movements include microsaccades [4], which are unconscious movements that occur when focusing a point and appear in the image as a horizontal shift of several image lines. Blinking, which is only partially avoidable, interrupts the scanning of the retina and appears as a dark horizontal stripe in the image.

A MHz OCT system developed by our research group covering a FOV of $60°$ has already been used in a clinical en-

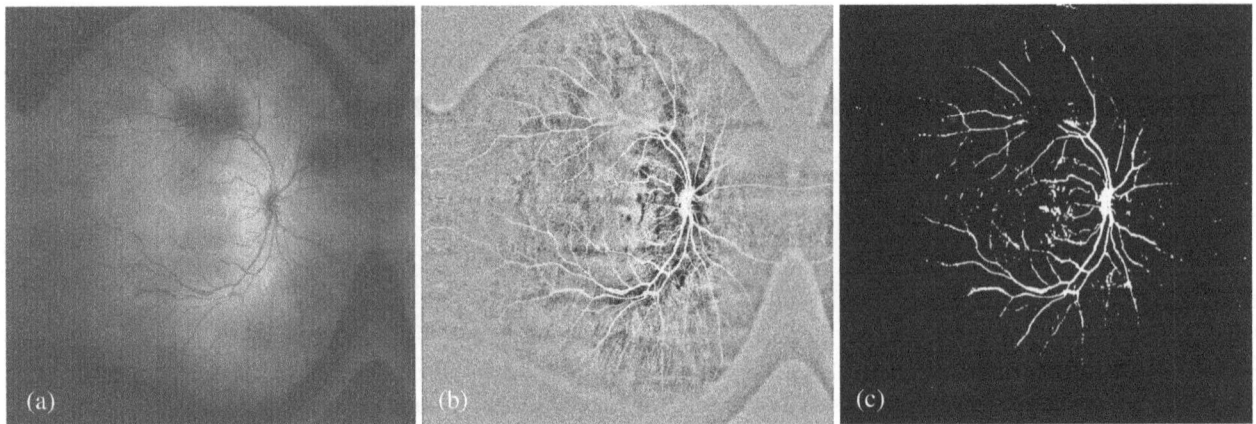

Figure 1: This figure shows the preprocessing steps discussed in Section 2.1 applied to an example for a retinal en face image. The original image (a) shows irregular illumination of the background as well as shadows in the central area. The periodic stripe pattern is clearly visible. After removal of the stripe pattern and the bright vessel centers, illumination equalization and contrast enhancement were performed on the image complement (b). After thresholding and median filtering the final binary image of the segmented vessels (c) is obtained.

vironment [5]. This work uses en face images with a FOV extended to 85° which are created by averaging the scan volume in the axial direction. Since relatively long acquisition times are necessary, motion artifacts inevitably occur. The main focus of this paper is to develop methods to detect, classify and correct the artifacts and to create an artifact-free image by registering intact image segments.

2   Material and Methods

The swept source OCT setup used for the acquisition of the images uses a FDML laser with a central wavelength of 1060 nm. With a resulting A-scan rate of 1.68 MHz, the recording time for an image size of 2088x2088 pixels covering a FOV of 85° is 3.6 s [3].

The processed images are en face images. Several images of the patient's eye were captured. By averaging this data set, an artifact-free image of the eye is to be created. For this purpose, the images must be registered and artifacts must be eliminated. The registration is necessary, since even the smallest spatial changes of the position of the patient's eye relative to the patient interface shift the structures and the instability of the patient adjustment leads to distortions between the single images. Artifacts are misinformation and should be removed so that they do not contribute when averaging the images.

Feature-based registration was performed. The most suitable structures for finding features are the blood vessels, as they stand out quite clearly from the background, which is very irregular due to noise, the tissue structure, shadowing or lighting artifacts. The vessels are segmented, a binary image is created and can then be registered. After registration, a transformation matrix is obtained and applied to the original image. Especially because of the microsaccades, a global transformation of the images is not possible. The transformation has to be performed locally on artifact-free image segments. The detailed steps are described hereafter.

2.1   Image Preprocessing for Vessel Segmentation

The image is prepared for segmentation by the following methods as displayed in Fig. 1.

The images contain a visible, periodic pattern of vertical stripes whose origin has not yet been clarified. To remove the stripes, the intensity variation caused by the stripes is calculated and subtracted from the image.

In the OCT en face images the vessels appear dark with a slightly brighter center. Since a binary image is to be generated, those lighter centers should be removed. This can be achieved by a morphological opening operation which suppresses the local differences in brightness.

At the complement of the image, an illumination compensation is performed: Symmetrically around each pixel a neighborhood is chosen. The average intensity in this neighborhood is calculated and the intensity of the central pixel is reduced by this mean value.

To enhance the contrast between the vessels and the rest of the fundus, histogram equalization is performed using contrast enhanced adapted histogram equalization (CLAHE) [6]. CLAHE performs local contrast adjustments. As a result, less pronounced vessels are displayed comparatively well. At the same time, this algorithm prevents excessive amplification of noise in homogeneous regions.

The binary image is created by choosing a suitable threshold. Here, a compromise must be found in which the vessels remain as connected as possible, but as few background pixels as possible are segmented. Remaining falsely segmented pixels are further reduced by applying a median filter. This filter even closes small gaps in the vessel tree, but it also eliminates small vessel fragments.

2.2   Artifact Detection

Due to the direction in which the eye is scanned, the artifacts appear horizontally. Consequently, the image should be divided horizontally into segments at the points where artifacts are found. After classification of the segments, the individual image parts should be registered to a master image. Detection of the artifacts was realized as follows.

The scanned images show a dark outer area that contains no relevant information. Due to clipping of the beam at the iris or eyelashes, the area of interest might also be limited by shadows . Thus, the relevant area is cut out by manually fitting an ellipse.

For the detection of the artifacts, the mean intensities of the individual lines within the ellipse are calculated. This is done on an image, which has been downsampled with a suitable factor. The mean intensities of successive lines are compared. The presence of a possible artifact is determined by means of a threshold, which is calculated individually for each image from the mean of the amount of intensity change of successive image lines. Since this also yields false positive results, criteria (cf. Section 3.2) must be found to decide if the image segment quality is not sufficient for inclusion in the registration (e.g. in case of blinking).

2.3 Registration

Due to the curvature of the retina in the image caused by the optics of the setup and the scanning geometry as well as image distortions which can arise due to patient movement, a rigid transformation of the image is not sufficient, but an affine transformation is necessary. Since in some cases there are considerable distortions between the individual images, the images are first aligned globally on the basis of the binary images. For this purpose, the speeded-up-robust-features (SURF) algorithm [7] is used, which is characterized by a high recognition rate of features and a rotational and scaling invariance.

3 Results and Discussion

The images were captured with a FOV of 85°. As discussed in Section 1, this leads to clipping and shadowing artifacts, which do not occur to such an extent for commercially available OCT devices with a much smaller FOV of around 20°. In addition, in the series of images used for this work, we tried to simulate the range of possible motion artifacts as good as possible by performing certain movements of the eye. We applied the processing as described above to test registration accuracy and artifact detection.

3.1 Vessel Segmentation

In general, the illumination compensation applied to the images followed by contrast enhancement works well. If, however, the vessels are located in shaded areas, so that their grey levels are very similar to those of the environment, the contrast cannot be significantly improved. This can be seen in Fig. 1b. Therefore, especially in the peripheral area only a few vessels can be segmented (cf. Fig. 1c).

3.2 Artifact Detection and Classification

The detection of strong artifacts, such as those caused by blinking, works reliably with the method described in Section 2.2 (cf. Fig. 2). Since the calculation of the threshold for each image is affected by the changes of the mean intensities from line to line, the threshold is lower for images without strong artifacts than for those with strong artifacts. Thus, in the former there is a higher susceptibility to false positives. In particular, horizontal vessels and vessel branching are easily detected as false positives.

Microsaccades are rarely detected. Here, a direct comparison of pixels of successive lines or working on a gradient image has not led to an improvement. These approaches were even more prone to misdetections. As a result, the image was split into too many and thus too small image segments. Too narrow segments, however, complicate the subsequent registration.

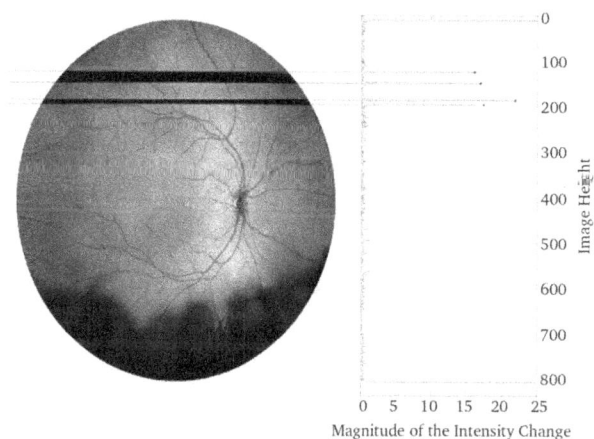

Figure 2: The graph on the right indicates the amount of intensity change of two consecutive lines used for the detection of artifacts. The blinking artifacts of the corresponding image, which is shown on the left, are reliably detected.

So far, for some artifacts no reliable criteria could be found yet to make a statement as to whether the image segments are of sufficient quality for registration. Due to the varying intensity throughout an image as well as between the images, finding reliable criteria proved to be difficult. Artifacts caused by blinking can be reliably excluded due to their low intensity. Other image segments of poor quality could possibly be sorted out by registering the segments and subsequently using the registration error to make a statement about the quality of the segment.

3.3 Registration

The coarse global registration based on the binary images works well in the region of the nerve head and in the areas where dominant vessel structures are present (cf. Fig. 3). However, due to the microsaccades causing horizontal shifts of successive image lines, global registration is not sufficient to exactly align the structures. Therefore, the image must be divided at lines where saccades occur. Based on the global transformation the individual segments are registered finer. An optimal solution, which is suitable even for very narrow image segments or segments from the edge area, is still being sought. The segmentation of the vessel towards the edges of the image should be improved to lead to a satisfactory registration even in the peripheral area.

4   Conclusion

This work focused on processing of 2D en face images. Over a wide area of the central region the registration of the images worked well. The very large FOV compared to conventional OCT systems facilitates the occurrence of motion artifacts. Particularly the presence of microsaccades as well as shadowing, which complicates good segmentation in the periphery, affect the registration. Even in images without strong artifacts, a global registration is not sufficient, as the microsaccades would have to be corrected first. Until now, no suitable criteria could be found to reliably detect them in the en face images. One way to address this problem might be to correct the microsaccades before creating the en face image by registering successive B-scans [8]. It may also be possible to record the eye movement with a pupil camera to correct motion artifacts. Correction of microsaccades could greatly simplify the registration.

Acknowledgement

The work has been carried out at the Institute of Biomedical Optics, Universität zu Lübeck.

5   References

[1] D. Huang et al., *Optical coherence tomography.* Science, vol. 254, no. 5035, pp. 1178–1181, 1991.

[2] R. Huber, M. Wojtkowski, and J. G. Fujimoto, *Fourier Domain Mode Locking (FDML): A new laser operating regime and applications for optical coherence tomography.* Opt. Express, vol. 14, pp. 3225–3237, 2006.

[3] J. P. Kolb et al., *Ultra-widefield retinal MHz-OCT imaging with up to 100 degrees viewing angle.* Biomedical Optics Express, vol. 6, pp. 1534–1552, 2015.

[4] S. Martinez-Conde, S. L. Macknik, X. G. Troncoso, and D. H. Hubel, *Microsaccades: a neurophysiological analysis.* Trends in Neurosciences, vol. 32, pp. 463–475, 2009.

[5] L. Reznicek et al., *Megahertz ultra-wide-field swept-source retina optical coherence tomography compared to current existing imaging devices.* Graefe's Archive for Clinical and Experimental Ophthalmology, vol. 252, pp. 1009–1016, 2014.

[6] K. Zuiderveld, *Contrast limited adaptive histogram equalization.* Graphics gems IV, Paul S. Heckbert (Ed.).

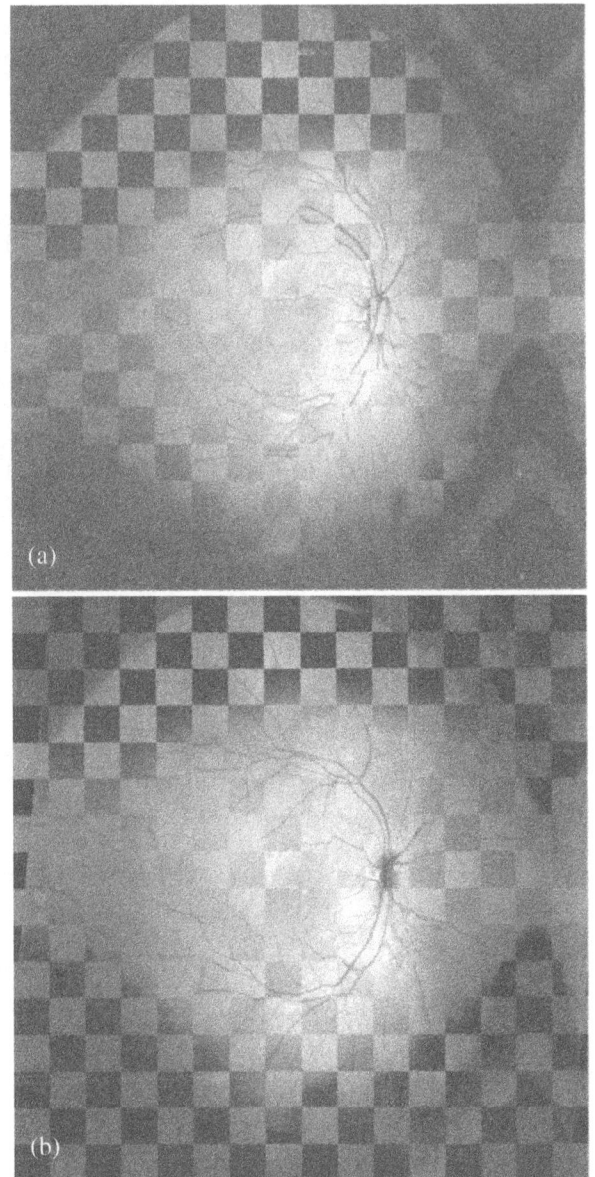

Figure 3: The figure shows two images to be registered, which are shown alternately using a checkerboard pattern. It can be seen in (a) that before registration the two vessel trees are strongly shifted from each other in the vertical direction. The checkerboard view shows that after registration (b) the vessels blend into each other very well, especially in the central area (FOV of $40°$), i.e. the registration of the images in this area has worked very well.

Academic Press Professional, Inc., San Diego, CA, USA, pp. 474–485, 1994.

[7] H. Bay, A. Ess, T. Tuytelaars, and L. Van Gool, *SURF: Speeded Up Robust Features.* Computer Vision and Image Understanding (CVIU), vol. 110, no. 3, pp. 346–359, 2008.

[8] R. J. Zawadzki et al., *Correction of motion artifacts and scanning beam distorsions in 3D ophthalmic optical coherence tomography imaging.* In: Proc. of SPIE, vol. 6426, 2007.

Spectral Characteristics of KINEVO® 900 from ZEISS

A. Britten [1], S. Meinkuss [2] and P. Reimer [2]

[1] Medizinische Ingenieurwissenschaft, Universität zu Lübeck, anja.britten@student.uni-luebeck.de

[2] Carl Zeiss Meditec AG, Oberkochen, {stefan.meinkuss, peter.reimer}@zeiss.com

Abstract

KINEVO® 900 from ZEISS (K900) is a Robotic Visualization System™ which includes three different fluorescence filter systems: INFRARED 800 (IR800), BLUE 400 (B400) and YELLOW 560 (Y560).

To use the properties of different fluorophores, K900 has to perform special lighting characteristics.

Appropriate excitation wavelengths have to be provided and emission wavelengths have to be filtered for visualization.

We measured emission ranges and losses of K900 directly after the light source, after the fiber optics, in the operating field and in the oculars to figure out if K900 illumination fits to the spectra of the fluorophores and to get new insights in the amount of transmission lost for modification of the illumination system.

K900 and its fluorescence systems provide the desired wavelengths for excitation of the fluorophores used in surgeries. The absolute loss between the xenon source and the illuminating field is more than 55 percent and in the fiber optics more than 35 percent.

1 Introduction

Fluorescence guided resection is an innovative technique in neurological surgery which helps surgeons to control whether tissue is malignant tumor or healthy tissue. K900 contains three different fluorescence system to make three different fluorophores visible.

One of the fluorescence techniques uses 5-aminolevulinic acid (5-ALA), which results in accumulation of protoporphyrin IX (PpIX) after injection to the patient. 5-ALA is a naturally in human cells occurring product of heme biosynthesis. PpIX is a fluorophore which can be excited by characteristic wavelengths of light included in the fluorescence system B400. K900 detects the of the fluorophore emitted light and visualizes it in the oculars and on a 3D screen. Fluorescent areas under B400 appear magenta and the areas around dark blue.

As an alternative, sodium fluorescein (FL) can be used to make various brain tumors visible. Under the Y560 filter of K900 areas of the brain enriched with FL can be seen. Tumors appear light yellow while the surrounding tissue still can be seen. In contrast to 5-ALA FL is chemically produced by fusion of phthalic anhydride, resorcinal and a catalyst. Surgeries using FL have the big advantage that bleedings in the operative field can be observed while Y560 is activated because the illumination is bright. That is why it is not necessary to switch between white light and Y560 fluorescence.

The third fluorescence filter system IR800 uses the fluorophore indocyanine green (ICG) which is produced by condensation of indolinium using sodium iodine. IR800 is utilized to see the intraoperative blood filled aneurysms, bypasses or arteriovenous malformations. Blood filled vessels appear white on black background. With the additional function FLOW 800 it is possible to visualize intraoperative the increase of fluorescence in the blood vessels. FLOW 800 creates colored pictures in which the color is a benchmark for the fluorescence intensity.

Using a surgical microscope without fluorescence guided technique, tumor resection depends only on the experience of the surgeon, who learned how color and consistency distinguish the tumor from healthy tissue [1].

If a fluorescence system is used, it is more probable to resect the tumor aggressively or even completely compared to a surgery only with a white light microscope [2].

In addition to that some studies have showed the positive influence of the complete resection of the tumor on the survival of the patient [3]-[5]. It is the goal of every oncological surgery to resect the malignant tumor completely.

Nevertheless a complete resection is especially for Glioblastoma multiforme (GBM) a real challenge because of the danger of functional loss of the patient [6].

GBM is one of the most malignant tumors and has despite the treatment with chemotherapy, radiotherapy and surgical treatment a poor prognosis [7]. It is difficult to remove the tissue in this disease because the malignant tumor is usually strongly intertwined with the surrounding tissue.

However the resection using 5-ALA can increase the resected tumor volume and extend the time to tumor progression for GBM. In the B400 fluorescence mode the tumor shines magenta and that is why the surgeon can see the malignant tissue better [8].

To remove spinal cord haemangioblastomas it is safer to use a microscope with ICG angiography than a normal microscope [9]. Haemangioblastomas are vascular-rich benign tumors of cerebellum. With the IR800 and FLOW 800 system surgeons can see every little vessel and can avoid the risk of not detecting pathological tissue. The use of fluorescence systems has a benefit for the patient.

2 Material and Methods

For all measurements a laboratory microscope K900 was used which was equivalent to the devices in clinical use. K900 is equipped with two 300 Watt xenon light sources which allow due to a two way illumination system with a variable spotlighting, a lighting with less shadows in the depth and in narrow channels. Emitted by the source including filters the light passes through a fiber optic and gets thereby from the source to the head of the microscope. In the head it enters the illumination optic and lights the illuminating field where it is reflected, passes the observation optic and is visualized in the oculars.

If no fluorescence filter system is activated, a UVIR filter is coupled in the path of light after the xenon light source which blocks UV and IR radiation to avoid damage caused by ultraviolet or infrared light. High energy infrared radiation lets molecules oscillate which can cause thermal damage and ultraviolet radiation can damage genetic material.

The additional option FLOW 800 uses the same filters as IR800. The different pictures on the screen are just created by a software algorithm.

In a first step we measured the spectral irradiance properties of the light source. For this we used a laboratory xenon light source which was equivalent to the light source of the K900 in clinical practice.

We focused the light, emitted by the light source, through a pinhole on an integrating sphere so that the full intensity of the xenon source impinges at the opening of the integrating sphere. The xenon light source was operated with 100 percent of its intensity and the sphere was connected to a spectrometer which measured the spectral transmission using a special software. The aperture opening of the integrating sphere was 20 mm and we averaged the result over five measurements. This measurement is used as maximum value for the transmission curve, e.g. as 100 percent light intensity of the light source for the following analysis.

After that we measured the spectral irradiance of the light source with three filters (UVIR, B400 and IR800). In a full K900 system the filters are held in a rotatable filter wheel and the K900 software turns the wheel for every configuration of the user interface in the right position. For this measurement arrangement we held the filters by hand in the optical path.

In a second step the spectral irradiance was measured again after the passage through the fiber optics which guides the light from the source to the head of K900. For the measurements served a six meter standard light guide which was equivalent to the fiber optic installed in the full K900 system. A 2 mm aperture was fixed in front of the end of

the fiber optic and the complete power of the xenon light source was fed into the lightguide. A special mount was used to fix the light guide at the entrance of the integrating sphere. This mount sealed the complete 20 mm opening of the sphere. The spectral irradiance was again measured for UVIR, B400 and IR800 filters.

To measure the spectrum of the illuminated field with the different filter options full K900 was used. The illuminated field is the area in which the surgeon operates. A working distance of 200 mm was selected on the screen and a medium light field diameter was adjusted. The aperture of the integrating sphere was focused with the help of the auto focus system and the spectra of the light field were measured.

The spectral measurements in the oculars were made with the help of an Anritsu measuring adapter, which works like the previously used integrating sphere and measures the spectrum of the irradiance. The adapter was mount at the tube of the microscope like an eyepiece and was interfaced to the spectrometer system. The working distance of 200 mm and a medium field of light was maintained. The measurements were made with barium sulfate standard white in the illuminated field which reflects the light coming from the illumination optic. The light passes the observation optic and gets into the oculars. For B400 the measurements were repeated with a ZEISS fluorescence target in the illuminated field. The fluorescence target consists of PpIX and simulates the use of fluorophores during a surgery.

3 Results and Discussion

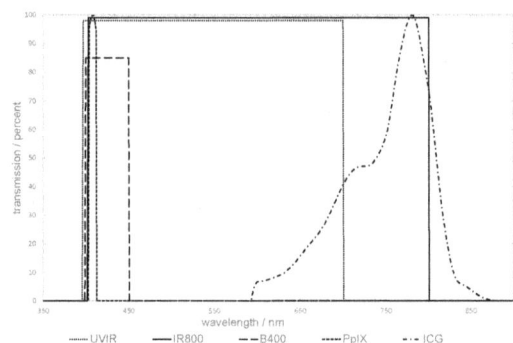

Figure 1: Emission region and losses of the light source for UVIR, B400 and IR800 filter in comparison to the excitation spectra of ICG and PpIX.

The light of the xenon light source passes one out of four different filters and gets not only filtered but also attenuated. We calculated for every graph the losses in relation to the irradiance after the xenon source. The following figures show the emission ranges which were extracted from the transmission spectra. In addition the spectra of ICG and PpIX are outlined.

Fig. 1 shows the emission region and losses of the xenon light source with the different filters. The UVIR filter lets light pass between wavelengths of 400 nm and 700 nm, which means that the destructive UV and IR radiations are removed. IR800 filter transmits between 400 nm and 800 nm which complies with the excitation wavelengths of ICG which is in a wavelength range between 600 nm and 900 nm with a maximum peak at 775 nm. B400 transmits between 400 nm and 450 nm which matches with the maximum excitation peak of PpIX which can be found at 410 nm.

The efficiency of UVIR and IR800 filter in above mentioned wavelengths range is nearly at 100 percent. Solely the B400 filter transmits only about 85 percent of the light after the source. This value can be explained by the maximum irradiance of the source which is gained without any filters only at wavelengths greater than 450 nm.

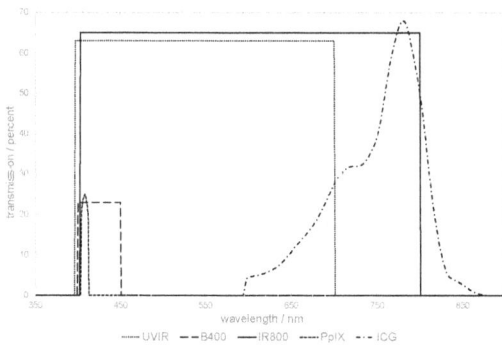

Figure 2: Emission region and losses of UVIR, B400 and IR800 filter measured after the transit through the fiber optics in comparison to excitation spectra of PpIX and ICG.

Due to the transit through the fiber optic the wavelengths range has not changed (Fig. 2). Only light intensity is reduced due to additional attenuation by the fiber. Using the fiber optic UVIR and IR800 intensity is reduced by about 35 percent compared to the intensity of the xenon source and B400 intensity is reduced by about 60 percent. That means that blue radiation is not as good transmitted through the fiber optic as greater wavelengths.

The emission region and losses of the illuminated field (Fig. 3), look very similar to the transmission after the light guide. The main difference can again be found in the amount of losses. Filtered with the UVIR or the IR800 filter only about 40 percent of the light after the light source is transmitted. That means that 20 percent of the light got lost in the optical path between the end of the fiber optics and the operating field. On this way the light filtered by B400 filter lost about 5 percent of its light intensity after the fiber optics. That means in the spectral range from 400 nm to 450 nm only 17 percent of original intensity after the light source are transmitted.

Compared to the initial intensities after the light source the light filtered with UVIR and IR800 filter lost on the optical path between xenon light source and operating field more

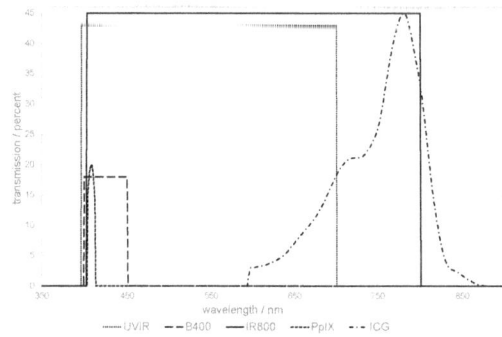

Figure 3: Emission region and losses of UVIR, B400 and IR800 filter measured in the by the head of the microscope illuminated field in comparison to excitation spectra of PpIX and ICG.

than 55 percent intensity and light filtered with B400 more than 80 percent. For every filter at least more than half of the initial light intensity got lost in this path. The ranges of the wavelengths of B400 and IR800 do not change in the first three measurements and that is why the excitation spectrum of PpIX and ICG fits.

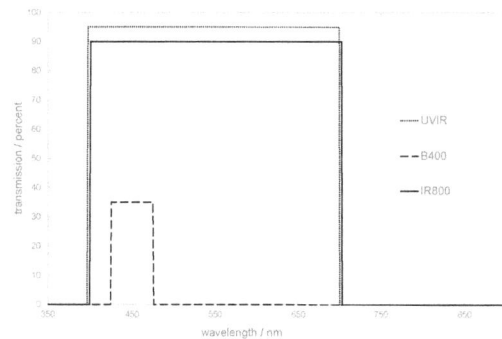

Figure 4: Emission region and losses of the whole microscope from the light source to the oculars by the use of UVIR, B400 and IR800 filter and with standard white in the illuminated field.

The light which is reflected in the illuminated field and captured in the oculars has not been measured with the help of the integrating sphere, but with the Anritsu adapter because the light in the eyepieces can not be coupled into the integrating sphere. The adapter and the sphere are two different measurement systems and that is why the weakening of the initial intensity of the light source can not be compared to the values of the oculars. Nevertheless the emission region of the measured spectra can be characterized.

Fig. 4 shows the spectra of UVIR and B400 filter with reflection on standard white which are similar to the spectra in the illuminated field. It is expected that the spectra look the same apart from the wavelengths which are taken away by the observation optic.

For UVIR and IR800 filter wavelengths between the range of 400 nm and 700 nm are transmitted to the eye of the user. A special filter in the observation optic blocks wavelengths greater than 700 nm to protect the eyes of the observer against thermal effects caused by infrared radiation. That is why wavelengths greater than 700 nm can not be ·found in the measurement with the IR800 filter.

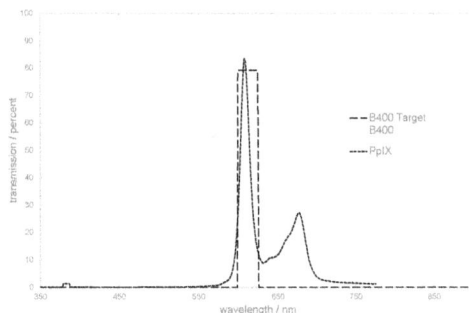

Figure 5: Emission region and losses in the oculars measured with B400 filter and a B400 fluorescence target in the illuminated field in comparison to emission spectrum of PpIX.

In Fig. 5 the emission of the ZEISS B400 fluorescence target in the illuminated field can be seen. The oculars visualize wavelengths between 600 nm and 625 nm. This range fits to the emission spectrum of PpIX which covers wavelength between 450 nm and 700 nm. The shift between the emission and excitation spectra is large enough to prevent spectral overlap. Thus reflected excitation light is effectively suppressed.

4   Conclusion

K900 fulfills the spectral requirements to use ICG and PpIX in clinical applications because the fluorescence filter systems deliver the right excitation wavelengths for the standard fluorophores ICG and PpIX. At least 55 percent of the light intensity got lost in the optical path between the light source and the operating field. Due to the transit through the light guide at least 35 percent of transmission got lost for UVIR and IR800, 60 percent for B400.

It is an objective to improve the efficiency and to reduce losses. One approach is to change the fiber optic to a liquid light guide. A liquid light guide transmits better than a glass fiber light guide. Especially in the range of blue wavelengths the amount of transmission is higher. Nevertheless it has to be taken in account that the handling of liquid light guides is more difficult and the optical properties are different. Short wavelengths in the UV range can destroy transmission properties and long IR wavelengths can overheat the fiber and cause bubbles. The effects depend on the power of illumination. To conclude we need a good fiber optic which looses less transmission so that we need less initial power of the light source.

Acknowledgement

The work has been carried out at Carl Zeiss Meditec AG, Oberkochen and supervised by Gereon Hüttmann, Institute of Biomedical Optics, Universität zu Lübeck.

5   References

[1] M. S. Berger, *The fluorescence guided technique*. Neurosurg Focus, vol. 36, no. 2, p. E6, 2014.

[2] W. Stummer, U. Pichlmeier, T. Meinel, OD. Wiestler, F. Zanella, HJ. Reulen et al., *Fluorescence-guided surgery with 5-aminolevulinic acid for resection of malignant glioma: a randomised controlled multicentre phase III trial*. Lancet Oncol., vol. 7, no. 5, pp. 392-401, 2006.

[3] F. Acerbi, M. Broggi, M. Eoli, E. Anghileri, C. Cavallo, et al., *Is fluorescein-guided technique able to help in resection of high-grade gliomas?* Neurosurg Focus, vol. 36, no. 2, p. E5, 2014.

[4] M. Lacroix, D. Abi-Said, D. R. Foruney, Z. L. Gokaslan, W. Shi, et al., *A multivariate analysis of 416 patients with glioblastoma multiforme: prognosis, extent of resection, and survival*. J. Neurosurg, vol. 95, no. 2, pp. 190-198, 2001.

[5] M. J. McGirt, K. L. Chaichana, M. Gathinji, F. J. Attenello, K. Than, et al., *Independent association of extent of resection with survival in patients with malignant brain astrocytoma*. J. Neurosurg, vol. 110, no. 1, pp. 156-162, 2009.

[6] M. A. Vogelbaum, *Does extent of resection of a glioblastoma matter?* Clin. Neurosurg, vol. 59, pp. 79-81, 2012.

[7] R. Stupp, M. E. Hegi, W. P. Mason, M. J. van den Bent, M. J. Taphoorn, et al., *Effects of radiotherapy with concomitant and adjuvant temozolomide versus radiotherapy alone on survival in glioblastoma in a randomised phase III study: 5-year analysis of the EORTC-NCIC trial*. Lancet Oncol., vol. 10, no. 5, pp. 459-466, 2009.

[8] S. Eljamel, *5-ALA fluorescence image guided resection of glioblastoma multiforme: A meta-analysis of the literature*. Int. J. Mol. Sc., vol.16, no. 5, pp. 10443-10456, 2015.

[9] N. Benedetto, *Use of near-infrared indocyanine videoangiography and Flow 800 in the resection of a spinal cord haemangioblastoma*. B. J. Neurosurg, vol. 27, no. 6, pp. 847-849, 2013.

Interferometric detection of laser induced nano- and micro bubble dynamics in water and tissue

K. Nadji[1], S. Freidank[2], A. Vogel[2] and N. Linz[2]

[1] Medizinische Ingenieurwissenschaft, Universität zu Lübeck, kevin.nadji@student.uni-luebeck.de

[2] Institute of Biomedical Optics, Universität zu Lübeck, linz@bmo.uni-luebeck.de

Abstract

An interferometric detection technique of laser-induced bubble formation was used to investigate bubble dynamics in water and transparent tissues. This technique enables to determine the bubble dynamics in a single-shot method, with 160 ps temporal and 70 nm spatial resolution. In the present work, this technique was used to investigate the bubble collapse in water and to compare the bubble dynamics in water and tissue. The bubble collapse can be characterized by the ratio of first and second oscillation and it was shown that the bubble dynamic is self-similar down to a bubble radius of 3 μm, but shows strong changes due to viscosity and surface tension for small nano bubbles. For bubbles in tissue, the technique requires additional flash photography of the maximum bubble radius to calibrate the radius-time-curve. Comparison of bubble dynamics in tissue and water yielded new insights regarding the visco elastic plastic properties and their influence on the bubble dynamics.

1 Introduction

The dynamics of laser induced cavitation bubbles has fascinated scientists since decades, because of its extremely high energy density in spherical bubble collapse. Besides the fundamental investigations in water, the medical application of laser induced bubbles in refractive surgery becomes ever more important. Laser-in-situ-keratomileusis (LASIK) is one of the most common surgery for correction of refractive errors [1]. During surgery, a corneal flap is created, folded to the side and subsequently corneal tissue is ablated to obtain a radius curvature change for vision correction [2]. The flap creation by focused femtosecond laser pulses is based on optical breakdown and the destructive expansion of the cavitation bubble inside the corneal stroma. Previous experiments on cavitation bubbles were often performed in water, but the dynamics in fibril structures like in corneal tissue remains almost unexplored. To investigate bubble dynamics both in water and in tissue, the working group of Prof. Vogel at the Institute of Biomedical Optics has developed a new interferometric detection technique based on backscattering of a cw probe laser beam. This technique is used to investigate spherical bubble dynamics, which is hard to realize. Limiting factors of spherical bubble formation are a compact form of the laser induced plasma, interacting interfaces close to the bubble wall, and buoyancy. Therefore cavitation bubbles have been produced by UV-microchip laser pulses at 355 nm wavelength (560 ps pulse duration), that are focused through a microscope objective (NA = 1.2) into the medium. The focusing conditions lead to very compact plasmas and formation of spherical nano bubbles. The interferometric detection for investigation of bubble dynamics has been used in water and in different tissues like corneal epithelium, stroma and crystalline lens of porcine eyes. By combining the interferometric detection technique with flash photography of the bubble size, the dynamics in tissue have been determined with unprecedented high temporal and spatial resolution.

2 Methods

Investigations of laser induced cavitation bubble dynamics in water and tissue require reproducible, spherical bubble generation. Therefore, temporal smooth UV ns laser pulses are focused through high NA into the transparent medium. At very high irradiances optical breakdown occurs via non linear energy deposition. A high energetic plasma is formed and the vaporization at high temperature leads to the investigated bubble. In the following section the applied methods for detection of laser induced cavitation bubble dynamics in water and biological tissue are explained. The whole setup has been realized on the optical bench [3].

2.1 Interferometrical Detection

Interferometric bubble detection enables the determination of bubble dynamics with high temporal and spatial resolution. The interferometric signal originates by interference of a cw probe laser that is backscattered at the bubble wall. As soon as the cavitation bubble in the laser focus is formed, the confocal and collinear aligned probe laser is reflected at the bubble front and rear side. Fig. 1 illustrates schematically the signal formation.

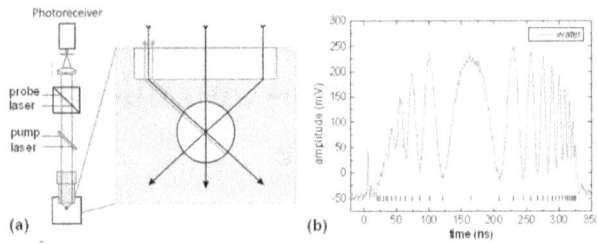

Figure 1: a) Interferometric signal formation by back reflection of the cw probe laser at front and rear side of the bubble, b) detected interferometric signal.

By confocal orientation of pump and probe laser and for bubbles larger than the beam waist, the cavitation wall forms an ideal mirror which reflects the probe laser. The path difference Δx between bubble front and rear side reflection corresponds to twice the bubble diameter D. For a path difference of an integer of the probe laser wavelength

$$\Delta x = 2D = n * \lambda, n \in \mathbb{N} \tag{1}$$

constructive interference leads to a maximum, which can be detected by the photoreceiver. As soon as the path difference is a half odd integer of half the wavelength

$$\Delta x = (n + 0.5)\lambda, n \in \mathbb{N} \tag{2}$$

destructive interference takes place and the interference signal has a minimum. Evaluation of the interference signal regarding maxima and minima enables to draw conclusion about the bubble dynamics, so that this method leads to radius-time-curves $R(t)$ with high temporal and spatial resolution via single-shot measurements. Spatial resolution is given by the probe laser wavelength. The distance between interference maxima and minima corresponds to a path difference of $\lambda/2$ at twice the bubble diameter, which results in a radius change $R(t)$ by $\lambda/8$. A 561 nm probe laser wavelength leads to an accuracy of 70 nm. Temporal resolution is affected by the bandwidth (BW) of the photoreceiver, that is used for detection of the interference signal in back reflection. A high sensitivity photoreceiver with 200 MHz (HCA-S-2001, 2x10^4 V/A transimpedance gain with 1.1 ns temp. res.) has been used for experiments with small nano bubbles while larger micro bubbles with faster dynamics have been investigated by a 2.2 GHz receiver (160 ps temp. res.). The photoreceiver signal was recorded using an 6 GHz oscilloscope with 25 GS/s. Corresponding to signal theory, a maximum velocity

$$u = \frac{\lambda}{8} \frac{BW}{0.22} = 0.568\lambda * BW \tag{3}$$

for the bubble wall of 63.75 m/s (200 MHz) and 701.25 m/s (2.2 GHz) can be detected with a probe laser at $\lambda = 561$ nm wavelength.

2.2 Interference Signal Evaluation in Water

Bubble dynamics in water for large bubbles is well investigated. In optical breakdown, a plasma is formed and the very high pressure and temperatures lead to an explosive expansion of the vaporized volume. The bubble expands with high velocity, slows down until it reaches its maximum radius and collapses again with high velocity. With the oscillation time T_{osc} deduced from the interference signal, the bubble maximum radius can be calculated via the Rayleigh equation [4]

$$R_{max} = \frac{T_{osc}}{1.83} \sqrt{\frac{p_\infty - p_v}{\rho_0}} \tag{4}$$

where p_∞ is the hydrostatic pressure (0.1 MPa), p_v the vapor pressure inside the bubble at 20°C (2330 Pa) and ρ the water density at 20°C (998.2 kg/m^3). However, for accurate calculations of the maximum radius for small bubbles one needs to chose the Gilmore model [5], which, in contrast to the Rayleigh model, takes surface tension and viscosity of water into account. Because of its high complexity, the radii in this work have been first determined by the Rayleigh equation and subsequently multiplied with a correction term $f(T_{osc})$ to obtain the radii predicted by the Gilmore model [6]. As bubble oscillation period T_{osc}, the period between bubble formation and its first collapse is identified (Fig. 2).

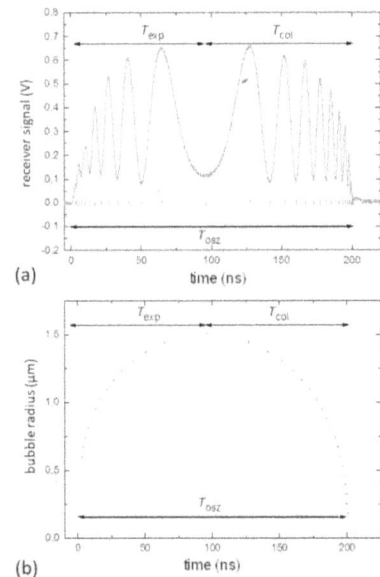

Figure 2: a) Interferometric signal of a laser induced bubble. The oscillation time is T_{osc}= 200 ns. This period can be divided in an expanding part T_{exp} and a collapsing part T_{col}. The expansion begins initially at optical breakdown and slows down until it reaches its maximum radius. The deceleration is characterized by increasing temporal distance between signal-maxima and -minima. After the turning point, the bubble collapses again. b) $R(t)$-curve determined by evaluation of the interference fringes and calculating R_{max}.

The evaluation of the interference signal for determination of the bubble dynamics $R(t)$ is according to the following scheme and is illustrated in Fig. 2: First the maximum radius R_{max} is calculated with the oscillation time T_{osc} by using (4). The maximum radius is given at the central-maximum or -minimum. The interference signal provides at each maxima and minima one measurement point for the $R(t)$-curve, where the bubble radius R is reduced by $1/8$ of the probe laser wavelength. In this work, a 561 nm probe laser wavelength is used that leads to a radius change of 70 nm.

2.3 Interference Signal Evaluation in Tissue

Interferometric determination of cavitation bubble dynamics in tissue requires additional information on the maximum bubble radius R_{max}. In water, the maximum bubble radius can be calculated by T_{osc} and (4). This equation can not be used in tissue, because the bubble dynamics strongly depend on different tissue parameters and reliable numerical models have not yet been developed. Therefore, the maximum bubble radius for calibration of the $R(t)$-curve must be determined by additional flash photography. Hence, the experimental setup has been extended by a flash lamp illumination and imaging beam path [3].

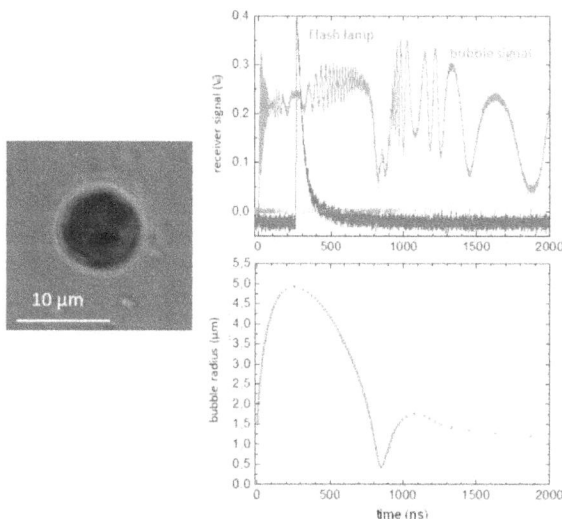

Figure 3: A flash photography of the bubble (left side) is used to calibrate the maximum radius. The flash lamp signal is shown together with the interference signal (top right) yielding a $R(t)$-curve (bottom right).

The time of the flash photography can be evaluated by the flash signal of an additional photodiode that is displayed on the oscilloscope (Fig. 3).The photographically determined bubble radius (in Fig. 3 exactly R_{max}) is used as starting point for the calculation of $R(t)$ by subtracting $\lambda/8$ at each time of an interference fringe in the bubble signal. The combination of flash photography for calibration of R_{max} and interferometric signal evaluation yields $R(t)$-curves in tissue with unprecedented temporal and spatial resolution.

3 Results and Discussion

The dynamics for bubbles in water between $R_{max} = 800$ nm and $R_{max} = 15.3$ μm is summarized in Fig. 4. The

radius-time-curves $R(t)$ show that bubbles with maximum radii smaller than 3 μm leave the regime of self-similar bubble dynamics and the second oscillations seems to disappear. However, the maximum radius of reoscillations is too small for interferometric detection, so that only Rayleigh backscattering occurs. The duration of the scattering signal matches the oscillation time of the second oscillation.

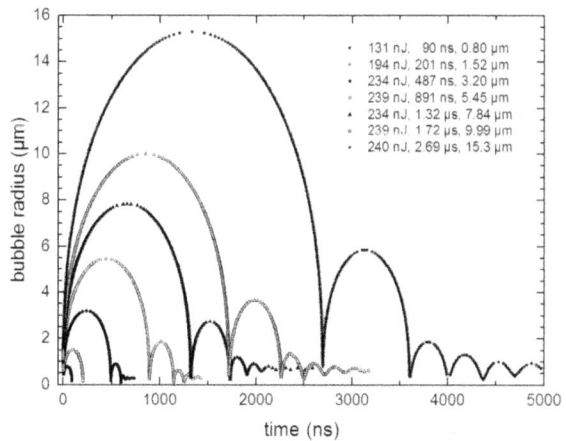

Figure 4: Summary of experimental determined $R(t)$-curves for laser induced cavitation bubble formation with maximum radii from 0.8 μm to 15.3 μm in water.

The size of the reoscillating bubble depends on energy losses in the bubble collapse. The stronger the collapse, the smaller the reoscillation. This enables a statement about the bubble collapse dependence on the bubble size by calculating the ratio of the first and second oscillation time T_{osc1}/T_{osc2}. This ratio in dependence on the maximum radius R_{max} is shown in Fig. 5.

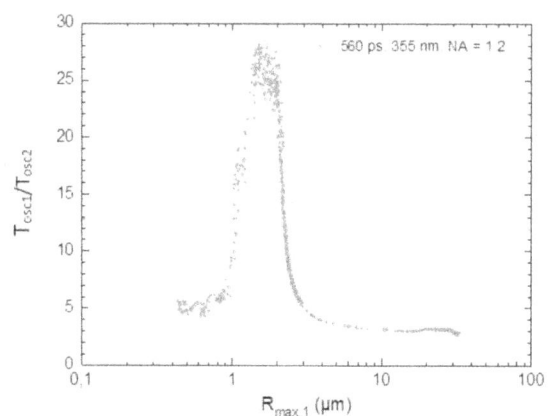

Figure 5: Ratio of the first and second oscillation times T_{osc1}/T_{osc2} as function of R_{max} for bubbles in water.

It was shown that bubble dynamics of bubbles > 3 μm have a constant ratio $T_{osc1}/T_{osc2} = 3$ (self-similarity). For smaller bubbles, the ratio strongly increases tenfold and reaches its maximum at $R_{max} = 1.5$ μm. This means that with decreasing bubble size, the energy loss in bubble collapse increases. The border of the self-similarity regime is defined by the interplay of surface tension and viscous damping. The surface tension leads to an additional pressure on the bubble wall with decreasing bubble size. Viscosity leads to an contrary effect, which slows the collapse down and dampens it. The increase of T_{osc1}/T_{osc2} for bubbles < 3 μm shows that surface tension has stronger effects than viscous damping, yielding higher collapse pressures and stronger dissipation.

The curve drop below R_{max}= 1.5 μm is a result of a stronger influence of viscous damping and possibly an elongated plasma shape close to the optical breakdown threshold, which limits spherical bubble formation.

Radius-time-curves R_{max} in water and tissue determined by interferometric detection and flash photography measurements are shown in Fig. 6. The $R(t)$-curves for $R_{max} = 5$ μm were normalized and plotted together. Measurements in water serve as reference. It is clearly visible that the bubble dynamics in water, epithelium, lens and stroma are very different. Comparing the collapse phase in water and tissue, one can see that the bubble wall velocity in water is significantly faster than in tissue. The collapse velocity exceeds the temporal resolution of the interferometric detection. In lens and epithelium, the collapse is much slower, such that interferometric detection enables the determination of the bubble dynamics through the collapse. In stromal tissue, the bubble wall velocity in the collapse is strongly dampened and is very slow.

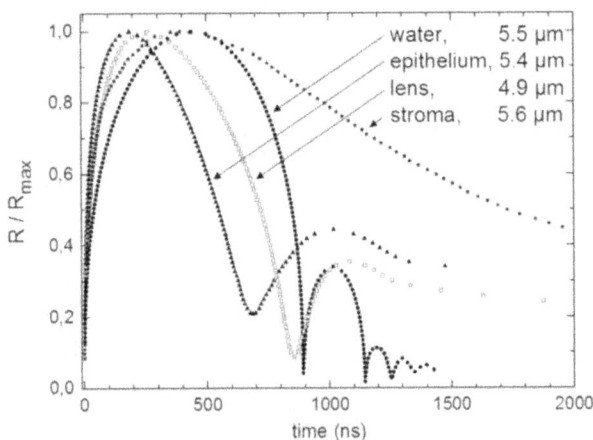

Figure 6: Normalized $R(t)$-curves of bubbles in water, epithelium, lens and stroma for $R_{max} \approx 5$ μm.

Regarding bubble expansion, it has been shown, that the expansion phase strongly depends on surrounding material conditions. In tissue, bubbles require higher pressures and stronger driving forces to overcome the elastic-plastic restoring forces. Thus, higher energy densities are needed and bubble expansion is initially faster than in water. Table 1 summarizes the maximum detectable velocities during expansion and collapse phase. Maximum velocities up to 360 m/s have been measured in epithelium, whereas bubbles in water expand with only 232 m/s. However, the initially very fast expansion in tissue is slowed down by elastic-plastic restoring forces. In total, the expansion phase T_{exp} of the 5 μm bubble in epithelium is 184 ns, whereas

Table 1: Maximum detectable bubble wall velocities during expansion and collapse for bubbles at $R_{max} = 5\mu$m

	Water	Epithelium	Lens	Stroma
$V_{max.exp}$	232 m/s	360 m/s	120 m/s	265 m/s
$V_{max.col}$	350 m/s	15 m/s	5 m/s	0.3 m/s

the expansion phase in water is two times larger (445 ns). The bubble collapse in tissue is accelerated by the elastic response of the deformed tissue matrix. However, the initially acceleration will be slowed down by viscous damping. Thus, the total collapse phase in tissue is longer than in water. For epithelium, a collapse time of 500 ns, lens up to 598 ns, and water as reference with 446 ns was measured. In stroma, no final collapse time has been determined.

4   Conclusion

An interferometric detection technique in combination with flash photography for determination of bubble dynamics has been used in water and different tissues like corneal epithelium, stroma and crystalline lens of porcine eyes. The bubble dynamics in tissue yielded new insights regarding the visco-elastic-plastic behavior under high strain rates. For the first time, the bubble collapse in tissue has been fully resolved with 70 nm spatial and 160 ps temporal resolution. Based on its high precision, the bubble interferometry has the potential to become the gold standard for the determination of bubble dynamics in water and tissue.

Acknowledgement

The work was supervised by the Institute of Biomedical Optics, University of Lübeck.

5   References

[1] T. Kohnen, *Basiswissen refraktive Chirurgie.* Deutsches Ärzteblatt 105:163–173, 2008.

[2] D. T. Azar and Koch, *LASIK.* Taylor & Francis Group, pp. 1–2, 2002.

[3] K. Nadji, *Interferometrische Detektion der Dynamik lasererzeugter Nano- und Mikroblasen in Wasser und Gewebe.* Bachelorthesis, Universität zu Lübeck, 2016.

[4] L. Rayleigh, *On the pressure developed in a liquid during the collapse of a spherical cavity.* Philos Mag, 34:94–98, 1917.

[5] F. R. Gilmore, *The growth or collapse of a spherical bubble in a viscous compressible liquid.* Calif. Inst. Techn. Rep., pp. 26–4, 1952.

[6] N. Linz, *Controlled nonlinear energy deposition in transparent dielectrics by femtosecond and nanosecond optical breakdown.* ISBN: 978-3-86247-076-1, 2010.

Investigation of Influences on Measurement Accuracy of Glucose with a Faraday Modulated Polarimeter

C. A. Carvajal Arrieta [1,3], C. Stark [2,3], R. Behroozian [2,3], and S. Müller [3]

[1] Biomedical Engineering, University of Applied Sciences Lübeck, cesar.andres.arrieta@stud.fh-luebeck.de

[2] Graduate School for Computing in Medicine and Life Sciences, Universität zu Lübeck, {christian.stark, reza.behroozian}@fh-luebeck.de

[3] Medical Sensors and Devices Laboratory, University of Applied Sciences Lübeck, stefan.mueller@fh-luebeck.de

Abstract

Optical blood glucose monitoring is an important medical affair worldwide. With a broadband Faraday modulated polarimeter, glucose is measured due to its optical rotatory dispersion (ORD) as a chiral molecule. However, its distinction from similar optically active impurities like proteins requires high setup stability. This paper reports influences on glucose measurements due to changes in currents' drivers. Thus, for extracted intensities, $I(w)$ and $I(2w)$ from the sinusoidal modulation, fitting polynomials were built and error propagation methods were applied when varying the LED, Faraday rotator amplitude, and offset currents. For simplicity, only glucose was measured at the LED central wavelength. Consequently, current changes explained around 10 % of the actual approximated error of ±1 mg/dl for a sample with 500 mg/dl. Also, glucose determination with the ratio $\frac{I(w)}{I(2w)}$ compensated these deviations by 26 % compared to $I(w)$. Finally, up to 2000 times higher variations gave acceptable errors representing high setup stability.

1 Introduction

To optimally measure blood glucose, an accurate and long-term stable optical setup is widely desired. Absorption spectroscopy is an option. However, it is limited in sensitivity by the low absorption of glucose in the near-infrared range and expensive for its required instruments [1]. With a broadband Faraday modulated polarimeter, optical rotatory dispersion (ORD) is extracted in the visible spectrum. Thus, glucose is measured due to its particular optical activity as a chiral molecule, i.e. polarized light is specifically rotated in proportion to its concentration [2]. However, its distinction from other optically active impurities with similar ORD demands high setup stability. This molecular rotation in millidegrees affects the transmitted light intensity, and its extracted components, $I(w)$ and $I(2w)$ from sinusoidal modulation at frequency w, required for glucose prediction.

Recent studies focus on birefringence or scattering effects when measuring glucose in the eyes with non-invasive polarimetric approaches [3]. However, little is known about the influences of electrical currents driving a setup measuring glucose with sample cells. This paper presents error estimations of glucose concentration prediction due to variations in the LED (i_L), Faraday rotator amplitude (i_A) and offset (i_O) currents during the measurements. Two determination methods were tested. First, using intensity $I(w)$ and second, with the ratio $\frac{I(w)}{I(2w)}$. Also, the stability against influences and theory compliance of components $I(w_y)$, with w_y the respective frequency w or $2w$, were evaluated. Consequently, current deviations represent a portion of glucose determination errors, partially explaining their source.

2 Material and Methods

To investigate effects on glucose measurement accuracy, currents driving the setup were measured and related to the extracted amplitudes $I(w_y)$ required for glucose determination. The measurement setup, procedures to identify influences and estimate errors are detailed as follows.

2.1 Measurement Setup

The polarimeter setup can be divided into four blocks. The first is the power sources for the current drivers of the LED (a) and Faraday rotator (c) in Fig. 1. The second block has the currents drivers' board with their electronic stages. The

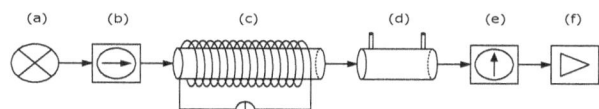

Figure 1: Configuration: (a) Light source LED, (b) Glan-Thompson polarizer, (c) Faraday rotator solenoid with Lead Silicate SF59 glass rod, (d) Flow through cuvette, (e) Modifiable Glan-Thompson polarizer and (f) Photo-detector.

third is a measurement box with external light isolation and temperature control, which contains an automatic sample generator with stock solutions and neMESYS syringe pumps, and the configuration shown in Fig. 1. Finally, the last block is the data acquisition card with its software. For simplicity of data analysis, the setup was evaluated with only glucose at the LED central wavelength $\lambda = 528\,\text{nm}$.

The detected light intensity I of setup in Fig. 1 is proportional to the electrical field as

$$I \propto \mathbf{E}^2 = (-\sin(\Theta_m \sin(wt) + \Phi))^2, \qquad (1)$$

where Φ is the sum of the sample angle Φ_s and the angle Φ_f between filters (b) and (e) on Fig. 1. Θ_m is the Faraday modulation depth, w its frequency and t a certain time [4]. With the approximation $\sin(\alpha) \approx \alpha$ for (1), equation (2) expresses the transmitted intensity as the sum of components $I = I(DC) + I(w) + I(2w)$ [2], [4].

$$I = \frac{\Theta_m^2 + 2\Phi^2}{2} + 2\Phi\Theta_m \sin(wt) - \frac{\Theta_m^2}{2}\cos(2wt) \quad (2)$$

2.2   Determination of Influences

The modulation depth Θ_m changes with i_A as in (3), where ν_λ is the SF59 verdet constant at wavelength λ, μ_r its relative permeability and N the number of solenoid turns [5].

$$\Theta_m = \left[\frac{90\mu_0\mu_r N\nu_\lambda}{\pi}\right] i_A = [k]\, i_A \qquad (3)$$

Changes in offset i_O modify the ORD symmetry given by the Faraday rotator, adding an angle Φ_{i_O} to the initial Φ. Also, variations on i_L generate a proportion K_{i_L} on amplitudes $I(w_y)$. Thus, intensities $I(w)$, $I(2w)$ and ratio R can be expressed theoretically as in (4), (5) and (6) respectively.

$$I(w) \propto 2\Phi\Theta_m = 2k[\Phi + \Phi_{i_O}][i_A][K_{i_L}] \qquad (4)$$

$$I(2w) \propto \frac{\Theta_m^2}{2} = \frac{k^2}{2}[i_A]^2[K_{i_L}] \qquad (5)$$

$$R = \frac{I(w)}{I(2w)} = \frac{4\Phi}{\Theta_m} = \frac{4}{k}\left[\frac{\Phi + \Phi_{i_O}}{i_A}\right] \qquad (6)$$

To assess currents i_A, i_O and i_L influences on glucose prediction, intensities $I(w_y)$ were measured in mV. Since the blood sugar concentration can rise up to $600\,\text{mg/dl}$ for patients with hyperglycemia [4], random order concentrations from $[0 - 500\,\text{mg/dl}]$ in steps of $125\,\text{mg/dl}$ were used. For each of them, inputs i_x, with x the current type $\{A, O, L\}$, were independently changed in a large driver's tolerable range. The set initial conditions were $i_A = 2.56\,\text{A}_{\text{pp}}$, $i_O = 0.7\,\text{mA}$, $i_L = 400\,\text{mA}$, third block at $32\,°\text{C}$ and modulation frequency $f_0 = 1.009\,\text{kHz}$. Moreover, polynomial regression was used to find the relation between currents and measured intensities. Their order was selected according to (4) or (5), and a determination coefficient $R^2 \approx 1$. The resulting fitting curves $I(w_y, C, i_x)$ are defined as in (7), with coefficients $a_{j,C,x}$, $j \in \{0, 1, 2\}$, which represent influences on amplitudes $I(w_y)$ by currents i_x in A, for concentrations C under typical measurement conditions.

$$I(w_y, C, i_x) = a_{2,C,x}[i_x]^2 + a_{1,C,x}[i_x] + a_{0,C,x} \qquad (7)$$

For getting the modulation depth, $I(w_y)$ was characterized with filter (e) of Fig. 1. By setting the previous initial conditions and $C = 0\,\text{mg/dl}$, leading to $\Phi_s = 0\,°$, filters' angle Φ_f was varied in $\pm 6\,°$. Since the filter's datasheet provides the conversion from turns in mm to degrees, angle Φ_f was known. Then, similar polynomials as in (7) were built, but $I(w_y, \Phi_f)$ depending on angle Φ_f. By using (6), the measured (M) modulation depth Θ_M can be calculated as

$$\Theta_M = 4\Phi_f\left[\frac{I(2w, \Phi_f)}{I(w, \Phi_f)}\right]. \qquad (8)$$

To evaluate the setup's theory compliance, the frequency components were measured with $\Phi_s = 0\,°$ and initial $I(w) = I(2w)$, leading to $\Phi_f = 0.25\Theta_M$ with Θ_M as in (8). Then, currents i_x were individually varied to compare amplitudes $I(w_y)$ and R with (4), (5) and (6).

2.3   Estimation of Prediction Errors

To estimate absolute prediction errors $\xi(i_x)_M$ for each concentration C, the differential method in (9) was applied. $D_M(i_x)$ is the determination method with $I(w, C, i_x)$ or with the unit less $R(w_y, C, i_x)$ from curves in (7), and Δi_x is the current variations in A. Function $F(\xi)$ was built to convert errors from mV, or unit less, to mg/dl by a fitting curve of initial amplitudes measured for each concentration.

$$\xi(i_x)_M = F\left(\Delta i_x\left[\frac{\partial D_M(i_x)}{\partial i_x}\right]\right) \qquad (9)$$

The absolute prediction errors $\xi(i_x)$, expected for the measured $D_M(i_x)$ were estimated as in (10). $D(\Theta_m, \Phi)$ is the theoretical determination method with $I(w)$, or the ratio R, and $\Psi(i_x)$ their influenced parameter Φ or Θ_m.

$$\xi(i_x) = F\left(\frac{\Delta i_x}{D(\Theta_m, \Phi)}\left[\frac{\partial D(\Theta_m, \Phi)}{\partial \Psi(i_x)}\right] D_M(i_x)\right) \quad (10)$$

Considering all currents i_x as independent error sources for $I(w_y)$ on each method D_M, their influence on total prediction error of glucose concentration, $\xi_T(C)$, was calculated with the sum in quadrature shown in (11). Finally, to evaluate the currents drivers' reliability, all inputs i_x were monitored when varying a specific current x.

$$\xi_T(C) = \sqrt{[\xi(i_A)_M]^2 + [\xi(i_O)_M]^2 + [\xi(i_L)_M]^2} \quad (11)$$

3   Results and Discussion

With the defined influences and theory, and applying the error propagation methods, the results are as follows.

3.1   Theory Compliance

The solenoid (c) from Fig. 1 has $N = 375\,\text{turns}$, glass SF59 a $\nu_\lambda = 43.90\,\text{rad/mT}$ and $\mu_r \approx 1$ [6]. With the initial current i_A in A peak to peak, the theoretical modulation depth from (3) was $\Theta_m = 1.92\,°$. Moreover, from characterization of angle Φ_f and equation (8) with $\Phi_f = 0.48\,°$

expected from condition $I(w) = I(2w)$, the measured modulation depth was $\Theta_M = 1.62\,°$. This results in $15.6\,\%$ of deviation. However, (3) is an approximation for long coils depending on literature parameters. Also, this difference can be caused by the assumption of homogeneous magnetic fields and inaccuracies of the micrometer filter (e). With small angles Φ_f, as in approximation (2), Θ_M should remain relatively constant. However, it varied only $0.5\,\%$ for the large testing range of Φ_f, proving the approximation reliability and modulation depth stability required for glucose prediction.

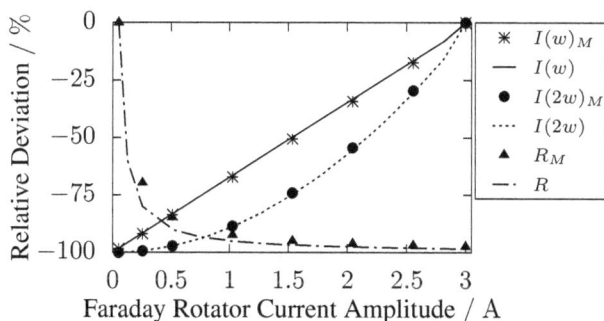

Figure 2: Relative deviations in % from maximum values of $I(w)$, $I(2w)$ and ratio R, measured (M) and theoretical, for variations only on Faraday rotator current amplitude i_A.

For the theory compliance of intensities $I(w_y)$ and R when varying only i_A, the percent of amplitude deviations with respect to their maximum measured values are illustrated in Fig. 2. There, $I(w)_M$ changed linearly as expected from (4), $I(2w)_M$ quadratic as in (5), and R_M inversely as in (6). On the other hand, when measuring i_O variations, $I(w)_M$ and R_M behaved linearly, while $I(2w)_M$ remained constant. Thus, the setup fulfilled appropriately the theory described for inputs i_A and i_O. Whereas with i_L, a more quadratic behavior was observed for $I(w_y)_M$ and R_M. However, theory for this effect is still in development.

3.2 Influences on Prediction Accuracy

With the measured and theoretically verified amplitudes $I(w_y)$, fitting polynomials $I(w_y, C, i_x)$ as in (7) were calculated. Typical deviations Δi_x were smaller (up to 12000 times) than the initial values of i_x. This suggested the use of partial differential error propagation methods in (9) and (10). To apply these estimations, the condition number $\kappa = i_x [D'_M(i_x)/D_M(i_x)]$ was evaluated, resulting in well-conditioned equations. Consequently, the estimated glucose concentration prediction errors, for measured and theoretical determination methods with $I(w)$ and R, in total and due to currents i_x are illustrated in Fig. 3.

Measured $\xi(i_x)_M$ and theoretical $\xi(i_x)$ prediction errors, with $D = I(w)$, are depicted by a linear response to Δi_x in Fig. 3a. The slope for currents i_A and i_L is a result of the partial derivatives, (9) and (10), dependency with Φ_s. Instead, the curve for i_O depends only on the constant Θ_M. For the sum in quadrature of measured errors $\xi_T(C)$, the expected square root behavior increases

for sample angles Φ_s of higher concentrations. When measuring glucose concentration, typical fluctuations on currents i_x were $\Delta i_A = \pm 0.217\,\mathrm{mA}$, $\Delta i_O = \pm 0.016\,\mathrm{mA}$ and $\Delta i_L = \pm 0.08\,\mathrm{mA}$, with $\Delta i_x = \sigma/\bar{i_x}$, σ the standard deviation and $\bar{i_x}$ the average current. Hence, by using fitting polynomials in (7) with a concentration of $500\,\mathrm{mg/dl}$, the estimated error $\xi_T(500\,\mathrm{mg/dl})$ with (11) was $\pm 0.112\,\mathrm{mg/dl}$, which corresponds to $10.6\,\%$ of actual determination error ($\pm 1.05\,\mathrm{mg/dl}$) when using intensity $I(w)$.

The estimated prediction errors with $D = R$ are shown in Fig. 3b. Their response to changes Δi_x was similar to the previous behavior explained for Fig. 3a. Nevertheless, the curve for $\xi(i_L)_M$ had a considerable reduction in its amplitude. This demonstrates that ratio R, as in (6), compensates considerably effects on concentration prediction by variations on i_L. For the test concentration, the estimated error $\xi_T(500\,\mathrm{mg/dl})$ was $\pm 0.083\,\mathrm{mg/dl}$, which is $8.9\,\%$ of $\pm 0.94\,\mathrm{mg/dl}$ determination error when using the novel ratio R. Thus, $\xi_T(500\,\mathrm{mg/dl})$ was reduced by $26\,\%$ compared to $I(w)$. Those results suggest that the influence K_{i_L} is very similar for $I(w)$ and $I(2w)$ as in (4) and (5), characteristic taken by the ratio, with more information by using also $I(2w)$, to cancel influences as presented in (6).

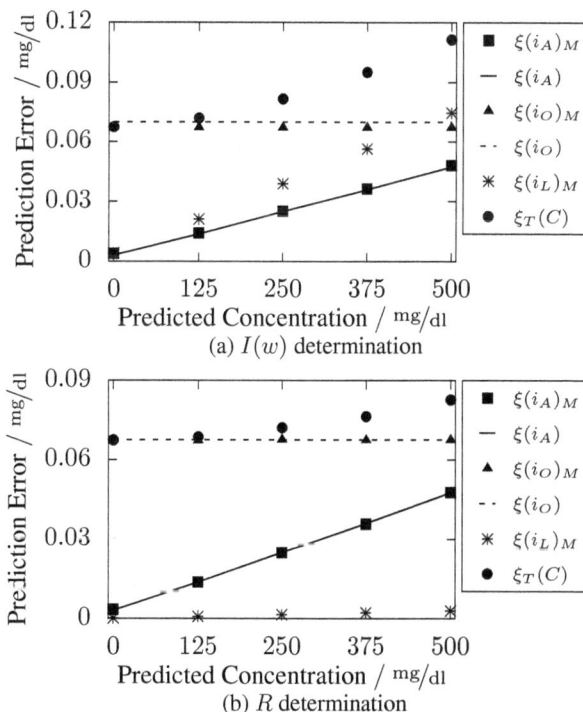

(a) $I(w)$ determination

(b) R determination

Figure 3: Partial and total prediction errors ξ of glucose determination methods with $I(w)$ in (a), and R in (b). The values are for typical variations on the LED i_L, Faraday rotator amplitude i_A and offset i_O currents during measurements with sample concentrations C.

Despite the very small prediction errors given by the setup, parameters with their biggest influence were found. For the ratio R, an improvement on Faraday rotator current offset will reduce the error $\xi_T(C)$ shown in Fig. 3b, and therefore improve the concentration prediction even more, which is already more accurate than other reference devices. On the other hand, for the determination with intensity $I(w)$, more effort is required to reduce the errors, since i_L will also have to be stable to get a positive effect on glucose prediction. This is avoided by using the ratio R as proved in (6) and Fig. 3. However in this setup $I(w)$ also predicts concentration with very high accuracy.

The remaining percentage of actual prediction errors can be related to dosage and mix inaccuracies from syringe pumps of the automatic sample generator. These effects explain 56 % of errors with $I(w)$, and 62 % with ratio R. The still residual percentages can be associated with concentration remains on cuvette (d) of Fig. 1. These mixture disturbances are reflected directly in the final glucose concentration prediction. However, their influence is also small, since the setup is still very accurate.

To investigate the setup's stability for large current changes Δi_x, prediction errors were estimated as illustrated in Fig. 4. The Clarke's grid gives an admissible glucose prediction error up to 20 % of the measured concentration [7]. Here, when measuring $500\,\mathrm{mg/dl}$, simultaneous variations up to 18 % of inputs i_x gave an acceptable error with R of $\xi_T(500\,\mathrm{mg/dl})_R = 100\,\mathrm{mg/dl}$, when assuming currents as the main source of errors. Since this range is 2000 times higher than the typical variations, and Δi_x is different for each current type x, it is very unlikely to get such errors in practice. This proves the high setup stability to current variations required to predict glucose concentration. Also, Fig. 4 shows that i_A becomes a critical variable for large variations Δi_x. Again, $\xi(i_L)$ is considerably reduced with R compared to $I(w)$, as explained before for (6).

Finally, a modulation of current i_L at f_0 was observed when varying currents i_x. This signal mixture changes the measured intensities inducing errors in the concentration prediction. The issue was solved by mounting the LED driver in a different driver's board. Thus, power supplies and ground references separation was needed. Consequently, currents i_A, i_L and i_O reliability was improved, and prediction errors were reduced to the presented values.

4   Conclusion

The Faraday modulated polarimeter setup for glucose concentration determination fulfills appropriately the theory and is stable for driver currents deviations. Since these perturbations represent a small portion of the total prediction errors, electronics regarding the first two blocks can be excluded from their biggest source. Also, considering possible mixing inaccuracies from syringe pumps in the third block, a large portion of the actual approximated prediction error of $1\,\mathrm{mg/dl}$ for a sample concentration of $500\,\mathrm{mg/dl}$ was explained. Since no reference devices can achieve such accuracy on glucose concentration determination, this paper reports the first characterization of influences on concen-

tration prediction by current drivers changes explaining the actual measurement errors. On the other hand, determination of glucose by using the novel ratio $\frac{I(w)}{I(2w)}$ compensates the LED current deviations considerably, which reduces the error on the final prediction compared to $I(w)$. Finally, with the verified stability, further measurements with the broadband setup could be performed with the distinction of glucose with a large concentration of impurities.

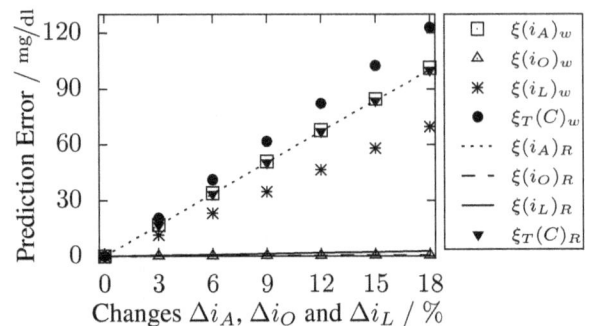

Figure 4: Measured prediction errors ξ with amplitude $I(w)$ and R determination methods, for different changes on the LED i_L, Faraday rotator amplitude i_A, and offset i_O currents with a sample concentration $C = 500\,\mathrm{mg/dl}$.

Acknowledgement

The work has been carried out and supervised by the Medical Sensors and Devices Laboratory, Lübeck University of Applied Sciences.

5   References

[1] K. Hazen, M. A. Arnold and G. W Small, *Measurement of Glucose and Other Analytes in Undiluted Human Serum with Near-Infrared Transmission Spectroscopy.* Analytica Chimica Acta, vol. 371, pp. 255–267, 1998.

[2] J. S. Baba, B. D. Cameron, S. Theru and G. L. Cote, *Effect of Temperature, pH, and Corneal Birefringence on Polarimetric Glucose Monitoring in the Eye.* Journal of Biomedical Optics, vol. 7, no. 3, pp. 321–328, 2002.

[3] B. Malik and G. Cote, *Advancement in Polarimetric Glucose Sensing: Simulation and Measurement of Birefringence Properties of Cornea.* In: Proceedings of SPIE, vol. 7906, 2011.

[4] B. D. Cameron, *The Application of Polarized Ligth to Biomedical Diagnostics.* Texas A&M University, pp. 9–65, 2000.

[5] E. Hecht, *Optics.* 4th ed., Addison-Wesley Publishing Company, pp. 366–368, 2003.

[6] H. Bach and N. Neuroth, *The Optical Properties of Glass.* Springer Science, pp. 116–120, 2012.

[7] W. Clarke, *The Original Clarke Error Grid Analysis (EGA).* Journal of Diabetes Technology & Therapeutics, vol. 7, no. 5, pp. 776–779, 2005.

Characterization of 3D Printed Scaffolds with Microscopic Optical Coherence Tomography (mOCT)

C. Stehmar[1], H. Schulz-Hildebrandt[2,3], T. Jüngst[4], M. Ahrens[2], T. Eixmann[3], M. Münter[2], J. Groll[4] and G. Hüttmann[2,3]

[1]Medizinische Ingenieurwissenschaft, Universität zu Lübeck, charlotte.stehmar@student.uni-luebeck.de
[2]Institut für Biomedizinische Optik, Universität zu Lübeck, {schulz-hildebrandt, ahrens, münter}@bmo.uni-luebeck.de
[3]Medizinisches Laserzentrum Lübeck GmbH, Lübeck, Germany, {eixmann, huettmann}@mml.uni-luebeck.de
[4]Lehrstuhl für Funktionswerkstoffe der Medizin und der Zahnheilkunde, Universität Würzburg, {tomasz.juengst, juergen.groll}@fmz.uni-wuerzburg.de

Abstract

Biofabrication by a 3D printing technique of synthetic materials together with biological components like cells holds an enormous potential for the fabrication of individual 3D human replacement tissue or for pharmaceutical tests of active ingredients. As the field of biofabrication has been grown rapidly over the last decades, there still remains one challenge - the non-invasive and label-free in-process-quality-control of 3D printed scaffolds. This work will focus on the question if optical coherence tomography (OCT) is the adequate method to characterize these specimen and to monitor cell interactions. For this investigation, microscopic optical coherence tomography (mOCT) is used, a tomographic imaging method that combines high axial and lateral resolution by using a supercontinuous source and a high numerical aperture objective. In this study we were able to show that this mOCT can not only be seen as a promising tool for future in-process control but also for the examination of the connectivity between scaffold architecture and its biological functionality.

1 Introduction

The emerging field of *biofabrication* uses 3D printing methods to merge cells with engineered biomaterials, such as polymers. These materials in combination with cells are also called bioinks. To mimic biological growth and to create a natural habit for cells, these inks have to meet specific conditions. An ideal bioink offers a high shape fidelity and a sufficient mechanical strength post-printing. Currently used bioinks are either natural materials, such as alginate, collagen or gelatin, or synthetic polymers [1]. As bioinks come together with cells, it is important to guarantee their viability, migration and proliferation. In addition, for the use of the scaffolds *in vivo*, degradation must allow gradual replacement of the artificial construct with the cell produced matrices. Therefore, waste products and the bioink should not cause inflammatory host response when placed in the human body [2]. As fabrication of these 3D printed materials is growing rapidly, one important challenge is still not addressed adequately - the non-invasive and label-free inprocess-quality-controll of the fabricated specimen.

Previously released work shows the attempt to visualize biofabricated scaffolds, but with little success. MicroCT, for example, allows high imaging depth but goes along with relative high costs, long data acquisition rates and high-energy radiation [3].

Optical Coherence Tomography (OCT) is an imaging technique, which uses Low Coherence Interferometry to obtain cross-sectional images *in vivo* and in *real time*. Its functionality can be compared with ultrasound imaging. The difference lies in the fact that OCT uses near infrared light instead of sound waves to generate images [4]. Still, OCT does not allow imaging single cells. However, previous work shows that microscopic optical coherence tomography (mOCT) using high numerical apertures together with a supercontinuous light source, provides high enough lateral and axial resolution to visualize cellular structures [5]. In contrast to other imaging modalities, like confocal or multiphoton fluorescence microscopy, this non-destructive technique allows the study of unstained cells in their natural habit. This makes mOCT interesting for imaging of living organisms. With its high resolution and the absence of contrast agents, mOCT offers optimal condition for characterizing present biofabricated specimen.

The aim of this work is to examine 3D-printed specimen with a mOCT system, in order to answer the question if OCT is an appropriate method to analyze these kinds of samples. Analysis in this case means the differentiation of the various morphologies and dimensions that may appear on a single OCT-image. As the microarchitecture determines the cell activity on a scaffold, it is of high importance to guarantee their quality for the later implantation in a human body.

2 Material and Methods

2.1 Microscopic optical coherence tomography (mOCT)

The main part of the experimental setup used for the present measurements is a Michelson Interferometer, shown in Fig.1. A supercontinuum light source (SuperK EXTREME supercontinuum laser, NKT Photonics A/S, Denmark) with a broad bandwidth of 400 nm and a center wavelength of 750 nm provides an axial resolution of below 1 μm. Passing a filterbox (SuperK Split; NKT Photonics), the beam is collimated onto a galvanometer scanner (6210H; Cambridge Technology, Bedford, MA, USA). A custom-made spectrometer (SD-OCT spectrometer, Thorlabs GmbH, Lübeck) and a high-speed camera form the detection arm. For high resolution images and in order to resolve single cells, high numerical apertures of 0.3 and 0.5 were used. A spatial resolution of 1 μm was reached.

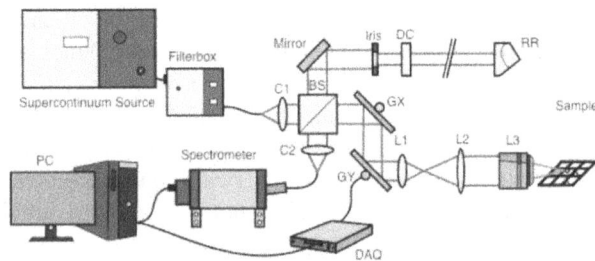

Figure 1: Schematic mOCT setup used for this work.; C1/C2, collimators; BS, 50:50 beam splitter; GX/GY, galvanometer mirror; L1/L2, beam expander lenses; L3, microscope objective; DC, dispersion compensation; DAQ, data acquisition; RR, retro-reflector.

2.2 Polymer scaffolds

In order to favor cellular interactions, a special biodegradable and thermoresistant polymer is used as a scaffold in this work. The polymer PCL (Poly-ε-Caprolacton) should mimic the natural habit of cells und thus enhance their migration and proliferation. As the microstructure of the scaffolds influences cell activity, the strand diameters were chosen to be relatively small, 5 μm or 10 μm. All specimen were fabricated at the *Lehrstuhl für Funktionswerkstoffe der Medizin und der Zahnheilkunde (FMZ), Universität Würzburg* with the 3D-printing method melt electrospinning writing (MEW). Polymer fibers were pulled out of a polymer melt by electrostatic forces which enables precise and stable drawing with the cooled and solidified bioink, as shown in Fig. 2. This emerging technique is described in more detail by Brown et al. [7]. Later on, the polymer scaffold were colonized with murine fibroblasts (cell line L929). The specimen were cultivated in a medium which supplied essential nutrients for the cells and cultured in an incubator.

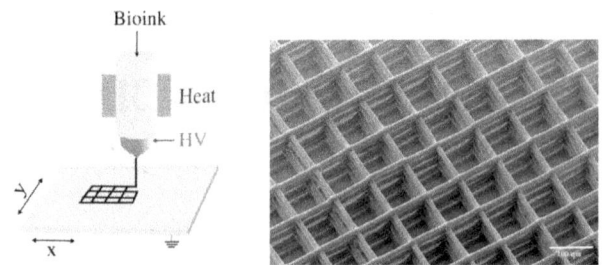

Figure 2: Bioink is melted and a electrical field (HV) between the container and a movable plate pulls the polymer into the desired shape. The fabricated 3D-scaffold is shown in the right scanning electron microscope (SEM) image.

2.3 Characterization of scaffold parameters

In medical use, each fabricated scaffold is an individual product. Therefore it is important to guarantee a precise and undamaged shape. Precise in this context means, that the desired strand diameter and distances in between the strands are kept constant. Broken strands or foreign matter attached to the scaffolds are not acceptable. Moreover it is known that the geometry of the scaffold, in particular the strand size and curvature, as well as the porosity influence the migration and proliferation of cells [8, 9]. For the characterization of the synthetic scaffolds, the area between the strands was calculated from processed OCT *en face* image sequences. Following four image processing steps were applied to a 2D image using Matlab:

1. 5×5 median filter for the reduction of speckle noise,

2. binarization of the image to emphasize its morphologies,

3. application of morphological operations such as opening to remove small holes,

4. calculation and labeling of the area between the strands.

Fig. 3 shows the sequence of these processing steps. The calculated area between the strands quantifies the regularity which should be preserved over the whole sample in order to guarantee high quality of the fabricated scaffold.

2.4 Calculation of speckle variance

OCT allows functional imaging by visualizing capillaries without the use of exogenous contrast agents. Often referred to as speckle variance OCT, this imaging modality calculates the interframe intensity variance from a sequence of structural images. In order to differ living cells from the scaffold and to visualize cell motion, speckle patterns are investigated. Interframe speckle variance $(V(x, z))$ of a stack of 150 time-shifted B-scans was calculated using the following equation

$$V(x,z) = \frac{1}{n-1} \sum_{i=1}^{n} |I(x,z,i) - \frac{1}{n} \sum_{i=1}^{n} I(x,z,i)|^2, \quad (1)$$

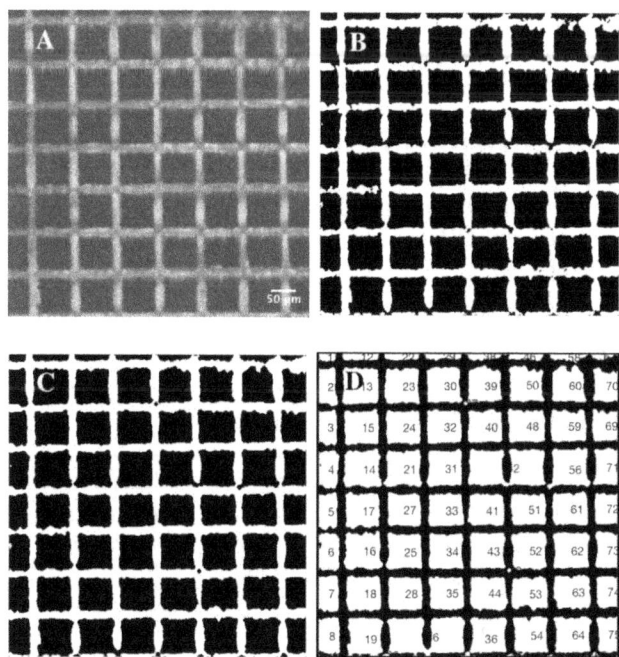

Figure 3: Images *A*, *B*, *C* and *D* each show one step of the Matlab algorithm: *A* shows the OCT median filtered OCT image, *B* depicted the image after a binarization, following, *C* depicts an opened image and *D* reflects the same picture with numbered, well-defined areas.

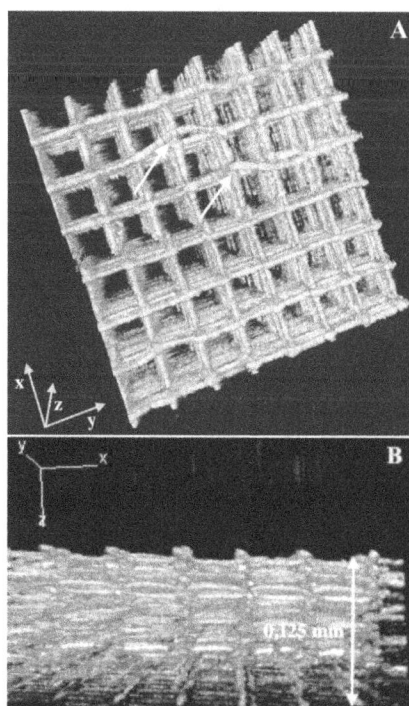

Figure 4: 3D image of the 0.6×0.6 mm section of a polymer scaffold (*A*) and the corresponding cross section (*B*). Arrows are indicating irregular structures. The imaging depth over 125 μm was reached.

where n is the number of B-scans and $I(x, y, i)$ displays the intensity of a pixel at x, y, and i. Additionally a intensity threshold was set, to substract the background and only compare regions where a signal was detected.

3 Results and Discussion

3.1 Depth information with OCT

In contrast to other microscopic imaging techniques like digital microscopy, mOCT provides imaging depth of up to several hundred micrometers. Fig. 4 reveals the obvious benefit of the here used imaging technique. A single volume acquisition allows inspection of all scaffold layers over a depth of 125 μm and discloses irregularities of the structure.

3.2 Quantification of the scaffold microarchitecture

To characterize a layer of a 3D scaffold using the previously described Matlab algorithm, the areas of the meshes in different depth were calculated. Results were then compared to the data calculated by a human observer, using the line selection tool of ImageJ. The average and the standard deviation of all areas in each layer were calculated (Fig. 5). Automatic and manual calculation agree within the standard deviation of the measurements. The decrease of the values outside the focus in layer 0 are due to signal loss of the mOCT, a low lateral resolution and signal-to-noise ratio.

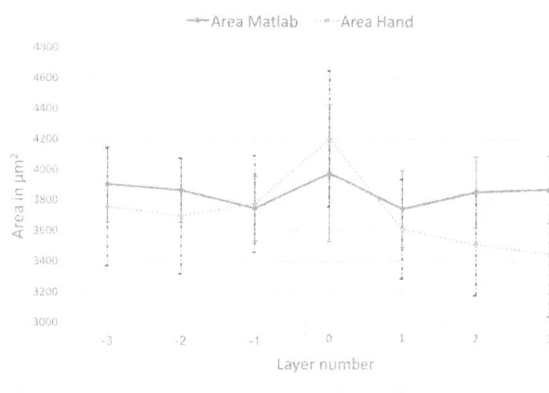

Figure 5: Measurements of the areas between the scaffold in seven different layers. The solid line depicts the average areas calculated using Matlab. The dotted line shows measurements by a human observer. Higher standard deviations outside the focus in layer 0 are due to a low lateral resolution and signal-to-noise ratio.

3.3 Colonized polymer scaffolds

We also investigated the question, whether it is possible to gain insight into cell-scaffold interaction using mOCT. The results of initial experiments are shown in Fig. 6. The mOCT images show cells adhering to the scaffold strands. As the polymer scaffold is mainly hydrophobic, these structures build a unfavorable nature for the fibroblasts. This explains the observation of large cell clusters and makes it hard to distinguish single cells.

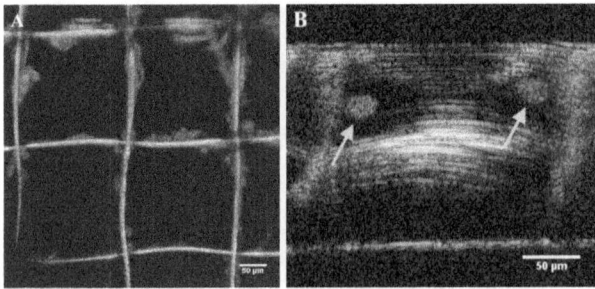

Figure 6: *En face* image of a scaffold with colonized fibroblasts (*A*). A cross section image (B-Scan) with arrows pointing to single cells (*B*).

3.4 Detection of cell activity by speckle variance

In order to differ cells and cell clusters from the scaffold, speckle variance was used for investigations. A high variance in the pattern indicates microstructural motion of living cells attached to the sample. Fig. 7 *B* shows speckle variance according to Eq. 1, compared with the original data in 7 *A*. Highlighted regions show high variation in the speckle pattern, displaying a significant proof of the viability of the attached fibroblasts. As a result, mOCT allows the monitoring of cultivated cells on polymer scaffolds.

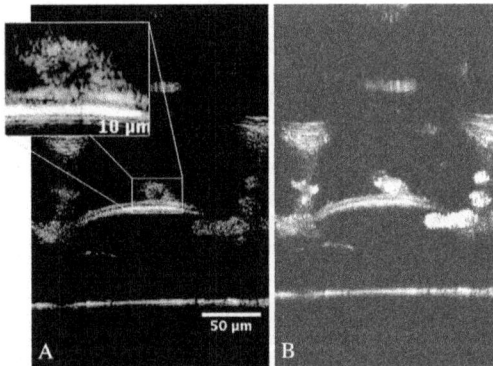

Figure 7: Visualization of fibroblast micromotion. Speckle variance in B-Scans was investigated of a cell cluster (*A*). The highlighted segments of *B* show high speckle variance calculated with Matlab.

4 Conclusion

In this paper, we demonstrate a successful way to characterize biofabricated scaffolds with mOCT. High numerical aperture imaging was used to display fine structures and cell clusters while still maintaining depth information. The latter distinguishes mOCT from other microscopic techniques. Future work will address the characterization of scaffolds-cell interaction. Migration of the cells shall be observed to gain information about their favorable scaffold habit. In order to improve 3D bioprinting techniques and biofabrication of functional engineered tissues in general, it is indispensable to assess the internal architecture of the polymer scaffolds and the vitality of the cultivated cells.

Acknowledgement

The work has been carried out at the *Institut für Biomedizinische Optik (BMO), Universität zu Lübeck* in cooperation with the *Lehrstuhl für Funktionswerkstoffe der Medizin und der Zahnheilkunde (FMZ), Universität Würzburg*. Work was funded by the BMBF (Photon Control, 13N14205).

5 References

[1] A.S.P. Lin, T.H. Barrows, S.H.. Cartmella and R.E. Guldberg, *Microarchitectural and mechanical characterization of oriented porous polymer scaffolds*. Biomaterials, 2003; 24(3):481-9.

[2] S. Ji, K. Schacht and M. Guvendiren, *Recent Advances in Bioink Design for 3D Bioprinting of Tissues and Organs*. Frontiers in Bioengineering and Biotechnology, 2017; 5: 23.

[3] E. DeSimone, K. Schacht, T. Jüngst, J. Groll and T. Scheibel, *Biofabrication of 3D Constructs: Fabrication Technologies and Spider Silk Proteins as Bioinks*. Pure and Applied Chemistry, 2015; 54(9):2816-20.

[4] J.G. Fujimoto and W. Drexler, *Introduction to Optical Coherence Tomography*. In: Optical Coherence Tomography - Technology and Applications, Springer, pp. 1-40, 2008.

[5] J. Horstmann, H. Schulz-Hildebrandt, F. Bock, S. Siebelmann, E. Lankenau, G. Hüttmann, P. Steven, C. Cursiefen *Label-Free In Vivo Imaging of Corneal Lymphatic Vessels Using Microscopic Optical Coherence Tomography*. Investigative ophthalmology and visual science, 2017; 58(13), 5880-5886.

[6] C. Wei and J. Dong, *Direct fabrication of high-resolution three-dimensional polymeric scaffolds using electrohydrodynamic hot jet plotting*. Journal of Micromechnanics and Microengineering, IOP Publishing Ltd, 2013; 23 025017.

[7] T.D. Brown, P.D. Dalton and D.W. Hutmacher, *Direct writing by way of Melt Electrospinning*. Advanced Materials, 2011; 23, 5651-5657.

[8] B.L. Farrugia, T.D. Brown, Z. Upton, D.W. Hutmacher, P.D. Dalton and T.R. Dargaville *Dermal fibroblast infiltration of poly(ϵ-caprolactone) scaffolds fabricated by melt electrospinning in a direct writing mode*. Biofabrication, 2013; 5 025001.

[9] J.M. Sobral, S.G. Caridade, R.A. Sousa, J.F. Mano and R.L. Reis, *The effect of geometry on three dimensional tissue growth*. In: Acta Biomaterialia, Elsevier Ltd, pp. 1009-1018, 2011.

3

Biochemical Physics

The Dynamics of Mutant Keratin 14
in Response to Thermal Stress in Human Keratinocytes

M. Göb [1], Y. Z. Ng [2], H. Itoh [2], Y. Miura [3], and E. B. Lane [2]

[1] Medical Engineering, Universität zu Lübeck, madita.goeb@student.uni-luebeck.de
[2] Institute of Medical Biology, Agency for Science, Technology and Research (A*STAR), Singapore
[3] Institute of Biomedical Optics, Universität zu Lübeck, miura@bmo.uni-luebeck.de

Abstract

Epidermolysis bullosa simplex (EBS) is an autosomal dominant inherited skin blistering disease characterized by increased skin fragility after minor mechanical stress. The majority of mutations within keratin genes *KRT5* and *KRT14* are associated with EBS and it is believed that the disintegration of keratin filaments and the formation of aggregates play an important role in the increased fragility of the skin. In this study, we analysed the dynamics and distribution of aggregate formation in mutant N/TERT-K14-R125P human keratinocytes in response to heat stress during live-cell imaging. We found out that a maximum number of aggregates is seen after five minutes of heat stress followed by an unknown cell response attenuating aggregate formation. Aggregates tend to cluster at cell corners and we observed immediate disappearance of aggregates during recovery. Further findings using 3D cultures suggest that aggregate dynamics are also related to the differentiation status of the keratinocytes.

1 Introduction

Epidermolysis bullosa simplex (EBS) is an autosomal dominant inherited skin blistering disease characterized by increased skin fragility after minor mechanical stress. In most cases EBS blistering occurs in the basal layer of the epidermis due to mutations in keratin genes keratin 5 (*KRT5*) and keratin 14 (*KRT14*) [1]-[3]. These proteins expressed in the basal layer form dimers that assemble intermediate filaments providing structural strength in healthy epithelial tissues [4]. Consequently, mutations within *KRT5* and *KRT14* lead to misfolded intermediate filament proteins and thus to structural fragility. In the most severe cases of EBS (severe-generalized EBS - previously known as Dowling-Meara type), keratin aggregates can be seen by electron microscopy in EBS skin [5].

Anecdotally, EBS patients noticed more blisters occurring at higher temperatures in summer. *In vitro*, thermal stress easily reduced stability of keratin filaments and resulted in formation of aggregates described in previous studies. Morley et al. observed a maximal number of aggregates at 15 minutes of 43°C heat shock and the filament network structure was completely reversed after 60 minutes of recovery [5]. However, since these experiments were conducted on fixed cells, observations are limited in spatial resolution and the keratin dynamics in mutant K14 cell lines and the biological link in causing blistering in EBS patients is still unclear.

To model EBS phenotypes *in vitro* under heat stress, we exposed keratinocytes to warm media perfusion. In this study, we focused on live-cell imaging of wild-type (WT) and mutant keratin cell lines tagged with green fluorescent protein (GFP) subjected to heat stress to further understand the biophysical mechanisms of tissue fragility in EBS patients.

2 Material and Methods

2.1 Cell Culture

In this study N/TERT immortalized human keratinocytes were used as stable model cell lines with keratin mutations mimicking severe EBS [6]. Since the substitution of p.Arg125 is the most severe mutation hotspot related with EBS, the GFP-tagged keratin mutation K14 R125P was introduced into the WT K14 sequence [7]. Using a pLVX-EF1α-AcGFP1-C1 lentiviral expression vector, WT (N/TERT-K14-WT-GFP) and mutant (N/TERT-K14-R125P-GFP) pathomimetic cell lines tagged with enhanced GFP were created for live-cell imaging [7].

For imaging 2D cultures, WT and mutant keratinocytes were cultured on coverslips in GIBCO keratinocyte serum-free medium (K-sfm), containing 25 µg/ml bovine pituitary extract, 1% penicillin/streptomycin, 0.2 ng/ml epidermal growth factor (EGF), and 0.4 mM calcium chloride (Thermo Fisher Scientific, #10724-011, USA) [6]. Cells were cultured for a few days to post-confluence. For 3D culture, 100,000 primary human fibroblasts were seeded on coverslips followed by 1,000,000 N/TERT-K14-WT-GFP or N/TERT-K14-R125P-GFP and grown in modified cFAD without EGF [6] for 2.5 weeks until a stratified multilayer epidermal equivalent is formed.

2.2 Heat Shock

N/TERT-K14-WT-GFP and N/TERT-K14-R125P-GFP cells were subjected to 43°C heat stress in order to induce filament breakdown and produce aggregates as seen in EBS patients. Cells were heat-stressed with warm medium using a perfusion system during live-cell imaging. The perfusion setup included a peristaltic pump (Longer Precision Pump Co., Ltd, BT100-2J, China), waterbaths for heating media, insulated silicon tubings and a magnetic perfusion chamber (Live Cell Instrument, Chamlide CM-B25-1PB, South Korea).

During imaging there was constant media perfusion while cells were kept at standard conditions at 37°C for 10 minutes, heat stressed at 43°C for 30 minutes followed by standard conditions again for 80 minutes of recovery. Temperature of the cells was monitored in real-time concurrently using a temperature sensor (Graphtec Corporation, midi LOGGER GL10-TK, Japan) placed directly into the chamber.

2.3 Cytotoxicity Assay

Cell viability was determined on cells that were heat-stressed for 30 minutes and after 24 hours recovery using a Pierce LDH cytotoxicity assay kit and following their standard protocol (Thermo Fisher Scientific, #88953, USA).

2.4 Imaging

All experiments described in this study were performed using a spinning disk confocal microscope system. It includes a 3D-FRAP Nikon Ti inverted microscope, stage-top incubator and a CO_2 control system that allows precise photokinetic live-cell imaging. A Nikon perfect focus system prevents z-drift in timelapse imaging.

Images were acquired using a 40x magnification plan fluor lens with oil immersion, 1.3 NA and 1.5x digital zoom. Fluorescence was excited with a 491 nm laser diode at 7.5 mW and 100 ms exposure time. The emitted light was detected using a Photometrics Evolve EMCCD and emission filter for 505-545 nm wavelengths. Depending on the sample thickness, two to four volumes per minute were acquired with 0.5 μm Z-resolution.

All data shown in this publication were processed using ImageJ 1.50e (Fiji), Imaris 9.1 (Bitplane), Microsoft Excel for Mac 2011, Version 14.6.7 (Microsoft Cooperation) and midi LOGGER GL 10 Software (GRAPHTEC).

3 Results and Discussion

The difference in appearance and distribution of intracellular keratin of wild-type and mutant keratinocytes is shown in Fig. 1. WT keratin is visible as a dense cytoplasmic filament network that forms a mesh around the nucleus, providing structural strength to the cell. In mutant cell lines, the keratin filament network is less developed. A weaker mesh around the nucleus and small peripheral aggregates can be observed. As seen in Fig. 1B, aggregates appear as small granules (<0.5 μm) near the cell membrane, which move towards the cell center and grow in size (<2 μm). Having reached a certain size, these clusters seemed to disassemble to form filamentous structures as described in previous studies [8].

The overall goal of this study was to examine the effect of heat stress on keratin and its dynamics. Initially, we analysed optimal conditions of different parameters that affect the cell behaviour.

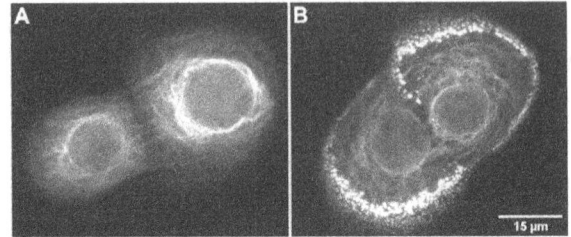

Figure 1: Fluorescent images of fixed keratinocytes. N/TERT-K14-WT-GFP (A) and N/T/TERT-K14-R125P-GFP (B). Cell lines with wild-type keratin show a filamentous network in the cytoplasm (A), whereas keratin aggregates are dominant in mutant cells (B).

3.1 Temperature Titration

In order to confirm that mutant cells behaved in similar manner to previous EBS models seen before, cells were exposed to different elevated temperatures from 39°C to 46°C. Appearance of aggregates was observed at around 41°C (data not shown). To ensure minimal stress without harming the cells, further experiments were performed at 43°C for 30 minutes.

To exclude shear force as stress factor, control experiments were conducted under constant perfusion. As expected, no increase in aggregate numbers was seen in WT and mutant cells. In addition, when WT cells were heat-shocked under constant perfusion, no aggregates were observed and the keratin filament network remained stable.

3.2 No Heat Shock related Cytotoxicity

After 24 hours of recovery no increase in cytotoxicity in heat-shocked cells was observed. Following the Pierce LDH cytotoxicity assay kit procotol, the determined cytotoxicity of both, WT and mutant was close to 0% (control data not shown).

3.3 Aggregate Dynamics

We observed an increase in numbers of aggregates in the N/TERT-K14-R125P-GFP cell lines during heat stress. Fig. 2A visualises the total amount of aggregates in relation to temperature changes. We achieved a very fast temperature increase within three minutes using constant perfusion. An immediate response in appearance of a greater number of aggregates was seen with a lag of less than one minute, correlating well with the temperature change. The maximum number was seen after five minutes of heat exposure (Fig. 2B). Unexpectedly, this peak number of aggregates started decreasing even though the temperature was still at 43°C. This suggests an ability of the mutant cells to adjust to the thermal stress. However, the number of aggregates was still elevated during heat stress. Furthermore, the number of aggregates returned to the initial levels within ten minutes of recovery. In contrast to normal behaviour [8], aggregates of all sizes disappeared quickly after reducing temperature.

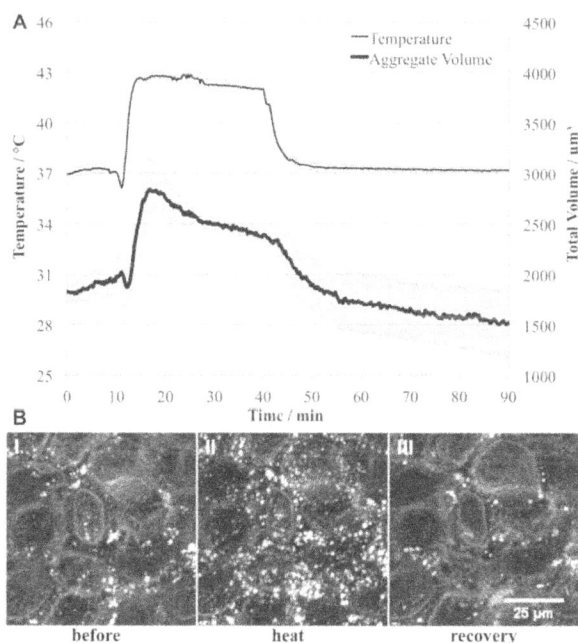

Figure 2: Temperature dependent appearance of aggregates. 2D mutant keratinocytes were heat-shocked and imaged as described above. (A) Average cell temperature and total aggregate volume of three experiments is plotted over time and indicates a correlation of temperature and aggregate formation. Error bars are standard error. Increased aggregate appearance during heat shock is also visible in the representative maximum z-projections (B) at timepoints 5 minutes (I), 15 minutes (II), and 60 minutes (III).

Aggregates tend to appear at specific locations within the cell volume visualized in a representative graph (Fig. 3A). Each grey dot of the scattered plot represents at least one aggregate at the specific Z-position of the 3D volume. Three layers of stratified cells in 3D culture can be seen as indicated by asterisks. Clustering of the grey dots into bands over time (Fig. 3A) suggests that aggregates are located at the top and bottom of the cells corresponding to the side view projections of the cell volume (Fig. 3B). All experiments were conducted on an overexpressing system in which K14-aggregates were present in all cell layers, but an increase in aggregate formation during heat shock was most obvious in the basal cell layer.

Similar observations are also demonstrated in top-view projections of the same experiment in Fig. 3C. During heat stress, aggregates that appear in top cell layers tend to be evenly distributed along the cell membrane (indicated by

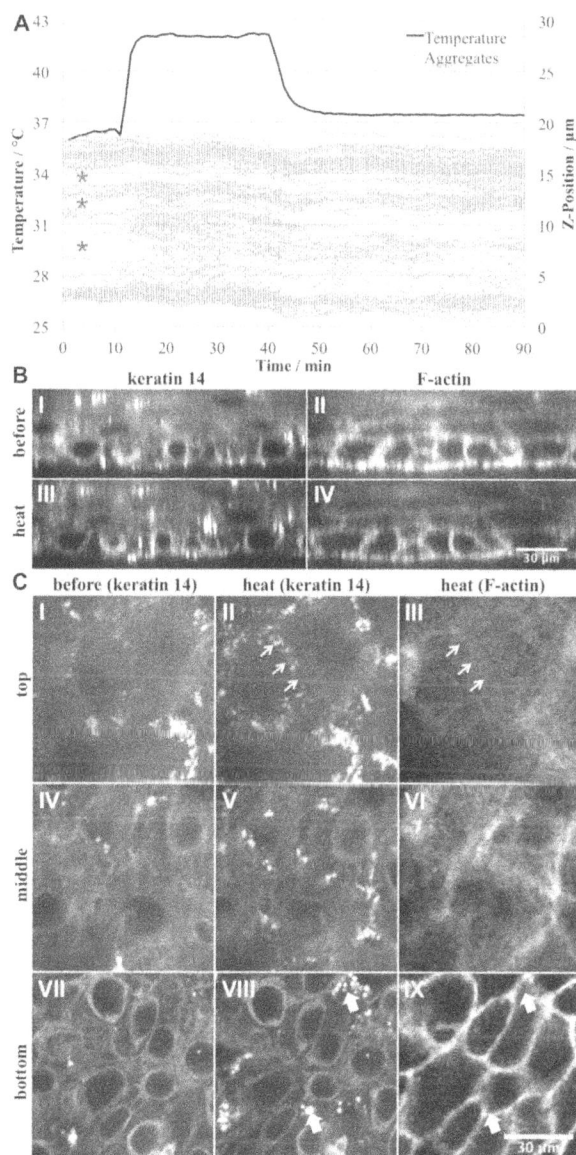

Figure 3: Aggregate distribution in different cell layers. 3D mutant keratinocytes were heat-shocked and imaged as described above. Aggregates indicated as dots are scattered according to time and z-position (A). They are predominantly located in bottom and top cell layers. Asterisks represent three different cell layers (***).

Similar dynamics are seen in XZ-slices (B) and XY-slices (C) of the same volume. (B I,III) and (C I,II,IV,V,VII,VIII) show GFP-tagged mutant keratin 14. (B II,IV) and (C III,VI,IX) display F-actin (B II,IV) (using RFP-tagged Lifeact) to visualize cell membranes. XY-slices at the top (C I-III), middle (C IV-VI) and bottom (C VII-IX) of the 3D keratinocytes are displayed. The localisation of aggregate formation during heat stress varies in different cell layers. (B) and (C) demonstrate conditions before (at 5 minutes) and during heat stress (at 20 minutes). Arrows indicate distribution of aggregates in different layers.

thin arrows), whereas clusters of aggregates in bottom layers were observed mainly in cell corners where at least three cells meet (indicated by thick arrows).

3.4 Discussion

Physiologically, we would expect the skin to adapt quickly to thermal stress. In WT keratinocytes we noticed that the filamentous network remained stable under heat stress. However, the cells with mutant K14 exhibited a dramatic response to heat in the form of transient appearance of aggregates.

Our observations that mutant cells were able to adjust to the thermal stress reveal that dynamics are more complicated than expected. The non-linear increase in aggregates at the beginning of and the drastic decrease after the heat shock indicate the possibility of an intrinsic biochemical response of keratin molecules to thermal stress. In contrast, the decline in aggregate numbers following the peak during heat stress suggests that a separate active process may be involved, such as possibly phosphorylation. Other forms of stress in cells that are known to induce aggregate appearance are mechanical or osmotic stress [9]. It will be interesting to see if the keratin dynamics for osmotic shock are similar to what we observed during heat stress.

Furthermore, the observation of higher numbers of aggregates in basal cell layers suggests that protein profile in undifferentiated keratinocytes may also affect mutant keratin dynamics.

Possible future experiments to dissect the rapid and transient appearance and disappearance of aggregates in response to thermal stress include western blot analysis of heat stress related proteins as well as soluble and insoluble fraction of keratin proteins. Additionally we suggest future projects to use deconvolution to improve image clarity and analyse in real time the dynamics of aggregates in relation to the filamentous network.

4 Conclusion

This work provided an insight into the dynamics of keratin aggregates in mutant N/TERT-K14-R125P cells in response to heat stress. Short-time exposure to heat stress less than 30 minutes does not result into obvious cell fragility, such as bursting or cell death. However, we saw a transient increase in aggregates in a temperature dependent manner while live-cell imaging. A maximum number of aggregates was observed after five minutes of heat stress followed by an unknown cell response that damps down aggregate appearance. During recovery, we observed quick disappearance of aggregates in all sizes.

Despite an overexpressing system, aggregate formation was most obvious in basal cell layers alluding to a positive correlation between aggregate appearance and differentiation status of the cells. Consisting findings in both, 2D and 3D cultured mutant keratinocytes revealed a tendency of aggregates to cluster in cell corners.

However the biophysical mechanisms are still not yet understood. Further research in this area will hopefully yield new therapeutic interventions for EBS.

Acknowledgement

This work was carried out at the Institute of Medical Biology of the Agency for Science, Technology and Research (A*STAR), Singapore and supervised by PD Dr. Yoko Miura, Institute of Biomedical Optics, Universität zu Lübeck.

I would like to thank the IMB Microscopy Unit for assistance and providing excellent microscopy equipment and Tong San Tan for providing the cell lines.

5 References

[1] J. M. Bonifas, A. L. Rothman, and E. H. Epstein Jr. *Epidermolysis bullosa simplex: evidence in two families for keratin gene abnormalities.* Science, vol. 254, pp. 1202–1205, 1991.

[2] P. A. Coulombe, M. E. Hutton, A. Letai, A. Hebert, A. S. Paller and E. Fuchs, *Point mutations in human keratin 14 genes of epidermolysis bullosa simplex patients: Genetic and functional analysis.* Cell, vol. 66 pp. 1301–1311, 1991.

[3] E. B. Lane et al., *A mutation in the conserved helix termination peptide of keratin 5 in hereditary skin blistering.* Nature, vol. 356, pp. 244–246, 1992.

[4] J. Schweizer et al., *New consensus nomenclature for mammalian keratins.* The Journal of Cell Biology, vol. 174, no. 2, pp. 169–174, 2006.

[5] S. M. Morley et al., *Temperature sensitivity of the keratin cytoskeleton and delayed spreading of keratinocyte lines derived from EBS patients.* J Cell Sci, vol. 108, pp. 3463–3471, 1995.

[6] M. A. Dickson et al., *Human Keratinocytes That Express hTERT and Also Bypass a p16(INK4a)-Enforced Mechanism That Limits Life Span Become Immortal yet Retain Normal Growth and Differentiation Characteristics.* Molecular and Cellular Biology, vol. 20, no. 4, pp. 1436–1447, 2000.

[7] T. S. Tan et al., *Chapter Nine - Assays to Study Consequences of Cytoplasmic Intermediate Filament Mutations: The Case of Epidermal Keratins.* Methods in Enzymology. M. B. Omary and R. K. H. Liem, Academic Press, vol. 568, pp. 219–253, 2016.

[8] R. Windoffer, S. Wöll, P. Strnad, and R. Leube, *Identification of Novel Principles of Keratin Filament Network Turnover in Living Cells.* Molecular biology of the cell, vol. 15, pp. 2436–48, 2004.

[9] M. Alessandro, et al., *Keratin mutations of epidermolysis bullosa simplex alter the kinetics of stress response to osmotic shock.*, Journal of Cell Science, vol. 115, pp. 4341–4351, 2002.

Characterization of the interaction between a lipid monolayer and nebulized substances by using a film balance

F. Mütel [1], C. Nehls [2], and T. Gutsmann [2]

[1] Medical Engineering, University of Applied Science Lübeck, fabian.muetel@student.uni-luebeck.de
[2] Division of Biophysics, Priority Area Infections, Research Center Borstel, {tgutsmann, cnehls}@fz-borstel.de

Abstract

To investigate the interaction between inhalable pharmaceuticals and the pulmonary surfactant a Langmuir film balance can be utilized. The drugs can be administered by injecting into the buffer or as shown in this work by nebulizing onto the monolayer. The reason for these measurements are the specific constitution of the pulmonary surfactant where pharmaceuticals mainly interact from the air phase and not from the aqueous phase. The measurements are performed at initial lateral pressures of 10 mNm^{-1} and 20 mNm^{-1} using ultrapure water and a salbutamol solution (1 mgml^{-1}) as active substances and 1,2-dioleoyl-sn-glycero-3-phosphocholine (DOPC) as the lipid for the monolayer. After analyzing the results there is a difference in the interaction of salbutamol and ultrapure water for the nebulization method. Further measurements with different parameters could consolidate the results and lead to the development of new pharmaceuticals.

1 Introduction

An essential component of cells and eventually of the human organism is the membrane system. This system is separating the processes in the organism from the surrounding environment. It features different functions which are essential for the viability of the organism. Some of these functions are the compartmentalization, the conversion of energy, the selective permeability as well as the signal transduction [1].

When the organism is affected by environmental conditions such as pharmaceuticals the first interaction happens with the membrane system. Therefor, the research of different interactions between drugs and lipid membranes is an essential component of developing new pharmaceuticals.
Due to the amphiphilic constitution (hydrophilic headgroup and hydrophobic tail) of the phospholipid molecules the process of self-aggregation occurs. The highly ordered lipid membrane is formed because the overall entropy increases due to the displacement of water molecules from the acyl chains [2][3][4].
Generally, the surrounding medium in the organism is aqueous whereby the phospholipid molecules form a phospholipid bilayer. However, there are organs in which two different phases come together. One of these organs is the lung in which the gaseous phase of the environment encounters the aqueous phase of the organism. Hence, the phospholipid molecules form only a monolayer instead of a bilayer. This monolayer is also called the pulmonary surfactant. The surfactant consists of phospholipids and the proteins SP-A to SP-D [5][6][7].

The aim of this work is to study the interaction between drugs (e.g. salbutamol) and the surfactant according to two different administrations of the active substances to the surfactant monolayer. The Surfactant is mimicked by a DOPC (Tab. 1) monolayer on a Langmuir film balance. For this purpose the active substances are either injected into the subphase/buffer or nebulized onto the monolayer. Therefor, the nebulization setup including the funnel, the containment, and the nebulizer is designed and experiments characterizing the interaction between salbutamol and DOPC are performed.

2 Materials and Methods

For the characterization of the usability of different methods to administer drugs on the surfactant monolayer the KSV Nima KN2001 Langmuir Blodgett Trough (Biolin Scientific, Gothenburg, Sweden) is used. The film balance consists of a trough for various buffers/subphases, two movable barriers and the Wilhelmy plate to measure the lateral pressure. To avoid falsification of the measurements due to temperature fluctuation, an external water circulation is installed to the trough which keeps the temperature constant at 21°C. Furthermore, to vary the liquid surface there are two movable barriers which can synchronously compress as well as expand the surface. For the acquisition of the changing lateral pressure induced by the addition of phospholipids a Wilhelmy plate is used. This Wilhelmy plate is made of a rectangular filter paper with a width of 10 mm and a height of 25 mm. Equation (1) describes the lateral pressure acting on the Wilhelmy plate.

$$\Pi = \gamma_w - \gamma_f \qquad (1)$$

Here, γ_w stands for the surface tension of pure water whereas γ_f stands for the surface tension of the surface with the film [6].

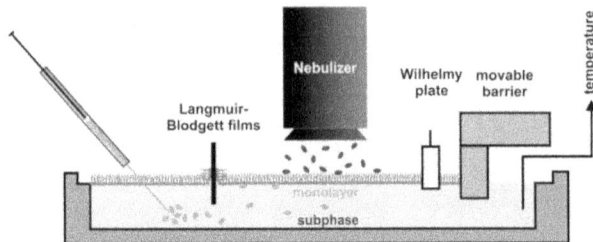

Figure 1: Schematic setup of a Langmuir trough. Here, the two different administration methods for the substances as well as the essential components (Wilhelmy plate, barriers, temperature control, and phospholipid monolayer) of a Langmuir trough are shown. The part of the Langmuir Blodgett film is not further used for the experiments.

The subphase consists of 150 mM NaCl and 5 mM HEPES (Tab. 1) solved in ultrapure water adjusted to a pH value of 7.4 by potassium hydroxide. For each measurement 59 ml of the subphase solution is poured into the trough. To prepare the lipid monolayer either 4 µl or 8 µl of a DOPC solution of 1 mgml^{-1} is spread onto the subphase by using a Hamilton syringe. By waiting at least five minutes the volatile solvent (chloroform) has evaporated resulting in a solvent free monomolecular layer [6]. The administered lipid volume per measurement depends on the used type of drug administration method.

Table 1: Substances used in the experiments

Substance	Molecular Weight	Short
Ultrapure Water	18.02	Aqua Ster. Dest.
Sodium Chloride	58.44	NaCl
2-[4-(2-hydroxyethyl)-piperazin-1-yl]ethanesulfonicacid	238.81	HEPES
Potassium Chloride	74.55	KCl
Chloroform	119.38	CHCl$_3$
1,2-dioleoyl-sn-glycero-3-phosphocholine (Avanti Polar Lipids, Alabama, USA)	61.00	DOPC
Salbutamol (Sigma-Aldrich, Darmstadt, Germany)	239.31	C$_{13}$H$_{21}$NO$_3$

For the measurements there are two different administration methods used to bring the substances (salbutamol solution and ultrapure water as control) in contact with the lipid monolayer. Before administering the substances, a certain lateral pressure has to be reached by compressing the monolayer. By moving the barriers towards each other the lipid molecules will be compressed and the lateral pressure increases. The two utilized lateral pressures are 10 mNm^{-1} and 20 mNm^{-1}. Thus, it is possible to consider differences in the interaction mechanism. At a certain lateral pressure the lipid monolayer becomes instable [4].

After reaching the desired lateral pressure the barriers are stopped, and the surface area stays henceforward constant. The substances are administered after waiting for another 10 minutes. By that it is assured that the lipid molecules have time to equilibrate. The measurements are done with two different solutions. On the one hand with ultrapure water as control and on the other hand with a salbutamol solution with a concentration of 1 mgml^{-1}. The lateral pressure is measured for 60 minutes and after that the barriers are moved to their initial positions and the measurement is completed. Each measurement is done in a sealed containment to avoid potential environmental disturbances on the experiment as well as keeping the aerosol inside the containment. Therefor, a HEPA-Filter H13 is mounted to the containment which provides a possibility to filter the air from inside the box before opening after the measurement is done.

2.1 Injection method

Using the administration method via injection the substance is injected below the lipid monolayer directly into the subphase. For that purpose a Hamilton syringe is used to inject 50 µl of the substance solution.

2.2 Nebulization method

With the second administration method the substance is administered by nebulization onto the lipid monolayer. Therefor, a funnel is designed in Solid Edge and printed with a 3D printer from the Fraunhofer EMB (Lübeck, Germany). This funnel is adapted to the Langmuir Trough (Fig. 2) and covers a surface area of 5 cm x 5 cm.

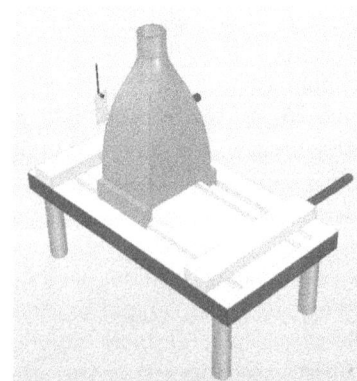

Figure 2: The KN2001 film balance setup with the designed and adapted funnel. The barrier on the right is movable, the barrier on the left is fixed.

Using this administration method only one barrier is moved

forward whereas the other one is locked into position. By reaching the desired lateral pressure, the movement of the barrier is stopped and also locked into position. After waiting for 10 minutes the Beurer IH18 nebulizer (Beurer GmbH, Ulm, Germany) is turned on to nebulize the substance solution onto the lipid monolayer. Therefor, the solution containment of the nebulizer is connected with the funnel via a pipe inside the sealed containment. The compressor of the nebulizer is placed on the outside of the box and is connected via another pipe with the solution containment inside the box. The nebulization time is chosen to be 7 minutes so that the nebulized volume corresponds to the injected volume approximately.

3   Results and Discussion

The following part shows the results of several measurements in the shape of lateral pressure vs. time diagrams. All experiments are performed at least two times and representative curves are shown in the following figures. These curves demonstrate the differences in the interaction of injected and nebulized drugs with the monolayer. The starting point of each curve is the point where the barriers reached the desired lateral pressure. This is followed by 10 minutes waiting time for the equilibration of the lipid molecules. After this period a decrease of the lateral pressure can be observed due to the administration via injection or nebulization. The decrease depends on the administration method. The lateral pressure is recorded after the administration for 60 minutes. The curves recorded during the movement of the barriers is not shown.

The results of the injection experiments which are illustrated in the following Fig. 3 and Fig. 4 show no major longterm differences between the interaction of either water or salbutamol with the monolayer. Nevertheless, there is a difference between salbutamol and water at 10 mNm^{-1}. Right after the injection the salbutamol leads to an increase of the lateral pressure which is followed by a slight decrease. The water curve does not show this manner. This effect will be further investigated with atomic force measurements by using Langmuir Blodgett films. There are no such differnces in the shorterm trend of the lateral pressure at an initial lateral pressure of 20 mNm^{-1}. Comparing the two measurement series at 10 mNm^{-1} and 20 mNm^{-1} it is obvious that there are different interactions occurring at 10 mNm^{-1} and 20 mNm^{-1}. This is indicated by the slight increase of the lateral pressure for both solutions at 10 mNm^{-1} (Fig. 3) whereas there is a slightly decreasing pressure at 20 mNm^{-1} (Fig. 4).

In Fig. 3 the lateral pressure decreases by 0.25 mNm^{-1} for the water solution as well as for the salbutamol solution in the first 10 minutes of waiting after reaching the initial lateral pressure. This decrease is followed by a further decrease of 0.75 mNm^{-1} due to the injection process. The lateral pressure of the water curve is increasing constantly. For the salbutamol curve the lateral pressure is decreasing in the

first 30 minutes but is then increasing similar to the water curve.

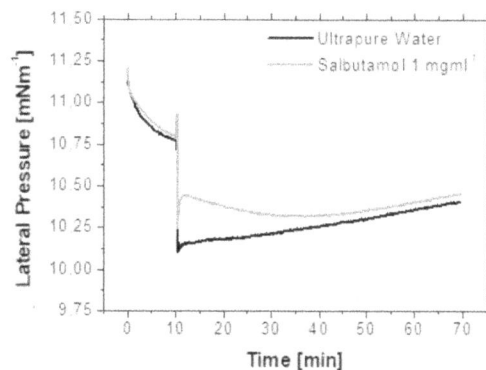

Figure 3: Lateral pressure of two different measurements while injecting substances at a pressure of around 10 mNm^{-1}.

In Fig. 4 the progression of the two curves is almost the same. At first the lateral pressure decreases by 1 mNm^{-1} in the first 10 minutes. After the injection the lateral pressure drops again by 1.5 mNm^{-1}. In the following progress the pressure decreases slightly by another 0.5 mNm^{-1}.

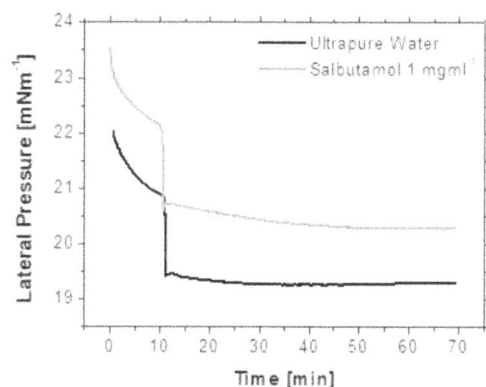

Figure 4: Lateral pressure of two different measurements while injecting substances at a pressure of around 20 mNm^{-1}.

The results of the nebulization experiments which are illustrated in the following Fig. 5 and Fig. 6 are showing major differences between the interaction of water and salbutamol with the monolayer. The lateral pressure decreases stronger for water than for salbutamol during the nebulization process. Comparing the two measurement series at 10 mNm^{-1} and 20 mNm^{-1} it is clear that there are different interaction processes taking place at 10 mNm^{-1} and 20 mNm^{-1}. This is indicated by the restricted exponential increase of the lateral pressure for both solutions at 10 mNm^{-1} whereas there is an overall constantly decreasing pressure at 20 mNm^{-1}.

In Fig. 5 the progression of the two curves is almost the same. At first the lateral pressure decreases by 0.5 mNm^{-1} in the first 10 minutes. After the nebulization the lateral pressure drops by 1.25 mNm^{-1} for ultrapure water whereas the pressure drops by only 0.6 mNm^{-1} for salbutamol. Then, in the following progress the pressure increases by 0.5 mNm^{-1}.

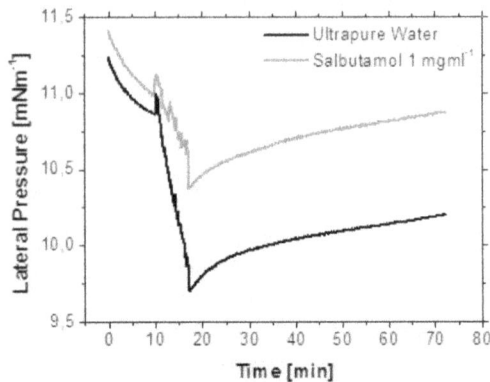

Figure 5: Lateral pressure of two different measurements while nebulizing substances at a pressure of around 10 mNm^{-1}.

In Fig. 6 the progression of the two curves is almost the same. At first the lateral pressure decreases by 1 mNm^{-1} in the first 10 minutes. After the nebulization the lateral pressure drops by 1.5 mNm^{-1} for ultrapure water whereas the pressure drops by only 1 mNm^{-1} for salbutamol. Then, in the following progress the pressure decreases by 0.5 mNm^{-1}.

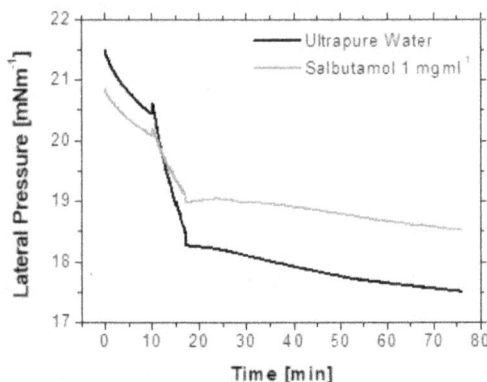

Figure 6: Lateral pressure of two different measurements while nebulizing substances at a pressure of around 20 mNm^{-1}.

The major difference between the two administration methods is the changing of the lateral pressure during the nebulization process. Compared to the injection process in which the lateral pressure decreases in the same manner the lateral pressure decreases different during the nebulization. Here, the salbutamol effects the monolayer and prevents the lateral pressure from decreasing as strong as for the water curve. That could be the result of the salbutamol integrating into the monolayer and thereby increasing the lateral pressure slightly.

4   Conclusion

This work proofs that the results obtained by the designed nebulization system are different compared to those obtained by the conventional injecting method. Therefor, there is now a valid model to investigate the interaction of inhalable pharmaceuticals with a pulmonary-like surfactant in a more realistic way. Nevertheless, the designed nebulization setup and the surfactant model offer a wide range of modifications. To provide an almost exact copy of the pulmonary surfactant system it is possible to use a more detailed composition of lipids and proteins than in the performed measurements. In addition to that, it is also possible to measure at several temperatures such as 37°C. Regarding the nebulization system there are chances to modify the nebulization methods such as using different types of funnels as well as using a different nebulizer with another operating pressure.

Acknowledgement

The work has been carried out at the division of biophysics, Research Center Borstel, supervised by Thomas Gutsmann, head of the division of biophysics and Christian Nehls, postdoc at the division of biophysics.

5   References

[1] A. Hädicke, *Interactions of Cationic Peptides with Anionic Lipid Bilayers and Monolayers - Influence of Peptide and Lipid Modifications on Binding*, Martin-Luther-Universität Halle Wittenberg, 2016.

[2] F. M.Goñi, *The basic structure and dynamics of cell membranse: An update of the Singer-Nicolson model*, In: Biochemica et Biophysica Acta 1838, pp. 1467–1476, 2014.

[3] C. Nehls, *Charakterisierung der Wechselwirkung zwischen dem bakteriellen Protein VapA und der Phagosomenmembran an Modellmembransystemen*, Forschungszentrum Borstel, 2016.

[4] J. Mertens, T. Gutsmann, C. Nehls, *Characterization of the interaction between the antimicrobial peptide LL-32 and a lipid monolayer using a film balance.*, In: Student Conference on Medical Engineering Science 2017, Infinite Science Publishing, Lübeck, pp. 65–68, 2017.

[5] H. Halliday, *Surfactants: past, present and future*, In: Journal of Perinatology 28, Nature Publishing, 2008.

[6] G. T. Barnes, I. R. Gentle, *interfacial science: an introduction*, Second Edition, Oxford University Press Inc., New York, 2011.

[7] J. Goerke, *Pulmonary surfactant: functions and molecular composition*, In: Biochemica et Biophysica Acta 1408, pp. 79–89, 1998.

Development of an Immunoassay for the Detection of Tetrahydrocannabinol and Methamphetamine

A. Kruse [1], K. Lettau [2] and C. Hübner[3]
[1] Medizinische Ingenieurwissenschaft, Universität zu Lübeck, alien.kruse@student.uni-luebeck.de
[2] Drägerwerk AG & Co. KGaA, Lübeck, kristian.lettau@draeger.com
[3] Institut für Physik, Universität zu Lübeck, huebner@physik.uni-luebeck.de

Abstract

Psychoactive substances cause 20% of all deadly road accidents. Australia is facing this problem with nationwide road blocks where persons are tested through the car window in less than five minutes. Therefore, this paper examines a simple drug testing system based on a Lateral Flow Assay (LFA) for the substances Tetrahydrocannabinol (THC) and Methamphetamine (MET). To receive a test system that provides fast results a substance for the control line has to be found which correlates with the dynamics of the test substances. In this work Fluorescein isothiocyanate (FITC) is investigated as one possible option. Results show, that FITC is a promising candidate to match the time and performance requirements which is subsidized by its similar characteristic to the test substances THC and MET. To assure, a reliable system is developed, long term studies concerning the stability are recommended.

1 Introduction

Point-of-care testing is a fast-growing field, becoming more important in primary care every day [1]. Besides diagnosis of pathogens, it can be used to detect many substances in different matrices e.g. blood, urine and saliva. The demand for a test system that detects psychoactive substances fast and easy is growing due to increasing number of deadly car accidents caused by drug abuse [2]. The developed drug test is tailored to particular needs of the Australian market. Australia is the only country testing every person through the car window for Tetrahydrocannabinol (THC) and Methamphetamine (MET) nationwide in so called "Road Blocks". By law, these tests are allowed to take a maximum amount of time of 5 minutes.

The test system is based on saliva samples to detect the current amount of THC and MET in the blood. Saliva is particularly suitable, since it is an ultrafiltrate of the blood and samples can easily be taken [3]. The chosen test set up is a Lateral Flow Assay (LFA). Small molecules like THC often lack additional binding sites, for which reason a competitive assay format is used. Negative test results will cause a disappearing test line on the test strip.

The aim of this work is to investigate the possibility to design a LFA based test system to detect the psychoactive substances THC and MET in less than five minutes of time. Further necessary requirements are availability, low cost of all components and stability for at least 14 months. To be able to compete with other products on the market reliable results in a temperature range from 5-40°C are needed. The focus of this work will be on the chosen substance for control line. The dynamics of the control line determines the velocity and sensitivity of the test system. As soon as the control line shows up, an interpretation of the test is possible. Therefore, it is important to have similar kinetics for test line and control line to avoid wrong results especially at the edge of the temperature range. Fluorescein isothiocyanate (FITC) is evaluated as a possible c- line substance. Its size is similar to the size of THC and MET with a molecular weight of 389,4 g/mol [4] with the adventage of similar kinetics. FITC is also available in large amounts, can be conjugated without additional chemical synthesizing [5], and is a low cost substance.

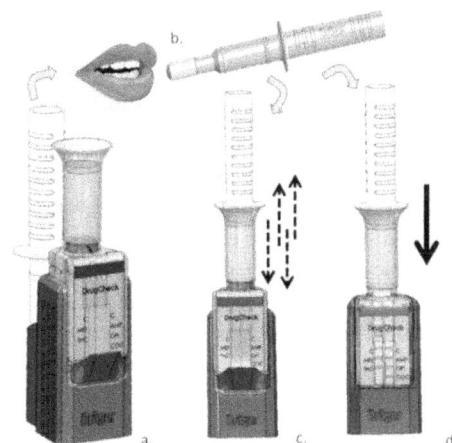

Figure 1: Schematic layout of the DrägerDrugTest. a) drug Test, b) sample taking, c) starting the drug test by shaking for 20 sec, d) pressing the test cassete down and wait for test lines to develop

2 Material and Methods

In this section, the selected test set up is explained as well as the substance chosen for the control line to meet the described requirements.

2.1 Lateral Flow Assay

The core of every Lateral Flow Assay (LAF) is formed by the test strip, which consists of a membrane, a sample pad and an absorbent pad. To receive a functional test strip, the components have to overlap. To provide the required stability, all pieces are glued on a backing card. The test lines composed of the to be recognized molecular compounds are striped on the nitrocellulose membrane as well as the control line. The antibody-gold conjugate which is needed for detection, is placed on the sample pad. Driven by capillary forces, an on the sample pad applied sample of saliva and running buffer will resolve the antibody-gold conjugate and travel over the membrane to the absorbent pad [6]. A competitive assay format as shown in Figure 2, which is most suitable for detecting small molecular compounds is applied. If the sample does not contain an analyte, the antibody binding sides are able to bind on the test line and a red line appears. In contrast, a THC or MET positive sample will cause a disappearing test-line. The control line appears independently from the test result. A missing control line leads to an invalid test result.

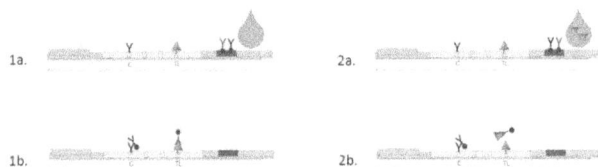

Figure 2: The design of typical LFA test strip. 1a negative sample is applied, 1b control line and test line are visible-negative test result, 2a positive sample is applied, 2b only control line becomes visible - positve test result

2.2 Control Line Design

The design of the control line determines if the described requirements above can be forfilled. When a negative sample is tested, the test line of the test strip has to develop before the control line becomes visible, otherwise the test can be interpreted as false positive. For a positive sample the result can be interpreted as soon as the control line appears. Considering the change of antibody binding affinity with the change of temperature, it is important to find a molecular compound for the control line that changes affinity in an equal way to avoid false positive results.

2.3 Fluorescein Isothiocyanate

FITC is a widely used fluorochrome for applications involving fluorescent labeling for example flow cytometry with

the excitation maximum at 495 nm and the emission maximum at 519 nm. For that reason, it is available in large amounts and is fairly cheap. FITC has a similar molecular weight to THC and MET which is expected to lead to a similar reaction to temperature changes. For functionalizing FITC as a polyhapten that will stick to the nitrocellulose membrane, it has to be linked to a protein. Bovine Serum Albumin (BSA) is a protein with high stability and strong affinity to the membrane and most suitable for immobilizing substances on the test strip. For conjugating FITC and BSA, a reactive group is necessary. Since FITC is already functionalized with an isothiocyanate reactive group, no further chemical synthesizing is needed for the conjugation process. The reactive group will react with N-terminal amino groups of the BSA when the coupling reaction is performed at a pH value of about nine [7].

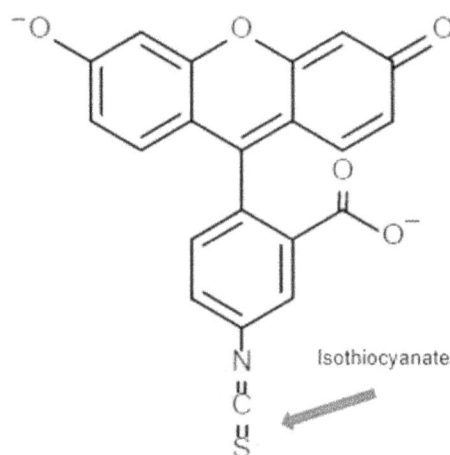

Figure 3: Fluorescein Isothiocyanate and its reactive group

2.4 Calibration of Test System

In this section, the calibration procedure is explained. For a reliable test system, a calibration method has to be developed. For comparable results, a color standard was established to estimate the intensity of the appearing lines on the test strip. Lines are rated from Color Master (CM) value 0 (no line) to CM value 7 (deep red line). The visual boundary where the operator recognizes a test line is set at a CM value of 1.5. The test parameters have to be adjusted in such a way that the THC line disappears at a concentration of 25 ng/ml in the sample while the cutoff value for the MET line is set at a concentration of 35 ng/ml. The CM value for negative tests depends on the polyhapten concentration on the test strip, as well as on the applied OD of antibody-gold conjugate in the system. To reach the cutoff values, the amount of gold in the system is important. The system gets more sensitive the less gold is in the system. After some pretesting, the polyhapten concentration was set to 1.5 mg/ml for all three substances. For calibration, five negative samples were taken and the CM value was notated after 120 sec for three different gold concentrations. Also,

Table 1: Example for OD variation for calibration

Variante	A	B	C
THC	OD 0,15	OD 0,25	OD 0,35
MET	OD 0,12	OD 0,20	OD 0,28
FITC	OD 0,12	OD 0,20	OD 0,28

three samples were prepared with the cutoff concentration for THC and MET and the CM values notated after 120 sec. From the taken measuring points with three different gold concentrations the OD was plotted against the CM value. The needed OD to reach a CM value of three for the FITC line and a four for the THC/MET line was than interpolated. For a positve sample an OD value was chosen to meet a CM value less than 1.5. After these steps another calibra-

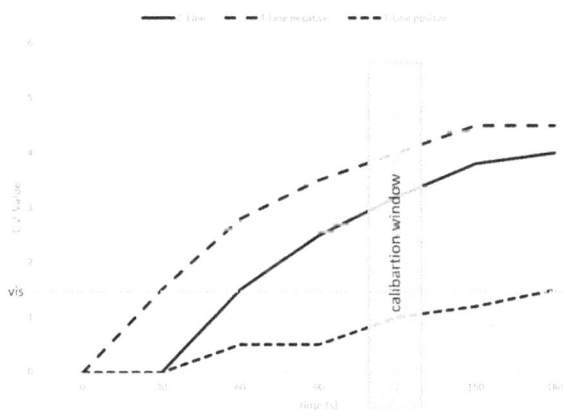

Figure 4: Example of a test calibration where the CM values meet the requierements. After 120 sec the control line has a CM of 3.5 and the test line a CM of 4 for negative samples and a CM of less than 1.5 for a positve sample

tion procedure was started to verify the estimated OD. In the following, the calibrated system was tested for its performance in different temperature settings.

3 Results and Discussion

In this section, the performance results of the calibrated test system for the application in different tempertures is presented and discussed.

3.1 Results

While working with lots of different test strips and gold conjugates, several different calibrations demonstrated that the system can be adjusted reliable to the desired CM values with the chosen method. Measurements at 5°C with at least 10 negative individual samples showed a slower development of the test and control line. When placing the test lines in different heights on the test strip a placement of THC/MET at the lowest position (position 1) and the control line right above on position 2, in some cases the CM value for the control line was stronger than the value for the

test line. Especially the slow development of the THC-line in combination with a faster developing c-line was critical and caused a couple of false positive results.

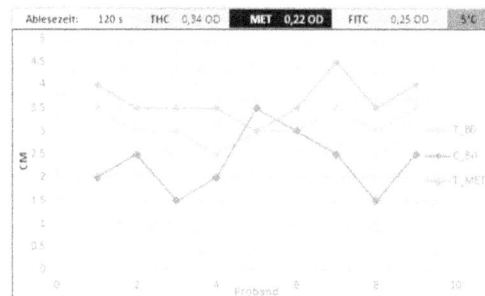

Figure 5: FITC-System at 5°C with line position T1C2 shows CM values after 120 sec for ten different tested probands. In case of sample number five it can be seen, that the control line has a stronger CM value as the test lines which can be a problem at low temperatures.

To minimize the likelihood of false positive results, the control line was moved up to position 3 on the test strip. Due to flow characteristics, the 3rd position on the test strip is reached by the liquid front with a noticeable time shift. Especially when working in low temperatures, the viscosity of the saliva sample buffer mixture changes and slows the process additionally down. The measurements showed a positive effect, no false positive effects were detected any further. The test system with line T1C3 positions was tested at 35-40°C afterwords. At high temperatures, the kinetic of the antibody binding rises and the lines on the test strip developed extremely fast. In this range, false positive results are not a problem, but the minimum cutoff value changes extremely. Samples with different cutoff concentrations were measured to determine the best possible cutoff at 35°C and 40°C. For THC, the representable cutoff value rised from 25 ng/ml to 75ng/ml at 35°C and to 87 ng/ml at a temperture of 40°C. The cutoff for MET changed from 35 ng/ml to 122.5 at 35°C and 140 ng/ml at 40°C. In both cases, it is a factor between three and four. Positive to remark is the fast development of the control line, which enables a positive detection of test results. The test has to be interpreted as soon as the control line gets visible, otherwise the test line will appear soon and the test gets interpreted false negative. The following figure 6 shows the cutoff values exemplary for the analyte THC.

First, storage studies for two weeks at 60°C were made to estimate the stability of FITC. The Hoffsche law implies a relation between aging and temperature. A temperature rise of 10°C will cause a doubling in reaction rate [8]. Measurements were taken after two weeks and compared to tests stored at RT. The FITC signal was extremely stable and did not loose any of its CM-value, while the signal for THC/MET dropped to 0.5 CM.

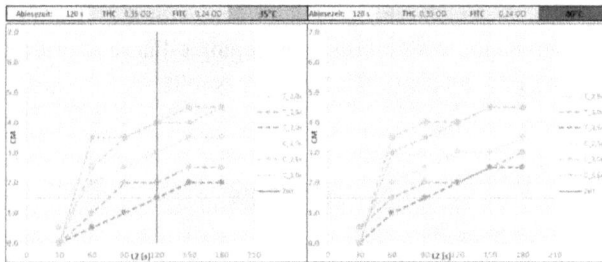

Figure 6: Cutoff determination of the FITC-System at 35 and 40°C shows, that with rising temperture the test system gets less sensitve.

3.2 Discussion

The FITC System showed a good performance at room temperature. Through measurements at different temperatures a similar response characteristic between THC/MET and the control FITC could be demonstrated. Performing the measurements at RT, a detection window of 60 sec can be reached in which the test can be interpreted as positive or negative safely. By dropping temperatures under 10°C, the false positive likelihood could be minimized by moving the c-line from position 2 up to position 3 on the test strip. Temperatures above 35°C provoke a strong loss in sensitivty. The reached cutoff value triples at least.

Furthermore, the detection window in which the test can be interpreted is by maximum 20 sec. To guarantee the discussed cutoff values above well-trained personnel for the testing procedure is necessary. The test has to be interpreted as soon as the c-line shows up, particularly for high temperatures, to avoid false negative results. Before a blood sample is taken in Australia, positive test results are verified by a second tempered drug test system, which has a higher accuracy. For that purpose, the small detection window seems acceptable. For a better performance at high temperatures the buffer system could be optimized. The buffer system is the main factor that can be adjusted further while sticking to the chosen antibodies and polyhaptens. Otherwise there will always be a tradeoff. If the customer is testing mainly in a certain temperature range, it makes sense to optimize the test for these characteristic temperatures.

Considering the aging of the test, stability data must be collected in long term experiments at RT since the 60°C data gives only a rough idea on processes taking place. Referring to old data, the CM value of THC/MET is expected to drop from 4 to 3.5/3 over the period of 14 months at 21°C. If FITC is not aging in the same way as first studies imply, this will cause problems, especially at low temperatures. Setting a fixed time from three minutes for interpreting the test in temperatures under 10 °C could solve this problem. For high temperatures, a possible aging effect would even come in handy and cause more sensitive cutoff concentrations.

4   Conclusion

In this work, a drug testing system for the Australian marked, testing for the substances THC and MET, was developed. The main focus was on the control line design to obtain a fast test that works in a wide temperature range. It was proven, that for designing a fast-developing test, a small molecular compound such as FITC has performance advantages to commonly used control lines consisting an antibody. The use of the test in a wide temperature range still causes losses in sensitivity which have not been improved so far.

The aim of future work will be to design assay components that will solve the performance problems in varying temperatures in the long run. LFA's of all kind have issues on reliability when it comes to a not tempered environment outside the laboratory. An approach might be to design a newly buffer system that inhibits the effects.

Acknowledgement

The work has been carried out at Drägerwerk AG & Co. KGaA, Lübeck and supervised by the Institute of Physics, Universität zu Lübeck.
I would like to thank everyone, who helped me while working on the project and who supported me with this paper.

5   References

[1] J. Verbakel, P. Turner, M. Thompson, and A. Plueddemann, "Common evidence gaps in point-of-care diagnostic test evaluation - a review of horizon scan reports," *BMJ Journals*, vol. 7, pp. 1–10, 2017.

[2] L. Vandam, B. Hughes, and P. Griffiths, *Drug use,impaired driving and traffic accidents*. Publications Office of the European Union, 2014.

[3] Y. Caplan and B. Goldberger, "Alternative specimens for workplace drug testing," *Anal Toxicol*, vol. 7, pp. 396–399, 2001.

[4] W. Luttmann and K. Bratke, *Der Experimentator Immunologie*. Elsevier, 2006.

[5] D. Wild, *The immunoassay Handbook*. Elsevier, 2005.

[6] H. Hsieh, J. Dantzler, and B. Weigl, "Analytical tools to improve optimization procedures for lateral flow assays," *Diagnostics*, vol. 29, pp. 1–14, 2017.

[7] N. Barbero, C. Barolo, G. Viscardi, and A. Plueddemann, "Bovine serum albumin bioconjugation with fitc. world journal of chemical education," *World Journal of Chemical Education*, vol. 4, pp. 80–85, 2016.

[8] J. Reece and N. Campbell, *Campbell Biologie*. Pearson Studium, 2015.

Hypotonic swelling as a procedure of encapsulating fluorescence labelled superparamagnetic iron oxide nanoparticles into human red blood cells

A. Steuer [1], M. Klinger [2], R. Pries [3], and K. Lüdtke-Buzug [4]

[1] Medizinische Ingenieurwissenschaft, Universität zu Lübeck, annkathrin.steuer@student.uni-luebeck.de
[2] Institute of Anatomy, Universität zu Lübeck, klinger@anat.uni-luebeck.de
[3] Department of Otorhinolaryngology, University Hospital of Schleswig-Holstein, Campus Lübeck, Ralph.Pries@uksh.de
[4] Institute of Medical Engineering, Universität zu Lübeck, luedtke-buzug@imt.uni-luebeck.de

Abstract

Superparamagnetic iron oxide nanoparticles (SPIONs) are used as tracers in medical imaging, e. g. for magnetic particle imaging (MPI) or magnetic resonance imaging (MRI). Since the half-life time of the SPIONs in the bloodstream is quite short and they are quickly absorbed by the reticuloendothelial system, the particles are introduced into human red blood cells (RBCs) to increase their half-life time in the blood circulation. The hypotonic swelling procedure is used to incorporate the particles into the RBCs. Before the SPIONs are introduced into the RBCs, they are fluorescent labelled. To evaluate the result, a transmission electron microscope, a magnetic particle spectrometer and a fluorescence microscope are used. Fluorescein isothiocyanate and rose bengal were chosen as fluorescent dyes because their biocompatibility is guaranteed. The results suggest that the method hypotonic swelling can be used to successfully introduce the nanoparticles into RBCs and that the magnetic properties of the particles that are important for imaging are not lost.

1 Introduction

Superparamagnetic iron oxide nanoparticles (SPIONs) are used, for example, as tracer material or contrast agent for medical imaging. They are injected into the patient for this purpose. The distribution of SPIONs in the tissue is strongly influenced by surface properties and particle size. The hydrodynamic diameter of the nanoparticles used here varies between 80 and 120 nm.

Particles with a diameter of more than 50 nm are quickly trapped in the bloodstream by macrophages and transported to the liver. These are therefore particularly useful for medical images of the liver. However, the SPIONs quickly disappear from the bloodstream. Large particles are detected faster as foreign bodies than smaller particles [1].

To increase the half-life time of the SPIONs in the bloodstream and in other organs, there is a method [5] of introducing the particles into the patient's own erythrocytes and then reintroducing these particle-loaded red blood cells to the patient. This not only increases the time for the nanoparticles to circulate in the bloodstream, but it also enables the particles to overcome the blood-brain barrier.

This would also enhance the possibility of taking images of the blood vessels in the brain using MPI or SPION-based MRI. Such a method is already used to keep different drugs or substances in the bloodstream or to allow them to overcome the blood-brain barrier.

The aim of this work is to introduce the SPIONs, which were previously coupled with a biocompatible fluorescent dye, into the human red blood cells (RBCs) and then evaluate them using a transmission electron microscope (TEM), a magnetic particle spectrometer (MPS) and a fluorescence microscope.

2 Material and Methods

Different methods for fluorescence labelling and the introduction of nanoparticles into red blood cells are used, which will be explained in the following subsections. The first subsection describes the synthesis process followed by the labelling process. In the second subsection, the hypotonic swelling is explained in more detail.

The fluorescence-labelled SPIONs are subsequently referred to as f-SPIONs.

2.1 Synthesis of fluorescence-labelled superparamagnetic iron oxide nanoparticles

To produce unlabelled nanoparticles, iron (II) and iron (III) salts are precipitated with a base, like sodium hydroxide or ammonia. Here ammonia was used. The reaction occurs in the presence of the coating material, here dextran, and under influence of heat.

Ammonia is highly soluble in water and forms hydroxide

ions (OH^-). The following reaction takes place:

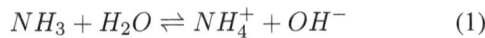

$$NH_3 + H_2O \rightleftharpoons NH_4^+ + OH^- \qquad (1)$$

The iron salts react with the hydroxide ions at a target temperature of 0 degrees Celsius:

$$Fe^{2+} + 2Fe^{3+} + 8OH^- \xrightarrow[cooling]{} Fe(OH)_2 + 2Fe(OH)_3 \qquad (2)$$

In this phase, the temperature must not exceed 15 degrees Celsius. After completely draining the ammonia, the solution is heated to 80 degrees Celsius for 30 minutes. Within the 30 minutes magnetite (Fe_3O_4) coated with dextran is formed in a condensation reaction:

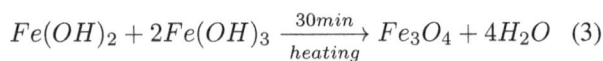

$$Fe(OH)_2 + 2Fe(OH)_3 \xrightarrow[heating]{30min} Fe_3O_4 + 4H_2O \quad (3)$$

After the nanoparticles have formed, they are dialysed against distilled water to remove the ammonia.

Two methods are available for coupling the nanoparticles with a fluorescent dye. In the first method, the particles are produced as described above and then coupled with a dye. The second process first couples the dye with the coating material, then uses this labelled dextran to synthesize the nanoparticles. For the here presented particles the second method was used.

Two dyes with biocompatible abilities have been selected for fluorescence labelling. The dyes used here are fluorescein isothiocyanate (FITC) and rose bengal.

Two different methods were used for coupling the dyes.

For FITC labelling, the dextran (molecular wight 70.000) and FITC are dissolved in dimethyl sulphoxide and the reaction is accelerated with two catalysts, namely Dibultyltin dilaurate and a few drops of pyridine. This process is called transesterification and is performed at 95 degrees Celsius [2]. The result is a stable thiocarbamoyl linkage and shown in Fig. 1.

Marking the dextran with the dye rose bengal is done by means of Steglich esterification. A carboxylic acid (in this case, the dye) forms a carboxylic acid ester with an alcohol when water is removed [3].

The so-called Steglich catalysts dicyclohexylcarbodiimide (DCC) and 4-(dimethylamino)-pyridine (DMAP) are used, with dichloromethane or chloroform serving as solvents.

2.2 Encapsulation of f-SPOINs in human RBCs by using hypotonic swelling

The erythrocytes are loaded with nanoparticles produced by the synthesis method are described in subsection 2.1.

To introduce these f-SPIONs into the RBCs, the hypotonic swelling method is applied. This method is performed in three steps.

The materials required for all steps are listed in table 1.

The first step is to isolate the RBCs from freshly extracted blood by centrifugation at 1400 g for 10 minutes. The serum is removed and the RBCs are washed three times with hepes buffer, consisting of hepes, sodium chloride and glucose.

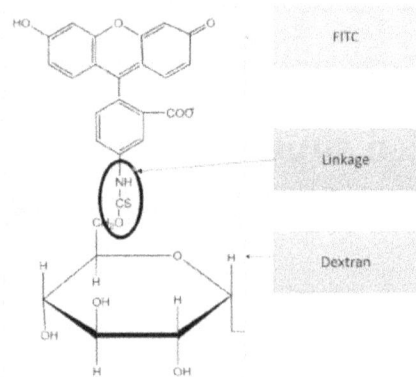

Figure 1: FITC dextran structure [4]

In order to achieve a hematocrit level of 70 %, 0.3 ml of hepes buffer is now mixed with 0.7 ml of centrifuged and washed RBCs. Then 1:1 f-SPIONs (1 ml) are added to this mixture [5].

The second step involves dialysis, in which the RBCs swell to an extent that allows the nanoparticles to enter the cell through the pores of the cell membrane. For this purpose, the nanoparticle-RBC-mixture is filled into a dialysis tube with 12-14 kDa-Cutt-off and dialyzed in the dialysis buffer for 75 minutes [5].

RBCs are reclosed in the third and final step of the method. For this purpose, 0.1 volume PIGPA per volume of dialyzed RBCs is added to the solution and incubated for 45 minutes at 37 degrees Celsius.

To remove excess unencapsulated nanoparticles, the cells are centrifuged at 400 g and washed four times with hepes buffer, as already done in step one [5].

The various steps of the applied procedure are shown schematically in Fig. 2. From left to right, this representation describes an unprocessed red blood cell (a) to which the f-SPIONs are added (b). Through hypotonic swelling, the particles can enter the blood cell through the open pores of the cell membrane (c). In the next step, the PIGPA is added and the cells are incubated, causing the pores of the membrane to close again (d). There should be no free f-SPIONs left after four washes (e).

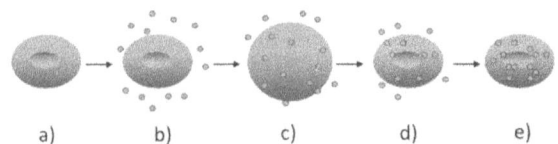

Figure 2: Schematic representation of hypotonic swelling [5]. a) shows a normal erytrocyte, b) indicates the addition of f-SPIONs, c) reveals hypotonic swelling, d) represents the resealing of the RBCs and the result of the washing process can be seen in e).

Table 1: Materials for introducing f-SPIONs into RBCs

Method step	Material	Amount
Step 1:		
Hepes Buffer	Hepes	10 mM
	$NaCl$	140 mM
	Glucose	5 mM
Step 2:		
Dialysis Buffer	$NaHCO_3$	10 mM
	NaH_2PO_4	10 mM
	Glucose	20 mM
	$MgCl_2$	4 mM
	ATP	2 mM
	red. Gluthatione	3 mM
Step 3:		
PIGPA	Adenine	5 mM
	Inosine	100 mM
	ATP	2 mM
	Glucose	100 mM
	Sodium pyruvate	100 mM
	$MgCl_2$	4 mM
	$NaCl$	194 mM
	KCl	1,606 M
	NaH_2PO_4	35 mM

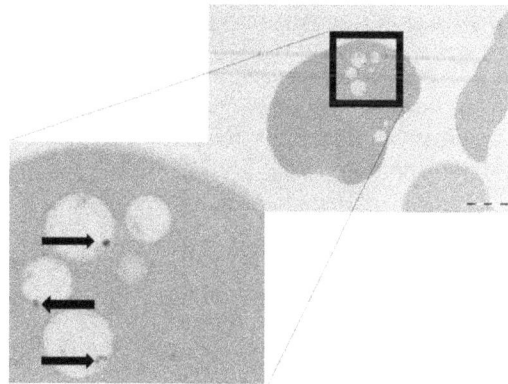

3 Results and Discussion

The results of the insertion of f-SPIONs into the RBCs are shown below and have been evaluated with three different instruments, consisting of the transmissions electron microscope (JEM-1011, Jeol), the magnetic particle spectroscope (MPS, Folklabs) and the fluorescence microscope (Carl Zeiss Microscopy GmbH).

3.1 Transmission electron microscopy and magnetic particle spectrometry

The Fig. 3 and 4 show TEM images of RBCs loaded with FITC-SPIONs (Fig. 3) and rose bengal-SPIONs (Fig. 4).

To be able to see more precisely where the particles are, an enlargement of the respective areas has been made and the arrows point to the f-SPIONs.

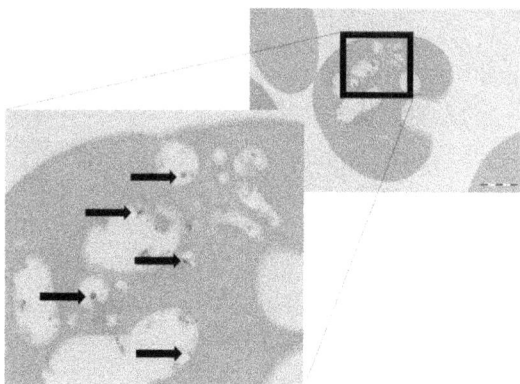

Figure 3: TEM image of RBCs loaded with FITC-SPIONs

Figure 4: TEM image of RBCs loaded with rose bengal-SPIONs

The images reveal that the erythrocytes are loaded with a few f-SPIONs. However, they are not randomly distributed in the cytoplasm, but are located in vacuoles. This fact indicates that the pores of the blood cell membrane have not opened far enough and the particles have entered the cell through an invagination process.

However, since SPIONs are in the RBCs, encapsulation can still be considered successful. And as measurements with the magnetic particle spectrometer (MPS) indicate, the loaded RBCs show a magnetization behavior that makes them suitable for imaging (Fig. 5). The first picture in Fig. 5 shows the harmonics of the nanoparticles. The steeper and smoother the curve, the better the quality of the SPIONs.

The magnetization curve is shown in the second image, which performs a sigmoidal curve typical of superparamagnetism.

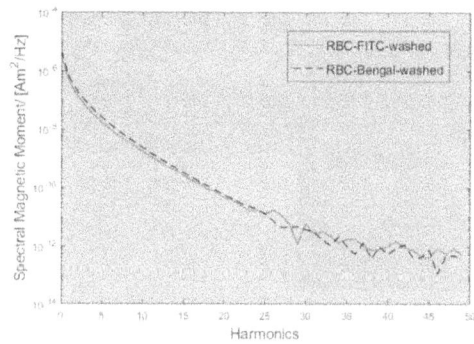

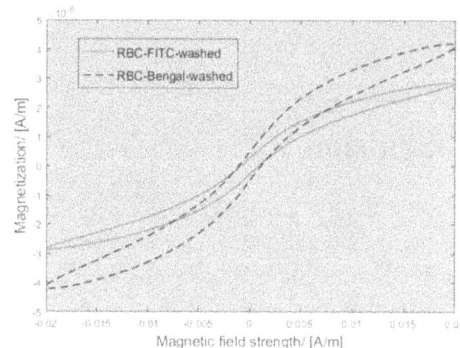

Figure 5: The harmonics (top) and magnetization (below) curves of RBCs filled with FITC- or rose bengal- SPIONs.

3.2 Fluorescence microscopy

Fig. 6 is a fluorescence image taken with a fluorescence microscope on which the RBCs loaded with FITC-SPIONs fluoresce clearly in front of the black background.

The loading with rose bengal-SPIONs was also successful, according to Fig. 7. Here, too, the charged erythrocytes are clearly fluorescent on a black background.

The RBCs were applied to the slides after sufficient dilution with cytospin, so that individual cells actually become visible.

The images of the fluorescence microscope show that the particles with the coating, and thus also the fluorescence dye, have entered the red blood cells.

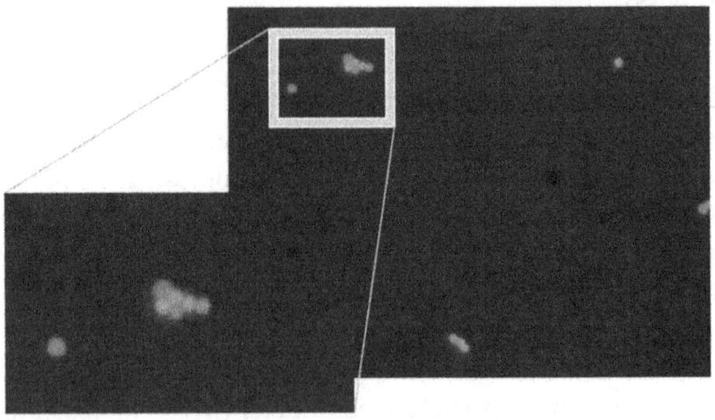

Figure 6: Fluorescence image of RBCs loaded with FITC-SPIONs

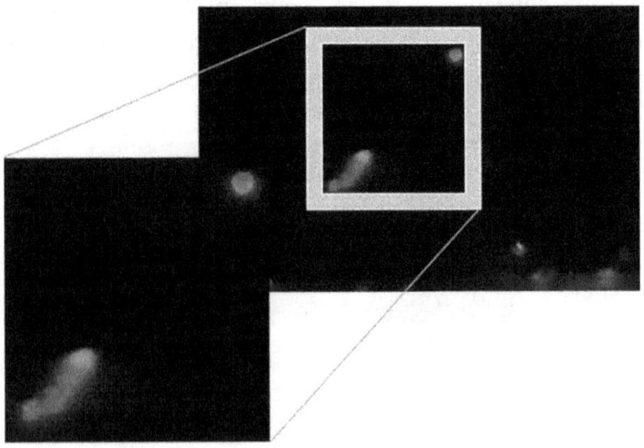

Figure 7: Fluorescence image of RBCs loaded with rose bengal-SPIONs

4 Conclusion

Hypotonic swelling is a method of encapsulating nanoparticles into red blood cells. The aim is to swell the cells in a dialysis fluid and open their pores so that the superparamagnetic nanoparticles can enter them. In an incubation cabinet, the pores are to be closed again in the next step and the particles are to be encapsulated.

The images of the TEM and the fluorescence microscope showed that the method was successfully applied, but the nanoparticles were too large to directly cross the membrane, but were encapsulated only by invagination.

Nevertheless, the RBCs carry good magnetisation properties, measured in an MPS. The results show that they would be suitable for MPI or MRI.

Future experiments should aim to reduce the particle size or increase the pores of the cell membranes and in addition, the osmolarity has to be adjusted to this experiment.

Acknowledgement

The work has been carried out at the Institute of Medical Engineering at the Universität zu Lübeck. The nanoparticles were produced in the institute's own laboratory.

The TEM images were taken in the Institute of Anatomy, in this respect a thanks to Kerstin Fibelkorn, and the fluorescence images in the Department of Otorhinolaryngology at the University Hospital Schleswig-Holstein.

The research team would also like to thank Kristin Müller, who helped to carry out the hypotonic swelling and Kirstin Plötze-Martin for the help with the fluorescence microscope.

5 References

[1] J. Maier, K. Mäder, T. Groth, and M. Antonietti, *Synthese und Anwendung von FERR-b-PEO stabilisierten SPIO Partikeln als Kontrastmittelsystem für die Magnetresonanztomographie*, Naturwissenschaftliche Fakultät I, halle-Wittenberg, 2008.

[2] S. Hauptmann, *Organische Chemie*. Deutscher Vlg f. Grundstoffindustrie, Frankfurt am Main, 1985.

[3] A. Holleman, and E. Wiberg, *Lehrbuch der anorganischen Chemie*. De Gruyter, Berlin, 1995.

[4] A. de Belder, and K. Granath, *Carbohydrate Research - Fluoresceinisothiocyanate-Insulin (Preparation and properties of fluorescein-labelled dextrans)*. Elsevier, Amsterdam, 2011.

[5] A. Antonelli, C. Sfara, L. Mosca, E. Manuali, and M. Magnani, *New Biomimetic Constructs for Improved In Vivo Circulation of Superparamagnetic Nanoparticles*. Journal of Nanoscience and Nanotechnology Vol. 8, 1-9, 2008.

[6] K. Müller, and K. Lüdtke-Buzug, *Incorporation of Superparamagnetic Iron Oxide Nanoparticles into Erythrocytes for MPI*. IWMPI accepted, 2018.

[7] M. Schuhmacher, *Superparamagnetische Eisenoxidnanopartikel für Magnetic Particle Imaging und Fluoreszenzmikroskopie*, Bachelorthesis at University to Lübeck (IMT), 2014.

4

Image Processing

Occlusion Estimation in 3D Point Clouds using Visual Data from Home Care Scenarios

D. Laule [1], J. Diesel [2], M.P. Heinrich [3]

[1] Medizinische Informatik, Universität zu Lübeck, david.laule@student.uni-luebeck.de
[2] Drägerwerk AG & Co. KGaA, Lübeck, jasper.diesel@draeger.com
[3] Institute of Medical Informatics, Universität zu Lübeck, heinrich@imi.uni-luebeck.de

Abstract

Today, home care monitoring systems are implemented more frequently, which is mainly due to the increasing number of elderly people and the reduced number of medical staff [1] [2]. This makes it all the more important that these systems are reliable and safe for its users. In this paper, we propose a risk management feature in form of an Occlusion State Index (OSI) based on the overall occlusion within an observed scene that can be integrated into an in-house developed smart home care monitor prototype. To implement and evaluate the proposed feature, visual data is acquired within a home care test scenario using two *Kinect 2.0* depth cameras. After preprocessing, the recorded depth information is merged into a point cloud from which a scene occlusion map is computed based on the ray box intersection algorithm of [3] and the fast voxel traversal algorithm of [4]. Finally an OSI is computed depending on the amount of occluded voxels in a 3D point cloud.

1 Introduction

The fall rate of people in their own houses dramatically increases with their age. One in every three adults older than 65 years is estimated to suffer a fall at least once a year [5]. To minimize these risks, an increasing number of households are equipped with specific camera systems to monitor elderly people in their home, mostly for safety and care reasons. Such a camera-based monitoring system that can detect a fall of a person in their home and send an related alarm notification to an authorized person like a caregiver or any relative of the affected person, could potentially reduce the number of fall-related deaths or long-term injuries. While observing a home care scenario, it can be important to know if an object in the monitored scene is invisible for a camera due to occlusion or because of missing depth information. In the home care setup, occlusions can be caused by multiple objects in the scene such as persons, furnishing (sofas, tables, etc.) or movable objects (wheelchair, drying rack, etc.). In general, the knowledge of all the pixels in a scene that are occluded by other objects and located between the given pixel and the sensor, can primarily serve as an important risk management feature for a camera-based monitor system.

Different research groups have already considered the problem of occlusion estimation in visual systems but in different field of applications and rarely in 3D. While [6] reflect the problem in stereo vision in combination with stereo matching, [7] and [8] propose more complex and costly solutions also based on 3D point clouds and voxelization.

In this work, we address the problem of occlusion estimation in 3D point clouds generated from visual data from multiple sensors. To accomplish this, we build a system on the work of [3] and [4], which efficiently computes an OSI based on estimated occluded voxels in a monitored 3D point cloud.

2 Material and Methods

2.1 System Architecture

The system architecture consists of two *Kinect 2.0* depth cameras and an internal developed smart home care monitor prototype which is based on the *Robot Operating System* (ROS) and mainly implemented in C++. The proposed feature can be divided into four main components which are shown in Fig. 1.

Frameworks. The different components of the prototype are implemented as ROS nodes to allow a loosely coupled architecture using ROS messages for the internode communication. The *Point Cloud Library* (PCL) and the *OpenCV* framework are used due to the provided high-performance implementations of multiple algorithms to solve point cloud specific computer vision tasks. The transfer of the RGB, NIR and depth images and the registration of the RGB and depth images are handled by the *libfreenect2* driver for *Kinect 2.0* devices and the *iai-kinect2* library. For the calibration of the *Kinect 2.0* sensors the external ROS package *OpenPTrack* is used.

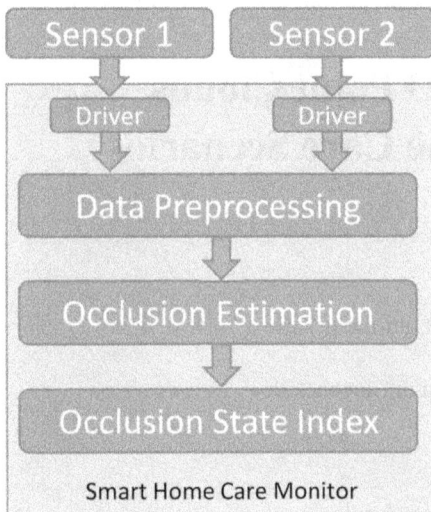

Figure 1: Architecture of the proposed occlusion estimation with Smart Home Care Monitor integration.

2.2 Data Acquisition

Two *Kinect 2.0* depth cameras were used to gain visual data as a stream of registered and rectified RGB, NIR and depth images at a frame rate of 30 Hz, provided by its integrated cameras. The *Kinect 2.0* were installed at the ceiling of the test room defining the observed scene. The cameras were connected via USB to a workstation with a Linux 16.04 LTS operating system, 16 CPUs (Model: Intel(R) Xeon(R) CPU E5-1660 v4 @ 3.20GHz), 32GB of memory and a *nVidia* GeForce GTX 1060 6GB graphic card with a cuda capability of 6.1. The sensor calibration as well as the recording of the required test data as ROS bag files is performed with an in-house developed setup tool. The hardware setup is illustrated in Fig. 2. For development and evaluation, a dataset with five minutes recording time of a test person operating in the test room is acquired.

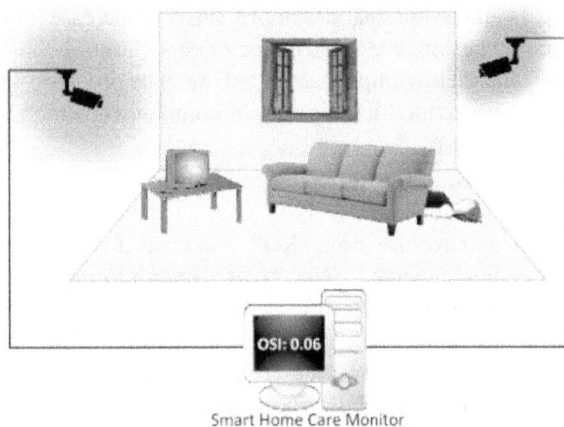

Figure 2: Setup of the Smart Home Care Monitor System.

2.3 Data Preprocessing

Synchronization. After recording the available RGB, NIR and depth data, image frames of each sensor are synchronized using an approximate time filter, so that all corresponding frames have the same time stamp.

Point cloud generation. In the next step, all the images gathered up to this point are converted into one point cloud $P = [x, y, z, r, g, b]$ for each sensor, whereby x, y, z stand for the point position and r, g, b specify the colour information. Both point clouds are then down-sampled by a voxel grid approach, provided by PCL and further transformed in the same world coordinate system using the transformations that were determined during the calibration and registration step of each sensor. This point cloud serves as the input of the occlusion estimation.

2.4 Occlusion Estimation

Ray box intersection. In the very beginning, a voxel grid box (from now on called "box") based on the input point cloud is generated. The box has to be axis aligned with the coordinate system and can be described by two points (b_{min}, b_{max}) representing the minimum and maximum extent of the box in the 3D space. Then, for each "empty" voxel in the box its occlusion state is estimated iteratively, sending a ray from the sensor origin O to the centroid of the target voxel. The ray can be expressed with the equation $O + D \cdot t$. The direction D of the ray is represented by a vector between the two considered points in the 3D space. By changing t (any positive or negative value) each point on the ray can be defined based on the ray's position and direction.

Next, the intersection of this ray into the box is determined using the slab method introduced by [3]. A slab is defined by a set of two lines that run in parallel to the considered axis of the coordinate system. In fact, there are three slabs for the bounds of the box. To determine the intersection of the ray with the box, the intersection of the lines for each dimension (x, y and z) is computed by calculating the ray equation with each line equation independently. The result is a set of six values indicating where the ray intersects with the box planes parallel to the axes of the coordinate system. The next step is to find out which of these six values corresponds to the ray box intersection, if the ray does not miss the box. The ray intersects the planes defined by the minimum extend of the box at two places (e.g. $t0x$ and $t0y$), but it does not necessarily mean that these intersection points lie on the box. Therefore, these two intersection points are compared to determine the point t_{min} located on the box. In fact, t_{min} is the point for which the value for t is the greatest. The second intersection point t_{max} of the ray with the box can be determined by comparing the values t of the planes defined by the maximum extend (e.g. $t1x$ and $t1y$) of the box, choosing the smallest one of them. As shown in Fig. 3, it is also possible that the ray does not intersect the box at all, which must also be checked, and occurs when for example $t0x > t1y$ or $t1x > t0y$. Depending on the position of the sensor origin it might be possible, that t_{min}

is greater than t_{max}. Then t_{max} has to be used as t_{min} to describe the intersection with the box.

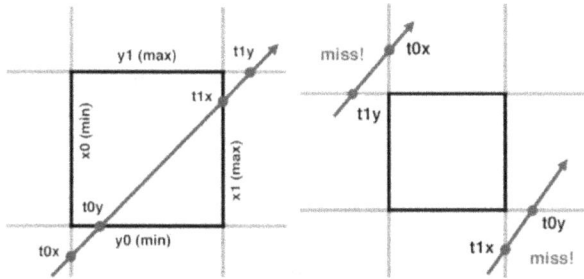

Figure 3: Visualization of the ray box intersection algorithm.

Ray Traversal. After determining the entry point of the ray in the box, the ray traversal algorithm of [4] is performed for each voxel under consideration returning its occlusion state. Tracing the ray from the sensor to the target voxel, every interjacent voxel is checked on whether it contains depth data. If so, the target voxel is marked as occluded and the traversal algorithm stops. Thereby, the start voxel is the centroid coordinate of the voxel located at the boundary of the box. The required step size in each dimension to the next voxel passed by the ray is computed with the ray direction and the chosen leaf size of the voxel grid. If the target voxel is reached without passing a voxel having depth information in the point cloud, it is not occluded and marked as such. The traversal of the ray and its components is illustrated in Fig. 4.

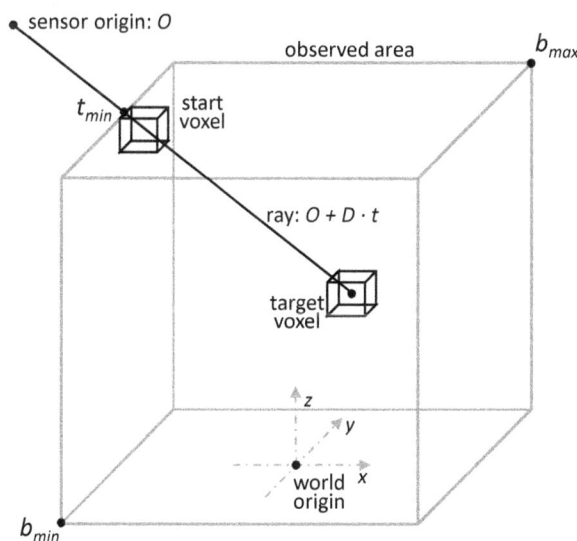

Figure 4: Visualization of the ray traversal algorithm.

2.5   Application Integration

To integrate the functionality, the existing PCL implementation needs to be adapted to the given problems. For that a new ROS package is created providing a node factory and nodelet plugin interface to integrate and run the occlusion estimation within the Smart Home Care Monitor System. The ROS node subscribes to the full point cloud topic which is limit to the observed scene in order to reduce the amount of voxels for the occlusion estimation and therefore shorten the algorithm run time by focusing on a scene of interest. The final output of the algorithm is a voxel grid point cloud including "not empty" voxels at all voxel coordinates that were estimated as occluded. In the initialization phase the sensor names are obtained from the ROS parameter server. Further, the observed scene (position and orientation) is loaded from a corresponding file to crop the box of the scene of interest from the full point cloud using the *OpenCV* FileStorage class and store it to an *OpenCV* matrix. In the next step, a transformation is computed to center and axis-align the input point cloud with the grid of the world frame, so that subsequently an axis-aligned voxel grid can be used. A callback on the subscribed full point cloud within the observed scene is used to perform the occlusion estimation for each frame received by the node. Then the following algorithm steps are performed.

1. Generate a voxel grid (PCL class) based on the input point cloud of the observed scene with a predefined leaf size to down sample the amount of voxels for which the occlusion is estimated. Each voxel represents the occlusion state of the points located in it. One voxel in the grid is represented by the centroid coordinate and the given leaf size of each voxel.

2. Initialize the voxel grid and set the sensor origins and orientations for each sensor.

3. Define a vector of tuples which includes the x-, y- and z- indices of the voxels of the voxel grid box, related to its minimum and maximum extend.

4. For each "empty" voxel in the voxel grid:

    (a) Compute the ray box intersection for each sensor independently using the slab method.

    (b) Estimate the occlusion state for sensor independently using the fast ray traversal algorithm.

    (c) Add the x-, y- and z- indices of the voxel to the in 3. declared vector, if the voxel is occluded for both sensors.

5. Create a new point cloud with centroid coordinates of the occluded voxels based on the vectors with the voxel indices.

6. Transform the new point cloud back to the sensor origin and orientation in the world coordinate system.

7. Publish the final point cloud including the occluded voxel centroids.

2.6   Occlusion State Index

Based on the amount of occluded voxels in relation to all voxels within the observed scene an OSI between 0 and 1 is computed and published as a ROS message. The OSI can then be used to send alarm notification, e.g. based on simple thresholds, and alert help.

3   Results and Discussion

We examined the mean estimation algorithm runtime in relation to different leaf sizes of the voxel grid as well as the mean occlusion reduction by using two sensors instead of a single one. The results are shown in Fig. 5.

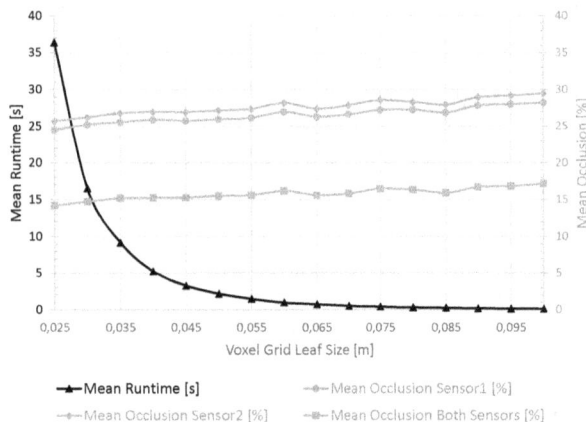

Figure 5: Results of the algorithm runtime and occlusion analysis depending on the leaf size of the voxel grid.

Reducing the leaf size of the voxels exponentially increases the runtime of the occlusion estimation algorithm. Based on our experiments and current implementation a larger leaf size with reduced occlusion estimation accuracy would have to be chosen to obtain low enough processing times for specific cases such as fall detection. Additionally, it could be shown that the amount of occluded voxels in the monitored scene can be reduced significantly by 10% using two sensors instead of one, which furthermore results in a general relative increase of visible scene content up to 40% with regard to any available depth information in the 3D point cloud. In general, the proposed system provides information about the reliability of its integrated system based on the OSI and can further be used to provide additional information on observed objects or persons that are potentially occluded by other objects in the scene.

However, the system also comes with some limitations. In the current implementation, no angle of view of the cameras is considered. Furthermore, a maximum of two sensors is supported and therefore the implementation is strongly dependant on the application. Computational costs could be reduced by considering voxels located in the line of a ray behind an occluded voxel in the first place as well as performing a parallel occlusion estimation for both sensors and not sequential.

Problems with the existing implementation occur when the ray is exactly parallel to the box which makes the computation of the ray box intersection impossible. Besides, there is no possibility to detect occlusion caused by objects that are closer than 0.5m to the sensor causing invalid depth data indistinguishable from empty depth data.

4   Conclusion

Occlusion is one of the greatest challenges in home care video monitoring and patient surveillance. In this paper, an approach for occlusion estimation and localization of occluded regions within 3D point clouds acquired by using multiple *Kinect 2.0* depth cameras is proposed. The estimated occlusion in a three dimensional space can be used to provide an OSI based risk feature for visual monitoring systems to measure its reliability. It allows an occlusion estimation with a frame rate of approximately 10 frames per seconds depending on the leaf size of a generated voxel grid. It could also be shown, that the occlusion within the observed scene is significantly reduced using two depth cameras instead of one resulting in a relative increase of visible scene content up to 40 %.

Acknowledgement

The work has been carried out at Drägerwerk AG & Co. KGaA, Moislinger Allee 53, 23558 Lübeck and was supervised by the Institute of Medical Informatics, Universität zu Lübeck.

5   References

[1] World Health Organisation, *World Report On Aging And Health*. World Health Organisation, 2015.

[2] A. Afentakis and T. Maier, "Projektionen des Personalbedarfs und -angebots in Pflegeberufen bis 2025," *WISTA - Wirtschaft und Statistik*, pp. 990–1002, Nov. 2010.

[3] A. Williams, S. Barrus, R. K. Morley, and P. Shirley, "An efficient and robust ray-box intersection algorithm," tech. rep., University of Utah, 2004.

[4] J. Amanatides and A. Woo, "A fast voxel traversal algorithm for ray tracing," tech. rep., University of Toronto, 1987.

[5] J. Stevens, M. Ballesteros, K. Mack, R. Rudd, E. De-Caro, and G. Adler, "Gender differences in seeking care for falls in the aged medicare population," *American Journal of Preventive Medicine*, vol. 43, pp. 59–62, July 2012.

[6] C. L. Zitnick and T. Kanade, "A cooperative algorithm for stereo matching and occlusion detection," *IEEE Transactions on Pattern Analysis and Machine Intelligence*, vol. 22, May 2000.

[7] K. Salvaggio, C. Salvaggio, and S. Hastrom, "A voxel-based approach for imaging voids in three-dimensional point clouds," in *SPIE Defense+ Security*, 2015.

[8] S. Hagstrom and D. Messinger, "Line of sight analysis using voxelized discrete lidar," *Proc. of SPIE*, 2011.

Automated landmark refinement in 3D ultrasound images of the aortic root

P. Merks [1], J. Hagenah [2], and A. Schweikard [2]

[1] Medizinische Ingenieurwissenschaft, Universität zu Lübeck, pascal.merks@student.uni-luebeck.de
[2] Institute for Robotics and Cognitive Systems, Universität zu Lübeck, hagenah@rob.uni-luebeck.de

Abstract

During the last years, effort was taken to plan valve-sparing aortic root reconstruction pre-operatively. The proposed planning methods are based on 3D ultrasound image data of the dilated aortic root and are capable of predicting the optimal prosthesis size for the individual patient, aiming on the reconstruction of his original healthy aortic root geometry. However, the approaches rely on the manual identification of specific landmarks. This user interaction is prone to individual errors. In this work, we present a model-based refinement of the annotated landmarks based on the morphology of the surrounding tissue as seen in the ultrasound volumes. Our first results indicate that an automatic refinement could be possible. Additionally, we could show that the underlying model of the valve geometry might not represent the real geometry sufficiently. This implies that the proposed planning methods relying on this geometric model should be revised concerning their underlying valve geometry description.

1 Introduction

The dilatation of the aortic root is one of the major causes of aortic valve insufficiency [1]. In this case, the valve diameter is increased through an enlargement of the root, which leads to the insufficiency because the leaflets cannot adjoin each other. There are a few possible treatments, whereas a valve replacement is the standard even though this method suffers from the risk of different complications such as thromboembolism, endocarditis, and anticoagulant-related hemorrhage [2]. Another promising treatment is the valve-sparing aortic root replacement, in which a prosthesis is attached to the intact leaflets of the patient to remodel the dilated root. One major challenge is the choice of the right diameter for the prosthesis, since it critically influences the closure of the valve and thus the long-term success of the surgical procedure [3]. This is especially problematic because the aortic root is not under pressure during surgery and therefore the structure and original diameter of the valve is inaccessible. One possible way to tackle this problem was introduced in [4]. Here, 3D ultrasound pictures of the aortic root and valve are produced with a transesophageal echocardiogram (TEE) and the data is measured. A program uses four landmarks, which are manually selected out of the 3D data, to predict the individual prosthesis size for the aortic root. In these previously published methods for planning valve-sparing aortic root reconstruction, the coaptation lines, i.e. the areas where the leaflets collide, are approximated as lines [4]. Hence, the whole valve geometry (see Fig.1) can be described by three commissure points (v_1, v_2, v_3) and the coaptation point (v_4).

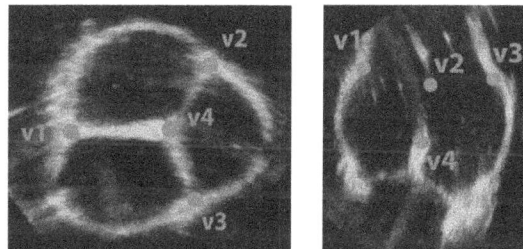

(a) Horizontal slice of the aortic root in the coaptation plane. The dots show the annotated landmarks projected on the displayed slice.

(b) Vertical slice of the aortic root. The dots show the annotated landmarks projected on the displayed slice.

Figure 1: Slices through the aortic root.

The surgery planning routine is based on these four landmarks that are identified manually in 3D ultrasound images of the aortic root. However, different studies showed a high inter-observer-variability by the manual measurement of the landmarks. In this paper a way to refine the manual selected landmarks is introduced. The basic idea is a model-based search for morphological structures in the area in between the landmarks to refine the landmark positions.

2 Material and Methods

The starting point for the program are the 3D ultrasound images and data which are taken with TEE ex-vivo and under physiologically realistic conditions (as seen in [5]). The 3D data is converted into a 318 x 318 x 217-matrix and used to create a point cloud (with threshold segmentation) which is necessary for further segmentation steps. In this work, 3D ultrasound volumes of 10 aortic roots were examined. The landmarks were already manually identified by experts with a technical background. The first step is the segmentation of the aortic root with the circle hough transformation (see section 2.1). This is possible because the aortic root can be approximated by a cylinder. In the same way the coaptions are approximated by lines which are detected with the hough transform. After that the intersection between the three lines and the circles are determined to refine the landmarks which are used in [4]. As these steps depend on the understanding of the hough transform a short introduction into this topic will be given first.

2.1 Hough Transform

In this section the ideas of the hough transform will be explained in short, as the following steps are based on it. The hough transform is a feature extraction technique used in computer vision and was proposed in 1972 by Richard Duda and Peter Hart [6]. Its aim is to find objects within a certain class of shapes by a voting process. The simplest application of the hough transform is the detection of straight lines in images. For reasons of clarity, the following description is for the 2D case, while the method is similarly applicable in higher dimensional spaces as well. An example is presented in Fig. 2. The input image with a straight line is pictured on the top left (note: the origin of the image lies in the top left corner). In general a straight line $y = mx + b$ can be represented as a point in the parameter space (m, b). Vertical lines pose a problem, because they give rise to unbound values of the parameter m. In [7] Duda and Hart proposed the use of the Hesse normal form

$$\rho = x \, cos\theta + y \, sin\theta \qquad (1)$$

to overcome this problem, where ρ is the distance from the origin to the closest point on the line and θ the angle between the x-axis and the line connecting the origin with the closest point of the line (illustrated in Fig. 2 at the bottom). Thus, each line can be represented in the parameter space or hough space (ρ, θ) by their corresponding tuple value. The hough space for the input image can be seen in Fig. 2 on the right. The algorithm looks at each pixel of the input image. If there is enough evidence that this pixel is part of a straight line it calculates the parameters (ρ, θ) and increments the corresponding bin in the hough space. After the examination of every pixel the voting process is completed and the element with the highest value in the hough space indicates the straight line that is most represented in the input image.

By changing the dimensionality of the hough space, it is also possible to detect circles. In this case, the hough space is three dimensional (a, b, r), containing the coordinates of the center point and the radius.

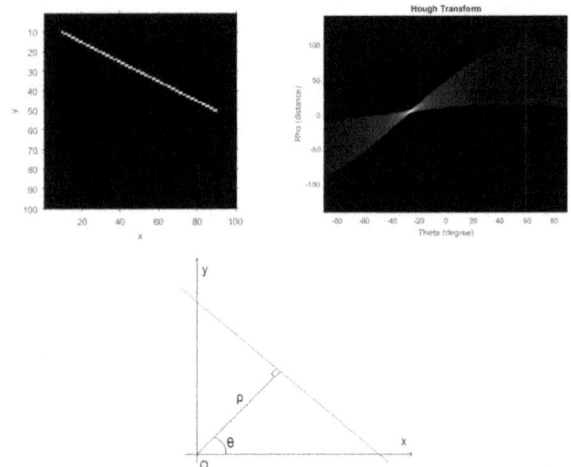

Figure 2: Principle of the hough transform. The input image of a line is to the top left and the hough space to the top right. The image at the bottom illustrates how a line can be represented by the angle θ and the distance ρ in hough space.

2.2 Aortic Root Identification

The implementation is separated in three parts. In the first step the aortic root is approximated by circles from the beginning of the root until it merges into the aortic valve. After that the coaptations are approximated as straight lines to optimize the landmarks for each data set which were determined by experts. Finally the intersection and consequently the optimal landmarks between the straight lines and the circles are identified .

For the computation of the circles the circle hough transform is used. The starting point is the first y-plane of the 3D image and each layer is searched for the optimal circle utilizing a Hough transform as described above. As the aortic root doesn't start in the first y-plane some parameters have to be defined to find its beginning. For one the radius of the found circle in a layer has to lie between two specified values. If this condition has been met the most likely circles in the next two layers must have a radius in a certain range to the first one and their centers have to be almost identical. When these conditions are met the beginning of the aortic root is identified. Like this, a smooth shape of the aortic root is provided and the algorithm is stabilized against noise. The subsequent layers are also searched for circles which meet the given terms with respect to the layer directly above the one examined. This process is continued until the circles in two consecutive planes don't meet the conditions. Such a point is reached when the aortic root crosses over into the aortic valve, because the radii of the found circles are way bigger compared to the preceding radii. These layers mark the end of the aortic root and thus the beginning of the aortic valve. The first part of the implementation is finished.

2.3   Coaptation Line Identification

As stated above, the coaptations are approximated as lines, crossing the coaptation point and one commissure point, respectively. Hence, a regular hough transform was used to identify these lines. As the aim of the method is a refinement of the already manually identified landmarks, the search radius was set up around the lines defined by the landmarks. The goal is to optimize these landmarks with the hough transform. The three lines are between the points $\vec{v_1}$ and $\vec{v_4}$, $\vec{v_2}$ and $\vec{v_4}$, $\vec{v_3}$ and $\vec{v_4}$. For each line a separate search is initiated, but the crucial point is to improve the chosen landmarks. To archive this the starting point $\vec{v_1}$ and the end point $\vec{v_4}$ of the line are constantly changed. This change occurs around each point in every direction with a chosen maximal deviation. For illustration the case of the line $\vec{v_1}$ and $\vec{v_4}$ is used: The new point $\vec{q_1}$ is the difference of $\vec{v_1}$ and $\Delta \vec{v_1} = (\Delta v_{1x}, \Delta v_{1y}, \Delta v_{1z})^T$, which is the deviation vector. The values of the elements of the vector are $-d \leq v \leq d$, where d is the predefined maximal deviation. The same procedure is applied for $\vec{v_4}$, which leads to six variable values and thus a six dimensional hough matrix. Each possible combination for the six values are tested and finally the most likely line is chosen by the hough transform. Therefore the straight line between the two points is defined as

$$\vec{x} = \vec{v} + s \cdot ||\vec{u}|| = \vec{v_1} + s \cdot ||(\vec{v_4} - \vec{v_1})|| = \\ (\vec{v_1} + \Delta\vec{v_1}) + s \cdot ||(\vec{v_4} + \Delta\vec{v_4}) - (\vec{v_1} + \Delta\vec{v_1})|| \quad (2)$$

where s is a scalar. After the three lines are calculated the four landmarks are also updated and improved. Now they should be located closer to their real position.

2.4   Landmark Refinement

To optimize the points further the intersection between the calculated circles (Step 1) and the three straight lines are computed. This is done by finding the closest point on one of the circles in respect to each line. After this last step the landmarks can be used in other programs to calculate the diameter of the healthy aortic root for further steps like an operation.

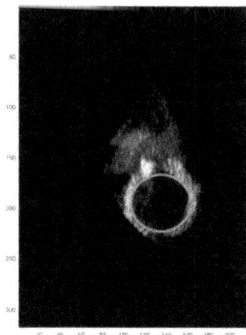

Figure 3: Example of the circle hough transform with an optimal circle on a plane inside of the aortic root.

3   Results and Discussion

The program was carried out for ten different data sets and to evaluate its functionality the line between two landmarks and the deviation of the manually selected points is computed.

3.1   Approximation of the aortic root

In the first step the aortic root was approximated. Fig. 4 shows the different circles for each plane and the resulting structure whereas Fig. 3 shows the optimal circle for one plane. The position of the center and the radii varies between the optimal circles for each plane to match the structure of the aortic root. The implementation identified the beginning of the aortic root which corresponds to the first circle in Fig. 4. The transition of the aortic root into the aortic valve can be seen by the slightly bigger circles in the middle of the image which is the case because of the anatomical structures.

All in all the data shows that it is possible to find a good approximation of the aortic root and part of the aortic valve as well as to find specific planes and structures with the use of the circle hough transformation.

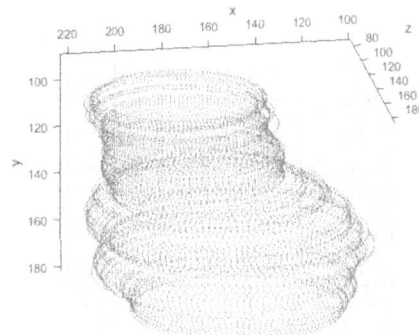

Figure 4: The approximated aortic root with different circles for each plane. The broader circles in the middle signify the slow transition of the aortic root into the aortic valve.

3.2   Approximation of the coaptations and refinement of the landmarks

The implementation for the second step, the detection of a straight line in three dimensional space and the refinement of the start and end point of the line was tested on simple examples (straight lines in three dimensions). These tests yielded results which confirmed the functionality of the implementation.

The refinement of the landmarks were performed on ten different data sets with a deviation of ± 7 pixels for every coordinate of the two points. The results can be seen in Fig. 5 in form of a boxplot. The y-axis indicates the deviation and the x-axis the specific coordinate. The first three are part of the starting point $\vec{v_1} = (v_{1x}, v_{1y}, v_{1z})^T$ and the last three the second point $\vec{v_4} = (v_{4x}, v_{4y}, v_{4z})^T$ of the straight line.

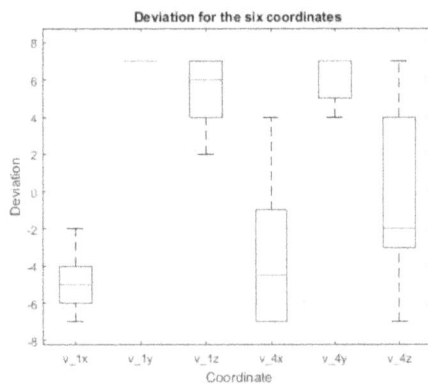

Figure 5: Boxplot of the refinement for each of the six coordinates. The first three are the deviations for the point $\vec{v_1}$ and the last three for the point $\vec{v_4}$. The maximal deviation was ± 7.

Fig. 5 and the deviation for each coordinate indicate a clear direction. Whereas the values for point $\vec{v_4}$ vary heavily for each data set indicated by long boxes, the deviation for the vector $\vec{v_1}$ shows a tendency in the same direction for each data set and therefore a low standard deviation. The coordinate $\vec{v_{1y}}$ is the most obvious case as the deviation for every data set is 7 (i.e. only one line in the plot), the maximal possible change for a single coordinate due to the settings. This means the implementation wants to shift the point and therefore the line in positive y-direction or down in Fig. 4 and thus closer to the aortic valve, which is illustrated in Fig. 6. The same is true for the second point $\vec{v_4}$ and the coordinate v_{4y}, only not as distinct as in the other case. This points to a wrong assumption in the basic model of the aortic valve. The approximation of the coaptations as a straight line doesn't seem to hold, as the results clearly point to with the adjustment in positive y-direction. This means that a linear approximation for this problem is not possible, at least not with the needed accuracy. A quadratic approximation for the coaptations and thus a curve between the two landmarks should yield better results. This would be also more accurate to the anatomical structure of the aortic valve. As the used model doesn't apply the last step to search for intersections between the circles and lines was not performed because the results would be to error-prone. First a new model has to be found and tested.

4 Conclusion

We proposed a model-based method to refine manually annotated landmarks in ultrasound images of the aortic root. A first qualitative analysis indicates that this automatic refinement could be possible. Furthermore, a quantitative analysis of the refinement revealed inaccuracies in the selected geometric model. Our results indicate that a further analysis of the geometric model should be subject of future work on aortic valve surgery planning tools.

Acknowledgement

The work has been carried out at the Institute for Robotics and Cognitive Systems, University of Lübeck.

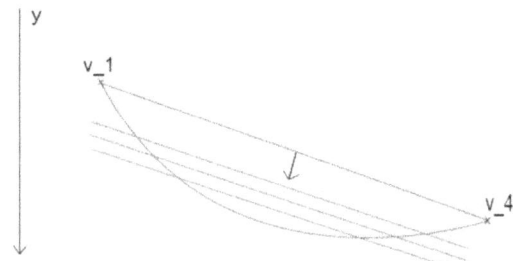

Figure 6: Scheme of the line shift done by the implementation. The top most line connects the manual selected points $\vec{v_1}$ and $\vec{v_4}$. The curve represents the actual anatomic structure of the coaptations. The algorithm tries to find an optimal approximation for this curve and therefore shifts the line and the two points in positive y-direction. This is illustrated by the three straight lines around the curve

5 References

[1] B. Iung, G. Baron, E. G. Butchart, F. Delahaye, C. Gohlke-Bärwolf , O. W. Levang, P. Tornos, J.L. Vanoverschelde, F. Vermeer, E. Boersma, P. Ravaud, A. Vahanian.*A prospective survey of patients with valvular heart disease in Europe: the Euro Heart Survey on valvular heart disease*. Eur Heart J 2003; 24:1231–43.

[2] J. Canádyová, A. Mokráček, R. Bush, *Aortic Valve Sparing Operation*, INTECH Open Access Publisher, 2011

[3] M. Scharfschwerdt et al., *Impact of progressive sinotubular junction dilatation on valve competence of the 3F Aortic and Sorin Solo stent-less bioprosthetic heart valves*, European Journal of Cardio-Thoracic Surgery, Oxford University Press, vol. 37, no. 3, pp. 631-634, 2010

[4] J. Hagenah, M. Scharfschwerdt, A. Schweikard, C. Metzner.*Combining Deformation Modeling and Machine Learning for Personalized Prosthesis Size Prediction in Valve-Sparing Aortic Root Reconstruction*. International Conference on Functional Imaging and Modeling of the Heart, pp. 461-470, 2017

[5] J. Hagenah, M. Scharfschwerdt, B. Stender, S. Ott, R. Friedl, H.H. Sievers, A. Schlaefer. *A setup for ultrasound based assessment of the aortic root geometry*. Biomedical Engineering/Biomedizinische Technik, 2013

[6] Duda, Richard O. and Hart, Peter E. , *Use of the Hough Transformation to Detect Lines and Curves in Pictures*, Comm. ACM, Vol. 15, pp. 11–15, 1972

[7] Duda, Richard O. and Hart, Peter E. , *Use of the Hough Transformation to Detect Lines and Curves in Pictures*. Artificial Intelligence Center. SRI International, 1971

Low-Rank Mask R-CNN

M. Sambale [1,2], M. Heinrich [3] D. Sciretti [2], and N. Trujillo [2],

[1] Medical Informatics, Universität zu Lübeck, mauriceraphael.sambale@student.uni-luebeck.de

[2] DENSO ADAS Engineering Services GmbH, Lindau, {n.trujillo,d.sciretti,m.sambale}@denso-adas.de

[3] Institute of Medical Informatics, Universität zu Lübeck, heinrich@imi.uni-luebeck.de

Abstract

This work provides a light weight version of the Mask R-CNN introduced by He et al. in 2017. This lighter model improves the run time on the same public dataset by 17%. Mask R-CNN is a state of the art neural network for multi object instance segmentation problems. It incorporates detection, classification and segmentation subnetworks. This high performance comes at the cost of high model complexity with 61 million trainable parameters. The light weight Mask R-CNN network reduces the computational cost using only 36 million parameters by applying low-rank approximation (LRA). This reduces the number of parameters greatly and preserves weights trained beforehand with only a precision loss of 5 percent points on the coco *minval* dataset. In the experiments the algorithm was compared with a preservation of 70%, 80%, 90% accuracy to the original Mask R-CNN and analyze the run time and computational cost.

1 Introduction

Instance segmentation is nowadays a common task in the field of computer vision. For cars, airplanes, robots and any agent that uses vision sensors, it is the first step to autonomous decision making and interaction with its environment. In the medical context it can be applied for interactive health-care agents as well as medical imaging, e.g. aid in histological examination, where there is the need to identify multiple instances of different cell types. Another example is lesion segmentation, where first attempts have been made by Rezaei et al. [1]. But especially in the medical sector the input sizes are very large, due to 3D data or very high resolution images. So, for applying the technology to these areas, additional requirements have to be met, such as computational cost and memory constrains. This work is tackling those problems, by lowering the computational cost on the use case of Mask R-CNN, a state of the art CNN, and speeding up its detection time.

Mask R-CNN by Kaiming He et al. was published in April 2017 and is the most promising instance segmentation networks of that time outperforming the winners of the COCO challenges 2016, which include bounding-box object detection, instance segmentation and key-point detection [2]. One of the drawbacks of this technology however is the size of the network during deployment time. For an image of 1024×1024 pixels the network needs 8GB of RAM for the detection in the tensorflow implementation by matterport [3]. This limits Mask R-CNN to high end GPUs. Furthermore, even though since the declared detection time of 5 fps is already impressive it is not enough for usage on larger input sizes.

The goal of this work is to reduce the computational cost and time consumption.

In the following an overview of the current state of the art in instance segmentation and optimization is given. Section 2 introduces Mask R-CNN and explains the approach. In section 3 experiments are presented, followed by a conclusion in Section 4.

1.1 Instance Segmentation

Multiple instance segmentation incorporates the tasks of detection and segmentation. One approach is to use a detector in a chain with a segmentation network, such as a previous explored FusionSeg network [4] together with Faster R-CNN [5]. A second approach combines the task in one network, such as Mask R-CNN [2].

1.2 Optimization

Since neural networks tend to get larger and more complex, their transformation to more efficient and smaller networks was explored. Two general categories of approaches are followed in that case. Either small architectures are used from the beginning, or large trained models are compressed in a second step. The latter can be done using low-rank approximation (LRA) [6, 7] Even though LRA is well known in the computer vision community, to the best of our knowledge, we are the first to apply it to Mask R-CNN.

2 Method

Since this work relies on Mask R-CNN the architecture is recapitulated first, followed by an explanation how the network was optimized.

2.1 Mask R-CNN

Mask R-CNN by Kaiming He et al. is a multiple instance segmentation and classification network that takes an image of arbitrary size as input. The output consists of a bounding box, a classification label and a mask for each detected object in the image. It is based on Faster R-CNN [5] but extends the idea by adding the mask branch to the network. The network can be divided in two parts: the backbone for feature extraction containing a Region Proposal Network (RPN) and two utility layers, and the network head with the classification and mask branch, as shown in Figure 1. The backbone network used by the authors is ResNet101 [8] without the classifier head. Instead a pyramid pooling at each stage, i.e. before downsampling, is applied, giving a total of 5 feature maps in different resolution stages. Next each stage is upsampled by a factor of two and added to the feature activation of the previous stage. Additionally a sixth feature map is created by maxpooling stage five. These six feature maps are now input for the RPN introduced in Faster R-CNN. The outputs of this layer are ROIs, i.e. bounding boxes, and the corresponding class predictions in the context of the different pyramid stages. The following proposal layer rearranges the outputs such that for each pixel bounding boxes of the different stages are associated. These ROIs are input for the detection layer which, in the training phase, learns rescaling and slight translation to the nearest ground truth ROI as well as cropping of wrongly proposed and duplicate bounding boxes. Its output are therefore the refined ROIs. During training it also passes belonging ground truth data. [2, 3]

The network head contains the classifying and segmentation branch. The input of both branches are the features from the pyramid pooling and the ROIs from the detection layer. First step for both branches is the processing by the ROIAlign layer which does a more precise interpolation of the features to the sub-pixel input bounding boxes.

The classifier branch passes the input through a convolution and two fully connected layers and predicts class probabilities and bounding boxes. The segmentation branch contains five convolutions and one deconvolution and gives back a binary class-independent segmentation of the size of the input ROI, saving space this way.

Notable is that the class and mask loss are decoupled, which leads to better results according to the experiments provided by Mask R-CNN. Specific parameter numbers and timings are provided in section 3. [2, 3]

2.2 Optimization through LRA

In this work a reduction of the computational cost is achieved by channel reduction in all convolutional layers of the network. It is assumed that a neural network often contains redundant model parameters that are not necessary and which can be reduced while preserving similar classification accuracy.

To remove those additional parameters a low-rank approximation (LRA) is performed on each convolutional layer of Mask R-CNN to optimize correlated weights and non-

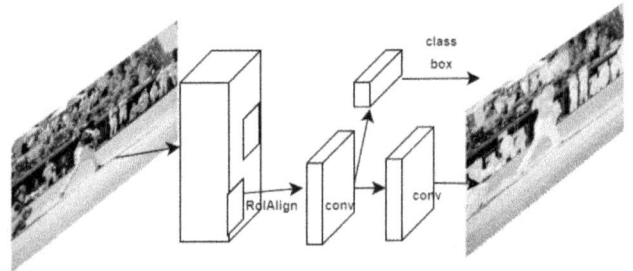

Figure 1: Mask R-CNN framework for instance segmentation [2]

informative channels. Next the outcome is sorted by information content and cut all channels not needed for 70%, 80%, 90% of information preservation. To fit the newly formed layers, the original convolutional layers are replaced by the linear combination of their truncated version and another convolution reassigning the original number of channels. This is done by following the spirit of [7] splitting $[N \times C \times M_1 \times M_2]$ convolutions into $[K \times C \times M_1 \times M_2]$ and $[N \times K \times 1 \times 1]$ convolutions with $K << C$. To be able to reuse already trained models, a LRA of the convolutions is used to estimate starting weights for the replaced layers. The Principal Component Analysis (PCA) and Singular Value Decomposition (SVD) to get a truncated convolution were compared.

The PCA relies on computing the eigenvalues and eigenvectors of the reshaped and zero centered weight matrix X $[CM_1M_2 \times N]$, where N is the number of filters, C is the number of input channels, M_1M_2 the kernel size. On the other hand the SVD computes the singular values and vectors of X.

The PCA is computed using the covariance matrix Σ which is the product of $X^T X$. Since this is symmetric, one can diagonalize the matrix and normalize the eigenvectors such that they are orthonormal, with:

$$\Sigma = WDW^T, \qquad (1)$$

where D is the diagonal matrix with eigenvalues and W is a matrix containing the eigenvectors. Thereafter the eigenvalues are used to compute the rank ratio which represents the information preservation and dictates how many eigenvectors are used in W_{trunc}. The new layer weights are the principal components given by XW_{trunc}.[9, 10]

In contrast, SVD uses that real matrix X can decomposed:

$$X = USV^T \qquad (2)$$

with U $[N \times N]$ containing the left singular vectors in its columns, V $[CM_1M_2 \times CM_1M_2]$ containing the right singular vectors in its columns and S containing the singular values. The correspondence $\Sigma = VDV^T$ is further shown in [9], with $D = S^2$ containing the squared singular values. Therefore the eigenvectors are equal to V and the eigenvalues are equal to D. [9]

It has to be noted that the SVD is more stable numerically, especially given matrices containing very small values, as it is the case with weights of neural networks. This can explain why the SVD needs less eigenvectors for the same information preservation.

2.3 Implementation Details

So far the authors of Mask R-CNN have not publish their code. Instead an implementation by an unassociated organization called "matterport" was used, which can be found here: [3]. Note that with this implementation a reproduction of the highly accurate results of the original paper was not possible, see table 1.

As a baseline the pretrained weights provided in this repository were used. A caffe framework implementation from [6] was used for the LRA and re-implemented for tensorflow making use of the same underlying libraries for the PCA and SVD. To regain lost accuracy the new network was fine-tuned on $MSCOCO_{2014}$ for 40k epochs, 2000 iterations per epoch and a batchsize of 2. It therefore results in fewer parameters for the same information content preservation.

3 Experiments

The approach was in three different versions. 70%, 80% and 90% of the remaining information after LRA were selected as test networks, which will be referred to as "L90", "L80", "L70" afterwards. For all models ResNet101 was used as backbone.

The datasets for fine tuning were the same sets used by Kaiming et al. : $MSCOCO_{2014}$ [11]. It is a challenging dataset differentiating 80 classes and multiple instances. It contains 117k images for instance segmentation with provided ground truth masks and labels for each instance. During the evaluation the "minival" set of Coco was used which contains a total of 4952 images.

As evaluation a part of the standard COCO evaluation metrics average precision (AP) was used. Average precision is computed by averaging the thresholded outcome of the intersection over union (IoU) with the ground truth for all instances and classes, e.g. AP_{50} means all bounding boxes with an IoU higher then 50% are true positives.

The number of network parameters was calculated by summing up the number of elements in all weight matrices of the convolutional layers. Parameters trained by other layers of the network were left out, because they were not targeted by the optimization and are a small percentage of the whole network size. The models was fine-tuned for 160K iterations on images of size 1024×1024 pixels with a learning rate of 0.001, a momentum of 0.9 and a weight decay of 0.0001. The baseline were pretrained weights of ResNet101 and all additional parameters correspond to the parameters used at the second release of [3]. Table 1 is showing the results for the bound box AP as well as performance metrics. Analysing Table 1 it becomes obvious that the resulting precision is on overall lower than expected. During training the

learning behavior was controlled via tensorboard and verified the convergence of the network after a given number of iterations. Then the visual performance was inspected. Results of L70, $original_{ft}$ and the ground truth are shown in Figures 2, 3 and 4. Here the problems can be seen. On the one hand the network shows false negatives, but produces false positive detections as well, which increase with LRA. On the other hand the ground truth images do not seem to cover all possible ground truth labels neither. Due to time constraint further experiments, like fine tuning with different hyper parameters such as smaller learning rate, was not attempted.

The L70 model compared with the $original_{ft}$ network achieved a speedup of $\times 1.17$ with a precision loss of 3% over mean of ten AP thresholds between 0.5 and 0.95. It also reduces the parameters of the convolutional layers by 40%. L90 is just 2% more precise at most and much slower then L70. Therefore the trade off is not good enough comparing speed and accuracy loss for this model. Nevertheless, in general there has to be an application-depended trade off between accuracy and model speed up.

SVD is needs less filters for the same initial information preservation. So it requires fewer parameters for fine tuning the network. The results show that this is a disadvantage since the performance is lower in those cases, even though the speed up is slightly better. In fact comparing PCA and SVD models with a similar amount of parameters, it seems that after fine tuning there is no preferable method. Rather a correlation between precision and number of parameters is detectable. Hence an analysis of the network performance without fine tuning is interesting in the future.

4 Conclusion

With the approach of LRA a significant reduction in model complexity by nearly half and computation time by nearly one image per second was achieved while only loosing 3% accuracy. Comparing different levels of information preservation, it was possible to show that less than 90% preservation is necessary to gain speed and hence making a trade off considerable. In the area of autonomous agents the ultimate goal for time optimization would be to reach real time performance, i.e. running on at least 25 fps. This research has shown, that there are additional steps necessary to succeed concerning computational complexity. Using this network on off-line prediction tasks such as segmentation of medical data, a speed up of $\times 1.17$ means a difference of an hour on 100 000 images.

4.1 Future Work

Research regarding the exchange of parts of the network, e.g. replacing ResNet101 with ResNeXt or another model will be tackled next. Using lighter networks before compression will already make the whole network smaller and is going to be explored in the next months.

Another topic is the reduction of memory usage, since Mask

Table 1: Evaluation of LRA on the bounding box AP metric. Mask R-CNN refers to the results provided in [2] while $original_{ft}$ stands for the results of the fine tuned unmodified implementation of [3].

@[IoU	area	maxDets]	Mask R-CNN	$original_{ft}$	fine tuned on Coco 160k iterations					
			L70	L70svd	L80	L80svd	L90	L90svd
AP@[0.50:0.95	all	100]	0.371	0.223	0.192	0.142	0.189	0.180	0.207	0.199
AP@[0.50	all	100]	0.600	0.349	0.323	0.250	0.322	0.310	0.345	0.337
AP@[0.75	all	100]	0.394	0.248	0.206	0.150	0.201	0.190	0.221	0.213
avg. time per Image (sec)	0.2	0.2432	0.2082	0.1977	0.2132	0.2054	0.2437	0.2163
network parameters	-	$61E6$	$37E6$	$29E6$	$42.7E6$	$34.3E6$	$50.9E6$	$43.4E6$

Figure 2: L70 segmentation   Figure 3: $original_{ft}$ segmentation   Figure 4: Ground Truth

R-CNN requires 8GB of RAM during deployment. One approach could be quantization of the network weights, since works in [12] show promising results on that topic.

Acknowledgement

The work has been carried out at DENSO ADAS Engineering Services GmbH, Kemptener Straße 99, D-88131 Lindau and supervised by the Institute of Medical Informatics, Universität zu Lübeck.

5 References

[1] M. Rezaei, H. Yang, and C. Meinel, "Deep learning for medical image analysis," *arXiv preprint arXiv:1708.08987*, 2017.

[2] K. He, G. Gkioxari, P. Dollár, and R. Girshick, "Mask r-cnn," in *Computer Vision (ICCV), 2017 IEEE International Conference on*, pp. 2980–2988, IEEE, 2017.

[3] I. Matterport, "Mask r-cnn implementation by matterport on github." https://github.com/matterport/Mask_RCNN, 12.01.2017.

[4] S. D. Jain, B. Xiong, and K. Grauman, "Fusionseg: Learning to combine motion and appearance for fully automatic segmention of generic objects in videos,"

[5] S. Ren, K. He, R. Girshick, and J. Sun, "Faster r-cnn: Towards real-time object detection with region proposal networks," in *Advances in neural information processing systems*, pp. 91–99, 2015.

[6] W. Wen, C. Xu, C. Wu, Y. Wang, Y. Chen, and H. Li, "Coordinating filters for faster deep neural networks,"

[7] A. G. Howard, M. Zhu, B. Chen, D. Kalenichenko, W. Wang, T. Weyand, M. Andreetto, and H. Adam, "Mobilenets: Efficient convolutional neural networks for mobile vision applications," *arXiv preprint arXiv:1704.04861*, 2017.

[8] K. He, X. Zhang, S. Ren, and J. Sun, "Deep residual learning for image recognition," in *Proceedings of the IEEE conference on computer vision and pattern recognition*, pp. 770–778, 2016.

[9] K. P. Murphy, *Machine learning: a probabilistic perspective*. MIT press, 2012. Chapter 12.1, 12.2.

[10] L. I. Smith *et al.*, "A tutorial on principal components analysis," *Cornell University, USA*, vol. 51, no. 52, p. 65, 2002.

[11] T.-Y. Lin, M. Maire, S. Belongie, J. Hays, P. Perona, D. Ramanan, P. Dollár, and C. L. Zitnick, "Microsoft coco: Common objects in context," in *http://cocodataset.org, European conference on computer vision*, pp. 740–755, Springer, 2014.

[12] S. Han, H. Mao, and W. J. Dally, "Deep compression: Compressing deep neural networks with pruning, trained quantization and huffman coding," *arXiv preprint arXiv:1510.00149*, 2015.

Multiple Landmark Localization in medical CT Scans using Deep Neural Networks with Heatmap Regression

J. Wessel [1] and M. P. Heinrich [2]

[1] Medizinische Ingenieurwissenschaft, Universität zu Lübeck, joeran.wessel@student.uni-luebeck.de

[2] Institute of Medical Informatics, Universität zu Lübeck, heinrich@imi.uni-luebeck.de

Abstract

This paper describes the localization of multiple medical landmarks using deep convolutional neural networks. We therefore extend the state-of-the-art CNNs architectures UNet and SegNet for image segmentation to a regression task. Our network enables high precision and captures context about the organ arrangement. Based of the idea detecting landmarks with heatmap regression, we developed a 3D network, which can be trained end-to-end. Our results show, that segmentation networks can be straightforwardly adapted to landmark regressions. Based on experiments using the VISCERAL dataset with a resolution of 7.2x7.2x11.5 mm, our network achieved an excellent localization error on average 9.45 mm for 20 different anatomical landmarks on thorax abdomen CT scans with minimal values of 6.59 mm. Furthermore this paper shows that the UNet architecture predicts landmarks with heatmap regression better than the SegNet architecture.

1 Introduction

The visual system enables humans to recognize objects quickly and accurately. This allows us to perform complex tasks like driving cars. Inspired by the human visual system, deep convolutional neural networks were developed in machine learning [1]. These neural networks are able to automatically learn to extract meaningful visual features by weak supervision in contrast to classic machine learning algorithms (such as random forests) that rely on handcrafted filters. Meanwhile, deep neural network approaches set astonishing new performances for many computer vision tasks. By increasing the number of weights that need to be optimized, the network model complexity overcomes increasingly complex problems, e.g. self-driving vehicles or achieving an equal level of diagnostic accuracies as radiologists or pathologists. However, this requires a large amount of data to prevent overfitting. Particularly, acquiring medical data is challenging, and therefore this limits the model complexity for medical landmark detection. Furthermore, working with 3D data increases enormously the required number of weights in the network to be learned and the filters have to be extended by one dimension.

In this paper we combine the two state-of-the-art models SegNet [5] and UNet [4] for segmentations. We apply this combined model to medical 3D CT scans to extract the landmarks with just one network. The localization of anatomical landmarks is an important task in automatic analysis and provides an initial position for many image registration and segmentation algorithms.

Based on the approach of Pfister et al. [3] to learn landmarks using heatmaps instead of direct coordinates, we evaluate different fully convolutional deep networks.

1.1 Related Work

In the paper by Mader et al. [9], anatomical landmarks of finger epiphyses are determined with a regression tree ensemble and a conditional random field. In a first step, each landmark localization was determined by an ensemble of decision tree regressors based on the local environment. In a second step, the positions of the landmark were encoded with a conditional random field, modeling spatial relations. Pfister et al. [3] introduced landmark detection using heatmap regression for pose estimation. They found that landmark detection using heatmap regression is easier to learn than landmark coordinates directly.

In the work of Payer et al. [2] deep convolutional neural networks have already been used for multiple landmarks localization in medical data. The idea of representing each individual landmark by a dense heatmap was also used. In their work, landmarks were determined in X-ray images (2D and 3D) of the hand in an end-to-end manner.

2 Material and Methods

2.1 Heatmap Regression

As described in [2] and [3], we also use heatmaps to determine multiple landmarks directly from the input images using a neural network. We predict a landmark for each organ. These landmarks represent the organ centroid, determined from given segmentations. Each landmark is represented

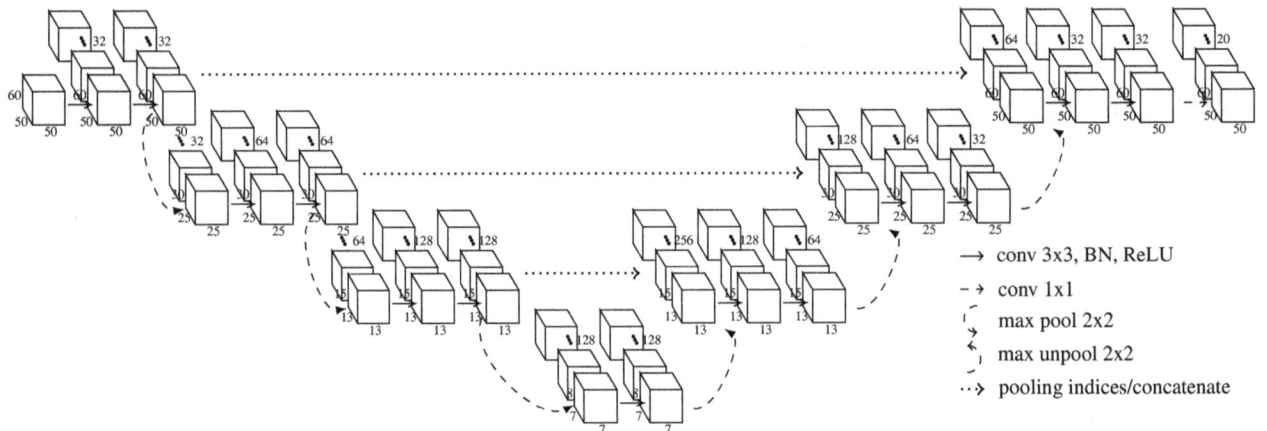

Figure 1: Schematic representation of the 3D USegNet architecture. The numbers at the volumes specify the dimensions of the data, the numbers between the volumes specify the channels. *Right Arrow:* Convolutional layer with 3x3 kernel size, batch normalization and ReLU activation. *Dashed Arrow:* Convolutional layer with 1x1 kernel size. *Dashed curved arrow down:* Max pooling layer with 2x2 kernel size. *Dashed curved arrow up:* Max unpooling layer with 2x2 kernel size. *Dotted Arrow:* Return pooling indices for the max unpooling layer and concatenate extracted features with the upsampled features.

in a 3D volume according to the image size by a Gaussian smoothed dot. To determine the predicted landmark position, we use the maximum response of each predicted heatmap.

2.2 Network Design

Fig. 1 shows our network architecture, which combines the UNet [4] and the SegNet [5]. The model is a fully convolutional neural network. The USegNet consists of 14 convolutional layers and therefore it is four layers smaller than the origin UNet or SegNet. This reduction ensures less memory requirement and a faster computation with approximately the same accuracy. With exception of the last convolutional layer, the network uses only 3x3 kernel sizes with padding. Each convolutional layer is followed by a batch normalization and ReLU activation. After every second convolutional layer, the features are downsampled with a max pooling layer with 2x2 kernel size and upsampled with a max unpooling layer according to the SegNet. In addition, the network contains the concatenate method of the UNet. The last convolutional layer with a 1x1 kernel size reduces the channel size to the number of landmarks to be determined.

By combining the two networks, the network gains a lot of information about the arrangement in the picture. The network can also cover large areas of the image with only small kernel sizes while maintaining high accuracy.

We have implemented three more models. Each of them is a modified version of our USegNet. The second model corresponds to the network in Fig. 1 without the concatenate method and is thus a reduced SegNet. The third model corresponds to the UNet, which upsamples without a max unpooling layer and pooling indices. The last model SequentialNet is the most simple network architecture without any skip connections and thus neither

concatenate nor unpooling.

The network downsamples the image data three times, so the smallest volume size is 7x7x8 pixels. After upsampling, the final output size of the network is 50x50x60x20 pixels.

2.3 Training

The network has been trained on the VIS-CERAL/Silvercorpus dataset [6]. From this dataset, the 64 CT scans from the thorax abdomen were used. We used anatomical structures to generate our ground truth 3D landmarks positions. The original average size of the volume is 512x512x420 pixels with a resolution of approximately 0.73x0.73x1.5 mm. Due to the very high memory requirements for 3D convolutional neural networks (CNNs), we downsampled the dataset to a size of 50x50x60 pixels. The resulting resolution is approximately 7.2x7.2x11.5 mm.

To avoid overfitting during training we performed a random affine transformation with scaling, shearing and translation. The data augmentation expands the 64 scans to 256 scans. The volumes were then split into 224 training data and 32 test data.

We evaluated the 3D dataset with eightfold cross-validation. In addition, we compared the performance of our network with the previously presented networks. All networks were trained with the same dataset and cross-validation. The networks were built from scratch using the Pytorch framework and not pretrained with existing image databases.

To optimize the parameters, the Adam optimizer and the mean squared error loss between predicted and ground truth heatmap were used. The learning rate was 0.001 and the network was trained over 40 epochs at a constant learning rate and a batch size of four. The bias were initialized with zero and the weights with the Xavier initialization. Using a NVIDIA GeForce GTX 1050 with 4096 MB RAM, the training time was approximately 35 min and the evaluation time of one test volume was less than 0.5 s.

3 Results and Discussion

Fig. 2 contains a 2D slice from a volume of the test dataset with the plotted landmarks from the surrounding frames. As shown in Fig. 2, the ground truth and predicted landmarks were detected close to each other or in the same pixel. It should be noted that the shown positions are depending on the x- and y-coordinates and can deviate in the z-direction. The resolution of the data is so much reduced that fine structures like the bronchial tubes are difficult to recognize. Despite this resolution, it is possible for our network to detect landmarks of small organs, e.g. the adrenal glands. The adrenal glands are partially represented in the 3D data only by a single pixel. This shows that the network not only learns information from the intensity values of the organs but also allocates organs to the environmental context.

Table 1 contains the localization errors from the different networks over all 20 landmarks. The first column lists the 20 organs corresponding to the landmarks. For each landmark, the mean standard deviation in mm is given for each network. These mean values and standard deviations have been averaged over the eightfold cross-validation. In addition, the localization errors of the untrained mean landmarks are given in column two for each organ. Furthermore,

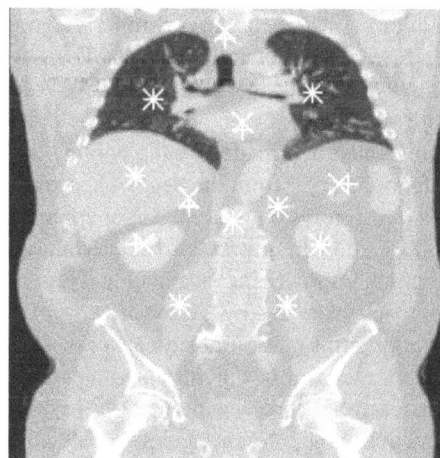

Figure 2: Representation of an upsampled (224x224 pixels) and contrast enhanced slice from a CT volume of the dataset. The '+'-symbols indicate the ground truth positions of the landmarks. The 'x'-symbols indicate the predicted positions of the landmarks. To visualize several landmarks in the 2D image, the landmarks within ±10 frames in anterior posterior direction were plotted.

the mean localization errors for each network is given for all organs. The last column of the table lists the total number of existing ground truth landmarks in the dataset. Our results show that the UNet has the lowest localization

Table 1: Localization errors on 3D dataset containing 256 images (including 3x augmented data) with 20 landmarks, detected with four different networks and the optimal naive prediction without training in image data. Each landmark corresponds to the main focus of one organ. The localization error is given by mean ± standard deviation in millimeters. Each localization error was averaged over the eightfold cross-validation. #LM describes the number of ground truth landmarks for each organ in the original dataset.

Landmark	Untrained (mean) prediction	USegNet	SegNet	UNet	SequentialNet	# LM
Liver	49.22 ± 27.06	9.40 ± 3.18	10.51 ± 3.93	9.11 ± 2.86	9.60 ± 3.28	64
Spleen	59.84 ± 26.85	10.19 ± 6.58	11.30 ± 5.87	9.19 ± 6.49	11.94 ± 6.75	64
Pancreas	56.36 ± 32.24	10.79 ± 5.66	13.07 ± 8.20	10.56 ± 4.46	13.14 ± 9.21	61
Gallbladder	54.47 ± 33.55	9.304 ± 5.60	12.79 ± 10.21	8.83 ± 4.78	11.67 ± 10.65	51
Urinary bladder	62.13 ± 30.68	$11.28 + 9.73$	11.93 ± 10.67	10.55 ± 9.61	11.78 ± 9.96	64
Aorta	43.03 ± 23.72	10.80 ± 4.66	12.36 ± 7.82	10.04 ± 4.09	11.48 ± 7.56	64
Trachea	36.11 ± 22.58	9.27 ± 3.90	9.33 ± 4.69	$8.89 \perp 3.77$	9.99 ± 4.32	63
Lung R	39.57 ± 24.15	9.37 ± 3.26	9.50 ± 3.24	9.12 ± 2.81	9.00 ± 3.01	64
Lung L	46.91 ± 24.33	9.33 ± 3.60	10.01 ± 4.07	8.65 ± 3.37	9.14 ± 3.29	63
Sternum	38.75 ± 22.60	9.35 ± 3.90	10.75 ± 5.17	9.21 ± 3.67	9.95 ± 4.00	64
Thyroid gland	36.13 ± 22.90	11.61 ± 4.80	12.03 ± 5.29	10.38 ± 4.18	11.67 ± 4.66	63
Lumbar vertebra	49.76 ± 24.31	10.09 ± 4.10	11.35 ± 5.45	9.81 ± 3.55	10.80 ± 4.29	64
Kidney R	54.37 ± 27.50	8.97 ± 3.26	9.97 ± 4.40	8.52 ± 3.04	10.73 ± 4.59	64
Kidney L	59.88 ± 26.76	9.90 ± 3.31	10.95 ± 4.32	9.23 ± 4.83	11.97 ± 5.22	64
Adrenal gland R	49.40 ± 31.14	7.26 ± 5.49	9.46 ± 6.81	6.59 ± 5.39	9.67 ± 7.65	48
Adrenal gland L	52.20 ± 30.90	9.66 ± 8.92	12.4 ± 9.86	8.36 ± 7.29	13.35 ± 12.84	50
Psoas major R	52.77 ± 25.77	9.79 ± 3.85	10.27 ± 4.45	9.13 ± 3.53	9.63 ± 4.13	64
Psoas major L	56.19 ± 27.18	9.85 ± 4.04	10.80 ± 4.72	9.70 ± 3.68	10.38 ± 4.10	64
Rectus abdominis R	59.33 ± 35.15	11.80 ± 10.06	17.37 ± 17.85	12.21 ± 13.49	16.17 ± 21.31	64
Rectus abdominis L	60.38 ± 31.38	11.12 ± 5.03	14.94 ± 12.50	10.94 ± 4.82	12.06 ± 6.69	64
Mean	50.84 ± 27.54	9.96 ± 5.15	11.55 ± 6.98	9.45 ± 4.98	11.21 ± 6.88	

error for almost every organ. Similarly good results are achieved by our combined USegNet. The SegNet and the SequentialNet provide worse localization errors than the other two networks. From this, it can be concluded that the additional information provided by the pooling indices and the max unpooling layer from the SegNet approach impairs the localization. It turns out that the UNet approach, which includes information from previously determined features after resampling, improves the localization. Networks without this approach are worse by a error of 1-2 mm.

The localization error of the trained networks deviates significantly from the error of the untrained mean landmarks. This means, each network was trained on the heatmap regression and does not represent random localizations.

Noticeable in the results are the high standard deviations, which vary from 30 - 100 % of the mean values. One reason for that could be the relatively small number of image data and test data for each cross-validation. A few mislocalizations would significantly affect the standard deviation.

The results show that the landmarks of organs with the same size and shape, e.g. the right and left adrenal gland, are determined with different accuracy. In this example, the right adrenal gland is determined up to 3 mm more accurately than the left adrenal gland. Another example are the landmarks for the right and left rectus abdominis. In this case, the right rectus abdominis is learned worse. In addition, the landmark for the rectus abdominis is the worst trained landmark by any network.

The total number of ground truth landmarks in the dataset indicate that landmarks for specific organs, e.g. the gallbladder or the right adrenal gland, are significantly less present. However, this does not appear to affect the localization error because the error of these landmarks is below the mean error of UNet and USegNet.

Considering that the resolution of the image data with 7.2x7.2x11.5 mm is low, the trained networks UNet and USegNet provide excellent results. The landmark positions were determined partially in a subpixel resolution. In the future, it would be a promising direction of work to address the memory limitations while training the 3D networks, which would consequently lead to further improved results when using higher resolution data.

4 Conclusion

We introduced a new network architecture USegNet, a model to detect multiple landmarks. The USegNet uses state-of-the-art segmentation networks by combining the methods of UNet and SegNet. The network predicts landmarks end-to-end using heatmap regression.

We demonstrate that our network can detect landmarks in medical scans. We found that the UNet is better suited for heatmap regression than the SegNet. It also turned out, that the UNet is slightly better than our USegNet. Our results show that the networks predict landmarks in low resolution with a subpixel accuracy using regional anatomical context. In a further work, an additional approach should be used to evaluate higher resolution 3D datasets with the networks.

This could lead to significantly better detections of the landmark positions.

Furthermore, other methods than a heatmap regression could be investigated for landmark detection. For this, the object detection systems of Redmon et al. [7], [8] are interesting approaches. These systems learn the landmark coordinates and also the width, height and the class probability of the objects directly as a regression task.

Acknowledgement

The work has been carried out at the Institute of Medical Informatics, Universität zu Lübeck.

5 References

[1] Y. LeCun, Y. Bengio and G. Hinton, *Deep Learning*. In: Nature 521, pp. 436-444, 2015

[2] C. Payer, D. Stern, H. Bishof and M.Urschler, *Regressing Heatmaps for Multiple Landmark Localization using CNNs*. In: MICCAI 2016, Springer, pp.230-238, 2016

[3] T. Pfister, J. Charles and A. Zisserman, *Flowing ConvNets for Human Pose Estimation in Videos*. In: CoRR 1506.02897, IEEE, 2015

[4] O. Ronneberger, P. Fischer and T. Brox, *U-Net: Convolutional Networks for Biomedical Image Segmentation*. In: MICCAI 2015, Springer, pp. 234-241, 2015

[5] V. Badrinarayanan, A. Kendall and R. Cipolla, *SegNet: A Deep Convolutional Encoder-Decoder Architecture for Image Segmentation*. In: CoRR 1511.00561, IEEE, 2016

[6] O. A. Jimenez-del-Toro, H. Müller, M. Krenn, K. Gruenberg, A. A. Taha et al., *Cloud-based Evaluation of Anatomical Structure Segmentation and Landmark Detection Algorithms: VISCERAL Anatomy Benchmarks*. In: IEEE transactions on medical imaging Vol. 35 (11), IEEE, pp. 2459-2475, 2016

[7] J. Redmon, S. Divvala, R. Girshick and A. Farhadi, *You Only Look Once: Unified, Real-Time Object Detection*.

[8] , CVPR, 2016 J. Redmon and A. Farhadi, *YOLO9000: Better, Faster, Stronger*. In: CoRR 1612.08242, 2016

[9] A. Mader, H. Schramm and C. Meyer, *Efficient Epiphyses Localization Using Regression Tree Ensembles and a Conditional Random Field*. In: Bildverarbeitung für die Medizin 2017, Springer, pp. 179-184, 2017

Performance enhancement of dictionary-based electrical properties tomography

N. Hampe [1,2], U. Katscher [2], and A. Neumann [3],

[1] Medizinische Ingenieurwissenschaft, Universität zu Lübeck, n.hampe@student.uni-luebeck.de

[2] Philips Research Laboratories, Hamburg, {nils.hampe, ulrich.katscher}@philips.com

[3] Institute of Medical Engineering, Universität zu Lübeck, neumann@imt.uni-luebeck.de

Abstract

Electrical properties tomography (EPT) provides clinically useful diagnostic parameters by deriving the electrical properties from standard magnetic resonance imaging data of a patient's tissue. In contrast to conventional EPT, in which numerical differentiation is used, in dictionary-based electrical properties tomography (dbEPT) local patterns are classified by comparing them to a dictionary. Although having several potential advantages over the conventional method the current implementation of dbEPT suffers from high reconstruction times in the range of hours. This paper introduces a method for reducing the dictionary size and thus the reconstruction time by exploiting the dictionary's symmetry. As a result the reconstruction time is successfully reduced by 83% without significant losses in accuracy.

1 Introduction

Electrical properties tomography (EPT) is concerned with mapping the electric conductivity σ and permittivity ϵ of a patient's tissue. Having first been mentioned in 1991 [1] EPT's non invasive character stands out from previous methods. Instead of requiring externally mounted electrodes, currents or radiofrequency probes, the derivation of electrical properties in EPT takes place on a complex image called B1 map, obtainable on a standard magnetic resonance (MR) imaging system. Belonging to the coils producing the radiofrequency (RF) pulse, the B1 map is a measure for the impact of the coil's electromagnetic field on the patient and reciprocally a map of the coil's sensitivity in receive mode [2]. Recently, first clinical benefits of EPT have emerged in specifying the diagnosis of breast tumours [3] and brain tumours [4]. Making use of certain assumptions (explained in 2), the mathematical problem can be reduced to calculating conductivity and permittivity separately from the B1 phase map and the B1 magnitude map respectively. Especially due to the relatively high duration of the magnitude map's measurement, the particular focus lies on the phase based conductivity reconstruction. This calculation raises two major technical issues. First, while the reconstruction formula requires knowledge of the transmit phase map, induced in the probe during the application of the RF pulse, the only quantity accessible to measurement is the combination of transmit and receive ("transceive") phase [2]. For asymmetrical objects these two can differ severely, disturbing the reconstruction in a non-trivial manner. The second challenge arises from the simplification of the central equations underlying the classic EPT algorithm under the assumption of locally constant electrical properties. This assumption's most common violation occurs at tissue boundaries, among others giving rise to the necessity of elaborate segmentation algorithms to exclude tumour borders for sufficiently precise measurements. While efforts have been made to address said challenges at the root, solutions go along with either a lengthier and costlier scan process or substantially higher computational costs due to iterative reconstruction algorithms.

An alternative way of coping with said issues is given by the so called dictionary-based EPT (dbEPT) [5]. Being inspired by MR finger-printing [6], a dictionary based modification of conventional MR imaging, the idea behind dbEPT lies in comparing small local patterns of the B1 map with entries of a simulated dictionary. Finding the best matching dictionary entry goes along with the assignment of the corresponding conductivity used to simulate this entry to the respective reconstructed location. The dictionary's simulation can include both transmit and receive stages, yielding conductivities unaffected from the mentioned transceive phase problem. Furthermore, this method opens up the possibility of including dictionary patterns derived from mixed conductivities, enabling in theory an exact solution at tissue boundaries. From a machine learning standpoint the described method corresponds to a k-nearest neighbour (k-NN) search with k equalling one. This classifier appears as a logical first choice as it is common practice in MR finger-printing [6]. While showing compelling first results in compliance with the classical EPT algorithm, its major downside pertains to long reconstruction durations in the range of hours. In order to obtain a handier algorithm for further research, this study introduces an approach for improving the k-NN matching time by reducing the dictionary size without introducing significant errors. In the outlook a brief demonstration of the applicability of the learning method random forest to the classification problem is given with the benefit of a highly reduced calculation time.

2 Theory

This section gives a quick summary of the physics and the resulting reconstruction technique behind the conventional, so called Helmholtz-based EPT (hhEPT). A more detailed description of the backgrounds is given in [7]. Furthermore, the realisation of dictionary-based EPT is presented.

Measuring a subject's electric properties with an MR scanner is intuitively accessible through the inspection of eddy currents induced by the RF pulse in the subjects body. These interact with the phase and magnitude of the resulting magnetization. In turn, the eddy currents depend on the subject's electrical properties, thereby linking the electrical properties to the measurable phase and magnitude of the RF field.

2.1 Helmholtz-based EPT

The basis for conventional EPT lies in the so-called Helmholtz equation which is a combination of Faraday's and Ampère's law with the constant magnetic permeability μ, the Larmor frequency of the utilised MR system ω and the magnetic field $\mathbf{H}$ [7]:

$$-\Delta \mathbf{H} = \mu\omega\kappa\mathbf{H} + \frac{\nabla\kappa}{\kappa} \times [\nabla \times \mathbf{H}]. \quad (1)$$

The quantity κ stands for the complex electric permittivity which is assumed to be isotropic. It is constituted by the electric permittivity ϵ and the conductivity σ as follows: $\kappa = \omega\epsilon - i\sigma$. The assumption of locally constant electric properties, i.e. $\Delta\kappa \approx 0$, yields the 'truncated' Helmholtz equation $-\Delta H = \mu\omega\kappa H$. Introducing the circularly polarized component $H^+ = (H_x + iH_y)/2 = |H^+|\exp(i\phi^+)$ as it underlies the RF excitation (transmit) and solving for κ produces:

$$\kappa = \frac{-\Delta H^+}{\mu\omega H^+}. \quad (2)$$

The assumption of small variations of $|H^+|$ compared to ϕ^+ yields 'phase-based conductivity mapping':

$$\sigma = \frac{\Delta\phi^+}{\mu\omega}. \quad (3)$$

Analogously, assuming that variations of $|H^+|$ are significantly larger than ϕ^+, 'magnitude-based permittivity mapping' is attained:

$$\epsilon = -\frac{\Delta|H^+|}{\mu\omega^2|H^+|}. \quad (4)$$

In conventional EPT the calculation of the Laplacians is achieved by utilising numerical differentiation kernels.

Thereby the problem arises in the transmit (TX) phase ϕ^+ not being directly measurable. Instead, the measured transceive (TRX) phase $\phi^0 = \phi^+ + \phi^-$ equals a superposition with the counterpart ϕ^-, the receive (RX) phase [2]. This problem is usually dealt with via the assumption $\phi^+ = \phi^0/2$, which theoretically only holds for symmetrical objects. The isolated measurement of the TRX phase is included in a variety of standard MR procedures as the simple known relationship of the real and imaginary part of the Fourier transform's result. As it is necessary to isolate the result from other phase influences, for example the influence of inhomogeneities of the static field B0, all spin echo sequences naturally qualify for EPT. This is due to the π-pulse, cancelling out temporally constant phase disturbing influences. Gradient echo sequences applicable to EPT either use very short echo times (TE) or balanced gradients to avoid a build-up of said disturbances. The most common sequence with balanced gradients is called steady-state free precession (SSFP). Due to the inherent influence of other contrast sources than B1 on an MR image's magnitude, the B1 magnitude's measurement has to follow a more elaborate procedure. The following formula describes the MR signal [2]:

$$S = V_1 M_0(\mathbf{r})\left|H^-(\mathbf{r})\right|\exp(i\phi^0(\mathbf{r}))\sin\left(V_2\alpha\left|H^+(\mathbf{r})\right|\right), \quad (5)$$

with system-dependent constants V_1 and V_2, flip angle α, spin density and relaxation effects M_0 and the transmit and receive sensitivities of the coil H^+ and H^- respectively. A unique sinusoidal dependency of the signal on a combination of the transmit sensitivity magnitude and the flip angle can be identified and exploited. In practice this implies the fitting of a sinusoidal functionality to the result of several scans with different known flip angles.

2.2 Dictionary-based EPT

Following a machine learning approach, dbEPT avoids numerical differentiation by comparing local patterns of the B1 map with a ground truth in the form of simulated entries of a dictionary, labelled with the corresponding electrical properties. To take into account the offset introduced by the second derivative in (3) and (4), a multitude of patterns related to a single combination of electrical properties is possible and must thus be included in the dictionary. As in hhEPT it is possible to simultaneously deduce maps for conductivity and permittivity by utilising the combination of the phase and the magnitude map or to solve for the quantities individually. So far the exploration of dbEPT has been conducted by searching for the one best-matching dictionary entry, being the equivalent of a k-nearest neighbour classifier with k equalling one. With each voxel in a pattern representing a feature for classification, the dictionary entry with the shortest distance (classification error) to this pattern, with regards to a chosen metric, has to be found. The shape of the patterns is arbitrary, e.g. 1D strings, or 3D objects like cubes or spheres can be used. Analogously to the size of the differentiation kernel in hhEPT the spatial extension of the patterns counterbalances the influence of noise and spatial resolution. Since matching patterns with a total number of n voxels provides n sets of properties to one voxel (each voxel can take each position in the pattern, apart from voxels at the edge of the image), the final value for each voxel gets evaluated by weighting the individual properties with the classification error.

3 Material and Methods

In this work the reconstruction of conductivity from B1 phase patterns (Fig. 1) is conducted using a transceive phase based dictionary derived from electro-magnetic field simulations with CONCEPT II (TU Hamburg-Harburg, Germany). Utilising an isotropic grid with voxel size 1 mm inside a quadrature head coil, the simulations consist of 20 homogeneous spheres with diameters of $20\,\mathrm{cm}$ at a Larmor frequency of $128\,\mathrm{MHz}$. The conductivity ranges from $\sigma = 0.1\,\mathrm{S/m}$ to $\sigma = 2.0\,\mathrm{S/m}$ in steps of $\sigma = 0.1\,\mathrm{S/m}$ with a constant relative permittivity of $\epsilon_r = 50$. Additionally, an entry with constant phase and thus $\sigma = 0\,\mathrm{S/m}$ is appended to the dictionary. By subtracting the mean from the entries as well as from the measured patterns, the dictionary is generalized concerning the constant phase offset from the first derivative. The shape of the entries is that of three dimensional crosses with 25 voxels consisting of lines with 9 voxels (middle voxel only counting once).

The in-vivo measurements stem from a scan of a healthy volunteer on a commercial 3T scanner (Ingenia, Philips Healthcare, Best, the Netherlands, equipped with a quadrature RF head coil). An SSFP sequence was used with $TR = 3.4\,\mathrm{ms}$, $TE = 1.7\,\mathrm{ms}$, voxel size $= 1 \times 1 \times 1\,\mathrm{mm}$, flip angle $= 25°$, 2 averages and a scan duration of 3:40 minutes. Informed written consent of the volunteer was obtained according to the local Institutional Review Board.

For the comparison of the measured pattern ϕ^{meas} with the dictionary entry ϕ^{dict} the Euclidean distance metric d is used $d = \sqrt{\sum_{k \leq N} \left(\phi_k^{dict} - \phi_k^{meas} \right)^2}$.

The weighting of N conductivity values σ for a voxel is calculated according to $\bar{\sigma} = \dfrac{\sum_{j \leq N} d_j^{-2} \sigma_j}{\sum_{j \leq N} d_j^{-2}}$.

All reconstructed images are filtered with a denoising median filter with its kernel being restricted to voxels with an SSFP signal magnitude of $\pm 10\,\%$ (Fig. 2).

In order to reduce the dictionary size and therefore the matching time, the symmetry of the spheres is exploited. With the phase courses of the octants of the simulated spheres being connected to each other via mirroring at the axes, each three dimensional entry has seven corresponding entries it can be transferred into by reversing one or more of the three phase lines. Thus, the orientation of the measured patterns has to be taken into consideration before matching with the reduced dictionary. This is done by comparing the sum of the two halves of each line. If the sum of the voxels preceding the middle voxel exceeds the sum of the voxels following the middle voxel the line is reversed (thus the top-front-right octant is used).

4 Results

Figs. 1 and 2 show the phase and magnitude map of slice 100 of 240 slices in transverse orientation as attained by the SSFP sequence. In Fig. 3 the result of the reconstruction with dbEPT with the full dictionary is depicted. The reconstruction time is 17h22'. Fig. 4 shows the reconstruction with the reduced dictionary with a reconstruction duration of 2h56'. All reconstructions were conducted on a 64-bit Intel processor (2.67 GHz, 12 GB RAM). As a measure of similarity between the map from the reduced dictionary f and the map from the full dictionary g the center of the cross-correlation is calculated as:

$$r = \frac{\sum_{m,n,l} \left(f(m,n,l) - \bar{f} \right) \left(g(m,n,l) - \bar{g} \right)}{\sqrt{\sum_{m,n,l} \left(f(m,n,l) - \bar{f} \right)^2 \sum_{m,n,l} \left(g(m,n,l) - \bar{g} \right)^2}}$$

(6)

with $\bar{f}$ and $\bar{g}$ as the means of the conductivity maps and m, n and l as the 3D voxel indices. A resulting value of $94.5\,\%$ ($100\,\%$ implying perfect agreement) confirms the visual impression of a high similarity. Corresponding to this impression a calculation of the mean absolute difference as in $\bar{d} = \sum_{m,n,l} |g(m,n,l) - f(m,n,l)|$ yields $0.022\,\mathrm{S/m}$. In both calculations voxels with a conductivity of zero were left out to constrain the calculation to voxels inside the brain. The number of thereby suppressed voxels inside the volume of interest is negligible.

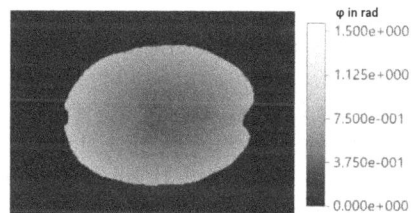

Figure 1: Phase map used in the reconstructions of the conductivity maps in Figs. 3, 4 and 5.

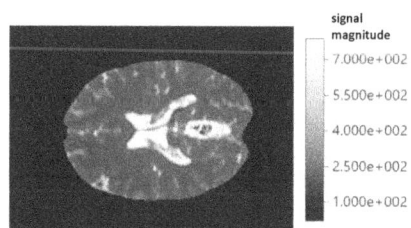

Figure 2: Magnitude map showing the underlying anatomical structures of the brain region.

5 Discussion and Conclusion

In this study the feasibility of the usage of the dictionary's symmetry for shortening the reconstruction time was presented. Utilising the sum of the partial lines from both sides

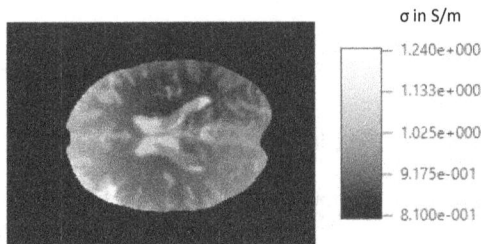

Figure 3: Conductivity map reconstructed by dbEPT using the full dictionary. The reconstruction time is 17h22'.

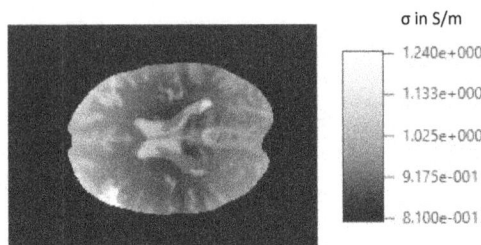

Figure 4: Conductivity map reconstructed by dbEPT with the reduced dictionary. The reconstruction time is 2h56'.

of the middle voxel as an indicator for reversing the order, the size of the dictionary can be reduced by the factor of 2 in each dimension and thus to a total of 1/8th of the original size. As it reduces the reconstruction time by 83 %, this serves as an important step for improving the practicability of the algorithm for further research.

Investigations showing uneven distributions in the radially resolved matching rates for each conductivity-related sub-dictionary imply potential for either further reducing the dictionary size or increasing the accuracy. The employment of labelled data from a simulated head as a separate dictionary or as an expansion of the current one will be tested. This would base the coverage of the dictionary on the rate of occurrence in vivo, maximizing the precision for the most frequently used patterns.

6   Outlook

The successful utilization of the nearest neighbour classifier provides a solid base for dbEPT. With reconstruction times still in the range of hours the method's current benefit is limited to academical purposes as it helps to better understand the general challenges of EPT. Alternative machine learning approaches to the underlying classification problem are currently being tested. In Fig. 5 a promising first result utilising a random forest regressor is presented.

With a training duration (training based on reduced dictionary) of approx. 1 minute and a matching time of 3 seconds this demonstrates the potential of machine learning for the practicability of dbEPT. Yet obvious differences (correlation 79.9 %, mean absolute difference 0.13 S/m) from Fig.

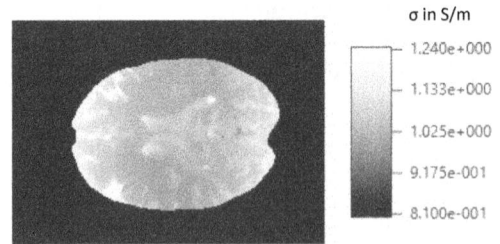

Figure 5: Conductivity map reconstructed by dbEPT using a random forest regressor. The reconstruction time is 3 seconds with a training duration of approx. 1 minute.

4, mainly appearing in a limited contrast especially in the central regions, indicate the necessity of future work in this approach or similar ones.

Acknowledgement

The work has been carried out at Philips Research Laboratories, Hamburg and supervised by the Institute of Medical Technology, Universität zu Lübeck. Furthermore, I would like to thank Thomas Amthor, Christian Findeklee and Mariya Doneva for their helpful support.

7   References

[1] E. M. Haacke, L. S. Petropoulos, E. W. Nilges, and D. H. Wu, "Extraction of conductivity and permittivity using magnetic resonance imaging," vol. 36, pp. 723–734, 1991.

[2] D. I. Hoult, "The principle of reciprocity in signal strength calculations - a mathematical guide," vol. 12, pp. 173–187, 2000.

[3] J. Shin, M. J. Kim, J. Lee, Y. Nam, and M. oh Kim et al., "Initial study on in vivo conductivity mapping of breast cancer using MRI," vol. 42, pp. 371–378, 2015.

[4] K. K. Tha, U. Katscher, S. Yamaguchi, C. Stehning, and S. T. et al., "Noninvasive electrical conductivity measurement by mri: a test of its validity and the electrical conductivity characteristics of glioma," vol. 28, pp. 348–355, 2018.

[5] U. Katscher, M. Hermrmann, C. Findeklee, M. Doneva, and T. Amthor, "Dictionary-based electric properties tomography," p. 3641, 2017.

[6] D. Ma, V. Gulani, N. Seiberlich, K. Liu, and J. L. S. et al., "Magnetic resonance fingerprinting," vol. 495, pp. 187–192, 2013.

[7] U. Katscher and C. A. T. van den Berg, "Electric properties tomography: Biochemical, physical and technical background, evaluation and clinical applications," vol. 30, p. e3729, 2017.

Automated defect recognition on X-ray images of aluminium castings based on change detection algorithms

J. Sauer [1], O. Schmidt [2], F. Kaiser [2], and T. M. Buzug [3]

[1] Medizinische Ingenieurwissenschaft, Universität zu Lübeck, julia.sauer@student.uni-luebeck.de
[2] VisiConsult X-ray Systems and Solutions, Stockelsdorf, info@visiconsult.de
[3] Institute of Medical Engineering, Universität zu Lübeck, buzug@imt.uni-luebeck.de

Abstract

Change detection in image series has been a well studied field in image processing. In this work, it is applied for automated defect recognition of aluminium castings using X-ray projections. An algorithm is derived to form a defect mask of a casting projection that requires a set of reference images. This algorithm consists of two basic steps: First, a pre-mask is formed. For this, the image is approximated block-wise by a polynomial picture model. Then it is tested if the model parameters have changed. Secondly, the pre-mask is combined with a difference image to estimate the change probability using Bayesian statistics. An extension of this test has been derived to reduce the detection of non-defect edge pixels. Therefore, an edge mask is created by the Canny edge detection algorithm. Subsequently, the a priori probability based on the pre-mask is modified assuming Gibbs distributed edge pixels. Detection masks of different defects are given and show reliable results.

1 Introduction

In the course of non-destructive material testing using X-ray technology, effective and reliable image processing is required to automatize inspection of large quantities of components. Turning to casting defects X-ray projections are acquired with the aim to visualize shrinkage defects, gas porosity, misruns, cold shuts, inclusions and other faults [1] that exclude the casting from safe usage. Obviously, tested components as well as occurring defects can be extremely diverse. Thus, development of detection algorithms is a complex task. Present techniques require costly parameterisations for each particular series of similar components. In addition to resulting inflexibility, certain defects are still not optimally segmented. Image processing techniques are presented that use multiple reference images for application of change detection algorithms. Especially large defects are supposed to be detected. However, the rarity of larger cavities causes problems: There is only few material to test developed methods. Thus, common defect detection techniques often miss them because of specialisation in detection of more often small porosities. Furthermore, human observers sometimes misjudge them for a correct part of the casting. Despite rare occurrence larger defects have to be detected extremely reliably because of high safety significance. Our aim is to find a sensitive and real-time capable detection method while minimizing parameterisation effort.

Many algorithms have been presented so far to detect changes in image time series (see e.g. [2]). We adapt these techniques to series of X-ray projections of equal compo-

nents. As the casting positions deviate to a certain extent, change detection strongly depends on quality of image registration [3], though this is not treated any further here. In the following an overview about change detection methods introduced by Hsu, Nagel and Rekers [4] is given. Based on the hypothesis test designed by Yakimovsky [5], they decide if a pixel value has changed. To estimate the variance of an image block, polynomial picture functions are used. Hsu, Nagel and Rekers further derived a method for automated threshold selection based on the used image model. This algorithm is applied to create a pre-mask, which is taken as prior knowledge for the adaptive change detection technique derived by Aach and Kaup [6]. Here, a change mask is estimated using a difference image and the a priori probability given by the pre-mask. This algorithm is adjusted in a way that the likelihood of classifying a pixel "changed" is reduced in the area of steep edges, where false-positives due to registration inaccuracies are more likely. These areas are determined using the Canny edge detection algorithm [7]. The final defect detection algorithm is tested on multiple castings with varying defects.

2 Material and Methods

The aim of defect detection is to assign each pixel with *okay* or *defect*. As we use change masks, this equals labeling the pixels either with "unchanged" u or "changed" c. In the field of object detection, Yakimovsky [5] assumed that grey values originating from different objects obey different normal distributions $N(\mu_1, \sigma_1^2)$ and $N(\mu_2, \sigma_2^2)$. Hsu, Nagel and

Rekers [4] extend this to decide, whether an image block has changed from one image to another by checking if the grey value distribution is different. This is achieved by maximum likelihood ratio

$$\lambda = \frac{\mathcal{L}(c)}{\mathcal{L}(u)} = \frac{P_1 P_2}{P_0}, \qquad (1)$$

where

$$P_k = \left(\frac{1}{\sqrt{2\pi}\hat{\sigma}_k}\right)^n \exp\left\{-\frac{1}{2\hat{\sigma}_k^2}\|\boldsymbol{f}_k - \boldsymbol{g}_k\|_2^2\right\} \qquad (2)$$

is the joint probability density function of image blocks $f_1(x,y)$ and $f_2(x,y)$ containing n pixels. The denominator

$$P_0 = \left(\frac{1}{2\pi\hat{\sigma}_0^2}\right)^n \exp\left\{-\frac{1}{2\hat{\sigma}_0^2}\sum_{k=1}^{2}\|\boldsymbol{f}_k - \boldsymbol{g}_0\|_2^2\right\} \qquad (3)$$

is the joint probability density of the image block combination, if the contained grey values follow equal distributions. The picture model vector $\boldsymbol{g}$ corresponds to the function $g(x,y) = \sum_{i=0}^{p}\sum_{j=0}^{p-i}\beta_{ij}x^i y^j$ and is determined by least square optimisation. Furthermore, [4] derived random variables for threshold selection. The maximum likelihood ratio is reformulated to

$$\lambda = \left(1 + \frac{q}{2(n-q)}z\right)^n, \qquad (4)$$

where z is the square of a t-distributed variable in the constant case and F-distributed for linear and quadratic picture functions. The factor q is the number of parameters of the respective picture function. Due to monotony of λ, the variable z is tested instead. Specifying a significance level α, the threshold z_α is calculated using the inverse of the respective distribution. If the related value of z, calculated from the image blocks, exceeds this threshold this region is labeled with c or else with u.

We use this method to determine a pre-mask. The pre-mask is used to calculate the a priori probability for the statistical test designed by Aach and Kaup [6]. They formulate the change problem in image series with Bayesian statistics. The provided data are the grey value differences between current and previous image in the sequence. As there is no temporal sequence available when testing castings, a reference image is used. Aach and Kaup start with the likelihood ratio

$$\frac{P(f_i = u|D)}{P(f_i = c|D)} \gtrless_c^u t \qquad (5)$$

that provides the decision rule, whether pixel i has changed or not using the threshold t and the difference image D. It is assumed that the labels of all pixels except i are known which is the reason why a pre-mask is required. With Bayes' theorem, the posteriori probabilities are replaced. The grey values of D in that region are further assumed normally distributed with zero mean and variance σ_u^2 and σ_c^2 for an unchanged or changed pixel, respectively. Inserting the joint probability density and taking the logarithm, the decision rule becomes

$$\overline{\Delta}_i^2 \gtrless_c^u -2\ln\left\{t\left(\frac{\sigma_u}{\sigma_c}\right)^n\right\} + 2\ln\left\{\frac{P(f_i = u)}{P(f_i = c)}\right\} \qquad (6)$$

with the normalized sum of quadratic difference grey values $\overline{\Delta}_i^2$. The decision threshold $t_s = -2\ln\{t(\sigma_u/\sigma_c)^n\}$ in (6) is modified by the a priori probability ratio on the right side. Aach and Kaup proposed to choose the threshold without estimating the variances. Instead, they assume $\overline{\Delta}_i^2$ to be a χ^2-distributed random variable and determine a significance level α. This is used to calculate the inverse chi-squared which is then used as threshold t_s. To improve results the a priori probability of the change mask is required. Markov random fields are used to describe properties of the local change mask around pixel i. Their joint probability follows a Gibbs distribution [8]. It is assumed that change regions are smooth which is especially accompanied with small Gibbs energy. Consequently,

$$P(f) = \frac{1}{Z}\exp\{U(f)\} \qquad (7)$$

is the a priori probability of change mask f using Gibbs energy U with the normalisation constant Z. Aach and Kaup count occurring border pixel pairs which means the number of pixels around i whose label differs from f_i. As less border pixels induce smoothness the energy is supposed to decrease simultaneously. Determining potential a, the energy

$$U(f_i = u) = an_i^a(c) \qquad (8)$$

is calculated, where $n_i^a(c)$ denotes the number of 8-connected neighbours of i with label c. Aach and Kaup finally derive

$$\overline{\Delta}_i^2 \gtrless_c^u t_s + 16a - 4an_i^a(c) \qquad (9)$$

as decision rule. The global threshold t_s and potential a are determined once for all pixels. Then the sum of squared difference grey values and number of border pixels is calculated for each pixel to decide between c and u. The algorithm starts with the pre-mask to count border pixels and updates the change mask continuously.

Both presented methods potentially detect casting defects as well as shifts that result from small deviations of the casting positions. Image registration [3] is applied to enable usage of change detection algorithms and to avoid false alarms. Still edge pixels might be assigned with c. See Fig. 4 as example. We present an extension of Aach and Kaup's test which reduces the a priori probability of $f_i = c$ when the pixel could be part of an edge. The Canny edge detection algorithm [7] is used to create an edge mask $\boldsymbol{h}$, where object pixels o and edge pixels e are labeled. It is assumed that unchanged pixels in the pre-mask always belong to an object because the method is highly sensitive referring to edge pixels. It follows

$$\begin{aligned} P(f_i = u, h_i = o) &= P(h_i = o|f_i = u)P(f_i = u) \\ &= 1 \cdot P(f_i = u) \qquad (10) \end{aligned}$$

as probability of joint events u and o. On the other hand, a changed pixel in the pre-mask can be a defect as well as an edge. The formation of pre-mask and edge mask $\boldsymbol{h}$ is assumed to be independent and resulting probabilities as well.

Consequently, the probability of i being a changed pixel and belonging to an object is

$$P(f_i = c, h_i = o) = P(f_i = c)P(h_i = o). \qquad (11)$$

To estimate the a priori probability $P(h_i = o)$, a Gibbs distribution of the edge pixels is assumed. This seems realistic as edges are rather smooth than pixelated. The probability distribution is defined similar to (7) with the Gibbs energy

$$U(h_i = o) = vn_i^v(e), \qquad (12)$$

where v is the potential and $n_i^v(e)$ is the number of edge pixel neighbours of i in h. The normalisation constant

$$Z = \sum_{n=0}^{8} \frac{8!}{(8-n)!} \exp\{-vn\} \qquad (13)$$

corresponds to the sum over all possible energy states which means possible arrangements of edge pixels 8-connected to i. Inserting this into (9) leads to

$$\overline{\Delta}_i^2 \underset{c}{\overset{u}{\lessgtr}} t_s + 16a - 4an_i^a(c) + vn_i^v(e) + \ln(Z). \qquad (14)$$

In (14), the resulting threshold value is raised with increasing number of edge pixels in close neighbourhood.
To apply the presented change detection algorithms a reliable reference image is required. It originates from multiple images of the current position. After image registration, the pixel-wise median is taken to avoid that occurring defects influence the resulting reference image. This image is used to calculate the difference whose grey values contribute to $\overline{\Delta}_i^2$ in (14). The reference is used to create the edge mask as well. The edge detection algorithm is applied on both reference and current casting projection image. The calculated masks are linked by the logical connective &. This ensures that no edges of defects are included.

3 Results and Discussion

The above-mentioned algorithms are combined to create a defect mask for the X-ray projection of a casting. In Fig. 1 the calculation progress is summarized. In this section, example results of the single steps are given for one casting projection, which is shown in Fig. 2. This defect is a usual gas hole which leads to exclusion. Results of the automated threshold selection using a picture model are presented in Fig. 3. Obviously, different polynomial orders (upper row of Fig. 3) lead to varying results. As to be seen their combination (bottom row) is an improvement with respect to false-positives resulting from noise and illumination inconsistencies. Secondly, the final defect masks with and without edge correction are calculated. Fig. 4 (left image) presents the results of the hypothesis test by Aach and Kaup [6]. Though the gas bubble has been detected sufficiently the mask contains many false-positive edge pixels. Finally, the resulting edge corrected defect mask is shown on the right side of Fig. 4. This method improves the defect mask according to false-positive edges. Still, there might be to many edge pixels detected. Other results of different occurring defects are demonstrated in Fig. 5a and 5b.

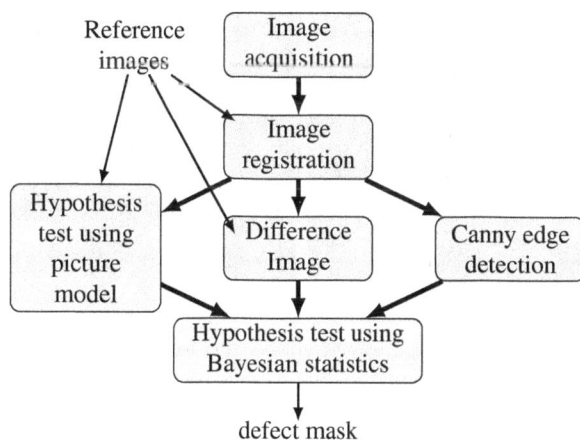

Figure 1: Schematic procedure of defect detection algorithm

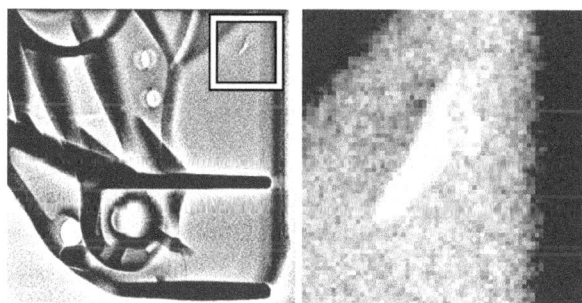

Figure 2: Casting with gas hole defect. The right image enlarges the framed section of the left image. Both have been processed with Fourier-based enhancement filter.

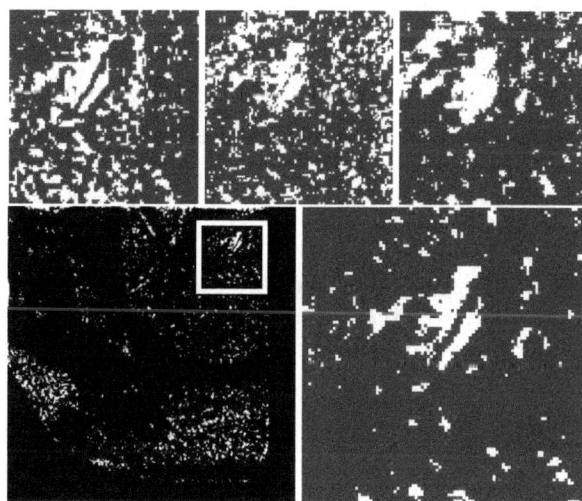

Figure 3: Results of hypothesis test by [4]. The change masks in the upper row depict the outcome of the framed defect from Fig. 2 with (from left to right) constant, linear and quadratic image model. These masks are linked by pixel-wise logical connective &. The bottom images show the resulting pre-mask. The complete mask is on the left side and the defect section on the right side.

Figure 4: The left image shows the final defect mask of Fig. 2 obtained from the change detection algorithm by [6]. The defect mask on the right side shows the result of the derived edge corrected hypothesis test. The amount of detected edge pixels has been considerable reduced. In both images the defect is framed.

4 Conclusion

The presented results verify that a defect detection is possible using the derived algorithm, though an improvement in edge correction is still required in some cases. Furthermore, the formation of a change mask depends on the choice of some parameters such as significance levels and potentials. It is further necessary to prove that one parameter set is reliable for complete projection series. Additionally, it is supposed to investigate, if the number of parameters can be reduced further, e.g. by normalisation. For automated defect recognition, real-time capability must be guarantied. This feature potentially applies for the presented algorithm.

Acknowledgement

The work has been carried out at Visi Consult X-ray Systems and Solutions in Stockelsdorf and supervised by the Institute of Medical Engineering, Universität zu Lübeck.

5 References

[1] *ASTM E155-00, Standard Reference Radiographs for Inspection of Aluminium and Magnesium Castings.* ASTM International, West Conshohocken, PA, 2000.

[2] R. J. Radke, S. Andra, O. Al-Kofahi, and B. Roysam, "Image change detection algorithms: a systematic survey," *IEEE transactions on image processing*, vol. 14, no. 3, pp. 294–307, 2005.

[3] L. König and J. Rühaak, "A fast and accurate parallel algorithm for non-linear image registration using normalized gradient fields," in *Biomedical Imaging (ISBI), 2014 IEEE 11th International Symposium on*, pp. 580–583, 2014.

[4] Y. Hsu, H.-H. Nagel, and G. Rekers, "New likelihood test methods for change detection in image sequences," *Computer vision, graphics, and image processing*, vol. 26, no. 1, pp. 73–106, 1984.

[5] Y. Yakimovsky, "Boundary and object detection in real world images," *Journal of the ACM (JACM)*, vol. 23, no. 4, pp. 599–618, 1976.

[6] T. Aach and A. Kaup, "Bayesian algorithms for adaptive change detection in image sequences using markov random fields," *Signal Processing: Image Communication*, vol. 7, no. 2, pp. 147–160, 1995.

[7] J. Canny, "A computational approach to edge detection," *IEEE Transactions on pattern analysis and machine intelligence*, no. 6, pp. 679–698, 1986.

[8] S. Z. Li, *Markov random field modeling in image analysis.* Springer Science & Business Media, 2009.

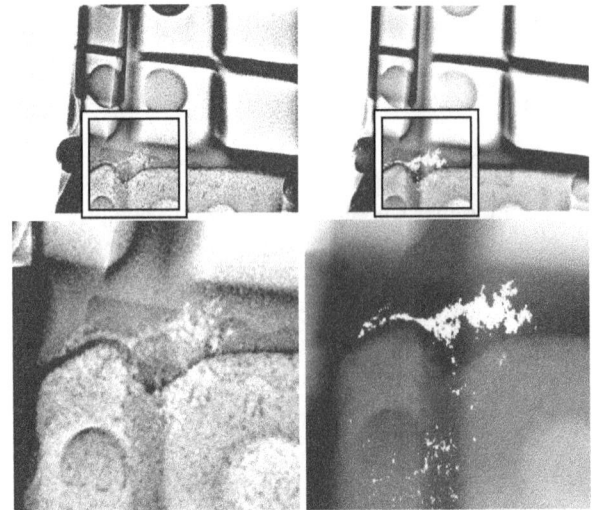

(a) Widespread gas porosity defect

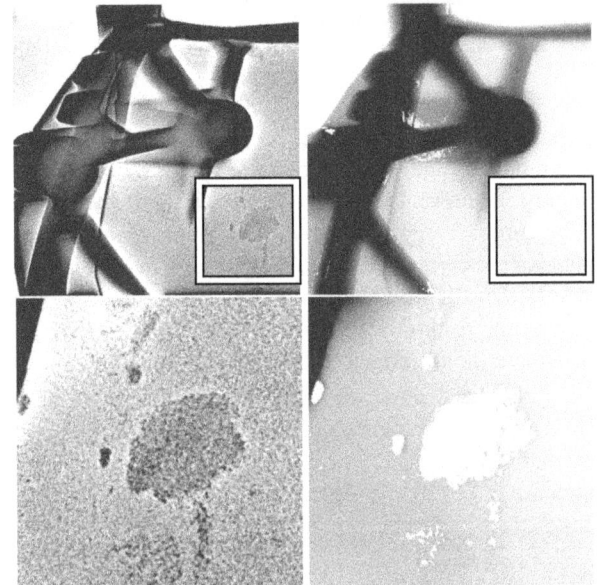

(b) Inclusion of sand or impure casting surface

Figure 5: Examples of different defects and corresponding detection result. For each example, the bottom images show the section containing the defect which is marked above. On the right side the resulting change mask is overlayed in white, respectively for Fig. 5a and 5b.

Development of a graphic user interface and cross manufacturer adaptation of a program for determining the pulse wave velocity in the aorta from phase-contrast magnetic resonance images

C. Schareck [1], T. H. Oechtering [2], A. Frydrychowicz [2], and M. A. Koch [3]

[1] Medizinische Ingenieurwissenschaft, Universität zu Lübeck, constantin.schareck@student.uni-luebeck.de
[2] Clinic of Radiology and Nuclear Medicine, Universitätsklinikum Schleswig-Holstein,
{thekla.oechtering, alex.frydrychowicz}@uksh.de
[3] Institute of Medical Engineering, Universität zu Lübeck, koch@imt.uni-luebeck.de

Abstract

Due to the periodic contraction of the heart, a pulse-shaped course of the blood velocity is caused in the aortic system. The speed at which this pulse wave moves is called pulse wave velocity (PWV). Since it depends, among other parameters, on the elasticity of the vascular wall, it can be used clinically as an indicator of atherosclerotic changes and, above all, as a measure of the risk of stroke for hypertensive patients. PWV can be determined non-invasively if the blood flow in the aorta is measured via flow-sensitive MRI. Within the scope of this work, existing phase-contrast (PC) image analysis algorithms, resulting in the PWV, were embedded in a graphical user interface (GUI). These algorithms were only able to evaluate PC images which were acquired on a Philips tomograph. Therefore they have been extended for the analysis of PC images acquired on a Siemens tomograph.

1 Introduction

The periodic contraction of the heart does not cause a constant volume flow through the vascular system but rather leads to a pulse-shaped course of the flow velocity, depending on space and time, which propagates along the aorta. The velocity of this propagation is called Pulse Wave Velocity. Due to the periodic behaviour of the heart a pulse wave is created on every heartbeat. The elasticity of the arterial vessel wall influences the speed of the pulse wave. Therefore the PWV is used as a clinical indicator for reduced elasticity of the vessel wall due to degenerative changes like atherosclerosis [1]. In such cases, the stiffness would increase, resulting in a higher PWV [2] [3].

The PWV can be calculated if the blood flow rate at two different vessel locations is known or rather can be derived from measured data. Comparison of the time-dependent flow rate at both locations yields the temporal shift, Δt, i.e. the time the pulse wave needs to travel between the first and second location. With given or rather derived course length, Δs, between both locations along the vessel, the PWV can be calculated by [4]

$$PWV = \frac{\Delta s}{\Delta t}. \tag{1}$$

Information about both time delay and vessel length can be obtained if the blood flow at least at two different locations in the aorta was measured with flow-sensitive MRI. Therefore the motion of the heart is used. The necessary data for an axial image at vessel location s is sampled over different heart beats. The acquisition starts in the same heart phase and lasts for a constant period of time. Thus the acquisitions are ECG-gated. This procedure is called CINE-Imaging and results in information about the velocity of blood, v_{blood}, within a vessel at location s at time t. By integration of v_{blood} over the cross-section, A, of the vessel at time t the volume flow rate (Fig. 3), Q, can be calculated with

$$Q(t) = \iint\limits_{A} v_{blood}(x,y,t)dx\,dy. \tag{2}$$

The temporal shift, Δt, can be determined by comparing the volume flow rates of the sections with respect to their temporal shift relative to each other. In order to calculate the volume flow rate of an PC-MRI image, flow velocity maps must be calculated from the PC images. This is necessary because PC-MRI images are usually stored normalized on an unsigned 12-bit interval, i.e. the stored pixel values do not represent the flow velocities. To handle this scaling, MRI manufacturers, here Philips and Siemens, have stored scaling parameters in the header file of each DICOM image.

2 Material and Methods

In this study existing algorithms for PWV-calculations of PC-MRI images were embedded into a graphical user interface. These algortihms were only usable for PC-MRI data measured with a Philips tomograph. In addition, those algorithms were unstable to wrong inputs, i.e the program was often aborted. Thus the main focus of this work was on the one hand the adaptation of the algorithms for Siemens PC-MRI data, and on the other hand improvement of the usability and stability. For this reason, the main functionality of the GUI is explained in this paper and usability improving features are discussed. The used datasets were provided by the clinic of radiology and nuclear medicine, Universitätsklinikum Schleswig-Holstein. Reference [6] shows the validation of the algorithms for a Philips MR-data study including 7 patients.

2.1 MRI

In order to test the functionality of the GUI, flow-weighted datasets, recorded on a Philips *Ingenia 3T* system and a dataset recorded on a Siemens *Magnetom Skyra 3T* system were provided. Each data set consists of at least 2 different T1-weighted phase contrast (PC-MRI) image series, in which on the one hand images of the ascending aorta (AOA) and on the other hand images of the descending aorta (AOD) were acquired. These image series were generated with throughplane velocity encoding and an ECG-gated acquisition scheme. Additionally each data set contains one anatomic T2-weighted sagittal (Sag) overview image stack. Each image in the stack shows parts of the aorta due to acquisitions at different slice locations. Thus the whole aortic course is indirectly imaged by means of iteration through the stack. Imaging parameters vary with weight, breath holding capabilities, patient size and heart rate (Table 1) [6].

Table 1: Parameters of the tested datasets

Parameters	Philips	Siemens
Matrix size (T2)	528x528	320x320
Matrix size (PC)	192x192	162x192
Unit of v_{blood}	ml/s	Unspecified
Repetition time	4.49 ms	37.12 ms
Echo time	283 ms	2.47 ms
Slice thickness	6 mm	6 mm
Voxel Size	1.56x1.56 mm^2	1.77x1.77 mm^2
v_{enc}	180 cm·s^{-1}	180 cm·s^{-1}
Rescale Intercept	−180	−4096
Rescale Slope	0.0879	2
Scale Slope	11.375	N/A

2.1.1 File Formats

The format of the files to be analyzed is DICOM. Thus each file of an image series contains two sections data. First the image data itself, second the corresponding image information, the DICOM header. The DICOM header is a list-like structure of entries that describe image parameters, such as those shown in Table 1.

2.2 PWV Computing Tool

The PWV Computing Tool is a Matlab based software tool. This tool is divided into 3 separate GUIs, the Main Task, the Pre-Computing Tasks and the Calculation Tasks. When the program is executed only the Main Task GUI is started. Then the Pre-Computing Task GUI and the Calculation Task GUI are executed one after the other.

2.2.1 Pre-Computing Tasks

The Pre-Computing Tasks primarily reads and sorts desired files within a folder at a direction defined by the user. The given reading and sorting algorithm works well for Philips but not for Siemens PC-MRI data. This is caused by different parameters in the respective DICOM headers. In addition, the way in which the data is stored differs. Philips stores all images (either AOD or AOA) within the same folder. Siemens stores images in acquisition specific folders, i.e. they generate more than one image folder. Therefore a new, Siemens data-specific reading and sorting algorithm was designed. The images of each folder are sorted within each acquisition with respect to the frame order.

2.2.2 Calculation Tasks

To succsessfully calculate the PWV the user has to determine a center line within the anatomic T2 overview series, such that the path length between the PC-MRI slice planes can be obtained. To this aim, the user places points in the middle of the vessel course. The points are then connected to each other via spline interpolation. Via the intersection points of the center line with the slice planes of the PC-MRI series, the start and end points of the path are determined. On this basis the course length Δs can be calculated (Fig. 1).

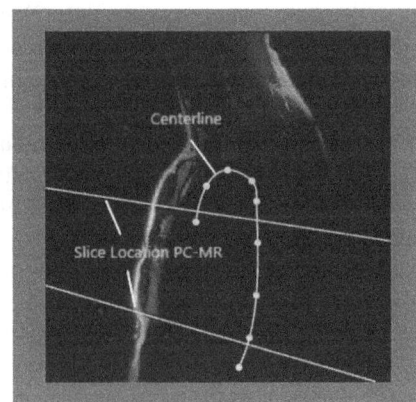

Figure 1: Determination of the center line and its intersection points with the PC slice locations by the user.

Due to a change of location and size of the aorta within the image series, a region of interest (ROI) has to be placed in an initial image. Thus two options are implemented. The ROI can be adapted interactively or automatically to the vessel changes in each image (Fig. 2). In case of the automatic adaptation an initial ROI is set in the first image at the vessel location. The ROI then adapts itself in size and location with respect to the vessel shape. This is done by a threshold based algorithm. First a median filter deletes outlier values. Then signal voids are closed and small intensitiy areas which connect between aorta and surrounding tissue are eroded by means of morphological operations. Two thresholds are set: one represents the maximum pixel value, the other represents the minimum pixel value inside the ROI. For each pixel inside the ROI, it is checked if pixel values of a 3x3 neighborhood are in between the two thresholds. For each pixel value where this is the case the pixel is added to the ROI. The adapted ROI is used as the initial ROI of the subsequent image ROI adaptation.

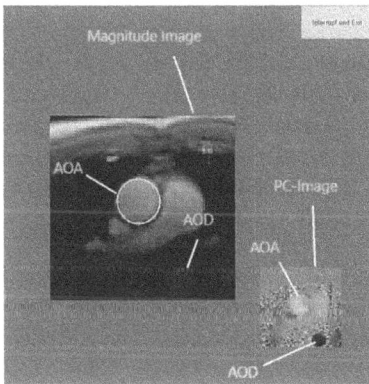

Figure 2: Visualization of the interactive ROI adaptation to the size and position of the vessel.

In order to derive the blood velocity from a PC series, each blood velocity, $v_{\text{blood,j}}$, contained in a pixel j must be multiplied by its corresponding pixel area A if the pixel to be considered is within the ROI, and zero otherwise. Mathematically this can be described by

$$Q(t) = \begin{cases} \sum_j v_{\text{blood,j}} \cdot A & \text{if } j \,\epsilon\, \text{ROI} \\ 0 & \text{else} \end{cases}, \qquad (3)$$

which is the discrete version of (2). Due to the fact that PC-MRI pixel values do not represent the velocity values but normalized intensities, these stored values (SV) must be converted. The conversion parameters Rescale Intercept (RI), Rescale Slope (RS) are stored in the DICOM headers (Table 1). In case of Philips PC-MRI another Parameter, Scale Slope (SS), is stored, thus velocity values can be calculated with

$$\frac{v_{\text{blood,j}}}{\text{cm} \cdot \text{s}^{-1}} = \frac{SV}{SS} + \frac{RI}{RS \cdot SS}. \qquad (4)$$

Due to a missing SS in the DICOM header of Siemens PC-MRI series this parameter has to be calculated. The shifting equation

$$\text{Shifted Value} = SV \cdot RS + RI \qquad (5)$$

yields results in the range $[-180; 180]$ for Philips PC-MRI data. In case of Siemens PC-MRI data, (5) yields results

in the range $[-v_{\text{enc}}; v_{\text{enc}}]$, i.e. $[-4096; 4095]$. These results show that SS is correlated to the phase encoding and therefore can be calculated by solving

$$-180 = \frac{RI}{SS \cdot RS} \qquad (6)$$

for SS. It is possible to calculate the velocity values with (4) for PC-MRI data from both manufacturers. Thus the volume flow curve (Fig. 3) of a PC series with (3) can be obtained.

To calculate the time shift Δt of the pulse wave between the chosen vessel locations, four different algorithms are implemented: time to peak (TTP, time to maximal flow), time to foot (TTF, time to intersection between a linear approximation of the 20-80% rising part of the curve and the zero line), time to upstroke (TTU, time to maximum in the rising part of the first derivation of the curve) [5] and the cross correlation (xcorr, measurement of the similarity of two series as a function of displacement) [7].

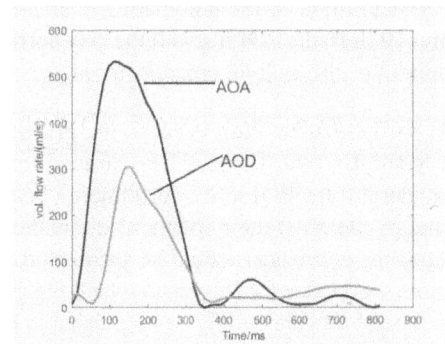

Figure 3: Visualisation of the calculated volume flow rate out of PC-MRI series from AOA and AOD of the same patient.

3   Results and Discussion

A GUI, based on already existing algorithms, was created. The previous tool was only usable for Philips 2D-MRI data. It was instable for wrong user input, in particular unexpected input. In addition, there was an impractical user guidance without instructions which informs the user about what to do next. In this respect, improvements have been implemented. The tool informs about incorrect inputs and provides an option to fix them. It provides information at all times about the current work step, as well as instructions on which inputs are expected so that potential sources of error can be minimized. In addition, an extension of the old and partly new algorithms was developed so that the tool can read and evaluate Siemens data. During the test runs the program ran stable. It remains to be seen whether the program remains stable for further Siemens data, which was not available during programming.

An obvious error source is the adaptation of the varying vessel size and location by the user. The user has to ensure that no values outside the vessel are within the ROI so that the calculation has a minimized noise error. Because the size of every ROI has to be checked and one series includes up to 60 frames it is a protracted task. Therefore the user can get negligent and sets vessel surrounding information inside the ROI, resulting in incorrect calculations. This can cause the flow curves to be erroneous. To ensure a precise adjustment of the ROI to the vessel, this task would have to be extended from a user-dependent adaptation to an automatic adaptation. In this respect, an attempt was made in this work. The resulting algorithm is threshold based and partially able to differentiate between aorta and surrounding tissue.

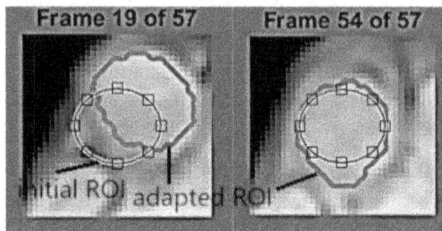

Figure 4: Visualisation of the automatically adapted ROI. At timepoint 19 (left) the ROI fits well the new aortic shape. At timepoint 54 (right) a slight error occurred.

This is not an optimal solution because the grayscale values are very similar near the heart. It can lead to the fact that no boundary of the ROI to the surrounding heart tissue can be found by means of the implemented threshold value method. Thus the entire heart could be segmented (Fig. 5). The algorithm works fine for images in which the grayscale values inside the aorta are clearly destinguishable from the surrounding tissue (Fig. 4). Because such conditions do not often occur, the algorithm remains unstable. For a stable condition it needs to be extended or in particular redesigned as a non-threshold based algorithm.

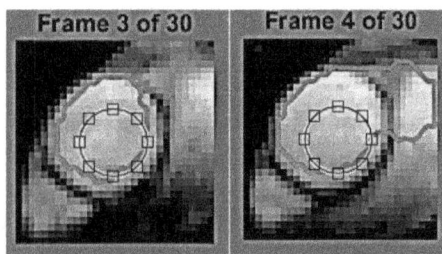

Figure 5: Visualisation of the automatically adapted ROI containing large errors. At timepoint 2 (left) the ROI fits well the new aortic shape. At timepoint 3 (right) a huge error occurs letting the ROI expand to the surrounding heart tissue.

4 Conclusion

A user friendly tool was developed to determine the pulse wave velocity based on Philips and Siemens PC-MRI data. Here, great emphasis was given to the demands and wishes of the physicians of the Clinic for Radiology and Neuro-radiology. The validity and stability for Siemens datasets

has not yet been properly verified. If there are no further problems, this tool will be used for clinical studies. In terms of stable automatisation, the tool remains in an extendable state. The basis for automatism has already been established.

Acknowledgement

The work has been carried out in cooperation of the Institute for Medical Engineering and the Clinic for Radiology and Nuclear Medicine, Universität zu Lübeck.

5 References

[1] G. Mancia et al., *2007 Guidelines for the management of arterial hypertension: The Task Force for the Management of Arterial Hypertension of the European Society of Hypertension (ESH) and of the European Society of Cardiology (ESC)*. European Heart Journal, vol. 28, pp. 1462–1536, 2007.

[2] T. J. M. Schlatmann and A. E. Becker, *"Histologic changes in the normal aging aorta: Implications for dissecting aortic aneurysm,"*. The American Journal of Cardiology, vol. 39, pp. 13–20, 1977.

[3] K. Sutton-Tyrrell et al., *"Elevated aortic pulse wave velocity, a marker of arterial stiffness, predicts cardiovascular events in well-functioning older adults"*. Circulation, vol. 111, pp. 3384–3390, 2005.

[4] M. Bock, L. R. Schad, E. Müller and W. J.Lorenz, *"Pulsewave velocity measurement using a new real-time MR-method"*. Magnetic Resonance Imaging, vol. 13, pp. 21–29, 1995.

[5] M. Markl, W. Wallis, S. Brendecke, J. Simon, A. Frydrychowicz and A. Harloff, *"Estimation of global aortic pulse wave velocity by flow-sensitive 4D MRI"*. Magnetic Resonance Imaging, vol 63, pp. 1575–1582, 2010.

[6] A. Timmermeyer, M. A. Koch and A. Frydrychowicz, *"Development and Validation of a Tool for Pulse Wave Velocity Measurements in MRI Phase Contrast Data"*. Student Conference on Medical Engineering Science 2014, pp. 3–4, Lübeck, March 12.-14, 2014.

[7] S. W. Fielden, B. K. Fornwalt, M. Jerosch-Herold, R. L. Eisner, A. E. Stillman, J. N. Oshinski, *"A new method for the determination of aortic pulse wave velocity using cross correlation on 2d PCMR velocity data"*. Magnetic Resonance Imaging, vol. 72, pp. 1257–1269, 1985.

Statistical Iterative Reconstruction Including Triple Coincidences for a Two-Layer Small Animal PET Scanner

P. Huß [1], M. Schaar [2] and M. Rafecas [2]

[1] Medizinische Ingenieurwissenschaft, Universität zu Lübeck, philipp.huss@student.uni-luebeck.de

[2] Institute of Medical Engineering, Universität zu Lübeck, {schaar, rafecas}@imt.uni-luebeck.de

Abstract

Small animal PET is a relevant imaging tool in biomedical research to visualize and assess metabolic processes in rodents. The small size of mice poses challenges to PET technology in terms of spatial resolution and sensitivity. Novel PET designs use small crystals to improve the resolution, and more than one crystal layer helps to enhance the detection efficiency. Our goal is to increase the sensitivity in software. To this end, we investigate to which extent triple coincidences can enhance image quality. This kind of coincidences are caused by inter-crystal scattering of photons. Using the simulation software GATE, we simulated a dual-layer scanner and an image-quality phantom. Reconstructions with and without triple coincidences were compared. Under the given simulation conditions, the inclusion of triple coincidences notably improved image quality in terms of contrast recovery. Future simulations should be performed to quantify the role of triples when other degradation sources are included.

1 Introduction

Positron emission tomography (PET) is an imaging technique widely used for diagnostic purposes in clinical routine. PET scanners dedicated to rodents have also become an important tool in biomedical research. One challenge faced by small animal PET scanners is the trade-off between spatial resolution and system sensitivity. To increase spatial resolution, some novel PET concepts use finely granulated detector crystals individually readout by photodetectors such as Silicon Photomultipliers. At the same time, two or more radial detector layers can be used to increase the sensitivity [1].

The goal of a PET detection system is to identify pairs of photons, each pair being simultaneously emitted after the decay of a positron and its subsequent annihilation with an electron. Each detected pair is assigned a *line-of-response* (LoR), which is the line connecting the two crystals involved in the detection (see line D in Fig. 1). On the other hand, individual readout allow inter-crystal scattering (ICS) events to be identified. The latter are characterized by three or more interactions (as represented in Fig. 1 by line T). Since the sequence of interactions in the detectors is unknown, a unique LOR cannot be assigned in such a case. Hence, ICS events are usually discarded. For multi-layer PET systems, the number of ICS events is non-negligible and could be thus used to enhance the system efficiency if included into the reconstruction. The present study aims to evaluate the impact of multiple coincidences on the reconstructed image for a dual-layer PET scanner for rodent studies. To this aim we have simulated such a system and

implemented data processing and image reconstruction algorithms, as described in the following sections.

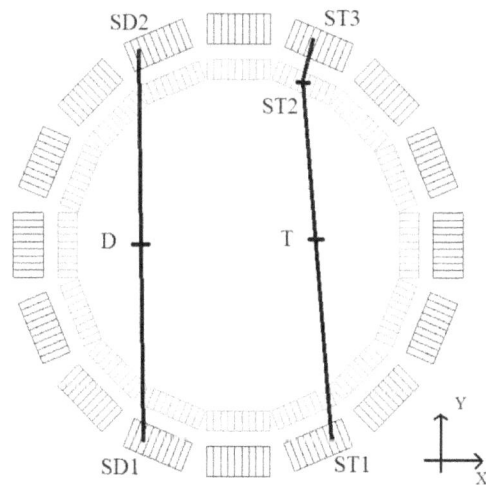

Figure 1: A schema of a double (D) and triple (T) coincidence in a dual-layer PET and their interaction locations in the crystals that form the signal.

2 Material and Methods

Simulation experiments were carried out using the Monte-Carlo simulation package GATE (*Geant4 application for tomographic emission*) [2], which is an open-source software especially conceived for emission tomography.

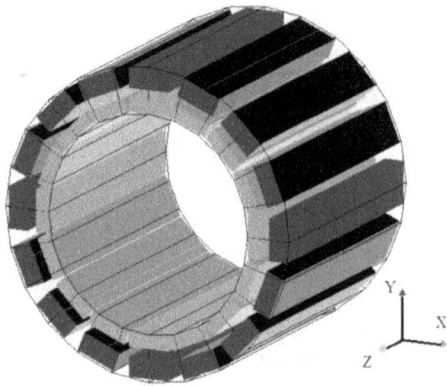

Figure 2: Small animal PET scanner with two stacked detection layers illustrated in grey.

2.1 Scanner

The simulated PET scanner is analogous to the one proposed by Zhang et al. [3]. As a key feature, it consists of two stacked detection planes (see Fig. 1 and Fig. 2), with 16 detection heads distributed on a ring. The inner and outer diameters of the scanner are 32.9 mm and 47 mm, respectively; the axial length is 77.22 mm. Each head consists of 6 detector blocks, each of one holding 9×9 scintillation crystals; this implies 7776 crystals per plane. The planes are separated by a 3 mm thick layer of air. The crystals are 4 mm long in the first plane, and 6 mm long in the second plane. The crystal cross-section is 1.422 mm $\times$ 1.422 mm, and the crystal pitch is 1.43 mm. The simulated scintillation material was cerium-doped lutetium oxy-orthosilicate (LSO).

2.2 Phantom

The simulated phantom is a modified version of the MicroPET IQ Phantom by NEMA [4]. The phantom consists of a cylinder with a diameter of 30 mm and a length of 48 mm which is divided into three regions, as shown in Fig. 3. Part A has a length of 14 mm and was filled with activity in our simulation experiments. In this region, two rods with diameter of 8 mm are placed; one rod contained no activity ("cold"); the ratio between the activity concentration within the other rod ("hot") and the background was set to 5:1. Part B only contained background activity, in the same concentration as in the background of A. In part C, 5 differently sized rods are placed in a cold background. The activity concentration in the rods was equal to the one in the background regions of A and B. The diameters of the rods differ by one millimetre, ranging from 1 mm to 5 mm. The rod length is 20 mm.

2.3 Simulation

To focus on the problem at hand, a system with perfect energy and time resolution was considered, and the radioactive decay was simplified by the emission of two antiparallel photons with an energy of 511 keV each. No atten-

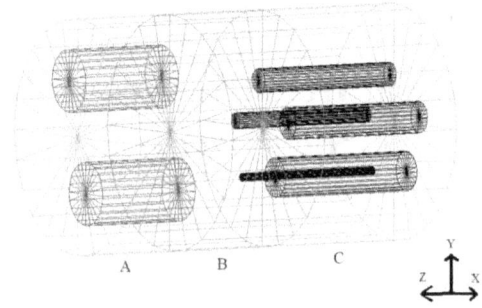

Figure 3: Simulated phantom. Part A consists of the hot and cold rods in uniform background, B is the uniform part and C consists of the differently sized rods in cold background.

uation media within the field of view was simulated. The total activity within the phantom was 4.255 MBq, distributed according to the aforementioned activity concentration ratios. The simulated scan time was 12 seconds to generate data of low statistical quality. Three independent simulations with these parameters were run to estimate statistical fluctuations.

2.4 Data Processing

The output of our GATE simulations was a list of single events ("singles"). A single arises when a photon interacts one or more times within a single crystal, either through Compton scattering or photoelectric absorption. If a photon deposits energy multiple times within the same crystal, the energy is added up and only one single is formed. To group those singles assumed to originate from the same emission, a coincidence sorter was implemented in MATLAB [5]: the first single of the list opens a time window of 1 ns, so that all singles that fall into this window are sorted together into a coincidence. The next single which lies outside the first window opens a new window with the same width, and so on. According to the number of singles within a window we can distinguish between "doubles", i.e., normal coincidences made of two singles, and multiple coincidences (more than two singles). Within the latter, we focused on those events containing three singles ("triples"). Higher-order coincidences were discarded. Both doubles and triples can be divided into two categories: Events whose photons originate from the same emission ("trues") or from different emissions ("random"). Fig. 1 shows a schema of a double (D) and triple (T): A double is accepted if the summed energy of two singles (SD1 and SD2) is equal to 511 keV. The corresponding LoR links SD1 and SD2, as shown in Fig. 1 (left). The triple coincidence is made of a pair, ST2 and ST3, and a remaining single (ST1). The event is further considered if the energy of ST1 is equal to 511 keV, and the sum of the other energy depositions is also equal to 511 keV. Given that the true sequence of interactions remains unknown, two possible LoR can be assigned: between ST1 and ST2 (true LoR) or between ST1 and ST3. In this work, the single with the highest energy was selected as an endpoint for the LoR.

2.5   Reconstruction

We have implemented the maximum likelihood expectation maximization (MLEM) algorithm, which is commonly used as reference algorithm in emission tomography. The iterative process of MLEM is described as follows [6]:

$$f_j^{(k+1)} = \frac{f_j^{(k)}}{\sum_{i=1}^{I} a_{ij}} \sum_{i=1}^{I} \frac{g_i}{\sum_{j'=1}^{J} a_{ij'} f_{j'}^{(k)}} a_{ij}, \qquad (1)$$

where f_j^k corresponds to the reconstructed activity in image voxel j after the k-th iteration, g_i corresponds to the number of coincidences detected in LoR i, and a_{ij} is a component of the so-called system matrix linking i and j. The measured data obtained by GATE were binned into a LoR histogram, and only non-zero entries were stored. The values a_{ij} were calculated on-the-fly using Siddon's ray-tracing [7]. Due to memory constrains, pre-calculation of the system matrix was not possible. Attenuation of photons by the first layer of crystals were also modelled into the sytem matrix. To this aim, attenuation weights were calculated based on the Beer–Lambert law. These weights were pre-calculated and included in the reconstruction. For each simulation, two data sets were generated: one set with only doubles, and another set with doubles and triples. Both data sets were reconstructed, and for each case MLEM was iterated until the difference between two consecutive images was very small. To quantify the difference, we used the sum of squared differences (SSD). The image grid was made of $65 \times 65 \times 125$ cubic voxels of length 0.6154 mm.

2.6   Figures of Merit

Various quality criteria were considered:

- The uniformity of the part B of the phantom:

$$\text{UNI}(\%) = (\sigma_B/\mu_B) \cdot 100, \qquad (2)$$

where σ and μ correspond to the standard deviation and the mean in the region of interest.

- The spill-over ratio (SOR) using the cold region (Co) with a volume of interest (VOI) with a diameter of 4 mm:

$$\text{SOR} = \mu_{Co}/\mu_B. \qquad (3)$$

- The contrast-to-noise ratio (CNR) between the hot region (H) and the surrounding background (HB) with VOIs with a diameter of 75 % of the physical diameter:

$$\text{CNR} = (\mu_H - \mu_{HB})/\sigma_{HB}. \qquad (4)$$

- The recovery coefficient (RC) of each rod from part C of the phantom with VOIs with a diameter of 75 % of the physical diameter:

$$\text{RC} = \mu_{rod}/\mu_B. \qquad (5)$$

Ideally, SOR = 0 and RC = 1. To reduce statistical fluctuations three data sets were used to calculate the mean and standard deviation of the aforementioned figures of merit.

3   Results and Discussion

First, the rate of misidentificated LoRs (ROM) for true triple events was determined which is possible because of the perfect time resolution. The number of trues and randoms and the ROM are shown in Table 1. It can be seen that the selection method yield a very poor identification rate.

	doubles	triples
trues	$12.69 \cdot 10^5 \pm 900$	$15.46 \cdot 10^5 \pm 1200$
randoms	7000 ± 110	8400 ± 120
ROM	-	49 %

Table 1: Number of true and random coincidences for reconstruction with only doubles and with triples included and the ROM.

The number of iterations used was 25 for doubles, and 35 when triples were also included. Further iterations did hardly change the SSD values.

The results for the uniformity, SOR and CNR are shown in Table 2. The SOR worsened when triples were included into the reconstruction. This is due to the high number of misidentified LoRs. The CNR of the hot region showed no significant differences in both reconstructions, while the uniformity improved when triples were considered. The latter effect might be due to the increase in the number of reconstructed events, which reduce the statistical noise amplification intrinsic to MLEM.

	doubles	doubles and triples
UNI (%)	73.3 ± 0.5	62.1 ± 0.3
SOR	0.171 ± 0.007	0.204 ± 0.014
CNR	5.67 ± 0.19	5.59 ± 0.34

Table 2: Mean and standard deviation of UNI, SOR and CNR for reconstruction of doubles and doubles with triples.

Fig. 4(a) shows a transaxial slice of part (C) of the simulated phantom. The reconstructed images for the same slice are shown in Fig. 4(b) for only doubles and Fig. 4(c) for doubles and triples. The smallest rod cannot be visually identified in both reconstructions. The second smallest rod is only visible in the reconstruction including triples. The other rods are also better reconstructed if triples are taken into account. This visual appreciation is supported by the quantitative evaluation of the RC (see graph of Fig. 5). In this graph, the change in RC as a function of the diameter of the rods is plotted. The rod with a diameter of 1 mm could not be reconstructed at all. In general, the smaller the rod diameter, the lower the RC. Including triples into the reconstruction had a large impact on the RC, which was significantly improved for all rods considered. The degradation of the RC as a function of the diameter is more significant when only doubles are reconstructed. The most significant observation is that the rod with a diameter of

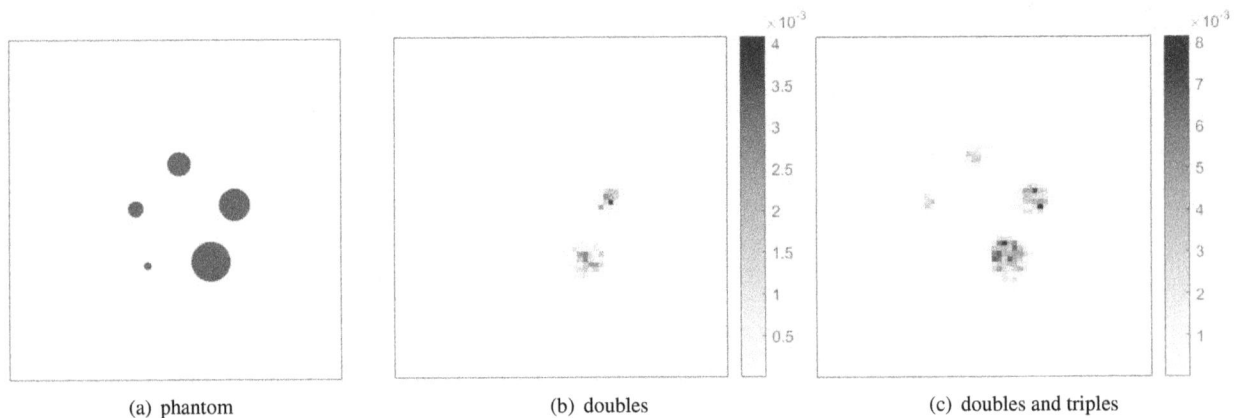

(a) phantom   (b) doubles   (c) doubles and triples

Figure 4: Emission phantom (a) and exemplary reconstructions using doubles (b) and using doubles and triples (c).

2 mm is only reconstructed if triples are included. This indicates that lesion detectability can be enhanced significantly thanks to the sensitivity increase resulting from the inclusion of triples.

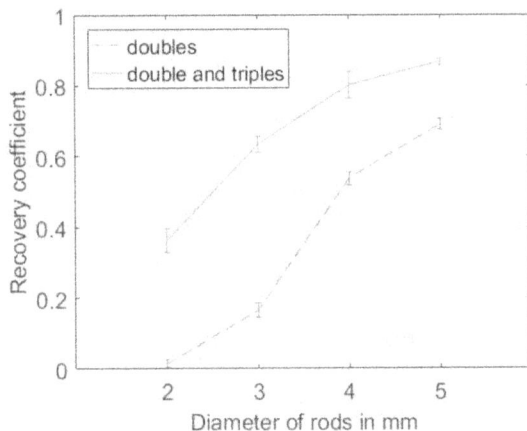

Figure 5: RC in part C of the phantom plotted against the diameter of the rods.

4 Conclusion

This investigation focused on evaluating the impact on the reconstructed image of inter-crystal scatter events for low-statistics scans. With dual-layer PET systems a higher number of ICS events in comparison to doubles are expected, so that triples resulting from ICS might offer a way to improve the sensitivity in software. The simulation study showed under perfect conditions that the detectability of small structures improves significantly under consideration of triples. The results are excellent if we consider that the rate of misidentification for triples was about 50 percent. Further improvements might be obtained if more sophisticated identification methods are used. On the other hand, the phantom did not contain any attenuation media. Further investigations should study the effect of triples under more realistic conditions.

Acknowledgement

The simulations were performed with resources provided by the North-German Supercomputing Alliance (HLRN). The work has been carried out at and supervised by the Institute of Medical Engineering, Universität zu Lübeck.

5 References

[1] M. Rafecas et al., *Inter-crystal scatter in a dual layer, high resolution LSO-APD positron emission tomograph*. Physics in Medicine and Biology, vol. 48, no. 7, pp. 821–848, 2003.

[2] S. Jan et al., *GATE: a simulation toolkit for PET and SPECT*. Physics in Medicine and Biology, vol. 49, no. 19, pp. 4543–4561, 2004.

[3] X. Zhang et al., *Development and evaluation of a LOR-based image reconstruction with 3D system response modeling for a PET insert with dual-layer offset crystal design*. Physics in Medicine and Biology, vol. 58, no. 23, pp. 8379–8399, 2013.

[4] National Electrical Manufacturers Association, *NEMA Standard Publication NU4-2008: Performance Measurements of Small Animal Positron Emission Tomographs*. Rosslyn, VA: National Electrical Manufacturers Association, 2008.

[5] MATLAB R2016a, The MathWorks, Inc., Natick, Massachusetts, United States.

[6] K. Lange and R. Carson, *EM reconstruction algorithms for emission and transmission tomography*. Journal of Computer Assisted Tomography, vol. 8, no. 2, pp. 306–316, 1984.

[7] R. L. Siddon, *Fast calculation of the exact radiological path for a three-dimensional CT array*. Medical Physics, vol. 12, no. 2, pp. 252–255, 1985.

Influence of Affine Image Registration on the Calculation of Diffusion Properties with a Kurtosis Model in Diffusion Weighted Imaging

K. Zantop [1], N. Kartalis [2], M. Heinrich [3], and R. Moreno [4]

[1] Medizinische Ingenieurwissenschaft, Universität zu Lübeck, karen.zantop@student.uni-luebeck.dc
[2] Department of Clinical Science, Intervention and Technology, Karolinska Institutet, nikolaos.kartalis@ki.se
[3] Institute of Medical Informatics, Universität zu Lübeck, heinrich@imi.uni-luebeck.de
[4] School of Technology and Health, Kungliga Tekniska Högskolan, rodrigo.moreno@sth.kth.se

Abstract

Diffusion weighted imaging (DWI) is a magnetic resonance imaging (MRI) technique which enables radiologist the measurement of tissue diffusion properties in the human body. Based on its diffusion coefficients tumorous tissue can be characterized or treatment response can be assessed or predicted. Motion, as for example due to respiration jeopardizes the computation of the diffusion properties. We registered series of diffusion weighted images before computing kurtosis coefficients of patients suffering from pancreatic cancer. This paper presents a comparison of diffusion weighted image series acquired at breath-holding, free-breathing, and respiratory-triggered modes. We found that affine registration mildly improves the calculation of kurtosis coefficients in the pancreas, but our findings would have to be validated in a larger cohort.

1 Introduction

Diffusion weighted imaging (DWI) is a magnetic resonance imaging (MRI) method. It is used for acute ischemic stroke detection [1], structural connectivity mapping of the brain [2], and neurosurgery planning [3]. In oncological imaging it has increasingly become a routine component in tumor detection, classification and treatment response in breast and liver cancers [4, 5, 6, 7].

DWI is sensible to diffusion at the cellular level. The objective of diffusion weighting in MRI is to detect diffusion movement by signal decay. This is interesting for oncological imaging, since malignant and benign tissues have different diffusion properties. Malignant tissues tend to have less extra- and more intracellular space. This restricts diffusion and leads to lesser signal attenuation in images with high diffusion weighting. On the other hand benign tissues have more extracellular space, which allows diffusion and thus results in a greater signal loss in images with high diffusion weighting. In general, diffusion is described by Fick's Law of Diffusion, given by

$$J = -D\nabla c. \qquad (1)$$

It connects the particle flux J to the diffusion coefficient D and the particle concentration c. The aim of DWI is to mea- sure the diffusion coefficient as in (1). Therefore, the image voxels are sequentially imaged with a varying diffusion gradient pulse that is described with the b-value. It is dependent on the gradient magnitude G of the diffusion weighting MRI pulse and given by

$$b = (\gamma\delta G)^2(\Delta - \delta/3) \qquad (2)$$

where δ is the gradient pulse duration of the DWI-sequence and Δ the diffusion time [9]. The obtained voxel intensities S_b with regards to the weighting b can then be modeled with an exponential function. It is in the simplest case given by

$$S_b/S_0 = e^{-bD}. \qquad (3)$$

S_0 is in this context the voxel intensity in the absent of a diffusion gradient ($b = 0$). The diffusion coefficient D is in practice also called Apparent Diffusion Coefficient (ADC). It models all occurring diffusion movements with the assumption of a Gaussian diffusion. A model not assuming Gaussian diffusion is the Kurtosis model, described with

$$S_b/S_0 = e^{-bD_K + b^2 D_K^2 \frac{K}{6}}. \qquad (4)$$

In this context, K is specified as the kurtosis coefficient that describes the grade of deviation from the Gaussian distribution, and D_K is the corrected diffusion coefficient [8]. It is crucial to provide aligned images of the diffusion weighted series to be able to compute valid diffusion coefficients. Therefore, the alignment of the volume of interest (VOI) in the image sequences obtained with different diffusion weightings is important. Motion of the pancreas while breathing is commonly observed [9]. Two alternative acquisition methods to circumvent free-breathing (FB) motion are breath-holding (BH) or respiratory triggering (RT).

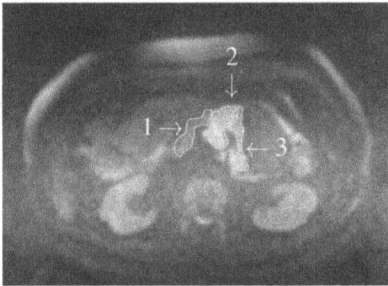

Figure 1: Slice of a patient's free breathing diffusion weighting ($b = 1000$) reference image. In this slice, the three sub-VOIs are visualized. The sub-VOI on the left is the upstream parenchyma (1, arrow pointing to the right). In the middle, the tumor was selected (2, arrow down) and on the right-hand side, the downstream parenchyma is shown (3, arrow left).

However, these methods also might be affected by motion. In this study we investigated the registration of diffusion weighted series with affine registration and compared them to the pre-registered data. Further, we calculated the diffusion coefficients in the kurtosis model.

2   Material and Methods

2.1   Study Population

The study included 15 patients. Of these, 7 were female, 8 male, with a mean age of 64 ± 7 years ($\pm$ standard deviation) and an age range of 54 to 77 years. The inclusion criteria of the study were: (a) Histopathological proof of pancreatic ductal adenocarcinoma (PDAC), (b) No history of previous chemo- or radiation therapy and (c) no contraindication for MR examination. The pancreas tumors had an average diameter of 3.2 ± 0.6 cm. One was in the body, 12 in the head and two in the tail of the pancreas. The pancreatic tissue to the right of the tumor is termed upstream (closer to the tail of the pancreas) and to the left of the tumor is termed downstream (closer to the head of the pancreas, see Fig. 1) parenchyma. In conclusion, the study has 15 whole volume regions of interest (VOIs), consisting of 15 tumor, 11 upstream parenchyma and 9 downstream parenchyma sub-VOIs. The study was approved by the local ethics committee.

2.2   MRI technique

A clinical 1.5 T scanner (Magnetom Avanto, Siemens Healthcare, Erlangen, Germany) was used. For all patients, the same single-shot spin-echo echo-planar DWI datasets were acquired (Table 1). The diffusion weighting gradients were applied in a tetrahedral scheme. Parallel imaging factor was 2 and the spectral selective fat saturation pulse was used. To maintain sufficient signal-to-noise ratio, 5 averages for all b-value acquisitions were chosen, except BH, where 2 averages were chosen. The DWI datasets contained

Table 1: DWI imaging parameters. Abbreviations: TE = echo time, TR = repetition time.

Imaging plane	Voxel size (mm)	
Axial	2.1x2.1x5	
Slice thickness/gap (mm)	TE (ms)	TR (ms)
5/0	75	2400

three images, taken breath hold (BH), free breathing (FB) and respiratory triggered (RT, with 3D PACE (Prospective Acquisition CorrEction, Siemens Healthcare)). Those in turn were serially imaged with five different diffusion weightings ($b = [0, 50, 300, 600, 1000]$). Additionally, a fourth FB image with $b = 1000$ was taken as a reference (REF). The three sequences were taken in the same session.

2.3   Image Processing

The pancreas VOIs were carefully drawn by hand, in REF, by a radiologist (NK) with 7 years of experience in pancreatic imaging. In the next step, the manually marked VOIs in REF were transferred to the $b = 0$ series of BH, FB and RT (see Fig. 2). For this, we used the MERIT rigid registration module of MeVisLab ([10], MeVis Medical Solutions AG & Fraunhofer MEVIS, Bremen, Germany). To be able to evaluate the outcome of pre-processing registration on the data, the BH, FB and RT images were duplicated. One copy stayed unchanged. The other was modified by registering the series of $b = [50, 300, 600, 1000]$ individually to the corresponding $b = 0$ image (see Fig. 3). MATLAB (Mathworks, Natick, Massachusetts, USA) was used for performing affine image registration with Mutual Information as similarity measurement and the step gradient de-

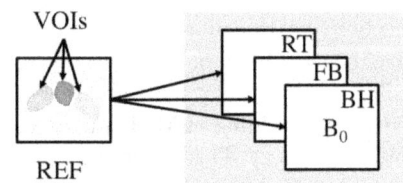

Figure 2: Registration of REF to the $b = 0$ series of BH, FB and RT. The obtained transformation was used to transform the sub-VOIs.

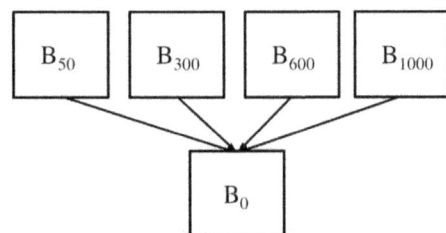

Figure 3: Scheme of the individual registration of the $b = [50, 300, 600, 1000]$ series to the corresponding $b = 0$ image.

Table 2: Mean R^2-error for the breathing acquisitions (BH, FB, RT), the tissue/parenchyma types before (b.R.) and after registration (a.R.).

	Tissue	$R^2 \pm$ std b.R.	$R^2 \pm$ std a.R
	Upstream	0.84 ± 0.19	0.84 ± 0.19
BH	Tumor	0.88 ± 0.14	0.89 ± 0.14
	Downstream	0.88 ± 0.14	0.89 ± 0.14
	Upstream	0.91 ± 0.11	0.92 ± 0.11
FB	Tumor	0.94 ± 0.08	0.95 ± 0.08
	Downstream	0.89 ± 0.11	0.90 ± 0.11
	Upstream	0.91 ± 0.11	0.91 ± 0.11
RT	Tumor	0.93 ± 0.09	0.94 ± 0.07
	Downstream	0.91 ± 0.10	0.92 ± 0.09

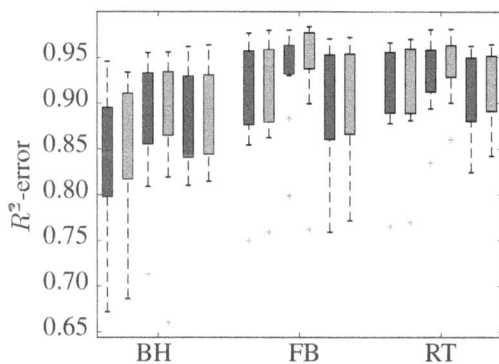

(a)

(b)

Figure 5: Boxplot of median corrected diffusion coefficient D_K (a) and the median kurtosis coefficient K (b). The y-axis is in $mm^2/s10^{-3}$ (a) respectively the unit-less value of K (b). The x-axis shows the breathing acquisition types BH, FB and RT. For each of the breathing acquisitions and parenchyma types, the patients median D_K (a) and median K (b) is plotted in boxplots. This results in six boxplots per acquisition. In these plots, the data is further subdivided in two boxplots per VOI: upstream (left), tumor (middle) and downstream (right). Dark grey is before, light grey after affine registration.

Figure 4: Mean R^2-error of the kurtosis fitting (y-axis). Each boxplot contains each patients mean R^2. The boxplots are divided in groups of six in the breathing acquisitions breath-holding, free-breathing and respiratory triggered. In these groups, the data is further sub-divided in two box- plots per VOI: upstream, tumor and downstream. Dark grey is before, light grey after affine registration.

scent algorithm as optimizer.

2.4   Evaluation methods

To evaluate the results of the model fitting to the acquired data, the coefficient of variation R^2 was used. In this method, the value of diffusion weighting is compared to the value of the calculated fit at the same b-value. R^2 is calculated with

$$R^2 = 1 - \frac{\sum(y(x) - f(x))^2}{\sum y(x)^2}, \qquad (5)$$

where $y(x)$ is the pixel intensity for all diffusion weightings and the model fit is described with $f(x)$.

3   Results and Discussion

Inspection of Table 2 and Fig. 4 shows that affine registration only mildly improves the curve fitting of the mean R^2-error. The best result is obtained by free-breathing acquisi-

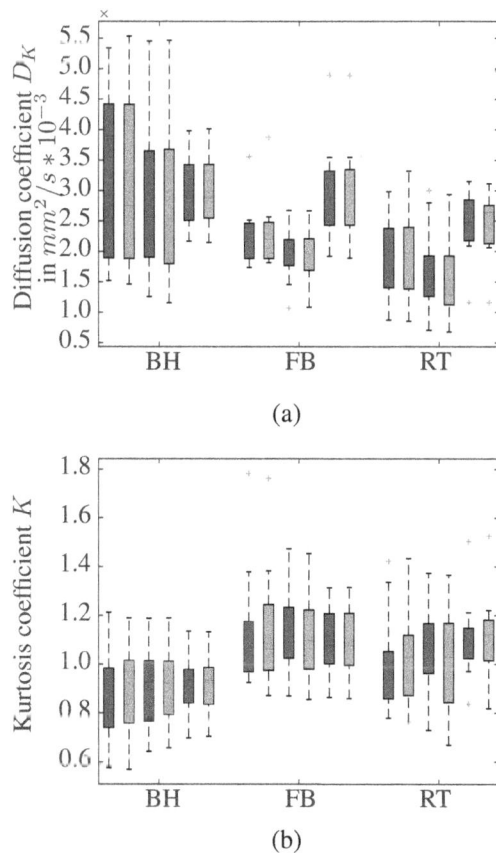

tion and the tumor VOI with $R^2 = 0.94 \pm 0.08$ before registration and $R^2 = 0.95 \pm 0.08$ after affine registration. Second best is RT acquisition, tumor, with $R^2 = 0.94 \pm 0.07$ and third FB, upstream with $R^2 = 0.92 \pm 0.11$ and RT, downstream parenchyma with $R^2 = 0.92 \pm 0.09$. This implies that data acquired with the FB-technique are sufficient and no substantial improvement is anticipated by registration of the image series upon which the calculation of kurtosis metrics are based. The good results for FB could be explained by different pancreatic tissue in patients with cancer compared to pancreas-healthy individuals, which results in diminished motion. The results for the kurtosis coefficients are plotted in Fig. 5 (a) (D_K) and Fig. 5 (b) (K). The median was used to eliminate outliers. Reference [11] showed, that there is a statistically significant difference in the D_K-value for the different tissue types upstream, tumor and downstream. As shown, for the breathing acquisitions FB and RT, we could obtain similar results. Still the distinction of the values and their median is limited to the breathing acquisition type.

4 Conclusion

Affine registration mildly improved the fitting of the kurtosis fit to the data. Breathing motion may cause local deformations to the pancreas, therefore nonrigid registration could further improve the alignment. Prior to the study, we assumed FB would obtain worse results than FB and RT image acquisition. However, the results might indicate that FB can obtain similar good results than respiratory triggered for the pancreas. This needs to be validated in a larger cohort. However, the results for the D_K-coefficient are distinct for the healthy/tumorous tissue types. The significance of these differences needs to be statistically investigated. Limitations of this study are the number of patients and b-values that were used. It has been shown that kurtosis fitting requires high b-value regions around $b = 2000$ [12]. The maximum value used in this study is $b = 1000$. Additionally, some literature states that six distinct gradient directions are insufficient to be able to compute valid fitting [12]. This implies that the study needs to be validated in a larger cohort and a greater range of b-values. In conclusion, we found that affine registration improves only slightly the results of the three different DWI acquisition modes, namely FB, BH, and RT in patients with pancreatic cancer.

Acknowledgement

The work has been carried out at Kungliga Tekniska Högskolan, School of Technology and Health, in cooperation with Karolinska Institutet and supervised by the Institute of Medical Informatics, Universität zu Lübeck.

5 References

[1] S. Warach, J. Gaa, B. Siewert, P. Wielopolski, and R. R. Edelman, Acute human stroke studied by whole brain echo planar diffusion-weighted magnetic reso- nance imaging, Annals of Neurology, vol. 37, no. 2, pp. 231–241, 1995.

[2] P. Hagmann et al. DTI mapping of human brain connectivity: statistical fibre tracking and virtual dissection, NeuroImage, 19(3):545 – 554, 2003.

[3] S. Dimou, R. A. Battisti, D. F. Hermens, and J. Lagopoulos. A systematic review of functional magnetic resonance imaging and diffusion tensor imaging modalities used in presurgical planning of brain tumour resection, Neurosurgical Review, 36(2):205–214, Apr. 2013.

[4] A. R. Padhani et al. Diffusion-weighted magnetic resonance imaging as a cancer biomarker: Consensus and recommendations, Neoplasia, 11(2):102–125, 2009.

[5] F. C. Schmeel et al. Diffusion-weighted magnetic resonance imaging predicts survival in patients with liver-predominant metastatic colorectal cancer shortly after selective internal radiation therapy, European Radiology, 27(3):966– 975, 2017.

[6] R. Rakow-Penner, P. M. Murphy, A. Dale and H. Ojeda-Fournier. State of the art diffusion weighted imaging in the breast: Recommended protocol, Current Radiology Reports, 5(1):3, 2017.

[7] L. Mannelli, S. Nougaret, H. A. Vargas, and K. G. Richard Advances in diffusion-weighted imaging, Radiologic Clinics of North America, 53, 2015.

[8] T. Ichikawa et al. High-b value diffusion-weighted MRI for detecting pancreatic adenocarcinoma: preliminary results, American Journal of Roentgenology, 188(2):409–414, 2007.

[9] V. S. Chandail, D. K. Bhasin, S. S. Rana. *The pancreas and respiration. Oblivious to the obvious!*, Journal of the Pancreas, 7(6), 2006

[10] T. Boehler, D. van Straaten, S. Wirtz, and H.-O. Peitgen. A robust and extendible framework for medical image registration focused on rapid clinical application deployment, Computers in Biology and Medicine, 41(6):340 – 349, 2011.

[11] N. Kartalis et al. Diffusion- weighted mr imaging of pancreatic cancer: A comparison of mono-exponential, bi-exponential and non-gaussian kurtosis models, European Journal of Radiology Open, 3:79–85, 2016.

[12] J. H. Jensen and J. A. Helpern. *MRI quantification of non-Gaussian water diffusion by kurtosis analysis.* NMR in Biomedicine 23.7 (2010): 698–710

Automated detection of vesicles in electron microscope images by using deep convolutional neural networks

N. Ghanad Poor [1], J. Lotz [2], M. Kleint [3],

[1] Biomedical Engineering, University of Applied Sciences Lübeck, niema.ghanad.poor@stud.fh-luebeck.de

[2] Fraunhofer MEVIS, Lübeck, johannes.lotz@mevis.fraunhofer.de

[3] Institute of Experimental and Clinical Pharmacology and Toxicology, University of Lübeck, maximillian.kleint@pharma.uni-luebeck.de

Abstract

The manual process of counting vesicles in the blood-brain barrier in order to investigate a hormonal connection to the sleep cycle is complex and time-consuming. An automated analysis of such EM images would increase the efficiency of this and comparable analysis tasks. We trained deep convolutional neural networks with two different approaches on an expert labeled dataset of real EM images of mouse brains. A pixel-wise network receives patches of $101 \times 101 \times 3$ pixels and evaluates the central pixels as either a vesicle or background. The second, fully convolutional network evaluates complete batches of $512 \times 512 \times 3$ pixels in order to distinguish vesicles from background by segmentation. Both networks showed difficulties differentiating vesicles when trained with similar objects. In order to make the problem easier to train, we will automate the process of finding the ROI as a next step.

1   Introduction

The blood-brain barrier (BBB), which consists of endothelial cells joined by tight junctions and forming a physical barrier to solutes [1], has been the subject of many publications [2]. Some macromolecules like proteins and peptides are transported through the BBB via vesicles [3]. The underlying project to this work is an investigation on the connection between that protein transport and the sleep cycle of mice. In order to prove the hypothesis that the protein transport is cyclic dependent on the point of the sleep cycle (more transport during the dark phase), electron microscopic (EM) images of mouse brains were taken at different times during one cycle. By counting the amount of vesicles it is assumed to get a direct indication on the protein exchange.

The vesicles of interest can be observed in Fig. 1. In general the vesicles have a round (sometimes a slightly oval) shape. During the transition phase, they can be observed as constrictions of the membrane. In total, the vesicles are distinguishable into luminal, abluminal and cytosolic vesicles, of which the first two are in formation phase and the last is passing through the endothelial cell's inside.

Furthermore, the vesicles of interest must be differentiated from other vesicles, which transport other substances or differ in size and location. While the location is supposed to be in or at the endothelial cells, the maximum size limit for our case is defined as 100 nm.

As it can be complicated and very time-consuming to manually evaluate such EM images, we aim to develop an au-

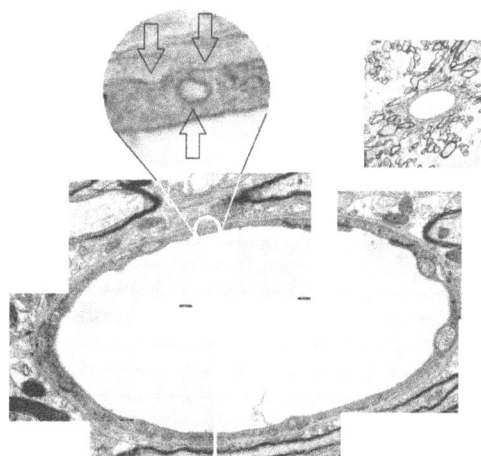

Figure 1: A collage with an overview of the BBB and a magnification of the endothelial cells. The magnification is superimposed from several images. The zoomed window shows two abluminal and one cytosolic vesciles.

tomated method. In this paper we examine the possibilities of detecting and counting these vesicles based on already existing methods used for comparable image recognition tasks. With the algorithmic capabilities of deep learning (DL) and the general computational power increasing, more and more medical applications make use of artificial intelligence [4, 5].

Previous work has shown success in detecting mitosis in multi-spectral images [6], counting cells in different types of microscopy images [7] or nucleus detection in 3D fluorescence microscopy [8]. While they all follow different approaches in how they detect the desired objects, most of them use a convolutional neural network (CNN) in their algorithm, since CNNs have been proven particularly efficient on image recognition tasks [4]. In particular, the so called AlexNet [9] won the the ImageNet Large Scale Visual Recognition Challenge (ILSVRC) of 2012 [10].

Unlike conventional neural networks, CNNs use convolution to learn the spatial topology of images [11]. Furthermore, the reduced number of weights leads to a less complex architecture [12].

Deep convolutional neural networks (DCNN) are often favored for working with images, because of the simple import of the images into the network and (in contrast to other networks) the inclusion of topological properties. Based on this investigation of neural networks and the amount of training data available, we chose to work with a simplified version of the AlexNet.

2   Material and Methods

We decided to reduce the complexity of the AlexNet, because it needs to be adapted to the quite small amount of data. Besides, it is easily adaptable to different input shapes and variables in its output principle.

The given dataset of 241 collages and a total amount of 4932 marked vesicles consists of EM images taken from slices of mouse brains with a thickness of 50 nm.

Every single collage of the dataset is combined from at least two images (compare Fig. 1), which have overlapping areas of different sizes. The collage's size can reach up to $5000 \times 6000 \times 3$ pixels. The collage is set on a large white background and has an additional overview of the surrounding area in the corner, which both are needless for the detection process. The collages are put together and the vesicles marked by a doctoral student with a background in biotechnology (B.Sc.) and toxicology (M.Sc.).

In a first step, we manually extended the expert annotations to a binary mask. The mask then mapped the full vesicle with ones and the background with zeros. With these new masks the quantity of the vesicles could be determined as it can be seen in Table 1.

Furthermore, the areal ratios between the actual image, its background (rectangular area around the patched images), a region of interest (ROI) and the marked vesicles were calculated in Table 2 and are visualized in Fig. 2.

Table 1: Frequency of vesicle occurrence

	Quantity	Min.	Max.
Size	419 ± 339 px	1 px	6179 px
Amount	20.46 ± 14.14	2	89

The large variance in size and amount per image of the vesicles is partly explainable by the different types (abluminal, luminal and cytosolic) and partly by the possible size range of the vesicles of 50 to 100 nm. This makes it very hard

Table 2: Areal ratios between (empty) background (B), image (I), ROI and vesicle (V). Here, the small percentage share of vesicles to image is illustrated.

Ratio	Value	Min.	Max.
I / B	0.7107 ± 0.1340	0.2214	0.9958
ROI / I	0.1132 ± 0.0613	0.0083	0.3736
V / I	0.0035 ± 0.0022	0.0001	0.0155
V / ROI	0.0330 ± 0.0123	0.0136	0.1059

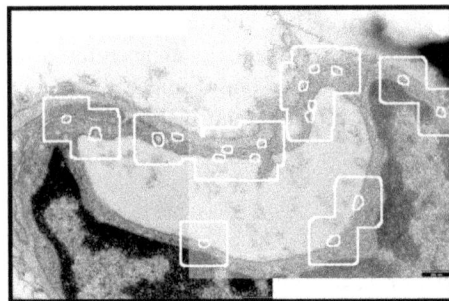

Figure 2: The marked areas exemplarily visualize the ratios, as given in Table 2.

for an untrained person to detect vesicles. Also, the great variance of ratios between the different collage subsections make it difficult to simply use the full image for the training process. With only 0.35% of the image marked as vesicles, the network could achieve high accuracy values with small error, if the weights in the network are trained such that the complete image is marked as non-vesicle. Additionally, possibly present vesicles outside of the ROI are not marked. To reduce this imbalance, we defined a ROI by convolving a patch with a size of 101×101 pixels with the mask, which leads to a ratio of 3.3%.

2.1   Data preparation

As the input size of the AlexNet can be adjusted to any desired value (it is only limited by the memory), we followed two approaches. In the first approach we generate patches of $101 \times 101 \times 3$ pixels for every pixel of the mask within the ROI. In a second approach we take larger cutouts of the image, but make use of the network as a fully convolutional neural network (FCNN) to directly classify the complete patch. This second approach benefits from the CNN's property of equivariance, which adjusts the output size according to the changing input size [12]. Both approaches return probabilities for the two classes *vesicle* and *background*. They only differ in the assignment of the area. The first version does a pixel-wise and the second an areal prediction.

In the first case, saving every patch locally would extend every reasonable storage requirement. The total number of over 70 million patches, each with approximately 0.25 MByte, would allocate more than 16 TByte of physical memory. As a solution we created lists for each image containing information about the position and state of each pixel in the ROI. The FCNN gets larger patches as an input:

512×512×3 pixels. It also needs a subdivision of the image, as the allocated memory for a complete image exceeds the available resources (NVIDIA GeForce GTX 1080 TI 12 GB memory). Here, the much smaller number of partly overlapping patches allows a direct generation during the training.

We implemented the network in Python in the Keras framework with TensorFlow as the backend. By using the environment Keras [13], it is possible to do a parallel precalculation of the patches during the concurrent execution of the training. Before the network is trained, the dataset is randomly subdivided into three batches: training batch (144 images), validation batch (49 images) and testing batch (48 images). This step assures that the validation during the training is independent of the testing afterwards.

2.2 Data training

We let the network train on the training set, while simultaneously performing a validation of the data on the validation set. This division of sets is also supposed to recognize overfitting.

To avoid local optima during the training [12], both approaches have been pre-trained. The major background share has a very large influence on the trained weights. The network would tend to adapt its weightings mostly influenced by the background (compare the ratios in table 2). Without pre-training both networks did not show a decreasing loss in the learning curve.

After pre-training, we trained the pixel-wise network with a balanced set, where positive and negative examples are in balance. In the pre-training step a balanced set of patches is randomly picked from several images and trained for ten epochs. As soon as we observed an increasing learning curve, we started the actual training.

The network for the fully convolutional approach was pre-trained repeatedly with random patches, until a temporary accuracy of 0.7 and an apparent gradient was reached.

With a pre-training accuracy of 84% the pixel-wise DCNN was trained for three epochs with 250.000 steps on unbalanced training batches of 64 and validation batches of eight patches each. As the accuracy converged, it was renounced to take more epochs. With a resulting training accuracy of 0.9924 and validation accuracy of 1, the 48 test images were predicted based on the trained network.

The attempt of repeating this experiment with a balanced training step during the actual training resulted in an accuracy of 0.5, meaning the network didn't reach the desired convergence and couldn't learn to differentiate the vesicles from the background.

The second approach, training with a FCNN, was executed for 48 epochs with one step for the varying number of patches of each image. It reached a training accuracy of 0.8687 and a validation accuracy of 0.9204. Again, we tested the network on several test images.

2.3 Data prediction

We used the test set for a independent evaluation of the networks. Similar to the training, the input of the pixel-wise predictor gets one patch per pixel from the area, which has to be evaluated. The obtained information on each pixel has to be reassembled to an image, resulting in a probability map. The FCNN gets the complete image as an input and directly returns a probability map.

After applying a probability threshold on this resulting map, we evaluated the results by doing a pixel-wise comparison with the masks.

3 Results and Discussion

We used a cross validation of the test images to calculate a confusion matrix. In addition to evaluating the results on the accuracy, we chose four more parameters: sensitivity, specificity, precision and the harmonic average (F1-score). The accuracy of 0.9456 indicates a good success rate, while hinting to the V/ROI ratio of 3.3%. The sensitivity of just 0.0958 shows that only one out of ten vesicle pixels were predicted correctly. Whereas, the specificity of 0.9734 gives evidence that by predicting most of the background pixels correctly, their major share in the image leads to the high accuracy. The precision of 0.1011 and F1-score of 0.0936 both underline this statement, as they show that most of the predicted vesicle pixels belong to the background. The reduction of the background and increased V/ROI ratio (Table 2) didn't achieve the desired result.

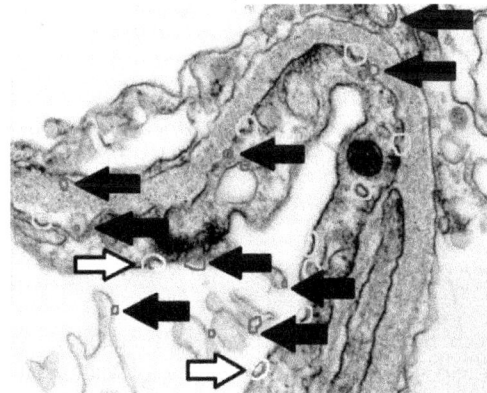

Figure 3: Cutout from a prediction by the pixel-wise DCNN. The white arrows point to correctly found vesicles, the black arrows to incorrectly found ones. Vesicles without an arrow are original expert label.

In the second approach, instead of distinguishing vesicles from the background, we performed an edge-detection. Therefore, we renounced to calculate a cross-validation.

While the FCNN seems to only have learned the characteristic of contrast between the bright interior and the dark edge, the pixel-wise CNN predicts formations similar to the vesicles. Most of these predictions are within the endothelial cell, but do not comply with the vesicle's requirements.

Then again, some predictions are outside of the endothelial cell, hinting that the ROI is not precise enough (compare Fig. 3).

For reasons of comparison, three images were relabeled. In these randomly picked images a high intra-observer variance on the labels became apparent. This shows the complexity of the task and raises the question, how accurate the labels are. Based on these findings we assume that both false positive and false negative labels lead to distorted weightings leading to difficulties in training the algorithm. It is also striking, that labeling cytosolic vesicles is easier for both the expert and the pixel-wise algorithm, as most of the correctly recognized or predicted vesicles were cytosolic.

4 Conclusion

Since the applied methods did not yield a network, that is able to automatically find vesicles, we try to understand, what needs to be changed. With the observations on the vesicle's shape and location, two possible adjustments arise: first, training on cytosolic vesicles only and second, learn to segment the endothelial cell to later use it as a ROI for training vesicle's positions. Since the number of cytosolic vesicles is quite low (<15%), the second approach appears to be a more promising task. Yielding a smaller and more accurate ROI, it can later on be used to return to the initial task. First experiments on automatic segmentation of the endothelial cell were promising.

We further aim to perform a re-annotation of the vesicles with multiple observers (because of the diverging expert annotation). Additionally, the segmentation of the endothelial cell itself yields useful information to the underlying project such as the endothelial cell's size, which enables the calculation of the areal and frequency distribution of the vesicles within these cells.

Acknowledgement

The work has been carried out at Fraunhofer MEVIS, Lübeck and supervised by the Institute of Mathematics and Image Computing, University of Lübeck.

5 References

[1] M. M. S. del Pino, R. A. Hawkins, and D. R. Peterson, "Neutral amino acid transport by the blood-brain barrier. Membrane vesicle studies.," *The Journal of biological chemistry*, vol. 267, no. 36, pp. 25951–25957, 1992.

[2] J. Greenwood, M. Hammarlund-Udenaes, H. C. Jones, A. W. Stitt, R. E. Vandenbrouke, I. A. Romero, M. Campbell, G. Fricker, B. Brodin, H. Manninga, P. J. Gaillard, M. Schwaninger, C. Webster, K. B. Wicher, and M. Khrestchatisky, "Current research into brain barriers and the delivery of therapeutics for neurological diseases: A report on CNS barrier congress London, UK, 2017," *Fluids and Barriers of the CNS*, vol. 14, no. 1, pp. 1–11, 2017.

[3] N. J. Abbott, A. A. Patabendige, D. E. Dolman, S. R. Yusof, and D. J. Begley, "Structure and function of the blood-brain barrier," *Neurobiology of Disease*, vol. 37, no. 1, pp. 13–25, 2010.

[4] H. Greenspan, B. van Ginneken, and R. M. Summers, "Guest Editorial Deep Learning in Medical Imaging: Overview and Future Promise of an Exciting New Technique," *IEEE Transactions on Medical Imaging*, vol. 35, no. 5, pp. 1153–1159, 2016.

[5] I. Kononenko, "Machine learning for medical diagnosis: History, state of the art and perspective," *Artificial Intelligence in Medicine*, vol. 23, no. 1, pp. 89–109, 2001.

[6] D. C. Ciresan, A. Giusti, L. M. Gambardella, and J. Schmidhuber, "Mitosis Detection in Breast Cancer Histology Images using Deep Neural Networks," *Proc Medical Image Computing Computer Assisted Intervention (MICCAI)*, pp. 411–418, 2013.

[7] W. Xie, J. A. Noble, and A. Zisserman, "Microscopy cell counting and detection with fully convolutional regression networks," *Computer Methods in Biomechanics and Biomedical Engineering: Imaging and Visualization*, pp. 1–10, 2016.

[8] F. Xing, Y. Xie, and L. Yang, "An automatic learning-based framework for robust nucleus segmentation," *IEEE Transactions on Medical Imaging*, vol. 35, no. 2, pp. 550–566, 2016.

[9] A. Krizhevsky, I. Sutskever, and G. E. Hinton, "ImageNet Classification with Deep Convolutional Neural Networks," *Advances In Neural Information Processing Systems*, pp. 1–9, 2012.

[10] O. Russakovsky, J. Deng, H. Su, J. Krause, S. Satheesh, S. Ma, Z. Huang, A. Karpathy, A. Khosla, M. Bernstein, A. C. Berg, and L. Fei-Fei, "ImageNet Large Scale Visual Recognition Challenge," *International Journal of Computer Vision*, vol. 115, no. 3, pp. 211–252, 2015.

[11] P. Simard, D. Steinkraus, and J. Platt, "Best practices for convolutional neural networks applied to visual document analysis," *Seventh International Conference on Document Analysis and Recognition, 2003. Proceedings.*, vol. 1, no. Icdar, pp. 958–963, 2003.

[12] W. Liu, Z. Wang, X. Liu, N. Zeng, Y. Liu, and F. E. Alsaadi, "A survey of deep neural network architectures and their applications," *Neurocomputing*, vol. 234, no. October 2016, pp. 11–26, 2017.

[13] F. Chollet *et al.*, "Keras." https://github.com/fchollet/keras, 2015.

A Convolutional Autoencoder for Motion Field Compression

M. Maus [1], T. Parbs [2], and A. Mertins [2]

[1] Medizinische Ingenieurwissenschaft, Universität zu Lübeck, maximilian.maus@student.uni-luebeck.de

[2] Institute for Signal Processing, Universität zu Lübeck, {parbs, mertins}@isip.uni-luebeck.de

Abstract

To compensate for arbitrary motion during MRI acquisition one has to know the motion fields for all sampled data points. One method to reduce the many degrees of freedom is the use of a motion model based on external sensor signals, which accurately describes the motion with much fewer parameters. We trained an autoencoder neural network using synthetic motion vector fields. We used this network to compress similar motion fields and showed that we can reconstruct them from only a few parameters. For the considered vector fields, we show that less than ten parameters are sufficient for vector field parameterization. This compressibility can then be used to approximate motion vector fields used in MRI image reconstruction.

1 Introduction

Patient motion during MRI scans can substantially degrade the quality of reconstructed images and can make an image unacceptable. In particular, respiratory and cardiac motion are major problems in long MRI acquisitions where breath holding is not a option. All methods for motion compensation can be broadly classified in prospective and retrospective techniques [1]. The former tries to correct rigid motion by modulating the magnetic field gradients and RF fields during acquisition. Retrospective motion compensation is motion correction during the reconstruction process. We focus on retrospective motion compensation and we are interested in correcting more complex elastic motion and not only rigid motion. In [2], a generalized reconstruction framework for compensation of arbitrary complex motion by solving a general matrix inversion problem, was proposed. However one needs accurate knowledge of the motion during the acquisition. In [3] the work was extended as an optimization framework for joint reconstruction and compensation of elastic motion called GRICS (Generalized Reconstruction by Inversion of Coupled Systems). But the problem is highly underdetermined, because one needs the knowledge of the motion field for each acquisition time. As a consequence one needs some form of regularization to cope with the large number of unknown motion parameters. In [3], the authors used a motion model as parameterization of the motion fields to make the problem more tractable and use alternating optimization and a multiresolution fixed point scheme. Concretely, they model the motion fields as a linear combination of a few external time signals which correlate with the motion, like respiratory bellows or navigator echoes. We want to expand this framework with a low-dimensional parameterization of the motion fields, which does not rely on additional sensor signals. Since it only works on the measured k-space data, the proposed approach

can be classified as a blind motion compensation. We assume that the motion fields, which for example describe respiratory motion, lie on a low-dimensional manifold which is embedded in high-dimensional space so that we can accurately describe the motion fields with much fewer parameters. Our final goal is to replace the external signals and to use an optimization algorithm on the basis of our low dimensional representation together with a quality measure to find the motion corrected image. In this work we use a deep autoencoder to learn a compressed representation from motion vector fields. Autoencoders are commonly used neural networks for unsupervised learning and their main applications are dimensionality reduction and feature learning. They use an encoder network to learn a code representation from high-dimensional data and a decoder network to recover the data from the code. It has been shown that autoencoders can be seen as a nonlinear generalization of principle component analysis [4]. We show that we can use an autoencoder to learn a low-dimensional representation of motion vector fields and that we can accurately reconstruct a motion field from few parameters. Our autoencoder learns the compression from the data and and not only applies a standard compression algorithm.

2 Material and Methods

In this section, the autoencoder neural network and especially our used network architecture will be described. At the end we further present the two quality metrics we used for evaluation.

2.1 Autoencoders

An autoencoder is an unsupervised neural network that can learn to encode an input into a compact representation and

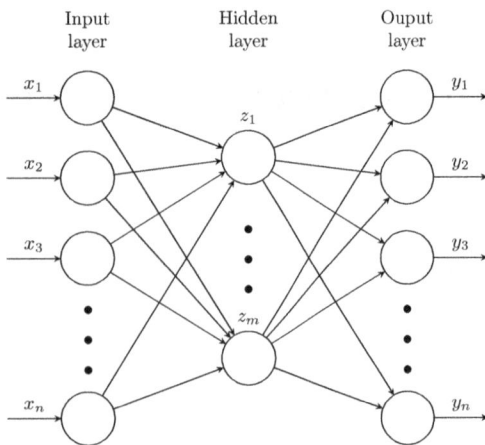

Figure 1: An autoencoder neural network with one hidden layer. By making $m \ll n$ we can force the autoencoder to learn a compression of the input.

decode this code to reconstruct the input as accurately as possible. Autoencoders have a long history and were used successfully in tasks like dimensionality reduction and feature learning [4][5][8]. Unsupervised means that we do not have labels but only a dataset $x_1, x_2, \ldots, x_N \in \mathcal{R}^n$ from which the network has to learn. A standard autoencoder with only one hidden layer, as shown in Fig. 1 takes an input $x \in \mathcal{R}^n$ and maps it to a hidden, or also called latent, representation $z \in \mathcal{R}^m$ using a deterministic function $z = f_\theta(x) = \sigma(Wx + b)$ with parameters $\theta = \{W, b\}$ and a (nonlinear) activation function σ, for example RELU or sigmoid functions. This mapping is called encoder. W is a $m \times n$ weight matrix and b a bias vector. Then the latent representation is reconstructed back to the input space, producing a reconstruction $y = g_{\theta'}(z) = \sigma(W'z + b') \approx x$ with $\theta' = \{W', b'\}$. This function is called decoder. So each training vector x_i is mapped to a corresponding latent vector z_i and then to a reconstruction vector y_i. The autoencoder parameters are optimized to minimize the reconstruction error $\mathcal{L}(x, y)$ between training samples and reconstructions

$$\theta^*, \theta'^* = \arg\min_{\theta, \theta'} \frac{1}{N} \sum_{i=1}^{N} \mathcal{L}(x_i, y_i) \quad (1)$$

where $\mathcal{L}$ is the loss function, e.g. the squared error $\mathcal{L}(x, y) = \|x - y\|^2$. In order to learn a compressed representation and not only the identity function we constrain the hidden layer to have a much smaller dimension than the input layer, as is shown in fig. 1. Consequently the autoencoder has to learn the most important features from the input to reach the best possible reconstruction. Training an autoencoder can be done just like in other feedforward networks, typically calculating gradients with backpropagation and learning by minibatch gradient descent [6].

2.2 Convolutional Autoencoder

Driven by the rapid progress in deep learning and the success of convolutional networks in learning high-level representations, the standard autoencoder was also modified, using deeper architectures with more hidden layers and using convolutional layers instead of fully connected layers [4][6][8].

A CNN is a hierarchical neural network with convolutional layers [6][7][8]. In the convolutional layers, the input is convolved with a set of filters (kernels), followed by some nonlinear activation function. For each filter we get a feature map z^k as output which is given as

$$z^k = \sigma(\sum_{\ell \in L} x^\ell * W^k + b^k). \quad (2)$$

Here z^k is the k-th feature map of the current layer, σ is the activation function, x^ℓ is the ℓ-th of L feature maps from the previous layer or image channels for the first convolutional layer, W^k and b^k are the weights (filters) and biases of the k-th feature map and $*$ is the 2D convolution operation. Convolutional layers are advantageous in many ways. First, the units in a feature map share the same filter weights, because the same filter is applied at every input position. This reduces the amount and redundancy of the parameters. Second, CNNs can account for the local structure in images because of the reduced number of connections. Often a convolutional layer is followed by a pooling layer, which downsamples the feature maps and makes the output invariant to small translations in the input. The most popular form is max-pooling where each output in a feature-map is replaced by the maximum output within a small neighborhood. The max-pooling layer operates independently on each feature map by taking only the maximum value over non-overlapping patches. The most common form is a filter size of 2×2 and a stride of 2, so that each feature map is downsampled by a factor of 2 . A max-pooling layer introduces sparsity over the hidden representations so that there is no need for further weight regularization [8].

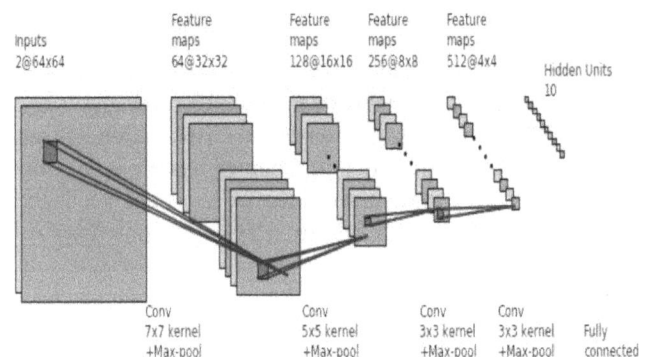

Figure 2: The encoder network. Convolutional layer, RELU and max-pooling are summarized in one block. The encoder compress the motion field from 8192 values to only 10 values. Then the decoder network (not shown) can accurately reconstruct the original motion field from the 10 values.

2.3 Data and Network architecture

The dataset was generated with a MATLAB program which simulates various breathing-like motions with gaussian modes and additionally adds small translations and rotations. We generated 80000 synthetic motion vector fields of size $64 \times 64 \times 2$, where the first channel describes the motion in x-direction and the second channel the motion in y-direction. We used 70000 motion fields for training, 5000 motion fields for validation and 5000 motion fields for testing and normalized the motion to the interval $[-1, 1]$. We used minibatches of 100 motion fields from the training set for training of the network. The weights in each layer were initialized from a zero-mean Gaussian distribution with standard deviation 0.02. Our model was implemented with the deep learning framework TensorFlow. The model was trained using stochastic gradient descent using the ADAM optimization algorithm [9]. The learning rate was initialized to 0.005 and was halved every 10000 iterations. Our loss function is the standard mean squared error. We used four convolutional layers with RELU activation function and 2×2 max-pooling in our encoder. In each convolutional layer the number of filters was doubled, starting from 64 filters in the first layer and reaching 512 in the last convolutional layer in the encoder. Each convolutional layer was followed by the RELU activation function and a 2×2 max-pooling layer which downsamples the feature maps by a factor 2. At the end of the last convolutional layer we had 512 feature maps of size 4×4. Then we used an additional fully-connected layer to reduce the dimension to the desired dimension of the latent space. The design of the encoder is shown in Fig. 2. The decoder has to reverse the operation of the encoder to bring the compressed representation back to original input size. For that we used four transposed convolutional layers with a constant stride of 2 for upsampling, RELU after the first three layers and tanh activation function at the end to squash the output back to the range $[-1, 1]$ of the input data. Further, we used batch normalization after each convolutional and fully connected layer, which normalizes the input to have zero mean and unit variance.

2.4 Quality metrics

The compression with the encoder leads unavoidably to a loss of information or quality, depending on the dimension of the latent space. Besides a subjective comparison of original image and reconstructed image, we also use the mean squared error (MSE) and the similarity index measure (SSIM) as quality metric to measure the similarity between both images. SSIM is considered to be correlated with the quality perception of the human visual system [10]. Given an input image x and the corresponding reconstruction y, the MSE is defined as the average of the squared intensity differences:

$$\text{MSE}(x, y) = \frac{1}{n} \sum_{i=1}^{n} (x_i - y_i)^2. \qquad (3)$$

The SSIM is a well-known quality metric used to measure the similarity between two images. SSIM measures the similarity between images through a combination of three comparisons, namely luminance l, contrast c and structure s [10]:

$$\text{SSIM}(x, y) = l(x, y) \cdot c(x, y) \cdot s(x, y), \qquad (4)$$

$$l(x, y) = \frac{2\mu_x \mu_y + c_1}{\mu_x^2 + \mu_y^2 + c_1}, \qquad (5)$$

$$c(x, y) = \frac{2\sigma_x \sigma_y + c_2}{\sigma_x^2 + \sigma_y^2 + c_2}, \qquad (6)$$

$$s(x, y) = \frac{\sigma_{xy} + c_3}{\sigma_x \sigma_y + c_3}. \qquad (7)$$

Here μ and σ are the mean and the standard deviation and c_1, c_2, c_3 are positive constants. The SSIM returns values in the interval $[-1, 1]$. If both images are equal, the SSIM would be 1, smaller values means less correlation between the images [10].

3 Results and Discussion

We trained the autoencoder for 50000 iterations. Fig. 3 shows the training loss and validation loss for a latent dimension of 32. The reconstruction error decreases as expected and the validation loss follows the training loss, so no overfitting occurs. Next we want to know how good our model can compress motion fields from the test set, which the autoencoder has not seen during training. Because the test motion fields are from the same data generating distribution, the autoencoder should also perform well on the test motion fields. Fig. 4 shows a motion field from the test set and the corresponding reconstruction from a 32 dimensional latent space. It is difficult to see any visual differences between original and reconstruction. For further eval-

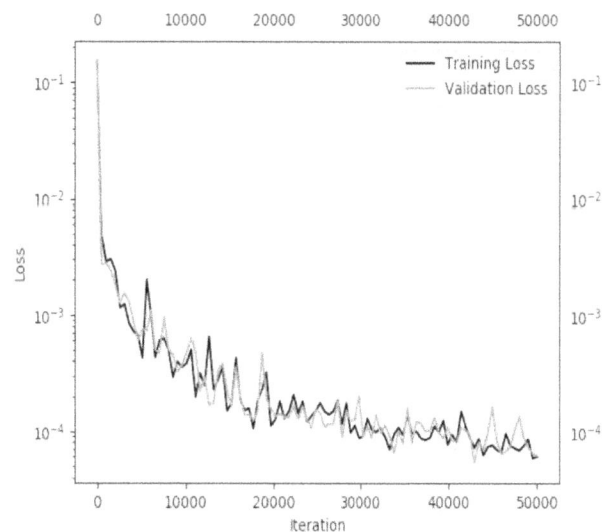

Figure 3: Error on training and validation set. The validation error follows the training error, so our network does not overfit.

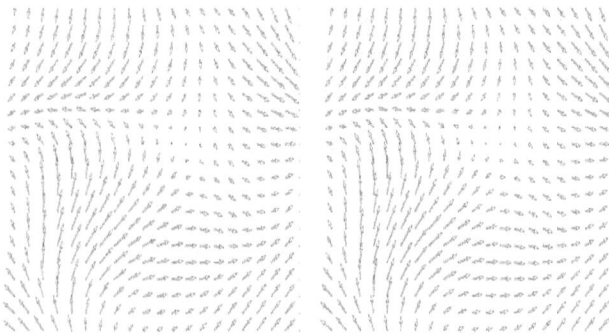

Figure 4: Left: A motion field from the test set. Right: The corresponding reconstruction from the autoencoder. The autoencoder has reconstructed the 8192 values from only 32 values. Differences are hardly visible.

uation we calculated the mean squared error and the SSIM for the test motion fields and take the mean over the test set. We did that for different dimensions of the latent space. The results can be seen in Table 1.

We see that the test error is only slightly larger than the training error. The autoencoder has learned to encode the data properly. He can reconstruct the input and generalizes to motion fields that are similar but which the autoencoder has not seen before during training. So the autoencoder has learned important aspects from which he can reconstruct the motion fields. Only for a mapping into a very low-dimensional latent space we see a much higher MSE and lower SSIM. But already for 8 dimensions we get accurate reconstructions and only slightly better results for more dimensions. However, one drawback of the autoencoder is that nothing forces the autoencoder to map motion fields which describe similar motions in input space to similar positions in latent space. But for our problem we want that small changes in high dimensional space also corresponds to small changes in the low dimensional space.

Table 1: MSE and SSIM for evaluation

Latent space dimension	MSE	SSIM
2	0.0076	0.68
4	0.0013	0.90
8	0.0007	0.91
16	0.0004	0.94
32	0.0004	0.94
64	0.0003	0.93
128	0.0003	0.94

4 Conclusion

We have shown that we can compress motion fields with our autoencoder network by a nonlinear mapping in a low-dimensional space. Next we will investigate how we can use spatial and temporal correlations between motion fields and modifications of our network to get a more controllable compressed representation.

Acknowledgement

The work has been carried out at the Institute for Signal Processing, Universität zu Lübeck and supervised by the Institute for Signal Processing, Universität zu Lübeck.

5 References

[1] A. Loktyushin, H. Nickisch, R. Pohmann and B. Schölkopf, *Blind retrospective motion correction of MR images*. Magnetic Resonance in Medicine, vol. 70, no. 6, pp. 1608–1618, 2012

[2] P. G. Batchelor, D. Atkinson, P. Irarrazaval, D. L. G. Hill, J. Hajnal and D. Larkman, *Matrix description of general motion correction applied to multishot images*. Magnetic Resonance in Medicine, vol. 54, no. 5, pp. 1273–1280, 2005

[3] F. Odille, P. Vuissoz, P. Marie and J. Felblinger, *Generalized Reconstruction by Inversion of Coupled Systems (GRICS) applied to free-breathing MRI*. Magnetic Resonance in Medicine, vol. 60, no. 1, pp. 146–157, 2008

[4] G. Hinton and R. Salakhutdinov, *Reducing the Dimensionality of Data with Neural Networks*. Science, vol. 313, no. 5786, pp. 504-507, 2006

[5] P. Vincent, H. Larochelle, Y. Bengio and P.-A. Manzagol, *Extracting and Composing Robust Features with Denoising Autoencoders*. In Proceedings of the 25th International Conference on Machine Learning, pp. 1096-1103, 2008

[6] I. Goodfellow, Y. Bengio and A. Courville, *Deep Learning*. MIT Press, 2016

[7] Y. LeCun, L. Bottou, Y. Bengio and P. Haffner, *Gradient-based learning applied to document recognition*. Proceedings of the IEEE, vol. 86, pp. 2278-2324, 1998

[8] J. Masci, U. Meier, D. Ciresan and J. Schmidhuber, *Stacked convolutional auto-encoders for hierarchical feature extraction*. In Artificial Neural Networks and Machine Learning–ICANN 2011, pp. 52-59, Springer, 2011

[9] D. Kingma and J. Ba, *Adam: A method for stochastic optimization*. arXiv preprint arXiv:1412.6980 , 2014.

[10] A. Hore and D. Ziou, *Image quality metrics: Psnr vs. ssim*. in IEEE International Conference on Pattern Recognition (ICPR), pp. 2366–2369, 2010

Deep 3D Encoder-Decoder Networks with Applications to Organ Segmentation

N. Bouteldja [1], and M.P. Heinrich [2]

[1] Medical Informatics, Universität zu Lübeck, nassim.bouteldja@student.uni-luebeck.de
[2] Institute of Medical Informatics, Universität zu Lübeck, heinrich@imi.uni-luebeck.de

Abstract

Deep learning approaches have been very successful in segmenting abdominal organs from CT volumes. Despite continuous progress, automated segmentation of the pancreas remains challenging due to its highly complex regional characteristics and large anatomical shape variability. To cope with these challenges, the incorporation of shape priors into neural networks for robust segmentation is an important area of current research. However, recent techniques which incorporate this prior knowledge, fail to train neural networks in an end-to-end manner. In this work, we propose a novel approach that enables us to train encoder-decoder networks simultaneously on manual segmentations and grayvalue CT volumes. The advantage is threefold: It is end-to-end trainable, improves robustness due to strong shape constraints and enables further applications such as coherent shape interpolation due to smooth transitions in the learned shape embedding space. We present high-accuracy segmentation and visual shape interpolation results on the NIH Pancreas-CT dataset.

1 Introduction

Image segmentation plays a key role in medical image analysis. For example, accurate organ segmentations are prerequisites for in-depth computer-aided diagnosis as well as successful radiotherapy treatment planning.

Despite remarkable advances in automated segmentation of various organs [1, 2], accurate approaches for segmenting small organs with large anatomical shape variability, e.g. the pancreas, are still missing. The difficulty arises especially from its highly complex regional characteristics, e.g. lack of strong boundaries, multiple nearby organs with similar appearance, as well as its large inter-patient variability. To cope with these challenges, model-based approaches have been frequently used in the last decade for robust segmentation. They often utilize statistical shape models (SFM) to provide compact shape spaces using principal component analysis. Therefore, spatially segmentation consistency is guaranteed [3].

However, due to the linear nature of the PCA-model, it fails in explaining shape variation which has not been seen in the training dataset. SFMs therefore seem to lack representation ability in contrast to the non-linear, powerful representation-learning and remarkably well-performing neural networks. In addition, incorporating shape priors into these networks can be of great importance, especially when tackling difficult segmentation tasks such as the segmentation of the pancreas. It strongly improves robustness leading to performance increases as shown in [4, 5].

Recent approaches [4, 5, 6] use convolutional auto-encoders (CAE) to learn a low-dimensional shape embedding (Fig.1).

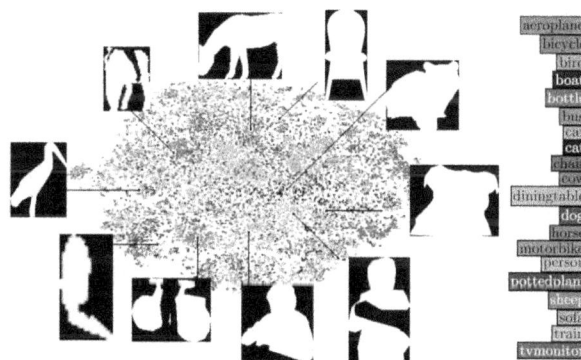

Fig. 1: t-SNE visualization of a learned shape space using a CAE on the Pascal VOC dataset. Similar objects are clustered in close proximity (best viewed in colour, see [6]).

In [4, 5], the authors projected both predictions and ground truth labels onto the shape space guiding models to follow global anatomical shape properties. Instead, Jetley et al. directly regressed input images to their shape encodings yielding a higher robustness as its predictions are constrained by the low-dimensional shape space [6].

However, these approaches fail to train models in an end-to-end fashion. Therefore, we present an end-to-end trainable, shape-constrained encoder-decoder network with applications to object segmentation, shape interpolation as well as object-mask registration. We train the network jointly on segmentation and CT volumes and modify the loss function to enable learning equal respresentations of both modalities. Visual and quantitative evaluation on the NIH Pancreas-CT dataset [7] demonstrate its efficacy.

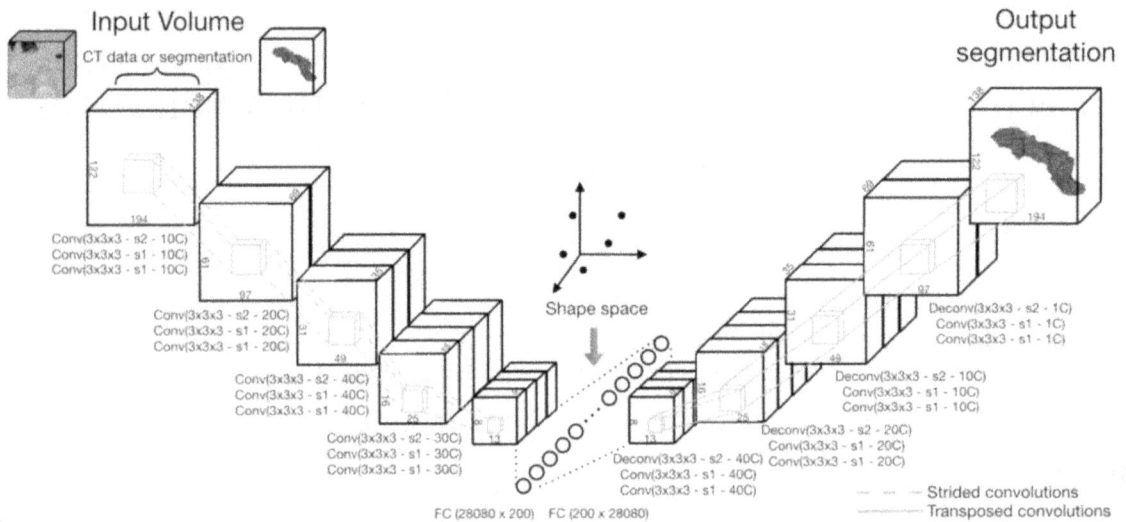

Fig. 2: Architecture of our proposed model. The abbreviation »conv(3x3x3 - s1 - 10C)« stands for a convolutional layer with 3×3×3 kernel size, 1×1×1 striding and 10 output channels. The first fully-connected layer (FC) produces low-dimensional shape codes (points in shape space) and is not followed by any activation function to avoid further constraints.

2 Material and Methods

The architecture of our proposed neural network is portrayed in Fig. 2. Special to our approach is the joint training on CT images I_i and segmentations S_i ($i = 1, ..., N$) as network inputs enabling end-to-end training.

Auto-encoder As far as segmentation inputs are concerned, the network is equivalent to a convolutional auto-encoder (CAE). CAEs are convolutional networks which optimise for an intermediate representation of the input that best reconstructs it (based on a low-dimensional code). The space of the intermediate representation is referred to as shape space (see Fig. 2) and is of low-dimensional nature to force the network to capture the most salient features of the underlying anatomy (e.g. shape structure). We aim to learn a manifold of valid shapes M as defined by the input shapes. Taken as a whole, the input is initially passed through the encoder E (contracting path of the network) which projects it onto M. The Decoder D (expanding path of the network) tries to further map shape space points (referred to as low-dimensional codes) to their corresponding input shapes as it attempts to achieve $S_i \approx (D \circ E)[S_i]$. It is important to note, that its predictions exhibit a relativly huge amount of smoothness due to its low-dimensional domain.

Image segmentation When individual CT images are considered as inputs, our approach resembles traditional segmentation networks, which aim at learning a highly discriminative, complex function f via stochastic gradient descent that transfers image into label information. In an analysis step, its encoder tries to extract meaningful and to some degree invariant features from the input, which are then passed to the decoder to synthesize segmentations. This is done by applying the softmax function to the final feature map in order to compute pixel-wise class probabilities, and therefore enables f to model the conditional distribution probability $P(S_i|I_i)$.

Joint training As our proposed network is trained simulta-

neously on CT data and segmentations, the aforementioned aspects are combined. In addition, the network is forced to learn features that are invariant across these two modalities and thereby result in an equal representation of both of them. However, to ensure the mapping into the same shape space, it is of great importance to avoid two seperate feature extraction paths throughout the network - one for segmentations and the other for CT scans. More precisly, the network could share half of its channels on every level for either analyzing CT or segmentation data. As this would lead to two seperate shape spaces in the embedding (see Fig. 3(a)), the application to reasonable object-mask registration would collapse.

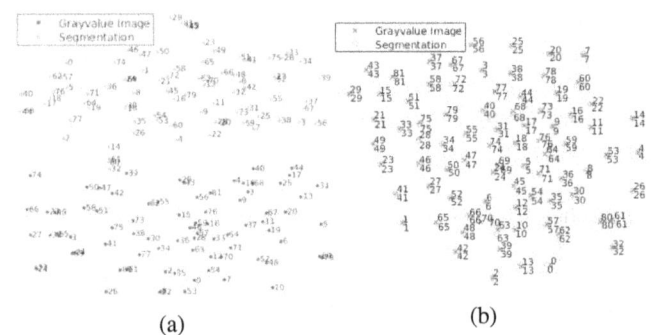

Fig. 3: t-SNE visualization of the embedding obtained by model training (a) without, (b) with the loss term L_C penalizing L1-distances between corresponding codes. Trained on the Pancreas-CT dataset (patient no. displayed).

We tackle this problem by adding the loss function L_C on the embedding to the objective function:

$$\min_\theta \left(L_{BCE} + \lambda_1 \cdot L_C + \lambda_2 \cdot ||\boldsymbol{w}||_2^2 \right). \quad (1)$$

Whereas θ denotes all trainable parameters, L_{BCE} refers to the binary cross-entropy loss to measure the similarity between predictions and ground-truth segmentations. The third term involves weight decay to reduce over fitting by reducing the complexity of the model when not needed. The key to solving the aforementioned problem is the introduction of $L_C = ||E(I_i) - E(S_i)||_1$ penalizing L1-distances between corresponding codes on the embedding. As a result (see Fig. 3 for its effect), the model is forced to map both data types onto a similar shape code.

Further applications Having learned the manifold of valid shapes, we have a natural distance measure between shapes since E maps non-linear shape deformations into an Euclidean space. We can thus linearly interpolate between the embedding of two shapes to obtain realistic shape interpolations. E.g., these can be used for upsampling cardiac cine MRI sequences to reduce their large slice thickness.

For a potential future application to object-mask registration, our approach can be of great support due to the equal embedding space for both data types and its smooth transitions. Fig. 4 illustrates the idea to project both inputs

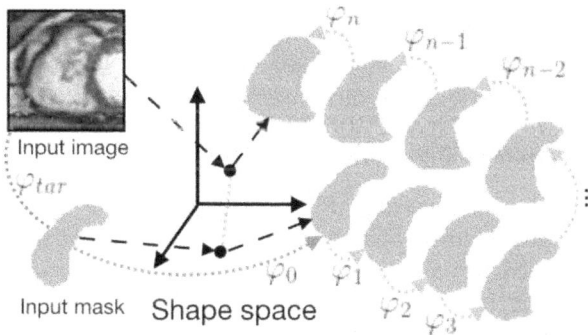

Fig. 4: Object-mask registration based on our approach. The idea is to compute consecutivly, simple registrations on shape interpolated versions of the inputs embedding.

into the shape space and then compute consecutivly multiple registrations between their shape interpolated versions. This breaks down a though registration problem (finding optimal φ_{tar}) into multiple, and due to low deformations, much easier ones: $\varphi_{tar} \approx \varphi_n \circ \varphi_{n-1} \circ \varphi_{n-2} \circ ... \circ \varphi_3 \circ \varphi_2 \circ \varphi_1 \circ \varphi_0$, whereas the amount of deformation can be controlled by the number of interpolation steps n. Especially in cases of low deformation, registration approaches based on geometry-constrained diffusion like [8] tend to be fast and accurate. When predictions of CT volumes are not accurate enough, one could still use this method as a strong pre-registration framework.

Implementation details The parameters of the model are initialised using the Xavier method and trained with Adam on random mini-batches of size 4 containing CT data and/or segmentations. The learning rate starts with 0.001 and is halfed after $400, 500, 570, 610$ and 750 epochs. Besides, every convolutional layer is followed by a LeakyReLU activation function (with $\alpha = 0.1$) and batch normalization except for the last layer which is finally followed by a sigmoid function. LeakyReLUs are also utilized for deconvolutional layers. Furthermore, we augment the data using affine transformations and empirically found suitable loss

weights as follows: $\lambda_1 = 0.11, \lambda_2 = 10^{-5}$.

Experiments Our evaluation is based on the NIH Pancreas-CT dataset from [7] which consists of 82 abdominal contrast enhanced 3D CT scans and manual ground-truth segmentations for the pancreas. The scans have a resolution of $512 \times 512 \times 181$ - 466 and a slice thickness between 0.5mm - 1.0mm. Our preprocessing pipeline starts with data resampling into isotropic voxel sizes of 1.0mm $\times 1.0$mm $\times 1.0$mm. We then crop bounding boxes with sizes of $194 \times 122 \times 138$ around its segmentations, a reasonable choice given that the mean size of the tightest bounding boxes is $135 \times 76 \times 84$. Finally, we apply a zero mean unit variance transformation on the cropped CT patches.

To measure the segmentation accuracy of our approach, we perform 4-fold cross-validation with each fold containing about an equal number of data, and report the mean Dice-Sørensen Coefficient (DSC). We further provide visual shape interpolation and shape embedding results.

3 Results and Discussion

Table 1 lists quantitative segmentation accuracies of our approach and the State-of-the-Art measured by the DSC.

Table 1: Segmentation results measured using the DSC.

Method	Mean DSC	Max DSC	Min DSC
Roth et. al [9]	71.42 ± 10.11	86.29	23.99
Roth et. al [10]	78.01 ± 8.20	88.65	34.11
Zhou et. al [1] CS	75.74 ± 10.47	88.12	39.99
Zhou et. al [1] MI	82.37 ± 5.68	90.85	62.43
Our approach	70.44 ± 5.69	80.84	54.42

We are able to achieve high-accuracy segmentation results (70.44 ± 5.69) even without incorporating the popular skip connections. Due to the low standard deviation of the mean DSC and the relativly high DSC-minimum, our approach provides stronger robustness in comparison to the better performing State-of-the-Art (comparing with [1] MI seems difficult since it uses multiple models and iterations following a coarse-to-fine strategy that could turn in our favor aswell). Besides, our approach is substantially faster since inference takes less than a second on a modern GPU in contrast to several minutes as in [1, 10].

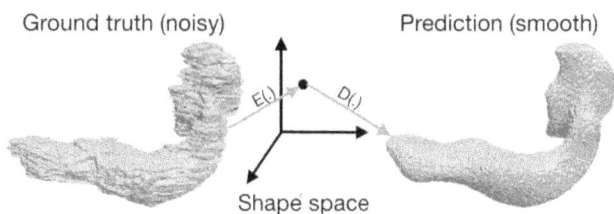

Fig. 5: Ground truth segmentation of NIH case 60 (left) seems to be quite noisy, whereas its prediction (right) provides a satisfying degree of smoothness.

Furthermore, Fig. 5 and 6 illustrate the properties of the learned shape space. As shown in Fig. 5, decoded shapes

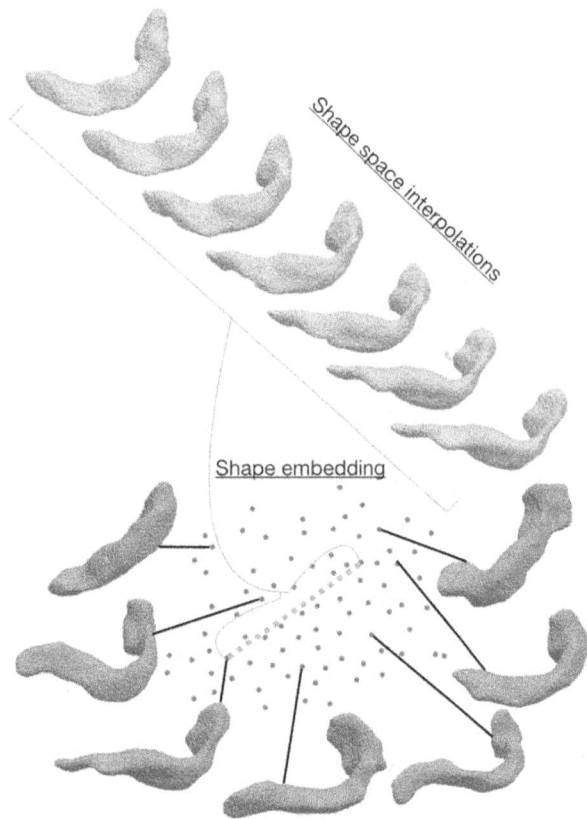

Fig. 6: Visualization of the shape embedding (dots), decoded shapes and shape interpolations along the dotted line. Besides the large inter-patient variability, note the smoothness of the shapes, the realistic shape interpolations as well as the smooth transitions in the compact shape space.

provide a desirable amount of smoothness despite rather noisy ground-truth segmentations. In addition, Fig. 6 also shows satisfyingly smoothed shapes decoded from a compact shape space with smooth transitions. The latter is illustrated by the shape space interpolations (dotted line), which exhibit considerably realistic shape deformations. On the whole, the shape space makes a well-structured impression.

4 Conclusion

In this work, we present a deep 3D encoder-decoder network with multiple applications to object segmentation, shape interpolation and object-mask registration. It is capable of tackling difficult segmentation tasks (e.g. absence of strong boundaries, large inter-patient variability) since its predictions are constrained by a learned, low-dimensional shape space leading to smooth output shapes. Special to our approach is that the network is end-to-end trainable, which recent techniques fail to provide. Our evaluation on the NIH Pancreas-CT dataset show high segmentation accuracies, robustness and consistent shape interpolations due to a compact shape space with smooth transitions.

In future work, we plan to expand our approach to a multi-label and multi-modality framework.

Acknowledgement

The work has been carried out and supervised by the Institute of Medical Informatics, Universität zu Lübeck. Furthermore, we gratefully acknowledge the support of NVIDIA Corporation with the donation of the Titan Xp GPU used for this work.

5 References

[1] Y. Zhou, L. Xie, W. Shen, E. K. Fishman, and A. L. Yuille, "Pancreas segmentation in abdominal CT scan: A coarse-to-fine approach," *CoRR*, vol. abs/1612.08230, 2016.

[2] M. P. Heinrich and O. Oktay, "Briefnet: Deep pancreas segmentation using binary sparse convolutions," in *Medical Image Computing and Computer-Assisted Intervention - MICCAI 2017*, (Cham), pp. 329–337, Springer International Publishing, 2017.

[3] N. Bouteldja, M. Wilms, H. Handels, D. Säring, and J. Ehrhardt, "Model-based 4d segmentation of cardiac structures in cine mri sequences," in *Bildverarbeitung für die Medizin 2017*, (Berlin, Heidelberg), pp. 18–23, Springer Berlin Heidelberg, 2017.

[4] H. Ravishankar, R. Venkataramani, S. Thiruvenkadam, P. Sudhakar, and V. Vaidya, *Learning and Incorporating Shape Models for Semantic Segmentation*, pp. 203–211. Cham: Springer International Publishing, 2017.

[5] O. Oktay *et al.*, "Anatomically constrained neural networks (ACNN): application to cardiac image enhancement and segmentation," *CoRR*, vol. abs/1705.08302, 2017.

[6] S. Jetley, M. Sapienza, S. Golodetz, and P. H. S. Torr, "Straight to shapes: Real-time detection of encoded shapes," *CoRR*, vol. abs/1611.07932, 2016.

[7] H. R. Roth, A. Farag, E. B. Turkbey, L. Lu, J. Liu, and R. M. Summers., "Nih pancreas-ct dataset," 2016.

[8] P. R. Andresen and M. Nielsen, "Non-rigid registration by geometry-constrained diffusion," *Medical Image Analysis*, vol. 5, no. 2, pp. 81 – 88, 2001.

[9] H. R. Roth, L. Lu, A. Farag, H.-C. Shin, J. Liu, E. B. Turkbey, and R. M. Summers, "Deeporgan: Multilevel deep convolutional networks for automated pancreas segmentation," in *Medical Image Computing and Computer-Assisted Intervention – MICCAI 2015*, (Cham), pp. 556–564, Springer International Publishing, 2015.

[10] H. R. Roth, L. Lu, N. Lay, A. P. Harrison, A. Farag, A. Sohn, and R. M. Summers, "Spatial aggregation of holistically-nested convolutional neural networks for automated pancreas localization and segmentation," *CoRR*, vol. abs/1702.00045, 2017.

5

Medical Imaging

Automated Lesion Detection with Neural Networks using Preprocessed Images in Transfer Learning

J. Sprenger [1], S. Reimers-Kipping [2], D. Schäfer [2] and T. Witter [2]

[1] Medizinische Ingenieurwissenschaft, Universität zu Lübeck, jo.sprenger@student.uni-luebeck.de

[2] Fuse-AI, Hamburg, {info}@fuse-ai.de

Abstract

Melanoma is the deadliest form of skin cancer, developed due to mutated DNA that leads the skin cells to multiply quickly. The earlier melanomas are diagnosed, the better are the chances for a succesful treatment. Skin cancer recognition is a very challenging task, due to images with low contrast and artefacts, reducing the quality. Many approaches have been made to improve automated lesion detection. In this paper the effect of image preprocessing, before training the neural network, is evaluated with a retrained Inception V3 network, using transfer learning with a limited amount of data. It is shown, that an improvement of performance can be achieved by combining several methods.

1 Introduction

Early detection of skin cancer reduces the risk of death significantly [6]. In order to support dermatologists, to improve the early melanoma detection and to enable nationwide skin cancer screenings, dermoscopy techniques were improved. Dermatoscopes are noninvasive and used to take well illuminated and enlarged images of the lesions. The manual inspection by a physician is time consuming and even a well trained dermatologist might miss a malignant tumor [3]. Automated lesion detection techniques, to recognize melanomas in early stages, can be fused with a dermatoscope, to generate a real time alert in case of a positive classification. This could help for example general practitioners in supporting the dermatologists.

Recently, Convolutional Neural Networks (CNNs) have performed very well in the task of automated melanoma detection, as well as in other medical imaging tasks [1, 3]. Automated melanoma detection is a very challenging task for a neural network, as lesion images contain lots of artefacts, such as hair, vignetting effects, gel bubbles, coulour charts in the images and differences in illumination and contrast. Furthermore the images show huge intraclass variation in colour, shape and texture.

In this paper the GoogleNet Inception V3 CNN architecure, which is pretrained on 1.28 million images, is investigated [7]. The Inception V3 was trained for the ImageNet Large Scale Visual Recognition Challenge 2014 [9] for 1000 object classes. In this work the CNN was retrained, using transfer learning. During the retraining step, only the final layer of the Inception V3 was retrained with the lesion images, since only a limited amount of data was provided. The Inception V3 performs quite well in skin lesion classification, given a large dataset for retraining as in [1]. As in most medical related tasks, the amount of labelled data is a problem. In this paper the effects of image preprocessing previous to the retraining step are investigated, to evaluate whether an enhanced performance can be achieved when the Inception V3 is retrained on a small dataset.

2 Material and Methods

2.1 Data

The data used in this task was retrieved from the website for the "ISIC 2017: Skin Lesion Analysis Towards Melanoma Detection" challenge. The images are from the ISIC archive, a publicly available collection of dermoscopic skin lesion images [2]. There are three different phases of the challenge, only the images for the classification phase were used. The dataset contains images of three different classes: melanoma, seborrheic keratosis and nevus. A groundtruth classification is provided.

The three classes are unbalanced. The dataset consists of 2000 images in total, 374 labelled as melanoma, 254 labelled as seborrheic keratosis and 1372 labelled as nevus. Fig. 1 shows a few examples from the dataset.

2.2 Retraining

Before the Inception V3 was retrained with the images, the data was equilibrated to avoid overfitting for a certain class. Therefore, the images of class melanoma and seborrheic keratosis were augmented using reflections. A new dataset was created, containing 1000 labelled images of each class. The neural network retrained on these data was tested on a balanced test dataset with 89 images per class. As retraining and evaluation took quite long, another neural network

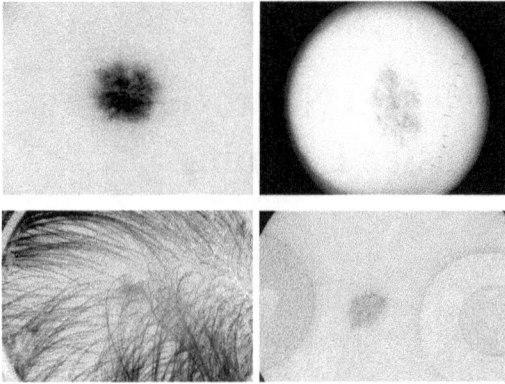

Figure 1: Different images showing the problems in automated lesion detection, such as vignetting effects, hair around the lesion and colour charts in the images

was retrained. The training dataset was balanced as well, but consisted of only 254 images per class. Both neural networks were trained with the same settings, 2000 training steps and a learning rate of 0.01 were applied. So far, no efforts were made to further improve the settings for the retraining. The network was retrained using the deep learning framework Tensorflow [5].

2.3 Image preprocessing

To improve the performance of the network for the given task, the attempt was taken to preprocess the images before retraining the network. To help the network extract better features, three different segmentation methods were evaluated to recognize the region of interest (ROI) in the image and enable a cropping. Thereby, only the important parts of the images were handed to the network and it was presumed that the Inception V3 achieves a better performance due to better feature extraction [3].

The methods were implemented in python. The segmentation methods were not evaluated before the retraining. As the segmentation is only used to crop the images, it is not necessary to segment every pixel exactly. The evaluation metrics of the neural networks are consulted to compare the different segmentation methods after retraining and testing.

2.3.1 Histogram based k-Means Clustering

The first segmentation method is the histogram based k-Means Clustering algorithm. The algorithm was applied on the RGB colour channels to find and reduce the number of similar colour regions. In this task, the images consisted of only five different colours after the algorithm was applied. The resulting image was converted to grayscale and the remaining colourvalues could easily be differenciated in the histogram of the grayscale image. As lesions are often dark, the darkest grayscale was chosen to be the threshold and a binary mask of the image resulting after applying k-Means Clustering was calculated. After calculating the binary mask from the image, morphological transformations

were used to reduce noise and prevent from oversegmentation [8].

2.3.2 Markerbased Watershed

Another method to segment images is the markerbased Watershed algorithm. The watershed transformation is applied on grayscale images and operates upon them like on topographic surfaces, whereat the different gray values represent the heights. The images were therefore transferred to grayscale. A variable threshold is calculated for each image, a combination of Otsu's threshold and a binary threshold. The marker contains information whether an object is foreground, background or a non-object. After performing morphological transformations, to reduce noise and small structures, the Watershed algorithm is applied to the image and a binary mask is created [4].

2.3.3 Simple threshold

As a third method, a simple adaptive threshold was chosen. The input image is converted to grayscale and afterwards a threshold is calculated by using a combination of Otsu's threshold and a binary threshold. After applying morphological transformations, the binary mask is created [4].

2.3.4 Cropping

After segmentation, a cropping method was applied to resize the images, to enable better feature extraction. As the lesion images often contain different areas and oversegmentation was an issue, only the largest connected region was considered in the cropping process. The cropping was performed by bounding boxes, whereat a so called pad was calculated to ensure the whole lesion is obtained in the cropped image. The pad was calculated to be 10% of the input image size.

Training and testing images were both preprocessed with the segmentation methods and cropping and used to retrain and evaluate the Inception V3.

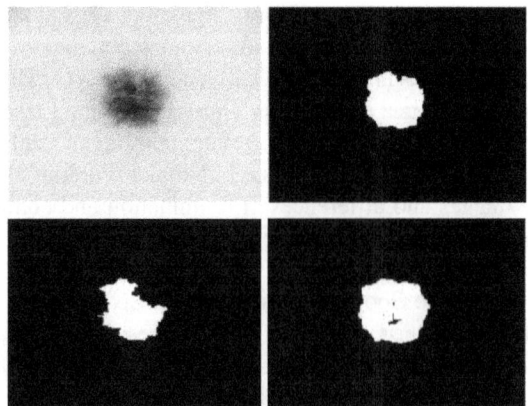

Figure 2: From top left to bottom right: original image, histogram based k-Means Clustering segmentation, Markerbased Watershed segmentation and the simple threshold segmentation

3 Results and Discussion

After retraining the Inception V3 with the original dataset using transfer learning, the performance of the network was evaluated, using the following different binary classification metrics. The accuracy is used as a statistical measure of the overall performance. The accuracy is defined as the number of true positives and true negatives divided through the total number of classifications. To evaluate how sensitive the network is for a certain class, the receiver operating characteristic (ROC) is consulted. Sensitivity and specificity of a varying threshold probability are computed. Examples are given in Fig. 3 and Fig. 4. Furthermore, a confusion matrix is calculated, which visualizes the classification decisions of the network. The horizontal axis shows the groundtruth classification and the vertical axis the classification decisions of the network, as can be seen in Tab. 1-5.

3.1 Performance

The evaluation of the retrained Inception V3 network with 1000 images per class yielded an accuracy of 67.04%. The retraining of the neural network with 254 images per class yielded an accuracy of 64.42%. As there is only a small difference and the retraining, image preprocessing and evaluation are much faster on small datasets, the neural network trained on 254 images per class was chosen for further comparison with the networks retrained on the preprocessed images. The confusion matrix and ROC curve of the network are visualized in Tab. 1 and Fig. 3. The networks trained on the preprocessed images yielded the following scores: the k-Means Clustering segmentation network got an accuracy of 64.42%, the Markerbased Watershed segmentation method an accuracy of 65.54% and the segmentation with a simple threshold led to an accuracy of 65.16%.

While comparing the accuracy scores there is no huge improvement, the confusion matrixes in Tab. 2, Tab. 3 and Tab. 4 show the difference in classification decisions. The k-Means Clustering preprocessing seems to lead to a network which performs better on the images of class nevus than before, but also performs worse for the remaining classes. The network trained on the images segmented by the Markerbased Watershed method seems to perform better for the melanoma class, equally for seborrheic keratosis, but worse for nevus. The simple threshold segmentation does not effect a certain class, but shows little improvement for melanoma and seborrheic keratosis.

As can be seen in Fig. 2, the segmentation methods perform quite differently on the images. While the Markerbased Watershed algorithm led to clear boundaries, but also a border loss, the simple threshold more often resulted in an oversegmentation and the results were strongly conditioned by the input image. The histogram based k-Means Clustering led to good results for all structures that are easily differenced in the histogram, but performed worse on images containing red lesions with low contrast. Fig. 2 only shows one example were all three methods led to a good cropping, even though the segmentation was not perfect. The performance of the segmentation methods on the other images varied highly.

Groundtruth:	melanoma	seb. keratosis	nevus
melanoma	54	20	15
seb. keratosis	11	74	4
nevus	21	24	44

Table 1: Confusion matrix of the network retrained on the original dataset (the horizontal axis shows the groundtruth)

Groundtruth:	melanoma	seb. keratosis	nevus
melanoma	50	25	14
seb. keratosis	7	73	9
nevus	17	23	49

Table 2: Confusion matrix of the network retrained on k-means clustering segmented images

Groundtruth:	melanoma	seb. keratosis	nevus
melanoma	61	20	8
seb. keratosis	9	74	6
nevus	29	20	40

Table 3: Confusion matrix of the network retrained on watershed segmented images

Groundtruth:	melanoma	seb. keratosis	nevus
melanoma	55	24	10
seb. keratosis	8	75	6
nevus	21	24	44

Table 4: Confusion matrix of the network retrained on simple threshold segmented images

Groundtruth:	melanoma	seb. keratosis	nevus
melanoma	58	23	8
seb. keratosis	5	81	3
nevus	17	25	47

Table 5: Confusion matrix of the ensemble network solution

The confusion matrixes and accuracy scores illustrate, that one segmentation method alone did not improve the performance as expected. As two of the segmentation methods seem to perform better for a certain class, an ensemble network solution was evaluated. Therefore, the output of the networks, the probabilities for each class, were combined and normalised. New probabilities for the classes were calculated and the results were evaluated. The results showed an improvement of the accuracy up to 69.66%. The confusion matrix in Tab. 5 shows an improvement of the classification for each class compared to the results of the network trained on the original image data in Tab. 1. While the methods lead to different classification results, all methods lead to neural networks which have a tendency to classify the images as seborrheic keratosis, rather than another class. This is visualized in Fig. 3 and Fig. 4 as well. The ROC curve for the class seborrheic keratosis converges more, which is due to the amount of images classified as seborrheic keratosis.

The segmentation methods need further evaluation. An incorrect chosen threshold can lead to a poor segmentation result. As the intraclass variations are high it is a very challenging task to find a threshold, which leads to a segmentation only containing the actual lesion. The sensitivity and specifity curves in Fig. 3 and Fig. 4 show the differences of the baseline and the best results with the ensemble network solution.

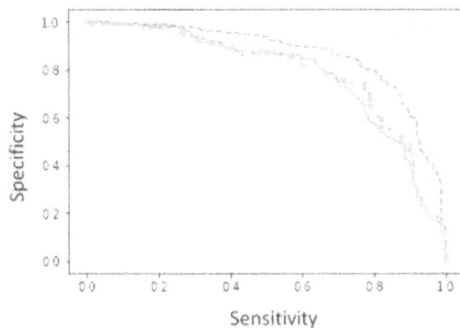

Figure 3: Specificity plotted against Sensitivity of the neural network trained on the original data (dotted line: seborrheic keratosis, dots: melanoma, line: nevus)

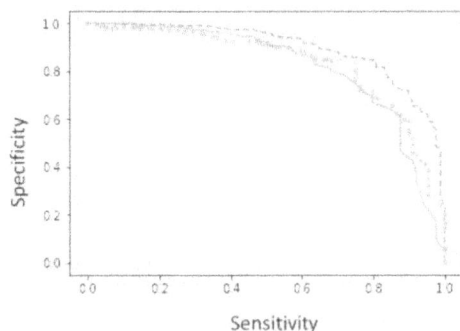

Figure 4: Specificity plotted against Sensitivity of the ensemble network solution (dotted line: seborrheic keratosis, dots: melanoma, line: nevus)

4   Conclusion

The segmentation methods did not lead to the expected improvement. As [3] already showed the positive effect of segmentation methods, it is most likely, the performance of the segmentation methods needs to be enhanced to improve the feature extraction. More tests with different settings could lead to better results. As the images show huge intraclass variation, it might be difficult to find settings that perform well on all images.

Another approach for enhancement is the usage of a neural network for segmentation, which has not been implemented

so far. A further improvement could be the retraining of the last few layers, instead of only retraining the last layer. More preprocessing steps, such as removing hair, adapting the illumination and removing artefacts such as vignetting effects, colour charts and gel bubbles could lead to further enhancements in the classification task, if they are performed autonomously. Previous to segmentation, it could as well lead to better segmentation results. Overall an improvement in the classification task could be measured using preprocessing and averaging over the resulting scores.

Acknowledgement

The work has been carried out at Fuse-AI, Hamburg and supervised by Prof. Dr. rer. nat. Thomas Martinetz from the Institute for Neuro- and Bioinformatics, Universität zu Lübeck.

5   References

[1] A.Esteva, B. Kuprel, R.Novoa, J. Ko, S. Swetter, H. Blau and S. Thrun, *Dermatologist-level classification of skin cancer with deep neural networks*. Nature, Vol. 542, No. 7639, pp. 115-118, 2017

[2] Codella N, Gutman D, Celebi ME, Helba B, Marchetti MA, Dusza S, Kalloo A, Liopyris K, Mishra N, Kittler H, Halpern A. *Skin Lesion Analysis Toward Melanoma Detection: A Challenge at the 2017 International Symposium on Biomedical Imaging (ISBI), Hosted by the International Skin Imaging Collaboration (ISIC)*

[3] Lequan Yu, Hao Chen, Qi Dou, Jing Qin and Pheng-Ann Heng *Automated Melanoma Recognition in Dermoscopy Images via Very Deep Residual Networks*. IEEE transactions on medical imaging, Vol. 36, No.4, pp.994-1004, 2017

[4] G. Bradski. *The OpenCV Librarys*. Dr. Dobbs Journal of Software Tool, 2000

[5] M.Abadi et al. *Tensorflow: Large-Scale Machine Learning on Heterogenous Systems*, 2015. Software available from tensorflow.org

[6] The Skin Cancer Foundation, at www.skincancer.org, (10.01.2018)

[7] C. Szegedy, V.Vanhoucke, S. Ioffe, J. Shlens Z. Wojna *Rethinking the inception architecture for computer vision*, 2015. Preprint at https://arxiv.org/abs/1512.00567

[8] S.L. Bill, K. Michael, S. Kalra, H.R Tizhoosh *Skin Lesion Segmentation: U-Nets versus Clustering*, 2017. Preprint at https://arxiv.org

[9] O. Russakovsky et al. *Imagenet large scale visual recognition challenge*, 2015. Int.J.Comput. Vis. 115, 211-252

A Comparison Study on MPI Reconstruction Methods for Multidimensional Lissajous-based Data-Acquisition Schemes

L. Bannoura [1], A. Cordes [2], and T. M. Buzug [2]

[1] Biomedical Engineering, University of Applied Sciences Lübeck, luay.bannoura@stud.fh-luebeck.de
[2] Institute of Medical Engineering, Universtät zu Lübeck, {cordes,buzug}@imt.uni-luebeck.de

Abstract

Magnetic particle imaging (MPI) is a new medical imaging technique that visualizes the spatial distribution of super-paramagnetic iron-oxide nanoparticles (SPIONs). In MPI, a method called x-space is used to reconstruct the images, it describes the recorded signal as a convolution of a nanoparticles distribution and a point spread function. However, the validation of x-space requires a linear and shift-invariant system (LSI). With a Lissajous based trajectory, standard x-space becomes challenging and ensuring the LSI of the system is not realized. A modified approximate solution to x-space was presented to counterbalance the problem. In this paper, we have qualitatively and quantitatively compared the quality of the different reconstruction methods for Lissajous trajectories using dice similarity coefficient (DSC). We show that the approximate x-space solution yield better DSC results in contrast to standard x-space. Additionally, we confirm that an exact reconstruction based on a Lissajous trajectory requires a system matrix.

1 Introduction

Magnetic particle imaging (MPI) is a new tomographic medical imaging technique that has been introduced by Gleich and Weizencker in 2005 [1]. MPI images the spatial distribution of superparamagnetic iron oxide nanoparticles (SPIONs) with high spatial and temporal resolution without using ionization radiation or any other harmful substances. SPIONs are clinically proven as a safe and biocompatible material [2].

The key to form MPI images is the nonlinear magnetization characteristics of the SPIONs when they are subjected to an oscillating magnetic field. A homogenous sinusoidal magnetic field, the so-called drive field is applied to excite the SPIONs and change their magnetization. The particles respond with harmonic signals at multiples of the excitation frequency that is measured using one or more inductive coils. Since a homogenous drive field is applied, all the particle will experience the same magnetic field. Therefore, a spatial encoding is needed, by applying an additional static gradient magnetic field, the so-called selection field, all the particles will be saturated except at a particular point called the field free point (FFP) or the dynamic region [3].

During data acquisition, the drive field moves the FFP along a closed trajectory, which causes the particles near the FFP to flip and induce a voltage in the receiving coils [3]. Various trajectories have been proposed and used. In practice, the most used MPI trajectories are either Lissajous or a Cartesian trajectory (see Fig. 1) [2].

In MPI, the measured signal does not give an image that is ready for direct visualization. For that reason, a reconstruc-

tion approach is needed to transform the measured signal into a spatial map of the particle concentration across the whole field of view (FOV). To date, there have been two reconstruction approaches that have been widely used, the system matrix-based reconstruction and the x-space reconstruction. The two approaches share the common assumption that the relation between the measured voltage $u(t)$ and the particle concentration $c(r)$ is linear [2].

$$u(t) = \int s(r,t)c(r)d^3r, \qquad (1)$$

where $s(r,t)$ is the MPI system function which describes the mapping between the measured voltage and the particle concentration at position r and time t [2].

For x-space it was shown in [4] that the measured voltage can be formulated as a convolution of the nanoparticle distribution and a point spread function (PSF). In this manner, the image can be directly reconstructed by simply gridding the measured voltage to the known location of the FFP. Afterwards, a deconvolution can be applied to increase the spatial resolution. The linearity and shift invariance (LSI) of the system is required in x-space, to guarantee that the reconstructed image is correctly reflecting the concentration and distribution of the nanoparticles samples being imaged [5]. The gradient field along the FOV should be linear, to ensure the unique location of the FFP in space. Similarly, the point spread function (PSF) should be isotropic along the FOV to ensure the shift invariance of the system.

Concerning 1-D line measurements, x-space showed promising results, where it was approved in [4] that the PSF is spatially invariant along the line when we normalize the

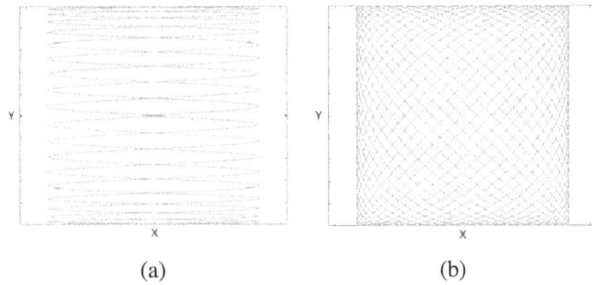

Figure 1: A graphical representation of possible MPI trajectories: (a) Cartesian, (b) Lissajous.

signal to the relative velocity of the FFP. As a result, the 1D x-space formulation can be described as linear and shift invariant.

For multidimensional systems, x-space has shown to be a suitable reconstruction method for Cartesian trajectories. However, for Lissajous trajectories, it turns out to be challenging whether one can ensure the LSI of the system. It was approved in [6] that the PSF does not only depends on the relative velocity of the FFP but also on the direction of the FFP velocity vector. And since the FFP velocity vector is not constant for Lissajous trajectories, this results in an anisotropic PSF that is a spatially variant for every FFP position, which abolishes the LSI of the system and makes the deconvolution in x-space a challenging task.

These difficulties were discussed in [7], where it was shown that, when x-space is based on Lissajous trajectory, the spatial distribution of the nanoparticles cannot be reconstructed by just gridding the voltage signal to a known FFP position. As a consequence, two different solutions have been discussed in [7]. First, an approximate solution to x-space, by assuming two diagonal reference frames to be used to grid the signal and a spatially invariant convolution kernel. Second, an exact solution by calculating a spatially variant deconvolution kernel for every position on the Lissajous trajectory based on a system matrix.

In this contribution, we have qualitatively and quantitatively compared the different reconstruction techniques for multidimensional Lissajous-based data acquisition schemes, in order to evaluate the quality of the approximate solution of x-space.

2 Material and Methods

2.1 Simulation Parameters

A simulation study was performed to compare the different reconstruction solutions for Lissajous trajectories. Data were acquired using a simulation tool written in C++ and loaded via a script written in MATLAB. The simulated MPI scanner uses an ideal magnetic field. The gradient field strength is chosen to be equal in both directions $G_x = G_y = 1.5$ T/m, a sinusoidal drive field is simulated with an amplitude of 0.075 T and excitation frequencies $f_x = 26.042$ kHz and $f_y = 25.253$ kHz in both directions. The magnetization behavior of the nanoparticles is simulated based on the Langevin theory of magnetization. A FOV of $[0.05 \text{ X } 0.05]$ m is used and sampled at 100 X 100 = 10000 positions. A 'P' phantom is used to acquire the measurements, and it is visualized in Fig. 2a.

2.2 X-Space Reconstruction

The x-space approach introduced by Goodwill and Conolly needs several assumptions to be realized. The approach ignores the relaxation effects and assumes that the particles are in thermal equilibrium, which means that the magnetization change of the nanoparticles can be described based on the Langevin theory of paramagnetism [4], [6].

For MPI data acquisition, the signal is recorded using an inductive coil. As a consequence, we can use the law of reciprocity to measure the received signal [3], this results in a measured voltage $u(t)$ that is formed as a convolution of the nanoparticle distribution $c(x)$ and a point spread function $h(x)$. For 1D x-space, the PSF is originated from the derivative of the nanoparticle magnetization with respect to the applied magnetic field strength [4]. For 2D x-space, the PSF becomes more complex, it consists of the summation of the nanoparticle magnetization and the derivative of the nanoparticle magnetization with respect to the applied magnetic field strength [6]. A simplified multidimensional formula of x-space is given by

$$\frac{u(t)}{|\dot{x}(t)|} = c(x) * h(x)\hat{\dot{x}}. \tag{2}$$

The complete formula with the detailed components of the PSF can be found in [6], where $\dot{x}(t) = [v_x(t), v_y(t)]$ is the FFP velocity vector and $\hat{\dot{x}} = \dot{x}(t)/|\dot{x}(t)|$ is the normalized velocity vector. The image is reconstructed in three steps. Firstly, normalization of the measured signal to the relative velocity. Secondly, gridding the normalized signal to the instantaneous position of the FFP. The gridding step is done using two techniques; the first technique is the standard x-space, where the signal is gridded instantaneously to the associated location of the FFP. The second technique is to assume two diagonal reference frames to be used to grid the measured signal, which is considered to be the modified approximate solution to x-space. To create the two diagonal reference frames the signal vector is divided into two vectors based on the direction of the FPP velocity. Each vector is gridded individually, and the resulting two gridded images are combined. To create the two vectors, the signal values that are measured when $v_x.v_y > 0$ are taken in one vector, and the values for $v_x.v_y < 0$ are considered in the second vector. Finally, a deconvolution of the blurred gridded image with a PSF shown in Fig. 3a is applied. The deconvolution here is done using a Wiener filter in the frequency domain.

2.3 System-Matrix Reconstruction

The linear equation Eq. (1) is discretized by forming a MPI system matrix and written in a matrix-vector notation.

$$u = Sc \qquad (3)$$

The system matrix S gives the connection between the measured voltage vector u and the spatial distribution of the nanoparticles c. The system matrix is composed of the simulated signals of a point source for every possible location on the volume grid. To create the system matrix, we place a simulated point source of nanoparticles at an exact position, and then a full MPI data acquisition is performed. Based on this system matrix a spatially variant deconvolution kernel can be calculated through matrix inversion [7].

The system of linear equation (Eq. 3) has to be solved through regularization and matrix inversion to reconstruct the spatial distribution of the SPIONs. Here we used the truncated singular value decomposition (TSVD) method to invert the system matrix and reconstruct the image [3].

2.4   Comparison of Methods

A quantitative comparison using the Dice Similarity Coefficient (DSC) is applied, to compare the Lissajous based reconstructed images from the three different solutions, standard x-space, x-space approximation and system matrix with the original phantom.

The Dice Similarity Coefficient is used to measure the spatial overlap between two binary segmented images; it can be applied to compare the reconstructed image and the original phantom image. It is defined as

$$DSC = \frac{2(A \cap B)}{(A + B)} \qquad (4)$$

where A and B are the two binary segmented images. The DSC values range within zero and one, where a zero value specifies no overlap, and a value of one specifies a perfect overlap [8].

3   Results and Discussion

Fig. 2 shows the reconstructed images based on a Lissajous trajectory. Fig. 2a shows the phantom used in the simulation. Fig. 2b shows the reconstructed image using the standard x-space method, where the measured signal is gridded instantaneously to the position of the FFP. The degraded quality and resolution of the reconstructed image can be seen clearly. The overall shape of the P letter can be visualized, but the small blocks that form the P letter are highly distorted without any particular related shape to the original P phantom, and false signal values are scattered across the whole image in incorrect positions. These artifacts appeared due to the fact that the signal depends on the direction of the FFP velocity vector (see eq 2). As a consequence, the MPI signal at the intersection points of the Lissajous trajectory is not unique, which results in a spatially variant PSF. Moreover, applying a deconvolution step using a Wiener filter with a simulated invariant PSF shown in Fig. 3a removed the blurring in the image, but resulted in a further quality degradation of the reconstructed image

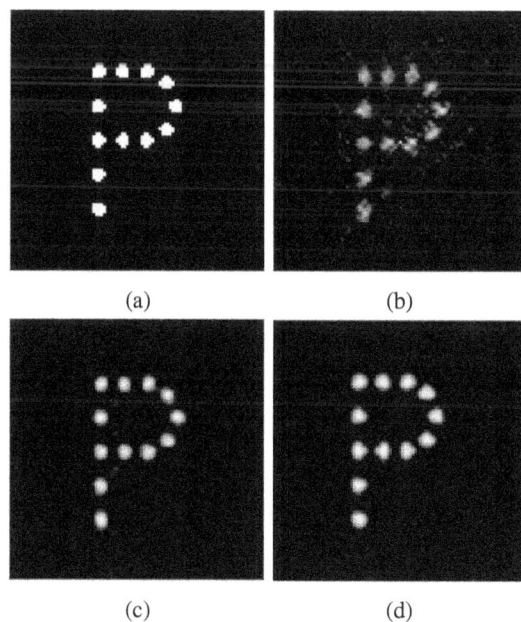

(a)   (b)

(c)   (d)

Figure 2: The reconstruction results of a P phantom image based on a Lissajous trajectory: (a) P Phantom, (b) reconstructed image using standard x-space, (c) reconstructed image using an approximate solution to x-space, (d) reconstructed image using the system matrix approach.

Table 1: Dice's coefficient results for reconstructed images based on a Lissajous trajectory.

Method	Dice's coefficient
System Matrix	0.90
X-Space Approximation	0.28
Standard X-Space	0.10

and in more scattered data points.

An approximate solution to this problem is to assume two diagonal reference frames to be used to grid the signal. The resulted reconstructed image of this technique can be seen in Fig. 2c. In contrast to the standard x-space, an enhanced quality and resolution can be observed, the shape of the P letter is more accurately reconstructed, and the small blocks are quite similar to the original P phantom. Evidently, remarkably less scattered points are visible, even though a few scattered points are still shown which is the result of applying a deconvolution step using a Wiener filter to improve the resolution and deblurring of the image. The better results of the approximate solution are confirmed by the DSC value shown in Table 1. The approximate x-space method gives a DSC of 0.28 compared to the standard x-space method of 0.10. At the same time, despite the considerably improved result, the small details in the structures blocks still cannot be sharply seen, where the blocks look approximately as small circles without any fine details, this can be explained due to two reasons. First, we assumed two diagonal reference frames, where two images are created for each frame, and the resulting two images are combined, this indicates that at various points along the trajectory wrong signals values are being considered. Second, we are assuming an

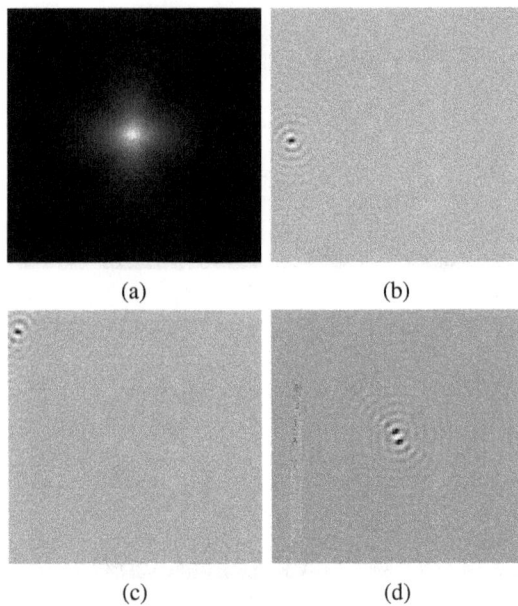

Figure 3: A graphical representation of different deconvolution kernels: (a) A simulated PSF, (b, c, d) A deconvolution kernel from the system matrix in different positions.

isotropic PSF to be used to deconvolve the image with, this assumption is incorrect, due to the fact that the PSF along the Lissajous trajectory is spatially variant for each position. The exact reconstruction of the SPIONs distribution using a Lissajous trajectory requires a system matrix as discussed in section 2.3. Since the system matrix gives the real deconvolution kernel for every position of the particles on the Lissajous trajectory. Fig. 3b- 3d shows the different shapes of the deconvolution kernel for various positions, the different shapes imply that the deconvolution kernel is spatially variant when MPI is based on Lissajous trajectories. The deconvolution kernels have negative values that subtract the smearing and compensate for the blurring. Moreover, the overall deconvolution kernels can be seen as similar to a high pass filter, in which it amplifies the high frequencies in the image that are responsible for the details and sharp edges. The reconstructed image shown in Fig. 2d supports these assumptions. It has improved quality and resolution compared to the previously discussed images. It can be seen clearly that the reconstructed image using a system matrix looks quite similar to the original P phantom. It is apparent that the structure blocks are very similar, especially with the small details and the sharp edges that can be observed. Scattered data points were not found. These overall results were confirmed by the DSC value of 0.90 shown in Table 1. The main limitation of the system matrix approach is that it suffers from the time required to create the system matrix. Furthermore, a large memory is required to store it. Moreover, in practice, different scanner configurations and any new SPION type requires constructing a new system matrix, which means it cannot be used as an instant reconstruction method like the x-space approach.

4   Conclusion

In this work, we have compared qualitatively and quantitatively the different solutions for MPI reconstruction methods based on Lissajous trajectories. We have shown that acceptable results can be obtained with the approximate solution for x-space. At the same time, for better reconstruction results, a spatially variant deconvolution kernel needs to be used for each position on the trajectory.

In summary, the system matrix reconstruction approach is considered the best to be used for MPI reconstruction based on Lissajous trajectories. Its main advantage that is it requires no additional deconvolution to be applied, since the deconvolution is involved in the system matrix inversion. The main limitation of this approach includes the time required for acquisition and inversion of the system matrix.

Acknowledgement

The work has been carried out at the Institute of Medical Engineering (IMT), Universität zu Lübeck, Germany.

5   References

[1] B Gleich and J Weizenecker. *Tomographic imaging using the nonlinear response of magnetic particles.* Nature, vol. 435, no. 7046, pp. 1214-1217, 2005.

[2] T Knopp, N Gdaniec and M Möddel. *Magnetic particle imaging: from proof of principle to preclinical applications.* Phy. Med. Biol, vol. 62, no. 14, pp. R124-R178, 2017.

[3] T Knopp and T. M. Buzug. *Magnetic Particle Imaging-An introduction to imaging principles and scanner instrumentaion.* Berlin: Springer 2012.

[4] PW Goodwill and SM Conolly. *The x-space formulation of the magnetic particle imaging process: one-dimensional signal, resolution, bandwith, SNR, SAR, and magnetostimulation.* IEEE Trans Med Imag, vol. 29, no. 11, pp. 1851-1859, 2010.

[5] PW Goodwill, K Lu, B Zheng and SM Conolly. *An x-space magnetic particle imaging scanner.* Review of Scientific Instruments, vol. 83, no. 3, pp. 033708-1-0337080-9, 2013.

[6] PW Goodwill and SM Conolly. *Multi-dimensional x-space magnetic particle imaging.* IEEE Trans Med Imag, vol. 30, no. 9, pp. 1581-1590, 2011.

[7] A Cordes, C Kaethner, M Ahlborg and T. M. Buzug. *X-space Deconvolution for Multidimensional Lissajous-based Data-Acquisition Schemes.* International Workshop on Magnetic Particle Imaging, pp. 74, 2016.

[8] K Zou et al, *Statistical Validation of Image Segmentation Quality Based on a Spatial Overlap Index.* Acad Radiol, vol. 11, no. 2, pp. 178–189, 2004.

Sinogram modeling for patient motion detection in dental Cone Beam CT

D. Wulff [1,2], S. Maur [2] and T. M. Buzug [3]

[1] Medical Engineering Science, Universität zu Lübeck, daniel.wulff@student.uni-luebeck.de
[2] Sirona Dental Systems GmbH, {daniel.wulff, susanne.maur}@dentsplysirona.com
[3] Institute of Medical Engineering, Universität zu Lübeck, buzug@imt.uni-luebeck.de

Abstract

In dental Cone Beam CT (CBCT) patient motion leads to artifacts which can complicate diagnosis. Existing methods for motion correction are computationally expensive. A method for motion detection helps to reduce this cost by limiting the correction needed region. Further, a local correction method makes sure that correct image information is not manipulated. In this paper, a model is presented which can be used for patient motion detection. The created model describes the trace of an object point in a 3-dimensional sinogram scanned by a CBCT on a circular trajectory. The accuracy of the model is evaluated by creating model instances for known traces and determining an error. The error is defined as the Euclidean distance between the model and the trace in the sinogram. It is presented that the model has a high accuracy with a maximum error of 2.24 pixels.

1   Introduction

In computed tomography (CT) a series of x-ray projections is acquired and used to determine a 3-dimensional image of a patient. In the backprojection process it is assumed that the scanned body has not moved. With the exception of cone-beam artifacts the resulting backprojection is a authentic approximation of the real object.

If the patient moves during image acquisition the back-projection process is inconsistent. The projections contain ambiguous information. Therefore, artifacts as streaking, blurring and double contours can occur and reduce the image quality significantly [1], [2].

Patient movements can be unconscious or deliberate. Especially agitated patients as children cannot stay as rigid as necessary to get images in a high quality [2]. The larger the movements the stronger the resulting artifacts [3].

A Cone Beam CT (CBCT) as used in dentistry has relatively long scanning times. Especially due to respiration and the heartbeat motion artifacts are unavoidable [4]. Reducing motion artifacts and improving image quality requires an image correction after acquisition. Projections in which the patient has moved have to be corrected. For this it is necessary to detect the time the patient moved during acquisition.

In the sinogram each angle can be associated with an acquisition point in time. Due to this it is obvious to assign the time the patient moved to a acquisition point directly in the sinogram.

In [5] polynomials of degree four are fitted to the shape of the scanned object in the sinogram. This fitting is used to compensate motion artifacts in CT. This approach is adapted to dental CBCT applications in this paper. A model is presented with which the trace of a scanned object in the sinogram can be described if no patient motion occurs. For this a sine function is used. Depending on the model, patient motion can be detected by looking for differences between the real traces and its ideal models.

2   Material and Methods

The method presented in this paper is based on the assumption that the trace of an object in a sinogram can be modeled by a sinusoidal function. For this, the used CBCT system is analyzed and a relation to the sine function is established.

2.1   System Geometry

The system assumed in this paper is a dental CBCT. It uses a conical x-ray beam to acquire one projection per angle in an angle range of $[0°, 200°]$. 200 projections are acquired in one system circulation and consecutively stacked as shown in Fig. 1.

This stack contains all acquired projections and is called *3-dimensional sinogram*. The x and y-axis characterize the

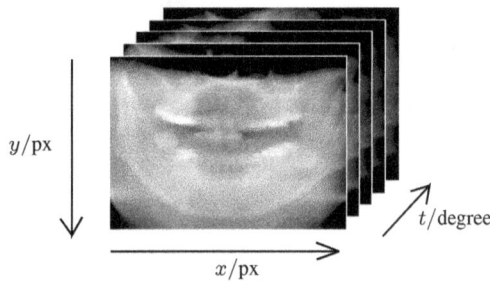

Figure 1: The projections are stacked to build the 3-dimensional sinogram.

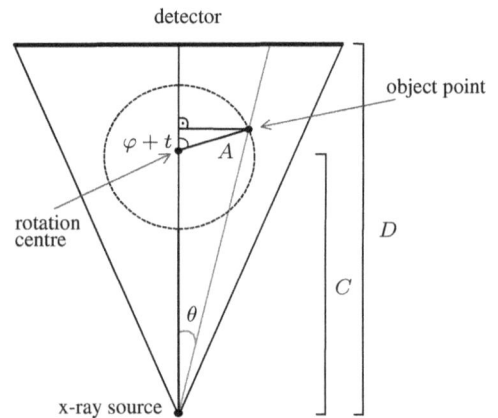

Figure 2: Topview on the main plane of the CBCT system. The polar coordinate system and an object point as well as its projection beam is illustrated.

projection pixel coordinates and the t-axis corresponds to the acquisition time. Each projection represents the same object volume from a different perspective depending on the circular system trajectory. Due to that, the unit of the time axis can be specified in degree.

Fig. 2 shows a topview of the system. The rotation centre is located between the x-ray source and the detector. C is the distance between the x-ray source and the rotation centre and D is the distance between the x-ray source and the detector. These parameters are fix and known from calibration.

A scanned object is located inside the system so a polar coordinate system is used to characterize its object points. With the coordinates $(A, G, \varphi + t)$ and the rotation centre as the origin, every object point in the volume can be characterized. Parameter A is the distance between the object point and the rotation centre in the main plane in mm. G is the object height in mm and $\varphi + t$ is the angle between the main beam and A in degree. This angle consists of the phase shift φ and the acquisition time of the projection t. The acquisition time corresponds to the system orientation in degree, in which the projection is acquired. These coordinates are marked in Fig. 2 and 3. As shown in Fig. 3 the main plane hits the detector at the bottom detector array.

Due to the circular system trajectory an object point follows a sinusoidal trace in the 3-dimensional sinogram. This trace depends on the object point position $(A, G, \varphi + t)$ in the scanned volume and can be split in two parts to reduce the dimensions. The first part is a sinusoidal trace in x-t-plane which only depends on the coordinates $(A, \varphi + t)$. The second part is a sinusoidal trace in y-t-plane which depends on all three coordinates.

2.2   2-dimensional Trace Modeling

First the 2-dimensional object point trace in x-t-plane is modeled. In this plane, the object height has no effect on the model. The general sine function used for this model is

$$f_x(t) = a_x \cdot \sin(\omega \cdot (t - \varphi)) + d_x. \tag{1}$$

Here a_x is the amplitude in pixel, ω the angular frequency in $\frac{1}{\text{degree}}$, φ a phase shift in degree and d_x is a height offset

in pixel. The angular frequency ω depends on the angle step size with which the projections are acquired. Here this acquisition frequency is one projection per degree so the angular frequency is constant, $\omega = 1 \frac{1}{\text{degree}}$.

The height offset d_x also is constant in x-t-plane, because it is independent of the object position. For every object the main beam goes through the rotation centre and hits the same detector element in the detector centre. The image origin is in the top left of a projection. So the sine function has a constant height offset of half a detector width.

Due to the cone beam geometry the amplitude a_x depends on the object distance A. As can be seen in Fig. 2 the opening angle θ depends on A and $\varphi + t$ so the projected amplitude a_x is influenced by these parameters. To determine A first the amplitude is used to get the opening angle θ by

$$\theta = \text{atan}(\frac{a_x}{D}). \tag{2}$$

A is determined by θ and the distance C by

$$A = \cos(\theta) \cdot C \cdot \tan(\theta) = \sin(\theta) \cdot C. \tag{3}$$

The angle $\varphi + t$ corresponds to the angle between the main beam and the distance A as can be seen in Fig. 2. If $\varphi = 0$ the resulting projected object point lies in the detector centre and the resulting sine function has no phase shift.

Due to the system geometry the resulting function does not have an ideal sinusoidal course. The cone beam arrangement leads to a shift of the maximum and minimum of the sine function and surrounding object positions. This effect leads to a shearing of the sine function in t-direction of the sinogram which can be determined by

$$s(t) = f_x(t) + t \cdot m. \tag{4}$$

The sine is shared along the time axis t depending on the function value $f_x(t) = x$ and a shearing parameter m. m is constant because it depends on the system geometry only.

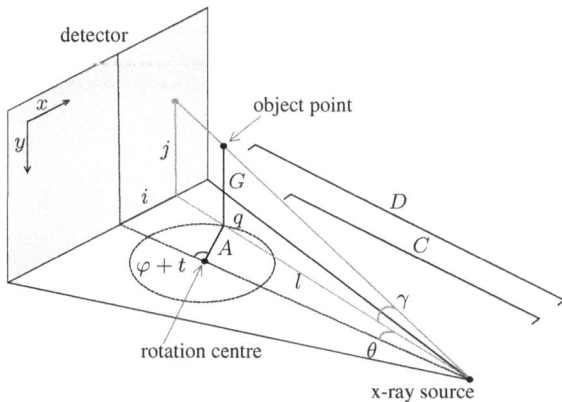

Figure 3: 3-dimensional view of the CBCT system. The object point is characterized by its coordinates $(A, G, \varphi+t)$ and its projection is illustrated on the detector.

With that the trace of an object point in the x-t-plane depends on the distance A and the phase shift φ. Due to the fact that the sinogram is 3-dimensional this 2-dimensional function is not adequate to model its trace completely. Therefore, a model for the y-t-plane is designed.

2.3 3-dimensional Trace Modeling

In y-t-plane also a sinusoidal trace of an object point is expected because of the circular system movement. But in this plane, as shown in Fig. 3, the trace further depends on the object height in the volume.

Furthermore, the function itself is different to the sine in the x-t-plane. The traces have a phase shift of 90° to each other. To compensate this phase shift the function in y-t-plane is adapted to a cosine so the general form of the function in this plane is

$$f_y(t) = a_y \cdot \cos(\omega \cdot (t - \varphi)) + d_y. \qquad (5)$$

Because the cosine in the y-t-plane also depends on the circular system trajectory the angular frequency $\omega = 1\ \frac{1}{degree}$ is constant. Due to the cosine the phase shift φ is equal to the phase shift in x-t-plane. The amplitude a_y and the height offset d_y are different than in x-t-plane because this variables additionally depend on the object height G.

From the known object point coordinates (A, φ) and the corresponding projection coordinates (x, y, t) the object height G can be determined. For this the distances i and j are needed which can be deduced by the maximum coordinates x_{max} and y_{max}. As can be seen in Fig. 3, j characterizes the height in the projection and i is the distance between the main beam and the projected object point. So they can be determined by

$$i = x - \frac{x_{max}}{2} \qquad (6)$$

$$j = y_{max} - y \qquad (7)$$

The distance l is the distance between the x-ray source and the point $(A, 0, \varphi + t)$ in the polar coordinate system as

illustrated in Fig. 3. It is determined by the use of the law of cosines by

$$l = \sqrt{A^2 + C^2 + 2AC \cdot \cos(t + \varphi)}. \qquad (8)$$

Furthermore, the distance q between the x-ray source and the point where the beam hits the detector in the main plane is determined by

$$q = \sqrt{i^2 + D^2}. \qquad (9)$$

These parameters and the distance j are used to get the object height G by using the Intercept Theorem by

$$G = \frac{j \cdot l}{q}. \qquad (10)$$

If the coordinates (A, G, φ) of an object point and its corresponding projected image coordinates are known it is possible to determine the height offset d_y and the amplitude a_y of the cosine in the y-t-plane.

A cosine has a zero-crossing if $\cos(\varphi = 90°) = 0$ so the height offset d_y corresponds to the cosine function value at this point. Due to that the following equations are used to determine d_y:

$$l = \sqrt{A^2 + C^2} \qquad (11)$$

$$q = \sqrt{\left(\frac{A \cdot D}{C}\right)^2 + D^2} \qquad (12)$$

$$d_y = \frac{G \cdot l}{q}. \qquad (13)$$

The distances l and q are the same as used before and illustrated in Fig. 3. Due to the known phase angle of $\varphi + t = 90°$ and $cos(90) = 0$ (8) can be formed to (11). Because i is unknown for $\varphi + t = 90°$ (9) is formed to (12). With these parametes the height offset d_y is determined with the Intercept Theorem by (13).

If the height offset d_y is known the amplitude a_y can be determined with $f_y(t) = y$ by

$$a_y = \frac{f_y(t) - d_y}{\cos(\omega \cdot (t - \varphi))}. \qquad (14)$$

Height offset d_y and amplitude a_y depend on the system dimensions and the object position. Furthermore, there is a relation between x-t-plane and y-t-plane. If one projection coordinate of a scanned object point is known all possible sinogram traces can be determined.

2.4 Evaluation of Model Accuracy

The model is implemented in Python 3.6 to evaluate its accuracy. A simulated spherical object with a radius of 0.1 mm is generated and forward projected as in a CBCT. The resulting projections are filtered so that in each projection only one pixel characterizes the object.

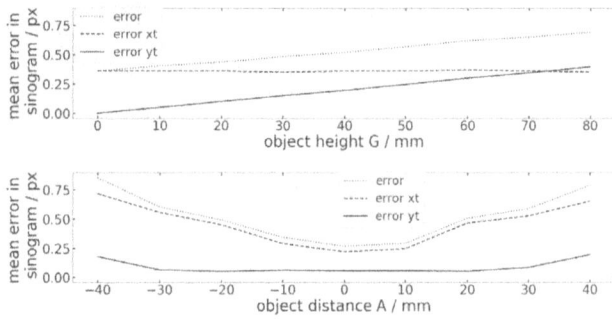

Figure 4: Mean error plots as a function of the object height
G (upper) and the object distance A (bottom). The mean
errors between the traces and their models are illustrated.

The object position in the CBCT system is varied in
distance A in a range [-40 mm, 40 mm] and object height
G in a range [0 mm, 80 mm]. For each object position a
sinogram is generated and evaluated.

The method is applied for all pixels of a trace in a sinogram. So several sets of possible model parameters are
determined. The best model parameters are determined by
looking for the maximum conformity in all parameter sets.

The accuracy is evaluated in each projection separately by using a distance metric to compare the model
$(x_{model}, y_{model}, t)$ and the trace $(x_{trace}, y_{trace}, t)$. The absolute differences between the coordinates for x-t-plane and
y-t-plane and the 2-dimensional Euclidean distance are determined. So three errors for each projection in a sinogram
are generated:

$$\text{error}_{xt} = |x_{model} - x_{trace}| \tag{15}$$

$$\text{error}_{yt} = |y_{model} - y_{trace}| \tag{16}$$

$$\text{error} = \sqrt{\text{error}_{xt}^2 + \text{error}_{yt}^2}. \tag{17}$$

3 Results and Discussion

To obtain an overall error measure for a sinogram the
means of error, error_{xt} and error_{yt} as well as its maximum
and minimum are determined. The results for the mean
errors are illustrated in Fig. 4. It can be seen that error_{xt} is
almost constant if the object height G is rising. That means
the model in x-t-plane is independent of the object height
G. However, there is a dependancy on A. The greater
the object distance A the larger is error_{xt}. The model in
y-t-plane is independent of distance A because error_{yt} is
almost constant if A is rising. Though error_{yt} depends on
the object height G.

In Fig. 4 can be seen that the mean errors are less than one
pixel. Furthermore, the maximum error ranges in pixels are
[0, 2.24] for error, [0, 3] for error_{xt} and [0, 1] for error_{yt}.
So the model has a maximum local inaccuracy of three pix-
els in x-t-plane and one pixel in y-t-plane. In addition, the
minimum error in each tested sinogram is zero. That means

every model has at least one part where the model fits perfectly to the real trace.

4 Conclusion

An object which is scanned by a CBCT leads to a characteristic trace in the 3-dimensional sinogram. The model
presented in this paper establishes a relation between its
trace in x-t-plane, y-t-plane and its position in the CBCT
system.

It is presented that the model provides an approximation of
sinogram traces in a high accuracy. The maximum distance
between the model and the trace is 3 pixels for x-t-plane
and 1 pixel for y-t-plane.

Due to the high accuracy of this method it can be used to
detect patient motion in CBCT. If a patient moves during
image acquisition traces in the sinogram are deformed. By
using a model the projections with patient movement can
be detected by the measurement for differences between the
real trace and its ideal model. So motion correction can be
performed much more specific.

Acknowledgement

The work has been carried out at Sirona Dental Systems
GmbH, Bensheim and supervised by the Institute of Medical Engineering, Universität zu Lübeck.

5 References

[1] R. Schulze, U. Heil, D. Groß, D. D. Brüllmann,
E. Dranischnikow, U. Schwanecke and E. Schoemer,
Artefacts in CBCT: a review. Dentomaxillofacial Radi-
ology, vol. 40, no. 5, pp. 265–273, 2011

[2] R. Spin-Neto, L. H. Matzen, L. Schropp, E. Gotfred-
sen and A. Wenzel, *Movement characteristics in young
patients and the impact on CBCT image quality*. Dentomaxillofacial Radiology, vol. 45, no. 4, 2016

[3] W. Lu and T. R. Mackie, *Tomographic motion detection
and correction directly in sinogram space*. Physics in
Medicine and Biology, vol. 47, no. 8, pp. 1267–1284,
2002.

[4] T. Hanzelka, R. Foltan, E. Horka and J. Sedy, *Reduction
of the negative influence of patient motion on quality of
CBCT scan*. Medical Hypotheses, vol. 75, no. 6, pp.
610–612, 2010

[5] D. Zerfowski, *Motion Artifact Compensation in CT*.
Medical Imaging: Image Processing, Proc. SPIE 3338,
1998

Classification of axial CT Images using Deep Learning for determining a Standard Coordinate System

M. Fleitmann [1], M. Seebaß [2], M. Westerhoff [2], D. Stalling [2], M. Heinrich [3], and M. Blendowski [3]

[1] Medizinische Informatik, Universität zu Lübeck, marja.fleitmann@student.uni-luebeck.de

[2] Visage Imaging, Berlin, {mseebass,mwesterhoff,dstalling}@visageimaging.com

[3] Institute of Medical Informatics, Universität zu Lübeck, {heinrich, blendowski}@imi.uni-luebeck.de

Abstract

A basic problem of DICOM data is that their internal coordinate system - the frame of reference - lacks an origin and therefore makes it difficult to locate corresponding slices in different CT volumes. Taking this localization problem as the starting point the classification of axial CT images using deep learning is implemented in this project.
Based on the results of this step a new standard coordinate system is introduced in order to obtain the shown body section of the image and thus an origin. Finally the successful embedding of an application in the Visage 7 Viewer is shown. The application will then have the ability to link a current and a prior CT study of one patient based on their classification results.

1 Introduction

One of the fastest developing areas competing in machine learning is deep learning [1]. Advantages of the neural networks like the end-to-end characteristic with respect to the feature extraction and the similarity measurement as well as the versatile applications are remarkable.

This project seeks to bring the embedding of the classification by means of deep learning in a medical and application-related context. To realize this, the classification of annotated axial CT image data similar to [2] is implemented. Alongside various experiments are conducted in which different configurations on the compilation of a training set are tested for optimizing the network's outcome.

In a second step a linking application based on the first step's results is created within the clinical viewer Visage 7. Visage 7 provided by Visage Imaging includes the Visage Server which centrally collects all image data from the hospital's image modalities. The user can then access the images trough the integrated viewer. The viewer itself comes with numerous integrated tools to perform image analyses.

The fetched DICOM data contains a coordinate system that defines the axis orientation but does not specify an origin. This prevents that one patient's current and prior study (Fig.1) could be linked without extensive registration or manually intervention. With the help of the classification results the application should be able to transfer a CT volume into a new standard body coordinate system such that studies can be aligned by means of a simple translation in the axial direction. As a result the process of manual linking can be accelerated to facilitate the practitioner's effort during diagnostic or control work.

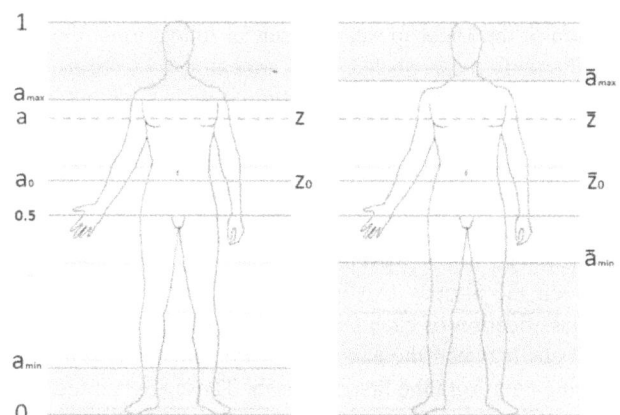

Figure 1: The figure shows a current and a prior study of the same patient. Any position $\bar{z}$ corresponding to the position z can be calculated by determining z_0 and $\bar{z}_0$ and the resulting translation factor t_z

2 Material and Methods

This section is providing information about the underlying network, the data partitioning and a description of constructing the standard coordinate system.

2.1 Background Network Information

Artificial Neural Networks (ANN) were developed to prescind the behavior of the human brain. Therefore, it contains interconnected units called neurons which process the incoming signals to compute an outcome. An enhancement is the Convolutional Neural Network (CNN) [3] . The novelties are the contained convolutional layers with self-trained filter kernel. They are amongst other suitable for classification or regression objectives for 2D data. To facilitate the tasks around the training of CNNs the open-source training platform DIGITS by Nvidia [4] is used in this project. It provides data management and helpful visualizations both at training and testing time. The open-source framework Caffe [5] by the Berkley Vision and Learning Center serves as the back-end. Caffe provides ready-made layers with implementations for CPU as well as GPU. Within DIGITS various pretrained models are given. For this project the GoogLeNet [6] was used. The network contains 22 layers. Some layers represent the network's core idea: The inception layers. Within these are 5×5, 3×3 and 1×1 filter kernel as well as dimension reducing elements included. This allows to use both local and high abstracted features without exploding parameter numbers.

2.2 Developing of Training Data Sets

To train a classification network labeled CT images are needed to create a training and validation set. The annotations required are set in the Visage 7 Viewer as text annotations in the axial slices of a volume. Meaningful labeling should consider only concise anatomical structures. Therefore, among other the bifurcation of the bronchi or the cross section of the kneecap were chosen as annotations. For the training only these labeled slices are extracted from the CT volumes. The number of annotations equals the number of classes leaving the rest of the volume unlabeled.

Considering the next step of constructing a standard coordinate system and classifying all the slices contained in the CT volume another approach to data partitioning was examined. The aim is to evenly distribute a number of classes between the already existing annotations. It results in a preclassification of each slice of a CT volume. The mean distances between the annotations are calculated from the training data from the first approach. These are summed up to an average patient length. Based on this and a fixed number of classes the width of a class is computed. Afterwards, the mean distances are used to calculate how many classes are inserted in each annotation interval and all slices were placed in their corresponding class (Fig.2).

During this project further extensions will be tested based on these two main configurations which will be further explained in the later sections.

2.3 Standard Coordinate System

As indicated in the beginning, results of the network are applied to compute a standard body coordinate system to overcome the lack of an origin of the DICOM coordinate system. The new standard coordinate system ranges from 0 at the feet to 1 at the top of the head as shown in Fig.1. Using a linear regression based on the classification algorithm the displayed CT volume section can be determined and transferred into the coordinate system. Based on this, a simple translation t_z in the z-direction (increasing from the

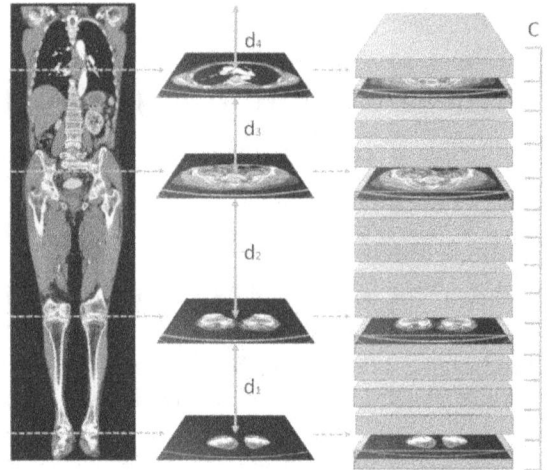

Figure 2: Based on the mean distances d_{1-4} of the set annotations the whole CT volume is divided in C classes.

patient's feet toward the head) is calculated to link corresponding slices of the current and prior study of a patient. To calculate the fitted regression line which is needed to obtain the body section, the study to be classified first requires a forward pass through the network. This results in the estimated class $c_i \in [0, C-1] \subset \mathbb{N}$ and associated score $s_i \in [0,1] \subset \mathbb{R}$ for each slice $i \in [0, N-1] \subset \mathbb{N}$. Here, N denotes the accumulated number of slices of the contained volumes in the study and C is the number of used classes. To ensure the accuracy of the regression some constraints are applied prior to it. A slice i is only included if $s_i > 0.8$ to preemptively exclude possible misclassifications. When the resulting number of slices N_r meets the criterion $N_r \geq 0.6N$ the regression is performed. By means of the slice distance the z-position of each slice z_i is then calculated in the corresponding DICOM frame of reference.

Through the z-positions and the corresponding class labels the equation

$$\begin{pmatrix} c_0 \\ \vdots \\ c_{C-1} \end{pmatrix} = \begin{pmatrix} z_0 & 1 \\ \vdots & \vdots \\ z_{N_r-1} & 1 \end{pmatrix} \begin{pmatrix} m \\ b \end{pmatrix} \quad (1)$$

can be formed. The equation is solved by using the method of least squares to gain the slope m and the intercept b of the fitted line equation $c = zm+b$. Based on z_{min}, a_{min} can be calculated as follows (a_{max} is calculated correspondingly):

$$a_{min} = (z_{min}m + b)/C. \quad (2)$$

With a_{min}, a_{max} of study g and $\bar{a}_{min}, \bar{a}_{max}$ obtained from the prior study $\bar{g}$ a fixed

$$a_0 = (min(a_{max}, \bar{a}_{max}) + max(a_{min}, \bar{a}_{min}))/2 \quad (3)$$

can be calculated as the midpoint of the overlapping area as displayed in Fig. 1. Based on this fixed value z_0 in the DICOM frame of reference can be determined:

$$z_0 = (aC - b)/m. \quad (4)$$

After $\bar{z}_0$ is obtained the same way the sought translation factor $t_z = \bar{z}_0 - z_0$ can be determined.

3　Experiments

In this section the conducted experiments on variant training and validation sets are explained in the first part. Subsequently the experiments for determining the regression's quality are described.

3.1　Training Data Set Configurations

The training set consists of 434 CT studies. It includes among other head/abdomen-thorax scans, run-offs and cutouts (e.g a single foot). For the testing/validation set additional data about 10 % of the training set's size were used. Unless otherwise stated the images were extracted with a fixed window = 500 HU and level = 80 HU.

Lm18　In total 18 annotations were distributed over the entire body. For training the network the slice images from each volume were taken in which an annotation was set.

Lm18bw　In addition to the fixed window the network was also given the window from the corresponding DICOM tags. This is due to the circumstance that for some images the fixed window changed them beyond recognition.

Lm18as　To raise the safety of the class transitions the spatial context of the slices was increased. The beneath and above laying slices with a fixed inter space of 3 mm where taken. They were embedded in the red or blue channel of a RGB image with the original image lying between them.

Lm54　A number of classes were distributed throughout the whole body with each image slice assigned to a class. For this and the following data sets 54 classes were used.

Lm54bw　Corresponding to Lm18bw both windows were combined in a 3-Channel image. The slice images were taken from the respective window, converted into gray images and stored in the red or green channel of a RGB image. The blue channel was kept black.

Lm54redu　Because the marginal classes partly contain all black images at the head as well as beneath the food the classes 53 and 52 as well as classes 0 to 2 were merged. The aim is to reduce the uncertainty of the network's estimates.

Network parameters　The GoogLeNet runs 50 epochs and each 0.5 epochs the validation set is passed forward through the network. The stochastic gradient descent acts as the solver. The learning rate base is 0.01 with a step size of 33.4% and 0.5 as gamma. The images where cropped to 240×240 at random places to fit the input specifications of the GoogLeNet. The training is executed on Titan X (pascal) or GeForce GTX 1080 Ti.

3.2　Regression

In another experiment the quality of the regression-based linking is measured. In order to conduct this 35 additional studies were selected with the requirement of possessing at least one prior study. Within the Visage 7 Viewer the linking is performed on a study and one of its priors. The remaining distance in millimeters between to CT image slices is measured by choosing a fixed a in the standard coordinate system and determining the corresponding slices from each fitted regression line (see Fig. 3). Taking this state as a starting point a registration is carried out by means of mutual information. As before, the distance is determined. The difference between those two values gives an indication of the quality of the regression based on the network classifications against a classical variation of registration based on Mutual Information.

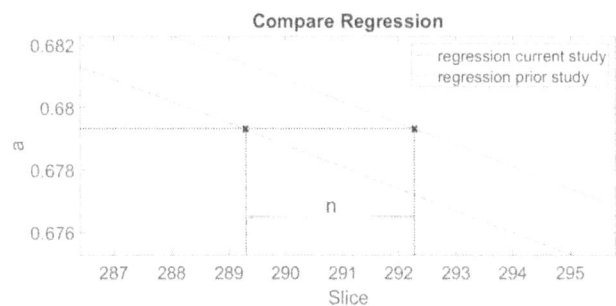

Figure 3: For a fixed value of a in the standard coordinate system compute the corresponding slices. The slice number n times the slice distance results in the measured distance.

4　Results and Discussion

In Table 1 different measures for determining the quality of each data set configuration are displayed. The column accuracy describes the proportion of images where the network's highest-rated estimated class matches the class of interest. The "Top 3" value provides information on the proportion of images in which the correct class was among the first three estimates. The last column shows a value that is relevant for the following regression. It shows the proportion of absolute class deviations greater than two.

Comparing the accuracy and the "Top 3" value of the experiments of the data configurations Lm18/bw/as the high values are noticeable. However, no statement can be made as to whether the use of both windows/level (Lm18bw) had a positive effect on the entire test data. The last column shows that Lm18as was not able to achieve the desired effect of securing the class transitions.

Considering the accuracy of the data configurations Lm54/bw/redu, no experiment reaches a proportion greater than 80%. Taking into account the second column it is noticeable that the values are as high as comparing to Lm18/bw/as. It can be stated that Lm54bw had no positive effect on the network's classification result. In addition Lm54redu as well could not outrun the accuracy and

"Top 3" value of Lm54 but shows a minimal improvement in class deviations.

Regarding determining the translation factor it's necessary to classify all slices of a CT volume. Fig.4 shows a CT run-off. Above and below are the results of the classification of a network trained based on Lm18 and Lm54. To compare both plots the y-axis shows the z-positions of an average patient instead of the class corresponding to the slice on the x-axis. It is noticeable that Lm18 results in more outlier. Therefore, Lm54 was used for the following regression.

For the comparison of the linking based on the network's regression and the subsequent registration by Mutual Information, an average difference of 4.61 mm with a standard deviation of 6.40 mm was found. Depending on this value and a visual control, the regression alone is already verified as sufficient for a majority of CT images. However, the registration can be used if the sole translation in z-direction is not sufficient in case e.g. a rotation is required.

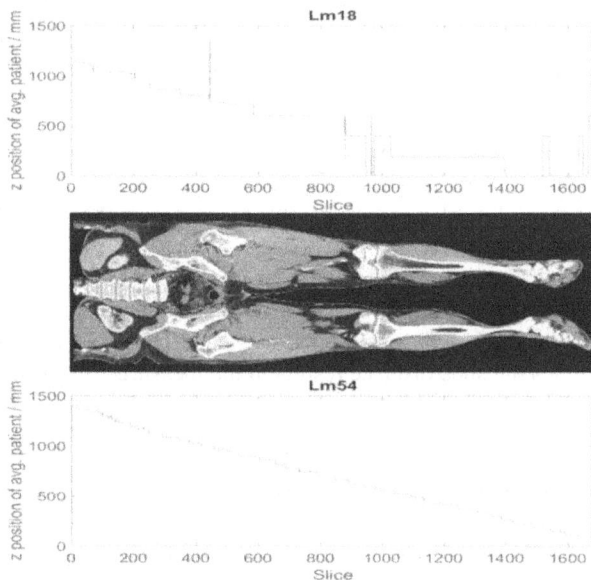

Figure 4: The figure shows a CT run-off. The classification results of a Lm18/Lm54 network are plotted additionally. The vertical axis is chosen so both plots are comparable.

Table 1: For each data set configuration, the mean and standard deviation of the accuracy (first network estimate), the "Top 3" accuracy and the proportion of absolute deviation greater than two of the sought class are given.

Config.	Accuracy	Top 3	deviation > 2
Lm18	0.98 ± 0.04	1.00 ± 0.01	0.001 ± 0.006
Lm18bw	0.98 ± 0.05	1.00 ± 0.00	0.001 ± 0.006
Lm18as	0.96 ± 0.08	0.98 ± 0.08	0.019 ± 0.078
Lm54	0.80 ± 0.13	0.98 ± 0.03	0.002 ± 0.005
Lm54bw	0.76 ± 0.16	0.98 ± 0.03	0.003 ± 0.006
Lm54redu	0.78 ± 0.14	0.97 ± 0.06	0.001 ± 0.003

In general, the equally high values in the second columns indicate that the Lm54/bw/redu performed as well as the others but by using many neighboring classes the class boundaries got blurred. This effect could also have occurred with Lm18as. Lm18/bw/as also have more outliers for the whole body classification explainable by the network not learning certain classification for each slice.

5 Conclusion

In this project the classification of axial CT images using deep learning was performed and different approaches for optimizing the training data to achieve a higher accuracy were tested. The focus there was on the contrasting juxtaposition of a few spatially distant classes and the division of the entire volume into many neighboring classes. Both delivered good results for the actual classification task but for the following objective of constructing a new standard coordinate system the approach with many classes was more suitable. Subsequently, the successful embedding of a linking application in the Visage 7 Viewer based on a regression using the classification network's results was demonstrated. In a direct comparison of the application to a registration by means of Mutual Information it has been shown that the simple linking is completely sufficient in most cases. Nevertheless, the application is not yet accomplishable for all CT volumes e.g. arms in different postures and orientations that cannot be classified in the coordinate system without automatically generated prior knowledge.

Acknowledgement

The work has been carried out at Visage Imaging, Berlin and supervised by the Institute of Medical Informatics, Universität zu Lübeck.

6 References

[1] M. I. Razzak, S. Naz, and A. Zaib, "Deep learning for medical image processing: Overview, challenges and the future," in *Classification in BioApps*, pp. 323–350, Springer, 2018.

[2] H. R. Roth and et al., "Anatomy-specific classification of medical images using deep convolutional nets," in *Biomedical Imaging (ISBI), 2015 IEEE 12th International Symposium on*, pp. 101–104, IEEE, 2015.

[3] J. Schmidhuber, "Deep learning in neural networks: An overview," *Neural networks*, vol. 61, pp. 85–117, 2015.

[4] L. Yeager, J. Bernauer, A. Gray, and M. Houston, "Digits: the deep learning gpu training system," in *ICML 2015 AutoML Workshop*, 2015.

[5] Y. Jia and et al., "Caffe: Convolutional architecture for fast feature embedding," in *Proceedings of the 22nd ACM international conference on Multimedia*, pp. 675–678, ACM, 2014.

[6] C. Szegedy and et al., "Going deeper with convolutions," *CoRR*, vol. abs/1409.4842, 2014.

Towards PET-CT Imaging of Fish: Development of a Dedicated Holder and a Digital Phantom

S. Seeger [1], M. Zvolsky [2], C. Schmidt [3] and M. Rafecas [2]

[1] Medizinische Ingenieurwissenschaft, Universität zu Lübeck, steven.seeger@student.uni-luebeck.de
[2] Institute of Medical Engineering, Universität zu Lübeck,{zvolsky, rafecas}@imt.uni-luebeck.de
[3] Isotopenlabor, Universität zu Lübeck, christian.schmidt@isolab.uni-luebeck.de

Abstract

Fish, in particular zebrafish, are becoming increasingly important as model organisms in medical and biological research. Although small animal PET-CT is a widely functional imaging tool commonly used to investigate metabolic processes, fish studies have been seldom carried out. One challenge faced by in-vivo fish imaging with PET-CT is the lethality of scans in air. We have developed an experimental setup which will enable long measurements of fish. For this purpose, a dedicated tank to be inserted in a small animal PET-CT was designed and constructed. Emphasis was put on the fixation of the fish and the oxygen supply. The additional attenuation of the radiation due to the materials used was studied. For this purpose, CT scans were performed. In addition, the watertightness of the structure was investigated. Furthermore, a digital phantom of a zebrafish was created, which can be used to simulate PET-CT measurements.

1 Introduction

Fish models have recently gained importance in biological medical research, with some fish species such as zebrafish (*Danio rerio*) being of particular interest. This species plays an important role in the investigation of gene functions and the systematic study of development processes. Their suitability results from a fully sequenced genome, the small body (3-5 cm), the transparency of the embryos and a good reproducibility and comparability of the experiments; the latter have been carried out due to the optimised husbandry conditions of the fish. For example to study gene functions in the whole organism, non-invasive imaging methods are frequently used [1]. In comparison to other imaging methods, nuclear imaging offers the possibility of visualizing biological, biochemical and functional processes very precisely. Depending on the application, different methods such as scintigraphy, single-photon emission computed tomography (SPECT) and positron-emissions tomography (PET) can be used. The latter is often combined with computer tomography (CT) in one single device [2]. The major advantage of a PET-CT system is the combination of functional information of the PET image with the good anatomical information provided by CT. Given the small size of the considered fishes, clinical PET-CT scanners are not well suited due to the insufficient spatial resolution ($\sim$4 mm). However, modern small animal PET-CT scanners are a promising alternative. For measurements with small mammals such as mice or rats, the animals are usually placed in a closed imaging chamber that can be easily plugged into the scanner. These chambers provide anaesthetic supplies, temperature control and, if necessary, ECG monitoring. There-

fore, long measuring periods without movement of the animal are possible. However, the measurement of fish places much higher demands on the measurement setup. Since fish can only survive for a very short time without water, the placement within a water-filled vessel is necessary [3]. To our knowledge, PET measurements of fishes in water tanks have been carried out so far only with larger species such as goldfish [4]. In that work, the size of the aquarium was chosen to exactly match the size of the fish, but had no further fixation or adjustment possibilities. Despite anesthesia, such a setup does not fully prevent small movements of the fish, which might seriously deteriorate the image resolution. The present project aims at paving the way for fish imaging using PET-CT. For this purpose, we have designed and constructed a dedicated fish holder to be inserted into a small animal PET-CT. Its properties have been studied in view to its future use. Given that commercial small animal scanners are designed for rodents, their spatial resolution might still be insufficient to accurately image very small structures such as those to be found in zebrafishes. Therefore, Monte-Carlo simulations of dedicated fish PET concepts are planned. The latter should provide better spatial resolution. To this end, we are developing digital phantoms reproducing the anatomy of fishes, to be included into the simulations to provide realistic data. These phantoms consist of a voxellized 3D dataset, which can be provided with a certain activity distribution and attenuation coefficients per voxel. To calculate the attenuation coefficients to be assigned to each voxel, measurements with a Micro-CT were performed.

2 Material and Methods

In order to fulfill the aforementioned criteria for an experimental setup, it is necessary to select suitable materials; furthermore, the construction should be kept as simple as possible, to be produced using common manufacturing methods.

2.1 Material

When selecting the materials, our main focus was good compatibility with water. This means that the material should be neither brittle nor permeable to water, nor should it swell due to the permanent exposure to water, as is the case with many polymers. No metal parts were allowed in the whole construction, since they would severely attenuate the radiation. The surface structure also plays a role; small amount of radioactive tracer might be excreted by the fish and deposit on the material surface, if it consists of a rough structure. This effect might lead to image artifacts or hinder decontamination of the holder. Another feature to be observed is the biocompatibility of the plastic. Since in-vivo imaging is to be carried out, the plastic must not exhibit toxic properties. Due to these requirements and the possible processing methods of the respective materials, three different polymers have been used for various parts of the construction, as shown in Table 1. These materials fulfil the requirements in the best possible way. The materials polymethyl methacrylate (PMMA) and polylactides (PLA) have a low density, which suggests that the radiation attenuation is only slightly higher than that caused by water. The rather higher density of the polyoxymethylene (POM) does not play a significant role, as this material is only used in areas which are not in the regions of interest. The water absorption of PMMA and POM is relatively low compared to other polymers. There is no reliable value for PLA so far. The values for the water uptake are determined according to DIN EN ISO 62 and show the water uptake after 4 days in gram.

Table 1: Used Materials [5]

Material	Density / (g/cm^3)	Water uptake / g
PMMA	1.18	0.4
POM	1.41	0.3
PLA	1.24	/

2.2 Construction

Our early designs focused on fixing the fish directly inside a watertight vessel. However, we finally opted for dividing the volume into a watertight case and an internal component for fixation. Figure 1 shows the final prototype with both components. All components were constructed using SOLIDWORKS Cad Software.

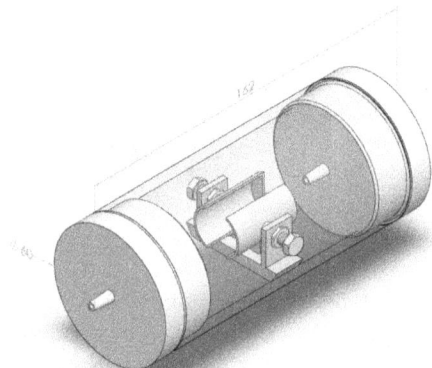

Figure 1: Final prototype with watertight cylinder and internal component for fixation of the fish. Dimensions in mm.

2.2.1 Watertight case

First of all, the dimensions of the holder and the whole setup to be built are important. Since not only zebrafish, but also slightly larger fish should be measured in the future, the maximum dimensions are based on a commercially small animal PET-CT system, the *nanoScan* PET/CT (P82S) from Mediso. The imaging chamber of nanoScan P82S allows for a construction with a maximum diameter of 70 mm. Hence, a cylinder with a diameter of 60 mm was chosen to leave room for supply lines within the imaging chamber. Since at least one of the two sides had to be opened in order to place the fish in the cylinder before the measurement, lids were planned for each side. The latter can be screwed into the cylinder with a thread. Since water supply and drainage are required, connections for hoses have been integrated in the lids. The water exchange is performed by using a peristaltic pump that pumps the used water out of the cylinder, allowing fresh, oxygen-rich water to flow out of a separate tank into the cylinder at the same time. On the inside of one of the lids there is a conical cavity in which the fish's mouth can be pushed so that the fresh water can reach the gills directly. The other lid has an internal hose connection that allows for the flexible drainage of used water near the gills, depending on the length of the used fish. This system can help to reduce the distribution of possible radioactivity released by the fish. The cylinder consists of a PMMA tube with a wall thickness of 4 mm, into which a thread was leathed. The lids are both made of POM and were leathed from one piece. Due to the 15 mm long thread a watertightness should be achieved, but additional o-rings or a teflon tape could be used.

2.2.2 Fixation of the fish

In previous tests, the possibility of enclosing the fish up to the head (leaving the operculum free) in low melt agarose was investigated, but this procedure is very time-consuming and very stressful for the fish. Agarose, however, seems to be ideal for fixing a fish without damaging the skin due to its high water content. Therefore, a fixture was sought that fixes the fish using an agarose "protective coating". In order to make this possible, a mounting bracket has been developed that allows two half-shells poured with agarose to be moved towards each other by means of threaded screws. This allows a slight pressure to be exerted, which fixes the fish between the two plates. The trays and the holder are made of PLA with an Ultimaker 2 3D-printer. M5 polyamide threaded screws, which are shortened to fit, as well as matching nuts which are glued in fitting notches of the holder have been used. Figure 2 shows this setup.

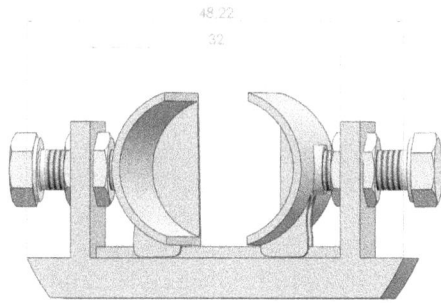

Figure 2: Internal mounting bracket for fixation. Dimensions in mm.

2.3 CT Attenuation measurements

In order to determine the attenuation coefficients of the prototype, CT images of the water-filled cylinder were measured in a clinical Siemens system (SOMATOM Definition AS+). For these images, the corresponding Hounsfield units (HU) of the individual pixels were determined from the data, i.e., using the DICOM header provided by the system during a measurement. These HU values can be converted into single attenuation coefficients for each pixel using known coefficients of water and bone at different energies. This is done by means of the following equation [6], where CT corresponds to the current HU value:

$$
\mu^{PET} = \begin{cases} \mu_{H_2O}^{PET}\left(CT+1000\right)/1000 & CT \leq 0 HU \\ \mu_{H_2O}^{PET} + CT\frac{\mu_{H_2O}^{CT}(\mu_{Bone}^{PET}-\mu_{H_2O}^{PET})}{1000(\mu_{Bone}^{CT}-\mu_{H_2O}^{CT})} & CT \geq 0 HU \end{cases}
$$
(1)

To determine the attenuation coefficients for the individual components and regions, the image data were segmented using Otsu's method [7]. This method uses a specific number of thresholds to provide segments with uniform values, which are representative attenuation coefficients in this case. An automatic determination of the thresholds using the matlab own function `multithresh`, searches for a given number of well-fitting thresholds in the image, did not work very well for this data set. Therefore the limit values were selected by histograms in such a way that a good segmentation of the individual areas was possible.

2.4 Phantom

To create an attenuation phantom, images of two euthanized zebrafish of similar size were measured with a Bruker Skyscan 1172 Micro-CT. To avoid movement during the measurement, the fishes were completely embedded in 1 % agarose in a 10 ml scintillation vial. The image data sets were further gauss-filtered and segmented with Otsu's method. The thresholds were, as before, obtained using `multithresh`. Since the aforementioned Micro-CT images do not provide HU values, the attenuation was estimated as follows: regions identified as filled with agarose were assigned the attenuation coefficients of water. Previous tests supported this approximation, since agarose consists mainly of water. The areas in the image that have been identified as air served as a further reference. The attenuation of the fish again be obtained by (1).

3 Results and Discussion

The components of the experimental setup were tested separately. Following the production and assembly of all components, we verified that the specified dimensions of the two assemblies were well implemented with the selected production methods. The caps for the cylinder fit into the inner thread with millimetre accuracy and thus cause tightness without the need for additional sealing agents between cylinder and cap. Shortly after the first filling of the cylinder there were still small leaks, but these did not occur during further filling. A long test over 4 days also confirmed the tightness. Furthermore, no deformation or swelling of components could be detected due to the influence of water. Also the inner mounting made of PLA, whose water absorption is not exactly known, does not show any deviations. However, a manufacturing problem has arisen in this segment. In the lower part of the holder, larger air pockets appear to have been trapped during the printing process. This can be recognized by the fact that the holder, which should actually be located at the bottom of the cylinder, floats around 180° rotated at the top of the cylinder (Figure 3). Due to the symmetrical design, this does not pose any problem for the actual positioning of the fish. But there is a risk that the holder in the inside of the cylinder will move before and after the actual measurements, when the body is shaken or moved. The air inclusions are also clearly visible on the CT image.

The segmentation of the CT images of the cylinder resulted in attenuation coefficients. For the energy of 511 keV applied in PET measurements, the individual components have the following values: The lids made from POM attenuate with a value of $0.1106\,\mathrm{cm^{-1}}$. The cylinder made of PMMA attenuate with $0.1046\,\mathrm{cm^{-1}}$ and the inner support made of PLA with a maximum of $0.1017\,\mathrm{cm^{-1}}$. The latter values vary greatly due to the air inclusions in the material. It should be noted that the attenuation values for the conversion from HU to attenuation coefficients found in the literature correspondent to 140 kV; on the other hand the CT scans were performed with a tube voltage of 100 kV. Never-

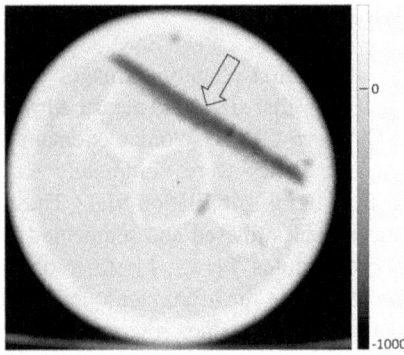

Figure 3: CT image of the cylinder with internal holder. The arrow marks the air inclusion in one part of the holder. The bar shows the gray-levels corresponding the HU values.

theless, these values allow a good estimation of the overall weakening of the radiation during real measurements. Especially in the main measuring direction, i. e. perpendicular to the axis of symmetry of the cylinder, the attenuation is mainly due to the water and the material used deviates only by a maximum of $0.005\,\text{cm}^{-1}$ or 5,2% from the values for water.

The CT images for creating the fish phantom were taken with a resolution of $16\,\mu\text{m}$ per pixel. Thus, at least for the elements of the fish visible in the CT, a very detailed data base is available (see Figure 4). The visible elements mainly comprise the skeletal structure, but also organs such as the swim bladder. The further processing of the images has shown that ring artifacts are contained in the images, which make segmentation and determination of the attenuation coefficients slightly more challenging. Another difficulty in creating the phantom is the separation of the fish from its environment, in this case agarose. Since the attenuation in the CT is almost identical in both cases, the automatically determined threshold values can only provide a first approach for good segmentation, manual correction was absolutely necessary.

Figure 4: Volumetric CT image (inverted gray scale) of the fish as a basis for the attenuation phantom.

4 Conclusion

Overall, the developed fish holder fulfills the requirements for measuring fish in small animal PET-CT. Also allows for longer measuring times, as the fish can be fixed and supplied with fresh water. In addition to measuring ze-brafish, the newly designed version also allows for the use of larger species, whereby only the inner support needs to be adapted. The selected materials attenuate the radiation only to a minor extent and attenuation effects could be compensated for in the image reconstruction process.

With regard to the production of the attenuation phantom, the CT images have created an excellent basis for the precise inclusion of components of the skeleton in particular. In order to obtain information about the organs and soft tissues of the fish, MRI images are currently being evaluated with the aim of further improving the phantom.

Acknowledgement

This work was developed in cooperation with the Institute of Medical Engineering, the Isotope Laboratory and the Fraunhofer EMB. Special thanks to Dr. Sebastian Rakers (EMB), who provided the fish for the measurements, Maren Bobek (IMT), who carried out the CT measurements, Prof. Martin Koch (IMT), who carried out the MRI measurements and Moritz Schaar, who helped with the segmentation.

5 References

[1] European Society for Fish Models in Biology and Medicine e.V., *Haltungsbedingungen für Zebrabärblinge*. Available: https://www.eufishbiomed.kit.edu/59.php [last accessed on 2018-01-10].

[2] S. Cherry, J. Sorenson and M. Phelps, *Physics in Nuclear Medicine*. Saunders Elsevier, Philadelphia, 2012.

[3] E. Fine, L. Herbst, L. Jelicks, W. Koba, D. Theele *Small-Animal Research Imaging Devices*. Seminars in Nuclear Medicine 44, pp. 57–65, 2014.

[4] W. Koba, L. Jelicks, E. Fine *MicroPET/SPECT/CT Imaging of Small Animal Models of Disease*. The American Journal of Pathology, vol. 182, no. 2, pp. 320–324, 2013.

[5] B. Schröder, *Kunststoffe für Ingenieure*. Springer Vieweg, Wiesbaden, 2014.

[6] C. Burger, G. Goerres, S. Schoenes, A. Buck, A. Lonn, G. Schulthess *PET attenuation coefficients from CT images: experimental evaluation of the transformation of CT into PET 511-keV attenuation coefficients*. European Journal of Nuclear Medicine, vol. 29, no. 7, pp. 923–927, 2002.

[7] N. Otsu *A Threshold Selection Method from Gray-Level Histograms*. IEEE Transactions on Systems, Man and Cybernetics, vol. 9, no. 1, pp. 62–66, 1979.

6

Signal Processing

Detection of Acoustic Alarms in Industrial Environment

A. Wiggers [1], L. Fornasiero [2] and H. Botterweck [3]

[1] Biomedical Engineering, University of Applied Sciences Lübeck, anne.wiggers@stud.fh-luebeck.de
[2] Drägerwerk AG & Co. KGaA, Lübeck, livio.fornasiero@draeger.com
[3] Department for Applied Natural Sciences, Fachhochschule Lübeck, henrik.botterweck@fh-luebeck.de

Abstract

Industrial workers are exposed to various dangers. One aspect to prevent them from damage to health is a reliable automated detection of alarms in situations of acute danger. When radio-transmission is impossible this might be performed acoustically. To keep hardware simple the method used should be as simple as possible. Three methods are tested whether a reliable detection of an alarm with known acoustic features during high background noise is possible with them: Euclidean distance of peak vectors with a comparison vector, autocorrelation of single frequencies with control of the repetition rate and cross-correlation with a known alarm signal. All methods are found to be insufficient for practical applications. There is no doubt that it is generally possible to detect an alarm also with high background noise levels. Further research is to be done on more advanced methods to find the simplest method to perform a reliable detection.

1 Introduction

Industrial workers are often exposed to various dangers during their work. Employers need to minimize these dangers wherever possible. One aspect of safety is, to be able to detect alarms in case of acute danger as early as possible to ensure safety for industrial workers and prevent them from damage to health. This includes a reliable automated detection of alarms from mobile warning devices. Difficulties in detection may occur, if a device cannot send radio signals or in places where radio transmission is impossible. Optic methods fail as well, if there are many corners or windings. A possibility to detect alarms in such situations is acoustic detection as acoustic signals are less affected by these restrictions.

The acoustic situation in industrial environments is quite diverse. It can include high background noise levels and the kind of noise varies. There can be stationary, fluctuating or impulse noise. In many cases there are several acoustic sources mixing different kinds of noise. This diversity makes it difficult to automatically detect a certain acoustic signal like an alarm.

The aim of this project is to find a method to automatically detect an alarm with known acoustic features in different kinds of noise especially with high noise levels, ideally even with negative SNR between the alarm and the background noise. The method should be as simple as possible, such that it works with simple hardware. As simple as possible in this case means different aspects. It should be implemented with tools of digital signal processing like e.g. Fourier transform. It should also need as few calculation steps as possible and therefore work as quick as possible. A third criterion is that the training of the algorithm should

also be simple. If an algorithm needs a lot of different data for training the effort of collecting the data can be quite high. In this case there is a high risk that the amount of data is insufficient and the algorithm then works only for certain situations but alarms are missed in others.

Prevalent detection and recognition methods as they have been used e.g. in speech recognition often fail in acoustic environments with high noise levels, especially if the noise is not stationary. Also they require large data bases for comparison. Those speech recognition methods, that are stable also in noise, are much more complicated than the method that should be found in this project. [1], [5], [6]

An interesting method to detect and classify impulsive sound in noisy environment is found in [2]. The impulsive sound is detected by finding sudden increases of the sound energy within the sound signal. The kind of impulse is then determined with either Gaussian Mixture Models or Hidden Markov Models. This algorithm could possibly work for alarms that consist of short tone impulses with high sound pressure levels. Yet it is limited to stationary background noise, while in this project the alarm signal should be detected in all kind of background noises. Also, both Gaussian Mixture Models and Hidden Markov Models need data bases for training, the risks of which being already discussed. So, a different approach must be found.

It is important that there are approximately no false alarms even with new background noises that the method has not been trained with. A real alarm occurs rarely in safe industrial environment. A false alarm might lead to disturbance of the production process and in worst case to unnecessary shutdown of a plant. To examine how often a false alarm or a missed alarm occurs, there are statistical values like sensitivity, specificity and positive predictive value used. The sensitivity in this case shows the percentage of alarms correctly classified as "alarm". The specificity against that shows how many records without an alarm are correctly classified as "no alarm". The positive predictive value shows the likelihood that there truely is an alarm when a record is classified as "alarm". The test device used in the project is assumed by the manufacturer to play 1 min of alarm in 8 hours operating time a day [3], including daily functionality tests of the device. Even if there was 1 min of true alarm within 8 hours and assuming that the method provides a sensitivity of 100%, a specificity of approximately 100% is required to receive an appropriate positive predictive value. Assuming that in reality there are a lot more acoustic situations than what can be simulated in test situations a criterion for the tests shall be that the method provides a specificity of 100% in test situations. This should also work if the microphone is changed no matter, if it is a better or worse quality one.

So, to sum this up, existing recognition methods are not sufficient for the aim of this project but an algorithm is needed which fulfills the following criteria:

- no large data bases needed for comparison

- detection works in all kinds of background noise, also with high noise levels (negative SNR)

- approximately no false alarms, i.e. specificity shall be 100% in test situations

- as simple as possible

2 Material and Methods

Three methods are tested in different kinds of noises whether it is possible to detect an alarm with known acoustic features. The methods are further described later.

2.1 Test set-up

The set-up of the hardware is demonstrated in Fig. 1. The hardware used for testing are a speaker (Genelec M030 Studio Monitor), a microphone (Earthworks Audio M30 30kHz Measurement Microphone, omnidirectional), an external soundcard (RME Fireface UC) and a laptop with MathWorks MATLAB R2016b. The alarm is given by a Dräger Pac 5500 O2 which is a gas measurement device to measure the oxygen concentration within the air. The alarm of the Pac starts if the oxygen concentration is either too low (below 19 Vol.-%) or too high (above 23 Vol.-%). Given these thresholds an alarm can easily be activated by breathing on the sensor.

The speaker and the microphone are connected with the external soundcard, which is connected with the laptop. This way both the speaker and the microphone can be controlled with MATLAB.

Background noise is played with the speaker. The detection works in a process of first recording for approximately 6s and then checking the recorded signal. If there is no alarm

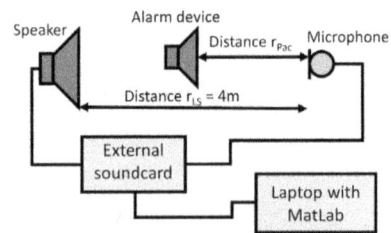

Figure 1: Set-up to test the detection methods for performance: Background noise is played by the speaker and the alarm is started. The microphone records all and the record is analyzed with MATLAB. Both speaker and microphone are connected with an external soundcard and controlled with MATLAB. The alarm device is activated by breathing on the sensor.

detected, a new record is started and checked afterwards. This process goes on until an alarm is detected or the background noise stops. If there is an alarm detected, the background noise is automatically switched off and "ALARM! " appears on the screen of the laptop.

There are 10000 tests performed for each method with present alarm and noise and also 10000 without alarm.

2.2 Description of the Alarm

The alarm to be detected consists of four tone impulses. The tone frequencies alternate between approximately 2775 Hz and 3150 Hz with respective harmonics, starting with the lower frequency. Each tone impulse has a length of 100ms. After the second tone there is a break of 50ms before the third tone starts. This signal is repeated after a break of 400ms, so the complete time from the beginning of the alarm signal to the beginning of the next one is 850ms and the repetition rate of the alarm is approximately 1.2Hz. The spectrogram of the alarm is demonstrated in Fig. 2.

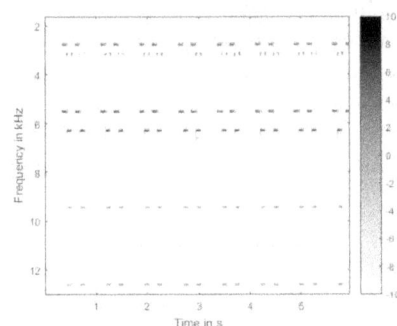

Figure 2: The spectrogram of the alarm. One single alarm consists of four tone impulses with alternating frequencies of 2775 Hz and 3150 Hz with respective harmonics. Each tone lasts for 100 ms with a break of 50 ms. The repetition rate of the alarm is approximately 1.2 Hz.

2.3 Background Noises

The background noises used for testing should present as many different kinds of noise as possible. Therefore, there are both artificial noise (white noise and pink noise) and records of different acoustic sources, e.g. an angle grinder, hammer strikes on metal or a car's engine, used. All noises are recorded in established quality with a sampling rate of 44.1 kHz and 16 bit quantization. Over all there are three noises, which are classified as stationary, three impulsive noises, one fluctuating noise and two which consist of a stationary mixed with a fluctuating part. A list of noises used in test situations and their classification can be found in Table 1.

Speech is also to be treated as noise for the detection. As a speech sample the International Speech Test Signal (short: ISTS) is used. This signal is a mixture of speech records from six different languages (American English, Arabic, Chinese, French, German and Spanish). It holds all typical features of spoken language including the average long term frequency distribution of the six languages without being comprehensible.

Table 1: Classification of background noises from tests

Noise	Classification
Car engine	stationary/fluctuating
Axe during wood chopping	impulsive
Angle grinder	fluctuating
Hammer strikes on metal	impulsive
ISTS	Speech
Machine hall	stationary
Pink noise (artificial)	stationary
Drum-set	impulsive
Vacuum cleaner	stationary/fluctuating
White noise (artificial)	stationary

3 Results and Discussion

The three methods tested are described and discussed in this section.

3.1 Euclidean Distance of Peak Vectors

There is a Fast Fourier Transform (in the following FFT) with FFT length of 1024 samples and an overlap of 512 samples performed for the whole record. It is then counted for each frequency how often a local maximum at this frequency occurs in the magnitude of the FFT spectra. Random peaks due to noise or turbulences on the microphone membrane are cancelled out by setting all values smaller than a threshold zero. The threshold is set by taking the mean value of peaks at all frequencies plus $\{0, 1, 2, 3\}$ times the standard deviation. For comparison the method is also performed without thresholding.

Denoising the signal does not affect the results in test situations but increases the processing time and is therefore considered unnecessary.

After cancelling out random peaks a vector is build covering the number of peaks in 25 frequencies of which 7 are part of the alarm signal. The Euclidean distance to a comparison vector containing the alarm is determined. If the distance is below a set threshold the signal is classified as "alarm ".

The sensitivity and specificity obtained in test situations with a threshold of 0.9 and different correction values can be found in Table 2. Calculating the Positive Predictive Value from these sensitivity and specificity values emphasizes why a specificity of 100% in test situations is necessary: The positive predictive values that are achieved here are between 0.18% and 16.28%, with the best value achieved for the mean value plus three times the standard deviation as correction value. The second best result is significantly lower giving a positive predictive value of 2.15%. These values are not acceptable for practical applications. In case of a sensed alarm it is far too likely with this method that it is a false alarm which can then lead to unnecessary disturbances of the production process.

Table 2: Statistics for Alarm Detection with Euclidean Distance of Peak Vectors for different correction values

Correction Value	Sensitivity	Specificity
None	100.00%	0.00%
Mean	87.93%	30.36%
Mean + SD	86.49%	82.14%
Mean + 2 SD	84.68%	91.96%
Mean + 3 SD	82.88%	99.11%

A threshold higher then 0.9 leads to better sensitivity but decreases specificity. Changing the threshold to lower values until the specificity is 100% for at least one correction value, moves the sensitivity to below 10% for all correction values. This is insufficient for practical purposes.

3.2 Autocorrelation of Single Frequencies with Control of the Repetition Rate

This method also starts with a FFT with a length of 1024 samples and an overlap of 512 samples. All FFT values are written in a matrix where a line contains all FFT magnitude values of one section of 1024 samples. The matrix is denoised with a two-dimensional Wiener filter. After that an autocorrelation is performed for 10 frequencies, resp. 10 columns of the FFT matrix. The formula for the autocorrelation according to [4] is

$$r_{xx} = \int_{-\infty}^{\infty} x^*(t) \cdot x(t + \tau) dt. \qquad (1)$$

The autocorrelation signal is normalized and checked for local maxima with a peak prominence given by a set threshold. The maxima are then used to determine the repetition rate of the frequency. Five of the frequencies are part of the alarm, i.e. they are one of the two base frequencies or their harmonics, and should occur with the associated repetition rate of approximately 1.2 Hz. The others do not occur within the alarm and should therefore not have the same repetition rate. Otherwise it can be assumed that there is some impulsive noise occurring with the same repetition rate. The signal is assumed to contain an alarm if there are at least two alarm frequencies occurring with the associated repetition rate and none of the other frequencies or if there are all five alarm frequencies and no more than one of the other frequencies occurring with the alarm repetition rate.

This detection method meets the requested specificity of 100% in test situations with all thresholds. Unfortunately, the sensitivity obtained with this method is too low for all kinds of noise. The best sensitivity is given with 6.25% at peak prominence thresholds between 0.08 and 0.095. This sensitivity is far too low to be sufficient for practical purposes.

3.3 Cross-correlation with Alarm Signal

For this method there is a cross-correlation of the signal and a single alarm signal performed. The cross-correlation signal is normalized. The cross-correlation signal is further examined by finding peaks that exceed a set threshold. From these peaks the distance between the seven highest peaks is measured. If there are less than seven peaks exceeding the threshold all of them are used. Less than two peaks exceeding the threshold lead to the assumption that there is no alarm. The distance between the peaks can be treated as the periodic time of the repetition frequency. If it is too small (below 800ms) it is assumed that there is no alarm within the record. If it is too big it cannot necessarily be assumed that there is no alarm because in fluctuating noise it is possible that parts of the alarm are masked in a way that there is no peak or the peak is below the threshold. In that case there are probably less than seven peaks above the threshold. Setting the threshold lower is no solution for this because a too low threshold leads to taking peaks into account that do not correspond to the alarm. In that case the distance between the peaks becomes too small or irregular and alarms are missed. So, if the distance between the peaks is too big, it is necessary to take into account that parts of the alarm may be masked and the corresponding peaks in the cross-correlation signal are missing. To compensate this the peak distances are checked whether they are multiples of the periodic time of the alarm repetition rate. If all peak distances are, it can be assumed that there is an alarm.

Tests with different peak thresholds show that a threshold below 0.4 leads to a specificity lower than 100% in test situations, so the threshold should be chosen higher than that. The best result with this method is reached with a peak threshold of 0.405. The sensitivity is then 20.00%. For higher thresholds the sensitivity decreases.

This method works for stationary or fluctuating noise but fails for all kinds of impulsive noise. If only stationary noise is used as background noise the sensitivity exceeds 80% for some thresholds and between 70% and 80% for fluctuating noise. Against that the sensitivity is 0% for impulsive noises. Because in practice impulsive noises are likely to occur, this method is also considered insufficient for practical purposes.

4 Conclusion

The simple methods tested in this project appear to be insufficient for the aim of reliably detecting an acoustic alarm in industrial environment. All three methods do either not meet the criteria that there cannot be false alarms in test situations or – with different parameters – the sensitivity of the method is too low to provide a reliable detection of the alarm.

Considering what is possible in prevalent acoustic detection and classification methods, (e.g. speech recognition techniques or music recognition) there is no doubt that it is generally possible to perform an acoustic alarm detection. Further research is currently done on methods that are more advanced than the presented ones.

As soon as a reliable method is found it should then be tested for requirements for hardware and restrictions in functionality. Also further tests have to be made with more than one alarm device to ensure reliability of a method independently from the number of alarm devices.

Acknowledgement

The work has been carried out at Drägerwerk & Co. KGaA, Lübeck and supervised by the Department of Applied Natural Sciences, Fachhochschule Lübeck.

The authors would like to thank Michael Brodersen from Dräger Safety AG & Co. KGaA for his support during the project.

5 References

[1] J. Li, L. Deng, Y. Gong and R. Haeb-Umbach, *An overview of noise-robust automatic speech recognition.* In: IEEE/ACM Transactions on Audio, Speech, and Language Processing, vol. 22, no. 4, pp.745-777, 2014.

[2] A. Dufaux, L. Besacier, M. Anorge and F. Pellandini, *Automatic sound detection and recognition for noisy environment.* In: Signal Processing Conference, 10th European, IEEE, Tampere, 2000.

[3] Dräger Safety, *Dräger Pac 3500/5500 Co, H2S, O2, Instructions for Use.* Edition 11, Dräger Safety AG & Co. KGaA, Lübeck, 2017.

[4] A. Mertins, *Signaltheorie.* Vieweg+Teubner, Wiesbaden, 2010.

[5] D. Stowell, D. Giannoulis, E. Benetos, M. Lagrange and M.D. Plumbley, *Detection and Classifications of Acoustic Scenes and Events.* In: IEEE Transactions on Multimedia, vol. 17, no. 10, pp.1733-1746, 2015

[6] A. Mesaros, T. Heittola, A. Eronen and T. Virtanen, *(Acoustic Event Detection in real life recordings).* In: Signal Processing Conference, 18th European, pp.1267-1271, 2010

Automated analysis of ear canal geometries

V.-M. Gerant [1], F. Gassenmeyer [2], A. Mertins [3], and H. Husstedt [2]

[1] Medizinische Ingenieurwissenschaft, Universität zu Lübeck, victoriamarie.gerant@student.uni-luebeck.de
[2] Deutsches Hörgeräte Institut, Lübeck, {f.gassenmeyer,h.husstedt}@dhi-online.de
[3] Institute for Signal Processing, Universität zu Lübeck, mertins@isip.uni-luebeck.de

Abstract

Soundwaves entering the ear are guided through the ear canal before reaching the eardrum. Thus, the transmission of sound is influenced by the geometrical shape of the ear canal. In audiological practice such as hearing aid fitting this acoustical transformation is of interest. Therefore, geometrical features of ten human ear canals are investigated. First, a centerline along the ear canal is calculated by an iterative procedure (extended version of [1]). With the area, from slices perpendicular to this axis, an area function is plotted against the length of the ear canal at different positions. The area function displays similarities in rise, length and thickness between the ear canals even if they all have a different shape. Further, this area function is used to calculate an acoustic transmission matrix. The entries of the transmission matrix are plotted over the frequency where the plots take similar courses but with different pronounced minima.

1 Introduction

In many audiological applications acoustical quantities are known at the entrance of the meatus. Still as a reference for further sound perception, measures in direct vicinity to the tympanic membrane are more convincing. Yet, the soundfield depends on the individual shape, which differs in length, thickness and curvature. This raises the questions of how to describe a geometry in a meaningful way and how to limit this geometry with defined boundary conditions.

Stinson and Lawton's paper on the specification of the geometry of an ear canal [1] provides the most frequently mentioned work on this topic. In their study they used imprints of 15 human temporal bones, for which they calculated a curved centerline and slices perpendicular to this axis, to derive the area function and to predict soundpressure distributions.

In the context of this paper, the method of Stinson and Lawton is used to characterize ear canals by means of imprints of 10 subjects in-vivo. Molds of the ear canals are created and handed over as a standard tessellation language (STL) file. Centerline and areas were created whereby area functions and an acoustical transmission matrix is derived. Using these methods, the ear canals are evaluated and future methods are discussed.

2 Anatomy and function of the ear

The human ear is divided into three main parts named outer, middle and inner ear. In Fig. 1 the outer ear lettered with the parts pinna, ear canal and eardrum are shown. These are the main parts of the outer ear and important during further remarks. The pinna passes incoming soundwaves along the ear canal towards the eardrum. The ear canal is a slender curved tube with a length of approximately 25 to 35 mm and varying cross sectional areas [2]. In most chanals two bendings can be distinguished, one near the pinna and one near the eardrum. The eardrum is a thin membrane which vibrates to transmit the soundwaves. The movement is transmitted to the auditory ossicles of the middle ear and is amplified further (Fig. 1) [3]. Since the eardrum forms a limit in the description of an ear canal, this study focusses on the hearing process associated with the outer ear.

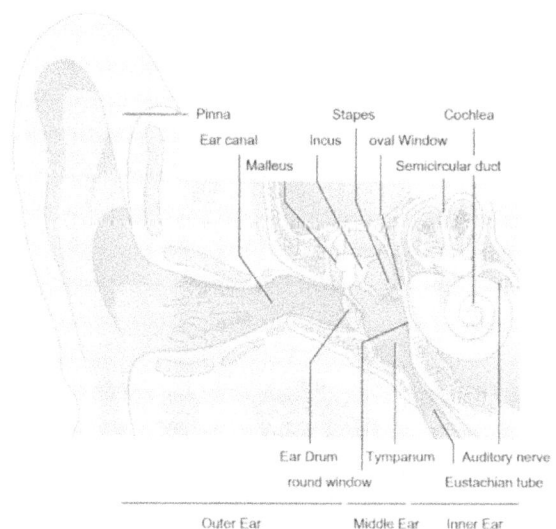

Figure 1: Illustration of the human ear. Only the outer ear with the Pinna, the ear canal and the ear drum is further considered [4].

3 Materials and Methods

Impressions of ten outer ears were made by injecting a special modeling clay into the ear canal. An optical 3D scanner (smart optics duo Scan) was used to digitize the molds with an accuracy of 20 micrometers. The data was saved as a standard tessellation language file. It consists of a list of triangular facet data, where each facet is uniquely identified by a unit normal and three vertices [5]. Further computations and implementations are done in Matlab. As all further computations are based on a centerline through the unknown, arbitrary geometry, a definition for such a centerline is given. Along the centerline parallel slices can be derived which are used to determine an area function and ensuing to calculate the transmission matrix to describe the ear canal in a mathematical way. This is achieved with the following methods.

3.1 Centerline and cross-sectional slices

Creating the centerline is an iterative process. The centerline s is defined by compound of center points c of the areas perpendicular to this line, whereby $s = 0$ defines the position of the ear drum. The boundary of the plane that makes it an area is the intersection of the ear canal border. The areas will further be named as slices where forty slices will be made for each ear canal. This leads to results, that are the best compromise between calculation efficiency and accuracy. Since the centerline is initially unknown, a first estimate of this is done by creating an initial line along the medial axes from the eardrum to the pinna. Parallel slices along this initial line created by a point, a normal vector and the information of the STL file are derived as shown in Fig. 2. At every point of this preliminary centerline the Frenet-Serret equations will be applied. The equations give an orthonormal basis of three vectors (tangent vector, normal vector, and binormal vector) that describe the local behavior of the curve [6].

In a new iteration, slices perpendicular to the tangential vector (concerning the initial center line) will be determined and the center points are calculated. As soon as the resulting center points do not change anymore, their connection will represent the final centerline. The criterion for stopping the iteration was defined as

$$||c_{\text{new}} - c_{\text{prev}}|| < \xi \tag{1}$$

where c_{new} represents the new center point and c_{prev} the previous one. It is checked if the new center point lies within the radius of the old center point with a selected tolerance. The tolerance was chosen as $\xi = 0.5$mm for all 10 ear canals.

An ear canal with adapted slices, initial centerline and final centerline is shown in Fig. 3.

The creation of the adapted slices does not occur for all previously created parallel slices, since the ear canal is previously limited by boundary conditions.

3.2 Condition for ear canal entrance

Since the mold of an ear canal consists conditionally parts of the pinna, the ear canal must be divided into different areas to get an impression of the geometry. There is no universal definition of an entrance to the ear canal, since not all impressions have the same shape. During this work one aim was to find a criterion for separating the ear canal and the pinna which can be applied equally to all molds. The derived criterion

$$\max(|A_{j+1} - A_j|) \wedge \max(d_j) \tag{2}$$

consist of two parts. During the first step the areas A of the parallel slices are calculated using

$$2A = |\sum_{i=1}^{n} x_i y_{i+1} - x_{i+1} y_i| \tag{3}$$

which is known as Gauss's area formula (3) [8]. A plane is defined by i many points, where $x(i)$ and $y(i)$ represent the coordinates of a point. Beginning at $s = 0$, the area of a slice A_{j+1} is compared with the area of the previous slice A_j. All differences are stored in an array. After all slices have been compared, the algorithm searches for the maximum value within the array. The highest difference occurs as a result of the transition from ear canal to pinna.

In some cases, the end of the mold can contain a bigger slice which mislead the algorithm to pass the wrong point. In this case a second condition to underpin the first one has to be implemented. Therefore, the Euclidean distance d (4) is calculated which gives the length between two points and based on the Pythagorean theorem. The algorithm is looking for the maximum of all distances d_j between the points of the centerline. The location where maximum area difference and maximum point distance are equal, the entrance to the ear canal is defined.

3.3 Determining of an area function

With the help of the area function, common properties about the geometry of the ear canal, such as changes in the thickness of the ear canal at different positions or the length of the ear canal can be made. To determine the area function, the areas of the slices were calculated with the help of (3) previously mentioned.

To use Gauss' formula (3), the area has to be transformed in the cartesian basis by a standard rotation matrix. Until now, only the areas of the initial parallel slices are used to determine the area function. The calculated areas are plotted against the current length of the ear canal at the position of the slice. To determine the current position the center point of the considered slice is used which is given by the centerline. The length of the ear canal is calculated continuously along the points of the centerline and summed up afterwards beginning at $s = 0$. Distance d between two consecutive centerpoints $x_{i-1}, y_{i-1}, z_{i-1}$ and x_i, y_i, z_i is calculated using

$$d = \sqrt{(x_i - x_{i-1})^2 + (y_i - y_{i-1})^2 + (z_i - z_{i-1})^2} \tag{4}$$

which is known as the Eucleadian distance. The calculated areas and lengths are plotted against each other and compared with regard to thickness, gradient and length (Fig. 4.

3.4 Extraction of ear canal models

The derivation of the following equations can be found at reference [7]. The calculated lengths l_i and areas A_i can be used in addition to the area function to express the ear canal in a mathematical description. As a first approximation, the ear canal can be devided into discrete sections, where each section is modeled as a uniform cylindrical tube. For each section one can set up a transmission matrix

$$K_i = \begin{pmatrix} e_{11} & e_{12} \\ e_{21} & e_{22} \end{pmatrix} = \begin{pmatrix} \cos(kl_i) & i\frac{Z_0}{A_i}\sin(kl_i) \\ i\frac{A_i}{Z_0}\sin(kl_i) & \cos(kl_i) \end{pmatrix}$$ (5)

where the wavenumber k is calculated with the angular frequency divided by the speed of sound $k = \frac{2\pi f}{c_0}$. Here the regarded frequency f range is 200 to 10000 Hz. The speed of sound in air is set to the constant value of $c_0 = 343$ m/s . The specific acoustic impedance in air Z_0 is calculated with the air density $\rho_0 = 1.2$ kg/m³ multiplied by the speed of sound c_0. The entries of the transmission matrix e_{11} to e_{22} are plotted against the frequency in Fig. 5. The product of all matrices

$$\begin{pmatrix} p_{\mathrm{a,in}} \\ q_{\mathrm{a,in}} \end{pmatrix} = \prod_{i=1}^{N} K_i \begin{pmatrix} p_{\mathrm{a,N}} \\ q_{\mathrm{a,N}} \end{pmatrix}$$ (6)

represents the connection between the input parameters $(p_{\mathrm{a,in}}\ q_{\mathrm{a,in}})$ and the output parameters $(p_{\mathrm{a,N}}\ q_{\mathrm{a,N}})$. With the help of the acoustical transmission matrix the sound pressure and the sound velocity can be calculated later on.

4 Results and Discussion

In Fig. 2 one can see the impression of an ear canal with parallel slices and the previous centerline. In comparison, Fig. 3 shows the ear canal with adapted slices, the previous centerline and the final centerline. In both figures the x and y axis give the dimensions of the mold. In Fig. 3 the condition for limiting the ear canal was applied.
A future task is to limit the ear canal after adapting the slices to get more suited results, regarding the acoustic relevant entrance area. At this point of work the adapted slices spread all over the geometry if it was not previously bounded.

In Fig. 4 the area functions for five ear canals are shown. The length of the ear canal in millimeter on the abscissa is plotted against the area in square millimeter on the ordinate. For clarity, only five out of ten ear canals were used. All of them approximately reach the ear drum, so the length can be reliably compared. They reach a length of 25 to 35 mm which is in good accordance with literature [2]. Even if the

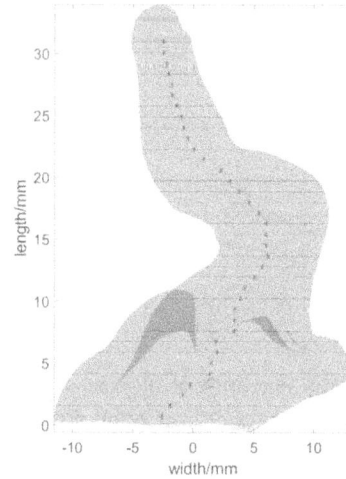

Figure 2: Impression of an ear canal with parallel slices and previous centerline.

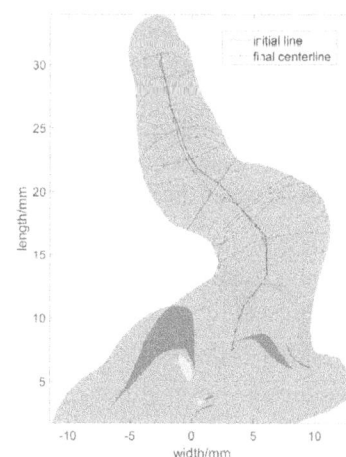

Figure 3: Impression of an ear canal with adapted slices, the previous centerline and the final centerline.

area functions differ at many points, characteristic behavior can be found.

Up to a length of about 10 mm, all ear canals show a positive slope of the area, which suggests that the ear canal gets steadily thicker at the beginning from the eardrum to the pinna. In the range of 15 to 25 mm the area of canal 4 and 5 increases and decreases noticeably which displays the prominent bends of the ear canals. If the ear canals reach a length of 25 mm to 30 mm a huge increase of the area can be seen, this is where the geometry expands and slowly passes over into the pinna. All curves have maxima and minima, which can be seen particularly well on ear canal 5. It shows that the geometry becomes very thin shortly before the transition into the pinna. On canal 1 and 2 the maxima and minima are not very pronounced which indicates that the ear canal is straight and not very curved.

A future task is to compare the area functions of the parallel

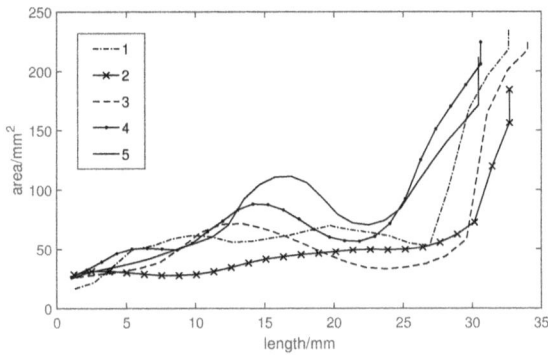

Figure 4: The area functions of five ear canals are shown. The areas of slices along the ear canal are plotted against the length of the ear canal at the position of the slice. Even if every ear canal has a different shape, analogies of the area functions in slope, length and behavior of the graphs can be seen.

slices with that of the adapted slices. It would be of interest to use more ear canals, fully scanned up to the ear drum with other methods than creating the mold by hand. An optical method would be conceivable.

In Fig. 5 the entries of the transmission matrix are plotted against the frequency in a range of 200 to 10000 Hz for the same five ear canals on which the area function was derived.

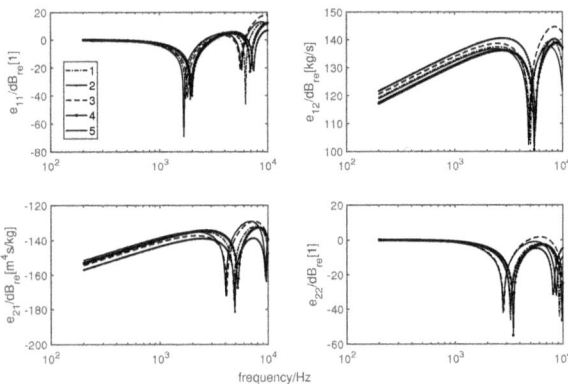

Figure 5: Entries of the acoustical transmission matrix for the mathematical description of five chosen ear canals. Entries of the transmission matrix e_{11} to e_{22} are plotted against the frequency in a range of 200 to 10000 Hz. The minima of the ear canals are frequency shifted among each other but always take the same course.

The plots always take the same course but with different pronounced minima at different frequencies.
Ear canals with a high minima and maxima at the area function also show the highest resonances during the matrix entries.
With the help of the transmission matrix one could calculate the sound pressure and velocity and derive many different statements about the ear canal by changing certain input parameters like the frequency.

A future task is to calculate the acoustical transmission matrix with the help of the areas of the adapted slices and to compare these with the areas of the parallel slices. The sound pressure and velocity can be derived and compared in the following from both type of slices.

5   Conclusion

In this study, methods were developed to make reliable statements about the geometry of an ear canal. Therefore, the centerlines of 10 given ear canal models were calculated. Perpendicular to this centerline, slices were created which were adapted to the ear canal geometry. A condition which defines the entrance of the ear canal is proposed. The adjusted slices were used to derive area functions which were compared afterwards. Despite very different geometry, similarities could be found among the area functions. Finally, the areas were used to determine the transmission matrix to descibe the ear canal in a mathematical way.

Acknowledgement

The work has been carried out at Deutsches Hörgeräte Institut, Lübeck and supervised by the Institute for Signal Processing, Universität zu Lübeck.

6   References

[1] M. Stinson and B.W. Lawton, *Specification of the geometry of the human ear canal for the prediction of sound-pressure level distribution*. In: The Journal of the Acoustical Society of America, vol. 85, no. 6, pp. 2492–2503, 1989.

[2] H. Loeweneck, *Sinnesorgane* Diagnostische Anatomie: Eine Hilfe zum ärztlichen Handeln, pp.19–34, 1981.

[3] T. Lenarz and H.-G. Boenninghaus, *Hals-Nasen-Ohren-Heilkunde*. p. 488, 2012.

[4] B. Kollmeier, *Anatomy, Physiology and Function of the Auditory System*. In: Handbook of Signal Processing in Acoustics, pp. 147–158, 2008.

[5] Y.H. Chen, *Generation of an STL File from 3D Measurement Data with User-Controlled Data Reduction*. In: The International Journal of Advanced Manufacturing Technology, vol. 15, no. 2, pp. 127–131, 1999.

[6] W. Kühnel, *Differentialgeometrie*. 5. Auflage, 2010.

[7] M. Kaltenbacher, J. Kolerus and J. Metzger, *Akustik für Ingenieure*. Technische Universität Wien, 2016.

[8] R. Pure and S. Durrani, *Computing Exact Closed-Form-Distance Distributions in Arbitrarily Shaped Polygons with Arbitrary Reference Point*. In:The Mathematica Journal, 2015.

Pulse detection in video sequences acquired with a thermographic camera using MIT's Eulerian Video Magnification

H. Siebert [1], M.-F. Uth[2], and A. Mertins[3]

[1] Medizinische Ingenieurwissenschaft, Universität zu Lübeck, hanna.siebert@student.uni-luebeck.de
[2] Drägerwerk AG & Co. KGaA, Lübeck, marc-florian.uth@draeger.com
[3] Institute for Signal Processing, Universität zu Lübeck, mertins@isip.uni-luebeck.de

Abstract

Monitoring vital data is essential in neonatal intensive care units. Contact-free measurement of the pulse rate increases the patient's comfort and facilitates the workflow in hospitals. This paper outlines a possibility of extracting the pulse rate from video sequences acquired with a thermographic camera. The presented approach uses *Eulerian Video Magnification* developed by the Massachusetts Institute of Technology (MIT) to enhance the pulse signal in video data and extract the magnified information with the help of several different methods [1]. We point out an approach that is capable of calculating vague values for the pulse rate from thermographic data and is worth further optimization in future.

1 Introduction

The pulse rate is one of the essential vital parameters to be monitored in neonatal intensive care. Amongst neonates, the pulse rate varies between 70 and 170 beats per minute (bpm) depending on the patient's activity [2]. The pulse is generated inside the heart when the ventricles contract and blood flows out of the left ventricle into the aorta. Different mechanical processes contribute to the propagation of the pulse. Therefore, the pulse can be described by the velocity of the blood flow, blood flow rate, and blood pressure [3], [4]. Due to heat exchange caused by convection and conduction between blood vessels and surrounding tissue, the temperature of the tissue is modulated with the pulse-shaped blood flow. This leads to a modulation of the skin temperature which can be detected with the help of a highly sensitive thermographic camera and an appropriate signal analysis method [3].

In this paper, we examine the possibility of detecting pulse frequencies in present data of a thermal imaging camera using *Eulerian Video Magnification* developed by the Massachusetts Institute of Technology (MIT) [1].

2 Material and Methods

Eulerian Video Magnification is a computational technique developed by the MIT and presented by Wu et al. in 2012 [1]. It is a method revealing subtle temporal motions and colour changes in video sequences that normally are impossible or very difficult to see with the human eye. Input of this method is a video sequence that is processed by a spatial decomposition and temporal filtering. The perceived signal is then amplified and thus hidden information is displayed [1].

The following section will outline the video data to be processed and will describe the approach of *Eulerian Video Magnification* more detailed. Furthermore, the methods used for extraction of the pulse rate from the output of *Eulerian Video Magnification* are explained.

2.1 Video data

In order to test the method of *Eulerian Video Magnification* for the available dataset of thermographic video sequences, different exemplary sequences are selected. These video sequences comply with the following criteria: The video material is acquired with a frame rate of 25Hz and reference data from patient monitoring is available. Furthermore, the chosen data has to be recorded from different patients in different bedding positions with little movement.

Based on these criteria, nine different video sequences are selected for further processing. Table 1 shows the chosen video data with the corresponding bedding positions and the minimum and maximum pulse rate monitored by a patient monitor during video acquisition. The used video data is uncompressed and each video sequence has a duration of approximately ten minutes.

All video sequences contain gray values from $0 - 255$ that encode a temperature range from 28 to $40°C$. Additionally, every video sequence is created in RGB colour space and with gray values encoding a smaller temperature range from 35 to $38°C$. Using the same range of gray values, this leads to a higher resolution for temperature encoding within the considered smaller temperature range.

Table 1: Video data with type of bedding position and monitored range of pulse rate.

Video ID	Position	Pulse Rate [bpm]
1	dorsal	163-178
2	lateral	121-172
3	abdominal	164-171
4	lateral	144-164
5	lateral	137-146
6	dorsal	131-138
7	dorsal	117-156
8	abdominal	146-164
9	lateral	143-175

2.2 Eulerian Video Magnification

The approach of *Eulerian Video Magnification* [1] considers the variation of pixel intensity values over time. Therefore, temporal and spatial processing methods are combined. In the first instance, the video sequence is decomposed into different spatial frequency bands. On each spatial band temporal processing is performed by observing the time series corresponding to the value of a pixel in a frequency band. A bandpass filter is applied to extract the frequency bands of interest. The resulting signal is then multiplied by a magnification factor α specified by the user. Finally, the magnified signal is added to the original signal and an output video is generated. In the following, the processing steps are described in detail.

Temporal processing is based on a first order Taylor series expansion. To explain the method, a one-dimensional signal undergoing translational motion is considered. This one-dimensional case can then be generalized to locally translational motion in two dimensions.

The image intensity $I(x,t)$ determines the intensity of the image pixel at the spatial position x at time t. After translational motion, the resulting intensities can be expressed by $I(x,t) = f(x+\delta(t))$ and $I(x,0) = f(x)$, where $\delta(t)$ is the displacement function. With regard to our application, motion means a spatial variation of the recorded temperature values.

If the image intensities can be approximated by a first order Taylor series expansion about x, the image can be expressed by

$$I(x,t) \approx f(x) + \delta(t)\frac{\partial f(x)}{\partial x} \tag{1}$$

at time t.

Applying a broadband temporal bandpass filter to $I(x,t)$ at every position x results in $B(x,t)$. For our application, the bounds of the bandpass filter are selected regarding the minimum and maximum pulse rate (see Table 1). Assuming the displacement function $\delta(t)$ is within the passband of the broadband temporal bandpass filter,

$$B(x,t) = \delta(t)\frac{\partial f(x)}{\partial x} \tag{2}$$

is obtained.

The bandpass signal is amplified by the magnification factor α and added to $I(x,t)$ which results to

$$\tilde{I}(x,t) = I(x,t) + \alpha B(x,t) \tag{3}$$

as the processed signal.

Inserting (1) and (2) into (3) leads to

$$\tilde{I}(x,t) \approx f(x) + (1+\alpha)\delta(t)\frac{\partial f(x)}{\partial x}. \tag{4}$$

By assuming the first order Taylor series expansion is a good approximation for larger perturbation $(1+\alpha)\delta(t)$, the amplification of the temporally bandpassed signal can be related to motion magnification. Thus, the resulting output is

$$\tilde{I}(x,t) \approx f(x + (1+\alpha)\delta(t)). \tag{5}$$

The spatial displacement $\delta(t)$ of the local image $f(x)$ at time t has therefore been amplified to the displacement $(1+\alpha)\delta(t)$ [1].

2.3 Extraction of the pulse rate

To extract a pulse rate from the video sequences being the output of the method described above, we consider three different approaches, which are described in the following section. The extracted pulse rate is then compared with the available reference data from patient monitoring.

Each of the approaches takes the framewise varying mean gray value as a basis. This determined variation of intensity values over time is processed further.

Computation based on local maxima To determine the pulse rate of the recorded patients, the first approach is to observe the local maxima of the extracted intensity value modulation. Therefore, the number of appearing local maxima within a defined time interval is determined. This results in a frequency that directly leads to the searched frequency that has been magnified by *Eulerian Video Magnification*. For the purpose of avoiding to involve local maxima with low values in the calculation of the pulse rate, a correction step is included in the approach. Therefore, the range of intensity values between the maximum and the minimum value occurring in the considered time interval is extracted. Only the local maxima within the upper 75% of the range of values are selected for further processing. The approach is illustrated in Fig. 1.

Computation based on FFT The second approach to determine the pulse rate is based on the Fast Fourier Transform (FFT). It is illustrated in Fig. 2. By performing the FFT, a digital signal is decomposed into its frequency components [5]. When applying the FFT on the extracted intensity value modulation, a spectrum is obtained that shows the different frequencies occurring in the signal. The frequency which is emphasized the most by *Eulerian Video Magnification* is the frequency component with the highest amplitude in the obtained spectrum.

Computation based on weighted FFT The third approach to extract the pulse rate is a method that combines the FFT with an amplitude threshold and a weighting of frequency components. In this approach, not only the frequency component with the highest amplitude is included in the computation, but all frequency components with an amplitude higher than a threshold being half of the maximum amplitude. Based on the resulting frequency components and their amplitude, a weighted mean value is calculated. This frequency value is then associated with the pulse rate. This approach is illustrated in Fig. 3.

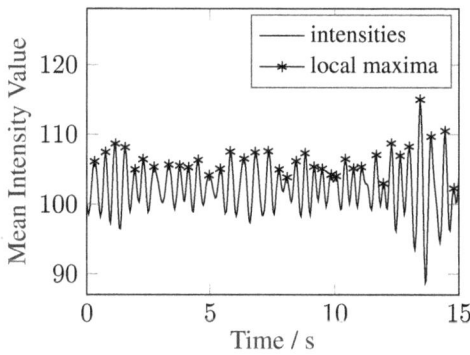

Figure 1: Detection of local maxima within the intensity values over a certain time interval.

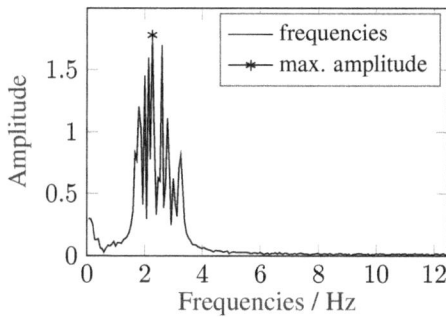

Figure 2: Detection of the frequency with the highest amplitude.

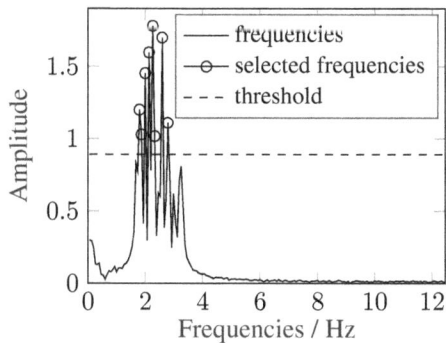

Figure 3: Detection of a frequency value by weighting frequency components with amplitudes higher than a certain threshold.

3 Results and Discussion

In order to compare the reference pulse rate from patient monitoring with the pulse rate extracted with the help of *Eulerian Video Magnification* and the described methods for pulse rate extraction, we evaluated the data over different time intervals. Therefore, we determined the mean value of the patient monitoring data over the chosen time interval and compared it with the extracted pulse rate within the same time interval.

In the following, the results are presented that arise from variation of the methods to extract the pulse rate, the variation of gray value coding or colour coding of the video sequences, and the considered time interval for data comparison.

3.1 Comparison to data of patient monitor

Table 2 shows the mean value of the absolute differences between the pulse rate calculated with the help of *Eulerian Video Magnification* and the reference data from patient monitoring. The results are presented for different time intervals for evaluation, for every method used for extracting the pulse rate, and for the different temperature coding possibilities. Table 3 shows the results for the same modalities considering the maximum absolute difference occurring in the compared data. For Table 2 and Table 3, the results for all video sequences presented in Table 1 are taken into account.

Table 2: Mean value in [bpm] of the absolute differences between the pulse rates recorded with patient monitoring and calculated with *Eulerian Video Magnification*. Values are presented by method for pulse rate extraction, time interval, and temperature coding.

method		total	28-40°C	RGB	35-38°C
local	$5s$	16.018	15.799	15.891	16.364
maxima	$10s$	14.026	13.679	13.876	14.522
	$15s$	13.679	12.804	13.876	14.522
FFT	$5s$	29.297	29.575	29.575	28.734
	$10s$	32.694	32.016	32.074	33.991
	$15s$	32.622	33.099	33.835	32.933
weighted	$5s$	18.790	18.730	19.127	18.405
FFT	$10s$	19.408	19.214	19.294	19.294
	$15s$	20.310	19.570	19.874	21.488

When comparing the presented results of the different methods to extract the pulse rate from magnified video sequences, the method that has calculated the pulse rate based on the number of local maxima within the time interval showed the best results with mean differences around 15bpm, followed by the approach that has performed a weighted FFT with differences around 20bpm. Simply using the FFT and considering the maximum amplitude led to differences up to 34bpm. Regarding the maximum differences for the same data, the ranking of pulse rate extraction methods was identical. Encoding a smaller tempera-

Table 3: Maximum value in [bpm] of the absolute differences between the pulse rates recorded with patient monitoring and calculated with *Eulerian Video Magnification*. Values are presented by method for pulse rate extraction, time interval, and temperature coding.

method		total	28-40°C	RGB	35-38°C
local	$5s$	60.000	60.000	60.000	60.000
maxima	$10s$	54.000	42.000	43.000	54.000
	$15s$	43.800	38.000	37.000	43.800
FFT	$5s$	150.000	82.000	82.000	150.000
	$10s$	166.000	81.000	81.000	166.000
	$15s$	163.000	78.000	78.000	163.000
weighted	$5s$	150.000	64.354	81.000	150.000
FFT	$10s$	151.100	63.824	60.057	151,100
	$15s$	151.733	54.580	54.945	151.733

ture range did not lead to improved results. The best results were also achieved with the method having used the local maxima approach. Here, differences with maximum values between 44bpm (evaluation over $15s$) and 60bpm (evaluation over $5s$) were reached. We attained the best results for both mean and maximum differences when evaluating over a time interval of $15s$. Comparing the different available temperature coding possibilities, the least differences including all examined video sequences were achieved with gray values encoding temperature values from 28 to 40°C. Fig. 4 and Fig. 5 outline the comparison of the different video sequences. The best results regarding the mean absolute difference was the result of video sequence 8 with 5.87bpm. Observing the maximum absolute difference, video sequence 4 showed the best results with a maximum difference of 15bpm.

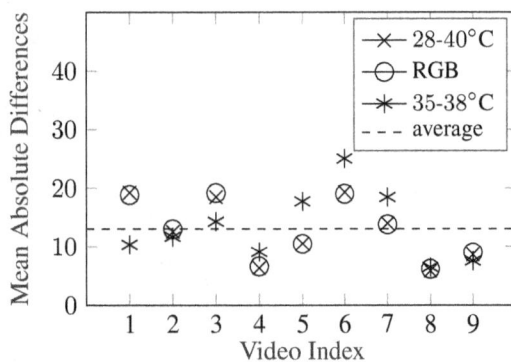

Figure 4: Comparison of the mean absolute differences of the different video sequences, using an evaluation time interval of $15s$ and examining the local maxima.

3.2 Discussion

The available thermographic video sequences combined with the method of *Eulerian Video Magnification* and the chosen methods for pulse rate extraction did not provide a reliable alternative to conventional patient monitoring.

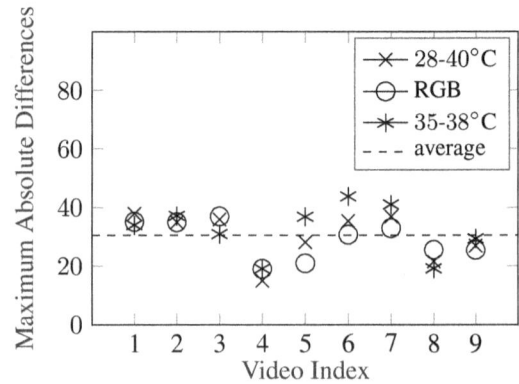

Figure 5: Comparison of the maximum absolute differences of the results of the different video sequences, using an evaluation time interval of $15s$ and examining the local maxima.

Nevertheless, when optimizing the utilized methods, the presented approach is worth further examination.

4 Conclusion

We presented an approach that has the potential of being a contact-free alternative for conventional pulse measurement. By using the described methods, we were able to achieve a vague calculation of the pulse rate. To provide a reliable procedure to extract the pulse rate from thermographic data, further optimization has to be done in future work.

Acknowledgement

The work has been carried out in cooperation with Drägerwerk AG & Co. KGaA, Lübeck and supervised by the Institute for Signal Processing, Universität zu Lübeck.

5 References

[1] H.-Y. Wu et al., *Eulerian Video Magnification for Revealing Subtle Changes in the World*. In: ACM Trans. Graph. 31.4 (July 2012), New York, USA, pp. 65:1–65:8. Code available: people.csail.mit.edu/mrub/evm/.

[2] M. Obladen and R. F. Maier, *Neugeborenenintensivmedizin*. Springer, Heidelberg, 2006.

[3] N. Sun, M. Garbey, A. Merka, and I. Pavlidis, *Contact-Free Measurement of Cardiac Pulse Based on the Analysis of Thermal Imagery*. In: IEEE Transactions on Biomedical Engineering 54.8 (Aug. 2007), pp.1418–1426.

[4] P. Salvi, *Pulse Waves - How Vascular Hemodynamics Affect Blood Pressure*. Springer, Milan, 2012.

[5] J. W. Cooley and J. W. Tukey, *An algorithm for the machine calculation of complex Fourier series*. In: Math. Comp. 19 (1965), pp.297–301.

Sampling and Interpolation of Sound Fields

H. AbdelRahman [1,2], F. Katzberg [2], T. Parbs [2], and A. Mertins [2]

[1] Biomedical Engineering, University of Applied Sciences Lübeck, hesham.abdelrahman@stud.fh-luebeck.de

[2] Institute for Signal Processing, Universität zu Lübeck, {katzberg, parbs, mertins}@isip.uni-luebeck.de

Abstract

As much as studying sound fields in closed rooms is interesting, it is complicated for how the room affects the sound waves. At a single point, the receiver would receive the direct wave and reverberations from the walls. To understand the behavior of sound waves in a certain room and how this room affects sound, we measure the spatially dependent room impulse responses of this room. This paper focuses on the conventional stationary sampling of room impulse responses and investigates the accuracy of their spatial interpolation. The study was carried out in the labs of the Institute of Signal Processing, Universität zu Lübeck. A hexapod robot was used to hold the microphone to provide accuracy and stability in positioning. Results showed low interpolation error values and they can be even lower by solving a hardware malfunction.

1 Introduction

For source-receiver scenarios in closed rooms, the transmission of sound waves involves the direct path of waves, and, additionally, successively reflected paths from the room walls. Due to the reflections, the listener receives reverberated sound. This may impair the sound experience for the listener, and, moreover, may decrease the performance of acoustic devices, such as telecommunication and speech recognition systems. However, the knowledge of the sound field inside the volume of interest allows for compensating room effects. Methods for listening room compensation are given in [1].

Sound transmission from one position to another position inside the room is described by room impulse responses (RIRs). The entire set of spatially dependent RIRs inside the volume of interest provides the sought sound-field information. Methods for measuring RIRs with stationary microphones are commonly based on correlation techniques exploiting specially designed excitation sequences, such as maximum-length sequences [2], [3] and exponential sine sweeps [4].

The spatio-temporal sampling of RIRs by using stationary microphones is not practical for higher audio frequencies, since the uniform sampling grid in space requires an extremely large number of sampling points, in order to fulfill the spatial Nyquist-Shannon sampling theorem. Sound-field sampling and reconstruction has been investigated in [5].

For reducing the effort for spatial sampling, compressed sensing (CS) based methods have been proposed, allowing for sub-Nyquist sound-field sampling and interpolation [6], [7], [8]. Recently, a method for sound-field recovery was introduced, that combines the theory of CS with a dynamic measurement procedure [9]. Here, samples taken by moving microphones are used to set up a system of linear equations. The solution of the system is the uniformly sampled sound-field belonging to the time-invariant sound source configuration. For practical applications, the microphone trajectories will lead to ill-posed and even under-determined problems considering given volume of interests. Therefore, the principle of compressed sensing is used for reconstruction.

The CS based measurement of sound-fields proposed in [9] relies on spatial interpolation of RIRs at positions on a uniform grid. The accuracy of the interpolation method used for modeling the linear system directly affects the quality of the recovered sound field. Since spatial interpolation is the key for solving the sampling problem in [9], we investigate in this paper how accurate the sound field may be actually interpolated given uniformly sampled sound-field data. In cotrast to the simulations in [9], we use measured real-world data. For measurements, we use a hexopod robot that precisely moves to the points of the spatial grid. On top of the robot, a microphone is mounted that takes stationary measurements of the grid. RIRs are based on maximum length sequence (MLS) excitation.

2 Material and Methods

In this work, we first designed a setup for measuring RIRs on a uniform grid in space. Then interpolation was tested by measuring RIR in the intermediate grid and comparing the measured results with the interpolated data. To do so, a MLS signal was designed and used as source signal. This signal was sent through speakers and recorded by a single microphone held by an arm of a hexapod robot that moves it through points of the uniform spatial grid. RIRs were sampled at these points and estimated at the intermediate points through interpolation. For the baseline of the interpolation accuracy, we also measured the RIR at the intermediate points of the original grid (ground truth) to compare them with the interpolated data. These steps are described in more details within this section.

2.1 Hardware

To perform the experiment, the hexapod microrobot H-820.D1, from Physic Instruments (PI), was used to hold the receiver microphone. The use of the robot provided a high level of stability, high accuracy for positioning and finer gridding, and reduced measuring time. The robot was programmed in MATLAB. MATLAB was also used for generating the source signal and communicating with all hardware including the audio interface and the robot. The audio interface used is FIREFACE 800, from RME, which served as input and output server. It was connected to the speaker transmitting the source signal, at the same time, it was connected to the microphone recording the received signals. Such a system included a hardware delay which was compensated using a feedback channel. The source signal is sent out through K+H O110 speakers, from Klein+Hummel.

2.2 Sampling of Sound Fields

The spatio-temporal sound pressure field $X(x,y,z,t)$ can be defined as: "the sound pressure recorded at location (x,y,z) and time t, given an acoustic event in a room." [5]. For the general case of single point source, generating source signal $s(t)$ being a Dirac impulse at time $t=0$, the sound field simplifies to the RIRs at locations (x,y,z), denoted as $h(x,y,z,t)$. Accordingly, the observed signal can be described as:

$$X(x,y,z,t) = \int_{-\infty}^{\infty} s(\tau)h(x,y,z,t-\tau)d\tau. \quad (1)$$

Given that the sound field is band-limited, it can be reconstructed through equidistant sampling in time and space. To perform such sampling, the Nyquist-Shanon Theorem (NST) must be considered. The NST determines sampling rates in time and space that ensure aliasing-free reconstruction.

For temporal sampling, to avoid aliasing, the sampling frequency should be set according to the following equation:

$$f_s \geq 2f_c, \quad (2)$$

where f_s is the sampling frequency and f_c is the cutoff frequency.

For the spatial sampling, the hexapod robot moves through points on a uniform grid in space. For this grid with regard to the NST, the spatial sampling is done according to the equation:

$$\Delta_\xi < \frac{c_0}{2f_c} \forall \xi \in \{x,y,z\}, \quad (3)$$

where c_0 is the speed of sound, which is equal to $343\,\mathrm{m/s}$ in air at $20°C$, and Δ_ξ is the spatial sampling interval.

In the following, the sampled sound field is denoted by $h(g,n)$, using the discrete spatial-grid variable g and the discrete time variable n.

2.3 Maximum Length Sequence

Measuring RIRs by use of periodic MLS as excitation is based on deconvolution [10], [2]. It exploits the fact that the MLS possess a perfect periodic autocorrelation function. The recorded signal may be simply correlated with the appropriate inverse signal, in order to extract the room impulse response. The inverse signal of the MLS is its time-inversed version. Let $s'(n)$ be one period of the MLS, and let $s(n)$ be the repeated MLS. Then, we have the helpful property

$$s(n) * s'(-n) = \sum \delta(n+m), \quad (4)$$

where $\delta(n)$ is the unit pulse. For the experiments in this paper, an MLS $s'(n)$ of length 8191 was used ($2^N - 1$ with $N=13$). The MLS was repeated twenty times to create the source signal $s(n)$.

By using the property (4) and regarding (1), the deconvolution of the RIR $h(n)$ is performed through the cross-correlation:

$$h(n) = X(n) * s'(-n). \quad (5)$$

2.4 Interpolation

For the spatial interpolation of RIRs, we use the Lagrange interpolation. In the experimental part, we test the worst case for interpolation. From the original sampled grid, all intermediate points being exactly in the middle between grid points shall be interpolated. These points also form a uniform grid, shifted from the original one by one half of the sampling interval.

In interpolation in general, if the interpolation points are distinct, then finding a polynomial passing through the data point is similar to solving a system of linear equations. The Lagrange interpolation of RIR $\tilde{h}(g_i,n)$ from RIRs $h(g_j,n)$ on a grid can be written as:

$$\tilde{h}(g_i,n) = \sum_{j=0}^{K} h(g_j,n)\mathcal{L}_j(g_i), \quad (6)$$

where g is a one-dimensional grid variable and K is the order of the polynomial. Note that the interpolation for a multidimensional uniform grid is separable into multiple one-dimensional interpolation problems. The Lagrange polynomials $\{\mathcal{L}_j\}_{j=0}^K$ must satisfy

$$\mathcal{L}_j(g_i) = \begin{cases} 1, & \text{if } i=j \\ 0, & \text{if } i \neq j \end{cases} \quad (7)$$

for the interpolation. This property is fulfilled by the polynomial

$$\mathcal{L}_j(g_i) = \prod_{k=0,k\neq j}^{K} \frac{g_i - g_k}{g_j - g_k}. \quad (8)$$

For the simplest case with $K=0$, the Lagrange interpolation is equivalent to a simple linear interpolation.

3   Experiments and Results

We measured the sound field in an office-sized room. The volume of interest was a 2D plane of size $10\,\mathrm{cm} \times 10\,\mathrm{cm}$. The sampling frequency of the microphone was $32\,\mathrm{kHz}$, accordingly, the temporal cutoff frequency is $16\,\mathrm{kHz}$. In order to satisfy the spatial NST (3), the spacing of the sampled grid must be $1\,\mathrm{cm}$ maximum. For the interpolation, we tested two spatial sampling intervals, $\Delta = 1\,\mathrm{cm}$ and $\Delta = 0.5\,\mathrm{cm}$. The last one corresponds to twofold spatial oversampling. For the stationary RIR measurements, we used 20 repetitions of an MLS of length 8191 as excitation, deconvolved the RIR by autocorrelation and averaged over the repetitions.

3.1   Measurement Setup

Additionally to the robot taking RIR measurements on the uniform spatial grid, we placed a fixed reference microphone next to the volume of interest. From this reference microphone we could determine if there were some time-variant effects, e.g. different delays due to changing of the room temperature, and, thus, the speed of sound. We could not observe any time-varying delay of this reference RIR, so the temperature and speed of sound can be assumed to be nearly constant during measurements. However, during the measurement process, we could observe that the amplitudes of the reference RIR scale differently over time. Experiments and analysis showed that the reason for this behavior comes from the audio interface, the FIREFACE 800. By simultaneously measuring the channel impulse response of the FIREFACE, i.e., directly connecting one input channel with one output channel via cable, we saw that the measured amplitude scales over time. The snippets of the measured reference RIR at the beginning of the experiment and after $i = 267$ iterated measurements are depicted in Fig. 1. In addition, Fig. 2 shows how the amplitude of the direct sound peak scales for the reference RIR over the iterated measurements. In that experiment, the amplitude oscillates around 0.065 and drops at $i = 267$. From Fig. 3, it can be seen that at the same time, the amplitude of the internal impulse response of the FIREFACE also drops drastically. In other experiments we also observed linear and random changes of the amplitudes. We still investigate the reason for this effect. In the following shown results, the scaling effect is not compensated.

3.2   Interpolation Results

We let the robot sample the region of interest on a virtual spatial grid. Then, we interpolated any intermediate positions being exactly in the middle between four grid points. As interpolation method, we used the Lagrange interpolation with $K = 9$. For having a ground truth for comparison, we also sampled the intermediate positions by the stationary MLS method. For evaluation, we define the interpolation error

$$\frac{\|\boldsymbol{h}_i - \tilde{\boldsymbol{h}}_i\|_2}{\|\boldsymbol{h}_i\|_2}, \qquad (9)$$

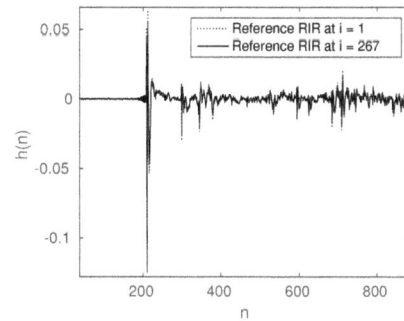

Figure 1: Time variant RIR at reference microphone for measuring interval $i = 1$ and $i = 267$.

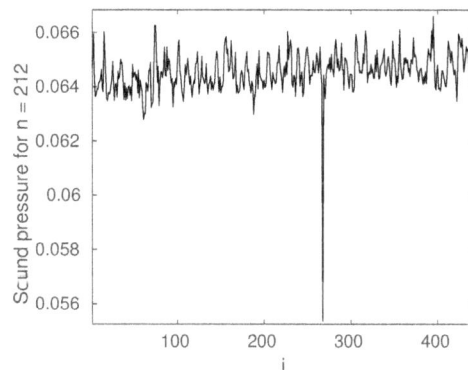

Figure 2: Amplitude changing of the direct peak of the reference RIR over the measurement intervals i

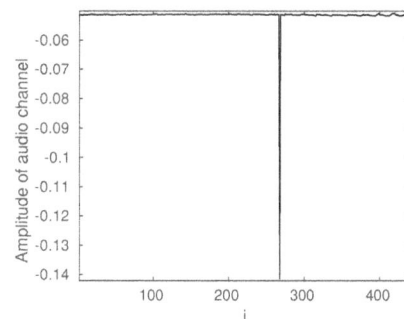

Figure 3: Mean amplitude changing of the internal impulse response of the audio interface.

which gives the normalized error signal comparing the ground-truth intermediate RIR in vector $\boldsymbol{h}_i$ with the interpolated RIR in $\tilde{\boldsymbol{h}}_i$. For $\Delta = 0.5\,\mathrm{cm}$, the region of interest is spanned by a spatial grid of size 21×21. The interpolation error for all 20×20 intermediate positions is shown in Fig. 4. Comparing the interpolated data with the ground-truth data, the interpolation error is found to be in the range between $-22\,\mathrm{dB}$ and $-32\,\mathrm{dB}$ for the case of twofold spatial oversampling with $\Delta = 0.5\,cm$, as shown in Fig. 4. For spatial sampling at the Nyquist frequency with $\Delta = 1\,\mathrm{cm}$, the interpolation error is between $-10\,\mathrm{dB}$ and $-20\,\mathrm{dB}$. The reason for the varying interpolation er-

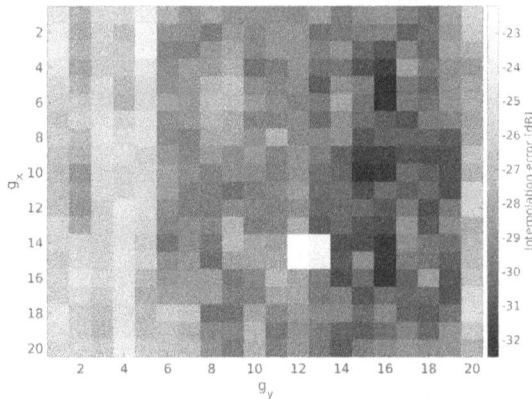

Figure 4: Interpolation error at positions of the intermediate grid for twofold spatial oversampling.

rors is given in Sec. 3.1. The used audio interface acquires inconsistent data over the measurement process. The amplitude drop of the internal impulse response of the audio interface shown in Fig. 3 happened in that moment, when the robot was measuring the original grid points close to the intermediate positions where the highest error (-22 dB) is present in Fig. 4 ($g_x = 14/15$, $g_y = 12/13$). This disturbed data used for interpolation is then inconsistent with the measured ground-truth data. Nevertheless, we see that the Langrange method for spatial RIR interpolation is quite accurate. We also tested the simple linear interpolation for the same setup as for Fig. 4. Here, the error was between -18 and -22 dB.

Even with some hardware malfunction present, we could test the interpolation method. It was clear that the Lagrange interpolation showed better results than linear interpolation. Also, as expected, the spatial over-sampling caused better results with the lowest error value. Solving the hardware problems would get us much better results and will decrease the error value with a huge factor.

4 Conclusion

To measure the RIRs inside a closed room and use it to interpolate the sound field in a region of interest, a MLS signal was created and used as source signal. The source signal was sent through a single source and captured through a single microphone fixed at an arm of a hexapod robot. Using a robot arm provided higher stability and accuracy for positioning. Microphone signals were measured at points of a uniform spatial grid with sampling interval fulfilling the Nyquist-Shannon sampling theorem. The received signals were then deconvolved to get the spatially dependent RIRs. Knowing RIRs at the grid positions, the sound field was interpolated between them. Lagrange interpolation was used to determine the RIRs at the intermediate points. Despite inconsistent data caused by the used hardware, the Lagrange method came out to be a good choice for sound-field interpolation. This study was the first step to studying sound-

field measurements using moving microphones, where spatial interpolation plays the key role for recovery. The next step will include these dynamic measurements with moving microphones.

Acknowledgement

The work has been carried out and supervised by the Institute for Signal Processing, Universität zu Lübeck.

5 References

[1] S. Spors, *Active Listening Room Compensation for Spatial Sound Reproduction Systems*. (University Erlangen-Nuremberg 2006) Ph. D. Diss. thesis http://www.lnt.de/lms/publications.

[2] M. Thomas, *MLS Theory*. Available: http://www.commsp.ee.ic.ac.uk/~mrt102/projects/mls/mls%20Theory.pdf. [last accessed on 15.12.2017]

[3] O. H. Bjor, *Maximum length sequence*. Norsonic AS, 28, 2000.

[4] A. Farina. *Advancements in impulse response measurements by sine sweeps*. In Audio Engineering Society Convention 122. Audio Engineering Society, 2007.

[5] T. Ajdler, L. Sbaiz, and M. Vetterli, *The Plenacoustic Function and Its Sampling*. IEEE transactions on Signal Processing vol. 54, no. 10, pp. 3790–3804, 2006.

[6] A. Benichoux, L. Simon, E. Vincent, and R. Gribnoval, *Convex regularizations for the simultaneous recording of room impulse responses*. IEEE Transactions on Signal Processing, vol. 62, no. 8, pp. 1976–1986, 2014.

[7] R. Mignot, G. Chardon, and L. Daudet, *Low frequency interpolation of room impulse responses using compressed sensing*. IEEE/ACM Transactions on Audio, Speech and Language Processing (TASLP) vol. 22, no. 1, pp. 205–216, 2014.

[8] R. Mignot, L. Daudet, and F. Ollivier, *Room reverberation reconstruction: Interpolation of the early part using compressed sensing*. IEEE Transactions on Audio, Speech, and Language Processing vol. 21, no. 11, pp. 2301–2312, 2013.

[9] F. Katzberg, R. Mazur, M. Maass, P. Koch, and A. Mertins, *Sound-field measurement with moving microphones*. The Journal of the Acoustical Society of America, vol. 141, no. 5, pp.3220–3235, 2017.

[10] M. Vilkko and T. Roinila, *Designing Maximum Length Sequence Signal for Frequency Response Measurement of Switched Mode Converters*. In Nordic Workshop on Power and Industrial Electronics (NOR-PIE/2008), June 9–11, 2008, Espoo, Finland. Helsinki University of Technology, 2008.

Evaluation of the Simple Open EtherCAT Master for the communication in a modular medical device

L. Kleinhans [1]

[1] Medical Engineering Science, Universität zu Lübeck, Lukas.Kleinhans@student.uni-luebeck.de

Abstract

The goal of modularization is to reduce the complexity of a system and to parallelize the development. This approach is used to face the growing complexity of modern medical devices. For the purpose of communication between modules a bus-system can be implemented, which fulfills requirements concerning latency and cycle times. In a prior work EtherCAT was chosen because of its good performance and availability. In this work the Simple Open EtherCAT Master (SOEM) was implemented on different systems with two different network setups, evaluating its real-time capabilities for different cycle times. The work focused on the measurement of the operational latency, in order to develop a lightweight and inexpensive embedded real-time EtherCAT master. The evaluation presented here shows that the SOEM can be real-time capable on a Raspberry Pi 3 with a configured real time kernel. Advantages of using the SOEM are its flexibility and the low costs of the implementation.

1 Introduction

The complexity of modern medical devices is ever growing, resulting in prolonged development times and costs, delaying innovation to be used in clinics. One way to deal with those issues is modularization. It allows to reduce time and costs of development and to integrate new functionality in existing products. Both customer and manufacturer can benefit from this approach [1].

A key aspect is the communication between the modules of a system. Therefore, a bus-system can be implemented that handles the exchange of data. The term real-time means that the system is capable to perform tasks within specific time constraints. There are as well requirements on safety and reliability as on real-time performance for the bus-system. The time constraints are defined by the process [8]. Therefore, maximum values for transmit-rates and operational latency for medical devices have been evaluated in [2]. Further details will be discussed in section 2.1.

There are different bus-systems used in automotive and automation industries but in previous works [2]. EtherCAT was selected due to its good performance and availability. EtherCAT is a real-time capable Ethernet technology originally developed by Beckhoff Automation. Since the goal during development was to minimize cycle times for automation applications requiring hard real-time, the protocol is optimized for short cyclic process data with cycle times less than $100\,\mu s$ [3], [4].

The EtherCAT master can be implemented on any hardware with a standard network interface controller (NIC) and an Ethernet port. It transports date directly within standard Ethernet frames (according to IEEE 802.3) using Ether-type 0x88a4 specified for EtherCAT. An EtherCAT network consists out of at least one master and one or multiple slave devices in which the EtherCAT master is the only device allowed to actively send an EtherCAT telegram. A telegram passes all slaves without being interrupted. The slaves read the data addressed to them on the fly while the frame passes through the EtherCAT Slave Controller (ESC) as shown in Fig.1. Furthermore, the ESC inserts the data as the frame moves downstream [3, 4]. This is the reason why EtherCAT is able to achieve the short cycle times with low jitter.

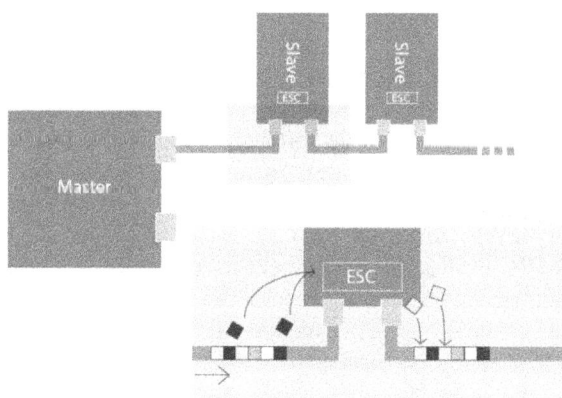

Figure 1: Operating mode of an EtherCAT bus-system. The squares represent parts of the frame processed by the slave [5].

There are several master implementations available, commercial and open source. The Simple Open EtherCAT Master (SOEM) was implemented on two different platforms with three different operating systems (OS). It was evaluated on which OS the SOEM can fulfill the requirements for the real-time constraints in a modular medical device as specified in [2]. The OS were Microsoft Windows 10, Raspbian 8 with a mainline kernel running on a Raspberry Pi 3 and Raspbian 8 with a patched kernel also running on a Raspberry Pi 3.

The rest of the paper is organized as follows: In section 2, the features of the slave and master implementations are described. Also, the requirements and the test setups are specified. In section 3 the measurement results are presented and discussed. The paper is concluded in section 4.

2 Material and Methods

2.1 Requirements

There are different types of bus-systems which could be used for therapy devices. The decision which type of communication infrastructure should be used requires the knowledge about the demands on the system. The following hard real-time requirements are based on the current communication in therapy devices as they were evaluated in [2]:

- Cycle times less than 1 ms
- Operational latency of less than 300 µs

EtherCAT is fast enough, reliable and developed for distributed control algorithms of safety relevant systems. It has also a low bit-error rate and is able to recognize bit-errors. This is why EtherCAT was chosen in the first place.

2.2 Slave

The slaves were implemented on XMC4800 Relax EtherCAT Kit evaluation boards. To use a micro-controller as an EtherCAT Slave it needs an ESC to process the passing frame. The software consists of three layers: The application, the slave stack code and the hardware-code. The application layer implements the behavior of the slave, for example control algorithms. The slave stack code implements the functions for EtherCAT communication, as for example addressing or reading of the EtherCAT frame. The hardware code assigns the signals from the code to the pins on the relax kit.

For this evaluation two different slave applications were implemented. The first application was implemented to simulate communication in a modular device. It sends 150 bytes of data every cycle on the bus. The second application directly echoes back the received data to the master without alteration or delay. The purpose of this slave is explained in subsection 2.4.

2.3 Master

The SOEM is an open source project and allows functionality enhancement using the C programming language. However, it uses the network-drivers of a host system which

is not necessarily real-time capable. In [2] the SOEM as it is available as freeware was adjusted to the application. The master was running on a Microsoft Windows 7 OS. Some minor adjustments were made to adapt the SOEM for the operational latency analysis on Microsoft Windows 10 Home running on an Acer Aspire VN7-571G Laptop to reproduce and confirm the results.

For comparison the SOEM was implemented on a Raspberry Pi 3 model B with Raspbian GNU/Linux 8 (jessie) OS using a mainline kernel. This OS is much more lightweight than Microsoft Windows. Windows is a commercial OS developed to run on many platforms. It has a lot of background processes that cannot be interfered with. This compromises the real-time capability of the OS. Raspbian on the other hand is optimized for the Raspberry Pi hardware and is just a set of basic programs and utilities that make the Pi run. The user can individually configure the system to its own needs [6].

Furthermore, the Linux mainline-kernel was patched with the PREEMPT-RT 4.4.9-rt17-v7+ patch. The kernel's primary function is to mediate access to the systems resources. By patching the kernel the distribution of resources and priorities is changed according to real-time needs [7]. To use the patched kernel and evaluate its performance advantages some bigger adjustments to the SOEM software had to be made. A POSIX thread (pthread) was used to ensure the priority of the EtherCAT main thread that controls transmitting and receiving the cyclic data.

2.4 Operational Latency Analysis

The term operational latency is defined as the time it takes from sending a message by the master application until it is received in the slave application or vice versa. In the performed tests, the time was estimated by dividing the round-trip time of a message by two. This time is an upper bound for the time a signal needs to be sent from the slave to the master.

To test the performance of the EtherCAT bus-system two system were implemented. One system using six slaves and a master and one system just with one slave and a master. Fig. 2 shows a schematic setup of the systems.

The five slaves in Fig. 2(a) in between the master and the last slave are the slaves that are simulating the communication within the modular device. The last slave – the echo-slave – echoes back the data from the master. It was placed at the end of a daisy-chain topology to simulate a worst-case scenario. In total 150 bytes were sent every cycle in one frame. Assuming a measurement was performed at the slave, the worst case would be that this update has to wait for the next passing frame for a full cycle time until it is sent to the bus.

To measure the operational latency, a time-stamp with the current system time was sent by the master. This time-stamp was immediately echoed back to the master once it reached the echo-slave at the end of the daisy-chain topology. When the master received the message, the time difference between the current system time and the time-stamp

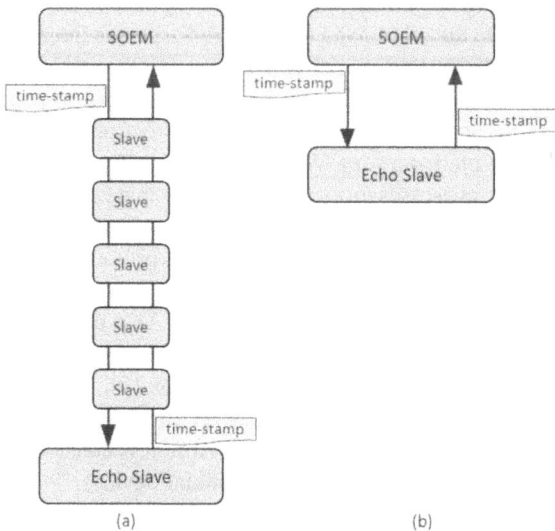

Figure 2: (a) Setup of the measurement with five slaves to simulate data on the bus (b) the time-stamp is the only byte on the bus.

Table 1: Average operational latency

Cycle-time	Windows	Raspbian	Raspbian RT
0.1 ms	0.99 ms	0.10 ms	0.07 ms
0.2 ms	1.04 ms	0.10 ms	0.07 ms
0.3 ms	1.05 ms	0.16 ms	0.09 ms
0.5 ms	1.09 ms	0.11 ms	0.07 ms
1 ms	3.02 ms	0.11 ms	0.15 ms
2 ms	4.55 ms	0.17 ms	0.09 ms
3 ms	6.13 ms	0.13 ms	0.12 ms
5 ms	9.10 ms	0.13 ms	0.13 ms
Average	3.37 ms	0.13 ms	0.10 ms

Table 2: Maximum operational latency

Cycle-time	Windows	Raspbian	Raspbian RT
0.1 ms	35.50 ms	0.50 ms	0.17 ms
0.2 ms	31.00 ms	0.22 ms	0.16 ms
0.3 ms	32.00 ms	0.28 ms	0.17 ms
0.5 ms	33.00 ms	0.25 ms	0.15 ms
1 ms	32.00 ms	0.37 ms	0.24 ms
2 ms	33.00 ms	0.57 ms	0.22 ms
3 ms	72.50 ms	0.47 ms	0.23 ms
5 ms	37.00 ms	0.52 ms	0.25 ms

was evaluated. The result corresponds to the time the message needed to run through the system to the slave and back including the processing by the slave's and master's software. If this time gets divided by two, the result is a worst-case estimate for the operational latency of the message.

There are two setups to evaluate the influence of additional slaves in the network occupying the bus with data. The measurements were performed with cycle times between 5 ms down to 100 μs. This covers the requirements of a cycle time of 1 ms as specified in 2.1. For each cycle time 5000 operational latency values were taken in 5 different measurements.

3 Results and Discussion

Measurements were performed according to the previously defined scenarios. The results of the average operational latency with just one slave and the time-stamp as only byte on the bus are summarized in Table 1.

The results of the worst case operational latency with just one slave and the time-stamp as only byte on the bus are summarized in Table 2.

The measurements yielded different operational latencies for every measurement which is why the average time of all measurements was compared. The average operational latency over all cycle times with Microsoft Windows 10 was 3.37 ms with the shortest measured operational latency of 0.99 ms at a cycle time of 100 μs. This is a strong deviation to the required 300 μs threshold. Even worse were the highest measured values of 72.5 ms. The results are similar but not completely consistent to the results in [2] with the same test setup. The differences in the results imply that the SOEM running on a Windows OS is also dependent of the used hardware.

Also, it could be observed that the influence of the additional five slaves in the network is almost not recognizable. Therefore this test setup was not considered in the further comparison. However, the SOEM running on a Microsoft Windows OS is without major adjustments not an option for a real-time master implementation for a modular device with the aforementioned requirements.

The SOEM on a Raspberry Pi 3 with the Raspbian 8 OS yielded quite different results. The average operational latency over all cycle times dropped down to 0.13 ms. This is an strong improvement to the prior results of the Windows OS. The shortest measured latency was 100 μs at cycle times of 0.1 ms and 0.2 ms. The latency at a cycle time of 1 ms was 110 μs, which is almost a third of the requirement of 300 μs. However, the maximum operational latency measured was 52 ms and the maximum latency at the required cycle time of 1 ms is 370 μs. The definition of real-time specifies that a system is only considered to be a real-time system if *all* tasks are performed within defined time intervals. This means even tough the average operational latency fulfills the requirements it cannot guarantee an operational latency of 300 μs for all data transferred on the bus.

3.1 Performance on Linux with real-time kernel

The comparison with the results of the mainline kernel shows the influence of the real-time patch on the performance of the SOEM on the Raspberry Pi 3 model B. The average operational latency was reduced by 30 μs to 100 μs. But more important is the maximum measured latency which is now 250 μs at a cycle time of 5 ms. At the required cycle time of 1 ms the maximum operational latency is 240 µs, thus this setup seems to fulfill the requirements and confirms this setup to be real-time capable.

3.2 Distribution of measured values

The graph in Fig. 3 shows the distribution of the measured operational latencies. The solid line shows the values of the mainline kernel peaked around an operational latency of 75 µs. There is another peak between 140 µs and 180 µs which causes the average latency to drop to 130 µs. The smaller peak is located at almost twice the operational latency than the first peak. The standard derivation is 26 µs. The few values above 350 µs causing the mainline kernel not to be real-time capable are not shown in this plot.

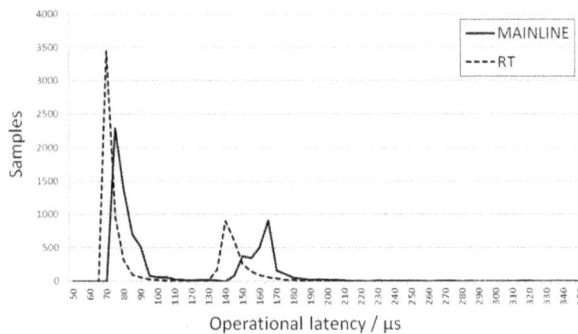

Figure 3: Distribution of the measured values of the operational latency

The dotted line shows the values of the patched kernel. Its main peak shifted to 70 µs and increased by 1200 samples. The peak is also noticeably narrower. The smaller second peak also shifted to 140 µs, which is exactly twice the operational latency of the main peak but its height did not change. Its standard derivation though reduced to 16 µs.

4 Conclusion

This paper focused on the implementation of the SOEM on different OS and has presented a performance comparison of them. The comparison was made by measuring the operational latency at different cycle times. The results show that the SOEM on a Windows OS in not real-time capable. The SOEM on the Raspbian OS on the other hand was able to reduce the operational latency within the required time intervals but only the setup with the patched kernel was able to reduce also the maximum latency to ensure real-time performance.

There are different EtherCAT master implementations available. In [2] the SOEM was compared to the TwinCAT3 master. TwinCAT3 provides real-time functionality regardless of the OS but its implementation is more complicated. For a master application on a Windows OS, TwinCAT3 is the better choice regarding real-time requirements. Nevertheless, the results achieved with the SOEM on the Raspberry Pi 3 model B with the patched kernel can compete with the results of TwinCAT3. The advantages of the setup

of this work are on the one hand the high flexibility of the SOEM. On the other hand the financial aspect is interesting. The Raspberry Pi 3 model B is commercially available for about 35 euros. This makes the implementation favorable compared to a solution with TwinCAT that needs a more expensive platform for a Windows OS.

EtherCAT possesses more useful features that can be implemented with the SOEM that were not investigated closer. The File Safety over EtherCAT (FSoE) protocol which is optimized for safe communication might be very important to transport safety critical information. Also the File over EtherCAT (FoE) feature was not tested, this could be useful to transport larger buckets of information [9]. These considerations make clear that EtherCAT provides a lot more features than just performance, therefore it is considered as a reasonable choice for medical modular devices.

Acknowledgement

The work has been carried out at the Institute for Electrical Engineering in Medicine, Universität zu Lübeck and was supervised by G. Männel, M. Sc.

5 References

[1] F. Börjesson, *Approaches to modularity in product architecture.* Master's thesis, Royal Institute of Technology (KTH), Stockholm, 2012.

[2] S. Gareis, *Development and Concept Evaluation of a Modular Therapy Device Using EtherCAT an Communication Protocol.* 2016

[3] EtherCAT Technology Group, https://www.ethercat.org/, [last accessed on 2018-01-08].

[4] BECKHOFF, http://www.beckhoff.de/, [last accessed on 2018-01-19].

[5] EtherCAT B110: Handbuch, 1.2, BECKHOFF, 2016

[6] C. Immler, *Schnelleinstieg Raspberry Pi.* Franzis Verlag GmbH, Haar, 2014.

[7] Elektronik Praxis, https://www.elektronikpraxis.vogel.de/echtzeit-mit-dem-raspberry-pi-a-630497/index2.html, [last accessed on 2018-01-18].

[8] G. Kang, *Shin and Parameswaran Ramanathan. Real-time computing: A new discipline of computer science and engineering.* In Proceedings of the IEEE, volume 82, 1994.

[9] EtherCAT Technology Group, *EtherCAT – Der Ethernet-Feldbus*, 2017

7

Safety and Quality

Problem Solving Methods in Medical Quality Management

S. Schonebeck [1,2], A. Rupp [2], S. Hauttmann [2], and M. Leucker [3]

[1] Medizinische Ingenieurwissenschaft, Universität zu Lübeck, sara.schonebeck@student.uni-luebeck.de
[2] Philips Medical Systems DMC GmbH, Hamburg, {sara.schonebeck,anna.rupp,stefan.hauttmann}@philips.com
[3] Institute for Software Engineering and Programming Languages, Universität zu Lübeck, leucker@isp.uni-luebeck.de

Abstract

Problem solving methods are techniques used to decrease performance gaps efficiently and permanently. This paper gives an overview of problem solving methods which are nowadays used for solving problems in industry. The paper focuses on the methods used in operative-orientated quality management in x-ray tubes manufacturing. All here described methods share that they are based on the PDCA cycle, which is a fundamental concept of the continuous-improvement processes. The presented methods are 3C, 8D, PRIDE, DMAIC and Shainin. These methods can be distinguished according to the characteristics complexity, business impact and methodical skills which are presented in a developed methodical selection matrix. Some applications are presented in examples from the field of medical engineering (imaging systems production).

1 Introduction

The Philips Medical Systems Development and Manufacturing Centre (DMC) GmbH is a leading company in the healthcare sector. Medical imaging techniques like x-ray tubes and components are produced at the location Hamburg [1]. Since medical devices are produced there, regulatory conditions must be strictly adhered to and problems have to be solved sustainably and contained quickly.

The operative-orientated quality management has the task of ensuring the fulfillment of quality requirements and to achieve quality improvements. The FDA (Food & Drug Administration) 21 CFR (Part 820.100) and the ISO 13485 (chap. 8) formulate requirements for the quality management systems of medical device manufacturers. These require that processes have to be continuously monitored, that appropriate methods have to be used to identify any quality issues that arise, and that procedures have to be in place to implement appropriate corrective and preventive actions in order to maintain a system permanently [2], [3].

There are a lot of methods and tools for systematic problem solving, which serve to identify root causes efficient and sustainably. These methods help to identify problems at the product or in the process, to eliminate them quickly and implement permanent countermeasures. The advantage of a problem-solving method is that there is a structure that guides what steps to take to have a uniform language, a comparable format, transparency and standardization.

Problem solving methods are of supreme importance since without these methods one is willing to jump to an explanation immediately with a superficial analysis and inappropriate countermeasures may be introduced. The root cause of the problem might not or insufficiently recognized and the real issue is not resolved [4].

2 Material and Methods

In this section, different problem solving methods used in operative-orientated quality management are presented.

2.1 PDCA cycle

As developed in the 1940's by William Edwards Deming, the PDCA cycle (Deming circle) is shown in Fig. 1. The four-letter acronym PDCA stands for Plan-Do-Check-Act. These sequence describes the main process steps of a problem solving method and can be used as a fundamental concept for all other methods described here [5]. The

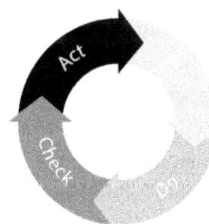

Figure 1: Schematic illustration of the PDCA cycle.

first step ("plan") is analyzing and describing the common state and studying the problem itself. Root causes have to be identified, new concepts and potential solutions developed, tested and prioritized and countermeasures defined. In the "do"-phase, solutions and countermeasures found in the plan phase, are implemented. Then ("check"-phase) the processes and results of the implementation are checked and evaluated whether they lead to success or whether they have to be improved. With a positive outcome, the process can be used across a broad front in the next phase ("act"). The next step is the implementation and standardization of the new general specification [6]. This problem solving method contains continuous improvements of the standard. Once the process has changed, the process is standardized and started with the next phase of planning [4].

2.2 3C-method

The 3C-method is a simple table-based method and important for fast problem solvings and less complex problems. A 3C-list, like the one presented in Fig. 2, is used for documenting the three Cs concern, cause and countermeasure as well as whom is by when responsible for the respective item and the state of the item. Causes are first identified for the

Concern	Cause	
		KW
Countermeasure	Who?	
	When?	

Figure 2: Possible representation of a 3C-list

problem with the highest priority. For every cause at least one countermeasure has to be defined. Then, countermeasures are implemented and documented. They are checked after a certain time if there is need for improvement [7].

2.3 8D-method

8D is a strategy-based problem solving method and stands for eight disciplines (process steps). In D1 ("Form a Team"), members for the team are selected. D2 ("Describe the Problem") is used for an exact definition of the problem and for collecting all the facts. D3 ("Implement Containment Actions") is an important step to protect subsequent processes from the problem. It doesn't solve the problem, but it buys more time for finding the root cause and solving the problem permanently. In D4 ("Identify Root Cause"), a root-cause-analysis is performed to eliminate the issue sustainably. An often used simple and fast technique is the 5x Why method. The root-cause is identified by repeating why-asking, starting with the most superficial cause question and iterating until the root-cause of the problem is found. Corrective actions are determined in D5 ("Choose Permanent Corrective Action"). These actions should target the root cause itself, not the symptoms. Once the corrective actions have been successfully implemented in D6 ("Implement and Verify Corrective Action"), the containment action is replaced by it. In step D7 ("Prevent Recurrence"), it should be ensured that the implemented actions are sufficient and that same or similar defects can no longer occur in the future. The last step D8 ("Congratulate Team") is to formally complete the problem solution. In this context, the knowledge gained will be made available for future processes [4].

2.4 PRIDE

PRIDE problem solving stands for the four phases *P*roblem definition and *R*ationale description, *I*nvestigate

root causes, *D*evelop, test, implement countermeasures and *E*nsure to sustain and reflect [8]. In the first phase, the task is to identify the perceived problem and summarize the reasons for the need to solve the problem. The next phase starts with the identification of the few critical causes with a 5x Why (see 8D-method) analysis with gemba evidence. The Japanese word gemba means "the real place" and describes the place and time of the actual occurrence of the problem. Then, hypotheses are tested, validated and reflected, corrective actions are implemented and its influence is verified. In the last phase, the implemented solutions are standardized to ensure sustainability. In addition, it must be ensured that the standards are established, continuously improved and used, and that the results are according to their goals.

2.5 DMAIC

The DMAIC-method is a principal item of Six Sigma which is a systematic method for optimizing processes [5]. DMAIC is suitable for major and complex problems where a large amount of data has to be processed in a sensible way [4]. The five steps are *D*efine, *M*easure, *A*nalyse, *I*mprove and *C*ontrol. The first step ("Define") is to define the problem and understand the consequences. The next step ("Measure") is to quantify the problem and collect data on it. In the "Analyse"-phase, the collected data have to be analyzed in order to identify the causes of the problem and possible improvements. In the "Improve"-phase, solutions are generated and evaluated. One is selected and the implementation is planned. The final phase ("Control") realizes the implementation plan of corrective actions and verifies the effect after implementation. If successful, the improvements or new procedures will be standardized.

2.6 Shainin method

The Shainin method also contains all steps of PDCA problem solving methods, but the focus is on quickly identifying the root cause. Important for the Shainin method is the

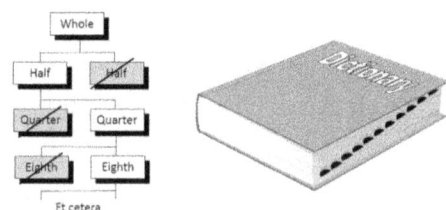

Figure 3: Principle of Dictionary Search.

Pareto principle which implies that among many influencing factors, normally only a few causes dominate, are relevant and produce the preponderant percentage of an effect [9]. So, it makes sense to address the biggest, most important problem first. This is called the "Red X" [6]. This dominant and most important factor for controlling the process result has to be identified [9]. To find the "Red X" a decision tree is used, for example, in which all relevant parts of the process are displayed. In turn, the individual processes are examined and compared with a is-not-comparison, so that in the best case one process step remains from which the further comparisons can be made. An example of this is Dictionary Search: the first question is whether the word is in the first or second half, then in which quarter, eighth, etc (cf. Fig. 3). In this way one finds the wanted root cause, has to check it and then confirms the "Red X".

3 Results and Discussion

3.1 Comparison of methods

In Fig. 4, a methodological selection matrix is presented which we developed by comparing the characteristics of the different presented methods. This can be used in industrial day-to-day life to decide which solving method to use for problem solving. It depends on the complexity and the business impact of a problem and the diagonal shows the methodical skills that are necessary to solve the problem. These increase with the diagonal.

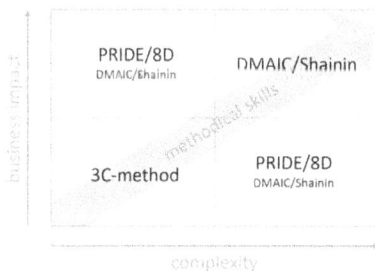

Figure 4: Methodological selection matrix.

For less complex problems with a low business impact (cf. Fig. 4) and where the root cause is often known, the 3C-method is used, which is simple and therefore suitable for every user [7]. This fast method doesn't use sophisticated tools or strategies, but jumping is avoided and actions are taken and documented which is the intention of problem solving methods in general.

PRIDE and 8D are suitable methods for problems with a high complexity and a low business effect or a low complexity and a high business effect. The 8D-method is based on the consistent execution of the individual tasks [4]. It describes formally and strictly what to do in each phase, but there are hardly any recommendations of tools to be used. It is used to respond to quality notifications from the customer. For solving the problem with PRIDE, there are several tools (5x Why, pareto, time plot) that can be used to highlight the causes of each problem and focus on them. In addition, it is relevant to PRIDE, which is a strategy based method, whether the problem really needs to be solved and what the benefits are (business impact) .

For highly complex problems with a high business impact, DMAIC and Shainin are well suited. When looking for solutions, the Shainin method focuses on a clear strategy and logical sequence [9], while DMAIC attaches great importance to using tools, statistics and is highly focused on data analysis. In DMAIC, there is a divergent root cause analysis which is based on possible causes. In contrast, the analysis at the Shainin method is convergent and the research is starting from the problem. In both methods expert knowledge is essential, because of the high complexity of the methods and problems. Overall, these two methods can be used side by side because they complement and support each other.

3.2 Methodology application

3.2.1 3C-method

One current manufacturing example, for which the 3C method can be used, is the problem that the measuring device for measuring the focal spot of a tube is not fixed strongly enough, it can be displaced, which could shift the exposure (concern). The reason is that the screws that attach the measuring device are too long (cause). To fix the problem (countermeasure), the screws are trimmed and it is checked if the problem has been solved. Then the case can be closed. All steps are listed in a 3C-list (cf. Fig. 5). The

Concern	Cause			KW
Measuring device not fixed strongly enough	Screws are too long			
Countermeasure Screws are trimmed	Who?	Person A		6
	When?	04.01.2018		

Figure 5: Representation of a 3C-list for a current example.

3C method was used in this case because cause and solution could be found quickly, no additional experts were needed to find a solution and the business impact was low. The other presented methods would have been too complex and time-consuming for this problem.

3.2.2 PRIDE

Figure 6 shows a section from the problem solving report of a general example ("Increasing complaints") of the PRIDE method in order to illustrate the first steps and the structure of this method. It can be seen that particular emphasis is placed on the business impact, which is captured right at the beginning (Rationale) and which is an important feature of the PRIDE method. The current numbers are compared to the target value and the gap is determined. These are presented in a time plot to illustrate the increasing trend of complaints of tubes. Next, the Pareto for the different tube types is set (Total Complaints) to determine the tube type with the most complaints. From this, a second Pareto is set up, in which the location of findings are presented in order to localize the problem. The next step, which is not shown anymore, is the 5x Why method to determine the root cause.

3.2.3 8D-method

A current issue in x-ray tube production are tubes inserts that have failed with the problem of "leakage". In this case, the problem of leakage was recognized because the tube insert under test failed the vacuum test. The vacuum within the tube is of supreme importance, in order that the electrons released from the cathode can not collide with other

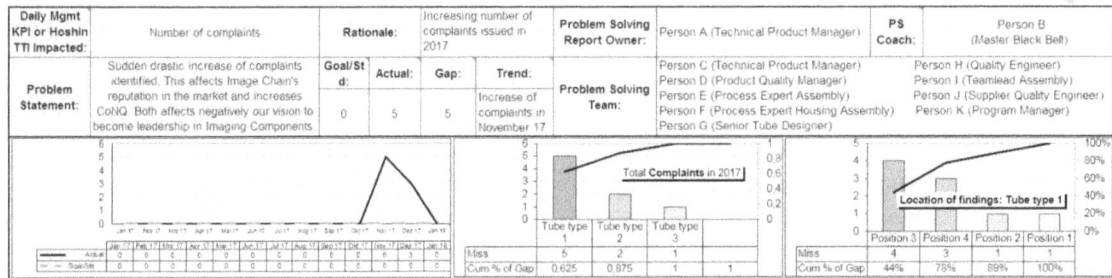

Figure 6: Section from the problem solving report of a general example for the PRIDE method.

gas molecules on their way to the anode [10]. This would affect the function as it increases the probability of gas discharge. To choose a suitable method, first the characteristics of the problem are collected. In the present problem, individual leakages have been identified during tube manufacturing process. In addition, the root cause is unknown, there is no recurring problem, subsequent processes are affected and expert support is needed. For this reason the 8D-method is used for problem solving because of complexity and business impact of medium size. The 3C method is not appropriate for this case scenario because there is too much information missing and the problem is more complex than could be handled with this method, so it would need to go through several iterations without reaching the goal. Thus, it makes more sense to choose a more complex method from the first in order to save time and solve the problem efficiently. DMAIC and Shainin are too complex and consuming for this case. This problem is currently in the "Root Cause" phase of 8D as it is contained, but root cause still under investigation.

4 Conclusion

Different problem solving methods were presented and compared to each other regarding complexity, business impact and methodical skills, so that it is possible to choose a suitable method for the problem at hand. The comparison has shown that it is important for a company to have a portfolio of problem solving methods that can be applied to a wide variety of problems so that it is possible to react appropriately to the problems. The benefit of the paper lies in the development of the methodological selection matrix that is disseminated and used within the company to choose from many methods the appropriate method for the respective problem. It was shown that the matrix can be used to select the appropriate method and it became clear that problem solving methods are useful and important in quality management to be able to act fast, to find the right root cause and to solve problems efficiently and permanently, complying with regulatory requirements.

Acknowledgement

The work has been carried out at Philips Medical Systems DMC GmbH, Hamburg and supervised by the Institute for Software Engineering and Programming Languages, Universität zu Lübeck.

5 References

[1] Philips Healthcare, *Philips Healthcare: Innovationen für eine verbesserte Gesundheitsversorgung.* Available: https://www.philips.de/healthcare [last accessed on 2018-01-11].

[2] FDA, *Mini Handbook: The Code of Federal Regulations Titel 21 - Food and Drugs.* GMP Publications, USA, 2014.

[3] ISO 13485:2016, *Medical devices - A practical guide.* © ISO, Switzerland, 2017.

[4] G. Linß, *Qualitätsmanagement für Ingenieure.* Fachbuchverlag Leipzig im Carl-Hanser-Verlag, München, 2011.

[5] A. K. Bergbauer, B. Kleemann and D. Raake, *Six Sigma in der Praxis: Das Programm für nachhaltige Prozessverbesserungen und Ertragssteigerungen.* Expert-Verlag, Renningen, 2006.

[6] T. Pfeifer, *Qualitätsmanagement: Strategien, Methoden, Techniken.* Carl-Hanser-Verlag, München, 1996.

[7] M. Radley, *Policy Deployment 3Cs: Principle Document.* GENEO Consulting Limited, Warwickshire, Available: http://www.geneo.co.uk/wp-content/uploads/2015/07/3Cs-Principle-Document-Lean-Model.pdf [last accessed on 2018-01-09].

[8] J. Shook, *Managing to Learn: Using the A3 Management Process to Solve Problems, Gain Agreement, Mentor and Lead.* Lean Enterprise Institute, Cambridge, 2008.

[9] K. R. Bhote and A. K. Bhote, *World Class Quality: Using Design of Experiments to Make it Happen.* AMACOM American Management Association, New York, 2000.

[10] L. Spieß, R Schwarzer, H. Behnken and G.Teichert, *Moderne Röntgenbeugung: Röntgendiffraktometrie für Materialwissenschaftler, Physiker und Chemiker.* Springer-Verlag, Wiesbaden, 2015.

On the Way to Realize a Zero Defect Production with a new Quality Gate Concept at the Electro Manufacturing Service Prettl Electronics Lübeck

T. Aldag [1] and G. Buntrock [2]

[1] Medical Engineering, Universität zu Lübeck, tobias.aldag@student.uni-luebeck.de

[2] Institute for Software Engineering and Programming Languages, Universität zu Lübeck, buntrock@isp.uni-luebeck.de

Abstract

With this project Prettl intend to reduce the defective production rate of 2500 parts per million (ppm) to nearly zero. That should be achieved with the management system Six Sigma. This management-system, developed by Motorola in 1986, combines more than one method in 5 phases (Define, Measure, Analyse, Improve, Control). At the begin the project and customers were defined and then the production is analysed and optimized on this basis. If this is applied correctly, the company can reduce the defective production rate from 2500 part per million (ppm) to 3,4 ppm. This is compared with 0.000 34 % defective parts. Prettl Electronics produces circuit boards on request for customers in the medical technology or aviation industry.

1 Introduction

The steady rise and growth of company's forces Electro Manufacturing Services (EMS) provider to ensure a high quality level, shorter production time and lower costs as another provider. An EMS provider is a company, who produces equipped printed circuit boards. Quality is defined as *the degree to which a set of inherent characteristics of an object fulfills requirements* [1]. The customer requirements must be known by the company. If this isn't possible, the quality management has to change or optimize their process. The company should establish a quality management system. The ISO DIN 9001 describes how you can build such a system [2]. Also, the 9001 requires that the company develops their own system.

This Project was created to analyse the Quality Gates: Automatic optical inspection (AOI) and the test field (combined test center and mechanical final inspection V2). If a customer makes an inquiry to Prettl Electronics (PEL), the company has to ensure that the customer requirements are 100 % fulfilled. At first the Production Support (PS) analyses the customer offer and requirements. Now the PS produces, based on the data, a production schedule and a material list. Furtheron the customer specific materials are ordered [3].

The next step is, that the Receipt, checks the ordered materials. If the materials pass the tests they go to the production. In the production, the circuit boards (CB) are equipped with surface-mounted devices (SMD) or Through Hole Technology (THT) components. All SMD components are checked by the AOI.

This contains a camera, which makes high-resolution pictures of the CB. With those pictures the AOI checks algorithmically if the CB is correct. If the outcome yields a defect, it will be checked again by an employee to ensure or reject this defect. After that station, the CB go to the manual assembly, because the through-hole technology (THT) components are put by employees. They have an optical pick and place. In the computer, on this table, is a program implement, that gives the right position of the right components by a laser point. After the assembly, the THT components are soldered with a shaft (wave soldering). At least the soldered components are checked by the employee and go to the next step. In this station the CB are equipped with components, which must be soldered by hand. These components are, for example plugs, connectors or cable.

If the customer wants an electric check, the CB goes to the Test Center. The employees make a functional test, an in-circuit test, which tests the conducting paths of the CB, or a high voltage test. After that the CB goes to the mechanical check V2, which examines the CB optical [4]. Because of the amount of data, in this paper only the mechanical final inspection is analysed.

2 Material and Methods

This project were described as a Six Sigma project. Six Sigma is developed from Motorola in 1986. This management system will be implemented with the DMAIC-cycle. DMAIC stands for *D*efine, *M*easure, *A*nalyze, *I*mprove and *C*ontrol. In the Six Sigma management-system distinguishes basically between 4 levels. The lowest is the *green belt*. This employee has basic knowledge about Six Sigma and works only about 20 % of the time on Six Sigma projects. The next higher level is the *black belt*. This belt has sound knowledge about Six Sigma and the tools. He works 100 % of the time on Six Sigma projects and works up to 4 projects per year. The next is the *master black belt*. He knows the same as the black belt. In addition to the work, training of the another belts. The last level is the *champion*. He know more about Six Sigma as the green belt but less than the black belt. The champion has the task to define and initiate projects [5].

2.1 Define

At the first step, the belt describes the project. An efficient tool for this is a projectplan (Figure 1). Here the belt has to think about the tasks and milestones. Which tasks need to be mastered, which employees are important to fulfil the project and what are possible barriers.

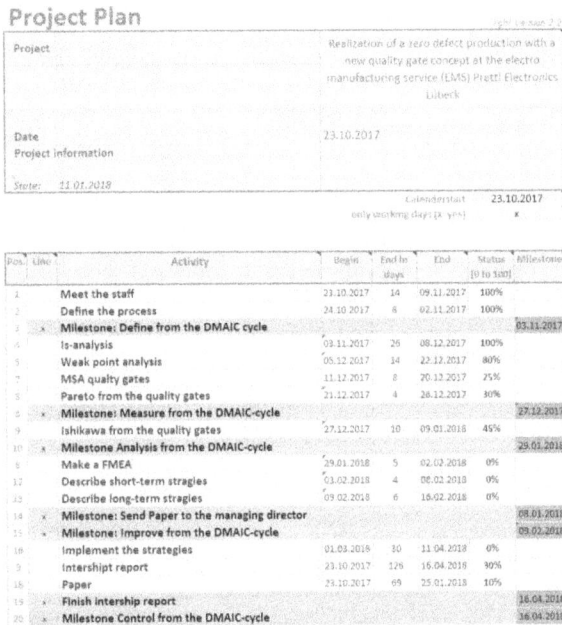

Figure 1: Projectplan about the intership. Goal of this graphic is, that the Six Sigma project belt have an overview over the project. Additional, on the first day the leader has to think about the tasks and possible troubles in the project.

Subsequently, the belt examines the process. What about the input of the process? What about the output? Which are the risks in the process steps? If we want a overview about the process, a Process-Flow-Chart is a good tool for this. But if the belt wants to look closer, he generates a *Turtle-Diagram* (Figure 2)

2.2 Measure

If we described the project and defined possible problems, we have to collect all data. This project should optimize the Quality Gates. Therefore we have to access the databases. In the V2, the employees has a excel-mask. Therein they

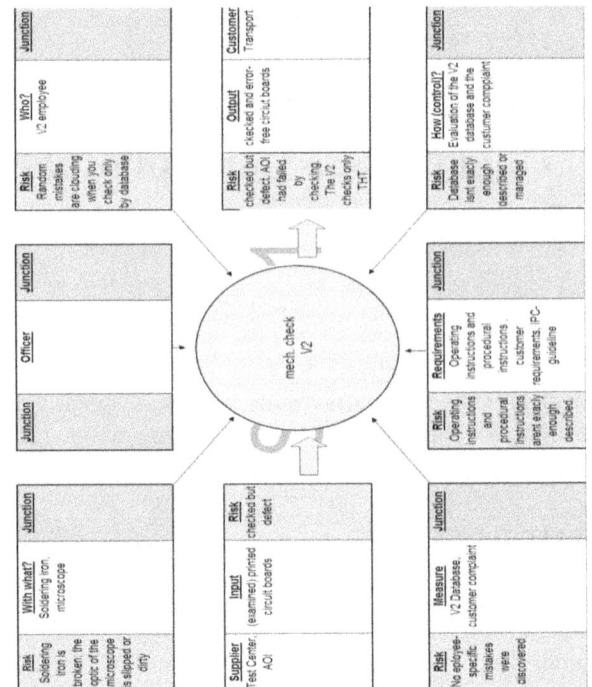

Figure 2: Turtle Diagram for the V2-check. The body of the turtle is the process. The head is the input and tail the output. The legs stands for how is the process works. With this graphics the belt intends to illustrate one process-step in the whole production chain and finds possible weakening (Because of the data protection reasons, the process owner was not named).

make entries about the CB, the work order and the error. This is then written and stored in a large Excel table.

The intership started in October 2017, so data between October 2016 and October 2017 was collected. In this database were over 2800 entries. So we make a *Pivotable*. This is a smart tool, which recap the data. We have to know, if the problem occures in the production or in the Quality Gate. So the monthly ppm-rate, arithmetic mean and the standard deviation is calculated. Additionally, we define action limits (arithmetic mean plus/minus three time the standard deviation) and draw this in the graphic and we get an *Original Data Chart* (Figure 3). If there is an outlier value, higher or lower as the action limits, we have to look for the reasons [6]. That would be an indication for an possible error in the production.

Additionally, the database contains the information why the CB is defect. The *Pareto-Diagram* is ideal to analyse this attribute (Figure 4). This chart contains a bard and a line graph. The bard graph describes the frequency of the errors or error codes and the line graph describes the accumulated ratio error. The Pareto is ideal to analyse how much effort is required to achieve an improvement. If we want an achievement about 80 %, we have to spend only 20 % on the effort. That means, if we reduce 20 % of the errors, we can get an success about 80 %.

After looking for errors in the production chain, we need to examine the Quality Gate. For this we can make a *Mea-*

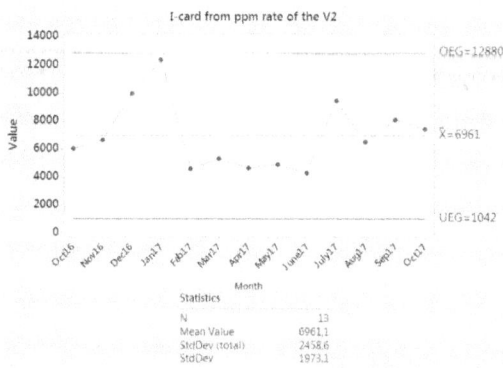

Figure 3: Original data chart for the V2. Here, from October 2016 to October 2017, the ppm rate for each month, the arithmetic mean and the standard deviation is calculated. The belt defines action limits (OEG equals arithmetic mean plus three times the standard deviation and UEG equals arithmetic mean minus three times the standard deviation). This graphic is a tool to examine the dispersion of the production (made by Minitab).

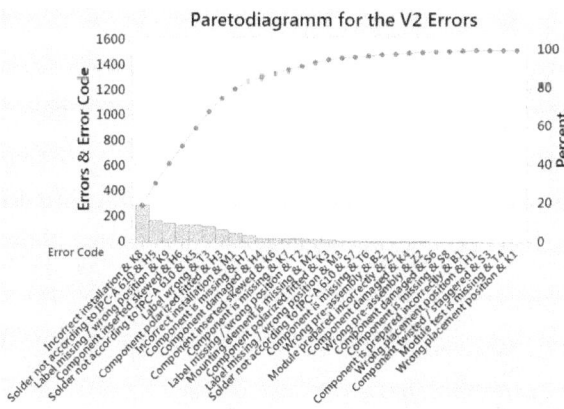

Figure 4: Pareto-diagram for the V2. Here the number of all error codes at the V2 were calculated. The number of the errors are illustrated in descending order by the bars and the cumulative total by a line. The belt can see which faults makes which percentage and know how many errors need to be reduced for a improvement of 80 % (Pareto principle). The letter in the error codes stands for the respective process step (see at section 1).

surement System Analysis (MSA). An employee (not working at the V2) manipulates 16 CB out of a total of 30. Now we give this CB to the V2-employees and they check this. They know what mistakes are made, but not where they have been installed. They checked them on a Monday before christmas and on a Monday after New Year. If they find all errors on a CB, we classify them as bad. But if they miss only one error, the CB is classified as good and this is a minus for the employee (Figure 5). The possible errors are: component is missing, component polarized fitted, mechanical damaged, component inserted skewed, tombstone effect, solder bridge. This is a measurement to examine the human factor in a quality gate [6].

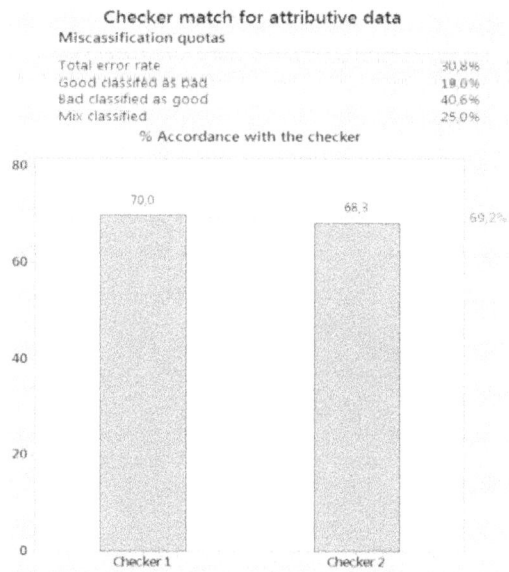

Figure 5: The MSA shows the eligibility of the V2-employees. The belt wants to show how good are the employees.

2.3 Analyse

After collecting and rework all data, we begin with the analysis. The main question here is, why are the customer requirements not fulfilled. In the section Measure we have get some threads. Now we have to select the right thread and this is possible with the *Ishikawa* or *Fishbone-diagram*. We look, why isn't the V2 100 % right and why we get the errors (see in Figure 4). For this we have to make an Ishikawa for each quality gate and their errors. Figure 6 shows an Ishikawa für K8 (incorrect installation at the completion). The Ishikawa-diagram has 6 limbs: People, materials, measurement, machines, methods and environment. On each limb, we write the possible causes for the main problem. We will find more than one possible causes. The next step is to develop strategies to avoid these problems. If there is no problem for one of the limb, we can delete this limb.

3 Results and Discussion

In Figure 1 we had think about the tasks and milestones. In Figure 2 we have the risk, that a CB is defect, even though it was controlled. And later we had see, that is true. A CB defect is detected through the V2 with probability about 70 %. The Figure 3 gives us to possible considerations. The production or the Quality Gate didn't work solid. The graphics doesn't show any outlier value, so we haven't to look clearly into the past, but the dispersion is very high from January to February and from June to July. It would be interesting, why is this happened, but not necessary.

Figure 4 is interesting, because of the error codes and where they come from. We can see the highest code K8 and the lowest K1. Here we can look when we achieve the 80 % mark. We have to reduce all error codes between K8 and

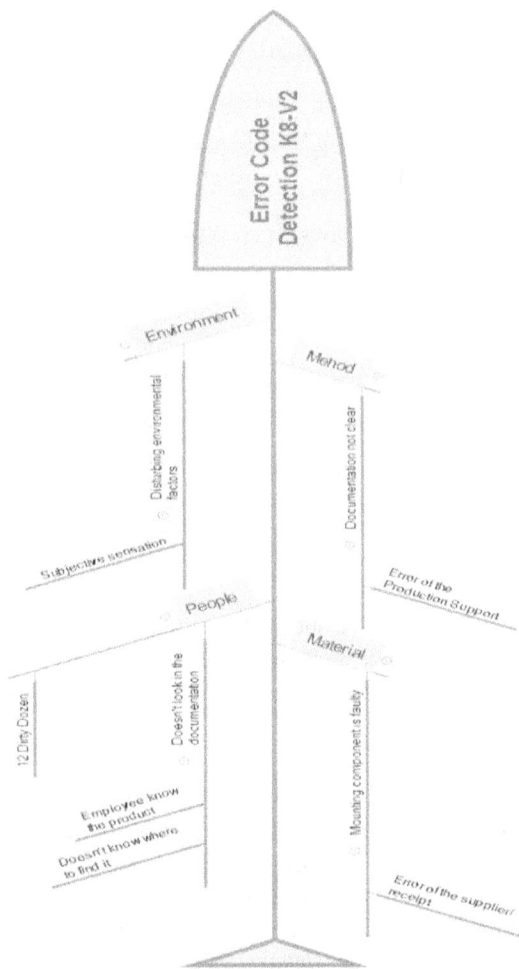

Figure 6: Ishikawa-diagram for the error code K8 from the quality gate V2 (see at Figure 4). From each of the 5 limbs (people, materials, measurement, machines, methods and environment) goes the question "Why". And if there is no answer, the main problem is found. You can delete the limbs with no ideas.

M1. Also we can see, that this isn't 20 % (it is 28 %). So this data don't follow the Pareto principle.

The MSA (Figure 5) shows us more than one detail about the V2. In subsection 2.2 is described that the employee checked the CB at two different days. For 25 % of the CB a defect was detected on a single day and on another day the defect was not detected. Such results may be caused by stress before Christmas, the recovery after the holidays or something completely different. It would be a good idea, for more expensive CB, to check them by Checker 1. Since Checker 2 will cause more costs for Prettl. Also from Figure 5 we adopt that defects were detected for 19.6 % of the CB although there was none. This causes unnecessary costs. On the other way around 40 % of the defect CB are sent to the customers. Again, unnecessary costs arise. In Figure 6 we present some reasons for a error code. The production works with the documentation, made by PS. If therein an error occurs, the production will work wrong. All companies have long term employees, since they represent

a high value. Nevertheless they work so long in the company, that they know all about the products. Therefore they believe that they know all about all documentations. But sometimes in the documentation something is changed. In this way even the long term employees can cause defects, when they don't update their knowledge about the product. If the supplier produce a defect component and the employee at the receipt and at the work station don't see this.

4 Conclusion

Since we are on the way to a zero defect production, the conclusion of our investigation consists in the explanation, how we use our results for the next steps. At first we have to find a good strategy to implement, such it will be economically justifiable and quick to adopt. We have seen that there are several different possibilities. Therefore we will create a "Failure Mode and Effects Analysis". Herein we declare every idea for possible optimizations and its evaluation.

Acknowledgement

This work was carried out at Prettl Electronics GmbH and supervised by the Institute for Software Engineering and Programming Languages. At first the first author thank Prettl Electronics in Lübeck, for the chance to make his placement in their company. Moreover, we want to thank our supervisor in Prettl, Karl-Jürgen Schneider, Dennis Wegner and Sebastian Praxl and outside from Prettl, the Black Belt in Six Sigma Kouross Kashanchi for their help and support. They gave us this topic, several hints for literature and a thread that we could follow.

5 References

[1] DIN EN ISO 9000:2015

[2] DIN EN ISO 9001:2015

[3] All procedural instructions of the company Prettl Electronics Lübeck

[4] All process description of the company Prettl Electronics Lübeck

[5] K. Magnusson, D. Kroslid and B. Bergman, *Six Sigma umsetzen: Die neue Qualitätsstrategie für Unternehmen*, (2003)

[6] J. Wappis and B. Jung, *Taschenbuch Null-Fehler-Management. Umseztung von Six Sigma*, 2010

ChessNet
– Learning to play chess by seeing grand master games –

S. Keser [1] and A. Madany Mamlouk [2]

[1] Medizinische Informatik, Universität zu Lübeck, seves.keser@student.uni-luebeck.de

[2] Institut für Neuro- und Bioinformatik, Universität zu Lübeck madany@inb.uni-luebeck.de

Abstract

In this paper we present the results of our project which aimed to learn the game chess by using grand master games. We used a freely available database which contained 2.2 million games and trained several different neural networks with data extracted from this database. The first goal was to prove if a convolution neural network will be able to extract the rules by only seeing the board and the move made by the grand master. This goal was partly reached, some parts of the rules were understood and could be found in filter kernels of the network. The network was connected to a correction framework which filtered the output to use the best valid move predicted. Since the network played on a very low level we tried to use recurrent structures to learn long term strategies. The Network did not train well using these recurrent structures but it was able to learn some rules of the game. A high level chess skill was not reached.

1 Introduction

This paper tries to continue the work done by Malte Lindenau for his master thesis [3]. The first step aimed to reproduce his results and then we continued to improve the quality. Since he did not use any recurrent structures, this work uses them.

2 Material and Methods

The networks were trained using the Keras framework for python[6]. Keras is a libary which needs a large scale tensor framework like Theano [7] or TensorFlow [8]. For this work the TensorFlow for Python backend was used since it supports CUDA, was easy to install and we had a well performing gaming GPU available. This improved the training significantly and allowed us to use larger data sets.

2.1 Related Publications

In 2017 DeepMind presented their work about the Alpha Zero Algorithm [1]. In comparison to this work, they used the method of reinforcement learning where the network plays millions of games against itself and improves its skills over time. They played 100 games against Stockfish [5] which was the strongest available chess engine at this time. 28 of these 100 games were won by Alpha Zero, the other 72 ended in a draw. Alpha Zero is also capable of playing Go. After only 8 hours of training it was capable of defeating its previous version AlphaGo in 60 of 100 games. Back in 1995 S- Thrun also made an attempt to train neural networks to play chess. His publication shows that it is possible to learn chess by feeding expert games to the network [2].

2.2 Data Format and Interface

The Data was formatted into sequences of three moves. Each move contains the chess board as an array of 768 boolean values and the moved pieces. There are 6 different types of pieces for each color. This means that a 384 element vector can represent each piece type on a 8x8 field, in total a 768 element vector was used to represent the place where pieces were taken from and where they were moved to. The Input array consists of 12x64 chessboard representations, where each of the 12 boards contains only one kind of chess piece. A value was set to true at a specific index if a piece of the correct type was present on this location of the chessboard. The format is shown in Fig. 2 (Input/Board) and Fig. 3 (Output/Move).

2.3 Playing against the Network

The trained networks were then connected via UCI(universal chess interface) protocol [10] to compete with regular chess engines or human players. Therefore we used the program Arena [11] which allows to see how the engine plays against other engines. The engine of choice was Stockfish, one of the strongest engines available. The framework for this connection was received from Malte Lindenau and just slightly changed to match new format specifications for the network I/O format. Since the move predicted by the network is not always valid - meaning that

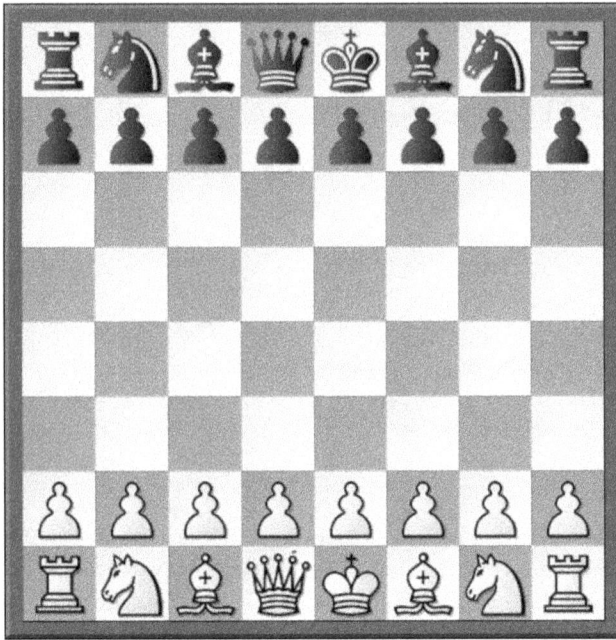

Figure 1: The initial piece position of the chess game. White has to do the first move.

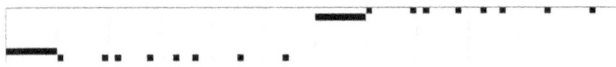

Figure 2: Start position converted to the 768 element wide vector and arranged in shape of chess fields. Black = true (piece present) and white = false (no piece) [3].

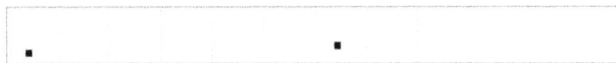

Figure 3: The format used for coding the moves into a 768 element vector. The move shown is d3-d4 (pawn)[3].

the rules of the game would allow this move - the interface contained a small engine which was capable of proving if a prediction was valid. This engine was connected between the neural network and the UCI interface to pass only commands which will be accepted.

2.4 Data Augmentation

We used the "millionbase" chess database which contains about 2.2 million games played by grandmasters [9]. This means that the moves contained in the database are very close to an optimal move. The first 10 million moves were converted from the Database (PGN Format) to the propietary format used for the networks. The set was then split into a 9 million moves training set and a 1 million moves validation set containing only white moves. the first million moves were taken to be the validation set. The networks were not trained well, since the test error reached its minimum mostly after 1-2 million single moves seen by the network. To make sure there is no overfitting, we removed duplicate moves from the database. This aims to the

openings and the first moves in the game. There are exactly 18 possible moves on the start position. the start position should be well known and is shown in Fig. 1. From this position every rook can be moved either one or two fields formward and each knight can be moved to one possible location. all other pieces are locked in place. Another idea is to filter the database to only contain moves from a game which was won by white. This has been made by Malte Lindenau in his Thesis but there was not measurable improvement to the networks playing skills. In total about ten models were build and trained. This paper introduces four of them. For all models the standard learning rates and the Adam optimizer were used.

2.4.1 Model 1

This was the first attempt to combine the convolution layers with the GRU (Gated Recurrent Unit) [4]. The 3x768 element input was split into 36 Slices of 64 elements, each representing the board for a specific piece. They all were zero-padded and convoluted to receive a 8x8 vector. They were flattened and partly concatenated into 3 different dense layers. Each of those dense layer constists of 768 neurons. The output of those dense layers was then concatenated and fed into a GRU layer which outputs a 3x768 vector. The sigmoid activation function was used.

2.4.2 Model 2

Compared to model 1 this model has 2 layers of convolution instead of 1. The zero-padding has been extended to get an output also fitting in an 8x8 vector. The dense layer between the convolution and the GRU was removed here.

2.4.3 Model 3

This model is based on model 2 and received a third convolution layer

2.4.4 Model 4

This model has 2 Layers of convolution followed by a 50% dropout layer. The GRU layer was also doubled and followed by a dense layer.

2.5 Network Construction

The Network design was inspired by the networks used in [3]. But there, the network was not able to learn any strategic skills due to the fact that it always just got a board and had to predict a move - without any preview of upcoming moves or history of moves which could help it to develop a strategy over multiple moves. The main idea of this work is to use Gated Recurrent Units (GRU) [4] to learn a kind of strategy over multiple moves. Therefore the network was constructed to accept sequences of three boards and predict three moves. The moves were chosen to be a sequence in the game flow and the board was fed into the network at the time t, t+1 and t+2. The network was formed into a three strand construction which were only connected widely in the GRU layer. Here we only compare 4 of those networks. Each of the networks has an input of 3x768 vector shape. this was reshaped and sliced into blocks of 64 values each representing one figure on the board. Each of those 64(8x8) fields was zero padded and passed through a 2-dimensional convolution layer. Depending on the network model it had 1 to 3 convolution layers followed by a densely connected layer which was splitted into 3, which means that each input board was passed to a different strand. The GRU Layer then connected all the outputs and delivered the final predictions.

2.6 Evaluation Methods

The ideal output vector consists of only zero values except for two elements having the value 1. This makes it really difficult for common accuracy measurements to be precise. Therefore the precision was measured by the total sum of value differences as shown in (1). P represents the prediction made by the network and R represents the reference which is taken from the database.

$$\sum_{i=1}^{768} |P_i - R_i| \qquad (1)$$

This means that a trained network which only gives a 768 vector of zeros will have a test error of 2. This would indicate that the network is not learning in a way as it is wanted. Having a validation error below that should indicate that the network has learned to reproduce the games from the database and chooses moves which are close to the moves in the validation set.

3 Results and Discussion

The Networks were all trained using the nine Million sequences long training set and were validated on a one million sequences long validation set. Table 1 shows the performace of the different network types. The validation error was calculated as shown in (1) and the valid moves count specifies the number of moves out of 10.000 where the highest rated move was valid.

Table 1: Trained Networks

Network	Validation Error	Valid Moves/10k
Model 1	1.47	1244
Model 2	0.86/2.45	1060
Model 3	1.14	1088
Model 4	4.14	573

3.1 Training Speed and Performance

The networks were all trained with the GPU supported TensorFlow backend. The average processing speed was about 6000 sequences per minute, depending on the net size. Each network was trained on the full set of nine million sequences on one epoch only. The difference between training and validation error was <0.01. therefore this work only contains informations about the validation error. The development of the validation error stopped mostly after around one million sequences seen by the nets. This is a sign of the nets not training well. But adding dropout and normalization layers did not change this issue. The progress of validation is shown in Fig. 4. the most interesting network here was model 2. It reached a local minimum of about 0.84 after learning one million sequences. After 4.6 million sequences the validation error suddenly jumped to 3.14 and decreased from there to 2.45 after nine million trained sequences. This behavior is usually a sign of overfitting but the net was trained on a duplicate free data set and passed only 1 epoch. due to this occurrence this part of the experiment was repeated using a different data set which contained all black moves instead of white moves. There no such effect appeared. A explanation for this effect was also not found.

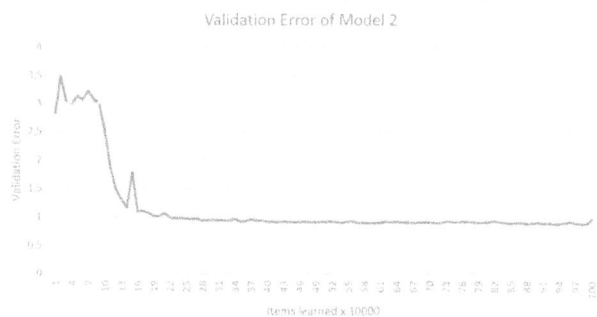

Figure 4: Decrease of validation error during training of Model 2

3.2 Playing against other engines

The validation results from the model 2 were the best so it was challenged to play against an engine. The Stockfish engine was used for this, but due to the fact that Stockfish is very strong the net got beaten very fast (7-12 moves). To see how strong the net is the search depth for Stockfish was decreased to one half move - but with no significant changes. The games ended mostly with a mate after 10-20 moves. The game of model 2 after seeing five million moves vs Stockfish v8 Modern is shown in Fig. 6 and Fig 7. For better understanding Fig. 7 shows which moves were made by the network in the sample game. The rating dropped at move 7 when white lost a bishop. The chess toll used for this work was Arena [11]. It also has an integrated method of evaluating the board and give a visual feedback which player is in the lead. As it can be seen in Fig. 5 the model 2 network was able to keep the game in a neutral balance for the first seven moves. Then the voting dropped down when white lost its first knight. After this move the rook was moved several times between h1 and g1 until the king was endangered and needed to be saved. This may be a cause that in early states there are often well-known positions which may be learned better than unknown strategies which can be applied only in very few games.

Figure 5: Evaluation over the game between model 2 and Stockfish. Better values for white (model 2) are above zero, for black (Stockfish) better values appear below zero.

```
1.d4 d6 2.d5 e5 3.c3 Bf5 4.e4 Bd7 5.Qf3 h5 6.Qe2
Ne7 7.Bf4 exf4 8.Nf3 h4 9.Rg1 c5 10.Rh1 Ng6 11.Rg1
Ne5 12.Rh1 b5 13.Rg1 Be7 14.Rh1 c4 15.Rg1 Kf8
16.Rh1 Nd3+ 17.Kd1 Kg8 18.Ne1 Qb6 19.Qd2 Bg4+
20.Qe2 Nxf2+ 21.Kc1 Bxe2 22.Bxe2 Qe3+ 23.Nd2
Qxe2 24.Rb1 Qd1#, 0-1 Mate
```

Figure 6: The moves made by the Mode 2 net (white) and Stockfish (black). black won.

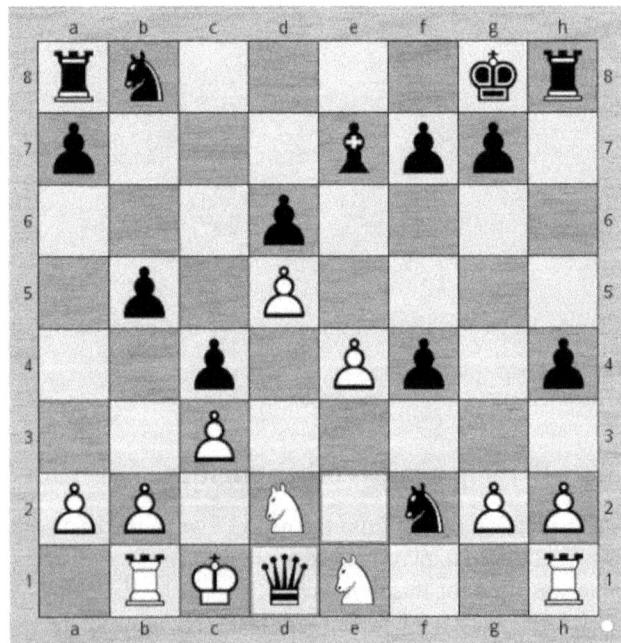

Figure 7: The board at the end of the game Stockfish vs Model 2

4   Conclusion

The main idea of this work did not give the expected output of a network beeing capable of learning long-term strategies in chess. The nets are not playing well and are only barely adapting the rules of the game. Due to this issue, it was not possible to figure out if the usage of the GRU-layer allowed it to learn a long-term strategy. In some expiriments it was recognized that changing some values like optimizers or activation functions has a huge impact on the outcome. More

expiriments with other network types could also bring up a new skill level.

Acknowledgement

The work has been carried out at the Institute for Neuro- and Bioinformatics, Universität zu Lübeck.

5   References

[1] D. Silver et al. *Mastering Chess and Shogi by Self-Play with a General Reinforcement Learning Algorithm*. DeepMind, London, 2017.

[2] S. Thrun, *Learning To Play The Game of Chess*. University of Bonn, Bonn, 1995.

[3] M. Linbenau, *Artificial Neural Nets Learning Chess by Imitation*. Universität zu Lübeck, Lübeck, 2017.

[4] Cho et al. *Learning Phrase Representations using RNN Encoder-Decoder for Statistical Machine Translation* EMNLP, 2014, arXiv:1406.1078

[5] *Stockfish Chess Engine* Available at https://stockfishchess.org/

[6] *Keras - The Python Deep Learning Libary* Available at https://keras.io/

[7] *Theano Libary* Available at http://deeplearning.net/software/theano/

[8] *TensorFlow - An open-source software library for Machine Intelligence* Available at https://www.tensorflow.org/

[9] *Million Base 2.2* Available at http://www.top-5000.nl/dl/millionbase%202.2.exe?forcedownload

[10] *UCI Protocol Specifications* Available at http://www.shredderchess.com/schach-info/features/uci-universal-chess-interface.html

[11] *Arena Chess GUI for Linux and Windows* Available at http://www.playwitharena.com/

Head-to-Steering-Wheel Distance Estimation Based on a Monocular RGB Camera with Convolutional Neuronal Networks

E. Vothknecht[1], F. Coleca[2], S. Klement[2] and H. Handels[3]

[1] Medical Informatics, University of Lübeck, erik.vothknecht@student.uni-luebeck.de
[2] gestigon GmbH - a Valeo brand, Lübeck, {foti.coleca; sascha.klement}@gestigon.com
[3] Institute of Medical Informatics, University of Lübeck, handels@imi.uni-luebeck.de

Abstract

The head-to-steering-wheel distance is an important piece of information for an airbag deployment system. If this information would be available in real-time, a safety system could adjust the force of the airbag deployment, resulting in fewer injuries during a car crash. In this paper we propose a way to track the 3D head position of the driver through a monocular RGB camera. Two Convolutional Neuronal Networks, based on the GoogLeNet[9], were trained using 2D images as inputs and 3D head positions as ground truth labels. The ground truth was automatically annotated using depth data coming from a Time-of-Flight camera. For our purpose, 11 volunteers were recorded with a Time-of-Flight camera and a RGB camera. The first network mean Euclidean error was 21.77 centimeters, and the second approach produced a lower 18.88 cm error. In both cases the networks failed to learn useful features about the position of the head.

1 Introduction

Airbags are nowadays available in every car. They have improved the safety of drivers and passengers significantly in years past. Ever since these systems were available, side effects like injuries to the head and the chest could occur because of the deployment of an airbag [1]. The deployment is in general triggered by car crashes and is performed in the same sequential order without taking under account the driver's actual physique in the vehicle compartment. If the airbag deployment would instead be customized based on an individual's height, weight and seating position, it could decrease the risk of injuries caused by these systems.

The head-to-steering-wheel distance of the driver is an important factor that can be considered in this context [2]. If the head position is tracked while driving, the deployment procedure of an airbag can be changed in real-time, based on the distance to the steering-wheel. Therefore we propose a way to predict the 3D head position of a driver in a motor vehicle from one monocular RGB camera via Convolutional Neuronal Network approaches.

Head pose estimation and head tracking have been studied in numerous studies in the literature [3]. The currently most common used face detection algorithm is the Viola-Jones face detection algorithm [4]. The algorithm is efficient and detects faces via a machine learning approach for object detection, based on 2D features in an image. The problem in driving scenarios is that the detection is aborted when the face is no longer in the field-of-view of the camera. For instance when the driver is looking over his shoulder. An alternative would be to follow an approach from J. García et al. [5]. They find locations of heads in 2D images based on circular patterns and use a tracking procedure, based on the Kalman filter. There is also a variety of different head pose estimation approaches[6][7][8], which can be used to detect the direction of a head in 2D images. The disadvantage of all these mentioned approaches, in our use case scenario, is the missing third dimension to measure the distance between head and steering-wheel.

Time-of-Flight cameras are the current state-of-the-art in real-time depth measurements and provide a depth map that can be used for distance measurements in our use case. The disadvantage of Time-of-Flight cameras, in contrast to RGB cameras, is their significantly higher price.

For these reasons we propose an approach based on Convolutional Neuronal Networks to predict the 3D position of the head of a driver with a monocular RGB image. We explore two different approaches with such networks. The aim is to provide the position of the head throughout the whole driving period, so that the head-to-steering-wheel distance can be calculated at all times.

2 Materials and Methods

To learn the 3D head position we are using two cameras. A Time-of-Flight camera is used to extract the ground truth head position based on a body part segmentation algorithm. The ground truth head position and the 2D images, from a separate RGB camera, are fed into a Convolutional Neu-

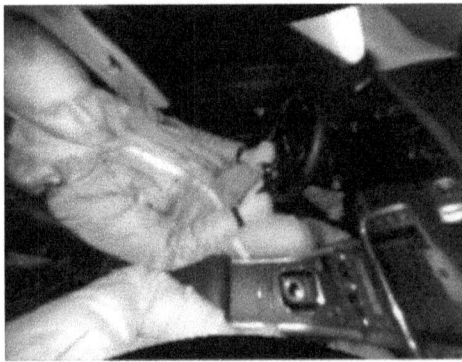

(a) Amplitude image of Time-of-Flight camera   (b) Monocular RGB camera

Figure 1: Images of the two cameras in the car. (a) is an amplitude image of the Time-of-Flight camera attached at the dome module in the car. (b) is an image of the monocular RGB camera attached below the air vents at the center console. The black square indicates an 224x224x3 input image which has been fed into the networks.

ronal Network to train the head position on 2D image features.

In this section the data acquisition for training and testing is described, as well as the CNN architectures.

2.1   Data Acquisition

To train a Convolutional Neuronal Network to estimate 3D head positions from 2D image features, a controlled data acquisition environment is needed. We recorded 11 drivers with a Time-of-Flight camera and a monocular RGB camera at the same time. Each volunteer was recorded in the driver's seat, performing normal driving activities for one minute. The drivers were also instructed to move their head around at various distances from the steering-wheel. We let the Time-of-Flight camera run for one hour prior to the recordings to achieve a stable working temperature.

2.1.1   RGB Camera

The RGB camera is an off-the-shelf Logitech C920 webcam (2MP CMOS sensor) which has been modified to accept a S-mount fisheye lens from Lensagon (BT2120). The lens and webcam are enclosed in a self printed camera case. The use of the fisheye lens enables a 170 degrees wide-angle few over the vehicle interior. The RGB sensor is capable to capture data with a resolution of 1920x1080 pixels at 30 frames per second.

2.1.2   Time-of-Flight Camera

The Time-of-Flight (ToF) camera is able to record depth data and amplitude images at a resolution of 320x240 pixels with 30 fps. The field-of-view is 109x81 degrees. The sensor is able to measure distances from 200 to 1500 mm.

2.1.3   Recording Environment

We used a Range Rover Evoque SUV for the recordings, not modified in any way, except for the roof module which was replaced by a 3D printed holder for the ToF camera. The

ToF camera was pointed at the driver lengthwise, in order to capture the full body images (Fig. 1a). The RGB sensor was placed on the center console, just below the air vents. This enables a good view of both, the driver and passenger, from the same location (Fig. 1b).

2.1.4   Data Pre-Processing

The head position needed to be extracted from the ToF data. We used a framework, supplied by the gestigon GmbH, that is able to provide real-time segmentation of different body parts. For each frame of the recorded streams, the segmented head of one person was extracted. The segmented pixels were used to compute the 3D coordinates of the head's center. The calculated centroids were defined as the ground truth head positions for further training.

We synchronized the ToF streams with the RGB streams by comparing the global timestamps of both streams, with respect to a threshold in milliseconds. Furthermore we cropped the RGB data to 672x672 pixels and resized the images to 224x224 pixels, centered on the driver due to CNN input size requirements (Fig. 1b).

2.2   Convolutional Neuronal Networks

All of the mentioned CNN pose estimation approaches [6][7][8] are based on 2D image features, while the predicted head positions are in 2D image coordinates. Our goal is to predict the head position with its three degrees of freedom based on 2D images.

In this context we have considered the GoogLeNet network topology by C. Szegedy et al.[9]. The GoogLeNet is an image classification and detection network. Its main improvement is the use of so called "inception modules". One module combines filters of different sizes and concatenate them to one output of a layer. This type of architecture allows to decrease the depth and the width of a network without significant performance penalties. Furthermore, with a decreased network size, the number of hyperparameters decreases which makes this network, in contrast to lager ones, less prone to overfitting.

Box Plots of Euclidean Distances in cm

(a) Coordinates learning approach

(b) Grid learning approach

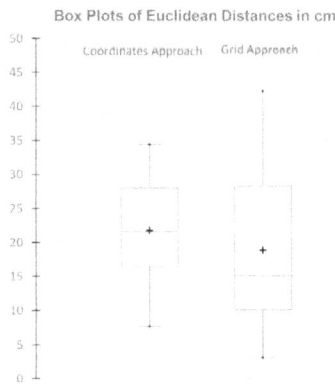

Figure 2: Box plots of the Euclidean distances of the coordinates- and grid learning approaches over the two test streams.

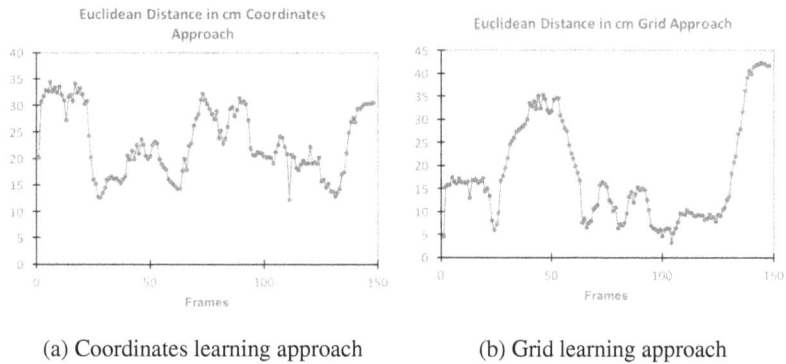

Figure 3: Euclidean distances between ground truth head positions and predicted head positions for the first 150 frames of a test stream for coordinates learning - and grid learning approach.

2.2.1 Learning the Coordinates

To predict the 3D head position based on 2D image coordinates, the GoogLeNet needed to be modified for our regression approach. We replaced the Softmax layer at the end of the GoogLeNet with an Euclidean-Loss layer. This layer computes the Sum-of-Square-Differences for its two inputs p^{pred} and p^{head}:

$$\frac{1}{2N} \sum_{i=1}^{N} \left\| (p_i^{pred} - p_i^{head}) \right\|_2^2$$

We supply a 224x224x3 input image to the network and use a vector of 3 coordinates (x, y, and z position of the head center) as the output.

2.2.2 Learning via Voxel-Maps

As a second approach we modified the input data to treat our use case in a three dimensional way. Based on the 3D coordinates of the ground truth head positions, we generated a 3D binary voxel grid covering the maximum and minimum head moments of all recorded drivers. The translation of the 3D coordinate into a bin of the grid was performed by:

$$bin_{x,y,z} = \frac{N_{bin} \cdot (p_{x,y,z} - min_{x,y,z})}{max_{x,y,z} - min_{x,y,z}}$$

, with $bin_{x,y,z}$ as the resulting bin, N_{bin} denotes the number of bins, $p_{x,y,z}$ the head position and $min_{x,y,z}$ and $max_{x,y,z}$ the minimum and maximum values of all head movements in each direction. The resulting 28x28x28 space is fed into the network as a vector of 21952 features for learning in combination with the 224x224 RGB images. As a network output we are receiving a tree dimensional heat-map which describes the probability of a head - position at a specific bin. The remapping is performed vice versa with a precision-loss of $\pm 2cm$.

3 Results and Discussion

The streams of the 11 recorded passengers have been divided into 9 for training and 2 for testing. The training of both networks was done using backpropagation for 20 epochs, with an ADAM optimizer (LR-0.0001, with applied step decay by reducing the learning rate every 6 epochs). After 20 epochs, the minimum loss was 0.014 for the coordinate learning - and 0.52 for the grid learning approach.

The results of the applied test streams on the trained networks can be seen in Fig. 2. The box plots are showing the overall Euclidean distances between the ground truth and the predicted head position for both approaches. The mean Euclidean distance of the coordinate learning approach is $21.77cm(\pm 6.84cm)$. The mean of the gird learning approach is $18.88cm(\pm 10.8cm)$. The interquartile range (IQR) for the grid learning approach is at 18.125 higher than the coordinates learning approach at 11.64.

We can note that both networks did not learn what they were supposed to learn. Both approaches are not near a Euclidean distance of $0cm$. In Fig. 3 the Euclidean distances between ground truth and predicted head positions are visualized. We can notice that the local minima and maxima are at the same frames, but with a different distance between the points for both networks. It indicates that both networks have learned fixed head positions. This assumption has been confirmed by visualising the test data in Fig. 4.

In the amplitude image of the Time-of-Flight camera the predicted head position of both networks and the ground truth head position are visualized during a forward leaning phase of one participant. The predicted head positions of both approaches are not moving during this phase over 19 frames for both networks. So both networks learned a fixed position. The difference of these positions is that the predicted position of the grid approach is more likely oriented towards the normal seating position of one passenger. In the coordinate learning approach, the fixed position is more oriented towards the average head position over all head movements that had been made.

These results can have a variety of reasons. On the one hand the recorded data can have an over representation of specific head positions during recording. The majority of all participants returned to their normal seating position after one movement was performed. Therefore it is very likely that head positions corresponding to normal seating positions occur more often than other positions. On the other hand the network design that we used might not be the optimal one for this use-case, and other topologies might need to be researched.

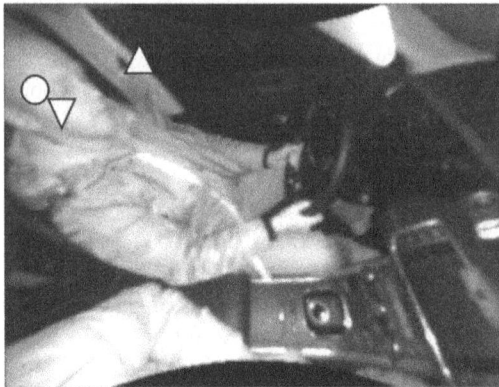

(a) Driver in normal seating position

(b) Driver in forward leaning position

Figure 4: Visualisation of the network outputs in the amplitude image of the Time-of-Flight camera. White point indicates the ground truth head positions. The upright triangle indicates the predicted coordinate of the coordinate learning approach. The downwards oriented triangle indicates the predicted coorinate of the grid learning approach.

4   Conclusion and Outlook

In this paper we tested two Convolutional Neuronal Networks to predict the 3D head position of one person from 224x224 RGB images. We recorded 11 volunteers for training and testing purposes with a ToF camera and a RGB camera in a vehicle compartment.

In conclusion, the coordinate approach, as well as the grid approach failed to learn the 3D head positions. The output of both networks are fixed head-positions. Further research

has to be focused on improving data pre-processing algorithms and the network architecture. Moreover the recording procedure should be executed with a higher diversity of poses and positions, in contrast to the normal seating position of one passenger. Additionally a higher amount of data should be recorded. Other approaches to infer depth from a monocular camera should also be considered.

Acknowledgement

The work has been carried out at the gestigon GmbH, Lübeck and supervised by the Institute of Medical Informatics, University of Lübeck.

5   References

[1] L. A. Wallis and I. Greaves, *Injuries associated with airbag deployment*. Emergency Medicine Journal (pp. 490-493), 2002.

[2] M. A. Manary, C. A. Flannagan, M. P. Reed and L. W. Schneider, *Predicting proximity of driver head and thorax to the steering wheel*. In Proceedings of the 16th International Technical Conference on Experimental Safety Vehicles, Paper (pp. 98-S1), 1998.

[3] E. Murphy-Chutorian and M. M. Trivedi, *Head Pose Estimation in Computer Vision: A Survey*. IEEE Transactions on Pattern Analysis and Machine Intelligence. Vol. 31, 2009.

[4] P. Viola and M. Jones, *Rapid Object Detection Using a Boosted Cascade of Simple Features*. Computer Vision and Pattern Recognition, 2001. CVPR 2001. Proceedings of the 2001 IEEE Computer Society Conference on. Vol. 1. IEEE, 2001.

[5] J. García, A. Gardel, I. Bravo, J. L. Lázaro, M. Martínez, and D. Rodríguez, *Directional people counter based on head tracking*. IEEE Transactions on Industial Electronics, Vol. 60, 2013.

[6] N. Ruiz, E. Chong, J. M. Rehg, *Fine-Grained Head Pose Estimation Without Keypoints*. arXiv preprint arXiv:1710.00925, 2017.

[7] J. Carreira, P. Agrawal, K. Fragkiadaki, J. Malik, *Human Pose Estimation With Iterative Error Feedback*. IEEE Conference on Computer Vision and Pattern Recognition, pp. 4733-4742, 2016.

[8] A. Bulat, G. Tzimiropoulos *Binarized Convolutional Landmark Localizers for Human Pose Estimation and Face Alignment with Limited Resources*. IEEE Conference on Computer Vision and Pattern Recognition, 2017.

[9] C. Szegedy, W. Liu, Y. Jia, P. Sermanet, S. Reed, D. Anguelov, D. Erhan, V. Vanhoucke, A. Rabinovich, *Going Deeper With Convolutions*. The IEEE Conference on Computer Vision and Pattern Recognition, 2015.

Regulatory requirement for Point-of-Care Testing (PoCT) Devices - Comparison of major regulatory systems and implications for an international regulatory strategy for market access of PoCT devices

O. Amasiatu [1] and F. Spitzenberger [2]

[1] Biomedical Engineering, University of Applied Sciences Lübeck, ogechukwu.godwin.amasiatu@stud.fh-luebeck.de
[2] Regulatory Affairs, University of Applied Sciences Lübeck, folker.spitzenberger@fh-luebeck.de

Abstract

Point-of-care testing (PoCT), also called "Near-Patient Testing" (NPT), refers to any in vitro diagnostic test administered outside the central laboratory at or near the location of the patient. Technological advances have made it possible to conduct many conventional laboratory tests at the point of care allowing rapid access to test results and therefore to increase efficiency in medical care. However, concerns about the overall utility and reliability benefits to patient care have accompanied the advancing technologies related to PoCT. This has led to the aim of this work as the comparison of regulatory systems for PoCT devices in major international markets was done using a table and discussed and also, developing a strategy for a virtual European manufacturer expanding to international markets with regards to PoCT of different risk classes.

1 Introduction

Over the past decades, the availability and use of PoCT have steadily increased in Europe and worldwide. Outside the hospital setting, PoCT provides laboratory examination services to under-serviced areas and general practitioners especially in Africa, South-America and Asia. However, concerns about the overall utility and reliability of the benefits to patient care have accompanied the acknowledgement of the advancing technologies related to PoCT. For example, important organizational and quality assurance challenges are addressed with the implementation of PoCT in any health environment and have led to the development of international standards for quality and competence in PoCT [1]. Examples of PoCT devices include as most prominent devices, glucose monitoring systems, blood gas testing devices and even high risk infectious disease testing devices as HIV test kits. Fig. 1 shows the use of a PoCT device. In view of regulatory affairs of medical devices, the specific characteristics of PoCT devices with regard to usability aspects, the higher level of complexity (especially in comparison to self-testing devices) and special quality requirements have recently been leading to new or increased regulatory requirements for safety and performance of these devices in various regulatory systems. Since these different regulatory requirements represent a significant challenge to the industry, manufacturers need to develop a valid and, practical regulatory strategy for international market access of PoCT devices. It is therefore the scope of this project is to

undertake a comparison of regulatory systems for PoCT devices with relevance to the major and growing international markets by identification and detailed analysis of regulatory market approval requirements and procedures. Countries and region included are: EU with a special focus on the differences between the requirements of the IVDD and the new EU IVDD, Canada, U.S.A, Brazil, India, China, Saudi Arabia and Nigeria. Based on this analysis, a regulatory strategy for international market access of PoCT devices of intermediate and high risk is developed, starting from a virtual European manufacturer expanding to the international markets listed above.

Figure 1: Patient testing with microINR from iLine Microsystems.

2 Material and Methods

For the comparison and critical analysis of the regulatory systems and the development of a strategy to access international markets, international and national regulatory documents, agreements, protocols, and standards were researched by a comprehensive web-based database search. Further available documents, reports, and articles are reviewed by online database searches. International agreements either directly addressed PoCT regulation or addressed certain aspects of PoCT quality, safety and efficacy. For the overview and comparison of the regulations and standards, seven countries and one region (European Union) were selected. These countries vary in their income, approach towards regulating laboratory diagnostic testing and, the availability of resources among other differences. The countries and region are as follows: EU, U.S, Canada, China, India, Saudi Arabia, and Nigeria. The following criteria are taken into consideration when performing the analysis:

- Major regulatory framework

- Structure of the regulatory authorities

- Hierarchy of regulation

- Definition of PoCT

- Risk classification of PoCT devices

- Comparison with regulatory requirements of self-testing devices

- Conformity assessment/marketing authorization procedures.

Overall, the consideration and reflection of the specific characteristics of PoCT devices within the regulatory framework is evaluated by these criteria.

Based on these criteria, a comparative and gap analysis on the levels of legislation/regulation on these countries is performed.

3 Results and Discussion

3.1 Hierarchy, transparency, and stability of regulation

Regulatory systems in the countries/regions reviewed in this project are of hierarchy which are identified as the following:

- *Primary or First level legislation* This is an executive law for medical devices and refers to binding and enforceable legislation, usually adopted at the level of individual countries by their respective legislatures and/or executives. For instance EU directive, statutory law and Act of parliament [2].

- *Secondary or Second level legislation* This is a form of law for medical devices, referring to written instruments that are binding and enforceable and are issued by the regulatory (executive) authority. Examples are EU Implementing Regulations [2].

- *Third level legislation or Guidelines* These are guidance documents that generally refers to non-binding normative documents issued by the regulatory authority, which offer guidance on recommended practices. They allow for scientifically-justified, alternative approaches and translation of regulatory, general acceptable guidance. They relate to the design, production, labelling, promotion, and manufacturing, testing of regulated devices, the processing content, and evaluation of submissions. Examples are technical standards and recommendations [2].

In a region and countries like the EU, U.S, China and Canada, there is high level of stability in the regulations although, with minimal difference amongst them. The importance of manufacturers understanding the application and conformity assessment requirements according to the three hierarchies of the medical device regulation for these target markets cannot be overemphasized. Usually, the primary legislation which is seen as the most important of the three tends set the directive, principles, definitions, framework and acts. While the regulations which are the secondary level legislation are the regulations which explains these first level principles. Most times it is quite difficult to differentiate between the two levels as much as they are both legally binding. In most cases, the regulations are used because the Acts are seen to be cumbersome or complex to follow. Most of the regulatory bodies of developed countries have these levels of legislation, for example the U.S has a primary level legislation in the Food, Drug and Cosmetics Act, which includes empowerment of power, recognition of medical devices, placing on the market and withdrawal, principles of safety and performance, quality management system requirement, reporting incident and medical device listing. The secondary level in the U.S is the Code of Federal Regulations (CFR Title 21) which sheds further light on the FD and C Act, with regulations, classification rules, the responsibilities and authorized representative and criteria for recalling of medical devices. In the case of EU, the first level legislation due to its complexity is also difficult for the parliament or the council of member states to pass or change, as it could take more than a decade to be issued, hence the use of Implementing Regulations for the amendment of the directives. The third level of legislation, also called "guidance documents", is composed of documents created to help stakeholders interpret and apply the directives and regulations in view of the manufacturers. They are recommendations related, for instance, to specific labelling requirement, good laboratory and clinical practices. Guidance documents are not legally binding. An example is the U.S FDA Guidance document and, in the EU, the MEDDEV guidance document. Standards as third level regulatory documents have a special relevance in the EU (and also in some other countries), because their application assumes for conformity with legally binding requirements from the primary and secondary legislation level. Some regulatory systems do not have three levels of legislation as shown in Table 1. For instance, Nigeria, does not yet have advanced guidelines and regulations to control the medical device market and taking a hint, one factor influencing regulation policies development is the region where a country belongs to. Nigeria being a member of the Economic Community Of West African States (ECOWAS), where also the members do not have strict or developed policies on medical devices makes the country susceptible to any kind of medical device quality, which has led to the deterioration of the health sector status. Saudi Arabia on the other hand, do not have much focus the medical device regulations, but has set its standard to any medical device seeking registration and market placement, must meet the conformity assessment requirements of either the regulatory bodies of the U.S, Canada, EU, Japan or Australia. This has helped bridge the gap in the absence of complex directives. India has recently developed new standards and regulations due the fast technology and manufacturing growth in other to meet up with the elites.

3.2 IVD and PoCT device regulation in the EU

Currently, the European IVD directive dating back from 1998 governs the in vitro diagnostic medical devices (IVDMD) in the European Union. This directive does not cover the devices related to new techniques and applications in current in vitro diagnostics testing, and also lacks conformity with current international guidelines and regulation systems with regards to the risk-based classification of IVDMD. To overcome these problems, the EU Commission has recently published the new regulation (EC) 2017/746 on IVDMD, called "IVDR". One of the aims is to include and regulate PoCT devices. The regulation defines NPT/PoCT as "any device that is not intended for self-testing but, is intended to perform testing outside a laboratory environment, generally near to, or at side, the patient by a health professional" [3]. According to the IVD directive, PoCT devices are currently found among all IVDMD categories and most of them are "other". The classification of PoCT devices according to the new IVDR follows the seven rules of classification criteria (Annex VII) on the basis of new risk-based classification system including four risk classes of A (lowest risk), B, C and D (highest risk) [4]. According to the IVDR, the regulation of PoCT devices is overall similar to self-testing ("home use devices"), except for devices classified as A. The minimum requirements include the "assessment of the technical documentation" of the device by the Notified Body according to Chapter II of Annex IX in the regulation [4]. This new procedure will increase the involvement of Notified Bodies in the assessment of PoCT devices in the future. Specific requirements for the language and format of the product information and specific safety and performance requirements including the labelling of devices for near-patient testing will be required. Of note, manufacturers of the higher risk PoCT devices (Classes C and D) will be obliged to make publicly available a "summary of safety and performance" including key elements of the supporting clinical data of the device. This will increase the level of information of the device performance data that

Table 1: The three legislation levels of the study countries in the respective rows order of The EU (EU Commission), U.S.A (FDA), Canada (TPD), China (CFDA), Brazil (ANVISA), India (CDSCO), Saudi Arabia (SFDA) and Nigeria (NAFDAC).

1st Level	2nd Level	3rd Level
EU MDD and MDR 2017/745	Commission Implementation Regulation (EU) No. 920/2013	EU MED-DEV Guidance document on Directive 2005/50/EC
Food Drug and Cosmetics Act, Title 21 [5]	CFR Title 21-175.300 [5]	FDA Guidance Document-FDA-2015-D-0025[5]
TPD Food and Drugs Act [6]	MD Regulations SOR/98-282 [6]	HC and TDP Guidance Document Master Files (MFs)-Procedures and Administrative Requirements[6]
Drug Admin Law of PRC No. 45 [7]	Regulation for the supervision and Administration of MD, order and measures No. 650 [7]	CFDA Regulatory Guide and standards - Data No.9 Medical Manuals [7]
ANVISA Law No. 9.782/1999 [8]	Resolution RDC decrees No. 185/2001 [8]	ANVISA guidance documents-Law No.6.360/1976 [8]
Drug and Cosmetics (Amendment) Act No.26 of 2008 [9]	Gazette Notification G.S.R 78(E) [9]	Guidance for Industry, Guidance- Common Submission Format for Registration/Re-registration-MD/GD/RC/01/00 [9]
	SFDA MD Interim Regulation decrees No.1-8-1429 [10]	MDS-G Guidance documents [10]
ACT CAP F33 LFN 2004 [11]		NAFDAC RR/007/00 Guidance document [11]

has been lacking from the perspective of IVDMD users. Table 1 describes the legislative levels in different parts of the world.

3.3 Regulatory Strategy development

The local regulatory bodies vary in different countries in terms of regulatory demands and even costs, and penetrating these markets will require that a potential manufacturer has a good overview and knowledge of the details of the medical device regulations which control each market. This is one of the major reasons why harmonization of the regulatory requirements through the Global Harmonization Task Force (GHTF) followed by the International Medical Device Regulators Forum (IMDRF) and the development of common economic trading regions were supported by major stakeholders of the medical device industry and representatives of the regulatory authorities [12]. Some parts of the world practice the common economic trading for instance the Mercosul in South-America and the Economic Community of West African States (ECOWAS). However, a lot of countries still impose their indigenous regulatory laws which requires analytical efforts to find their individual requirements. For a European manufacturer of PoCT devices of intermediate and high risks, there are certain steps to be followed for access to the different markets. The first important step is to obtain the CE-marking in EU and this is achieved by satisfying the standard level of EU conformity assessment. Further factors have then to be considered, such as a background assessment of the local regulations, product awareness or assessment, the PoCT classification or differentiation, authorized representatives necessity, documentation requirement, local medical device labels requirements, cost analysis and effect, time frame, regulatory steps and post-market surveillance [12].

4 Conclusion

The preliminary results of this study reveals that global regulation of PoCT devices increases with the acknowledged significance and increased use of these devices in in vitro diagnostic testing. The majority of markets reviewed as case studies include regulatory requirements for IVD medical devices and even special criteria for marketing authorization of PoCT devices. However, there is no international and unambiguous understanding of the term "PoCT" in details and regulatory levels are not the same. Manufacturing and competitive regions or countries tend to develop stiffer regulations while some take references from established directives and have recently modified their regulations like Saudia Arabia and India. Therefore, it is imperative to seek for globalization and harmonization of regulatory requirements for medical devices including PoCT device through the activities of IMDRF.

Acknowledgement

The work has been carried out at the Medical Sensors and Devices Laboratory, University of Applied Sciences, Lübeck, Germany. I thank Professor S. Klein for providing the facilities to carry out this work and discussions on regulatory affairs and medical device designs. A special thank you to Professor F. Spitzenberger for providing the topic and, the framework of this paper and all the encouragement and discussions.

5 References

[1] EN ISO 22870: Point-of-care testing – Requirements for quality and competence, *2016*. Available: https://www.iso.org/standard/71119.html [last accessed on 2018-01-22].

[2] WHO global model regulatory framework for medical devices including in vitro diagnostic medical devices, *WHO Medical Devices Technical Series*. World Health Organization, Geneva, 2017.

[3] The European Union Commission, *Official Journal of the European Union*. English edition, vol. 60, L 115, pp.188. 05-2017.

[4] The European Union Commission, *Official Journal of the European Union*. English edition, vol. 60, L 115, pp.149–150. 05-2017.

[5] The US Laws, Regulations and Guidance: https//www.fda.gov/forindustry/coloradditives/ guidancecomplianceregulatoryinformation/default.htm [last accessed on 2018-02-08].

[6] Legislation and Guidelines - Medical devices: https//www.canada.ca/en/health-canada/services/drugs-health-products/medical-devices/legislation-guidelines.html[last accessed on 2018-02-08].

[7] CFDA Laws and Regulations: eng.sfda.gov.cn/WS03/CL0758/ [last accessed on 2018-02-08].

[8] Brazil ANVISA Medical Device Regulations: https://www.emergogroup.com/resources/regulation-brazil [last accessed on 2018-02-08].

[9] India Regulators Publish New Medical Device Rules: https://www.emergogroup.com/blog/2017/02/indian-regulators-publish-new-medical-device-rules [last accessed on 2018-02-08].

[10] SFDA - The Medical Devices Interim Regulation: https://www.sfda.gov.sa/en/medicaldevices/regulations/Pages/default.aspx [last accessed on 2018-02-08].

[11] NAFDAC, Nigeria: Medical Devices Guidelines: www.nafdac.gov.ng/index.php/guidelines/medical-devices-guidelines [last accessed on 2018-02-08].

[12] Global regulatory strategy and country reports: https//www.emergogroup.com/service/worldwide/global-regulatory-strategy[last accessed on 2018-01-23].

Integration of a label printer into the software system of a cleaning and disinfection device for hygiene documentation

L. Preuße [1], J. Haase [2]

[1] Medizinische Ingenieurwissenschaft, Universität zu Lübeck, luise.preusse@student.uni-luebeck.de
[2] Institute of Computer Engineering, Universität zu Lübeck, haase@iti.uni-luebeck.de

Abstract

The cleaning and disinfection of reusable medical tools is an essential but critical step. Notably in the field of surgical operations, a small contamination can cause life-threatening complications with the next patient. Therefore, there are strict regulatory requirements to the quality assurance (QA) of the decontamination process. Modern disinfection systems can help the QA by creating process protocols. This work develops a software component that handles this step of the documentation process for a series of new disinfectors. This includes an analysis of the requirements, limitations and the conceptual design of the component. Ways to improve the assistance to the documentation process are also discussed in the last Section.

1 Introduction

An important part of today's medical treatment is maintaining a strict standard for the hygienic conditions. The contamination of a small wound can cause a life-threatening infection. To ensure the patient's safety there are strict regulatory requirements, for example the location, tools and workflow of the treatment. In the field of surgical operations, any reusable tool that came in contact with a patient needs to be cleaned and disinfected. In Germany, this process needs to follow the recommendation on the "Hygiene requirements for the reprocessing of medical devices" from the Commission on Hospital Hygiene and Infection Protection at the Robert Koch Institute and the Federal Institute for Drugs and Medical Devices [1]. According to this recommendation, a disinfection system needs a protocol that includes used chemicals, measured values, charge number, the approval decision and the approving person. Considering the amount of paperwork in a modern hospital, an automated documentation is probably the most effective choice here. Subject of this work is the development and implementation of a software component that handles the arrangement and printing of such protocols. The component is supposed to be used in a series of new cleaning- and disinfection systems, therefore an adaptable solution is needed.

2 Material

It was decided that the cleaning-protocols should be printed on stick-on labels. They can be adhered to the sterilized and packed tools or added to the documents easily. A reference printer was provided, whose features are introduced in the following Subsection.

2.1 The Opal OD9

The OD9 model by OPAL Associates GmbH is a versatile label printer with an ethernet, an USB and a serial port. It supports the Transmission Control Protocol/ Internet Protocol (TCP/IP). The OD9 can use two printing techniques. The Thermo-Direct mode uses heat-sensitive labels while the Thermal-Transfer mode needs an additional ink ribbon. Given that only the Thermal-Transfer technique creates labels that reliably last for years, it will be used to print the protocols. The OD9 uses the Eltron® Programming Language 2 (EPL2) as printer language, older EPL1-commands are also supported. The printer interprets binary data it receives as ASCII-text, any valid commands are executed. Invalid parts of the data are ignored. Fig. 1 shows a simple label and the corresponding EPL2-commands. The UTF-8 character encoding is unknown to the printer, therefore the encoding for the "Ü"-letter has to be corrected manually. The complete set of EPL-commands and their parameters is explained in the EPL-programming guide [2].

2.2 Software

The operating system for the new series of cleaning- and disinfection devices is a customized linux-based system. It was chosen because most developers are familiar with it and the great variety in drivers it offers. The customization reduces the memory occupied by the operating system, since only a necessary range of functions are included.

The programming environment used in the project is Rational Rhapsody® from IBM in version 8.0.61. Because the printer needs to be integrated into a larger software system it is helpful to use the same environment. Rational Rhapsody® is a visual programming environment that

```
I8,1,001
N
A700,680,2,5,1,1,N, "□BERSCHRIFT"
A200,680,2,2,1,1,N, "Seite 1"
LO36,620,750,4
A700,580,2,2,1,1,N, "Zeile 1:"
A700,560,2,2,1,1,N, "Zeile 2:"
A700,540,2,2,1,1,N, "Zeile 3:"
A450,580,2,2,1,1,N, "Sehr informatives Datenfeld"
A450,560,2,2,1,1,N, "Sehr informatives Datenfeld"
A450,540,2,2,1,1,N, "Sehr informatives Datenfeld"
LO36,510,750,4
P1
```

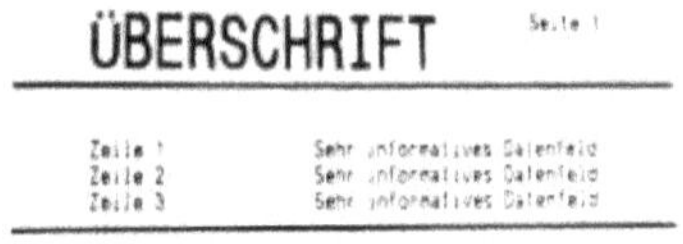

ÜBERSCHRIFT Seite 1

Zeile 1   Sehr informatives Datenfeld
Zeile 2   Sehr informatives Datenfeld
Zeile 3   Sehr informatives Datenfeld

Figure 1: The most used EPL2-commands for defining and printing a label. The I command sets the character encoding, N resets the image buffer of the printer. A and LO defines a text-field and a black line. The final command P starts the printing process. The "□" symbol in UTF8 corresponds to the letter "Ü" in DOS-LATIN 1.

supports automated code generation based on UML diagrams. This additional layer of abstraction simplifies the communication between developers. The functionality of another software component can be understood based on the UML models, so less time is spent on reading the source code. Additionally, the models are utilized in the documentation of the development process and can be reused in later projects. In addition, Rhapsody can be used to create test cases. Disadvantages of Rhapsody are the purchase costs and the high complexity which results in a steep learning curve for people unfamiliar with it.

The computers which are used for development run Microsoft Windows. Therefore, a virtual machine under Debian 8.5 was used to compile and test the implementation. The hardware of a disinfection system with its corresponding CAN-communication can be simulated. A prototype is also accessible.

2.3 Requirements

The concept for the printer component is based on a set of customer demands. The physical printer has to be connected to the ethernet port of the cleaning system directly. The support for two label sizes is demanded. This affects the variety in label material the customer can use with the printer. Each label contains a header with general information, a main field containing the cleaning protocol and a border. The design for the border is provided for both label sizes. The cleaning and disinfection system is intended for the international market. The header needs to be customizable since the regulatory requirements vary between nations. The support for multiple languages is also mandatory, therefore the difficulties concerning the character encodings must be addressed.

The development of the cleaning- and disinfection system is in an advanced state and some parts are already imple-

mented. The software architecture is divided into different layers. For example, there is a driver layer that contains all hardware-near drivers. The subsystem layer administrates and controls functional parts of the machine. These layers are sorted by hierarchy and communicate over a central control layer. The new printer will be part of the device layer and rely on the established TCP/IP-driver. Given the amount of different components that need to be managed, a consistent structure is the base of almost every software component in the system. As a consequence the printer component has to implement three basic classes: A manager, a connector and a builder. The manager inherits the statechart shown in Fig. 2 containing all major events necessary to control the component. This class controls the component and establishes connections with other parts of the system. The connector-class implements the functionality of the printer component. The builder creates an instance of the manager-class statically.

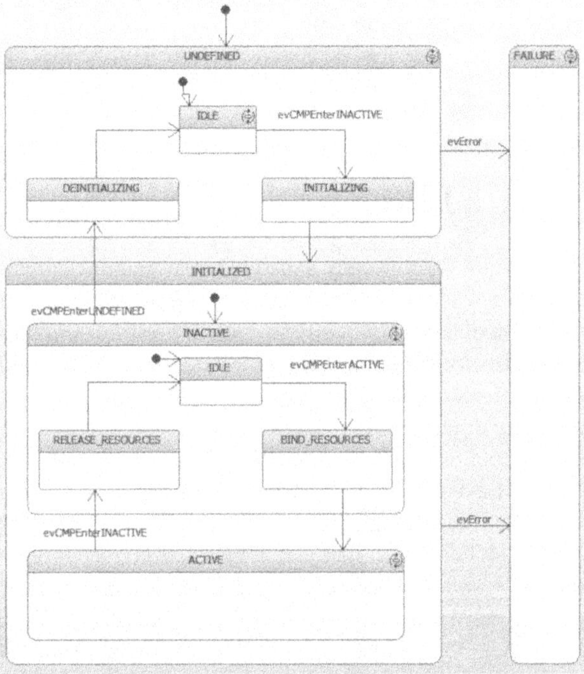

Figure 2: This statechart defines the basic behaviour of almost every component in the software system. The main functionality should be implemented in the ACTIVE-state. (De)-Initializations happen in the corresponding substates of UNDEFINED. The state INACTIVE can be used for preparations necessary for the functioning of the component. The transition to the FAILURE-state is only triggered in case of a critical error. Printed with friendly permission of ITK Engineering GmbH.

3 Concept and Results

The most basic functionality the new component needs to offer is the ability to connect to a physical printer and other components in the software architecture. Considering the demand of an ethernet connection with the OD9, the

TCP/IP protocol will be used. The OD9 does not support other transmission protocols and a TCP driver is already established. To access the driver the address of an instance of the connector class is requested from the manager. After a connector was received from the driver, the connection to the physical printer can be opened.

The printer component will be used by the component that manages the data of the cleaning process. As a consequence the manager has to implement a function that creates an instance of the printer's connector and returns the address. Theoretically the manager could provide an immense amount of different connectors. This would allow one machine to use multiple printers connected to it. In practice only one printer needs to be connected per disinfector and it is unlikely that the machine has many ethernet ports to spare. Usually only one document is printed per cleaning process. Even if a large number of stored protocols is printed, the process will not take much time on the OD9. With just one connector the implementation will be simplified. The manager has just one connection to administrate. Errors like the division of one label to multiple printers or the mixture of different labels can therefore be avoided. It can still happen that multiple commands are sent to the printer simultaneously, a protection with a method like mutex will be necessary and was implemented.

3.1 Printing Procedure

It is unlikely that parameters like port, IP address or character encoding change often after the cleaning device is installed at a location. Therefore, a set of default values will be imported from a configuration file during the initialization. As a result, the connection can be opened without specifying the parameters first. The main functionality is provided in the ACTIVE state. The connection will be opened a short time in advance in the state BIND_RESOURCES. This way the parameters can still be modified during runtime. A return to the previous state is not covered in the current statechart. Staying in BIND_RESOURCES or a new state leaves another problem. The loss of connection should not lead to the loss of the data already sent. All printing commands could be stored until the label is printed or a function to request data from another component could be added. In addition, the component triggering the transition from INACTIVE to ACTIVE is not necessarily the same component that uses the printer. To avoid new timing conditions between other components an easier solution is favored. The transition to ACTIVE is independent of the connection state. Before any text-commands can be transmitted, the component has to check if a label was initialized correctly. The state of the connection can be checked and modified at the same time. A change of connection parameters while the component is active is also imaginable. This way the printing process will consist of two phases. The first phase is a preparation phase that allows the change of parameters, the definition of the header and the selection of the label format. If all settings are valid the main phase allows the transmission of

strings to the printer. Parameters cannot be changed during that phase. After the print was executed or canceled a new preparation phase starts. Instead of a solution with substates and events, this behaviour is implemented with the use of attributes. The reason for this design decision is a more simple implementation of page breaks. Fig. 3 shows the basic concept for the functionality of the label printer.

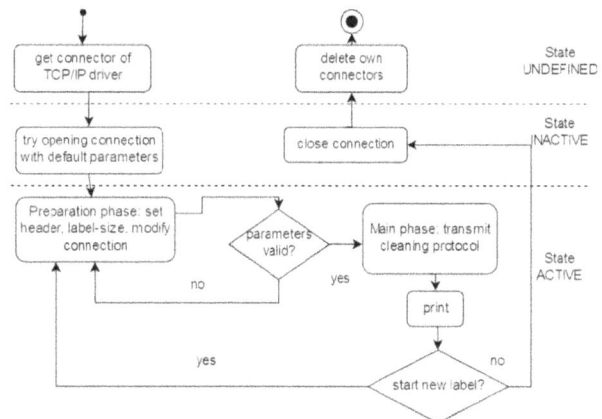

Figure 3: The basic functionality of the printer component with conditions and corresponding states of the manager-class.

3.2 Printing of Text

For a new label settings like character encoding and printing technique are required. After a label is initialized, text, figures, lines etc. can be defined in the image buffer of the OD9. They are printed as soon as the P-command is sent. The OD9 is able to set line breaks on its own, but in tests a few letters got lost from time to time. This might have occured due to slight differences in the positioning of the label material in the printer or an inaccurate defined label width. To increase this function's reliability, line breaks and page breaks will be implemented. Because label and font sizes are known in advance, the amount of characters per line and lines per page is also established. For the correction of the character encoding the iconv-implementation from the GNU C Library is used. This implementation supports a variety of character sets but loads only the conversion tables it needs for an encoding change. This way the only tables stored on the system are the ones used by the cleaning device and the printer. The correction of the encoding could as well be the task of the component managing the data. Charging the printer with that task makes its use more intuitive. The text sent to the component is the text printed on the label. This makes it also needless to store strings with different encodings.

3.3 Header and Label Format

Every label needs a header with general information. The provided border design already separates a portion of the label as header. Unfortunately this section is too small to fit all necessary information inside. Leaving it empty would result in an unpleasant looking protocol. Therefore, the following compromise was made: A small first section contains general information like page- and machine number and a second header is adjustable. The information in the first header is obligatory, therefore its contents must be checked before a main phase is started. A long second header could result in header-only labels and therefore trigger infinite page breaks. A maximum length must be defined depending on the size of the label. The headers are printed again after a page break occured with an incremented page number.

The component needs to support two label formats that only differ in label length. The goal of the implementation is an automatic printing process. Therefore, a method to detect the length of the current label material is desired. Some label printers including the OD9 have sensors for the label length. Reflectivity sensors can find the gap between labels, because of the highly reflective carrier material. Transmission measurements show the black lines between other labels. In case of the OD9 these measurements can only be started manually using the buttons on the printer. This information stays in the printer's storage even after power is removed. The EPL2-programming guide contains a command to read such data over the serial port. Fortunately the OD9 also transmits this data over other ports. Information on other printers concerning this command was not found. Given that a lot of user software uses this command and most printers are connected via USB the OD9 is probably not a singular case. The results are received as a string, there is no guarantee of a consistent formatting. Printers intended for the international market usally call the label length "Length" in their status page. Since the reliability of this function is questionable it remains a comfort feature of the OD9.

4   Conclusion

The goal of the project was the addition of a printer component to an existing software-system. The final implementation can create and print protocol labels like the one in Fig. 4. Two different sizes of label material are supported, on certain printers the size can be determined automatically. For general information a predetermined header and an optional one were implemented. The adjustment of character encodings secures the compability with different languages. Among the compatible character sets are Latin, Hebrew, Cyrillic, Greek, and Turkish characters. The tables also include additional characters for different languages. Due to the differences in character size and the implementation deadline the character sets for Chinese, Korean and Japanese are not yet supported by the software component. With them all character sets supported by the OD9 would be implemented.

Operability of the implementation with other printers with similar specifications to the OD9 is probable, since only basic EPL2-commands were used. Because there were no other label printers available the compatibility with other models remains to be tested.

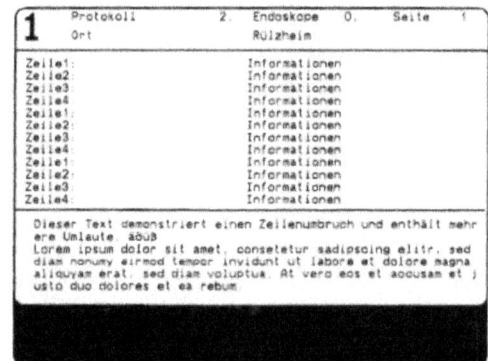

Figure 4: An example label created by the final implementation of the printer component. The second header has its maximum size.

During the development and implementation of the component some possible improvements became apparent. The built-in memory of the printer could be used to store additional conversion tables and allow the support of almost every character set imaginable. The amount of compatible printers could be increased by testing other EPL2-printers and including other printer languages. The user interface of the cleaning system could be updated to allow the modification of printing parameters. This needs changes in other components but the userfriendlyness of the printer would improve greatly. In the current implementation the printing component has no means to communicate with a digital hospital information system. Depending on the regulatory requirements protocols or parts of them might need to be filed. Including bar or qr codes on the labels would enable a fast distribution of information. The need for a solution like that depends on the local hospital information system and the regulatory requirements.

Acknowledgement

The work has been carried out at ITK Engineering GmbH, Rülzheim and supervised by the Institute of Computer Engineering, Universität zu Lübeck.

5   References

[1] Komission für Krankenhaushygene beim Robert Koch-Institut und das Bundesamt für Arzneimittel und Medizinprodukte, *Anforderungen an die Hygene bei der Aufbereitung von Medizinprodukten*. Available: https://www.rki.de/DE/Content/Infekt/Krankenhaushygiene/Kommission/Downloads/Medprod_Rili_2012.pdf ?__blob=publicationFile [last accessed on 2018-01-05]

[2] Zebra Technologies Corporation, *ELTRON® Programming Language*. Available: https://www.zebra.com/content/dam/zebra/manuals/en-us/printer/ep12-pm-en.pdf [last accessed on 2018-01-05].

Preparation of a concept for the provision of effective medical equipment maintenance in the emergency health cluster region of Syria by Dräger

P. Nama [1], F. Spitzenberger [2]

[1]Biomedical Engineering, Fachhochschule Lübeck, palak.nama@stud.fh-luebeck.de

[2] Fachbereich Angewandte Naturwissenschaften, Fachhochschule Lübeck, folker.spitzenberger@fh-luebeck.de

Abstract

The purpose of this paper is firstly to gain knowledge about the critical situation in Syria with regard to the healthcare system and the planning of a maintenance management process for medical devices in this war region. In order to gain the knowledge, a literature search method was defined. Therefore, various search engines were used and different NGO documents were analyzed. The preliminary results for the situational analysis are differentiated between the physical accessibility and the functionality of the hospitals of the region. In addition, major elements of a maintenance management process are introduced.

1 Introduction

Usually, in post crisis and developing regions, medical equipment and devices used in public health facilities are imported. The health sector in these regions faces challenges in ensuring that the equipment is adequately maintained and serviced during the life cycle of the devices.

Maintenance activities include proper installation qualification prior to usage and scheduled testing, calibration and other preventive, corrective and comparative measures to ensure that the devices continue to function properly (operational and performance qualification). The lack of, or inappropriate maintenance may jeopardize the safety and effectiveness of such devices and therefore often negatively impacts individual and public health in these regions.

Best-practice systems for maintenance management depend on adequate technical and regulatory conditions and requirements that build the framework for safe, effective and efficient use of medical equipment in post crisis and developing regions.

However, these conditions/requirements are mostly not well defined and established in these regions and differ a lot from higher income countries, where standards and harmonized regulations on minimum requirements for the maintenance system of medical devices are usually established.

In addition, maintenance management in the context of health relief covers at least two aspects:

1. to establish the necessary technical and regulatory conditions and requirements for effective maintenance by the affected region, and

2. to establish an efficient system for the provision of medical equipment maintenance by the provider/manufacturer of medical equipment that is specifically adapted to post crisis and developing regions.

1.1 Scope and aim of the internship

The first aim of the internship was to identify special needs, challenges and requirement related to the maintenance of medical devices in the emergency health cluster region of Syria: among these indicators are elements such as characteristics of devices relevant for the emergency support, number of people in need, situation and competence level of health and engineering staff, regulatory and technical requirements or the maintenance of relevant medical equipment in Syria.

The second aim was to introduce to and analysis of the currently established process at Dräger, related to the maintenance management for medical devices (manufacturers) are currently includes and covered by the maintenance system established by Dräger.

Due to the confidentiality obligation, companies' internal information were not detailed and listed in this paper. This mainly related to the maintenance process at Dräger. Instead, a general approach of an effective maintenance management system was researched and analysed.

2   Material and Methods

In order to achieve the predetermined results, listed in the introduction, a systematic and detailed literature search was carried out. In the following abstract, these are explained step by step:

The research was split into two parts according to the aims of the paper: the current situation in Syria and the implementation of a maintenance management system.

2.1   Situation analysis in Syria

In order to gain knowledge about Syria the search engines EMBASE, NCIB and SpringerLink were used. In addition to that, topic-related Non-government organization (NGO) documents and data were analyzed. For an analytical and understandable overview and not too exceed the limits of the paper, the following indicators were determined.

1) the documents shouldn't be older than year 2014

2) the document shall content information about: needs and challenge of people in Syria

3) the document shall content information about: accessibility and functionality of hospitals in Syria

4) if possible: already existing maintenance management concepts for medical devices in crisis regions

2.2   Maintenance Management of medical devices

The aim of the next step was to get a deep insight into the topic of maintenance for medical devices. The technical literature was used for this purpose. In order to get a comprehensive information about the subject, maintenance itself is described, followed by an overview of indicators which are necessary to build a maintenance program.

3   Results and Discussion

3.1   Situation analysis in Syria

The conflict in Syria has become one of the largest humanitarian catastrophes in the world. The death toll – most of them are civilians – is estimated at 500.000[1] people till now. An even greater number were wounded, tortured or ill-treated. By the beginning of the year 2018, the war is entering in its seventh year without any hope of an end of the war. Systematic attacks on doctors, staff, hospitals and patients are destroying the Syrian health care system and making even basic medical care for the civilian population almost impossible.

3.1.1   Accessibility of hospitals

Syria is divided into 14 different governorates[2]. Each of these regions are differently occupied and besieged. Due to the critical political situation, it is nearly impossible to get into some regions. The accessibility can be divided into three different parts: accessible, hard-to-reach and not accessible. In accessible regions, the Syrian population and hospital stuff have no disability to physically reach the hospital. Where in hard-to-reach regions, this is made more difficult due for example long distance or some special security situation. In regions that have been classified as inaccessible, residents have little to no chance of reaching the hospitals[3].

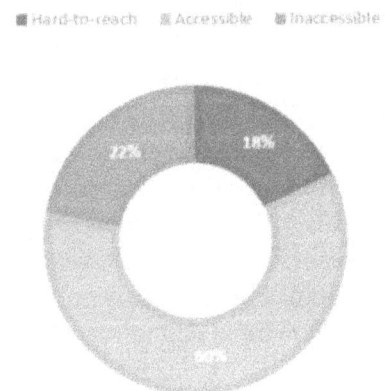

Figure 1: Accessibility Status Dec. 2015.

As illustrated in figure 1, 60 percent of the hospitals are very easy to reach, 18 percent are in hard-to-reach area, where 22 percent cannot be reached at all. The figure although does not provide information about the distribution of the hospitals. According to the HeRam

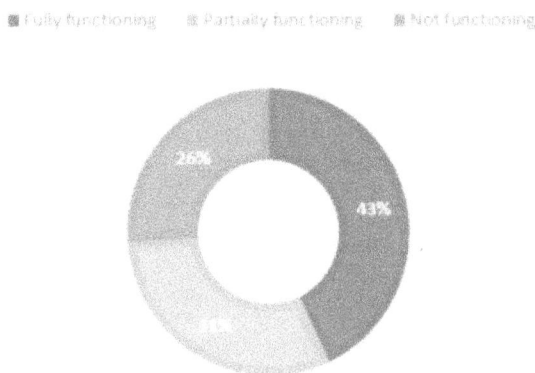

Figure 2: Functionality Status Dec. 2015.

Report, people in regions like Idleb, Dar'a and Rural Damascus have almost no access. Most of the accessible areas includes Damascus, Ar-Raqqa and Al-Hasakeh[2].

3.1.2 Functionality of hospitals

The functionality of hospitals, which includes infrastructure, staffing, equipment, specialties etc, influences the development of the concept. Like the accessibility there are three different types of functionality:

Fully functioning: Hospitals which are fully functioning can provide healthcare with full capacity.

Partially functioning: If the hospitals lacks in equipment or has to deal with shortage of staffing, it is considered as partially functioning.

Not functioning: These hospitals are not able to provide any kind of healthcare services. The reasons can be vary from infrastructure damages to political influences.

According to the HeRam reports 43 percent of the hospitals are fully functioning, 31 percent partially functioning, where 26 percent are not functioning at all (s. figure 2)[2].

Again, the regions Idlib and Dar'a are facing the most difficult situation. Up to 90 percent of all hospitals cannot be function in these regions[2].

3.2 Maintenance Management of medical devices

Medical devices are gaining more and more importance. To ensure that they can run for many years, an effective maintenance management system, which meets all criteria is required.

3.2.1 Definition

According to the EN 13306:2010 Maintenance terminology, maintenance is describes as follows: "Maintenance is the combination of all technical, administrative and managerial actions during the life cycle of an item intended to retain it in, or restore it to, a state in which it can perform the required function"[5]. Adding to above definition, the definition of an efficient maintenance management is "all activities of the management that determine the maintenance objectives, strategies and responsibilities,and implementation of them by such means as maintenance planning, maintenance control, and the improvement of maintenance activities and economics" [5].

3.2.2 Categories of Maintenance

Maintenance itself can be divided into two parts: Preventive and Corrective maintenance[6].

Preventive maintenance (PM):

The goal of the preventive maintenance is to increase the lifespan of a medical device and thus to prevent failure. This kind of maintenance involves scheduled activities like exchange parts, which have finite life (e.g. battery, tubing), cleaning special parts (filter) etc. The frequency of these actions is planned before and in specific intervals by the manufacturer[6].

Corrective Maintenance (CM):

Corrective maintenance is used after a failure has happened. It is also sometimes [6].

3.2.3 Framework of a maintenance management planning

As part of this research, many elements were found which influence the planning of a functional maintenance system. In order to summarize these points, they were categorized and listed as follows:

1) Human resources management:

This sub management focus on primarily on three aspects: number, type and task. The number of the stuff depends on the size and special facilities of the hospitals. Type and tasks depends on each other. In order to determine the people with different skills (technicians, manager etc.), the tasks have to be defined and if necessary, they have to be trained[7,8].

2) Financial resources management:

Another aspect is the financial planning of the maintenance management. This should not only include the human resource but also physical resources like workspace and tool, test equipment. The initial costs can be calculated very easily. The challenge is to estimate the cost during the whole process and to managing the estimated cost[7,8].

3) Operational management:

This part deals mostly with PM. Therefore, it includes on the one hand the scheduling of the maintenance and on the other the required steps, which are dependent on the devices. A very important part is to plan how to document the whole process and the results[8].

4) Performance monitoring and improvement:

In terms of performance monitoring a detailed documentation build the foundation. Based on this, one can measure the performance of the whole management. For a maintenance management to be complete, one should also consider to improve the performance, which is based on the analysis of the documentation. Any improvement is associated with changes, which may incur additional costs. A cost-benefit analysis can help to take the decision[8].

3.3 Discussion

As one can see from the results above, there are different regions of hospital and device accessibility with different influencing factors. In Damascus, for example, the occurring problems are lower than in Idlib. Additional to that, because of the unstable political situation, there is ongoing internal displacement of the people that hinders from hospital accession.

During this research, there is only few evidence for an existing maintenance system for medical devices. There is no specific information about the state of the devices, for example how many devices are in use, how many of them require maintenance, how many have to be repaired, what parts are missing etc. All this information is missing to plan an effective maintenance management. Assumption and speculation has to be made for further planning.

All in all, the concept for the development of a functional medical devices maintenance system has to be flexible enough to adapt to different circumstances. This includes the shortage of technicians as well as the condition of the equipment.

4   Conclusion

In this paper, the current situation of Syria was discussed as well as the major steps fro the development of a maintenance system for medical devices. As emphasized in the discussion, there is not much information about the installed base of medical equipment. In order to develop an efficient maintenance system, the following steps will be taken in the future:

A gap analysis will be performed between processes and practices currently established at Dräger and needs and requirements identified for the emergency health cluster region of Syria. The next step will be to evaluate and define a roadmap for effective medical equipment maintenance.

Acknowledgement

This work was carried out at Drägerwerk AG Co. KGaA and supervised by Silja Dennier.

I would like to express my gratitude and thanks to Prof. Dr. Spitzenberger as well as Prof. Dr. Oliver Rentzsch, who not only gave me the opportunity to be a part of this extraordinary project but also guided and supported me through the process. Secondly, I would like to thanks my supervisors Silja Dennier, Ann-Marie Baasch and Jan Hölterling at the company Dräger for their endless support, kind and understanding spirit. Without their immense knowledge this project would not have been possible.

5   References

[1] Bundeszentrale für poitische Bildung [2018-01-12].

[2] Zaki Mehchy, Rabie Nasser, Khuloud Saba, *Developing health centres and hospitals indices for Syria*. WHO, 2017.

[3] *Overview of hard-to-reach and besieged locations*. OCHA, Jan 2016

[4] *Availability of the Health Resources and Services at Public Hospitals in Syria*. WHO, 2013.

[5] Technical Committee CEN/TC 319 "Maintenance" *EN 13306:2010: Maintenance - Maintenance terminology*.

[6] Adolfo Crespo Marquez *The maintenance management framework*. Springer-Verlag London, 2007, p 69-72

[7] Riccardo Manzini et. al. *Maintenance for Industrial Systems*. Springer-Verlag London, 2010, p. 70

[8] *Medical equipment maintenance programme overview*. WHO, 2011.

Validation of software Stryker Anatomy Analysis Tool

N. Blum [1], A. Petersik [2], and B. Hofstätter [2]

[1] Medizinische Ingenieurwissenschaft, Universität zu Lübeck, nele.blum@student.uni-luebeck.de

[2] Stryker Osteosynthesis, Stryker Trauma GmbH, Kiel, Germany , {andreas.petersik,bernhard.hofstaetter}@stryker.com

Abstract

Stryker Anatomy Analysis Tool (SAAT) is a software tool designed to perform morphometric measurements on a large number of bone models utilizing a database of 3D bone models created from CT scans. It was developed to improve the knowledge of bone morphology and optimize the anatomical compliance of trauma implants. Due to major changes in code structure and library, the new version SAAT 5.3 had to be validated. Beside the validation with an older software version SAAT 4.5, an independent validation with two external software programs were performed. For the validation 3667 individual bones from six different bone types were used. The deviations between the new version SAAT 5.3 and the old version SAAT 4.5 were within an accep range often below 5 %. Only some results have higher deviation due to major changes of landmark definitions. The comparison with the independent software version showed nearly no difference.

1 Introduction

Stryker Anatomy Analysis Tool (SAAT) is part of "Stryker Orthopedics Modeling and Analytics"(SOMA), which is a system composed of different software tools and a 3D bone database containing over 17 000 bones from 63 bone types proprietary of Stryker Trauma GmbH. SOMA was designed to provide means to efficiently quantify bone geometry [1], evaluate, and optimize the anatomical compliance of osteosynthesis implants. Anatomically well contoured implants have been reported to have numerous clinical benefits [2][3]. SAAT was developed by the Technical University of Munich in cooperation with Stryker Trauma GmbH. It was designed to perform morphometric measurements on a large number of bone models utilizing the SOMA 3D bone database. The tool allows different kind of measurements based on geometric constructions on a generic bone shape, which is called "template bone". Employing a correspondence-mapping algorithm developed by Schröder and Gottschling [4], the constructions and measurements were mapped from the template bone onto the individual bone models of the 3D bone database. Additionally, SAAT provides different filter options which could be applied to the 3D bone database like age, gender, size or ethnic group of the patient. Due to some major changes and extension of the functionality of the software as well as changes of the 3D bone database, the new software version 5.3 of SAAT in combination with the new version 5.0 of the 3D bone database needed to be validated.

2 Material and Methods

The validation was basically divided into three parts. In order to proof congruence between past and future analyses, the old software version SAAT 4.5 with the old 3D bone database version 4.6 was compared to the new software version SAAT 5.3 with the new 3D bone database version 5.0. In the second part, the measuring accuracy was validated by the comparison with the independent software Geomagic Control 2014 (3D Systems, Rock Hill, South Carolina, USA), which is a comprehensive metrology software. In the third part, the statistical capabilities, which were newly introduced into SAAT 5.3, were validated by comparison with the independent statistical data analysis software IBM SPSS Statistics 20 (IBM Corporation, Armonk, New York USA).

The accuracy of measurement in SAAT was in general limited by the accuracy of the bone models of the 3D bone database. The 3D bone models were calculated from CT scans with an average resolution of 1 mm. Depending on the clinical CT scanner parameter including pixel size, convolution kernel, slice distance and thickness, interpolation method and the threshold value (or other segmentation parameters) the maximum achievable accuracy of 3D visualizations of CT scans was reported to be one tenth of the scanner resolution [5]. Consequently measurement values provided by SAAT were rounded to one decimal place. Output parameters from Geomagic Control 2014 and IBM SPSS Statistics 20 were also rounded to one decimal place when compared to results by SAAT 5.3.

2.1 Comparison between SAAT 4.5 and SAAT 5.3

The comparison of measurements performed with SAAT 5.3 and SAAT 4.5 was divided into two parts. The measurements were based on user-defined mapped points in one part, and in the second part, the measurements were based on pre-defined landmarks. Mapped points could be set by the user at every position on the surface of the template bone. They were automatically transferred to the corresponding position on every individual bone by employing the mapping algorithm of the SOMA 3D bone database. Landmarks were pre-defined points, lines or planes at specific, anatomical significant locations on the bone (see Table 2) , pre calculated based on constructions implemented in SAAT. They were bone-type specific and were stored in the 3D bone database for every individual bone. The landmarks were re-defined in the new version 5.0 of the 3D bone database, which is accessed by SAAT 5.3. The re-defined landmarks should led to more stable and anatomical relevant analyses of the bone morphology. The redefinitions of the three anatomical planes ("SOMA Transversal Plane ", "SOMA Sagittal Plane ", "SOMA Frontal Plane ") were especially important. They represent the orthogonal coordinate system for each bone, therefore were the reference for the correspondence mapping, and consequently determine the location of mapped points and landmarks. For these reasons, the assessment of the differences between measurements of SAAT 4.5 and SAAT 5.3 had to be performed separately for mapped points and landmarks. The measurements were applied on three different kinds of bones. Measurements based on mapped points were defined on humerus, pelvis and sternum ($N_{Humerus} = 285; N_{Pelvis} = 574; N_{Sternum} = 122$) (Fig.1). Measurements based on landmarks were defined on femur, pelvis and tibia ($N_{Femur} = 1301; N_{Pelvis} = 518; N_{Tibia} = 837$).

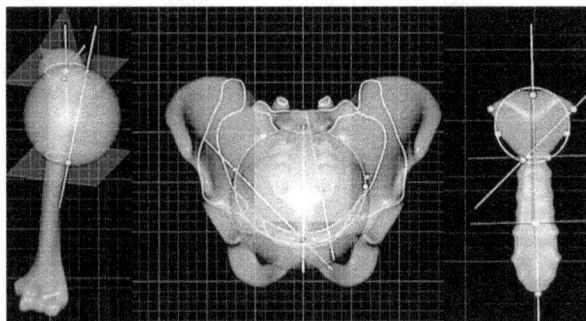

Figure 1: Constructions on humerus, pelvis and sternum for mapped point based comparison.

For the mapped point based comparison, the bone selection was based on two main arguments. Firstly, the chosen bones were frequently used bones for different kinds of evaluations. Secondly, bilateral bones, like pelvis and sternum, led to inconsistent constructions in software version SAAT 4.5. The correct mapping of these bones should be validated in SAAT 5.3. The bones, which were selected for the landmarked-based comparison, were bones with frequently used and important landmarks. The investigated populations were not restricted by age, gender, ethnic group or any other filter options. The constructions for the mapped point based comparison (Fig.1) as well as the landmark based comparison were build up in SAAT 4.5 to create as much inter-dependence as possible. Subsequently, all available

types of measurements from SAAT 4.5 for volumes, length, angles and areas were applied to these constructions as described in Table 1 and Table 2. The constructions and measurements described above were imported into SAAT 5.3. All measurement results of SAAT 5.3 and SAAT 4.5 were compared for each single bone.

Table 1: Measurements used for mapped point based comparison of Humerus (H), Pelvis (P), Sternum (S)

bone	Measurement	
H, P, S	Length	Point to Point Distance
S		Plane to Plane Distance
H, P, S		Circle Radius
H, P, S		Circle Arc Length
H, P		Circle Perimeter
H, P		Shape Contour Perimeter
H	Angle	Angle between two Lines
H,P		Projected Angle between two Lines
S		Angle between Line and Plane
H	Area	Shape Contour Area
S		Circle Area
H, P, S	Volume	Shape Volume
H, P		Sphere Volume

Table 2: Measurements used for landmarks based comparison of Femur (F), Pelvis (P), Tibia (T), landmarks in focus

bone	Landmark	Measurement
F	Femoral Neck Axis (FNAx)	(FNAx/FPAx) -CCD Angle
	Femoral proximal Axis (FPAx)	Point to Line Dist. (L30;FNAx) Anteversion: FNAx projected on TP
	Latera/ Mediall Epicondyle (L5/L6)	Point to Point Dist. (L5/L7,L6/L8)
	Lateral/Medial Condyle (L7/L8)	Point to Point Dist. (L7 to L8)
	Fronta/ Transversal/ Sagitall Plane (FP/TP/SP)	Angle between Line and Plane (FNAx/SP)
	Center Point Spongy 2/10 (L30)	
P	Left/ Right Acebtabular Sphere	Sphere Volume(LAS) / Radius (RAS)
	Left /Right Tubic Tubercle	Point to Point Dist. (L3L/L3R)
	Left Most Lateral Ilium Point	Point to Point Dist.
	Right Most Lateral Ilium Point	(L4L/L4R)
	Left Most Anterosuperior point on sciatic notch (L5L)	Point to Plane Dist. (L5L/SP)
	Right Most Anterosuperior point on sciatic notch (L5R)	Point to Plane Dist. (to SP) (L5R)
	Left Posterior Superior Iliac Spine	Point to Point Distance
	Right Posterior Superior Iliac Spine	(L6L/L6R)
T	Tibial Axis (TAx)	Angle between line and plane
	Transversal Plane (TP)	(TAx/TP)
	Medial Point on Tibia plateu	Point to Line Dist. (L1/TAx) Point to Point Dist. (L1/L2)
	Lateral point on Tibia plateu	Point to Line Dist. (L2/TAx)

2.2 Comparison between SAAT 5.3 and Geomagic Control 2014

The constructions were built up in SAAT 5.3 and reconstructed as similar as possible in Geomagic Control 2014. Because it was not readily possible to measure a large number of individual bones in Geomagic Control 2014, only ten individual bones of each femur, pelvis and the fourth lumbar vertebra were selected randomly from the SOMA 3D bone database. To rebuild the measurements in Geomagic Control 2014 the coordinates of the pre-defined mapped points and landmarks were imported into Geomagic Control 2014 together with the individual bone models from the 3D bone database. All measurements which were compared between SAAT 5.3 and Geomagic Control 2014 are listed in Table 3.

All constructions on the femur, which served as basis of measurements, are shown in Fig. 2.

Table 3: Implemented measurements in Geomagic Control 2014 and SAAT 5.3 for Femur (F) , Fourth Lumbar Vertebra (L04) and Pelvis (P)

bone	Measurement	
F,L04, P	Length	Point to Point Distance
P		Line to Line Distance
F,L04,P		Circle Radius
F, L04	Angle	Angle between Line and Plane
F,L04,P		Angle between two Lines
P		Angle between two Planes
F,L04,P	Area	Shape Contour Area
F,L04,p	Volume	Shape Volume

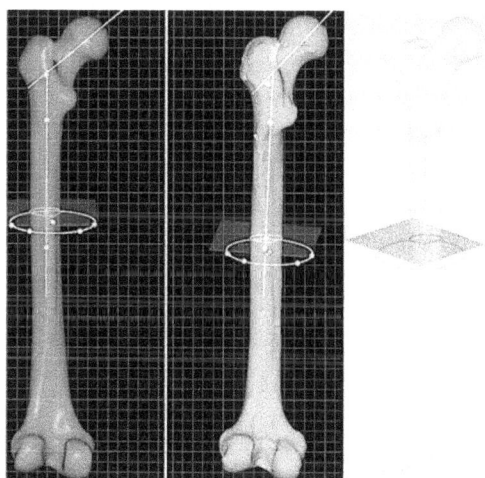

Figure 2: Constructions on template bone and individual femur bone in SAAT 5.3 (left) and on an individual femur bone in Geomagic Control 2014 (right).

2.3 Validation of statistic output

A new feature of SAAT 5.3 was a statistical evaluation of the performed measurements. Beside typical statistical values like mean, standard deviation, minimum and maximum, also percentile based figures (quartiles, median) and parameters resembling box plots were displayed. The measured values as described in section 2.1 were used for the validation of the statistical figures. The values measured by SAAT 5.3 as described in section 2.1 were imported into IBM SPSS Statistics 20. The statistical results calculated by SAAT 5.3 and IBM SPSS Statistics 20 were compared.

3 Results and Discussion

3.1 Comparison between SAAT 4.5 and SAAT 5.3

Comparing the measurements based on mapped point constructions in SAAT 4.5 and SAAT 5.3, the differences were often less than one millimeter, square centimeter, cubic centimeter or degree as shown in Table 4. Only few individual bones showed significantly different output values between the two software and database versions. Differences between measurements were caused for two reasons: Firstly, a few bone models in the new 3D bone database were resegmented from CT scans to improve their accuracy. This led to major changes in the dimensions of the said bones. To minimze the influence of this bones and because the distribution of the measurement results was not a normal distribution, the median was choosen over the mean as the statistical parameter for the evaluation. The second and more important reason, because it affected all bones, was the different correspondence mapping due to changes of the three reference planes described in section 2.1. In some cases especially constructions like shape contour intersections or best-fit sphere constructions based on mapped points led to noticeable differences (Fig.3).The minimization process of the best-fit construction based on a least sqaure solution of the mapped points to the plane. These types of constructions were inherently susceptible to small changes of the underlying mapped points.

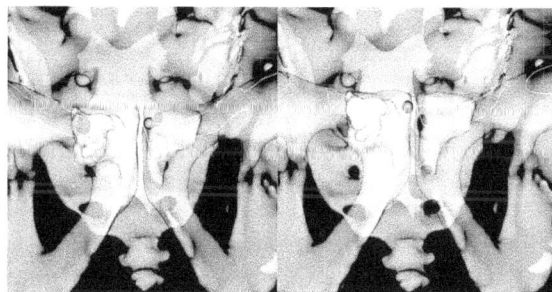

Figure 3: Intersection contour of a best-fit plane based on four mapped points with a pelvis shape. Left: anterior mapped point is located on the left pubic tubercle in SAAT 4.5. Right: the same mapped point is locate on the right pubic tubercle in SAAT 5.3 resulting in a different intersection contour.

Table 4: Differences between mapped point-based measurement results produced by SAAT 4.5 and SAAT 5.3.

Measurement	unit	Median	Interquartile Range
Point to Point Distance	mm	0.07	0.03 to 0.20
Plane to Plane Distance	mm	0.10	0.10 to 0.30
Circle Radius	mm	0.07	0.00 to 0.13
Circle Arc Lengths	mm	0.17	0.07 to 0.34
Circle Perimeter	mm	0.38	0.17 to 0.77
Shape Contour Perimeter	mm	0.63	0.23 to 1.67
Angle between two Lines	o	0.10	0.00 to 0.15
projected Angle between two Lines	o	0.10	0.00 to 0.20
Angle between Line and Plane	o	0.50	0.20 to 0.90
Shape Contour Area	cm^2	0.15	0.05 to 0.30
Circle Area	cm^2	0.00	0.00 to 0.00
Shape Volume	cm^3	0.00	0.00 to 0.00
Sphere Volume	cm^3	0.25	0.10 to 0.53

The deviations of landmark-based measurement results, as shown in Table 5, were notably higher between both SAAT versions compared to the differences of the mapped point-based measurements. The observed deviations between the landmark-based measurements resulted from the changes of the landmarks definitions. These were aimed to improve construction stability and the accuracy of landmark-based morphometric measurements.

Table 5: Differences between landmark-based measurement results produced by SAAT 4.5 and SAAT 5.3

Measurement	unit	Median	Interquartile Range
CCD Angle - (FNAx/FPAx)	o	1.2	0.6 to 2
Point to Line Distance (L30;FNAx)	mm	2.5	1.3 to 3.8
Anteversion: FNAx projected on TP	o	1.5	0.7 to 2.4
Point to Point Dist. (L5/L7)	mm	0.8	0.4 to 1.5
Point to Point Dist. (L6/L8)	mm	6.7	5.6 to 7.8
Point to Point Dist. (L7/L8)	mm	1.1	0.5 to 2.2
Angle betw. Line and Plane(FNAx/SP)	o	2.3	1.5 to 3.1
Sphere Volume(LAS)	cm³	2.8	1.5 to 4.2
Sphere Radius (RAS)	mm	0.4	0.2 to 0.6
Point to Point Distance(L3L/L3R)	mm	0.1	0.1 to 0.3
Point to Point Dist.(L4L/L4R)	mm	0.1	0 to 0.1
Point to Plane Dist. (to SP) (L5L)	mm	1.2	0.6 to 2
Point to Plane Dist. (to SP) (L5R)	mm	1.2	0.6 to 2
Point to Point Dist. (L6L/L6R)	mm	0.1	0.1 to 0.3
Angle betw. Line and Plane(TAx/TP)	o	4.8	3.5 to 6.6
Point to Line Dist. (L1/TAx)	mm	5.1	4.2 to 11.45
Point to Point Dist. (L1/L2)	mm	1.6	0.8 to 2.7
Point to Line Dist. (L2/TAx)	mm	4.3	2.9 to 5.8

3.2 Comparison between SAAT 5.3 and Geomagic

Comparing the values measured in SAAT 5.3 and Geomagic Control 2014 there were only a few small differences between the values and most were exactly zero. All results are listed in Table 6. Possible reasons for small deviations were rounding errors or differences in the calculation algorithms used by SAAT 5.3 and Geomagic Control 2014. For example, different resolutions of shape contours, defining cross section areas, were observed as shown in Fig. 4, where the contour drawn by Geomagic Control 2014 seems more finely resolved than in SAAT 5.3.

Table 6: Differences between measurement results produced by SAAT 5.3 and Geomagic Control 2014

Measurement	unit	Median	Interquartile Range
Point to Point Distance	mm	0.00	0.00 to 0.00
Line to Line Distance	mm	0.00	0.00 to 0.07
Circle Radius	mm	0.00	0.00 to 0.00
Angle between Line and Plane	o	0.00	0.00 to 0.00
Angle between two Lines	o	0.00	0.00 to 0.00
Angle between two Planes	o	0.00	0.00 to 0.00
Shape Contour Area	cm²	0.00	0.00 to 0.05
Shape Volume	cm³	0.00	0.00 to 0.00

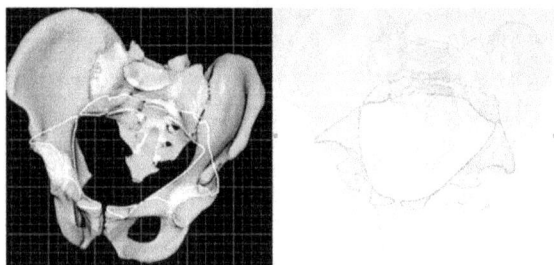

Figure 4: Intersection contour drawn in SAAT 5.3 (left) and in Geomagic Control 2014 (right).

3.3 Validation of statistic output

Comparing the statistical evaluation provided by SAAT 5.3 and IBM SPSS Statistics 20, there was no difference between the calculated values. This was expected, since it was a design requirement of SAAT 5.3 that the statistical output was identical to IBM Statistics SPSS 20.

4 Conclusion

The comparison with the older software version SAAT 4.5 yielded in small deviations due to the redefinitions of the new landmarks and consequently changing the working coordinate system of the mapping algorithm or in case of landmark redefinitions in more stable and anatomical more sensible results. Especially the comparison with the two independent software's Geomagic Control 2014 and IBM SPSS Statistics 20 showed none or negligible deviations from the output results in SAAT 5.3. Results provided by SAAT 5.3 fulfilled all measurement requirements as measurement accuracy and usability for development of osteosynthesis implants and scintific analyses.

Acknowledgement

The work has been carried out at Stryker Osteosynthesis, Stryker Trauma GmbH, Kiel, Germany and supervised by Prof. Handels , Institute of Medizinische Informatik, Universität zu Lübeck.

5 References

[1] M. Hartel, A. Petersik, A. Schmidt, D. Kendoff, J. Nüchtern, J. M. Rueger, W. Lehmann, L. G. Grossterlinden *Determination of Femoral Neck Angle and Torsion Angle Utilizing a Novel Three-Dimensional Modeling and Analytical Technology Based on CT Datasets*. PLoS ONE, Volume 11,2016.

[2] Park AY, DiStefano JG, Nguyen T-Q, Buckley JM, Montgomery WH, Grimsrud CD. *Congruency of scapula locking plates: Implications for implant design*.Am J Orthop.41(April):E53-6., 2012

[3] Ravindra, A., Roebke, A., Goyal, K. S. *Cadaveric Analysis of Proximal Humerus Locking Plate Fit: Contour Mismatch May Lead to Malreduction*. Journal of Orthopaedic Trauma, 31(12), 663-667, 2017

[4] M. Schröder, H. Gottschling, N. Reimers, M. Hauschild, R. Burgkart*Automated Morphometric Analysis of the Femur on Large Anatomical Databases with Highly Accurate Correspondence Detection*. Open Medicine Journal, Volume 1, pp 15-22, 2014.

[5] A. Pommert, U. Tiede, K. H. Höhne, *On the Accuracy of Isosurfaces in Tomographic Volume Visualization*. Medical Image Computing and Computer Assisted Intervention, Proc. MICCAI, Part II Lect. Notes Comput. Sci. 2489, 2002.

Evaluation of a calotte membrane as a part in the execution of system tests of a peritoneal dialysis machine

K. Hüvel[1], S. Sebesta[2], and P. Rostalski[3]

[1] Medizinische Ingenieurwissenschaft, Universität zu Lübeck, kerstin.huevel@student.uni-luebeck.de
[2] Fresenius Medical Care, Schweinfurt, sven.sebesta@fmc-ag.de
[3] Institute for Electrical Engineering in Medicine, Universität zu Lübeck, philipp.rostalski@uni-luebeck.de

Abstract

The peritoneal dialysis machine sleep•safe harmony is an automatic dialysis machine that is used in home dialysis as well as clinical dialysis [1]. The device contains two calotte membranes, which serve to position the device to the sleep•safe set, in order to ensure the coupling to the hydraulic system and thus the exact delivery of dialysis fluid. The sleep•safe set, which is a disposable item, comprises two hemispherical recesses in which the dialysis fluid is either displaced or sucked in by a diaphragm.

The aim of this project is to determine the suitability of the calotte membrane under the specified system conditions. For that purpose the properties of the calotte membrane component to be tested were defined. In order to test the lifetime of the calotte membrane, various system tests were developed and carried out.

1   Introduction

The function of the kidney within the human body is indispensable to life. Therefore people with partial or complete damage of the kidney face a life-threatening situation. The sector of peritoneal dialysis is essential, because as a result of ongoing research patients, despite serious kidney disease, regained major self-dependance during the treatments.

In peritoneal dialysis the elimination of substances obligatory excreted by urine, the balancing of the electrolyte and acid-base balance as well as the removal of fluid via the peritoneum is indispensabled of life. A peritoneal dialysis solution is regularly insert the abdominal cavity via a permanently implanted catheter and, after a certain dwell time in which the material exchange takes place, is discharged via the catheter and replaced by a fresh solution for the next dwell time [2]. The peritoneal dialysis machine sleep•safe harmony is an automatic system that performs the inflow and outflow of the dialysis fluid to drain and detoxify the patient's body. It is used in home dialysis and clinic dialysis [1].

A general and basic requirement of medical devices is the need to be designed in a way that allows resilience to ordinary operating conditions without impairing their characteristics. No user interaction shall lead to endangerment of patients health or safety [3]. Therefore, system tests are performed in order to detect possible errors in hardware or software as early as possible [4].

This project considers the determination of the suitability of the membrane under the specified characteristics. The System Requirement Documents (SRDs), which relate to the entire system and the Component Requirement Documents (CRDs), which specify the requirements for the individual components, are the requirement documents. For this purpose, the Component Requirement Documents (CRD) have been compared to the test parameters of the system test and the properties for the component have been iteratively adjusted according to the findings of the system test.

Within the scope of the requirements engineering process for the peritoneal dialysis system sleep•safe harmony were checked for their validity.

2   Material and Methods

2.1   System Test

When carrying out system tests the entire system is checked with regards to the performances. A system test serves, among other things, as an internal preparatory work of the manufacturer to deliver the system to the customer, the approval or the safety of the product. Furthermore, it supports the structured organization of the remaining project steps, as well as the possible extension of the project as a function of the results obtained [5].

The system was tested towards specification of the characteristics for the system and its architecture. By means of these measures mistakes could be identified. Evidence of the errors contributes to a continuous increase in product quality. For this reason, tests should be carried out at all levels of the system in the initial phases.

One of the main features of the system test is the early detection of errors without physical or economic damage. Such system tests are intended to demonstrate the functionality of the system with regards to certain properties [6]. Functional tests as well as non-functional tests are carried out to prove all functions of the product [5].

The "Alternative Components" project includes various test sequences for the peritoneal dialysis machine sleep•safe harmony, such as treatments, initial outflow, dry running of the calotte membrane and balancing. The various test sequences are carried out in order to test and document the function or answer of the sleep•safe harmony in different states. The test procedures are intended to provide information on the long-term stress of the calotte membrane, the balancing and dosing accuracy of the device, the performance data of the system regarding the heater in connection with the sleep•safe set, the durability of the sleep•safe set during long-term treatments (max. 24 h) and the trouble-free running of the system. In the system test 10 peritoneal dialysis sleep•safe harmony machines are checked for their function according to the defined characteristics.

2.2 Calotte Membrane

The pump is operated by a hydraulically driven two-piston diaphragm pump. This promotes the dialysis solution from the solution bags into the patient and from the patient into the drainage. The disposable sleep•safe set comprises two hemispherical recesses in which the dialysis fluid is either displaced or sucked by the movements of the membrane [4]. The two calotte inserts with the diaphragms are used to connect the hydraulic system to the sleep•safe set. By its use, the exact balancing of the dialysis solution is guaranteed [1].

The function of the calotte membrane is the correct coupling to the hydraulic system in order to ensure accurate delivery of dialysis fluid.

The coupling to the set interlayer is carried out during sleep•safe set pressing by air displacement between the set interlayer and the calotte membrane.

The procedure of coupling the membrane to the set interlayer of the sleep•safe set is as follows, the calotte membrane is fully deflected to the self-tensioning of the set interlayer on one side and by the deflected membrane on the other side. It is necessary to suppress the air between the two components and thus to establish a coupling. This procedure starts at the beginning of the treatment, as soon as the sleep•safe set is inserted into the panel.

Regarding the lifetime of the calotte membrane, it is inevitable that the large deflection has no negative effect on the internal stress of the membrane and also be within the limits specified in the characteristics.

The calotte membrane is required to have a plateau length of minimum 7 mm and maximum 29 mm (length sensor value) over the lifetime in a pressure range of -50 mbar to +50 mbar. The valid range of the calotte membrane for the balancing is shown in Figure 1. In Figure 2 the pressure profile and the plateau of the relevant calotte area are shown for the balancing.

The properties for the lifetime of the calotte membrane with regards to the load changes, taking account of the operating conditions, have to be greater than 2,9 M load changes for the sleep•safe harmony system.

This determination of the lifetime of the calotte membrane

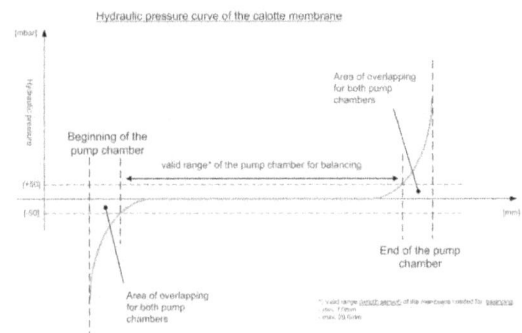

Figure 1: Deflection of the calotte membrane

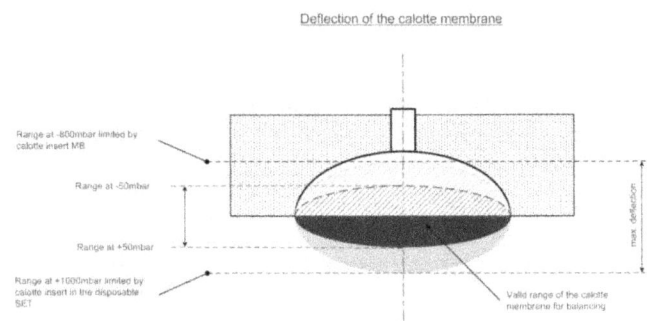

Figure 2: Hydraulic pressure curve of the calotte membrane

was calculated on the basis of the following data:
- Duration of treatment per day: 10 h
- Device Lifetime: 5 years
- Maximum Inflow Volume: 36 liter
- Number of pump cycles per treatment: 1600 strokes

Assuming the mean value at a length sensor value of 18 mm of the calotte membrane:
- Inflow: 22,5 ml per stroke
- Outflow: 22,5 ml per stroke

$$\frac{36000\ ml}{22,5\ \dfrac{ml}{stroke}} = 1600\ strokes \tag{1}$$

Corresponds to 800 strokes per chamber.

Calculation of the lifetime:

$$1600\ \frac{strokes}{treatment} \cdot 365\ \frac{treatment}{year} \cdot 5\ year \tag{2}$$
$$= 2920000\ strokes = 2,92\ M\ strokes$$

3 Results and Discussion

3.1 Length Sensor Values

The length sensors L1 and L2 are installed on the hydraulic pump. The length sensors measure the distance covered by the two pumps in the hydraulic system. They are used to detect the entire travel distance and the current position of the calotte membrane. With the help of the length sensor, the plausibility check of the values takes place as a function of the hydraulic pressure sensors.

By means of the length sensor values and the given cylinder surface of the hydraulic pump, the volume per pump stroke can be calculated. The length sensor values $L1_{start}$, $L1_{end}$, $L2_{start}$ and $L2_{end}$ are provided as log-files by the peritoneal dialysis machine sleep•safe harmony. From the difference $L1_{end}$ and $L1_{start}$, as well as $L2_{end}$ and $L2_{start}$, the length sensor value per pump stroke can be calculated for the respective pump chamber. By multiplying this value with the cylinder surface of the hydraulic pump, the volume per pump stroke can be calculated.

$$\frac{volume}{pump\ stroke}[\mu l] = \frac{length\ sensor\ value}{pump\ stroke} \cdot \frac{cylinder\ surface}{1000} \quad (3)$$

The mean volume per pump stroke is averaged from all treatments performed and is between 22.574 μl and 26.134 μl.

3.2 Initial Outflow

The initial outflow is the first step performed in a treatment with the peritoneal dialysis device sleep•safe harmony. In this case an outflow is initially started in order to determine whether there is some remaining fluid in the patient's peritoneal cavity. With this check overfilling is avoided.
The aim of the test sequence of the initial outflow is to test and document the conveying capacity. The calottes and pumps are stressed, but without heating the water.

3.3 Treatments

Treatments are carried out during the system test to determine the performance data of the system regarding the heaters in connection with the sleep•safe set, the stability of the disposable during long-term treatments (max. 24 h) as well as the smooth operation in the system during insert line and dismantling of the system. All delivered treatment volumes of the inflow and outflow are documented to be able to make statements about the system test.
The evaluation in Figure 3 (a) shows that seven of the system test devices have increased a volume between 25.000 liters and 30.000 liters. Three devices deviate from this volume. This is due to the fact that the devices 7, 9 and 10 are meant to be for the dry running procedure and thus hardly perform any standard treatments.
The total pump strokes per pump can be calculated from the delivered volume by means of the length sensors, from which a mean volume per pump stroke can be calculated. These are shown in Figure 3 (b). The evaluation in Figure 3 (b) shows that most of the devices performed around 500.000 to 600.000 pump strokes per pump.

3.4 Balancing and Dosing

The ratio between the inflow volume and the outflow volume is checked during the test sequence for balancing. The dispensing ratio between the actual value display and the measured value of the scale is controlled during dosing. A standard treatment with six basic cycles is set up for this purpose. A patient bag is used as a patient simulator. The patient line of the sleep•safe set is connected to the patient bag and the bag is placed on an aligned and calibrated scale. After the 4th cycle, data collection is started. During the inflow, the actual values of the scale are noted at the beginning of the inflow and at the end, as well as the target volume indicated by the device and the volume actually conveyed at the end of the inflow. In addition, the existing volume in the patient is noted at the beginning and end of the inflow in the device. During the outflow, the actual values of the scale are recorded again at the beginning and at the end, as well as the patient volume indicated by the device and the patient volume actually conveyed. Similarly, the volumes inside the patient were noted at the beginning and end of the outflow. The balancing and dosing deviation is calculated from the recorded values as described in the formulas below. According to the requirements the balancing must not deviate more than 1% and the defined limit value of the dosing deviation is ± 3%.
Formula for balancing:

$$Balancing\ [\%] = \frac{\frac{S_{End\ Outflow}}{S_{End\ Inflow}}}{100} \quad (4)$$

Formula for dosing:

$$Dosing\ [\%] = \frac{100 - (S_{End\ Outflow} + S_{Starts\ Outflow}) \cdot 100}{Aided\ Patient\ Volume_{End\ Outflow}} \quad (5)$$

$S = Scale\ Actual\ Value$

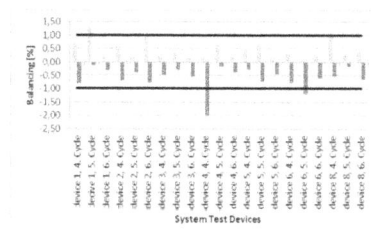

Figure 4: maximum value and minimum value per system test device of cycle 4 to 6 of the balancing over all measurements

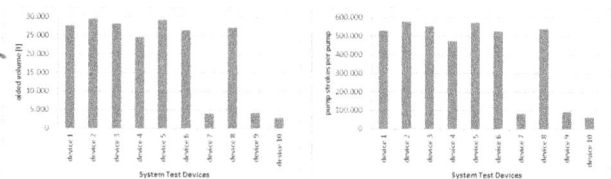

(a) delivered volumes of all treatments data (b) pump strokes per pump of all treatments data

Figure 3: delivered volumens and pump strokes of the treatments data

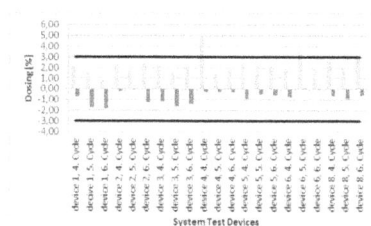

Figure 5: maximum value and minimum value per system test device of cycle 4 to 6 of the dosing over all measurements

Figure 4 shows the results of the balancing measurement. For each system test device for cycle 4 to 6 the average maximum and minimum value is shown in the diagram. The evaluation of the blancing chart in Figure 4 shows that the maximum balancing values as well as the minimum balancing values predominant observes within the balancing deviation. The measured values of the device 1 in the 5th cycle, the value of the device 4 in the 4th cycle and the value of the device 6 in the 5th cycle are outside the permissible balancing deviation. Since this behavior does not continue in the subsequent cycle, this error is most likely due to a measurement error.

Figure 5 shows the results of the dosage measurement. For each system test device for cycle 4 to 6 the average maximum and minimum value is mapped. The evaluation of dosage in Figure 5 also shows that the maximum dosing values as well as the minimum dosing values predominant observes within the dosing deviation of $\pm 3\%$. The value of the device 4 in the 4th cycle and the value of the device 6 in the 5th cycle are outside the permissible dosing deviation. Since this behavior does not continue in the subsequent cycle, this error is most likely due to a measurement error.

Since, apart from the measurement errors, all measurements are within the acceptable measured value deviations, it means that a strong stress on the calotte membranes and the hydraulic pump does not adversely affect the balancing and dosing of the sleep•safe harmony.

3.5 Dry Running of the Calotte Membrane

The dry run is carried out in order to simulate the long-term stress of the calotte membranes and to stress the pumps. Since a sleep•safe set is not suitable for treatment of more than 48 h and therefore no normal treatments can be carried out over the weekend, a dry run is performed. During this test sequence the calotte membranes are cycled.

At present, a test is carried out to check the lifetime of the calotte membranes. For this purpose, the geared motors of the hydraulic pump, which are connected to the calotte membrane via the hydraulic fluid, are operated under stress in dry-running operation. This means that the peritoneal dialysis machine sleep•safe harmony without the sleep•safe set is operated only with the setting of the maximum possible pump cycles and thus the calotte membranes are displaced or sucked by the hydraulic fluid. With the aid of this continuous run the durability of the calotte membranes are tested.

In Figure 6, the pump strokes per pump are shown for the dry-running process. When the pump strokes from the treatments are taken into account, it becomes clear, that the devices have reached the lifetime of 2,92 M strokes.

4   Conclusion

In conclusion, the project „Alternative Components" with its various test sequences for the peritoneal dialysis machine sleep•safe harmony shows that the calotte membranes sustain 5 years of regular usage. Thus a statement can be made about the long-term stress of the calotte membrane and its

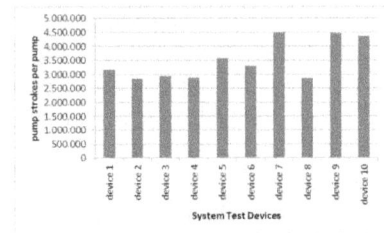

Figure 6: pump strokes per pump of all dry running data

durability for this period. With the additional measurements of the balancing and dosing, all within the scope of the measurement deviations, the calotte membrane holds up a lifetime of 5 years with a daily treatment duration of 10 h without losing their conveying accuracy.

It must also be taken into account that the membranes have been tested harder in the test sequences than they are used in reality, as in the system test the membranes are driven by the hydraulic pump under permanent stress. In reality, the peritoneal dialysis devices run about 10h a day here the dwell time in which the calotte is not claimed is already included.

Acknowledgement

The work has been carried out at Fresenius Medical Care, Schweinfurt and supervised by the Institute for Electrical Engineering in Medicine, Universität zu Lübeck.

5   References

[1] Fresenius Medical Care, *sleep•safe harmony Peritonealdialyse Service Manual.* Auflage 4A-2015. Schweinfurt: 2015.

[2] J. Roob, *Physiologische Grundlagen der Peritonealdialyse und Prinzipien der PD-Verfahren.* In: Nephro Script, Medmedia Verlag und Mediaservice Ges. m. b. H., Graz, p.6, 2013.

[3] N. Leitgeb, *Sicherheit von Medizingeräten, Recht-Risiko-Chancen.* Springer Vieweg, Berlin/Heidelberg, p.91, 2015.

[4] H. M. Sneed, M. Baumgarner and R. Seidl, *Der Systemtest, Von den Anforderungen zum Qualitätsnachweis.* Carl Hanser Verlag, München, p.8, 2012.

[5] G. Wirtz, *Systemtest [Internet].* Bamberg: 2016 [cited 2017 Mar 28]. Available form: URL:http://www.enzyklopaedie-der-wirtschaftsinformatik.de/lexikon/is-management/Systementwicklung/Hauptaktivitaten-der-Systementwicklung/Software-Implementierung/Testen-von-Software/Systemtest

[6] K. Sattler, *Methodik für den Systemtest in der integralen Fahrzeugsicherheit.* Magdeburg: Fakultät für Elektrotechnik und Informationstechnik der Otto-von-Guericke-Universität Magdeburg, 2015

General Client-Server Software Architecture for Medical Robotics (written in C++)

S. Mueller [1]

[1] Medizinische Informatik, Universität zu Lübeck, simon.mueller@student.uni-luebeck.de

Abstract

The Institute for Robotics and Cognitive Systems at the University of Lübeck conducts research projects in the field of medicine involving different robot platforms, and naturally each application requires the development of new software. In order to reduce the redundancy inherent in this task, the institute maintains a software library, sharing and standardizing the implementation of code across different projects. Common steps include connecting to a piece of hardware, communicating between computer and robot platforms, or transmitting messages via an interface. Described are the expansion of the aforementioned library with a general client-server architecture covering different forms of communication. Its features are modular design, asynchronicity, concurrency, optionality in operation, and extensibility.

1 Introduction

The development of a general client-server-robot software architecture was inspired by two specific projects, developed by the Institute for Robotics, University of Lübeck. It is probably a good starting point to briefly outline these two medical robotics applications.

One project involves a lightweight robotic arm intended for the precise manipulation of surgical instruments. The robot interface is running Java code; it opens a TCP port and accepts a list of commands to operate the arm, also sending replies in accordance to the results. The medical application using the arm is implemented in C++, mostly conducting kinematic computations, and is in turn controlled by a client application for text input, to be run locally or remotely, with the latter requiring an additional TCP connection. Commands are processed in sequence and can include additional data as arguments.

The second project always has two connections. The robot side is represented by a haptic robot arm; meaning that in addition to performing movement it can also measure it. The client side consists of a virtual reality simulation, viewed through VR goggles, and implemented in C#. Both sides constantly and simultaneously send messages – the robot side concerning its positioning and forces measured from human manipulation of the haptic arm, and the client side transmits movement directives triggered by certain events in the VR simulation. For example, if a virtual object was hit, the robot arm is supposed to counteract the physical manipulation exacted on it. Similar to the first application, the server program performs kinematic calculations based on the contents of the messages it receives and then sends the results to the other side. This technology is intended to serve as the basis for future implementations in the field of medicine, like training a surgical procedure.

2 Material and Methods

The software library (RobSoL) of the Institute for Robotics and Cognitive Systems at the University of Lübeck is written in C++. It is highly modularized, meaning that ideally all distinct pieces of functionality are separated into modules. Specific applications are then largely assembled from those. Modules of similar functionality, for example handling different types of connections, are supposed to be interchangeable in a derived context, like a server module. Modularization is supposed to make software easier to comprehend, reduce redundancy, and help prevent conflicting code modifications.

C++ is often chosen to implement applications that are performance-critical or operate physical hardware, like robots. C++ 11 introduced significant changes to the core language and standard library; most importantly *multithreading* capability. Other additions, such as *smart pointers*, *uniform initialization*, *lambda functions*, and *move semantics*, may help increase the succinctness and integrity of C++ programs. These terms are placed in italics and briefly explained whenever they were useful in the software implementation. [1] was consulted regarding C++ 11 changes. Online references were of help, in particular [2].

Boost is one of the most commonly employed collections of third-party C++ libraries and has even been influential to C++ core language changes. ASIO (Asynchronous Input/Output) is a Boost library for network and low-level I/O programming, especially asynchronous communication, modeled in an abstract and intuitive way. The implementation of the client-server-robot software architecture was inspired by the chat server example [3]. Specific components of Boost ASIO functionality are described where applicable, put in italics and capitalized, e.g: *I/O-Service*. [4] is a companion to official Boost documentation.

3   Results and Discussion

Basic Model: The medical robotics applications described in the introduction can be generalized to a client-server-robot model. Messages are transmitted through two connections, one from client to server, and another from server to robot.

Consider this example: Client is a terminal console taking input of commands in text form. The client connects to a remote server via an internet connection. Robot is a robotic arm and linked up to the server through a serial port.

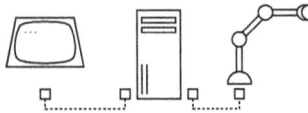

Figure 1: Client-server-robot model

From a software perspective, entering commands into the client console to move the robotic arm is comprised of the following operations. (C)lient, (S)erver, (R)obot. S: open serial port connection to R; open a TCP socket and wait for C. C: connect to S. S: wait to receive messages. C: input of move command and target coordinates; send text message to S. S: receive message, parse it, match command, convert coordinates, call move function, send instructions to serial port. R: Receive move instructions, conduct movement, send acknowledgment to S. S: send message to C that command was executed successfully.

Communication: The transmission of messages in the above examples is handled in a sequential manner – it is always initiated by C and then goes in sequence C–S, S–R, R–S, S–C. Asynchronous communication means that messages can be transmitted at any time; the opposite implies waiting between transmissions, which is a performance bottleneck. Moreover it is perfectly possible to execute synchronous communication within an asynchronous architecture if the server receives messages at any time but stores them in a queue and processes them in order. It is even conceivable that operations triggered by messages are executed concurrently (in overlapping time frames), if the application allows this and possibly with the additional requirement that responses are sent in the same order as initiating messages were received (i.e. pipelining). Consequently, communication ports within the client-server architecture are always operated asynchronously, as are typical network adapters.

This following class diagram depicts the way communication functionality is modularized. Com is an abstract class requiring the implementation of a "connect" function, and providing the means of attaching a message queue. Asyn-Com is another abstract class, this time requesting asynchronous read and write functions, while supplying wrappers of those functions for different kinds of byte arrays. AsioCom is a template class containing all the functionality to run communication ports with the Boost ASIO library. They all operate the same way procedurally, but the template class has to be bound to a specific ASIO *I/O Object*. The library can handle serial port connections as well as

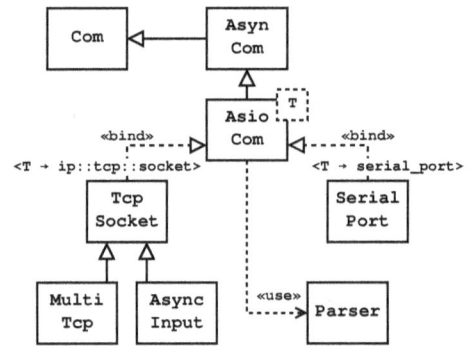

Figure 2: Communication Classes

internet connections using TCP. (UDP is also available but the protocol does not guarantee successful transmission of messages). TcpSocket and SerialPort bind their respective *I/O Object* to AsioCom, while also implementing constructors and a connect-method – the only other ways these two classes vary from one another. MultiTcp is a TCP socket for multi-client applications with two modes: token-based and free-for-all. The difference is, respectively, whether or not messages sent by all clients but the one owning a token are ignored. The server hands out the token if it is requested by a client and available, else the client holding the token possesses it until released, or until the client disconnects from the server. Depending on the message, server responses are sent individually or to all clients. To clarify, client programs would utilize one TcpSocket class object each and the server a MultiTcp class object. For communication between a single client and server, both would utilize TcpSocket class objects, which opens one TCP port and either accept a connection or tries to connect to an open port. Finally, AsyncInput sends user input as messages and prints delivered messages to the screen.

The core piece of any Boost ASIO library implementation is an *I/O Service* object, abstracting the operating system's input/output services and scheduling. *I/O Objects*, as bound and configured by TcpSocket or SerialPort, are linked to an *I/O Service* upon creation. ASIO's *Write_Async* and *Read_Async* methods take as arguments an *I/O object*, a completion handler, and either a message to send or a buffer to fill with a received message. These asynchronous method calls return immediately, but once the read or write operation actually finishes, the completion handler is processed (also termed callback functions.) The *I/O Service* object has a run method, which exits only if no completion handlers of previous asynchronous method calls are left to be processed. The connect methods of either class template binding class start accepting a connection or attempt to establish a connection, both of which are asynchronous method calls as well. So one way of running an *I/O Service* is to have the accept/connect completion handlers perform a *Read_Async* operation, which in turn executes another *Read_Async* call in its own completion handler. If an error occurs, the loop is broken, so the *I/O Service* automatically stops, until a new connection is established. Simultaneously to the execution of this *Read_Async* loop, *Write_Async* calls can be performed. Reading is done in one of two modes, either collecting messages or message fragments in a fixed-length buffer and sorting through those with a Parser object (explained in the next section), or filling a variable-length buffer until a given message delimiter has been transmitted. The second mode was necessary due to compatibility issues with non-ASIO TCP sockets. MultiTcp runs an Accept_Async loop, continuing to accept connection attempts by clients and grouping them in a Session subclass (not depicted), containing the Read_Async loop.

C++ 11 introduces *lambda functions*, also referred to as *anonymous functions* because they are unnamed. As usual however the accept parameters, called a capture, and may include return statements. Completion handlers can be defined as *lambda functions*, with the advantage that they can then be written within the asynchronous function call, instead of referring to another named function. So operations to be performed when such a call finishes are defined in the same context that they were started. Another C++11 feature, *Smart pointers* are mostly used to share ownership over a single instance of an object. The object is automatically destructed once no more references to it exist. Copying *smart pointers* only increases the reference counter. A class can create a *smart pointer* of itself. In the case of MultiTcp, if a Session object captures a reference to itself in the completion handler of the *Read_Async* loop, then the session automatically ends (is destructed), once all clients have disconnected.

Messages: Both TCP sockets and serial ports exchange messages in the form of encoded byte arrays – ASCII characters most commonly. Messages can be created as dynamically allocated arrays. They trigger function calls on the server, so they are arranged in lists of MessageFunctions. A MsgFunc contains the following information: 1) The command name as a character string; unique for each message. 2) A group ID the message belongs to; also unique per group. 3) A pointer to a member function called to handle a specific command. These functions and MsgFunc-lists are defined in the same class, as explained below. 4) A polymorphic wrapper of a static Parser object function. Parser objects transform message arguments expressed as a character string into data types used by the robot function. Parser has a static parse function, so no instantiation is required. By using the polymorphic wrapper, it is also possible to define an *anonymous function* for parsing. 5) A list of message options. 6) A binary representation automatically calculated from the previous list, which allows checking for message options in the form of a more efficient bitwise operation. 7) The length of the command string; calculated automatically.

The MsgFunc-list is included as a member variable of a so-called MessageFunctionIntermediary class (MFI), named for mediating between received messages and robot function calls. With *uniform initialization* this list can be created in the definition of its MFI class (all in one place), and additionally not all the parameters detailed above need to be set for any given message. Instead the initialization of a MsgFunc entry can stop after the group ID. (Option "pass-

through" sends the argument of a message directly to the other communications side without triggering a function). A MFI always includes an integer member variable identifying itself, which will be automatically used for the message group ID if it was not initialized. Last but not least, the MFI class must provide one member function for each message command (not "passed through"). These all have the same signature – the argument string of the message as a parameter – and return a MsgFuncReturn struct containing the reply string along with an error code.

Typically, the functionality of a given robot application is implemented in its own class. Thus the MFI may include this class as a member object, with the MFI functions solely calling their associated robot functions, with a conversion of the argument string to a data type required for the call (using the given Parser), and generating a MsgFuncReturn object from the result.

MFIs can be concatenated. For example, it might make sense to combine an MFI of generic robot commands with an MFI of application-specific commands. All MFIs are derived from a base class containing the handle message routine and an empty MsgFunc-list. *Move semantics* are used to transfer all MsgFunc-lists to the base MFI, which is much more efficient than copying them, but will leave the source list defunct. However, by using pointers to member functions, the base MFI will still execute the correct function of the source MFI, and inheritance between parent and child MFIs is still supported.

The syntax of MsgFunc-list entries is quite complex, and creating functions in MFIs for every message is repetitive. Thus MFI header and source files can be generated from a text file written in a much simpler format. This is done with a separate program, called within the RobSoL build process (using CMake) whenever the text file is modified.

In order to increase the speed of matching the entry in the MsgFunc-list to the command in a given message, the list is sorted by command length first and in alphabetical order second, when the server program is first started, and a look-up table is created referencing the starting index and number of MsgFunc-entries by command length to reduce the number of entries any incoming message is matched against. Binary search is utilized, helped by the fact that ASCII characters are also numbers. Even more efficient is naming commands after their index in the MsgFunc-list, which exists as a message option. This approach is only sensible if the list is never changed manually but extended with the program mentioned above.

Concurrency and Mutual Exclusion: Functions triggered by asynchronous messages may not always be executed concurrently, for example when access to a shared resource is mutually exclusive. This is also known as a race condition. C++11 enables concurrency via *threads*. A two-sided server handles incoming messages in two *threads*, one per AsynCom object. A *mutex* is required if any given MFI function presents a race condition. To enable shared access to a queue of messages between the server and a communication object, the template class MutexQueue was created, with every queue operation protected by a *lock*.

A function call receiving its own copy of data sent as an argument has unique access to this copy. Referencing data is quicker than copying it, but references are not thread-safe. *Move semantics* are useful here again, because they are as efficient as references in terms of passing data to functions, and in the context of a MFI function call the data is no longer needed outside of the function after entering it.

The AsioCom class can be run concurrently without requiring a *mutex*, because asynchronous function calls can be sent to the *I/O Service* via a post method, which always processes completion handler after already existing ones. This is only the case, however, if the *I/O Service* is run in a single *thread*. ASIO provides *Strand*, which is basically a *thread* within *I/O Service*. So reading and writing can be performed in two distinct *Strands*, making them thread-safe again. A server with two connections has four concurrent operations maintaining the communication, and can potentially process multiple robot functions simultaneously. Whether message functions are executed concurrently (in individual threads) is controlled by message options. It is possible that the same MFI function is triggered on the server by simultaneously arriving messages from different communication sides, so by default these calls are processed using the monitor pattern. Monitor is a software pattern introduced by Herb Sutter in 2012, based on multiple C++ 11 features [5]. As a template class, it can be applied to any given object, and basically what it allows is arbitrary *lambda functions* to be performed with that object, guaranteeing thread-safety via an internal *mutex* and *lock*.

RobotServer: The server module operates a chosen number of AsynCom objects, each one representing a communication side allowed to send "initiating" messages, meaning that they trigger MFI function calls on the server. Which communcation object's messages cause responses to be sent to what other objects also has to be configured. Received messages are placed in one MutexQueue per connection, filled by AsnyCom, emptied by RobotServer, and held as a *smart pointer* between the two. The server operates additional internal queues for scheduling as defined by message options, like different priorities or concurrent execution allowed between groups of messages. Since RobotServer must handle command messages regarding its own operation, it is itself an MFI.

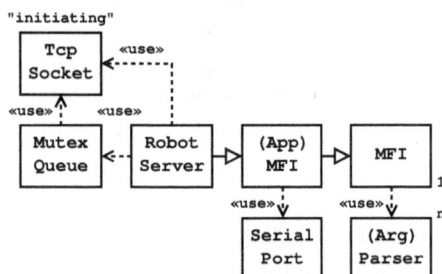

Figure 3: Example "pseudo-synchronous" RobotServer

Robot functionality addressed via the MFI can be as complex as desired. For example, it could operate additional Com objects. This is how the pseudo-synchronous com-

munication described in the basic model is implemented, as shown in the class diagram above. Only the client side maintains an "initiating" communication with the server, and the robot object addressed via the MFI in turn connects to the hardware on its own. AppMFI and ArgParser represent application-specific objects.

4 Conclusion

The client-server-robot software architecture was developed for the two projects described in the introduction. Future implementations by other developers depend on the comprehensibility and extensibility of the additions to the library. These two aspects contrast each other in a way, as it may not be obvious how to best adapt versatile modules to a particular application. Different ways of enabling concurrent operations also present different ways of messing up the correct flow of a program. Hopefully these issues are alleviated to some degree by extensive documentation that accompanied the work.

Software design is more of a craft than a science, but it is still beneficial to stay up to date on new developments in the field and learn better approaches to old problems. Considering this, the implementation presented here could never be deemed optimal. Concurrent processing of asynchronously transmitted messages has only been tested with one application and probably requires fine-tuning, particular the scheduling performed internally by the server. The freedom allowed in creating message functions has lead to confusing syntax, requiring a separate program to simplify the process. Message handling might be improved by using a different data structure like a hash map. Still, aspects of this implementation may hopefully be of interest to readers involved in the development of C++ software.

Acknowledgement

The work has been carried out at the Institute for Robotics and Cognitive Systems, University of Lübeck, and was supervised by F. Ernst and P. Jauer.

5 References

[1] B. Stroustrup, *The C++ Programming Language*. Addison-Wesley, Boston, 2013.

[2] *C++ Reference*. http://cppreference.com.

[3] C. Kohlhoff, *Boost ASIO Examples*. http://www.boost.org/doc/libs/1_66_0/doc/html/boost_asio/examples.html.

[4] B. Schäling, *The Boost C++ Libraries*. XML Press, Laguna Hills, 2014.

[5] H. Sutter presents C++ Concurrency, *C++ and Beyond 2012*. https://channel9.msdn.com/Shows/Going+Deep/C-and-Beyond-2012-Herb-Sutter-Concurrency-and-Parallelism.

Work instruction for replacement deliveries in case of commercial DEFective On Arrival

M. L. Wiegel [1], C. Oelze [2], J. Meyer [2] and M. Leucker [3]

[1] Medizinische Ingenieurwissenschaft, Universität zu Lübeck, maxi.wiegel@student.uni-luebeck.de
[2] Q&R DXR Operations, Philips Healthcare Hamburg, {Catherine.Oelze, jens-meyer}@philips.com
[3] Institute for Software Engineering and Programming Languages, Universität zu Lübeck, leucker@isp.uni-luebeck.de

Abstract

The purpose of this paper is to describe the process of replacement deliveries at Philips Healthcare Hamburg at the business unit Diagnostic X-Ray (DXR) in case of commercial DEFective On Arrival (DEFOA). This process needs to be defined for a standardized and efficient operation for all involved departments. Currently a standard work instruction for this process is missing. The aim of this work is therefore to establish a work instruction using the PDCA-Cycle and exemplify the logistical flow with the aid of a flow chart applicable for all employees at Hamburg DXR.

1 Introduction

Philips Healthcare Hamburg is manufacturer for a variety of X-ray systems, tubes and components. There are two main business units within Philips Healthcare Hamburg: GTC, which stands for generators, tubes and components as well as DXR, an abbreviation for diagnostic X-ray. This work focuses only on the processes and the procedures at DXR. In the following Philips Healthcare Hamburg DXR is denoted as DXR.

The quality assurance is an important part in the development and in the manufacturing from medical devices. To guarantee that the products do have high quality, the processes must continuously be supervised and procedures as well as work instructions need to be defined. Therefore, the Food & Drug Administration (FDA) as well as the International Organization for Standardization (ISO) are standards applicable at DXR to ensure safety and reliability for medical devices [1], [2], [3].

The ISO defines regulations for Quality Management Systems (QMS), medical devices and requirements for regulatory purposes. These requirements are applicable to all organizations, which develop, manufacture or install medical devices. According to the ISO 13485:2016 a quality management system describes how the organization directs and controls all activities to achieve the intended results. It consists of the planning, documents or records and processes. Additionally, the use of standards lead to the fulfillment of customer requirements [3].

The FDA is the US authority for food and drug safety. Regulation 21 CFR Part 820 of the FDA defines a current good manufacturing practice for medical devices. Due to the FDA each manufacturer shall establish quality system procedures as well as instructions. If one follows their regulations, medical devices can be exported to the US. According to the FDA, quality is defined as the totality of characteristics of a device which fulfill all needed requirements, including safety and performance. In case of nonconformity, the quality of a product is not met which means that a specified requirement is not fulfilled [1].

Deviation in quality can occur at any time of the process, like the manufacturing, delivery or installation. If this takes place during delivery or installation these defects are defined as DEFective On Arrival (DEFOA) [4], [5].

First priority of a DEFOA case is to provide the customer with the required material. It currently lacks in a standardized work instruction for the commercial DEFOA replacement delivery applicable at DXR.

Therefore, the focus of this work is the replacement delivery in case of commercial DEFOA, which is under examination and the creation of a work instruction by applying the PDCA-Cycle. Furthermore, the performance of the work instruction will be analyzed. It should be clarified if the implementation of a work instruction leads to time saving in the processing of commercial DEFOA cases.

2 Material and Methods

All nonconforming products (NCs) need to be controlled because of that each manufacturer shall maintain and establish procedures for these cases [1], [2], [3]. The *Control of Nonconforming Product* procedure at DXR describes the identification, documentation, evaluation, disposition and segregation of such NCs. Furthermore it explains the execution of DEFOAs, which is described below [5].

2.1 DEFective On Arrival

The abbreviation DEFOA is a basic element at DXR and had increasingly gained in importance in recent years. In the past defects on arrival were denoted as NCs but for a more exact evaluation of the root of a defect the term DEFOA has been established. A DEFOA component can be a missing screw, up to a defective X-ray system. The occurrence of DEFOA has an impact on reliability and leads to customer complaints.

There are two classes of DEFOA, the commercial DEFOA and the service DEFOA. Figure 1 illustrates the relation of these classes [4], [5].

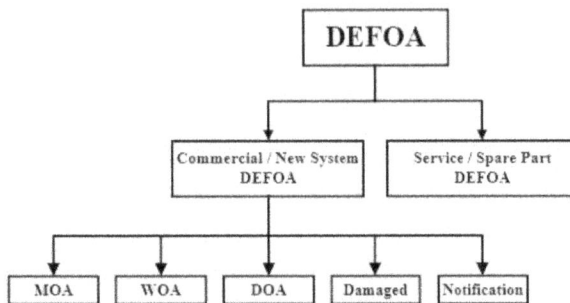

Figure 1: Tree diagramm of DEFOA based on [4].

A service DEFOA (or spare part DEFOA) is a medical component delivered that does not fix the customer's problem, which occurred due to a defect in the part delivery or part quality process. They are not related to installations but are related to installed base. This means that the X-ray system is already installed at the customer's but there is a problem with a component. A new part, also called spare part, is ordered and is defect during arrival or installation.

In contrast to this a commercial DEFOA (or new system DEFOA) describes an unplanned defective, missing, damaged or wrong item, delivered at the end-customer or warehouse. That kind of DEFOA occurs before (first) installation and does not meet the specification [4], [5].

There are five different categories defined for new system DEFOA. Missing on Arrival (MOA) describes a part from an order that is missing on arrival at warehouse or end-customer and is not on back order. The Defect On Arrival and/or Dead On Arrival (DOA) describes a part that does not function on arrival at warehouse or end-customer. These parts do not show any visual signs of damage. In comparison, there is the term Damaged on Arrival, an item that may not function on arrival at warehouse or end-customer and which shows visual signs of damage. A Wrong on Arrival (WOA) describes a item, which was not ordered from warehouse or end-customer. Another category of DEFOA is called Notification. The correct item was delivered and the problem was resolved on site. Nevertheless, there is a remark regarding quality like no return, no replacement needed, no credit and no claim.

A DEFOA can only be a commercial or a service DE-FOA. The separation point between both is the Customer

Acceptance Date (CAD). That date indicates the transfer of the whole system between the project manager and the customer itself. Before the CAD it is described as a commercial DEFOA and after that no commercial DEFOA can be claimed [4], [5].

There is an employee at DXR who processes commercial as well as service DEFOAs. This DXR DEFOA Manager prepares and reviews the DEFOA feedback, identifies DEFOA classes and forwards to responsible for analysis and finally coordinates rework activities on materials returned as DEFOAs [4], [5]. The term rework describes an action taken on a nonconforming product to fulfill the specified requirements [1], [3]. There are two different software tools for the processing of DEFOAs. The Customer Defect Report (CDR) tool is used in case of service DEFOAs and the One Event Management System (OneEMS) tool finds application for commercial DEFOAs [4], [5].

2.2 PDCA-Cycle

In order to optimize the processes in a company, there are various tools to achieve defined targets. The problem solving Plan-Do-Check-Act (PDCA) cycle, a tool for continuous quality improvement, is a basic element of the QMS, which is described within the ISO and was applied for the development of the work instruction (as shown in Fig. 2).

There is the need for a planning before the implementation of the process. The Plan-Phase includes the detection of potential improvements for a current process, the analysis of the present situation and the developing of a new concept. Knowing what type of output is desired helps to develop a plan. The Do-Phase allows the Plan to be enacted. During the Check-Phase the outcomes from the Do-Phase are analyzed. If the Check-Phases shows that there is an improvement for the current process the new process will be implemented and becomes a new standard of how the company should act. If it does not lead to an improvement the existing standard will remain (Act-Phase) or the PDCA-Cycle could be applied again [6].

2.2.1 Plan-Phase

Due to outstanding issues in performing various steps, it was decided to record the process of replacement delivery in case of commercial DEFOA. Therefore, meetings with employees from different departments involved in the process were prepared in advance and carried out. These departments were Production Control, Catalog Team, Quality and Regulatory (Q&R), Cost Engineering, Production, Shipping and Order Desk. The meetings were essential to find out which department is in charge of proceeding steps for replacement deliveries and in which order the steps are conducted. All employees decided to create a work instruction which will be implemented to the *Tagesarchive*, a database were only DXR documents are archived.

2.2.2 Do-Phase

A work instruction describes step-by-step how to perform and accomplish a specific task. It consists of a flow chart and corresponding short sentences with the main information. The flow chart itself is one of the *Seven Basic Tools Of Quality* (Q7). The Q7 are a fixed set of graphical techniques identified as being most helpful in the correction of issues related to quality. A flow chart is a diagram that represents a process, which shows the different steps with boxes and their order by connecting them with arrows. In a flow chart rectangular boxes stand for actions that need to be done and diamond-shaped boxes are used for logical decisions. The diamond-shaped boxes have therefore one arrow for input and two arrows for output with the decision yes or no. In order to create the flow chart the software tool *Microsoft Visio 2016* is applied [7].

Figure 2: Representation of the PDCA-Cycle based on [6].

3 Results and Discussion

An initial flow chart has been created by following the Plan- and Do-Phase of the PDCA-Cycle. This flow chart describes and visualizes the replacement delivery in nineteen major steps including five decisions to be made, which were set in the meetings. There are two main departments responsible for the logistical replacement delivery in case of commercial DEFOA.
The DXR DEFOA Manager, is responsible for the evaluation of the DEFOA case. An employee from Order Desk is responsible for the replacement delivery provision. The other involved departments of DXR also have an important role performing replacement deliveries.

Figure 3 shows the initial created flow chart. The replacement delivery starts when a new DEFOA case occurs in OneEMS, which is created from Keymarket and is reported to the DXR DEFOA Manager as replacement required (step 1). Next the DXR DEFOA Manager screens the DEFOA case for complete information (step 2) and has to check if the DEFOA case is a field replaceable unit (FRU) and therefore can be replaced on-site (step 3). In this case the Keymarket checks when the component is available via the service parts supply (SPS), a department off-side of DXR, which supplies all service parts (step 4). In case of a FRU and availability, the Keymarket needs to

order at SPS and a delivery from DXR is rejected (step 5). The DEFOA case can be closed afterwards.

If the component is not available via SPS (step 4) the Order Desk checks if the component is deliverable via DXR within a short time frame (step 6). If no step 5 is followed, otherwise Order Desk checks if the material is a material, which is handled in Production and therefore can be replaced directly (step 7). In case of no FRU (step 3) step 7 needs to be followed. If step 7 is answered with yes the Order Desk organizes the delivery of the needed component via the Shipping department to the customer (step 18) and updates the data in OneEMS (step 19). If step 7 is answered with no Production Control identifies the deliverable material (step 8) and Order Desk checks, if the material is an off-the-shelf component (step 9). If yes, step 18 and 19 are followed and the DEFOA case can be closed.

If the material is not an off-the-shelf component the Order Desk initiates and monitors following steps of producing the replacement delivery (step 10). First, Production Control defines a parts list for the missing component (step 11) and the Catalog Team creates a unique number for the material, to retrace this (step 12). Quality & Regulatory release the unique number, which was created by the Catalog Team (step 13). Cost Engineering is accountable for the pricing of the component (step 14). Production Control releases the production order (step 15) and Production manufactures the requested material and prepares the shipment (step 16). The Packaging department prepares the component for shipping (step 17), step 18 and 19 are followed and the DEFOA case can be closed.

The Check- and Act-Phase of the PDCA-Cycle could not be applied yet. There are still outstanding issues in some steps and therefore the work instruction is not implemented yet. Another meeting is planned to present and discuss the initial flow chart. The Check-Phase can be investigated when the last steps are clarified. The work instruction will be applied and the time by performing a replacement delivery will be under examination by measuring the time of single steps. If the new work instruction is time-effective the Act-Phase can be executed and the work instruction will be archived in the *Tagesarchiv* of DXR. Otherwise, the time-consuming steps need to be detected and analyzed.

4 Conclusion

It is shown that the replacement delivery in case of commercial DEFOA is an interdepartmental process and therefore a work instruction needs to be established. There are still some actions and steps inaccurate which need to be determined before releasing the final version of this work instruction to the *Tagesarchiv* of DXR. The question of which remains open is if the implemented work instruction leads to time saving in the processing of DEFOA cases. This will be examined in the near future.

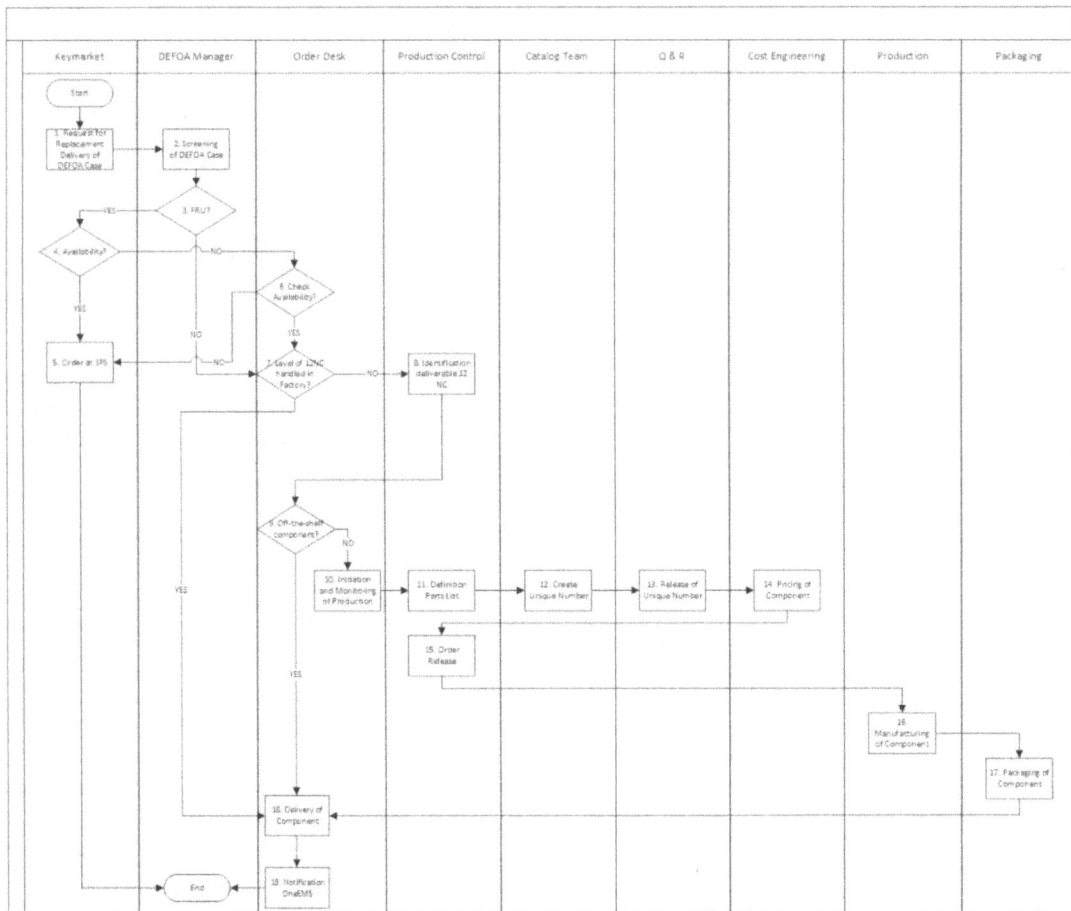

Figure 3: Initial flow chart for commercial DEFOA replacement delivery.

Acknowledgement

The work has been carried out at Philips Healthcare Hamburg and supervised by the Institute for Software Engineering and Programming Languages, Universität zu Lübeck.

5 References

[1] FDA, *Mini Handbook: The Code of Federal Regulations Titel 21 - Food and Drugs*. GMP Publications, USA, 2014.

[2] J. Ward, *The quality system compendium (CGMP Requirements and Industry Practice)*. 3rd edition, Association for the Advancement of Medical Instrumentation (AAMI), 2015. ISBN 1-57020-581-7.

[3] ISO 13485:2016, *Medical devices - A practical guide*. © ISO, Switzerland, 2017. ISBN 978-92-67-10774-5.

[4] Philips, *Commercial DEFOA Process*. YA116, Rev. 3, 2017. Internal document, Philips Healthcare. Available: https://share-intra.philips.com/:w:/r/sites/STS20140805104517/_layouts/15/WopiFrame.aspx?sourcedoc=%7BB21769D4-DEF4-4FA4-B45F-DFB145FACAE4%7D&file=YCO-050-XXX-SOP-Commercial%20DEFO A%20v1.0.docx&action=default&DefaultItemOpen=1 [last accessed 2018-01-10].

[5] J. Meyer, *Control of Nonconforming Products*. XI-DX R-090-2001, Rev. 8, 2017. Internal document, Philips Healthcare Hamburg. Available: http://qms.de.ms.philips.com/qms/visual.php?xml=qms&xsl=view_qms&block=3&qmsfd=2 [last accessed 2018-01-10].

[6] T. Pfeifer, *Qualitätsmanagement: Strategien, Methoden, Techniken*. 2nd edition, Carl-Hanser-Verlag GmbH & Co. KG, München, 1996. ISBN 10-3446185798, ISBN 13-9783446185791.

[7] V. Omachonu and J. Ross, *Principles of Total Quality*. 3rd edition, CRC Press LLC, Florida, 2004. ISBN 13-978-1574443264, ISBN 10-1574443267.

Improvement of the recall process at Olympus Winter & Ibe GmbH

D. Zahnow [1], Ch. Hübner [2], and D. Wladow [3]

[1] Medizinische Ingenieurwissenschaft, Universität zu Lübeck, daniela.zahnow@student.uni-luebeck.de

[2] Institute of Physics, Universität zu Lübeck, huebner@physik.uni-luebeck.de

[3] Quality and Regulatory Affairs - Complaint Management & Vigilance, Olympus Winter & Ibe GmbH, daniel.wladow@olympus-oste.eu

Abstract

The recall process of medical devices is a delicate subject in companies, that produce medical devices. It is closely connected to different extensive regulatory requirements. A considerable part is detailed in the respective quality system regulation; in Europe in the DIN EN ISO 13485, in the USA in CFR title 21 part 820 and in Japan the PAL. It is crucial to implement the recall process optimally, to meet the regulatory requirements and to ensure an invariant quality. During the internship at Olympus Winter & Ibe GmbH (short: OWI or OSTE for Olympus Surgical Technologies Europe, which is the umbrella organisation for European production and development centres in Europe with OWI as headquarter.) the current process was investigated and analysed. Certain problems were identified and possible solutions were evaluated. The analysis has shown that the main problem can be found inside the internal and external communication. Furthermore, the process is handled in different departments. For this reason, the standardisation is rather complicated.

1 Introduction

During the life cycle of a medical device and with it in the scope of the recall process, norms, laws and regulations have to be complied. In the quality system of the European Union, the DIN EN ISO:9001 (short: ISO:9001) is the most significant norm. This norm constitutes the minimum requirements on a quality management system [1]. The DIN EN ISO:13485 (short: ISO:13485) also deals with the requirements of a quality management system, especially with the individual phases of the life cycle of a medical device [2]. This implies the development, production, storage and distribution, installation, maintenance and definitive removal from servicing and disposal.

As well as the European regulations, the American regulations are also important. The competent authority in America is the Food and Drug Administration (FDA). Amongst other, the FDA regulates the approval and market surveillance of medical devices in human and veterinary medicine. The most significant law is the Code of Federal Regulations (CFR). Part 820 of the CFR represents the requirements concerning a quality management system, which is the counterpart to the ISO:13485 of the European Union. The recall process is described in part 810.

The rest of the world has norms, laws and regulations regarding the quality management system and especially the recall process. However, these also follow the FDA regulations. Consequently, it can be assumed that the FDA regulations are most important for the improvement of the recall process at OSTE.

A recall is a correction or removal of a medical device, which is damaged or faulty in a way that allegedly causes or increases the risk of a serious health effect or death [3]. Inside OSTE, a recall is regarded as a tricky but doable situation, which is called "final stage of escalation" within the Corrective and Preventive Actions (CAPA) process [4]. The CAPA is a process inside the quality management system of which the recall process is a consequence. The aim of a CAPA is to continuously monitor and optimise the quality system of a company. This process is part of the Current Good Manufacturing Process (CGMP) conforming work and results in systematic investigations of discrepancies, failures and deviations. Also, it serves as an escalation board [4].

2 Material and Methods

2.1 Brainstorming

A brainstorming is useful for collecting and sharing ideas, existing knowledge and associations regarding a specific topic or problem.

All individuals involved in the topic are invited. The initiator is also the moderator of the meeting and represents the individual central question or task. All meeting members are invited to report their experiences, ideas or problems they encountered. At the same time, approaches to solve a problem may be discussed.

2.2 Mind Map

A mind map is a cognitive technique for a visual representation of tasks, ideas, problems, solutions and all information someone wants to represent [5]. The main theme, problem or question is centred and the belonging tasks, ideas, problems and other information are arranged around it. It is possible but not necessary to add sub-ideas, sub-problems and so on (shown in Fig. 1).

Figure 1: Schematic figure of a mind map. In the centre, the main theme/problem/idea is located, of which problems, ideas, tasks or something else can be deduced. Also, further sub-problems, sub-ideas, sub-tasks and so on.

2.3 PROMOTE - Flow chart

The PROMOTE (**Pro**cess **M**anagement at **O**STE for more **T**ransparency and **E**fficiency) portal is an OSTE internal project to make business processes more transparent.
Flow charts are used for visual representation of processes and their execution. These diagrams enable an easy understanding of individual processes and how the single steps fit together [6]. It also helps to detect and correct weaknesses. Different symbols are defined by the PROMOTE portal and linked with arrows that indicate the flow direction. These individual symbols represent different functions (as visible in Fig. 2). The green rectangles with rounded corners (top left) show instructions or actions and yellow rectangles (top right) show the departments or individuals (the role) who are responsible (r), accountable (a), informed (i) or consulted (c) (bottom right, the lines with a, i, c or r). The red diamonds (top middle) show an event or status, which is attainable. Grey rectangles, with blue and underlined text inside (bottom left), show documents, which are necessary or a result of the instruction. Furthermore, there are grey rectangles with rounded corners (bottom central) that represent a decision possibility. These decision possibilities are AND (all subsequent instruction have to implement), OR (one or more of all subsequent instruction have to be implement) and XOR (one of all subsequent instruction has to be implement).
These defined symbols make the process flow easier to understand and more applicable.

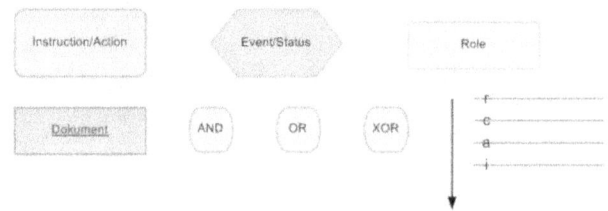

Figure 2: Schematic figure of the flow chart symbols inside the PROMOTE portal. First row (left to right): Instruction or Action you have to do, Event or Status that is reached and Role, that can be responsible (r), accountable (a), informed (i) or consulted (c) departments or individuals. Second row (left to right): Documents, needed or created, AND/OR/XOR which are decision possibilities and an arrow to see the direction of the flow and lines with the letter r, a, i or c to clarify which task the role have to do.

3 Results and Discussion

3.1 Result of the brainstorming with the resulting mind map

To find a good start for the topic, an initial brainstorming meeting was contemplated. Therefore, all departments involved in the latest recalls at OSTE were invited. The central question of the meeting was: "What are the problems during a recall?" The interactive exchange of the brainstorming meeting has identified several problems as well as ideas to resolve some of these problems. The information gained during the brainstorming was transformed into a mind map. Three main problems can be identified.
The first one is that the recall process occurs in different departments inside OSTE. Currently, some recalls are handled in the complaint department and other in the service department. Consequently, the process procedure is different. A Standard Operation Procedure (SOP) is a local business process description in which the implementation of the regulatory requirements respectively the requirements of the Global Operation Procedure are represented. It includes a graphically representation of the process and if applicable a description in text form and considered the local conditions [4]. The second problem is that the SOP, for a recall inside OSTE, is partly unknown in the various departments. To avoid this problem in the future, a target group for process training was identified. Both problems can be limited if the recall process is handled in one department only. As part of the complaint process, the defective devices are sent to OSTE's complaint department. It would be easier for the customer to handle recalls if the process is integrated in OSTE's complaint department. In this case, the customer knows what to do. Furthermore, the Medical Device Safety Officer (MDSO) is currently the head of the complaint department. The MDSO is the most important person during the recall handling. He is the owner of the recall process and responsible for the notification of all competent authorities, for the documentation during the recall and he keeps

the overview over the recall process. If the recall is handled in the complaint department, the MDSO has a better control of the returned devices.

On the other side, it would be straightforward for OSTE to handle the recall process inside the service department. Recall devices, which have to be repaired or exchanged, have to go through the service department whether or not the devices were previously handled by the complaint department. In general, the service department is responsible for repairs. In addition, it is their task to send the items back to the customer, or to distribute credit notes or to give the order to scrap the devices.

Currently, new and unused devices are handled by the customer support. This department can only handle new devices, but not those which were already at the end customer. In the future the customer support will have nothing to do anymore with the recall process. OSTE will integrate the process regarding new and unused devices into the other departments.

The third problem is that all employees feel that the internal and external communication is insufficient. That implies that the internal information flow is slow and insufficient. Not all involved parties have the same knowledge, which causes failures and process flow disruptions - especially cross departmental. Besides, exceptions are not adequately described. To prevent this in the future, it would be better to hold frequent meetings and to appoint one person of the recall team to take meeting minutes. Additionally, the documentation should be integrated into the internal Product Lifecycle Management (ProLIMA) portal. ProLIMA is an internal name of a product database management(PDM)/product lifecycle management (PLM) system. A PDM provided data and documents as a result of the product design and a PLM expands the PDM system with the lifecycle of the product. This way, all employees can also keep themselves informed about ongoing recalls. The integration in one department would help to improve the communication, too. Furthermore, decision trees or checklists could help during the notification.

These tools should help to improve the internal communication in the beginning. So the recall team can remember who needs to be informed and nobody will be forgotten.

While the recall is being in process, the meetings are, including the meeting minutes, a well-suited tool to keep up all involved employees up-to-date. At the same time, it is particularly important that the meeting minutes are available for all involved employees.

To improve the external communication, the Quality Information Letter (QIL) should be extended. Currently the QIL is a short and concise document with all relevant information for the customer. However, misunderstandings and problems still occur. To avoid this, the part with the Actions and Measures could be changed into a checklist, so that the customer can tick off the completed tasks. Only when all tasks are ticked off, the customer can complete the recall.

These three problems are the biggest problems, which affect all departments and which were directly approached.

Additional problems visible in the mind map refer to finances and the repairs. With regards to financial issues, it is crucial to estimate how much money the recall will require. Therefore, the total charges including scrap disposal-, replacement- and repair-costs have to be calculated. Besides that, additional costs for exceptions like unexpected repairs, goodwill cases or derogations also have to be considered. The repairs are strongly connected to the finances. The biggest problem regarding repairs is what to do when a repair regarding the recall is not sufficient for the defective item. It is possible that the investigation shows that an additional repair is necessary or that the investigator notices further problems. This will make the cost estimation more difficult and needs to be considered during the calculation of the total charge. The difficulty here is to decide who will fund the costs. This should be considered during the calculation of the total charges.

Fig. 3 shows a mind map with all referenced problems.

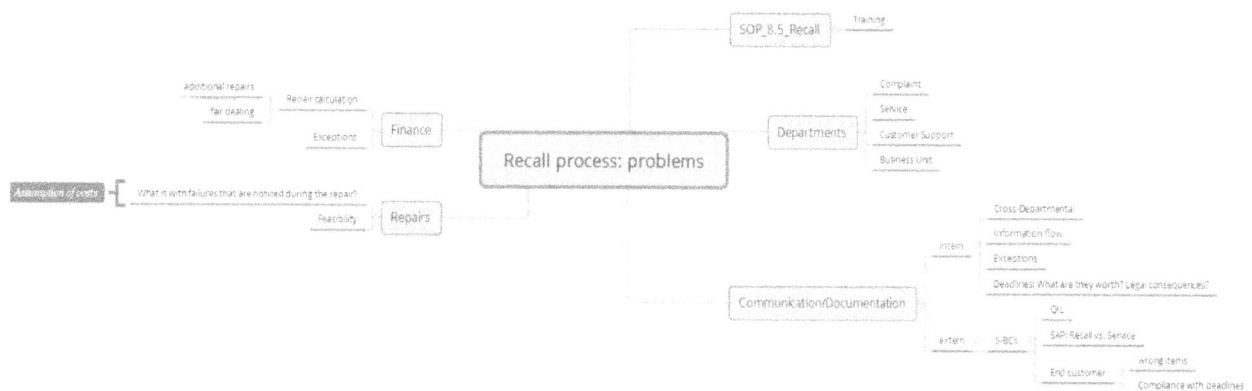

Figure 3: Resulting mind map. On the right side the three main problems (unknown SOP, different departments and communication) and on the left side further problems (finance and repairs).

3.2 Flow chart

For process analysis, the SOP of the recall was reviewed. This SOP is very detailed and comprehensible, but it is plaintext which increases the risk that employees may overlook something. Therefore, a flow chart of the recall process was created. With this, it is easier for the employees to see when what tasks have to be done and what tools have to be used. The resulting flow chart is shown in Fig. 4.

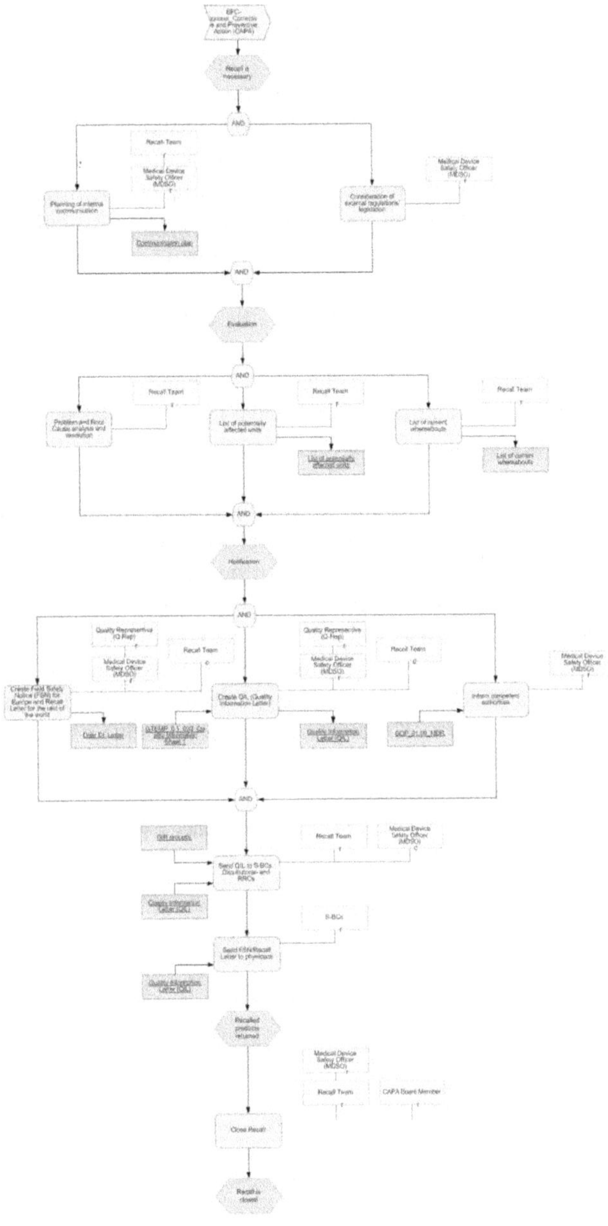

Figure 4: Resulting flow chart based on the current SOP.

During the creation and analysis of the resulting flow chart, it was found that a couple of steps are out of date. On the one hand some tasks are no more a part of this SOP and otherwise some of the processes, which were used in this SOP, were revised. These processes are out of date and can no longer used for the recall. It was an important step in the process analysis to be aware of these discrepancies. This way, it was possible to identify a probable root cause of some issues and failures during the process.

4 Conclusion

We demonstrated the discrepancies inside the recall process at OSTE and worked out possible strategies to improve the process.

In the end, it would be difficult to integrate the process only in one department, as both the service and the complaint department will always be involved to a certain extent. Only the customer support can be disregarded during the recall process. The service and also the complaint department need to be better involved and have to work together. For the customer, it is significant to know where to send the items. It would be advisable for OSTE to define rules, when the items have to be supervised by the service department and when by the complaint department.

To revise the SOP, it is necessary to initiate a change process. For this purpose a new flow chart should be created, which reflects the process optimally and includes explanations and important additional information.

In the next step we have to check whether the strategies are viable or not. The final step should be the integration of the feasible strategies into the process and after that inform and train the employees at OSTE accordingly.

Acknowledgement

The work has been carried out at Olympus Winter & Ibe GmbH and supervised by the Institute of physics, Universität zu Lübeck.

5 References

[1] DIN EN ISO, *9001: 2015-11: Qualitätsmanagement-systeme-Anforderungen*. Beuth Verlag GmbH, Berlin, 2015.

[2] DIN EN ISO, *13485, Medizinprodukte-Qualitätsmanagementsysteme-Anforderungen für regulatorische Zwecke*. Beuth Verlag GmbH, Berlin, 2016.

[3] U.S. Food and Drug Administration, *What is a Medical Device Recall?*. Available: https://www.fda.gov/MedicalDevices/Safety/ListofRecalls/ucm329946.htm [last accessed on 2017-11-07].

[4] Olympus Surgical Technologies Europe, *Product Recall SOP: 8.5-005-en-03*.

[5] Buzan, Tony and Buzan, Barry, *Das Mind-map-Buch: die beste Methode zur Steigerung Ihres geistigen Potenzials*. MVG-Verlag, 2013.

[6] Mind Tools Content Team, *Flow Charts - Identify and Communicate Your Optimal Process*. Available: https://www.mindtools.com/pages/article/newTMC97.htm [last accessed on 2017-12-12].

Development of a verification process
for software of medical devices

N. Razavirad[1], J. Lane[2]

[1] Medizinische Ingenieurwissenschaft, Universität zu Lübeck, naeimeh.razavirad@student.uni-luebeck.de

[2] HEYER Medical AG, Bad Ems, jlane@heyermedical.de

Abstract

This paper focuses on the verification of anesthesia device software. Software verification has a vital role in the medical technology, since errors can result in a device failure and consequently causes risk and threats to patients. Testing is a technique determines whether a software application performs correctly and to detect the possible errors. It is essential to understand which test methods are used and at which development level of the software the testing procedures takes place. In this work, first a systematic test procedure is defined. Furthermore, it is discussed whether the created test work flow can be integrated with test tools. For a better understanding of the function of this test tool, an example of a test case is given in Chapter 3. Finally, the result is analyzed, discussed and evaluated from the perspective view of a tester.

1 Introduction

The core task of an anesthesia device is to ventilate a patient during surgery, as well as to control and monitor their vital parameters. According to IEC 62304, a software system of an anesthesia device is categorized in safety class C and according to EN ISO 14971 in risk class IIb [1],[2]. Therefore, a failure of its software system can lead to significant, life-threatening risks to patients. In order to reduce the risk and thus to ensure a secure, stable, error-free, high-quality software system, it is necessary to test the software during the development process. A failure of a software system also causes costs. The costs incurred to correct an error will differ depending on the software development stage in which the error is detected. These can vary between 25% to 50% of the total project cost, and might be the largest share of costs [3]. All monetary test costs of a project comes due to the time required to execute test activities. Depending on which method is used to test the software system - manually or automatically - different efforts and times can be predicted. The cost, which is needed for the fixing of error should not be ignored. Figure 1 shows time line and context of error occurrence and resolution, and the corresponding costs incurrs per time. As in diagram a) is shown, the cost per error exponentially increases by a factor of ten in the course of development [4]. From diagram b) it can be seen that most of the errors occur during the implementation phase, and the number of errors discovered at each stage of the development decrease gradually. According to diagram c) most errors are corrected during the software and system integration test level.

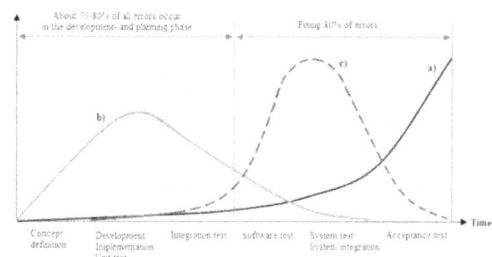

Figure 1: a) cost per error, b) error incurrence, c) bug fixing [5],[6].

Testing the software in each mentioned levels requires different testing techniques. These techniques describe which information is known during the execution of a test case and what to look for while checking the software. Because the internal programming structure is known during unit and integration level, the "white box"-technique can be used [7].

For subsystem and system tests, knowledge of the internal code structure is not a prerequisite for testing. The software is checked at this level using the "black box"-technique [8]. Here it is verified that all software requirements are met. With knowledge of the software structure and the requirement descriptions, the properties of these two above mentioned tests methods are used and compared, which is referred to "gray box"-technique [9].

2 Material and Methods

This chapter presents a closer understanding of the material and methods as well as the steps that were taken starting from the initial considerations up to the end of a testing process. Material and tools are described below.

2.1 Creation and application of a systematic, seamless test process work flow

Testing is an activity which should ideally start at the same time with the development phases and is done under certain conditions, e.g. based on a test case with a defined test environment. Followingly, a test work flow that was created for this work is explained. For better readability, the created work flow is shown in two sections (Fig. 2 and 3).

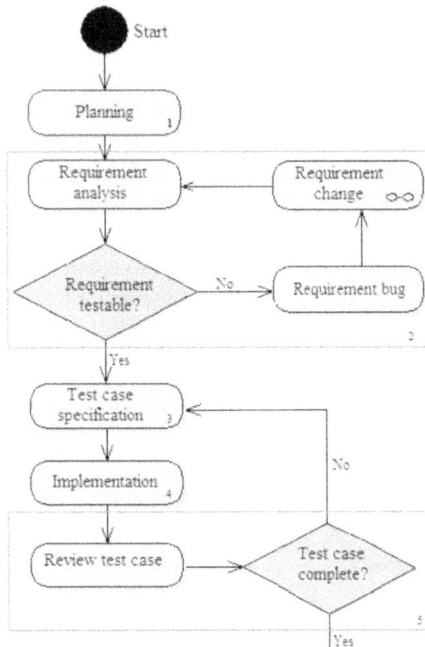

Figure 2: Planning, Requirement analysis (Own representation based on [8]).

During the planning phase (Fig. 2, No.1), it must be determined at which test level and with which test method the software is to be tested. In the second step, the test objects and the test conditions are defined more in detail. The requirements are to be analyzed first, and in case of , e.g. non-testable or incomplete requirement, this should be corrected by the Requirement Engineer in an iteration method (Fig. 2, No.2).

The test cases must be formulated in a structured, clear, testable, reproducible and in understandable form. In each step, the actions are described with necessary preconditions and expected results. In order to use automatic testing, all test cases must not only be formulated in textual form, but also be implemented in executable source code (Fig. 2, No.3 and No.4). A review is performed in order to check the completeness and accuracy. This serves to minimize errors in the test case execution (Fig. 2, No.5).

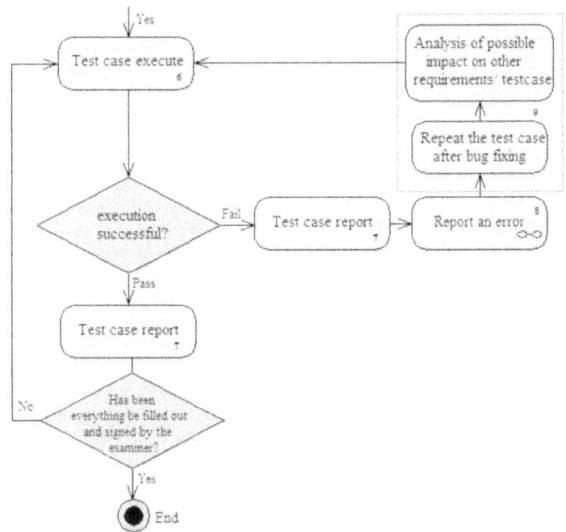

Figure 3: Test Case Execution (Own representation based on [8]).

After a successful accomplishment of the previous steps, the test case can be performed manually or automatically (Fig. 3, No.6). If the expected and observed results agree within previously specified acceptance criteria, then the status of the test case is "Pass". If an error occurred, the status of the test case changes to "Fail". A test report is then generated from the test results (pass or fail) (Fig. 3, No.7). The error found during test must be reported to a bug tracking system (Fig. 3, No.8). It is important to document exactly where and in which test step an error is found, the possible cause, and what has to be done from the viewpoint of the tester to redesign the test results most accurately. After fixing the bug, the detected defect is to be tested again, to verify that the bug is fixed or the change was successful. Therefore, it requires a retest on a new software version (Fig. 3, No.9).

The procedure described above can be used for manual and automatic testing.

2.2 Enterprise Architect (EA) as a test tool

While choosing a test tool, it is important to know which test method to manually or automatically test the software. The following tools are a few examples of manual and/or automatic test tools:

- Imbus TestBench supports test planning, test design, test automation, test execution and reporting [10].

- Rational Functional Tester Enables Automated Regression Testing [11].

- QF Test is used for performing endurance tests at the system level [12].

- FitNesse Framework is used for acceptance and integration testing and uses the programming interface for software validation [13].

- Enterprise Architect by SparxSystem.

In this work, Enterprise Architect version 13.0 was used as a testing tool [14] and investigated how and whether this tool is also suitable for a manual test. Enterprise Architect supports core features related to testing. These features are defined in Test case as Element, Build Run, Debug and Unit Testing, Attaching Test Documents. EA is also a development tool that provides:

- Design of software systems.
- System Modeling of business processes.
- Creation of the UML diagrams, for the planning and documentation of the development process.

Among other things, EA enables requirements and configuration management.

3 Results and Discussion

The workflow shown in the previous section was integrated in EA and implemented as an example on an anesthesia device. In this work, 64 test cases with 500 test steps in EA for integration and system tests were described and executed with Gray- and Black-Box test technique.

Figure 4 shows an example of these test cases. The marked parts in the figure are information that is necessary to create a work flow according to the description in chapter 2. This information has been realized in EA using testing, change and defect windows. Different attributes of these windows are used in generated documents. In the test window the actions, inputs and expected results can be defined. But for the documentation of the observed results in the testing, there is no possibility to save current results in this window without the last-performed test results being deleted from the database. However in order to get an accurate qualitative and quantitative result in the course of the development, the test cases have to be executed in different build versions. That means, a way needs to be found in EA for any repetition of the test cases without the old executed testresults being deleted and changed the original testcase. For solving this problem in EA, in any build version, all scheduled test cases that need to be done, are added as a new items in a new folder. Thus, all the test results are preserved in different build versions. And it is then also possible to create an analysis of all these test results over time.

Test case-Version	0.5.0
Build-Version:	1
Executed Date:	20.01.2018
Test Ergebnis:	**FAIL**

Execution Steps:

Precondition step
• Test Object = Anesthesia device AD
• Gasmixer-TestBoard(for simulation different setting)

Textual Step 01		**Status**
User Action	[SWA-GM-02][SWA-GM-03][SWA-GM-04][SWA-GM-06] (list of requirements) Select in GUI: Gas Combination = O_2/N_2O FG = 700ml/min, O_2 = 18% Then select a CMV (Continuous Mandatory Ventilation) ventilation mode.	
Expected Result	It should be possible to choose N_2O as second in additional to O_2. The requested setting for FG and O_2 must not be possible.	
Observed Result/ Defect ID	Adjusted O2/N2O as Gas Combination. It is possible to select FG value to 700m/min and O_2 value to 18 percent. (see Defect-120)	FAIL

Defect Reference:

Defect-ID	**Headline**
Defect-120	Possible to set a False setting for Gases

Executed by Naeimeh Razavirad Signature: NR	Approved by Jeff Lane Signature: JL

Figure 4: A section of document format (.doc or .pdf) generated in EA

To create a data analysis in EA, generate a custom SQL queries. The following SQL sample query shows the status of the test case (Fig. 4) of six different software build versions (see Fig. 5).

```
SELECT DISTINCT t_object.ea_guid AS CLASSGUID, t_object.Name as "Testfall",
t_object.Object_Type AS CLASSTYPE, t_objecttests.InputData
as "Version", t_objecttests.Status as "Status"
FROM t_object
LEFT OUTER JOIN t_package testf on testf.package_id = t_object.package_id
LEFT OUTER JOIN t_package packagename on packagename.package_id =
testf.parent_id
LEFT OUTER JOIN t_objecttests on t_objecttests.object_id = t_object.object_id
WHERE t_object.stereotype = 'testcase' and testf.name like 'SW-Build_\%' and
t_object.Name like 'SWTEST01\%' and t_objecttests.Test = 'TestErgebnis'
ORDER BY t_objecttests.InputData
```

Testfall	Version	Status
SWTEST01.02_Gasmixure	0.4.0.dev-01	Pass
SWTEST01.02_Gasmixure	0.4.1.dev-01	Pass
SWTEST01.02_Gasmixure	0.5.0.dev-0	Fail
SWTEST01.02_Gasmixure	0.6.0.dev-0	Fail
SWTEST01.02_Gasmixure	0.6.0.dev-01	Fail
SWTEST01.02_Gasmixure	0.6.0.dev-02	Fail

Figure 5: An example for Status of a test case represented in different software build versions.

Thus it can be analyzed, why the error appeared only once, why an error occurs permanently, in each version, or irregular intervals.

All executed test cases with the Pass or Fail results must be documented in each build version from a regulatory and qualitative viewpoint. In order to archive the documents, own templates were defined in EA, which can be exported as ".doc" or ".pdf" documents . An example document is shown in (Fig. 5).

4 Conclusion

Summarizing this paper answered the question, if and how the software verification of an anesthesia device is possible with the help of EA as a test tool. As discussed in the previous section, it can be seen that the testing work flow of an anesthesia device can be implemented and executed in EA. The advantages of EA include:

- Creation of test cases with all necessary conditions.

- Ability to reference existing requests and test cases to test pages.

- Tracing of the defects in the executed test case.

- Easy analysis by SQL query.

- Creation of custom templates for reporting

The disadvantage is that while performing the test cases, attention must be paid that without a copy of the test case as a new element, the stored data will be deleted.
From the perspective of a tester through early testing, the error can be found in time and can be eliminated more cheaply. But testing can only detect the mistake and that does not prove that a software system is error free. Also, the testing does not guarantee flawlessness, even with 100% coverage. Ideally, errors remaining in the system can be limited and estimated using empirical values and comparable development devices from the company. Using sufficient verification and documentation, quality and functionality can be ensured prior to the release of a software system so that no life-threatening error occurs.

Acknowledgment

This work has been carried out at HEYER Medical AG, Bad Ems and supervised by the Institute for Electrical Engineering in Medicine, Universität zu Lübeck.

5 References

[1] IEC 62304:2016-10+A1:2015(VDE 0750-101:2016-10), *Medical device software-Software life-cycle processe*, Beuth, Berlin.

[2] DIN EN ISO 14971:2013-04, *Medical devices - Application of risk management to medical devices*, Beuth, Berlin.

[3] M. Baumgartner; R. Seidl; H.M. Sneed, *Der Systemtest von den Anforderungen zum Qualitätsnachweis*, Edition 3, Hanser Verlag München 2012.

[4] *FMEA-Fehlermöglichkeitsund Einflussanalyse-Deutsche Gesellschaft für Qualität e.V.*, Edition 5, 2012. Available: https://www.dgq.de/dateien/Band_13-11_Auszug_Web.pdf, [last accessed on 2018-01-25]

[5] T. Pfeifer; R. Schmitt, *Qualitätsmanagement: Strategien–Methoden–Techniken*, Edition 5, Hanser Verlag, 2015.

[6] B. Regius, *Qualität in der Produktentwicklung: vom Kundenwunsch bis zum fehlerfreien Produkt*, Hanser Verlag, 2006.

[7] C. Kaner; J. Bach; B. Pettichord, *Lessons Learned in Software Testing-A Context Driven Approach*, John Wiley& Sons, 2011.

[8] *Certified Tester Advanced Level Syllabus, Testmanager International Software Testing Qualifications Board*, Austrian Testing Board, German Testing Board e.V. & Swiss Testing Board, 2012. Available: http://www.german-testing-board.info/wp-content/uploads/2016/07/CTAL_Lehrplan2012_TM_Final_Germ_V100.pdf[last accessed on 2018-01-25]

[9] D.W. Hoffmann, *Software Qualität*, Edition 2, Springer, Berlin, Heidelberg, 2013.

[10] *TestBench imbus*, Available: https://www.imbus.de/testbench/, [last accessed on 2018-02-09]

[11] *Rational Software, Rational Functional Tester 2004* Available: RationalSoftware, RationalFunctionalTester2004, [last accessed on 2018-02-09]

[12] *Quality First Software GmbH: QF- Test*, Available: www.qfs.de, [last accessed on 2018-02-09]

[13] *Ward Cunningham: FitNesse Test Framework,* Available: http://fitnesse.org/, [last accessed on 2018-02-09]

[14] *Enterprise Architect from SparxSystem*, 2017. Available: www.sparxsystems.com, [last accessed on 2018-01-25]

8

E-Health

Connecting MOLGENIS to HL7 FHIR: Transformation from Questionnaires to EMX

N. Deppenwiese [1], H. Ulrich [2] and J. Ingenerf [2,3]

[1] Medical Informatics, Universität zu Lübeck, noemi.deppenwiese@student.uni-luebeck.de
[2] IT Center for Clinical Research, Universität zu Lübeck, hannes.ulrich@itcr.uni-luebeck.de
[3] Institute of Medical Informatics, Universität zu Lübeck, ingenerf@imi.uni-luebeck.de

Abstract

Questionnaires from clinical trials or routine documentation can be represented in various formats. One interchange format for clinical questionnaires is HL7 FHIR. When corresponding meta data elements from questionnaires are mapped it becomes possible to interchange and reuse the collected data. MOLGENIS is a popular biobanking toolbox that supports meta data annotation and such a mapping. Since MOLGENIS can only import data available in the EMX format, in this work an application for converting HL7 FHIR Questionnaires into the EMX format was developed. The web based application can be used via an API or a graphical user interface. It fetches FHIR Questionnaires and generates corresponding EMX tables in which the general structure of the questionnaire is maintained. The Questionnaire items become entity attributes. Some information like identifiers or semantic annotations is however lost in the process due to the EMX format's inflexibility.

1 Introduction

To enable the reuse of data collected in clinical trials or routine documentation it is necessary to match corresponding metadata elements. These elements contain information like the question asked or the unit of measure used. They can be collected in metadata repositories [1], where they can be annotated with terminology codes and provided for reuse. This process is facilitated through using one standardized format for all data elements. Dugas et al. demonstrated how CDISC ODM can be used to build such a metadata repository [2], but ODM lacks flexibility due to its strict hierarchy [3] . Alternatively, HL7 FHIR, an emerging standard for health care applications, can be used as a more flexible interchange format for clinical questionnaires [4]. MOLGENIS is a web based software often used in biobanking, but in addition it supports metadata annotation and mapping. Unfortunately, it can only import data available in its own EMX format. Therefore, the objective of this work was to design and implement a web service which takes questionnaire data available on a HL7 FHIR Server and converts it into MOLGENIS EMX tables so it can be processed in MOLGENIS.

2 Material and Methods

While both FHIR and EMX can model questionnaires, there are some fundamental differences in the way these formats function that have to be considered when converting questionnaires from FHIR to EMX.

2.1 HL7 FHIR

FHIR stands for fast healthcare interoperable resources and is the newest standard published by Health Level Seven (HL7). It is based on so-called resources, objects with pre-defined attributes which can be represented in XML or JSON and are typically stored on FHIR servers from which they can be retrieved via a representational state transfer (REST) API.

The resource used to model clinical questionnaires is called *Questionnaire* and consists of two parts. The general part defines attributes for the *Questionnaire* as a whole, e.g. an unique *url* used to identify this *Questionnaire* regardless of the server it is hosted on.

The second part consists of *items* which can contain questions, texts or groups of *items*. Hence, *items* can have a variety of data types, e.g. *string*, *choice* or *group*. Every *item* needs an unique *linkId*. *Items* may also reference one or many terminology codes in their *code* field. *Items* can reference *ElementDefinitions* representing more general metadata elements via their *definition* attribute. Since *ElementDefinitions* are not resources but FHIR types, they can not be referenced or stored directly. Instead, the enclosing resource, e.g. *StructureDefinition*, is referenced by URL and the reference is complemented by the *ElementDefinition's* id. *ElementDefinitions* can be used to build a metadata repository enabling data element reuse and therefore improving semantic interoperability.

If an *item* represents a question, answers may be restricted by referencing a *ValueSet* in its *options* field. This resource can contain codes from different (external or self-defined) code systems directly in its *expansion* element which simply lists all contained codes. Alternatively, the *ValueSet* may define its codes in a more abstract manner via its *compose* element. FHIR Servers may be asked to generate the *expansion* for a given *ValueSet*, but not all servers support this operation for all *ValueSets*.

When a *Questionnaire* is answered, these answers are *not* collected in the *Questionnaire*. Instead, a new instance of the *QuestionnaireResponse* resource is generated. This allows a separation of metadata (*Questionnaire*) and data (*QuestionnaireResponse*). Fig. 1 shows the relationships between the FHIR schema definitions and the instances for *Questionnaire* and *QuestionnaireResponse*. This work focuses on the metadata in *Questionnaires*.

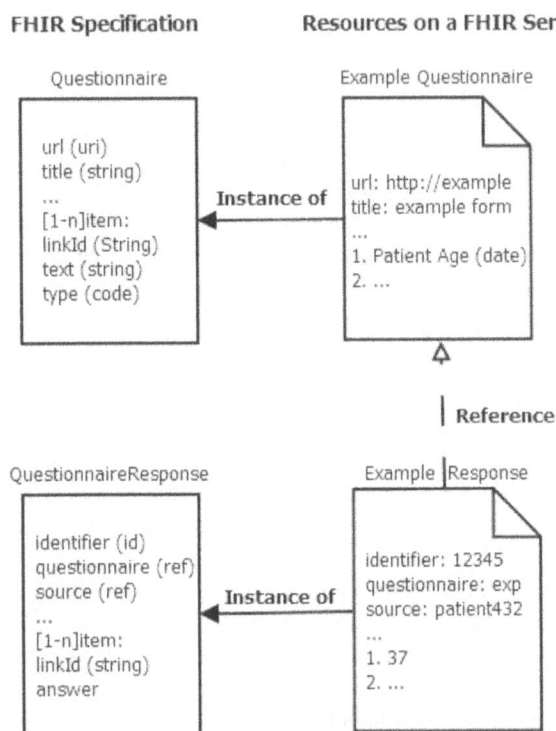

Figure 1: The FHIR specification defines schemas for resources. Concrete instances of the resources can be stored on FHIR Servers. The answers to a specific *Questionnaire* (an instance of the *Questionnaire* schema) are an instance of *QuestionnaireResponse* that references the *Questionnaire*.

2.2 MOLGENIS EMX

MOLGENIS stands for *mol*ecular *gen*etics *i*nformation *s*ystem, an open-source web application used to store and analyze data for biomedical research [5]. The system can be accessed either via graphical user interface or RESTful API. Preexisting data can be imported using a table-based data format called EMX. These data models usually consist of multiple tables containing data and metadata. In this context, the following terms are used:

- *Entity:* An entity corresponds to a FHIR resource or an object class in an object-orientated programming language. It can be contained in packages and defines attributes.

- *Attribute:* An attribute describes a field of an entity. Attributes can be nested or reference each other.

- *Package:* Packages are used to structure the data model. They can contain other packages or entities.

- *Instance:* Instances are the actual data objects. They contain values for the attributes for the entity they represent.

A valid EMX model has to contain at least two tables: One containing an entity and its instances (data) and one defining the corresponding attributes (metadata). For every column in the entity sheet, there must be a corresponding row in the attributes sheet containing at least the *name* and *entity* values. For example, the table *students* in 1 contains the data while its attributes are defined in the first three rows of *attributes*. Furthermore, every entity needs to specify one *idAttribute* which values must be unique for every instance of this entity. In table 1, *number* is the *idAttribute* for the *students* entity. All other objects, e.g. other entities or packages, must be specified in their respective tables. The permitted values for an attribute can be limited by using the data type *xref* for said attribute. Then, an additional column named *refEntity* needs to contain the name of another entity. The instances of this entity are the attribute's permitted values, the corresponding column in the referencing entity's table needs to contain the referenced instance's *idAttribute*. An example is the *students* entity's *course* attribute as seen in table 1. Furthermore, the data type *mref* enables an attribute to contain multiple references at once. The referencing column then contains a comma-separated list of the referenced instance's *idAttributes* values.

It is also possible to group attributes together by using the *compound* data type. Then, all attributes that are meant to be part of the group need to reference the compound's name in an extra column called *partOfAttribute*. A compound can contain other compounds.

The columns for the metadata tables like *entities* or *attributes* are given. It is not possible to add additional columns. However, in some cases it is useful to store further information on these objects. In newer MOLGENIS versions, *tags* can be used for this purpose. All tags must be defined in an extra table *tags*. Among other things, tags must can contain a *relationIRI*, an *objectIRI* and a *codeSystem* reference. Every *tag* needs an unique *identifier*. All tables allow these identifiers to be added in an extra *tag* column. Therefore, it becomes possible to use concepts from terminologies and ontologies to further describe EMX metadata.

EMX constrains identifiers like attribute or table/entity names. Only the letters A-Z and numbers as well as number signs (#) and underscores are allowed, the first character must be a letter. Furthermore, full entity names consist of all enclosing packages names joined by underscores, e.g. for the *students* entity in table 1 *base_uni_students*. Therefore, underscores can not be used in the actual entity names.

Table 1: Valid EMX example tables

(a) Table *students*

number	name	course
123456	Ada Lovelace	Mathematics
347036	Grace Hopper	Programming
459826	Alan Turing	Mathematics

(b) Table *courses*

name	abbr
Mathematics	MAT
Programming	PRO

(c) Table *attributes*

name	entity	dataType	idAttribute	refEntity
number	students	int	TRUE	
name	students	string		
course	students	xref		courses
name	courses	string	TRUE	
abbr	courses	string		

(d) Table *entities*

entity	package	description
students	uni	all students
courses	uni	offered courses

(e) Table *packages*

name	parent	description
uni	base	package for university data model
base		root package

The maximum length for identifiers is 30, even for full ones containing all the package names.

2.3 Data type mapping

FHIR items and EMX attributes use different data types, nevertheless many FHIR data types have EMX equivalents, e.g. *boolean* corresponds to *bool*. The mapping was developed while keeping in mind that while the developed application only converts metadata (*Questionnaires*), it should be possible to convert the actual data (the *QuestionnaireResponse*) into entity instances that can be saved in the tables generated by the converter. Therefore, for complex data types like *quantity* without EMX equivalents, a string representation was chosen.

2.4 Converter Design

The converter application was conceptualized as a web service providing both an application programming interface and a graphical user interface. In both cases, a FHIR server URL and *Questionnaire* ID serve as input parameters. Optionally, a MOLGENIS URL and login credentials can be submitted. This login information is used to connect to MOLGENIS during the conversion process. When the entity names are generated it is then possible to check whether they have already been assigned in this MOLGENIS in-

stance. The application also allows the user to disable the creation of EMX tags to ensure compatibility with older MOLGENIS versions. Errors that occur during processing are sorted into the following categories (adapted from FHIR error categories):

INFORMATION: Information about the conversion process. This may just be a success message.
WARNING: A problem that may result in lost information occurred. However, the generated EMX is still valid.
ERROR: Tables were generated but their validity can not be granted. Therefore no upload to MOLGENIS was attempted.
FATAL: The conversion process was aborted and no tables were generated.

When used via the API, the application first fetches the resource and translates it into an EMX model. One *Questionnaire* is transformed into one EMX *entity* with every *item* being translated into an attribute. Additional entities may be generated to represent *ValueSets* as described in section 3. If no ERROR or FATAL errors occur, the resulting tables are saved as tab-separated-value files and compressed into a ZIP file. This file is then uploaded to MOLGENIS if a URL and valid credentials have been provided. The converter application then responds with the status code *200*. If the program encountered ERROR or FATAL errors, an error status code (e.g. *422*) is send. Error messages are always listed in the response's JSON body.

The GUI workflow renders the generated tables and hence allows the user to verify the conversion results before uploading them to MOLGENIS or downloading the generated file. Again, the MOLGENIS upload is only possible if MOLGENIS credentials were provided and no ERROR or FATAL errors occurred. All error messages are displayed along with the generated tables.

2.5 Implementation

The application was implemented in Java 8. *Spark Java* (http://sparkjava.com/) was used as server framework, the *HAPI FHIR* library (http://hapifhir.io/) for processing the FHIR version 3.0.1 resources. Additionally, the *Apache HTTP Components* project (https://hc.apache.org/) was utilized to connect to MOLGENIS version 4.0.0.

3 Results and Discussion

The developed application can convert FHIR *Questionnaires* into valid EMX tables. Most of the information contained in *Questionnaires* can be represented in EMX, however, some information is lost due to the inflexibility of the EMX format.

Most of the general information contained in the *Questionnaire's* first part, e.g. *version*, *publisher* or *purpose* is lost because the EMX entity table has no columns that could be used to store it. The only free text attribute available is *description*, which in this work is used to store the *Questionnaire's description* attribute. While it would be possible to simply store all the extra information in this field, it would be difficult to retrieve single values from one string.

A *Questionnaire's items* can be nested and hence be used to divide the questionnaire into sections and group questions together. This structure can be modeled by using the *compound* data type in EMX. Therefore, the general structure of the *Questionnaire* is retained by the EMX model.

FHIR *Questionnaires* are identified by a unique URL. Since MOLGENIS *entities* need unique names as well, using the URL appears natural. But EMX identifiers like *entity* names are constrained as described in 2.2. FHIR URLs are typically too long and contain special characters so they can not be used here. In this work, they are hashed and the hashes a prefixed with the letter *A* to conform to the identifier restrictions. While this process generates suitable names for the generated entities, it is irreversible and makes it difficult to re-identify the *Questionnaires* they were generated from. Therefore the original URL is saved in the *entity's* description field. However, since this is a free text attribute and may contain other information as well, re-identification can not be guaranteed. The same problem may occur with *item ids*, these are transformed likewise.

If the *Questionnaire* references other FHIR resources like *ValueSets* or *ElementDefinitions*, these resources need to be fetched and processed as well. This may fail if the reference can not be resolved which may be the case if the URL is incorrect or describes an abstract resource, e.g. *StructureDefinitions* for LOINC Codes. In this case, the application creates a WARNING error and saves the unresolved URL in the *entity's* or *attribute's description*.

Questionnaires offer multiple ways to bind a set of possible answers to an *item*. First, all options can be included directly via the *option* attribute. These can be parsed into instances of a newly created entity, the *attribute* generated from the item then references this entity. Alternatively, the *item* can reference a *ValueSet*. If the reference can be resolved and the *ValueSet* contains an *expansion* (list all values), they are parsed likewise. Otherwise the FHIR server is asked to generate an expansion which is processed subsequently. If this fails the *ValueSet* is treated like an unresolvable reference and its URL is saved in the attributes description. While FHIR allows soft bindings (answers other than pre-defined options are also allowed), EMX only allows strict bindings. This may become problematic when actual data (*QuestionnaireResponse*) is being converted.

FHIR defines an optional *enableWhen* field for *items*. This attribute can contain a question the answer to which decides whether the *item* is shown, e.g. only ask for time of death if patient is deceased. This function can not be modeled in EMX, all related information is lost.

Questionnaires and their *items* can be annotated with terminology codes. This information is often central to mapping processes [2]. These *code* attributes are converted into EMX *tags* with the actual codes serving as identifiers. This is problematic when codes do not adhere to EMX identifier constraints, e.g. LOINC codes contain special characters and hence can not be used as tags. In those cases, a WARNING error is generated and the annotation is skipped.

4 Conclusion

Clinical questionnaires represented as FHIR *Questionnaires* can be converted into MOLGENIS EMX *entities* while preserving their general structure. Still, due to the inflexibility of the EMX format, some information is lost during conversion. This is especially problematic for FHIR identifiers like the *Questionnaire* URL, which can not be used as EMX identifiers due to naming restrictions and have to be irreversibly transformed. To still enable re-identification, the original URL is saved in the description. A conversion of *QuestionnaireResponses* into instances of the generated entities is conceivable. If the metadata extracted from the *Questionnaire* is enhanced using MOLGENIS it needs to be converted back to FHIR to allow its usage in metadata repositories. Since both FHIR and MOLGENIS are works in progress, the converter needs to be continually adapted to format changes, changes which may allow solving some of the problems identified in this work.

Acknowledgement

The work has been carried out and supervised by the Institute of Medical Informatics, Universität zu Lübeck.

5 References

[1] H. Ulrich, A.-K. Kock, P. Duhm-Harbeck, J. K. Habermann and J. Ingenerf, *Metadata Repository for Improved Data Sharing and Reuse Based on HL7 FHIR*. Studies in health technology and informatics, no. 228, p. 162, 2016.

[2] M. Dugas, A. Meidt, P. Neuhaus, M. Storck and J. Varghese, *ODMedit: uniform semantic annotation for data integration in medicine based on a public metadata repository*. BMC medical research methodology, vol. 16, no. 1, p. 65, 2016.

[3] J. Doods, P. Neuhaus and M. Dugas. *Converting ODM Metadata to FHIR Questionnaire Resources*. Studies in health technology and informatics, no. 228, p. 456, 2016.

[4] M. Baake, C. M. Drenkhahn, N. Deppenwiese, A.-K. Kock and J. Ingenerf, *Metadata Repository for Improved Data Sharing and Reuse Based on HL7 FHIR*. In: HEC 2016: Health – Exploring Complexity. Joint Conference of GMDS, DGEpi, IEA-EEF, EFMI. German Medical Science GMS Publishing House, Düsseldorf, 2016.

[5] M. A. Swertz, M. Dijkstra, T. Adamusiak, J. van der Velde, K. Joeri et al, *The MOLGENIS toolkit: rapid prototyping of biosoftware at the push of a button*. BMC bioinformatics, vol. 11, no. 12, p. 12, 2010.

Mapping of Laboratory Services included in an Internal Catalog to SNOMED CT using the UMLS

C. Drenkhahn [1], H. Ulrich [2], P. Duhm-Harbeck [2] and J. Ingenerf [2,3]

[1] Medizinische Informatik, Universität zu Lübeck, cora.drenkhahn@student.uni-luebeck.de

[2] IT Center for Clinical Research (ITCR), Lübeck, hannes.ulrich@itcr.uni-luebeck.de, petra.duhm-harbeck@uksh.de

[3] Institut für Medizinische Informatik, Universität zu Lübeck, ingenerf@imi.uni-luebeck.de

Abstract

Standardized representation of laboratory data is the basic requirement for semantic interoperability and computer-aided analysis, applying not only to test results but also to related billing information. For this reason a semi-automatic algorithm was developed and implemented that enables the mapping of laboratory services included in a hospital's internal catalog to an appropriate coding system. Due to syntactic and semantic variance in the source data SNOMED CT was chosen as mapping terminology instead of LOINC, using the UMLS to bridge the gap between German service terms and English target concepts along the way. By utilizing variances and substrings of the original search term the coverage of results could be improved so that finally three-quarters of the internal catalog's elements got mapped to a manually approved SNOMED CT concept or postcoordinated expression.

1 Introduction

Laboratory tests are a key factor of modern healthcare. They can provide valuable data for various purposes during diagnostics and treatment in a quick and economic way [1]. If the results are communicated in a standardized, machine-readable form, many possibilities arise for using them to improve medical care, for example in Decisions Support Systems or statistical analysis [2]. Although there are already medical terminologies available that can ensure semantic interoperability if used for laboratory data, standardization is still not implemented in most primary systems [3]. This issue applies not only to classical lab tests, but also to other data associated with laboratory procedures, in this project focused on the field of billing.

German hospitals often utilize a so-called 'Hauskatalog' (internal catalog) defining their local range of services to capture all performed medical procedures. As a result a detailed documentation and internal cost allocation are facilitated [4]. Besides services of other clinical departments the internal catalog contains many elements describing laboratory procedures as well. We worked together with a hospital interested in improving their internal catalog that is currently based on the tariff scheme *Deutsche Krankenhausgesellschaft Normaltarif* (DKG-NT) which is generally used to gain a tariff neutral representation independent from changing scales of fees. To improve the hospital's usage of their huge internal service data, for example by enabling better navigation and retrieval when it is analyzed in a data warehouse by the hospital's decision makers, the local textual descriptions of their internal catalog need to be annotated by suitable vocabularies. So, the project's aim was to map the internal catalog's elements to a reference terminology, preferably using automated means.

As a possible target system the *Logical Observation Identifier Names and Codes* (LOINC) terminology can be considered. This coding system addresses specifically the standardization of laboratory observations by summarizing six relevant data axis into one term and code for each test. Alternatively the *Systematized Nomenclature of Medicine - Clinical Terms* (SNOMED CT) provides the most extensive ontology of medical concepts [5].

For both terminologies there is currently no complete and up-to-date German translation available [6], [3], leading to linguistic difficulties when trying to find appropriate entries for the internal catalog's German terms directly. Fortunately the *Unified Medical Language System* (UMLS) integrates many health care-relevant vocabularies into one big, complex and multilingual database. Therefore it is a useful tool for cross-mappings between terminologies [2].

2 Material and Methods

The internal catalog used as source file for this project was provided by our partner hospital's controlling department in February 2017, containing all in all 7648 entries of clinical services. After restricting the contents to laboratory related elements in chapters M and N a total of 3833 items remained, composed of 935 normal DKG-NT services ('Standard') and 2898 locally specified procedures ('Hausleistungen'). All of these elements are grouped into eight different types of services, of which the category LABOR is the most important for this project and covers a majority of 1747 elements. The textual description in German that characterizes the respective service can be found exemplarily in the second column ('Text') of Table 1 and is the only part of the internal catalog relevant for mapping. As visible in this excerpt the catalog's terms aren't structured in a homogeneous way and vary largely in the range of information included.

Table 1: Example source data for the mapping: A part of the hospital's internal catalog. Services including the measurement of the substance creatinine with different additions.

Leistungsziffer	Text	SoH
13585000	Kreatinin	Standard
13585001	Kreatinin (Jaffé-Methode)	Hausleist.
23585001	Creatinin-Versand	Hausleist.
23585002	Kreatinin im Urin	Hausleist.
13615000	Kreatinin - Clearance (Zweimalige Bestimmung von Kreatinin)	Standard

The following data elements can be contained solely or combined in the procedure's description: the substance that was measured (e.g. 'Kreatinin'), the material in which it was measured (e.g. 'Urin'), the method or technique used (e.g. 'Jaffé-Methode'), plus any general conditions like location or circumstances (e.g. 'Versand').

2.1 Choice of Terminology

Before performing the actual mapping of the internal catalog the suitability of both reference terminologies in question - LOINC and SNOMED CT - was thoroughly evaluated, resulting finally in the selection of the latter.

Being the most widespread coding system for laboratory tests [7] the usage of LOINC would have been the obvious choice, facilitating the combined analysis of billing and lab data. But as each LOINC code is built from a unique set of six, mostly mandatory parts *Component, Property, Time, System, Scale* and *Method*, this fixed data structure produces ambiguities when trying to fit the heterogeneous information content of the internal catalog's items into it. Though most terms include data for the *Component* axis, the material for *System* isn't mentioned regularly, just as any further information needed to distinguish between disparate LOINC codes (Table 2). On the other hand the internal catalog focuses on techniques and ambient conditions that aren't in LOINC's scope of application [8].

Table 2: Some possible LOINC codes for the mapping of 'Kreatinin im Urin' (without axis *Method*). Currently 31 different options can be found for this combination.

Code	Comp.	Property	Time	System	Scale
2161-8	Creat.	MCnc	Pt	Urine	Qn
2162-6	Creat.	MRat	24H	Urine	Qn
20511-2	Creat.	Imp	Pt	Urine	Nom

SNOMED CT in contrast is an ontology with a hierarchi-

cal structure so that it offers concepts of different granularities, enabling a flexible adjustment for each service description. Although it is not a laboratory-specific terminology, SNOMED CT contains a large number of lab test describing concepts with 11931 *Evaluation Procedures* in version 07/2017. Furthermore, many of its concepts from other categories can be useful to incorporate partial data of an internal catalog's element. By making use of SNOMED CT's ability for postcoordination, different concepts can be combined via relationships in a machine-readable way (Fig. 1), so that the representation of information becomes even more flexible.

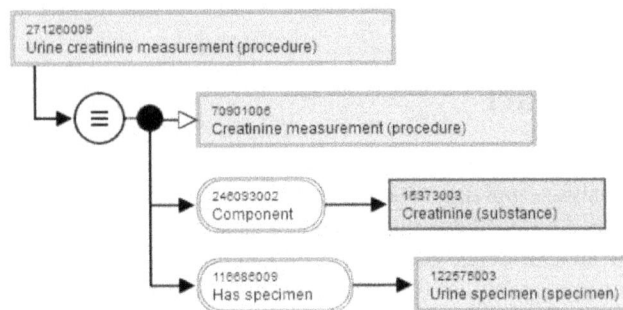

Figure 1: Representation of laboratory procedures in SNOMED CT, concepts in square boxes and relationships in rounded shapes. The precoordinated concept at the top is equivalent to the postcoordinated combination of the others.

2.2 Mapping with the UMLS

As mentioned before, a direct search for SNOMED CT concepts with the internal catalog's terms was not possible due to language issues. So, the idea to utilize the UMLS as an intermediate step of translation was developed. Similar to SNOMED CT the UMLS summarizes medical facts into discrete concepts, but for each of these it additionally collects all synonyms and representations in other terminologies with the same meaning. These references are called *atoms*, and besides SNOMED CT concepts, they include also translated terms from controlled vocabularies like *Medical Subject Headings* (MeSH). As a result the UMLS concepts can serve as a bridge between German lab procedure terms and the corresponding SNOMED CT concepts with English names.

So, the search function was divided into two steps: At first the UMLS was searched for concepts appropriate for the current internal catalog term. Afterwards the SNOMED CT atoms for each of the results previously found were retrieved and collected. Both types of queries were conducted using the UMLS *Representational State Transfer Application Programming Interface* (REST API) which is provided by the UMLS's publishing organization.

To execute the mapping in an automated fashion, a Java-based application was implemented whose schematic process is visualized in Fig 2. As input it receives the list of service terms extracted from the internal catalog. For each of its entries the UMLS is searched for appropriate

SNOMED CT concepts in the way described above. In the first query's reply a maximum of 15 UMLS concepts were returned and saved by their *Concept Unique Identifier* (CUI). The second request was conducted using this identifier and retrieved all SNOMED CT concepts that could be found as atoms in the UMLS's data structure. These results were written to a list including the code and the name of the respective SNOMED CT concept.

Because of the heterogeneous nature of the internal catalog's descriptions this simple approach of searching wasn't expected to yield results for every entry. To improve the amount and the quality of outcomes the search algorithm was extended by using variants and substrings of the original term. After analyzing problematic features of 200 randomly chosen terms a set of five different strategies was developed including: (1) removing spaces, (2) removing short filler words, (3) dividing into word groups, (4) dividing into words, (5) dividing the first word. With these variants the UMLS search was executed again in the same way as before, but for each adjustment separately and iteratively only if the previous attempt(s) failed in providing results.

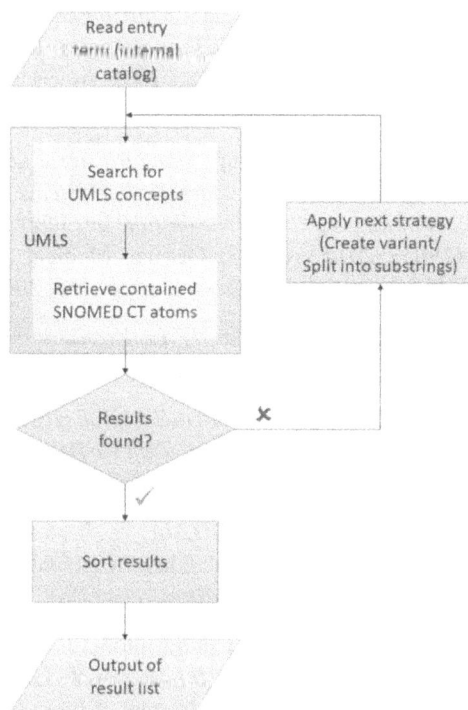

Figure 2: Schematic process of the application's algorithm.

Finally the obtained list of discovered SNOMED CT concepts was ordered to move better results to the front. As the internal catalog contains laboratory services, concepts of the fitting category *Procedure* were hereby rated as best options and those containing the words 'measurement' or 'count' as good ones.

2.3 Postprocessing

The UMLS-based mapping resulted into a tabular file with each row consisting of a service description from the in-ternal catalog and a varying number of SNOMED CT concepts. Because the task was too complex to be fulfilled automatically, the best fitting concept was chosen manually. If the most approximate representation of the term was given by a combination of concepts, all of these were selected, but no SNOMED CT concepts were added to the result list in this step. Along the way, a remark about the type of correlation for each mapping was included, describing if an exact representation of the internal catalog's service is achieved or if the SNOMED CT concept or expression is more general in its meaning.

After this manual examination the last step of final adjustments could be done algorithmically again: Where necessary a postcoordinated SNOMED CT expression was built. For those items, for which more than one concept was chosen in the previous step, the individual elements were concatenated by SNOMED CT intern relationships according to the official *Concept Model* in [9], as shown in Fig. 3.

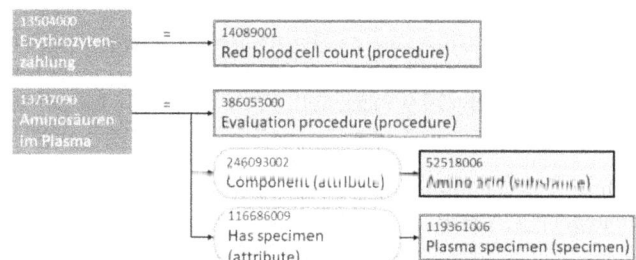

Figure 3: Exemplary mapping results with service descriptions on the left and exact SNOMED CT representations on the right, concepts and relationships as in Fig. 1.

3 Results and Discussion

For the project's aim of mapping laboratory services included in an internal catalog a total of 3833 terms were processed in a semi-automatic procedure. All in all, for 2892 of these an appropriate SNOMED CT concept or expression could be found, leaving a part of 24,5% without any result. Out of the successfully mapped terms slightly over a third was rated as an exact representation of the service, hence almost two thirds could only be mapped to some extent.

In a more detailed evaluation large differences in mapping quality and success can be found with regard to the elements' category and the search type used. As shown in Table 3 general laboratory procedures (LABOR) could be exactly mapped nearly three times more often than microbiological ones (MIKRO), caused by the service descriptions being more complex and specific for that topic.

In terms of search algorithm the usage of the unaltered full term proofed to be not very effective, producing results for less than 10% of most categories' items. Only for the major type LABOR a superior percentage of 22.5% could be determined this way, meaning that for 393 of 1747 queried terms at least one possible SNOMED CT concept was found in the automatic UMLS search. In the subsequent manual analysis

Table 3: Mapping results ordered by type of service and correlation.

Category	Terms	Exact	Partial	None
LABOR	1747	597 (34.2%)	729 (41.7%)	421 (24.1%)
MIKRO	793	95 (12.0%)	451 (56.9%)	247 (31.1%)
FREMD	635	288 (45.4%)	254 (40.0%)	93 (14.6%)
Other	658	63 (9.6%)	415 (63.1%)	180 (27.4%)
Total	3833	1043 (27.2%)	1849 (48.2%)	941 (24.5%)

363 of these findings were approved as containing an appropriate result (Table 4), indicating a good accuracy of 92.4%. Searching with substrings and variations on the other hand yielded results for almost every item (98.7%), but the results were a lot more inaccurate (74% approved). The portion of exact mappings adding up to 36.2% was lower than that for full term search (79.9%) as well.

Table 4: Evaluation of search methods for service terms of category 'LABOR'.

	Full term	Substring
Source terms	1747	1564
Results found (automated)	393	1544
Manually approved	363	1143
Exact mappings	290	414

Among the successfully mapped terms slightly less than half reached their best representation with one single SNOMED CT procedure. 51% could be mapped more precisely by postcoordination, creating expressions of up to four concepts with their corresponding relations.

In case of partial correlated mappings it should be noted that the grade of discrepancy varies a lot. While some of these results come really close to an exact representation, for others only a far more general SNOMED CT concept could be found, possibly lacking essential information of the internal catalog's procedure. Though the manual postprocessing was done with great caution and with the aid of other knowledge sources, its accuracy has yet to be evaluated by an expert of the laboratory domain. To enable aggregated analysis based on the mapping's results further software will be needed. For the combination with LOINC coded laboratory tests a good foundation is newly available by the official mapping between LOINC and SNOMED CT, established by their respective publishers [10].

4 Conclusion

This project's results show that German laboratory procedure terms included in a hospital's internal catalog can be mapped to a standardized terminology by a semi-automated approach to a large extent. While the LOINC coding system's structure was found to be too unflexible to represent heterogeneous source data, SNOMED CT appeared as a good alternative further improving mapping accuracy by postcoordination. For cross-mappings and linguistic issues the UMLS could be identified as an important and useful interface terminology. Though being more inaccurate, searching with the addition of substrings and variances made the mapping more successful on the whole.

Acknowledgement

The work has been carried out at and supervised by the Institut für Medizinische Informatik, Universität zu Lübeck.

5 References

[1] K. D. Pagana and T. J. Pagana, *Mosby's Manual of Diagnostic and Laboratory Tests*. Elsevier Health Sciences, 2013.

[2] K. W. Fung and O. Bodenreider, *Utilizing the UMLS for Semantic Mapping between Terminologies*. In: AMIA Annual Symposium Proceedings 2005, American Medical Informatics Association, p. 266, 2005.

[3] E. Pantazoglou and S. Thun, *Standardisierte Laboranforderung durch LP-Kodes des Kodiersystems LOINC zur Steigerung der Interoperabilität*. In: Jahrestagung der GMDS 2014, vol. 59, German Medical Science GMS Publishing House, Düsseldorf, p. 155, 2014.

[4] J. F. Debatin and P. Gocke, *IT im Krankenhaus: Von der Theorie in die Umsetzung*. MWV, 2015.

[5] T. Benson and G. Grieve, *Principles of Health Interoperability: SNOMED CT, HL7 and FHIR*. Springer, 2016.

[6] DIMDI, *LOINC/RELMA*. Available: http://www.dimdi.de/static/de/klassi/loinc/index.htm [last accessed on: 17.01.2018].

[7] D. J. Vreeman, M. T. Chiaravalloti, J. Hook and C. J. McDonald, *Enabling International Adoption of LOINC through Translation*. Journal of biomedical informatics, vol. 45, no. 4, pp. 667–673, Elsevier, 2012.

[8] C. McDonald, S. Huff, J. Deckard, S. Armson, S. Abhyankar and D. J. Vreeman, *LOINC Users' Guide*. Regenstrief Institute, 2017.

[9] SNOMED International, *SNOMED CT Editorial Guide*. International Health Terminology Standards Development Organisation, 2017.

[10] S. L. Santamaria, F. Ashrafi and K. A. Spackman, *Linking LOINC and SNOMED CT: A Cooperative Approach to Enhance Each Terminology and Facilitate Co-usage*. In: ICBO 2014 Proceedings, pp. 99–101, 2014.

A Testbed Design Approach for Teleoperative Medical Treatment

P. Prieß[1] and R. Allner[2],
[1] Medical Informatics, University of Lübeck, patrick.priess@student.uni-luebeck.de
[2] Institute of Telematics, University of Lübeck, allner@itm.uni-luebeck.de

Abstract

This paper presents a work in progress approach to enable teleoperative treatment. We are developing an testbed for a immersive environment, that allows medical practitioners to interact physically with patients without the need of being at the same place. For this attempt, we developed requirements and chose appropriate measurement instruments. The goal is to reach a high grade of immersion and enable new ways of telemedical treatment. This should guide the project through the whole development process.

1 Introduction

The need for telemedical solutions is often the result of an medical shortage especially in rural areas. Simple treatments and examinations often represent a hurdle and consume a lot of additional time for practitioners.

In practice, most of this telemedical solutions have set their focus on telecommunication. To expand the field of applications it is the next step to shift this scope to teleoperation. Related technologies like *Virtual Reality* (VR) have become very popular and the market size in this field of application increases significantly [1]. That technology enables additional sensory impressions normal telecommunication could not provide and allows to extend the common understandings of telemedicine by the term telepresence.

Therefore we are developing a testbed designed for teleoperation. The testbed is intended to be used by practicing doctors to get a better understanding of the conditions needed to provide sufficient medical care without the need of being spatially present. Thus, the goal is to develop innovative telemedical solutions for specific health care problems.

This paper provides information about this work in progress approach and its development stages. For this purpose important requirements were developed as well as appropriate instruments for measurement were selected. Further the difficulty of choosing hardware is discussed.

1.1 Procedure

Our approach consists of a physical test environment, using existing commercial technologies. A (user) testbed environment, an extendible middleware and several use case specific output devices will allow to research different approaches for all kinds of interaction. The project is separated into three different stages:

(1) Designing the *Local Area* (LA), the physical environment which provides certain input devices to track motions, sounds and other information from the user. In addition, the LA will be capable of giving back visual output, haptical feedback and acoustics. This represents the intended core mechanics of the LA and allows to fully interact with a virtual environment for test purposes.

(2) Providing corresponding output devices that could be accessed to transfer the user input to the real world. The virtual environment will be left behind and the user will be immersed at distant location called *Remote Area* (RA), shown in figure 1. In addition, the RA needs also to be equipped with input and output devices in order to record and emit all information and interactions to keep the both locations synchronized.

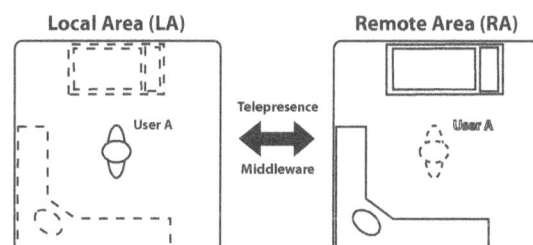

Figure 1: User A being telepresent at the RA.

(3) Developing the sophisticated testbed into a working prototype which will be used under realistic conditions. Furthermore, the system will be evaluated by the target audience.

1.2 Drivers and Enablers

The advance of VR technologies into the commercial market is a key enabler for this approach. The worldwide VR software and hardware market size nearly doubled from 2016 to 2017 and forecasts let assume that this behavior will continue [1]. The portfolio reaches from simple body tracking sensors to exoskeletal full haptical feedback gloves. *Head Mounted Display* (HMD) systems like the HTC Vive and Oculus Rift are a common choice for VR applications. Further, many medical devices possess interfaces to communicate with external computers or being controlled over distance. Common medical instruments like electrical stethoscopes are equipped with modern sensors and bluetooth, while standards for medical device communication, like the ISO/IEEE 11073, enable to combine several of those devices.

2 Basics

To get a better understanding of the researched telemedical approach, it is necessary to clarify what is meant by being telepresent at a RA to enable teleoperative treatment. In the following the terms telepresence, teleoperation and telemedicine are defined in a more precise way.

2.1 Telepresence

"Telepräsenz/Teleexistenz bezeichnet das Gefühl des Bedieners des Systems, sich in der entfernten Remote-Umgebung präsent zu fühlen..." [2]. Freely translated: Telepresence/tele-existence refers to the feeling of the system operator that he or she feels present in the remote environment.

2.2 Teleoperation

"Teleoperation ist die Erweiterung der sensoriellen und manipulatorischen Fähigkeiten einer Person für das Wirken an einem entfernten Ort (remote location)." [2] Freely translated: Teleoperation is the extension of a person's sensory and manipulatory abilities for working at a remote location.

2.3 Telemedicine

There are many different definations for *telemedicine* in use, but we will stick to the one favored by the *World Health Organization* (WHO):
"The delivery of health care services, where distance is a critical factor, by all health care professionals using information and communication technologies for the exchange of valid information for diagnosis, treatment and prevention of disease and injuries, research and evaluation, and for the continuing education of health care providers, all in the interests of advancing the health of individuals and their communities." [3]

3 Infrastructure and Requirements

In general the testbed will be a set of modular rooms in the real world that could be equipped like required. The infrastructure within this places allows to position all hardware as needed using existing commercial technologies. In stage

(1) the user will be inside the LA while being immersed at a virtual area and in stage (2) at the RA.

We developed requirements to give structure to the design process of the testbed. For this, we chose appropriate instruments, measuring relevant requirements. This will help us to evaluate existing technologies and to choose appropriate hardware and software components for the testbed. Further, this will form the basis for developing new teleoperative solutions and mixed reality applications.

3.1 Immersion

Achieving a high grade of immersion is very important for this project, because otherwise some intended tasks can not be accomplished. This might be specific for every use case. E.g. treatments like auscultation or measuring blood pressure require less complex and precise interactions then carrying out a brain surgery. Thus, the needed grade of immersion to successfully perform those tasks are different. In the course of this project, multiple applications for teleoperative treatment are conceivable and should be realized.

A persons feeling to be immersed at a distant location is very subjective. There are multiple aspects influencing this process like e.g. affinity for technology, susceptibility for motion sickness and more. Some of these might be user dependent and are hard to optimize in a general way. Many approaches for that kind of measurement in literature often evolved from existing usability evaluation methods.

We refer to the one published by Bowman, Johnson and Hodges [4], because it is a testbed evaluation approach that focuses on interaction. They separated interaction into navigation, selection and manipulation as well as system-control-based interactions. This method evaluates interaction outside the context of specific applications. It is separated in three major steps:

(1) A taxonomy is build, listing interaction tasks as well as corresponding techniques.

(2) Listing outside influences that might distort the results.

(3) Taking performance measures for evaluation.

Bowman, Gabbard and Hix discussed that this approach is considered to be an usability evaluation method for virtual environments [5]. Thus, we decided to exclude usability as a standalone requirement.

The taxonomy we want to use for selection and manipulation is the one that was proposed originally [4]. It focuses on the three general interaction tasks selection, manipulation and release, the user has to deal with in an virtual or immersed environment. Especially testing different feedback techniques will have impact on which hardware will be used. E.g. the taxonomy intends to test three different kinds of feedback for a selection tasks: graphical, force-/tactile and audio feedback [4]. In addition, we might get further information about how much immersion is needed to perform certain use case specific tasks.

For navigation tests Bowman, Johnsson and Hodges [4] proposed two search tasks, completed in a virtual environment including non transparent obstacles. Those two tasks were characterized by Darken and Sibert as naive and primed search [6]. In the first one the user has to search something with no a priori knowledge of the targets location. The second includes any search task where the environment is known. By using the mentioned testbed evaluation approach we have a performance measure that allows to optimize the testbed setup, as well as to compare different kinds of techniques and hardware.

3.2 Commercial off-the-shelf

Commercial off-the-shelf (COTS) products are sold in substantial quantities in the commercial marketplace by definition [7]. All products used for the testbed should be COTS products. This means hardware components used for the testbed have to be commercially available without being a custom-made solution. Designing and building hardware that fits the needs is a very time consuming task and would require domain experts from other fields as well as a significant amount of financial capital. We assume that out of the box solutions are easier, allow to access a wide spectrum of possibilities and there are certain sub requirements that do not have to be dealt with in detail. Two of them are reliability and extensibility. COTS components fulfill all essential conditions and government regulations needed for conventional use. This suggests that if we are using those products as intended, just in combination with other ones, this will provide reliable testbed hardware. In addition, most COTS products already support established interfaces. Unfortunately, there is also a downside for using commercial technology. There are certain regulations designers of consumer products mostly not have to deal with in terms of medical applications. Fortunately, this is no barrier for us right now, but it might be for future projects.

3.3 Safety

Safety of a system is defined as the absence of catastrophic consequences on the user(s) and the environment [8]. Typical error considerations are power failures, device failure, faulty operation, latency and communication errors. An example for this is dislocation of the operating output device e.g. robot arm caused by information loss. This also means to consider potential afflictions caused by the used technology like motion sickness in case of HMDs. It is important to determine what kind of hazards exist and how to eliminate or control them. There are several guidelines for designing safety-critical systems but we suggest a more scientific approach. Sojer, Buckl and Knoll [9] presented a method based on a more formal foundation. It needs to be considered that the following enumeration is only a very basic abstraction of their approach:

(1) Safety requirements and safety assurances are manually specified on a more abstract level e.g. actor model. All requirements will be iteratively back-propagated along the actor chain until they reach the input actors.

(2) Safety requirements are refined to the different hardware components on which the actor is executed.

(3) Selecting appropriate fault detection mechanisms.

It still needs to be tested if this approach could be fully or partially adapted for the purpose of this project. However, like Wears and Leverson stated: "Safety must be built into a system from the beginning" [10]. Thus, we think this is an appropriate first design approach of the safety requirement.

3.4 Testbed Hardware

This section gives insight about the hardware used for the LA. It will be discussed how components are classified to get a better understanding of potential hardware choices for different kinds of interaction.

The taxonomy mentioned in section 3 gives insight about what interaction we want to perform. Poupyrev, Weghorst, Billinghurst and Ichikawa [11] subdivide this between egocentric and exocentric interactions which could also be applied to VR input devices. The following two paragraphs contain information about how they approached this topic:

They described that egocentric interactions assume that the user is inside the virtual environment and performs actions in a known manner. For example, this includes basic object manipulation as well as virtual pointer selection tasks using a representation of their real hands. Choosing hardware as input devices is a major design factor regarding the mapping of the virtual hand and the real one.

Exocentric interactions concern influencing the virtual environment from the outside, also called God's eye viewpoint. An given example for this might be the World-In-Miniature technique. While immersed the user helds a virtual miniature model of the environment in his hands and is able to manipulate it that way.

A possible use case for this project might be the interaction with medical devices e.g. adjusting the patients bed height or setting up a heart rate monitor. Further administrative tasks like recording patient files are conceivable too. Manipulating the virtual environment using a regular computer could also be considered as exocentric interaction.

At this moment of time egocentric interactions receive more attention, but this might change in the future. Despite that, further considerations regarding navigation need to be made. Alternative movement methods like teleporting or fast forward movements have their own needs regarding hardware and will highly influence the process of choice too.

Further we need to choose medical devices that provide the needed sensory input at the RA. However, in stage (1) the focus is more set on ego- and exocentric input devices for the LA itself.

3.5 Input management

The collected and exchanged data from the LA as well as the RA needs to be organized. A dynamical input structure needs to be developed that allows to prioritize certain input depending on the situation. For example the visual motion capturing of the Kinect v2 might be more suitable to determine the users height and will be preferred to the measurement of the HTC Vive used as HMD. There will be several interfaces, regarding devices used by the testbed, that need to be implemented in this process.

Unfortunately it is not allowed to publish more detailed information at this point of time.

4 Discussion and Future Work

Within this work requirements for the development of a telepresence testbed and their evaluation methods were defined. It has to be shown that this is an accurate approach. The actual performance of many devices still needs to be testified. We assume that the defined requirements will help to make decisions in a less subjective way, but we are also aware of the weaknesses and possible failures of this attempt.

There is a lot of future work regarding hard- and software. The market is growing fast and a constant observation is needed to identify additional hard- and software for the testbed. Unfortunately in most cases there is no good way to evaluate the performance of hardware before purchase. Further, a lot of the VR devices are only available as development kits, which are quite expensive in most cases and do not fulfill our COTS requirement. Despite the fact that this excludes those products in most cases, we need to follow their development process to determine if they might be used for this project in future.

New available components need to be compared by specification against competitors in their field of application. Many of the modern medical equipment provides additional data and possibilities for interaction as mentioned and therefor needs to be considered too. Further the discussed safety evaluation approach needs to be elaborated in detail. But before that, there are several design decisions that need to be clarified to do so.

There might occur problems later on in the development process which have not been considered at the moment. Stage (1) still contains a lot of questions and obstacles that need to be overcome, leaving a lot of space for future work.

Acknowledgement

The work has been carried out at and supervised by the Institute of Telematics, University of Lübeck.

Special thanks goes to Prof. Dr. Stefan Fischer, director of the Institute of Telematics, for giving me the opportunity to do an internship at this institute.

5 References

[1] "Virtual reality software and hardware market size worldwide from 2016 to 2020." https://www.statista.com/statistics/528779/ virtual-reality-market-size-worldwide/. Accessed: 2018-01-24.

[2] "Wirklichkeitsnahe Telepräsenz und Teleaktion Begriffsdefinition Sonderforschungsbereich 453 der DFG ." http://www.sfb453.de/begriffsdef.html. Accessed: 2017-11-13.

[3] W. H. Organization, *Telemedicine: Opportunities and Developments in Member States*. World Health Organization, 2010. Report on the Second Global Survey on Ehealth 2009.

[4] D. A. Bowman, D. B. Johnson, and L. F. Hodges, "Testbed evaluation of virtual environment interaction techniques," *Presence: Teleoperators and Virtual Environments*, vol. 10, no. 1, pp. 75–95, 2001.

[5] D. A. Bowman, J. L. Gabbard, and D. Hix, "A survey of usability evaluation in virtual environments: classification and comparison of methods," *Presence: Teleoperators and Virtual Environments*, vol. 11, no. 4, pp. 404–424, 2002.

[6] R. Darken and J. Sibert, "Wayfinding behaviors and strategies in large virtual worlds," in *Proceedings of CHI*, pp. 142–149, 1996.

[7] "Federal acquisition institute definitions." https://www.acquisition.gov/far/html/Subpart%202_1.html#wp1158534. Accessed: 2017-01-10.

[8] A. Avizienis, J.-C. Laprie, B. Randell, and C. Landwehr, "Basic concepts and taxonomy of dependable and secure computing," *IEEE transactions on dependable and secure computing*, vol. 1, no. 1, pp. 11–33, 2004.

[9] D. Sojer, C. Buckl, and A. Knoll, "Propagation, transformation and refinement of safety requirements," in *Proceedings of the 3rd Workshop on Non-functional System Properties in Domain Specific Modeling Languages*, 2010.

[10] R. L. Wears and N. G. Leveson, ""Safeware": Safety-Critical Computing and Health Care Information Technology," in *Advances in Patient Safety: New Directions and Alternative Approaches (Vol. 4: Technology and Medication Safety)*, Agency for Healthcare Research and Quality (US), 2008.

[11] I. Poupyrev, T. Ichikawa, S. Weghorst, and M. Billinghurst, "Egocentric object manipulation in virtual environments: empirical evaluation of interaction techniques," in *Computer graphics forum*, vol. 17, pp. 41–52, Wiley Online Library, 1998.

Development of an Android Application to Monitor the Core Body Temperature

M. Ortac [1], T. Graßl [2] and P. Rostalski [3]

[1] Medical Engineering, Universität zu Lübeck, m.ortac@student.uni-luebeck.de

[2] Drägerwerk AG & Co. KGaA, Center of Competence Accessories & Consumables, Connect & Develop, thomas.grassl@draeger.com

[3] Institute for Electrical Engineering in Medicine, Universität zu Lübeck, philipp.rostalski@uni-luebeck.de

Abstract

The Dräger Tcore Monitoring System allows a non-invasive measurement of the core body temperature. As an additional monitor for the Tcore system, an Android application was desired for use by customers or for internal testing purposes. The current version of the application can monitor the results from up to four Tcore sensors. For each sensor a pbox (parameter-box: displays the value with unit) and a wbox (waveform-box: displays the graph) are added to the layout. During the measurement, the wbox adds all the values to a graph series and the pbox displays only the last received value. SQLite database on Android is used to save data from the sensor. The user interface can be adjusted. If the temperature value exceeds a limit, an audio-visual signal will be triggered. The application created in this work can be improved by separating the modules for writing and reading from the database and updating the views.

1 Introduction

The human body is able to keep its temperature constant. The temperature regulation is influenced by high energy expenditure and complex temperature regulation mechanisms such as vasomotor function, heat generation and sweat secretion. Vital organ functions and metabolic processes that depend on body temperature are impaired severely, if this mechanism is disrupted [2]. The temperature in the outer regions of the body and the extremities changes depending on environmental factors. Under normal circumstances the body's core temperature remains largely constant at $36.5°C$. The core of the body consists of underlying tissues, organs and brain. The core temperature can vary during the day; it is lowest in the early morning and reaches a maximum in the afternoon. Also, the female menstrual cycle causes a fluctuation of the core temperature. Physical fitness, acute stress, age, food intake and sleep behavior can additionally influence the individual core temperature [3]. A heat stroke can be the result of a disturbed cooling mechanism of the body. Strong physical activity at high environment temperatures, inappropriate clothes, high humidity and low airflow can promote a heat buildup in the human body. Persons at risk could monitor their body's core temperature continually by using a Tcore sensor (Fig. 1). The interaction of the particular components is described in the Section 2.

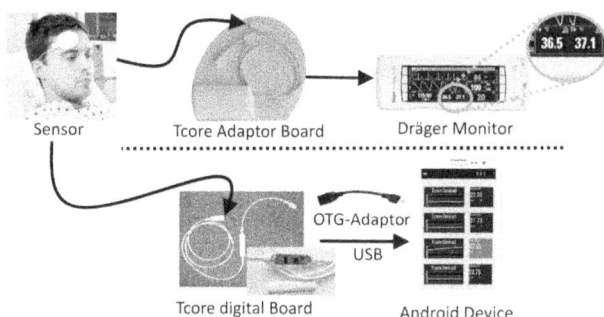

Figure 1: The existing Dräger Tcore Monitoring System and monitoring the core body temperature by using an Android device and Tcore digital board [1].

2 Material and Methods

For the core temperature measurement, the existing single-use Tcore sensor is used. The Tcore Monitoring system measures the heat flux between two temperature resistors. The analog signal from the sensor is converted to a digital signal which is processed by a microcontroller to calculate the core body temperature. A tablet with Android-Version 4.1.1, Software Development Kit (SDK)-Version 16 and a Universal Serial Bus (USB)-HUB with four ports were used for development of the Android application (Fig. 2).

2.1 Data Transfer Method

USB offers a transfer rate up to 480 Mbit/s [4] and power is also be supplied, so an extra power source is not needed. This type of connection is compatible with many terminal devices. A properly shielded USB cable offers a range of

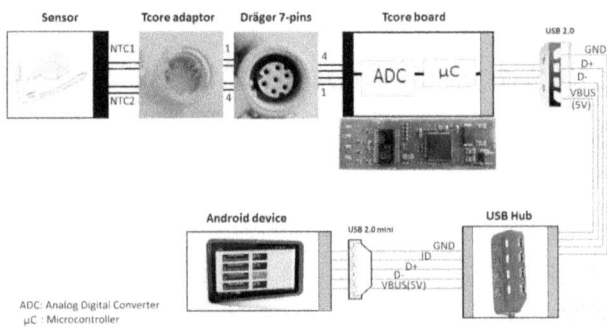

Figure 2: The interaction of the particular components of Dräger Tcore Monitoring System from sensor to Android application [1].

two meters between the Tcore sensor and the Android device.

2.2 Requirements and Demands

The application shall provide portable monitoring of patients. The measurement data of Tcore devices shall be displayed graphically in the application and shall be stored. A new connected device shall automatically be detected and started. For each sensor a pbox and a wbox are required which have to be updated automatically if new data are received. The pbox shall display a time signature of the last received value. It shall be possible to predefine limits. The user shall be alarmed audio-visually if the temperature value exceeds the limit. The user interface shall be adjustable for individual users. The graph in the wbox shall be scrollable and scalable. If a device has been detached, the system shall detect this action and the user interface shall be updated accordingly.

2.3 User Process

A flowchart was used to plan and realize the user process. During the design and evaluation process several loops had been necessary to support the usability with a failure-free plug-and-play functionality. Finally the result offers the possibility that the p- and wboxes can be added to a grid layout at runtime. To implement the user interface, pixels are converted to columns and rows to characterize the position, height, and width of each box. This information is necessary to properly display the p- and wboxes (avoid overlapping, objects outside of display range, etc.). Fig. 3 shows the simplified flowchart of the application. The *MainActivity* runs in the foreground and provides the graphical user interface. Menu items (e.g. buttons) are defined in this activity to enable editing views. Views can be dragged and dropped and the program detects any overlapping. The handler is a thread that runs parallel to the activity. *DeviceService* is started in the *MainActivity* and always runs in the background. The service retrieves a list of connected sensors and configures the connection for each sensor. An object creation from the class *DeviceBox* generates the p-

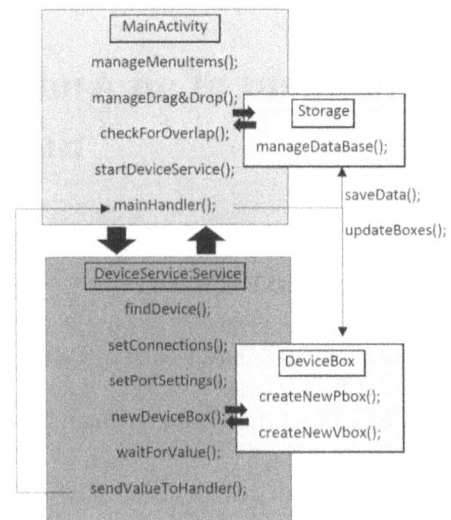

Figure 3: Simplified flowchart of the application: The *MainActivity* is responsible for the user-defined layout. It starts the *DeviceService* to configure a connection to the device and listens to the results of measurement. *Device-Service* generates an object for each device from the class *DeviceBox* to create p- and wboxes. The results will be send to the handler which saves the values and updates the views [1].

and wboxes for the sensor. A callback method listens to the data from the sensor and sends the information to the handler in *MainActivity*. Consequently, it is possible to update the views and save the results whenever the service provides a message. Android Studio was used to implement the application.

2.4 Android Studio

Android Studio 3.0 is a development environment for building applications on Android devices. The programming language is Java. Android SDK Tools version 26.1.1 is installed and the Android Platform version is API 26: Android 8.0 (Oreo). SDK version 16 and grandle version 4.1 (Jelly Bean) are the oldest versions which have to be supported, to make the application backward compatible. Android Studio provides debugging via WiFi (Wireless Fidelity) and releasing the USB-ports for connecting sensors during debugging. SQLite database on Android is used to save data from the sensors [5]-[7].

3 Results and Discussion

In this section, the results are listed and discussed. A graph view that displays the Tcore value over time was implemented to test the service class that connects to a sensor and listens to the results. Initially the application consisted of text views that could be moved and overlapping was prevented. The views and a drawer were generated in an XML-file and integrated into the layout. Programming for the

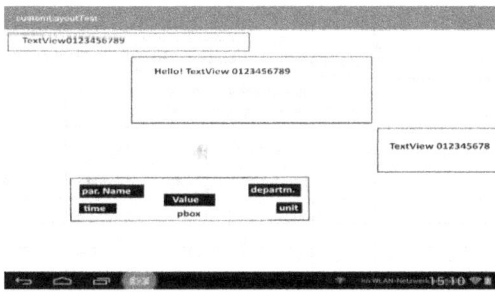

Figure 4: The application consists of text views that are created in an XML-file of the main layout [1].

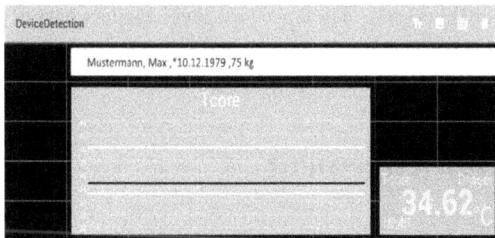

Figure 5: The application includes a USB-service and visualizes the core temperature from Tcore sensor with upper and lower limit [1].

views is shown in Fig. 4. One of the views (pbox) consists of five other text views.

For the next version of the application, the pbox (lower box), the programming of which was shown in Fig. 4, was retained, while the other text views have been removed. Additionally a wbox has been implemented to visualize the temperature values. For the simulation, the temperature data are generated randomly by an extra Java class. The pbox blinks if the related value exceeds a limit.

The USB-Service was integrated to implement the data connection between the Android device and the Tcore sensors; this is shown in Fig. 5. Upper- and lower-limit can be set to show normal range of temperature measurement. At this point in the development, the views were still generated by a XML-file and could only be used for one connected Tcore sensor. Up to four Tcore sensors can be connected via the four-port USB-IIUB and the application should include corresponding views for each sensor.

To use several sensors as plug and play devices, views defined in the XML-files were deleted and views are now generated at runtime when a sensor is connected. Generating the views at runtime allow for data from each sensor to be visualized in its own view. This was realized by sorting views in *linear layouts* as shown in Fig. 6.

With the linear layout, users are not able to drag and drop views. To integrate this feature, the *boxLayout* depicted in Fig. 4 is required. To add the views in the *boxLayout*, the height, width, horizontal-, and vertical-position are defined. Fig. 7a shows the views positioned in the *boxLayout*. Fig. 7b shows the user interface separated into the component layers and views.

Finally, to prevent an unwanted user action, the menu items

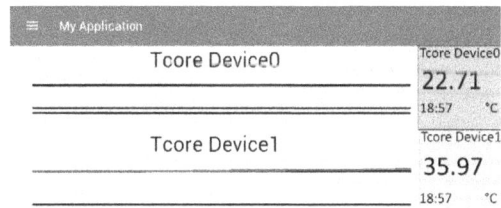

Figure 6: The views are generated at runtime when a sensor is connected and are ordered in linear layouts. Adjustments by the user are not supported [1].

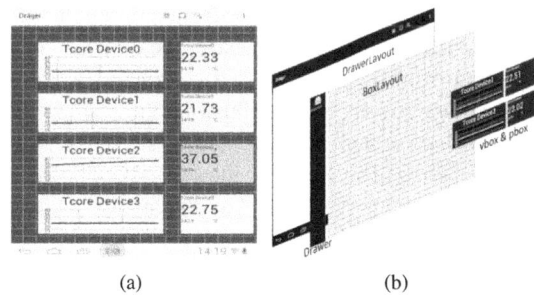

(a)   (b)

Figure 7: a)Views are generated for each connected sensor and added to *boxLayout*. A drag & drop function is implemented. b) The user interface consists of views on *boxLayout*, added to *drawerLayout*. A drawer is also placed on the *drawerLayout* [1].

are added to the application. The user interface will not be changed if the user touches the display of the Android device unintendedly. The desired user action has to be enabled by selection the menu item before using. If one item is activated, the other items will be deactivated automatically to allow only one user customization. The following items are implemented:

1. The first menu item allows to make views smaller.

2. The second menu item allows to make views larger. A "pinch-to-zoom in&out"-function (a multi-touch gesture for zooming in and out views or images) is not implemented, to change the size of the boxes only in a desired direction.

3. The third menu item must be clicked to scroll and zoom series of wboxes. So the graph can be considered more closely.

4. The fourth item allows to delete views if a medical device provides a value for which only a pbox is needed. Then the wbox can be deleted.

5. By using the fifth item, the user can drag and drop views. This function will be used to reposition the boxes on the layout. To create space on the layout, the views can be added to a drawer. To the contrary, the views can be pulled out of the drawer again.

The user is alarmed audio-visually if the temperature value exceeds a predefined limit. If the user missed an alarm, the

Figure 8: The data are saved in the SQLite database [1].

data can be fetched from the database. In Fig. 8 the database table is shown. A filter can be used to get data of a specific patient, from a particular sensor or from a time interval. To evaluate the results from the database, the database file has to be copied from the Android to a computer. A database browser software is needed, to illustrate the results. Alternatively, two separated implementation steps are needed; the first one stores the results and the second one reads from the database and generates boxes on the layout, which will be updated continuously.

The application was developed with a current SDK version (Version 26) and tested on an Android device with an old SDK version (Version 16) to verify the application runs on older devices. To allow for backwards-compatibility, the compiler should ignore differences in the SDK version when installing the application on a older device. This causes some errors; for instance, if four devices are connected via a USB-HUB to the Android, they are listed in the sensor list. Detaching the USB-HUB only removes one sensor from the list, leaving three other sensors listed as being connected. The size of the sensor list grows with each reconnection of the USB-HUB. The SDK version 18 does not cause this error. To solve the problem, all the listed sensors are copied in a temporary list and a connection is described for each sensor. If a connection is "null", the sensor is removed from the temporary list which, after this process, contains only connections which correspond with an actual, connected sensor. The temporary list is used for the next implementation steps.

The application also allows user accounts to save the defined user interface. The wboxes show all saved values and the pboxes show the last value from the sensor. The displayed time in the pbox can be compared to the current time. If the sensor does not send data for a defined time interval, the corresponding views can be removed automatically or manually from the layout.

4   Conclusion

The body core temperature is normally constant, but the temperature of extremities varies depending on outside influences. Some devices produce approximately results and some of them are not intended for long-term measurements or are invasive. The measurement with a Dräger Tcore sensor is realized non-invasively and can run for an extended time. The Dräger Tcore digital board can be connected to an Android device to monitor the core temperature. The

developed application can display and save the temperature data from up to four sensors at the same time. The user can reposition views on a layout by dragging and dropping. The size of the views can be changed by the user. The graph in wboxes is scrollable and scalable. When a sensor is detached or a new sensor is attached, the layout is automatically updated. Alternatively the view can be deleted manually. The data are saved in SQLite database on Android device under the filename *database.db*. The size of table *Results* will be extended if new data are added.

The application created in this work can be used for internal testing purposes. The app can be improved by separating the modules for writing and reading from the database and updating the views. The app was developed as part of an internship and developed as test software. It is not a medical device acc. MDD.

Acknowledgement

The work was carried out at Drägerwerk AG & Co. KGgA, department Hospital Accessories & Consumables, Lübeck and supervised by the Institute for Electrical Engineering in Medicine, Universität zu Lübeck.

5   References

[1] All Figures in this work are own representations.

[2] M. Flügel 2013 Lübeck, *Kontinuierliche Messung der Körperkerntemperatur bei Frühgeborenen mittels eines nicht-invasiven Sensors und Korrelation mit der durch Nah-Infrarot-Spektroskopie gemessenen regionalen zerebralen Sauerstoffsättigung*. Available: http://www.zhb.uni-luebeck.de/epubs/ediss1434.pdf [last accessed on 2018-01-04].

[3] Drägerwerk AG & Co. KGaA, *Die Bedeutung der Kerntemperatur - Pathophysiologie und Messmethoden*. Germany, 2016 Available; https://www.draeger.com/Products/Content/t-core-booklet-9067939-de.pdf [last accessed on 2018-02-05].

[4] Freescale, *USB 2.0 for Kinetis MCUs*, Available: https://www.nxp.com/docs/en/supporting-information/Universal-Serial-Bus-Training.pdf [last accessed on 2018-02-05].

[5] Google Play and Android Developers, Available: https://developer.android.com/index.html [last accessed on 2018-01-18].

[6] D. Felker and M. Burton and J. Muhr, *Android App Entwicklung für Dummies*, ISBN: 9783527711499, Wiley VCH Verlag GmbH, 2015

[7] D. Louis and P. Müller *Android: Der schnelle und einfache Einstieg in die Programmierung und Entwicklungsumgebung*, ISBN: 9783446451124, Carl Hanser Verlag GmbH & Company KG, 2016

Interaction Paradigms of a Ball-Shaped Input Device for Intensive Care Patients

A. Vandereike [1], S. Burgsmüller [1], J. P. Kopetz [2], M. Sengpiel [2] and N. Jochems [2]

[1] Medizinische Informatik, Universität zu Lübeck, {annkathrin.vandereike,svenja.burgsmueller}@student.uni-luebeck.de
[2] Institut für Multimediale und Interaktive Systeme, Universität zu Lübeck, {kopetz,sengpiel,jochems}@imis.uni-luebeck.de

Abstract

Mechanically ventilated intensive care patients have limited ability to communicate verbally. This problem is addressed in the ACTIVATE project by developing a „Ball-shaped Interactive Rehabilitation Device" called BIRDY. A (quasi-) experimental study was carried out with 40 participants in two age groups (young: M=23.45 years and old: M=67.25 years). The participants evaluated eight different objects regarding key characteristics and interacted with their favorite BIRDY-object within a typical clinical scenario. This paper focuses on participants spontaneous interaction to develop a universally usable BIRDY for intensive care patients. Participants spontaneous interactions with BIRDY indicate that option selection could be realized through rotating and rolling and execution through squeezing BIRDY. Implications for further user interface development will be discussed.

1 Introduction

The ACTIVATE („An Ambient System for Communication, Information and Control in Intensive Care") project is dealing with the development of a socio-technical system to support the communication in the postanaesthetic recovery period with mechanical ventilation. These patients have limited ability to communicate their needs. A system is being developed, to enable patients to communicate and obtain information.

As part of the project, a new „Ball-shaped Interactive Rehabilitation Device" called BIRDY will be developed. In the development process of BIRDY, a comprehensive requirements analysis is carried out. Part of the requirements analysis is a study in which relevant characteristics of BIRDY were determined for the potential user. To specify characteristics commercially available balls and eggs were evaluated by the participants. Second part of the study investigated the interaction with the favorite object chosen after the first evaluation. The participants should interact sponateously with the favorite object within a typical clinical scenario. Imaginable interactions of the participants were recorded and discussed during this process.

The following research questions are addressed in this study:
Key characteristics such as size, weight, shape, material, surface and „squeezability" and as well as preferred combinations of characteristics were determined and a favorite object were selected. This research question is addressed

in [1].
The focus of this paper is on the spontaneous interactions performed by the participants at the second part of the study. The interactions were carried out with the favorite object and recorded. Age-related differences in the interactions were identified.
In the following, the conduct of the (quasi-) experiments is described. A detailed description can be found in [2].

2 Material and Methods

This section contains the study procedure and material, the setting, the description of the sample and the data collection and data analysis.

2.1 Study Procedure and Material

Fig. 1 shows the process of the study. The study started with the welcoming of the participant, the presentation of the study and the recording of demographic data and experience in handling technical equipment by a questionnaire.

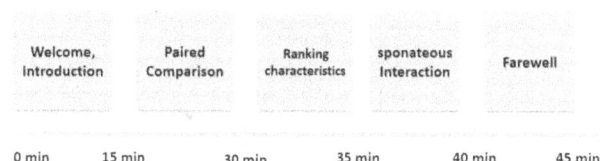

Welcome, Introduction	Paired Comparison	Ranking characteristics	sponateous Interaction	Farewell	
0 min	15 min	30 min	35 min	40 min	45 min

Figure 1: Process of the study.

This was followed by the introduction in which the participant was informed about the restricted mobility (shown in 2.2) and an audio file was then played back in order to put the participant mentally in the situation of a ventilated intensive care patient (see 2.2.1). This was followed by the paired comparison [3]. The participants first described their impressions of the eight objects seen in Fig. 2 are described in more detail in [1]. The paired comparisons were carried out with these eight objects.

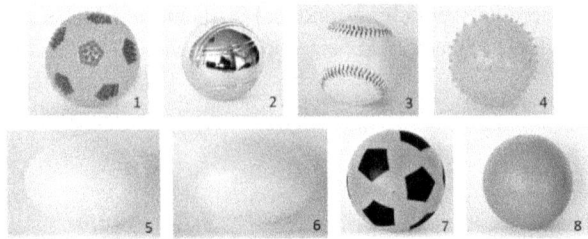

Figure 2: Eight objects used for the paired comparison.

Subsequently, a direct evaluation of the objects regarding to their key characteristics was carried out and then the participants selected their favorite object and gave reasons for their selection.

In the last part of the study, the participant had the possibility of free interaction with the favorite object. Meanwhile, the camera was aimed at the participant's arm and hand to record the interaction. Therefore a new scenario was introduced, which is described in 2.2.1. As a next step, the participant had time to carry out a first impulse in spontaneous interaction with the system. The interactions were recorded and subsequently discussed with the participant. After completion of the spontaneous interaction and discussion, the participant was debriefed.

2.2 Setting

In order to make the setting of the study as realistic as possible, the participants should lay down in a hospital bed in a simulated hospital room. In 30° upper-body position [5], the participants carried out the paired comparisons. Physical constraints such as swollen hands [6] of intensive care patients were simulated using gloves. Since the freedom of movement in the intensive care unit can be restricted by a fixation on the bed, the participants were not allowed to lift their elbows from the bed surface during the free interaction.

2.2.1 Scenarios

The first scenario at the beginning of the study was necessary so the participants could put themselves in the situation of an intensive care patient. Information and instructions were given on the scenario contained in the following text: «Please close your eyes. Try to put yourself in the shoes of an intensive care patient. After an accident, e. g. by car or bicycle, you wake up from anesthesia. You look around and suspect it's an intensive care unit in the hospital. Your whole body feels numb and you feel very weak. You are sensing that you have an object in your dominant hand and you can control something with it. How do you like this object? How does it feel? What do you like about it, what don't you like? »

The second scenario was needed to initiate the free interaction with the preferred object. The participant was instructed not to lift the elbows from the bed surface because of limited hand movement. Afterwards, the participants had the opportunity to interact in the following scenario: «You can't talk, but you want to tell the system that you are thirsty. How would you interact with the system? »

2.3 Sample

The target group of ACTIVATE are mechanically ventilated intensive care patients with an average age of approximately 64 years. To consider age differences two groups were recruited in the development of BIRDY. The aim was to create a sample balanced in terms of age and gender. The young group of 20 participants (M=23.45 years) with a gender distribution of 11 females and the old group of 20 participants (M=67.25 years) with a gender distribution of 12 females.

2.3.1 Inclusion Criteria

Inclusion criteria were an adequate level of health, the mobility to participate independently at the study and the ability to communicate verbally in German.

2.3.2 Recruiting

Participants were recruited in cooperation with the Institute for Multimedia and Interactive Systems (IMIS) via online and offline advertising for the target group and at places frequented by senior citizens.

2.4 Data Collection and Data Analysis

The data were collected in the form of questionnaires for the collection of demographic data and experience in handling technical equipment. Notes and audio recordings were made during the study for internal validation. Video recordings of the participants' hands and arms were recorded for a subsequent analysis of the interaction.

The quantitative evaluation of the interactions was carried out by calculating frequencies to determine the ranks, for example, which interaction was most frequently performed. The qualitative evaluation was used directly for the development of the BIRDY, the evaluation was based on content analysis according to Mayring [4] (Qualitative Content Analysis).

3 Results and Discussion

During the free interaction, the participants could interact with their favorite object, thereby different interactions were observed and quantified.

3.1 Observation

This section describes the interactions that were spontaneously performed by the participants and the related posture of the hand.

3.1.1 Posture

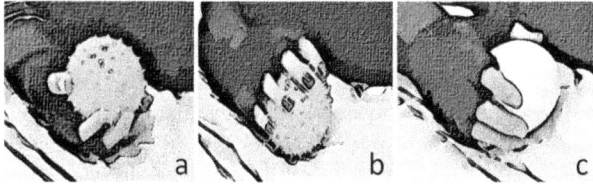

Figure 3: Hand postures without movement. a) Palm upwards. b) Palm downwards. c) Palm sidewards positioned.

During the interaction, three different basic postures of the hand were observed, whereby the changing of the hand posture was taken into consideration and recorded.
Fig. 3 shows the different postures. In a) the participant holds the palm upwards, in b) the palm of the hand is held downwards and in c) the participant holds the palm sidewards to the object.
Table 1 shows the frequencies of the different postures. It can be seen that the palm of the hand has been preferred to be oriented upwards and downwards. As far as the age group is concerned, it can be seen that the old participants often held the palm of their hands downwards.

Table 1: Preferred hand postures. Sums for the complete sample, the young group and the old.

Palm	Total	Young	Old
Downwards	26	11	15
Upwards	21	15	6
Sidewards	9	6	3

3.1.2 Selection

Two different interactions were observed for the movement of the object. 16 of the 40 participants rolled the object (Fig. 4a). During this interaction, the participants moved the object around themselves and moved it over the bed surface. The rolling movements were performed to the right, left, forward and backward.
15 of 40 participants (Fig. 4b) rotated the object in the hand. During this interaction, the object was moved around its own axis without any bed contact. Table 2 shows the frequencies of the used interactions.
Regarding to the hand postures, it can be seen that participants with the palm facing downwards have increasingly rolled the object over the bed surface.

3.1.3 Execution

For the execution of certain actions, the participants performed the following movements shown in table 2. The object was most frequently squeezed (Fig. 4c), this could be observed on 34 of 40 participants. The participants used different ways of applying pressure to the surface of the object, either with the whole hand or with individual fingers. Further actions were knocking (Fig. 4d) with the object by hitting the bed surface several times, this was observed on 7 of 40 participants.
Lifting (Fig. 4e) the object by elevating it away from the bed surface was observed on 12 of 40 participants and shaking (Fig. 4f) the object by moving it briefly and vigorously was observed on 6 of 40 participants.

3.1.4 Exploration

Exploration of the object has been performed in two ways. The conscious exploring (Fig. 4g) of the surface as well as the explicit searching for pressure points or buttons, which could be observed on 18 of 40 participants and the unconscious playful rotating or touching (Fig. 4h) with the fingers, which could be observed on 14 of 40 participants.
The young group (Table 2) showed a higher exploration unconsciously (11 out of 40 participants). Exploration is important for the later usage of BIRDY. If BIRDY is interesting for the user, he or she will explore it more and engage in interaction with it. Interaction could activate and support users' motoric and mental abilities.

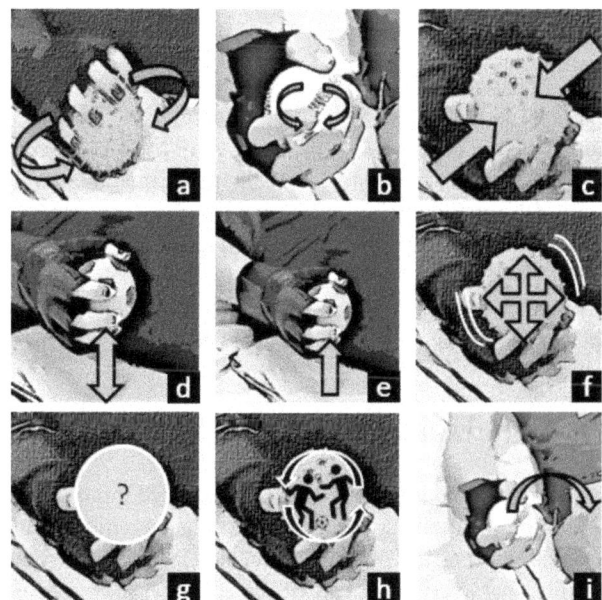

Figure 4: Representation of the various observations. a) Rolling the object over the bed surface. b) Rotating the object in the hand. c) Squeezing the object. d) Knocking the object on the surface of the bed. e) Lifting the object from the bed surface. f) Shaking the object. g) Exploring the object surface. h) Unconscious playful rotation of the object and/or sensing of the surface. i) Throwing the object out of the bed.

Table 2: Sum of the observed movements grouped by total, young and old. Group descriptions are split for selection, execution and exploration and sorted by the frequency of the observations.

Oberservation	Movement	Total	Young	Old
Selection	Rolling	16	6	10
Selection	Rotation	15	7	8
Execution	Squeezing	34	17	17
Execution	Elevation	12	3	9
Execution	Tapping	7	3	4
Execution	Shaking	6	5	1
Exploration	Conscious	18	10	8
Exploration	Unconscious	14	11	3

3.2 Self-Awareness of the Participants

The second part of the spontaneous interaction focused on the expressed thoughts of the participants in relation to the performed interactions.

3.2.1 Thought Concerning Identified Interaction Patterns

Squeezing the object was interpreted by 15 out of 40 participants as the execution of a selected action.
The movement of a possible cursor on a fictitious screen, introduced by the ivestigator, would be controlled by 10 out of 40 participants by rotating or rolling movements and by 10 out of 40 participants by movements in the air.

3.2.2 Free Association

12 of 40 participants associated throwing BIRDY out of bed (Fig. 4f) to attract the attention of a caregiver. The affordances of the object, i. e. throwing may have influenced the participants. One of them actually performed the movement. Observation shows that this behavior was more frequently mentioned by the young group of participants.
13 out of 40 participants spontaneously thought about using a computer mouse. 4 participants associated squeezing the object with the signalling of physical needs.

4 Conclusion and Outlook

In the development process of the ball-shaped input device for communication support, a comprehensive requirements analysis was carried out, which included the study to determine characteristics. Part of the study was focused on spontaneous interaction with the favorite object.
It was obvious that squeezing the object was spontaneously executed. The movement of the object was carried out as a rolling and rotating movement. Exploration was performed in a conscious and unconscious way by about half of the participants.
Regarding to the development of BIRDY, the following spontaneous interactions can be deduced from this study. A possible activation or execution of certain actions could

be realized by squeezing BIRDY. The selection in an interface of the system could be realized by rolling and rotating movements. Explorative hand movements shouldn't influence BIRDY's post-development functions to avoid sending unwanted signals to the system. On the other hand, these could also be taken up to show that BIRDY „is alive " by giving the user feedback, i. e. through vibration.
After finishing the complete requirements analysis, a demonstrator of BIRDY will be developed and integrated into an executable test system. The following phase evaluates partial aspects of the demonstrator. The results are used to optimize the system. After evaluation of the complete system, a field trial is carried out in the clinical environment to verify its applicability under field conditions. In the final phase, the overall system is optimized based on the findings of the field trial. The project shall followed by further product development and eventually market launch.

Acknowledgement

This work has been supported by the Federal Ministry for Education and Research (BMBF) and carried out at the Institute for Multimedia and Interactive Systems (IMIS) at the University of Lübeck.

5 References

[1] S. Burgsmüller, A. Vandereike, J. P. Kopetz, M. Sengpiel and N. Jochems, *Study of Desirable Characteristics of a Communication Device for Intensive Care Patients.* In: Buzug T.M., Handels H., Klein S. (eds.), Student Conference Proceedings 2018, Medical Engineering Science, Medical Informatics and Biomedical Engineering, Lübeck, Infinite Science Publishing, (accepted), 2018.

[2] D. T. Campbell and J. C. Stanley, *Experimental and Quasi-Experimental Designs for Research.* Ravenio Books, 2015.

[3] H. A. David, *The method of paired comparisons.* London, C. Griffin, 1988.

[4] Ph. Mayring *Qualitative Inhaltsanalyse. Grundlagen und Techniken.* (7. Auflage, erste Auflage 1983). Weinheim, Deutscher Studien Verlag, 2000.

[5] J. Rathgeber, *Grundlagen der maschinellen Beatmung: Einführung in die Beatmung für Ärzte und Pflegekräfte.* Stuttgart, Georg Thieme Verlag p.236, 2010.

[6] L. Ullrich, D. Stolecki, M. Grünewald, *Intensivpflege und Anästhesie.* Georg Thieme Verlag p.171, 2005.

Systematic Analysis of the Communication in Medical Care for the Development of Design Recommendations for Telemedical Applications

P. Kling,[1] and R. Allner [2]

[1] Student of Medical Informatics, University of Lübeck, philipp.kling@student.uni-luebeck.de

[2] Institute of Telematics, University of Lübeck, allner@itm.uni-luebeck.de

Abstract

Telemedicine is a growing market. The introduction of new technical applications to this field is often hampered by legalisation or a lack of acceptance by the users. In April 2017, a new eHealth law came into force. Inspired by this, a qualitative study has been performed to analyse the communicational needs of physicians and the possibilities presented by the law. For this study, eighteen health care professionals were interviewed about their opinion of the communication in their working environment. Ten hypotheses were extracted from the interviews and weighted by relevance. After a quantitative study to prove the results a guideline for developing telemedical applications that are better accepted by the users can be generated based on the hypotheses.

1 Introduction

Communication in medical working environments has a very high priority [1]. In Germany, the telecommunication in the health care sector or telemedicine still has much potential for improvement.

1.1 Telemedicine

The German Medical Association (GMA) defines the term telemedicine as a collective term for various medical care concepts, which have in common that medical services of the health care in the areas of diagnostics, therapy and rehabilitation as well as in the medical decision guidance are provided on spatial distances (or time delayed) [2]. The performed study concentrates on diagnostic and therapy in the working environment of physicians and medical assistants.

Telemedicine is one of the major topics in the development of future medical care concepts, especially in recent years [3]. In Germany the current applications in this area are limited mainly to the transmission of vital data [4].

This can be attributed to the legal situation, which prohibits remote treatment via telecommunication. Nevertheless, there are promising approaches and pilot projects that deal with telemedicine and try to evaluate and establish new forms of care using different types of telecommunication [4].

1.2 Legislation

In Germany, The legal situation is rigid and many applications that are already available in other countries are not legally possible. Telemedicine applications are still poorly established in international comparison [4]. A reason for this could be the legal situation and the attitude of the German Medical Association to new telemedical concepts. Most projects on this topic are concerned with telemedical monitoring in cardiology. They mainly include home-care, diagnostics, prevention and curative therapy [4].

Certain other functionalities like are simply forbidden. The main issue is the remote treatment ban [5]. In many parts of Germany, physicians are not allowed to diagnose or give initial treatment to patients from a distance. This could result in a medical undersupply, especially in rural areas.

1.3 New Opportunities for Telemedicine

An increase in the efficiency of communication would be important for the German healthcare sector, since local physicians treat significantly more patients per day in an international comparison and accordingly the available time per patient is lower [6]. A consultation via video could save the travel time for home visits. This would have an even higher impact in rural areas.

Currently the telecommunication in the medical sector is limited mostly to fax and telephone [7]. Other communication channels such as email are not frequently used due to a mistrust in its safety or the lack of accountability.

Although there are already established telecommunication

applications in the field of monitoring certain clinical pictures such as in stroke patients or patients with cardiac arrhythmia, other applications are excluded by the remote treatment ban [4].

The eHealth law "law for secure digital communication and applications in healthcare" released in April 2017 opened new opportunities in telemedicine [8]. Video consultation now allows a better exchange on a physical distance between physicians or a physician and the patient. The National Association of Statutory Health Insurance Physicians and the National Association of Statutory Health Insurance Funds have agreed on an adjustment of the billing catalog in an evaluation committee [8].

These video consultations are only possible for certain clinical pictures. The necessary indicators are currently limited to follow-up checks of wounds and movement restrictions as well as of dermatoses or to the assessment of the voice and speaking. According to the responsible authority the number of indicators will be expanded to cover more clinical pictures [4]. In addition, only general practitioners, paediatricians and specialists such as dermatologists, ophthalmologists or surgeons may offer this service. However, the video consultation may only be used as a follow-up inspection. Therefore, the initial diagnosis have to be performed in person [8].

Therefore the new eHealth law provides developers of telemedical sector with more opportunities to design new applications. However, some existing applications in this field that meet the law are still not accepted by users. A possible reason for this could be the omission to include the users into the development process and to sufficiently identify their needs and opinions.

Motivated by these circumstances, guidelines should be designed to help developers of new telemedical solutions. By using these ever-evolving guidelines, the end product should meet the needs and requirements of users and the relevant laws. This could lead to acceptance of users and facilitate the progress of medical telecommunication in Germany.

The first step to develop such guidelines is a qualitative study made in the form of interviews, in which impressions of the current situation of the telecommunication in hospitals and medical practices are recorded and key messages are extracted.

As in other countries, medical devices must be certified in Germany as well. Depending on the potential risk to the patient, this certification process can be very expensive. It is therefore difficult for small start-up companies to establish themselves in this market. The guidelines could eventually be helpful to lower the financial risks for the developers.

2 Methods

We designed a qualitative study in which semi-structured interviews with physicians and medical assistants were per-

formed. The results were extracted from the statements given in the interviews. The study is not finished yet. The presented numbers and results are preliminary.

2.1 Qualitative study

Qualitative studies consider a small number of participants and extract subjective opinions. The goal for this particular study is to generate hypotheses [9].
It is especially helpful if the researcher is not familiar with the field of his study. One major benefit of qualitative studies is the opportunity to see the area of interest from another perspective. Normally the participants have a much better understanding of the matter and so the researcher can profit from this knowledge [9].
In this study interviews were performed to collect the data. Qualitative studies are very different from quantitative studies, normally performed to detect the needs of a user group. For example, the questions in interviews are not completely fixed. The researcher should be the starting point of the communication but from that point on the participant should be the one leading. With this method the participant can decide what is important and the influence of the researcher is minimised.

2.2 Participants acquisition

For the recruiting of the participants, medical professionals were personally addressed. Interested parties got the information material sent to, as well as the interview questions to prepare them optimally. Participants were not getting paid or remunerated in other forms.
Eighteen health care professionals participated in the study. Sixteen of them were physicians and two were medical assistants. Thirteen participants were employed in a hospital and the other five in doctor's offices.

2.3 Interview guideline

In the qualitative study health care professionals from different working environments were interviewed. Therefore an interview guideline had to be created. This includes an introduction, a main section containing the interview questions and a closing part for the interview. The question segment was split in three thematic parts: General questions, questions about a messaging application and questions about teleconference.
The first part was designed to get a general overview of the knowledge and expectations of the participant regarding the telecommunication in there working environment. More precisely, participants were asked about preferences, barriers and enables of communication and what options developers could have to improve there applications.
The last two segments were created to evaluate ideas for telemedical applications that were in development. The questions about messaging applications refer to a bachelor thesis in which a messenger for medical use should be

developed. The questions about teleconferences relate to a master thesis for developing an application that simulate a doctor's office and conferences with physicians and patients. The focus in these segments was to get options about certain design features and to detect reservations about different functionalities.

All questions are designed to be as open as possible to avoid short answers like "yes" or "no". The goal was to get the participants to talk freely about the field of interest and beyond. When the participants speak freely much more information is given. With this style of questions we also tried to avoid manipulating the interview partners. Nevertheless it is impossible to completely prevent any influences to the communication partner during an interview [10]. Overall 18 questions were predefined for the guideline.

3 Results

Because it was a qualitative study the results alone are not proven statements but hypotheses that have to be verified. The number of questions in a single interview varied between 20 and 50. On average a question was answered with three statements.

For every interview, the given statements were gathered. Afterwards we compared the statements with each other. The hypotheses generated from the interviews are the statements that seem to be significant. This significance was concluded from the frequency of a certain statement in all interviews, the assumed importance of the statements and its comprehensibility.

The following list shows the ten most significant statements ordered by its significance:

1. The physical contact between a physician and a patient can not be replaced by the currently available technology.

2. If video conferences are offered, this can only be done with patients who are well-known to the physician.

3. The quick and easy sending and saving of pictures would make everyday work much easier.

4. Among other things, new telemedical concepts fail because most of the patients are unwilling or unable to use new technologies.

5. Fax is not considered impractical, but alternatives would be used if they are secure, easy and billable.

6. The outdated hardware and the lack of mobile devices in the working environment makes the introduction of certain telemedical applications almost impossible.

7. Frequently, the communication with a communication partner runs over several channels (telephone and fax at the same time). This is perceived as inconvenient.

8. The interface between digital and analog data management is poor and the data is often kept both analog and digital, which is too laborious.

9. There is a lack of training for the individual applications, which complicates the familiarisation and leads to the fact that these are used less often.

10. A joint infrastructure and data management of surgeries and hospitals within a region would be beneficial because test results are frequently requested to treat patients and to avoid duplication of procedures.

In order to generate guidelines out of the hypotheses they should be evaluated by a quantitative study first.

4 Discussion

The hypotheses can be divided in three main aspects: infrastructural problems, the interpersonal component and hurdles for new technologies to get accepted.

The infrastructural problems exist both in the small scale like outdated hardware at the work place and in a wider perspective such as the missing of a standardised data management and exchange system. The most significant hypotheses shows that the participants rate the physical contact between physician and patient very high and name it as a core aspect of there work. The participants named many hurdles new technology have to bypass. The reservations about new technology and the lack of training opportunities were mentioned repeatedly in the interviews and appeared to be significant.

It seems like the law position was not considered significantly good or bad by the participants.

In this study occurred different bias. Statistical bias are mistakes that can occur in any type of scientific study and have the potential to falsify the results [11]. There are different types of bias. Some of them are listed in Table 1.

The selection bias occurred in the study because the participation was optional and only interested health care professionals participated. It can be assumed, that on average they are more interested in the topic of telemedicine and new telecommunication options than it is normal in the healthcare sector. In addition, the participants were on average about 30 years old, which is also not representative.

Performance bias can not be prevented in this type of study. The interview as the way of obtaining information is too variable. In interviews like this, it is impossible to always have the same conversation with every participant. In addition, the discussions also resulted in questions that were asked only to specific participants but not to the others.

Whether a detection bias occurred, can not be decided. For example, errors in the analysis of the data may have occurred by calling certain statements irrelevant, although they should be considered as relevant.

The same goes for the reporting bias. It is difficult to say whether incomplete or unbalanced results reporting has occurred.

The attrition bias could be excluded because no participants have left the study.

Type of Bias	Source
Selection bias	Unbalanced selection of participants and thus a non-representative sample
Performance bias	Error in the execution; For example, by not asking every participants in the same way
Detection bias	Error while collecting the data
Reporting bias	Error that result from incomplete or unbalanced result reporting
Attrition bias	Error that occur when participants to drop out of the study and therefore the data must be eliminated

Table 1: A selection of different bias types in studies [12]

5 Conclusion

Especially for rural areas, telemedicine will be even more important in the future.

However there are many different barriers blocking its progress. We assume that some of them can only change with time like the access and knowledge about computer technology of patients but others like the legislation and the general opinion about telemedicine require a change in the mindset of the population.

The need for more efficient working cycles seems to exist and this study suggest that communication should be simplified.

6 Future Work

As already mentioned the study is not yet finished. More interviews and evaluation will be done. Further stages are planned and include the evaluation of the hypotheses through a quantitative study. For this purpose, a broad mass of participants should be asked about the hypotheses. This could be done by a questionnaire.

In addition, the results could also be evaluated by testing an application that uses the generated guidelines. However, this method is likely to give distorted results, because creating an application not only uses the guidelines, but also own ideas and skills. The quality therefore does not depend exclusively on the guidelines.

Another important aspect is the opinion of the patients. Due to the restrictions of the study the target group consists only of health care professionals but patients are also potential users of telemedical applications. For a proper user analysis every user group should be interviewed.

Acknowledgement

The work was supported by the Institute of General Medicine and the Institute of Telematics, University of Lübeck and supervised by Prof. Dr. S. Fischer, Institute of Telematics, University of Lübeck.

7 References

[1] S. Thun, *Digitalisierte Medizin*. Informatik-Spektrum, Springer, 2015.

[2] AG-Telemedizin der Bundesärztekammer, *Telemedizinische Methoden in der Patientenversorgung – Begriffliche Verortung*, 2015

[3] P. Kutscher and H. Seßler *Kommunikation–Erfolgsfaktor in der Medizin*. In: Teamführung, Patientengespräch, Networking & Selbstmarketing, Heidelberg, 2007

[4] N. van den Berg, S. Schmidt, U. Stentzel, H. Mühlan and W. Hoffmann, *Telemedizinische Versorgungskonzepte in der regionalen Versorgung ländlicher Gebiete*, 2015.

[5] W. Berg, *Telemedizin und Datenschutz*. In: Medizinrecht, 2004

[6] K. Koch, A. Miksch, C. Schürmann, S. Joos and P. Sawicki, *Das deutsche Gesundheitswesen im internationalen Vergleich*, In: Deutsches Ärzteblatt, 2011.

[7] R. Heinze and J. Hilbert, *Digitalisierung und Gesundheit: transforming the way we live*. In: Teilhabe im Alter gestalten. Aktuelle Themen der Sozialen Gerontologie, 2016

[8] Kassenärztliche Vereinigung Schleswig-Holstein, *Vergütung für Videosprechstunden geregelt*. Nordlicht, 2017.

[9] I. Steinke, E. Kardorff and J. Preece, *Qualitative Forschung - Ein Handbuch*, 2000.

[10] *Field Guide to Human-Centered Design*. Available: https://www.ideo.org/ [last accessed on 2017-9-12].

[11] G. Sica, *Bias in research studies*. In: Radiology, Radiological Society of North America, 2006.

[12] J. Higgins and S. Green, *Cochrane handbook for systematic reviews of interventions*, 2011.

Study of Desirable Characteristics of a Communication Device for Intensive Care Patients

S. Burgsmüller [1], A. Vandereike [1], J. P. Kopetz [2], M. Sengpiel [2] and N. Jochems [2]

[1] Medizinische Informatik, Universität zu Lübeck, {svenja.burgsmueller, annkathrin.vandereike}@student.uni-luebeck.de
[2] Institut für Multimediale und Interaktive Systeme, Universität zu Lübeck,{kopetz, sengpiel, jochems}@imis.uni-luebeck.de

Abstract

Mechanically ventilated intensive care patients are often unable to communicate verbally. Within the project ACTIVATE a ball-shaped interaction device is developed in addition to an interactive system to support communication. To determine requirements for the interaction device in the framework of a (quasi-) experimental investigation preferential judgments regarding design properties were recorded. A group of 40 male and female participants was divided into two age groups (18-40 years and over 58 years). The participants evaluated eight different objects regarding their characteristics and interact with their favorite object within a clinical scenario. This paper focuses on the user's preference regarding characteristics of the favorite object. It has been shown that a surface with a profile and a manageable size are the most important characteristics. The design of the prototype will be developed based on the results.

1 Introduction

The ACTIVATE project focuses on optimizing the communication of mechanically ventilated intensive care patients. These patients are often unable to communicate their needs in the recovery phase and to take meaningful contact with the environment. This places a strain on patients, relatives and nursing staff. The aim of the project is therefore to develop a system which supports the communication of patients with hospital employees or relatives to promote patients' quality of life and autonomy.

Within the project, a novel ball-shaped input device for use in bed will be developed. For the development process of the device named BIRDY (Ball-shaped Interactive Rehabilitation Device) a comprehensive requirements analysis is carried out which includes several usability studies. As a first step, needs and requirements for the use of the planned support system will be systematically identified and evaluated. The characteristics and their specifications were recorded based on commercially available balls and eggs, evaluated by the participants.

For BIRDY it is primarily important to ensure that it is accepted as an interaction object by the patients and to interact intuitively with it.

The first part of the study is about the investigation with regard to relevant characteristics for potential users. The second part focuses on interactions with the favorite object and was researched in [1].

This paper focuses on the first part of the study and discusses the results.

For recording the characteristics of BIRDY there was raised the following research questions:

There are some characteristics of size, weight, shape, surface texture, and softness. So the importance is which of these characteristics are frequently relevant for the selection of the favorite BIRDY. In addition, is to find out which of the study objects are preferred the most and what are the decisive characteristics.

2 Material and Methods

The study is intended to limit the characteristics of BIRDY. For this reason, eight objects were evaluated in a (quasi-) experimental investigation regarding their differences. A description for a (quasi-) experimental investigation can be found for example in [2]. The accomplishment is described below.

2.1 Study Design and Procedure

The study started with welcoming the participant and giving some information about the study and recording technical comprehension and demographic data by a questionnaire. Afterwards, the participant was informed about the restricted mobility (described in 2.2). When the participant was laying in bed an audio file was played to put him mentally in the situation of a ventilated intensive care patient (see 2.3). This was followed by a paired comparison [3] where the participants first described their impressions of the eight objects (Fig.2). Afterwards, they decided reasoned for an object in a direct evaluation with regard to the characteristics. Subsequently, they selected reasoned their favorite

BIRDY. The favorite BIRDY was then used for the second part of the study, the interaction with the object. This paper is limited to the choice of the favorite BIRDY. Fig.1 shows the order of the study process.

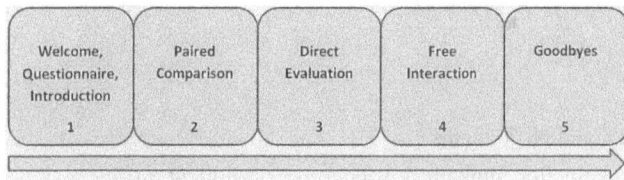

Figure 1: Order of study process

2.2 Setting

To make the study as realistic as possible a hospital room was simulated. The participants should lay down in a hospital bed with a 15-30° angle because ventilated intensive care patients usually have an upper body position [4]. Their hands are often tumid by medication caused by water retention. This prevents smooth and skillful motion sequence [5]. Gloves are used to simulate this limitation of mobility among the participants. Some of the ventilated intensive care patients are restricted in the radius of movement by a fixation of the hands. For this reason, participants should not lift their arms from the bed surface during the study. This is supposed to prevent an unplanned removal of incoming cables and tubes. For these reasons it is important to simulate such limitations and preconditions in the context of the study.

2.3 Scenario

To put the participant mentally into the role of an intensive care patient, an audio file was played at the beginning of the experiment, which was recorded before. This contains the following scenario:

The participant should close his eyes and empathize with the situation of an intensive care patient. After an accident, e.g. with the car or bicycle, he wakes up from anesthesia, looks around and suspects, that it is an intensive care unit in the hospital. He should imagine that his whole body is numb and feel very weak. Into the dominant hand, he gets an object with which he could operate or control something. The questions »*How do you like this object? How does it feel? What do you like about it, what you don't like?*«start the interview.

This should ensure that all participants have the same level of information.

2.4 Study Objects

The selection is based on results of a preliminary study. 30 different objects were tested on an exhibition with 12 participants. These had the free choice to choose an object and test it for its characteristics, to evaluate this and share their

experiences. The aim was to represent as many characteristics as possible. Fig.2 shows the eight selected study objects regarding differences in size, weight, shape, surface texture and softness.

Figure 2: Eight study objects with different characteristics in size, weight, shape, surface texture and softness

Different characteristics with regard to size, weight and shape (see table 1), material and surface texture (see table 2).

Table 1: Different sizes, weights and shape of the eight study objects

ID	Object	Size (in mm)	Weight (in g)	Shape
1	LED Ball	64	45	spherical
2	Bouleset Ball	73	722	spherical
3	Baseball	76	144	spherical
4	Hedgehogball	86	28	spherical
5	Syrofoamegg	80x62	5	egg-shaped
6	Syrofoamegg	120x86	12	egg-shaped
7	Softball	120	44	spherical
8	Foamball	89	30	spherical

Table 2: Different material and surface textures of the eight study objects

ID	Material	Surface
1	Flexible Soft Rubber	Rough and little, tight rubber stings
2	Stainless Steel	Even and crossed with rills
3	Synthetic Leather	Even and crossed with seams
4	Flexible Soft Rubber	Rough and big, far out rubber stings
5	Styrofoam	Even
6	Styrofoam	Even
7	Foamed Material	Even
8	Flexible Rubber Layer	Rough

2.5 Sample

There are different age groups because age-related differences in mobility and technical affinity are assumed and basically everyone can become an intensive care patient.

A young cohort (20 participants, 11 females) at the age of 18 to 40 and a cohort of old (20 participants, 12 females) over 60 years was planned. For experimental economic reasons, up to 58-year-olds were recruited during the process. In table 3 more detailed information about the age distribution of the participants is given.

Table 3: Age distribution of young and old participants

Age	Young	Old
Minimum	18	58
Maximum	31	84
Mean value	23,45	67,25
Variance	9,2	43,56
Standard Deviation	3,03	6,6

2.5.1 Inclusion and Exclusion Criteria

As an inclusion criteria, the participants should be healthy and mobile.
Not suitable for the study are participants who do not have sufficient ability for meaningful verbal communication in the German language.

2.5.2 Recruiting

The young cohort was recruited by students and staff of the University of Lübeck. The old cohort was recruited by flyers for retirement homes and pharmacies or personal response on the campus of the University Hospital of Lübeck and its environment

2.6 Data Collection- and Analysis

For data collection, the participants had to complete a questionnaire, which collected next to technical understanding also demographic data.
During the study, the whole interview with the participants was documented with notes and audio recordings for internal validation.
A quantitative evaluation of the collected data was made. For this purpose, the frequencies of the characteristics regarding the favorite BIRDY were calculated to put them in a ranking order. The qualitative evaluation of the collected data is based on the content analysis according to Mayring. A description can be found in [6].

3 Results and Discussion

In the following, the results regarding the favorite BIRDY are presented.

3.1 Evaluation of Favorite BIRDY's Characteristics

For the evaluation of the favorite BIRDY, characteristics mentioned about frequencies of their nominations were determined. The values in table 4 indicate how often the respective characteristics were named in relation to the participant's favorite BIRDY. This is how a ranking was created and where the frequency of the assessed characteristics can be seen.
Results of the sums of these characteristics show that especially the surface was relevant for the participants. With a

Table 4: Ranking of favorite BIRDY (FB) ordered by mentions based on 40 participants. Evaluated characteristics in this context are size (S), weight (W), shape (Sh), surface (Su) and softness (So).

Rank	ObjectID	FB	S	W	Sh	Su	So
1	1	13	11	4	1	10	5
2	4	9	6	1	0	8	7
3	8	7	3	2	0	6	4
4	3	4	4	5	0	5	2
5	5	3	3	1	2	0	0
6	2	2	2	1	0	2	0
6	6	2	1	0	1	0	0
7	7	0	0	0	0	0	0
Total		**40**	**30**	**14**	**4**	**31**	**18**

value of 31, the surface texture is the most important characteristic for choosing the favorite BIRDY, followed by the size with a value of 30.
Subsequently, the softness is on the third position. However, the frequency of these mentions is significantly lower than for the first two characteristics with a value of 18. 14 out of 40 participants choose their favorite object because of the weight. The shape has less influence on the choice of the favorite BIRDY.

3.2 Top 2 of Favorite BIRDY

As the results of the investigation show, object 1 is the favorite BIRDY chosen from 13 participants out of 40 followed by object 4 chosen by nine (see table 4). Object 1 is a LED Ball with a size of 64mm and a weight of 45g. The material consists of flexible soft rubber and the surface is rough with little, tight rubber stings (cp. table 1-2). The most reasons for choosing this object was mentioned by size and surface texture. Other relevant characteristics are softness and weight. These were less mentioned in relation to the first two characteristics and have therefore less relevance. The shape is only relevant for one participant.
The second rank of the favorite BIRDY is taken by object 4. It is an Hedgehogball with a size of 86mm and a weight of 28g. The material is the same as object 1 and the surface has big, far out rubber stings with even space (cp. table 1-2). The participants particularly preferred the surface, softness, and size. No essential decision criterion on the basis of nominations is weight and shape.

3.2.1 Requests and Comments to the Favorite BIRDYs by Participants

Object 1: A request for the object was an egg-shape. Besides, it could have a heavy weight or being like object 4 but smaller.
The size is manageable and all fingers can do something. Further comments are that the object has a smooth material and is light weight. Moreover, it was mentioned that the object is very catchy. Another aspect is that the temperature

of the material was not cold which was assessed as positiv.

Object 4: There was also the request for an egg-shape. The material was also described as smooth and in addition as pleasant. The object is easy to touch, through the nubs very manageable and not slippery. This makes you feel having something in your hand and tactile stimuli come into play. Besides the same aspect for the temperature was mentioned. In summary, it was rated as the best combination of all characteristics.

3.3   Result Discussion

Object 1 and 4 have the same material, which suggests that the prototype of BIRDY should have such or similar material. On the surface, the grip is in the foreground due to the nubs. Obviously, an object with a profile is an important aspect for the most participants and should be considered when designing BIRDY. It can give them a sense of security and in general to notice something. Consequently, the stimulus can arise intuitively to explore and try out the object. This is important for the patients who should recognize and use BIRDY.

A major difference between the two favorite BIRDYs is the size. Object 4 could be smaller regarding its manageability. Object 1 was chosen based on the size next to the surface, which means that the emphasis is placed on manageability. Thus, controlling elements such as buttons could be reached easier. Besides it is less straining to hold a small object in the hand, which is relevant for the patient, who should interact with BIRDY.

Another point is the softness, which has less mentioned, but nevertheless, it could be an interesting aspect. A good or easy flexibility could give the patient a feeling of confirmation of an action e.g. as a feedback function.

Both objects had the request for an egg-shape. Due to only a few or no mentions of the participants to the shape, this request is not primarily in focus.

Interesting is the remark of a good temperature of the material. A cold surface material should therefore be rather avoided.

No requests and comments were given to the weight and regarding the two favorite BIRDYs, there was not the focus of mentions with a value of 4 referring to object 1 and a value of 1 referring to object 4. But with regard to the weight of the two favorite BIRDYs a range from 28g to 45g could be feasible.

4   Conclusion and Outlook

For the development process of a ball-shaped interaction device to support the communication of intensive care patients a comprehensive requirements analysis was carried out. This included a study were characteristics of BIRDY should be limited. Investigated characteristics were size, weight, surface texture, softness, and shape. In summary, the following results can be mentioned. The participants' preferences for the favorite BIRDY clearly tend towards a surface with a profile and a manageable size. Furthermore, softness is the third most important factor. All in all, object 1 was able to achieve the best result across all characteristics. The results of the investigation regarding the favorite BIRDY should flow into the development of the prototype which will be integrated into an executable test system. Furthermore, partial aspects like the evaluation of interactions with BIRDY investigated in [1] will be used to flow in the development. After a study with the developed prototype in a clinical environment is carried out to use the BIRDY application under real conditions. Finally, the system will be optimized based on the clinical study results so that it can be produced and go to market.

Acknowledgement

This work has been supported by the Federal Ministry of Education and Research (BMBF) and carried out by the Institute for Multimedia and Interactive Systems at the University of Luebeck.

5   References

[1] A. Vandereike, S. Burgsmüller, J.P. Kopetz, M.Sengpiel, N. Jochems *Interaction Paradigms of a Ball-Shaped Input Device for Intensive Care Patients.* In: Buzug T.M., Handels H., Klein S. (eds.), Student Conference Proceedings 2018, Medical Engineering Science, Medical Informatics and Biomedical Engineering, Lübeck, Infinite Science Publishing, (accepted), 2018.

[2] S. Sonnentag, *Abschlussarbeiten und Dissertationen in der angewandten psychologischen Forschung.* Hogrefe Verlag, 2006.

[3] H.A. David, *The method of paired comparisons.* London, C. Griffin, 1988.

[4] J. Rathgeber, *Grundlagen der maschinellen Beatmung: Einführung in die Beatmung für Ärzte und Pflegekräfte.* Stuttgart, Georg Thieme Verlag p.236, 2010.

[5] L. Ullrich, D. Stolecki and M. Grünewald, *Intensivpflege und Anästhesie.* Georg Thieme Verlag p.171, 2005.

[6] Ph. Mayring, *Qualitative Inhaltsanalyse: Grundlagen und Techniken.* Beltz Verlag, 2015.

An Investigation on Automated Consolidation for Medical Database Schemata

J. Oehm[1], H. Ulrich[2], D. Simon[3], J. Ingenerf[4]

[1] Medizinische Informatik, Universität zu Lübeck, johannes.oehm@student.uni-luebeck.de
[2] IT Center for Clinical Research (ITCR), Lübeck, hannes.ulrich@itcr.uni-luebeck.de
[3] Agfa HealthCare GmbH, Trier, dirk.simon@agfa.com
[4] Insititue of Medical Informatics, Universität zu Lübeck, ingenerf@imi.uni-luebeck.de

Abstract

In this paper, we will investigate how the consolidation of two diverged relational database schemata and migrating them into a common one can be supported with *Data Exchange* technologies by using the *++Spicy* tool. We will evaluate this under the tightened constraint of being backward compatible to the old schemas version during a transition window. We came to the conclusion, that this is probably feasible, but would be to complex to realize for our specific task.

1 Introduction

In this paper, we will discuss the consolidation of two different database schemata, which diverged inside *Patient Administration System*, a module of ORBIS hospital information system. The code of version 8.4, which is currently used by many hospitals in Germany, France and Austria, was forked to version 8.5 in 2007 in order to develop faster tailored for the special needs of a customer, a large european hospital chain. We will introduce the term *codestream*, since new versions have been released for 8.4 as well as 8.5 ever since. The 8.4 codestream is still used by most hospitals, while 8.5 usage is currently limited to France and UK. To allow easy data exchange and statistics, one of the 8.5 customers is using a single database for over 36 hospitals at the same time, consisting of about 4 terabyte medical data. This is why update process must be made in a fast manner, since changes to the database schema decreases performance during that time. Especially expensive index rebuilds have to be avoided whenever possible.

Since there were initially no plans for merging the codestreams back together, there were bigger changes made in the database schema, which cumbers a consolidation of the code. In order to be able to consolidate the code, the consolidation of the database must be handled first.

Because hospitals must be operational around the clock, closing the software at the same time on all clients for an update is not an option. This means that the new database schema must be compatible to the old one.

2 Background

Consolidating the existing legacy systems is a big topic all areas of software development. The main reason for consolidation from a software company's perspective is the speed improvement in the development of new features, which saves resources and enables faster delivery to the end-user. A feature that should be available in both codestreams must be either developed in both codestreams independently, which doubles the costs, or must be developed in a codestream-agnostic way, so that it can be merged into the other one. Either way, the new feature must be tested in both codestreams.

Having the codestreams consolidated benefits also the customers: If there is a new feature requirement that has already been implemented for another customer (which is very common), it can be enabled in the settings. However, with multiple codestreams, they will not benefit if it is only available in another codestream, because merging a feature still makes it necessary to wait for the next release of the software. The release cycles usually take about 6 month from planning over testing over piloting phase until release, which means that the long waiting time with less efficient workflows results in increased operating costs.

Before it is possible to consolidate the code, it is indispensible to consolidate the database, since both codestreams need to have the same view on the data. Having the code to be aware of different schemata would lead to increased development complexity and harder maintenance, which completly reverses the benefits of a consolidation.

3 Material and Methods

3.1 Migration

Making changes to source code is pretty easy: The files can be managed using one of those mature *Version Control Systems* like SVN or Git. A project can be compiled and executed for any version at any time, since code does not depend on previous versions of itself.

Making changes to a database that contains production data is a lot more difficult. This process is usually referred to as migration. In order to reduce this issue, Agfa internally developed the *Agfa Database Updater* (ADUP), a special migration tool.

In normal projects, the desired database state is maintained using some SQL-Scripts containing many `CREATE TABLE` commands, which are kept with the source code. This is acceptable during the initial development phase and can be used for testing, but after the software is shipped, this SQL-Script may not be executed again, because using `CREATE TABLE` twice will lead the customers to lose all of their existing data in that table. Instead, for every change an `ALTER TABLE` statements must be used, which are usually created manually.

ADUP is built around its own SQL dialect called RSQL (Repository SQL). RSQL files look like simple `CREATE TABLE` SQL files for each table, which are managed using the *ClearCase* version control system and collectively describe the desired schema of the database. Every time a RSQL file is changed, ClearCase increases the version number of that file. From all RSQL files, a *database repository* file containing the history of all tables is created. In the update process, ADUP will check which version is currently installed on the database and use `ALTER TABLE` statements to convert the current schema into the latest one without losing data. To do so, it resolves the order of dependencies accross multiple tables by using a directed acyclic graph model. For complex migrations that can not be handled automatically, a set of comments can be applied to control the migration between versions using individually created SQL scripts.

3.2 Data Exchange

Because the problem of the so-called *data exchange*, which means a sensful transfer of the existing data in a source schema σ (called *instance* of σ) into a target schema τ, is very common in real life applications, the topic is well researched in the academic database community. There are two approaches to be distinguished [1]:

- **Materialization**: Construct a instance of τ using given instance σ. This process is closely related to migration.

- **Query answering**: Answer queries over τ by rewriting them into queries over σ. This approach is not an option for our case, since it will likely lead to performance losses by unintentional joins and complicates the further evolution of the schema.

A *relational mapping* is formally defined as a tuple $\mathcal{M} = (\sigma, \tau, M_{\sigma\tau}, M_\tau)$, were σ is the source schema, τ is the target schema with all relation symbols different from those in σ. $M_{\sigma\tau}$ is a set of first order logic formulae called *source-to-target tuple-generating dependencies* (st-tgds), which are of the form $\forall \vec{x}\vec{y}(\phi_\sigma(\vec{x}, \vec{y}) \longrightarrow \exists \vec{z}.\psi_\tau(\vec{x}, \vec{z}))$ with ϕ_σ and ψ_τ as conjunction of atoms over σ or τ respectively. M_τ describes the constraints on the target schema, which

can be usually expressed as *tuple-generating dependencies* (target-tgds) of the form $\forall \vec{x}\vec{y}(\phi(\vec{x}, \vec{y}) \longrightarrow \exists \vec{z}.\psi(\vec{x}, \vec{z}))$ and *equality-generating dependencies* (egds) $\forall \vec{x}(\phi(\vec{x}) \longrightarrow x_i = x_j)$. Note that the all-quantifiers are usually omitted [3].

With some constraint in the mapping rules (*weakly acyclic* target-tgds), which are almost always fullfilled in real life examples, a solution can be constructed. The solutions can be classified further into *universal solutions*, that only contain the information that can be derived from the source schemas. Among this solutions, there is the so-called *core*, which is the smallest among all solutions, meaning it contains no redundant data [2].

However, this approach assumes that we have a fixed target schema τ. In our situation, we have two source schemas, σ_1 and σ_2 and are supposed to transform (migrate) them into a common schema τ, s.t. all information from σ_1 as well as σ_2 must be available in τ. This is not directly handled by this methods.

3.3 Transistion Window

Since the impossibility to have downtime, we must use a *transistion window approach*. This means that we must arrange the migration in a way that the new database schema can be used together with the old versions of the software code, which means e.g. that directly dropping columns or tables is not allowed. Then, after we are sure the software on every computer in the hospital is updated, we can implement the breaking changes on the database. Since a new version of ORBIS (software and corresponding DB schema) is released with every quarter of the year, it is easiest to keep the transition window open for at least one version. Within this window, we must be able to assert that both versions operating in parallel are not producing incosistent data, e.g. both versions may not write the same information into different tables. This is not covered by the existing data exchange techniques.

3.4 Common Superschema

This leads us to the task of finding a common superschema τ for σ_1 and σ_2, which contains the union of all tables and all columns. During the transition period, both schemas must provide a consistent view for both software versions, which means, that if there are two different tables storing the same kind of information, they must be kept in sync.

We are not aware of any tooling for automatic generation of common database schemata. However, this would not be expediend, since we must be able to analyze, why the schemata evolved differently. Instead, we used *Oracle SQL Developer*, which provides existing schemata comparing features, for the manual creation of a common target schema. At this point, we made an analysis on the nature of differences. They can be categorized into three different types:

1. New introduced tables or columns for features that are not yet available in the other codestream.

2. New introduced tables or columns that were created for the same feature in both versions, but named differently.

3. Different table structures due to refactorings.

The first case is very simple, the new tables or columns can simply be added to the common superschema. The migration is already handled very well by the ADUP and adding columns is a cheap operation. The second case is fortunately seldom since most of the time new features that required new tables or columns were rather merged than implemented twice. The third case however, is more complicated, because at the beginning of the development of 8.5, the software was not in use, so major changes were made without a migration strategy.

3.5 Complex Data Migrations

An example of the third case is the handling of patients and physicians: A intrinsic question when creating data models for hospitals is the handling of natural persons, because a doctor can come into a hospital as a employee, a visitor, or a patient himself. The *Health Level 7 v3* standard introduces the concept of seperating the entity *Person* from a role as *Practictioner* or *Patient* to handle the commonalities, while the *Fast Healthcare Interoperability Resources* standard for example simply duplicates the common person-related attributes in the *Patient* and *Practicioner* resources [5]. The same issue can be found here: In 8.4, there is a table *Arzt* („Arzt" is the German word for physician) and a table *Patient*, which both store the basic administrative and demographic data. An address update or a name change due to marriage must be handled in both tables, which leads to potential inconsistencies.

Consider this simplified schema σ_1: It uses simply two different tables for each kind of person, ignoring the commonalities:

- Patient(<u>patid</u>: int, first: varchar, last: varchar, insurance: varchar)

- Arzt(<u>arztid</u>: int, first: varchar, last: varchar, station: int→Location.locid)

The 8.5 codestream's schema σ_2 looks like this (*TABLE_PER_CLASS*-inheritance in terms of Hibernate):

- Natperson(<u>persid</u>: int, first: varchar, last: varchar)

- Patient(<u>persid</u> → Natperson.persid, insurance: varchar)

- Physician(<u>persid</u>: int → Natperson.persid, station → Location.locid)

As you can see, in the σ_1 codestream the name information for a physician is stored twice, which leads to unwanted redundancy, while in σ_2, the information is only stored once, which means that the commonalities are handled far better. Because queries over patient and physician data will require

a join, there might be performance drawbacks. We still decided to use the σ_2 schema, since the benefit of having a *single source of truth* about a person's name data overweights possible performance drawbacks.

From the data exchange perspective, the following mapping/migration rules can be constructed:
st-tgds:

$$\text{Patient}(patid, first, last, insurance) \longrightarrow$$
$$\exists persid(\text{Natperson}(persid, first, last) \qquad (1)$$
$$\wedge \text{Patient}(persid, insurance))$$

$$\text{Arzt}(arztid, first, last, station) \longrightarrow$$
$$\exists persid(\text{Natperson}(persid, first, last) \qquad (2)$$
$$\wedge \text{Physician}(persid, station))$$

target-tgds:

$$\text{Patient}(persid, \_) \longrightarrow$$
$$\exists first, last(\text{Natperson}(persid, first, last)) \qquad (3)$$

$$\text{Physician}(persid, \_) \longrightarrow$$
$$\exists first, last(\text{Natperson}(persid, first, last)) \qquad (4)$$

In our example, target-tgds are always fullfilled by the st-tgds, for other cases, there are rewriting techniques. This is important, because st-tgds can be translated directly into SQL statements.

There exists an algorithm called *chase* to materialize an instance of the target schema. However, using external tools to perform row-wise manipulation over large database tables is not performant. Luckily, there are tools available, that can translate mapping rules into SQL scripts. We used *++Spicy* [4] for our experiments, which generated us the following (simplified) transfer script:

```
CREATE TABLE skolemTbl(
    id bigserial PRIMARY KEY,
    skolem text);

CREATE TABLE targetValuesArzt AS
    SELECT *, 'arzt'||arztid AS skolem
    FROM Arzt;
INSERT INTO skolemTbl(skolem)
    SELECT 'arzt'||arztid FROM ARZT;
CREATE TABLE targetValuesPat AS
    SELECT *, 'patient'||patid AS skolem
    FROM Patient;
INSERT INTO skolemTbl(skolem)
    SELECT 'patient'||patid FROM PATIENT;

INSERT INTO Natperson
    SELECT s.id, p.first, p.last
    FROM targetValuesPat p, skolemTbl s
    WHERE p.skolem = s.skolem
UNION
    SELECT s.id, p.first, p.last
    FROM Arzt a, skolemTbl s
    WHERE a.skolem = s.skolem;
```

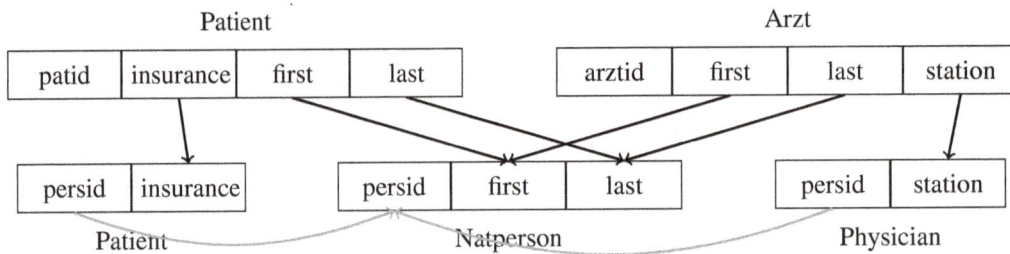

Figure 1: A visual representation of the mapping rules. Black arrows are st-tgds, target-tgds are noted with gray arrows.

```
INSERT INTO Physician (arztid, station)
   SELECT s.id, a.station
   FROM targetValuesArzt a, skolemTbl s
   WHERE a.skolem = s.skolem;
```

As you can see, Spicy uses an additional table to create the so-called *skolem terms* to generate IDs, the details of this process can be found in [3]. It is worth mentioning, that Spicy originally created intermediary tables to perform queries if there are target-egds violated. Because we have none, we can omit this step and copy the data directly into the target tables. We had to add an UPDATE statement to add the *persid* to the *Patient* table, since Spicy was not build for migration.

We have to keep the columns *first* and *last* of the table *Patient* for backward compatibility. During the *transition window*, we must also synchronize these columns of with the corresponding columns of *Natperson* by manually creating a trigger. The *arztid* must also be kept in order to be able to migrate the foreign key relationships of other tables to the correct new entry in the *Physician* table.

Since the table *Arzt* is still needed for backward compatibility, we created a mapping from the *Natperson* model back to the *Arzt* model, but modified the statement from copying data into creating a view. Oracle views are updateble, so old software versions will still be fully operational.

```
CREATE VIEW Arzt AS
   SELECT phy.arztid, pers.first,
          pers.last, phy.station
   FROM Physician phy, Person pers
   WHERE phy.persid = pers.persid;
```

We also had to create manually migration scripts to update the foreign key constraints of other tables pointing to the new created column *persid*. We must be aware of the transistion window approach there, too, which means creating new columns for the *persid* and triggers to keep them in sync with the old foreign key columns.

4   Conclusion

The usage of data exchange techniques for migration seems to have a lot of advantages. We are convinced that it is possible to create a generic version of the ADUP that can handle more complicated database schema migrations on its own. However, the transistion window requirement makes this process much more complicated, since we must be able to add the required triggers automatically. Further, ADUP can not be allowed to drop columns on its own, because migrating all the existing code to the newest schema can be quite time consuming and may take more than one release. Also, a new syntax must be introduced to annotate the existing RSQL-Files with the mapping rules, which weakens the benefits over manually writing migration scripts.

We finally decided against investing more effort into this, since for the current task of migrating σ_1 and σ_2 to the consolidated schema τ of about 200 tables, we had only 5 of these complex cases. We consider the manual migration resulting in faster migration scripts and being more safe compared to an early version of such an automatic tool. However, we suggest further looking into this topic since the migration is a big topic in all kinds of application development and a transistion window is not always required.

Acknowledgement

The work has been carried out at Agfa HealthCare GmbH, Trier, and supervised by the Institute of Medical Informatics, Universität zu Lübeck.

5   References

[1] Ö. Özçep, *Foundations of Ontologies and Databases for Information Systems*. (Lecture slides), Lübeck, 2017

[2] M. Arenas, P. Barceló, L. Libkin and F. Murlak, *Foundations of Data Exchange*. Cambridge University Press, New York, 2014

[3] B. Marnette, G. Mecca and P. Papotti, *Scalable data exchange with functional dependencies*. Proceedings of the VLDB Endowment Vol. 3, 2010, Nr. 1-2, S. 105 ff.

[4] B. Marnette, G. Mecca, P. Papotti, S. Raunich and D. Santoro ++*Spicy: an Open-Source Tool for Second-Generation Schema Mapping and Data Exchange*, Clio, 2011

[5] T. Benson and G. Grieve, *Principles of Health Interoperability - SNOMED CT, HL7 and FHIR*. Third Edition, Springer, London, 2016

Infinite Science
Publishing